The Wild Flower Key

How to identify wild flowers,
trees and shrubs in Britain and Ireland

D1340010

Francis Rose

Revised and expanded by Clare O'Reilly

Art Editor Delf Smith

Illustrations by Martine Collings, Delf Smith,
R.B. Davis, Lura Mason, Norman Barber
and Judith Derrick

FREDERICK WARNE

To Pauline

FREDERICK WARNE

Published by the Penguin Group
Penguin Books Ltd, 80 Strand, London WC2R 0RL, England
Penguin Group (USA) Inc., 375 Hudson Street, New York, New York 10014, USA
Penguin Group (Canada), 90 Eglinton Avenue East, Suite 700, Toronto, Ontario, Canada M4P 2Y3
Penguin Ireland, 25 St Stephen's Green, Dublin 2, Ireland
Penguin (Group) Australia, 250 Camberwell Road, Camberwell, Victoria 3124, Australia
Penguin Books India (P) Ltd, 11 Community Centre, Panchsheel Park, New Delhi 110 017, India
Penguin Group (NZ), cnr Airborne and Rosedale Roads, Albany, Auckland 1310, New Zealand
Penguin Books (South Africa) (Pty) Ltd, P O Box 9, Parklands 2121, South Africa

Penguin Books Ltd, Registered Offices: 80 Strand, London WC2R 0RL, England

First published by Frederick Warne 1981
This revised and expanded edition first published 2006
10 9 8 7 6 5 4 3

Design by Design 23
Cover design by Glen Bird

ISBN-13: 978-0-7232-5175-0
ISBN-10: 0-7232-5175-4

Printed and bound in China

While all the information in this book is believed to be true and accurate, neither the authors nor the publisher can accept legal responsibility for any errors or omissions that may have occurred.

Important notice: Conservation legislation and protected species designations are stated as at 1st March 2004.

CONTENTS

Dr Francis Rose, MBE (Original author)

Dr Rose is one of our best-known botanists, with an encyclopaedic knowledge of the lichen and bryophyte flora as well as of flowering plants, and of plant ecology and biogeography. His career as a botanist began at the age of six, learning to identify plants on country walks. After graduating in botany from London University, he spent most of his working life teaching there until retirement in 1981. *The Wild Flower Key* remains his most popular book. It took over twenty years to write and represents a lifetime's experience of plant identification.

Clare O'Reilly (Author of revised and expanded second edition)

Clare O'Reilly (previously Coleman) is a freelance botanist and writer who left her career as an environmental lawyer in order to revise this book. She recently gained an MSc in Plant Taxonomy but was a self-taught amateur botanist for many years. Her passion for plants began after winning a wild flower-in-a-vase competition, aged seven. Clare teaches beginners' courses on plant identification and this experience, coupled with Dr Rose's expertise in the original text, enables this revised edition to be even more useful for those new to field botany.

FOREWORD BY DR FRANCIS ROSE

The first edition of this book was published in 1981. Twenty-five years on, a new edition has now been prepared. We have incorporated three major modifications in this new edition.

1. The up-to-date nomenclature of Stace and Kent has replaced the Clapham, Tutin and Warburg nomenclature of the 1981 edition because there are quite a number of changes to both generic and specific names of plants. In the 1981 edition the nomenclature was based on Clapham, Tutin and Warburg's *Flora of the British Isles* which is now long out of date.

2. We have included in this new edition quite a number of new alien species, many of which are now well established in the wild.

3. In order to accommodate these additions we have decided to delete the plants of North-western Europe (France, Netherlands, Denmark etc) which were included in the first edition but are not or scarcely naturalised in Britain.

The whole object of these changes is to make edition two of *The Wild Flower Key* a more practical and efficient guide to the plants one is likely to find in the field in the British Isles.

The section of keys on plant habitats and the vegetative keys has not been altered because the magnitude of the task was too great at this time. Many new illustrations have been prepared for the new edition so that alien as well as native species are now represented. As in edition one, it was deemed not necessary always to illustrate the whole of the plant but only the essential parts. A few of the original illustrations were unsatisfactory and in such cases these have been replaced by more accurate pictures.

I would like to thank Clare O'Reilly for all her magnificent work on the text and the excellent artists who have prepared the illustrations. I would also like to thank my wife and family for their patient help during the preparation of the work. And finally I would like to thank users of the book for their continued enthusiastic support which has enabled *The Wild Flower Key* to stand the test of time.

INTRODUCTION TO SECOND EDITION

For 25 years this book has reigned supreme as the most useful field guide to British and Irish wild plants. No other field guide has full botanical keys as well as colour illustrations.

This new edition is in fact a wholesale revision. The book now concentrates on the flora of the British Isles, omitting the western European species contained in the first edition. As well as updating the plant classification and scientific names, it includes:

- over 150 additional plants and new colour portraits;
- new line illustrations drawn by an experienced botanist, comparing key features used in identification;
- new keys and 'ID tips' boxes;
- the most up-to-date keys available for certain groups, written by leading specialists especially for this book;
- additional field identification tips, often not in any other field guide;
- new features to assist those working in conservation, such as marking plants as endemic, BAP species, Red Data List species or with protected species status;
- a compilation of the latest research on ancient woodland indicator species;
- distribution details updated and expanded to follow the *New Atlas of the British & Irish Flora* (2002) by C. D. Preston, D. A. Pearman & T. D. Dines; and
- an expanded, illustrated glossary.

I have not amended or expanded the vegetative keys because this would be several years work in itself. I have also omitted the few species of grasses, sedges and rushes included in the first edition because there is a separate guide to these groups with keys: *Colour Identification Guide to the Grasses, Sedges, Rushes and Ferns of the British Isles and N.W. Europe* (1989) by F. Rose. If you have any comments or spot any errors, please send details to me care of the publishers, so that future editions can be improved.

The keys are the most important part of this book – please try them out. There are no short cuts to plant identification. Using a key, **together with** reading a text description **and** picture matching, is the **only** way to be reasonably certain of your identification of **any** plant, not just of look-alikes and plants in the more difficult groups. But this is fun and, above all, by learning to really look at wild plants, you will soon enjoy seeing the world through different eyes!

CLARE O'REILLY

How can you help save our wild plants?

Learn to identify them! We need more people with good plant identification skills, especially as most schools and universities no longer teach botany. This book shows you how to get started.

ACKNOWLEDGEMENTS

It has been a tremendous privilege to work with Dr Francis Rose and I would like to thank him for trusting me to revise his book. This project evolved into a much bigger task than anyone envisaged and I am very grateful to the many botanists who have worked with me. The following specialist botanists either reviewed my amendments or re-wrote the keys for certain difficult groups:

Paul Green – Onions (*Allium*)
Dr Ray Harley – Mints (*Mentha*)
Geoffrey Kitchener – Willowherbs (*Epilobium*)
Richard Lansdown – Water-starworts (*Callitriche*), Water-crowfoots
   (*Ranunculus* subgenus *Batrachium*) and Knotweeds (*Persicaria*)
Dr Chris Preston – Pondweeds (*Potamogeton* and *Groenlandia*)
Dr Anthony Primavesi – Roses (*Rosa*)
Dr Tim Rich – Fumitories (*Fumaria*) and Gentians (*Gentianella* and *Gentiana*)
Dr Fred Rumsey – Broomrapes (*Orobanche*) and Bladderworts (*Utricularia*)
Nick Stewart – Water-crowfoots (*Ranunculus* subgenus *Batrachium*)

Richard Lansdown also proof-read the whole text and provided many useful comments which have improved this book. Professor Clive Stace advised on several taxonomic issues, Keith Kirby (English Nature) on ancient woodland indicators, and Dr Chris Cheffings (JNCC) on the plant status lists. Eric Clement checked all of the new illustrations and Eric and Mike Grant assisted with several of the new species accounts. Mike Hardman advised on re-working the *Viola* key and Dr Barbara Pickersgill on the glossary. Paul Losse, Sue Cooper and Brenda Harold reviewed the new introduction. Thanks are also due to the artists, Martine Collings and Delf Smith, and to Gwynn Ellis, who produced the index. Particular thanks to Delf for going way beyond what he was contracted to produce in order to enhance this book. Delf also managed the lengthy plate design process and drew scales bars for the entire set of illustrations. I am grateful for the assistance of the staff of the libraries and herbaria at Reading University, Kew and the Natural History Museum, to those who tested the new keys in the field and to everyone who provided comments on the original text.

I would like to acknowledge the encouragement and teaching from many BSBI and WFS members over the years. It is not easy to single out individuals, but I must in the cases of Eric Clement and the late Franklyn Perring, both for their willingness to make the time to teach me and many others. Finally, a huge thank you to my husband John for suggesting that I gave up my job to do this project, without whom this book may well have gone out of print.

Illustrations acknowledgements

The London Natural History Society permitted use of line illustrations from the *Flora of London* by Rodney Burton. Line drawings of *Callitriche* fruits from the *BSBI Water Starworts Handbook* by Richard Lansdown are reproduced with permission of the Botanical Society of the British Isles.

ABBREVIATIONS, SYMBOLS AND CONVENTIONS

SCIENTIFIC DESCRIPTIONS

agg	aggregate
sp	species (singular)
spp	species (plural)
ssp	subspecies (singular)
sspp	subspecies (plural)
ann	annual
bi	biennial
per	perennial

PLANT FEATURES

alt	alternate, alternating
fl(s)	flower (s)
flg(d)	flowering (ed)
fr(s)	fruit (s)
infl(s)	inflorescence (s)
lf(lvs)	leaf, leaves
opp	opposite

OTHER TERMS & SYMBOLS

±	more or less
–	to (eg, oval–obovate denotes oval to obovate)
v	very
id	identification
◄	text description for this illustration is on the previous page
►	text description for this illustration is on the following page
▲	illustration for this text description is on the previous plate
▼	illustration for this text description is on the following plate

DISTRIBUTION

('NE' denotes Northeast,
'N, E' denotes North and East)

Br Isles	British Isles
C	Central
CI	Channel Islands
E	East, Eastern
Eng	England
Eur	Europe
GB	Great Britain
IoS	Isles of Scilly
IoM	Isle of Man
IoW	Isle of Wight
Ire	Ireland (without political division)
N	North, Northern
S	South, Southern
Scot	Scotland
W	West, Western

FREQUENCY

abs	absent
c	common
f	frequent
l	local
la	locally abundant
o	occasional
r	rare
vc	very common
vla	very locally abundant
vr	very rare

HABITATS

ar	arable	**hths**	heaths, heathland
calc	calcareous	**mds**	meadows
chk	chalk	**mt(s)**	mountain (s)
gd	ground	**rds**	roads, roadsides
gdn(s)	garden (s)	**wa**	wasteland or brownfield sites
gslds	grasslands	**wd(land)**	
hbs	hedgebanks	**(s) (y)**	wood, woodland, woods, woody
hds	hedges		

PLANT STATUS

AWI	ancient woodland indicator species
BAP	Biodiversity Action Plan priority species for GB and N Ire
End	species endemic to Br Isles (excluding microspecies)
introd	introduced or introduction
natd	naturalised

CONSERVATION & RARITY STATUS

EU	species protected by the Directive on the Conservation of Natural Habitats & of Wild Fauna & Flora (also known as the Habitats and Species Directive) (92/43/EU)
FPO (Eire)	the Flora Protection Order 1999
Sch 8	Schedule 8 of the Wildlife and Countryside Act 1981 (as amended)
WO (NI)	the Wildlife (Northern Ireland) Order 1985
EW	extinct in the wild
EX	extinct
CR	critically endangered
EN	endangered
VU	vulnerable
NT	near threatened

according to the Vascular Plant Red Data List for GB (2005) edited by C. M. Cheffings & L. Farrell

*	nationally scarce, found in only 16–100 different 10 x 10 km grid-squares since 1987 in GB & IoM
**	nationally rare, found in only 1–15 different 10 x 10 km grid-squares since 1987 in GB & IoM
**Ire	listed in the Irish Red Data Book: Vascular Plants (1988) by T. G. F. Curtis & H. N. McGough

BOTANICAL TEXTS & SOCIETIES

BSBI	Botanical Society of the British Isles
New Atlas	*New Atlas of the British & Irish Flora* (2002) by C.D. Preston, D.A. Pearman & T.D.Dines
New Flora	*New Flora of the British Isles* (2nd edition 1997) by C.A. Stace
Plant Crib	*Plant Crib 1998* by T.C.G. Rich & A.C. Jermy

THE SCOPE & CONTENT OF THIS BOOK

The area and plants covered
This book includes flowers, trees and shrubs found growing in the wild in the British Isles, which includes Eire, Northern Ireland, Scotland, Wales, England, the Isle of Man and the Channel Islands.

The flora of the British Isles totals over 3500 species and subspecies of flowering plant. Therefore a fully illustrated field guide with keys could never include all of these and remain portable. This book includes over 1600 of the flowering plants that you are most likely to find (details of how these were selected are at the end of this section).

How the plants are described
Plants may be:

- fully described and illustrated in the main text;
- described in the main text by comparison with a similar fully described and illustrated species;
- briefly described in the keys to genera;
- briefly described in the vegetative keys (keys to plants not in flower); or
- named in the keys, with both English name and scientific name, but not described elsewhere.

Where a plant is referred to in the text as the 'only species' in a genus with a certain character, this should be read to mean that this is the only species **found in the British Isles** with this attribute.

The main text of plant descriptions & illustrations
The main text contains full descriptions of the plant species, arranged by family. This sequence of plant families uses the plant classification system followed in the *New Flora of the British Isles* (2nd edition 1997) by C. A. Stace.

Species described and illustrated are distinguished both in the text and on the illustration pages by capital letters, A, B, C, D, etc.

The species **not** illustrated are described **by comparison** with the illustrated ones they most closely resemble and are labelled A.1, A.2, B.1, etc, the letter concerned denoting the described and illustrated species with which they are compared.

The form of the entries
Each entry in the main text follows a standard form:
- **English or common name**
- **Scientific name**
- **Description of the plant** including **measurements** which denote **length** unless otherwise stated.
- **Distribution and frequency of occurrence**. Distribution data are those of the plant as a **native**; if known to be introduced over part or all of its range, this is indicated by the abbreviation '**introd**'. A region given in brackets following mention of introduction is that of the plant's native place of origin, eg 'introd (Mediterranean)'.

- The plant's usual **habitat**.
- When a plant flowers (and, with certain trees or shrubs, fruits). Months are cited 1–12; so 'fl 4–6' shows that a plant flowers from April to June.
- Characters in **bold text** are key features that distinguish this species.
- The heading '**key differences**' introduces diagnostic characters which distinguish this species from look-alike plants.
- The heading '**Do not confuse**' introduces information about other, similar species that you could easily confuse with the plant described.

For **non-illustrated species** (A.1, A.2, etc) full details of a plant's description, distribution, habitat and flowering times are not always given. In such cases, you should assume that they are the same as for the preceding and illustrated species (A) with which they are compared.

'ID Tips' boxes

These boxes contain identification tips and hints to help you use the keys. The '**What to look at**' section lists the key characters of a plant that you should look at carefully. You could make a note of or take a photograph of each of these characters as you will need these details to identify many plants in this group.

The illustrations

This book does not always illustrate the **whole** of every plant. Where several plants are superficially very alike, **one** plant is illustrated fully, with only the details (often enlarged) of other similar species shown to demonstrate how they differ from one another.

Drawings labelled Aa, Ba, etc, are features of plants A or B respectively.

Scale bars

Most of the illustrations and details have a scale bar beside them; those **with no measurement** indicated represent **1 cm** length of the actual size of the plant concerned.

Plant names

Plant scientific names and English names generally follow those in the Botanical Society of the British Isles' standard list, published on its website, **www.bsbi.org.uk**. Scientific names used in the first edition of this book, which have now been changed, are given in brackets.

How the plants were selected for this book
Completely covered

- All the **native** and **long-established introduced** species of flowering plants;
- More recently **introduced** species which are widespread, having more than 300 10 km square records (counting all date classes) in the *New Atlas of the British & Irish Flora* (2002) by C.D. Preston, D.A. Pearman & T.D. Dines; and
- More recently **introduced** species with 50–299 10 km square records in the *New Atlas* which meet a set of criteria aiming to include species that are, for example, invasive, rapidly spreading or easily confused with native plants.

Partially covered
- **The so-called 'critical' species** among flowering plants, ie very similar species of, eg, *Alchemilla* (Lady's-mantles), *Rubus* (Brambles), *Sorbus* (Whitebeams), *Euphrasia* (Eyebrights), *Taraxacum* (Dandelions) and *Hieracium* (Hawkweeds);
- Some very common **Grasses, Sedges and Rushes** in the vegetative keys; and
- **Yew, Scots Pine and Juniper**. These, our only native conifers, are included in the vegetative keys, illustrated with line drawings.

Plants excluded
- Introduced flowering plant species with under 50 10 km square records which only appear on the *New Atlas* CD-Rom;
- Non-native (but widely introduced) conifers;
- **Grasses, Sedges and Rushes** in the main text;
- **Ferns, Horsetails, Clubmosses and Quillworts** – the vascular cryptogams; and
- **Mosses, Liverworts, Lichens and Algae** – the so-called 'lower plants'.

HOW TO USE THIS BOOK

The botanical keys are the most important feature of this book. This section explains why and how to use them.

What is a key?
A key is a series of questions that lead you to an answer: the name of your plant. This process is known as 'keying out'.

Why use keys?
Many people can **recognise** a buttercup, but keys help you to **identify** it because:
- keys make you focus on **diagnostic** (ie unambiguous distinguishing) plant characters;
- keys help you separate **similar-looking plants**; and
- keying out shows you what a plant is **not** and why, as well as what it is likely to be.

Keys also help you learn about plants because:
- you are more likely to **remember a plant** if you have spent time looking at it carefully while using a key; and
- keying out is **fun** (like solving a puzzle!)

Before you use the keys
You do need to be familiar with the basic parts of a flower. If you are not, read the section '**Guide to flower structure**' on p 15 and have a close look at the different parts of some common wild flowers, such as a buttercup and a daisy.

How to use the keys
The general key on p 24 is a key to all plant families in this book. Interspersed with the text descriptions are keys of two kinds, to genera within a family, and to species within a genus. There is an indexed list (p 562) of all keys within the main text.

All the keys in this book work in the same way and generally follow the dichotomous principle. At each numbered step you are confronted with a choice between two (or more) contrasting alternatives (known as a couplet). A key may start:

1   Petals 4...2
    Petals 5...6

This means that if your plant has four petals, you proceed to step **2** on the left hand side of the key. But if your plant has five petals, you go instead to step **6** on the left, lower down the key. After further choices you will eventually reach an entry which, together with the features established at earlier stages in the key (eg 'petals 4') seems finally to describe your plant and which will give its name (or that of its family and perhaps genus if you are using the general key).
The example 'petals 4' was very simple. In many cases, because plants may vary in so many features, more than one character may be given, for example:

1   Petals 4 – lvs opp – stem hairy...2
2   Petals 5 – lvs alt – stem hairless..6

Sometimes neither of the alternatives will agree fully with your plant. In this case, try the alternative where a **majority** of the characters agrees with your plant and proceed to the step indicated. Because some plants are so variable, and also because some features can be misinterpreted, some plants may 'key out' twice or even three times in the same key. If you, in effect, 'miss' a species at first (perhaps because your specimen is atypical) you therefore have a chance of reaching it again further on in the key.

If you have not used a key before, start by taking a plant that you know, like a buttercup, and see if you can get the right answer from the keys. You may be surprised to find that there are several, very similar species of buttercup, a fact that you might have missed if you had not used the key.

How to identify plants in flower
If you have no idea to which family or genus a plant belongs, go through the following steps until the keys lead you to an answer:

1   First go to the general key to the families for plants in flower (p 24).
2   Start with the Master Key, which outlines major, readily observed characters with line illustrations.
3   Choosing the major character that agrees with your plant, turn then to the sub-key indicated, and work through it (see **How to use the keys**, above). The sub-key should take you eventually to the right **family** (or **genus**).
4   Turn then to the page number given in the key. **First** carefully read the general description of the family; if it agrees with your plant, **then** work through the family key (if there is one) to find (or to confirm) the correct genus.
5   Then go through the genus key to find the correct species. If the family (or genus) is a small one of only one or a few species, go straight to the description of the genus, and then to those of the one or more species that follow.

6 Now read the species description **very** carefully, paying attention to the details of stem and leaves, as well as those of flowers and fruit.

7 If the description seems to agree with your plant, then (and **only** then) turn to the illustration to **confirm** that you have identified your plant correctly.

8 If the description and illustration do **not** match your plant, work backwards to find where you may have gone wrong. Provided you have observed your plant accurately, you are unlikely to be very far away, and you will probably find the correct species among adjacent text descriptions and illustrations.

If you think you know the family or genus to which your plant belongs:
- Use the index to find the main text entries on that family or genus.
- First carefully read the **general description** of the family or genus.
- If your plant seems certainly to belong there, use the keys to the family and its genera (if there are any large enough to have needed keys) to find the specific description of your plant. If the family or genus is a small one, go straight to the description of the genus and then to that of the species.
- Continue from step 6 above.

Tips on using the keys
- Read the whole of **both** alternatives for each couplet before attempting to choose between them.
- Always pay more attention to **shape** and **size** of flower parts than to their **colour** – colours of flowers can often vary, and therefore can be misleading.
- **Always** cross-check the text description and pictures after using the keys.
- Remember that the keys only cover the plants included in this book.
- If you are really stuck, try looking for an illustration that seems to match your species and then work backwards – read the text description and, if it seems to fit, try using the key to reach it.

EQUIPMENT

The only essential equipment needed to identify plants is a good quality metal **hand lens** of x10 magnification. Always carry a hand lens on a string round your neck, or you will inevitably lose it! A hard-backed **field notebook** and a **camera** are useful. An Ordnance Survey **map** or 'GPS' (Global Positioning System) is also useful, to record grid references if you find something unusual or that you cannot identify.

Hand lenses are available from ecology equipment suppliers and from BSBI Publications.

Using a hand lens
Hold the hand lens **very close** to your eye – so your hand touches your eyebrow – and hold it **still**, while you **slowly** bring the plant towards you until it comes into focus. Holding the plant and your hand lens up to the light often helps you to see fine details more clearly.

Using a microscope

Microscopes are **not** only for experts. One great advantage of using a microscope is that you have **both hands free** to dissect and measure a small structure such as a tiny flower. You can also see the parts of a plant much more clearly than through a hand lens. The type of microscope generally needed for flowering plant identification is known as a **low-power dissecting** or **binocular** microscope (rather than a high-power compound microscope with up to x400 magnification or more). You will find that you mostly use x10 to x30 magnification. These microscopes are no more expensive than a camera and are often much cheaper second-hand. You can learn how to use a microscope on introductory wild flower courses run by the Field Studies Council (see '**Where to find out more**' section p 22).

Measuring plant parts

In this book, measurements indicate either a range, eg 5–8 mm, or an average, eg 5 mm. All measurements denote length, unless otherwise stated. Ranges placed in brackets, eg (4–)5–8 (–10) mm mean that the plant part concerned is usually between 5 and 8 mm long but rarely it may be as short as 4 mm or as long as 10 mm.

When measuring plant parts, if possible look at several plants, as size is so variable. Population **averages** are much more reliable, so if ever in doubt, measure **10 plants** and use the average. When measuring **petal length**, choose a petal from a flower that is **fully open**, as petals in some species continue growing even after the anthers have shed their pollen. When measuring **fruit length**, check whether this **includes** any beak or persistent style.

GUIDE TO FLOWER STRUCTURE

Flowers are very varied in form, and it should be remembered that some of the parts shown in the drawing on p 16 may be missing in some species, or even in some of the flowers of a species. However, in most cases all the parts illustrated are present, though of varied form; so this illustration serves to show in a general way the relative position and sequence of the parts normally present.

Why are there different types of flower?

Flowers are the sexual reproductive organs of plants, and contain male and female organs; reproduction is the real *raison d'être* of flowers. The amazing variation that exists in flower form and structure is in fact connected with the diverse ways in which plants have evolved by natural selection to achieve cross-pollination, either by various insects, animals or by the wind.

About the parts of a flower

The flower-stalk normally has at its apex a swelling known as the **receptacle**, to which the parts of the flower are attached. This may be convex, or a concave cup.

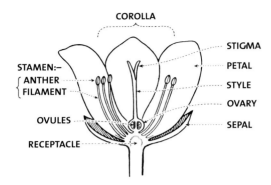

The outermost whorl of flower parts, known as the **calyx**, is composed of a number of usually leaf-like structures, individually called the **sepals**, whose normal function is to protect the inner parts of the flower in the bud stage. (In some species the calyx is absent, or, as in some members of the Daisy family, represented only by a ring of hairs, the **pappus**, which later becomes a parachute for seed dispersal.)

Within the calyx, the next whorl, the **corolla**, is composed of leaf-like parts, normally coloured, called the **petals**. Their function in most flowers is to make the flower visually attractive to pollinating insects. They may secrete nectar at their bases or in special spurs, and also may have scent-producing glands, usually at their bases. In wind-pollinated flowers the corolla may be inconspicuous or absent.

Sometimes the sepals are joined together into a **calyx-tube**, and the petals may be joined together into a **corolla-tube**. In such cases the number of the parts can, however, usually be counted from the number of lobes or teeth on the tip of the tube.

The numbers of sepals, petals and of the inner parts are of great importance in identification; it is also very important to note whether they are free or conjoined. For example, in the Buttercup family all the parts of the flower are free and separate, while in the Bellflower family the calyx and corolla are both in tubular form; in the Daisy family even the stamens (see below) are joined into a tube. Sometimes the sepals and petals are identical in form and colour, then they are collectively termed a **perianth**, as in the Lily family.

Regular and irregular flowers

In some families, such as the Dead-Nettle, Pea and Orchid families, the sepals and petals are not in regular symmetrical whorls, but are modified to form flowers of **irregular** form, symmetrical only about one axis from front to back. Such flowers usually have one or more petals enlarged to form a **lip** at the front and a **hood** or arched structure at the back. These flowers are highly adapted to visits by special kinds of insects, the lip forming a landing platform for the insects, while the hood at the rear forms a roof that protects the sexual organs of the flower from rain. The form of such flowers may even permit only certain kinds of insects to enter. Pea flowers have a **keel** of two lower petals in front, partially joined together, two **wings**, one on either side, and a **standard** at the back.

Inside a flower

Within the flower are its reproductive organs. Working from the outside towards the flower's centre, there are first the **stamens**, collectively called the **androecium**. These vary in number, but each stamen consists of a stalk or **filament** and a head or **anther**. The anther contains two or more pollen sacs. The **pollen** consists of microscopic grains – copious and dust-like in wind-pollinated flowers, less plentiful and sticky in insect-pollinated flowers – which contain the male sexual nuclei or **gametes**.

Next there is the female organ of the flower, the **gynaecium**. This comprises a number of parts called **carpels**, which are quite separate in Buttercups but fused together in most other families. In a normal fused gynaecium three parts can usually be distinguished. First, at the base, the **ovary**, a little case of one or more **cells**, which contains tiny **ovules**, each holding a single egg. Above the ovary is the stalk-like **style**, which may be single or multiple, and at the tip of each style is a **stigma**, the receptive surface for the pollen.

When pollen is transferred by insects or wind to a stigma at the right stage to receive it, each pollen-grain germinates, forming a microscopic tube which grows down inside the style to reach the ovules and their eggs. The male nucleus travels down the tube and fuses with the female nucleus in the egg. The fertilized egg can then grow into an **embryo**, which, with its coverings, becomes a seed. In time the ovary becomes a seed-box, or fruit.

Looking at carpels and the ovary

The **number** of carpels and **whether or not** they are **fused** together, is important for identification. In the Buttercup family the carpels are all quite separate, each with its tiny ovary, single short style and stigma on top. In this family they are easy to count, but are usually of a large, indefinite number. In most other families, where the carpels are fused, their number can be determined either by counting the number of stigmas, or of styles, if these are separate, or else the number of cells in the ovary. The simplest way to see these is to cut (with thumbnail or penknife) a cross-section of the ovary and examine it with a hand lens. In some plants, however, it is difficult to count the carpels; in such cases this feature is not used in this book.

Finally, the ovary may be **superior** or **inferior (or semi-inferior)**. A superior ovary is attached to the receptacle above the corolla-whorl and stamens; an inferior ovary has the calyx and the corolla arising from its top, so that it can be seen below the flower in side view. This character is important, for example, in distinguishing the Bellflower family, with inferior ovaries, from the Gentian family, with superior ovaries.

Groups of flowers on one plant

Flowers may be borne singly, that is, they are **solitary**, as in the Primrose and Daffodil, or they may be grouped together in various ways into an **inflorescence**. The various forms of inflorescence mentioned in the text (**cyme**, **raceme**, **spike**, **umbel**, **panicle**, **corymb**) are defined in the glossary, as are the leaf-like structures found in an inflorescence (**bracts** and **bracteoles**). In the Daisy and the Teasel families, the flowers are grouped into a head (**capitulum**) of unstalked tiny **florets** on a common receptacle. The whole may be mistaken for a single flower, especially if, as in the Daisy, there are symmetrical **ray florets** at its outer edge which look like petals.

PLANT STATUS

Native and introduced plants

This book makes no distinction between native plants and long established introduced plants. The classification of introduced plants as archaeophytes (naturalised before 1500) or neophytes (introduced post 1500), as used in the *New Atlas*, is not included in this book because it is controversial for certain species and currently subject to revision.

Rarity status

The asterisk star system in this book follows the interim plant status lists published by the Joint Nature Conservation Committee in 2004, based on the data in the *New Atlas*. These asterisks are only applied to native and long-established introduced plants and only to species distribution in Great Britain and the Isle of Man, excluding the Channel Islands.

*   Nationally **scarce** – found in only 16–100 different 10 x 10 km grid-squares since 1987 in GB & IoM
**   Nationally **rare** – found in only 1–15 different 10 x 10 km grid-squares since 1987 in GB & IoM

For Northern Ireland and Eire, the symbol **** Ire** indicates plants that are nationally rare, listed in the *Irish Red Data Book: Vascular Plants* (1988) by T. G. F. Curtis & H. N. McGough.

Not all 'nationally scarce' or 'nationally rare' plant species are legally protected species, although all wild plants are protected from uprooting without authorisation (see '**Wild plants & the law**' section, p 19).

Conservation status

The *Vascular Plant Red Data List for Great Britain* (2005) edited by C. M. Cheffings and L. Farrell (Red Data List) lists wild plants that are most in need of conservation action. Each plant is categorised according to not only its rarity but also the level of threat it faces. The most endangered plants are designated as follows: **EX**: Extinct **EW**: Extinct in the wild **CR**: Critically endangered **EN**: Endangered **VU**: Vulnerable **NT**: Near threatened

These categories are explained in the Red Data List, which is available from **www.jncc.gov.uk** and are in addition to the 'nationally scarce' and 'nationally rare' designations referred to above. No comparable list is available yet for Ireland.

This book cannot be used to provide a complete list of endangered, rare and scarce plants as many subspecies, microspecies and hybrids are not treated separately here.

Endemic plant species

This book generally follows the List of Plants Endemic to the British Isles compiled by T. C. G. Rich *et al.* (1999) *BSBI News* 80: 23, as updated by the Red Data List, for endemics in the UK, Isle of Man and Channel Islands and lists provided by Dúchas for Eire. Microspecies are not included. Endemic status is often controversial and may be changed in the light of new knowledge. For example, Boyd's Pearlwort, *Sagina procumbens* 'Boydii' (*Sagina boydii*) has been shown by genetic studies to be a variant, probably a cultivar, of Procumbent Pearlwort, *S. procumbens* and therefore is not marked as endemic in this book.

WILD PLANTS & THE LAW

All wild plants are the private property of someone, usually the landowner, and are subject to both the laws of property and the general criminal law, as well as laws in conservation statutes. The law surrounding wild plants is complex, and therefore this section inevitably contains **general guidance only** and cannot be relied upon as a definitive statement of the law. The precise legal position also differs according to jurisdiction; England & Wales, Scotland, Eire, Northern Ireland, the Isle of Man and the Channel Islands each have different laws relating to wild plants. European law is also binding either directly or through implementing legislation, but does not apply to the Isle of Man as it is not an EU member state.

All plant species

Under the Wildlife & Countryside Act 1981 (WCA) and its equivalent legislation in Northern Ireland and Eire, it is a criminal offence to intentionally **uproot** any wild plant without the landowner's permission.

Protected plant species

Some plant species are individually protected, under European legislation and/or the WCA or its equivalent. Protected plant species are marked in this book with abbreviations for the conservation legislation concerned, 'Sch 8, WO (NI) or FPO (Eire)' or 'EU' for the Habitats and Species Directive. These abbreviations are explained in the '**Abbreviations**' section on pp 8–9. It is a criminal offence for anyone (including the landowner) to intentionally **uproot**, **pick any part** of (arguably this includes collecting seed) or **destroy** these protected species. This list includes bluebell where this plant is collected for trade (including unlicensed collecting of seed). A list of these protected species is available in the Botanical Society of the British Isles' (BSBI) Code of Conduct for Field Botany downloadable at **www.bsbi.org.uk**.

Picking plants

It is not a criminal offence under the WCA to pick other wild plants. However, picking without permission is a civil wrong against the landowner, who could sue you for damages if they could prove that they had suffered a monetary loss as a result. In England & Wales, wild plants are excluded from the criminal laws of theft and criminal damage, with some exceptions, including if the plants are collected for commercial gain. In Scotland, picking wild plants without the landowner's permission is still covered by the criminal law of theft. In addition, sometimes local by-laws covering nature reserves prohibit unauthorised picking of plants.

Picking & identification

The general rule is always **take the book to the plant** in the field, not the plant home to the book. Use a field notebook to note the key features of the plant, make drawings or take photographs and avoid picking anything. In order to study the structure of any plant growing floating or submerged in water, you will need to collect a small piece of it. Return the specimen to the water as soon as possible, where it will usually continue to grow. However, to correctly identify wild plants, you will **sometimes** need to pick a single flower, floret or leaf. Use your common sense – only pick the part of the plant needed for identification and follow the BSBI's Code of Conduct for Field Botany, referred to above.

Photography

Plant photography should be done with great care as crushing plant leaves can kill plants and may do just as much, or even more, harm as picking the flowers. 'Gardening' before taking photographs may unnaturally expose a plant. Take care that nearby plants, including seedlings and slow-growing, inconspicuous mosses and lichens, are not inadvertently crushed. Remember that one or two photographs may not be sufficient for identification as they cannot show all of the important features. If possible, also use the keys and ID Tips boxes in this book to help you focus on the key characters needed for identification and make sure you have a full description of these.

ANCIENT WOODLAND INDICATOR (AWI) PLANTS

The nature conservation value of a wood is often related to the length of time that a site has been wooded. AWIs are one source of evidence used to assess whether a wood is 'ancient'. 'Ancient' woodland is defined as woodland that has been in continuous existence since before 1600 and may be much older. Woodland that has developed since 1600 is known as recent woodland and often has a much less rich ground flora.

The first detailed study of AWI plants was carried out by George Peterken in the 1970s in Lincolnshire. Most research into AWI plants has been concentrated in South and East England, although the species marked as AWI plants in this book are drawn from lists collated by English Nature from different parts of the country. Some species are only AWIs in certain areas, for example, Sweet Violet is considered an AWI in East Anglia but elsewhere it is a commonly introduced plant. Therefore, **where a plant is marked as an AWI, you should check the Table of AWIs on p 558** to find out in which areas of the country this plant may be regarded as an AWI species.

Important **limitations** to using flowering plant species as indicators of ancient woodland include:

- AWI plants differ from area to area so you need a list of species considered to be AWIs in your locality (some county floras have this information).
- There is no perfect AWI species as many also occur in recent woodland or non-woodland habitats.
- Presence of one (or even several) AWI species may have little or no significance. The concept is concerned with probability: the more AWIs present, the more the evidence builds up to support the argument that the wood may be ancient rather than recent woodland.
- The particular character and distribution of AWI plants in a wood must also be taken into account. For example, certain areas within a wood may have a higher number of AWIs due to localised soil types or vegetation communities, even though there is no difference in their history from the surrounding areas. The management history of a site, such as whether it has been thinned, coppiced, felled and replanted, may affect the number of AWIs present or their abundance.
- In trying to determine whether a wood is ancient or not, AWI plants should be used in conjunction with other source material, such as old maps, written records and archaeological evidence.
- In some woods, the bryophytes and lichens may be more valuable as evidence of ancient woodland than vascular plants.

FINDING & RECORDING WILD PLANTS

Why record?

We can't conserve wild plants, or their habitats, unless we know what we have got and where it is growing. This is the fundamental basis of plant conservation. Often we lack recent records for common plants, so don't assume that someone else will have already reported what you have seen.

Basic content of a botanical record

A botanical record must have 4 elements: a 'who, what, where and when'. '**Who**' is your name as recorder and the name of any person who checked the identification; '**what**' is the scientific name of the plant; '**where**' is a location description including a 6-figure grid reference (a location sketch map may be useful); and '**when**' is the day, month and year that the plant was found. Additional information is also useful, such as notes on the abundance of the species at the site, other plants present and the habitat.

Getting your records checked

If you are not reasonably certain of your plant identifications, ask an experienced botanist to check them, as it is crucial that records are correct. You should also ideally get your records checked for any plant that is difficult to identify or unknown in an area. To get started with plant recording, and get help with identification, join a local botany group (see '**Where to find out more**' section).

Where to send records

Send your plant records to all of the following bodies:

- the Botanical Society of the British Isles (BSBI) County Recorder (see '**Where to find out more**' section below). You can also submit records to the BSBI on-line at **www.reticule.co.uk/flora/content/record/recording.htm**;
- if your records are from a nature reserve, to the organisation(s) managing the reserve;
- to the local biological records centre or museum, if there is one for the area (see **www.nbn-nfbr.org.uk/nfbr.php** for a list), which collects data from all biological groups, including plants; and
- to English Nature, Scottish Natural Heritage or Countryside Council for Wales if the site is a Site of Special Scientific Interest (SSSI).

WHERE TO FIND OUT MORE

Botanical Society of the British Isles (BSBI) is the leading national society for everyone interested in British and Irish wild plants. The BSBI runs field meetings and publishes an academic journal, newsletter and other publications. Each county has a BSBI recorder, a volunteer responsible for keeping information on plants within their county. The BSBI also operates a network of referees for the more difficult plant groups. Contact details for all BSBI vice-county recorders and referees are in the BSBI Year Book, which you receive on joining. BSBI c/o Department of Botany, the Natural History Museum, Cromwell Road, London SW7 5BD 0207 9425002 (ans. phone) **www.bsbi.org.uk**

County Wildlife Trusts are the 47 local branches of the UK's leading wildlife conservation charity, which lobby government, carry out educational programmes and own and manage a large network of nature reserves. The Wildlife Trusts, The Kiln, Waterside, Mather Road, Newark, NG24 1WT 0870 0367711 **www.wildlifetrusts.org**

Field Studies Council (FSC) is an educational charity that runs wild plant identification courses for all levels from residential field centres situated around the UK. FSC Head Office, Montford Bridge, Preston Montford, Shrewsbury, Shropshire SY4 1HW 0845 3454071 **www.field-studies-council.org.uk**

Local natural history societies often have botanical groups. Ask at your local library or try an internet search. Examples of the most active societies are: Dublin Naturalist's Field Club **www.dnfc.net** Glasgow Natural History Society **www.gnhs.freeuk.com** London Natural History Society **www.lnhs.org.uk**

Plantlife is the UK's wild plant charity, that lobbies government on plant conservation issues, owns nature reserves and runs the 'Back from the Brink' programme aiming to conserve our rarest wild plants. Plantlife International, 14 Rollestone Street, Salisbury, Wiltshire SP1 1DX 01722 342730 **www.plantlife.org.uk**

Wild Flower Society (WFS) is a national society that promotes knowledge of the British flora by running field meetings, competitions and publishing a magazine. Keeping a WFS diary of all the wild plants you find in a year is an excellent way to learn plant identification. WFS, 82A High Street, Sawston, Cambridge CB2 4HJ 01223 830665 **www.rbge.org.uk/data/wfsoc**

FURTHER READING

*suitable for complete beginners
**improvers
***more advanced texts

Standard reference books & field guides
*****New Flora of the British Isles** by C. A. Stace (2nd edition 1997)
****Interactive Flora of the British Isles** DVD-rom by C. A. Stace and edited by Dr R van der Meijden and Dr I de Kort (2004). A DVD-rom version of the *New Flora*
*****Plant Crib 1998** edited by T. G. C. Rich & A. C. Jermy (1998)
***New Atlas of the British & Irish Flora** by C. D. Preston, D. P. Pearman & T. D. Dines (with CD-rom) (2002)
*****Field Flora of the British Isles** by C. A. Stace (1999)
***A New Key to Wild Flowers** by J. Hayward (1995)

Large-format illustrated books
***Cassell's Wild Flowers of Britain & Northern Europe** by C. Grey-Wilson & M. Blamey (re-issued 2003)
***The Wild Flowers of the British Isles** by I. Garrard & D. Streeter (1983)

Identification guides to plant groups
The Botanical Society of the British Isles publishes a series of **identification handbooks with keys and line illustrations, including the following groups: *Sedges* by A. C. Jermy, A. O. Chater & R. W. David (with a key to vegetative plants); *Umbellifers* by T. G. Tutin; *Docks & Knotweeds* by J. E. Lousley & D. H. Kent; *Willows & Poplars* by R. D. Meikle; *Crucifers* by T .C. G. Rich; *Roses* by G. G. Graham & A. L. Primavesi; and *Pondweeds* by C. D. Preston.
***Colour Identification Guide to the Grasses, Sedges, Rushes & Ferns of the British Isles** by F. Rose (1989) (with a key to vegetative grasses)
****Grasses** by C. E. Hubbard (3rd edition 1984) (with a key to vegetative plants)
***British Water Plants** by S. Haslam, C. Sinker & P. Wolseley (1982) (with a key to vegetative plants)
***The Fern Guide** by J. Merryweather & M. Hill (2nd edition 1995)
****The Ferns of Britain & Ireland** by C.N. Page (2nd edition 1997)

THE GENERAL KEY TO PLANT FAMILIES

MASTER KEY

1 Plants without fls, or submerged aquatic plants with minute, inconspicuous fls.....................
see Vegetative Keys, p 47
Plants with fls present **2**

2 Plants with the individual fls, or florets, **v small** (3 mm wide or less, or if larger with a strap-shaped corolla like a Dandelion floret), grouped together in **v dense** heads, catkins, spikes, or clusters, with the individual stalks to each floret either abs or scarcely visible **3**
Plants with the individual fls of varied size, but each fl possessing a separate and clearly visible (but sometimes v short) fl-stalk of its own..**5**

3 **Submerged** or **floating aquatic** plants with inconspicuous greenish or brownish fls ....**A**
Plants **neither** submerged **nor** floating in water (though they may be in marshy or boggy ground) **4**

4 Individual fls, or florets, massed together into close **heads**, which may be either rounded, button-, disc-, or shaving-brush-shaped – all the florets stalkless and seated on a common **disc** or receptacle, which has a collar or ruff of lf-like bracts around it at its base – each **fl-head** with a definite stalk, and itself often superficially resembling a single fl ....**B**
Individual tiny fls, or florets, massed into dense, simple, unbranched, cylindrical or ovoid **spikes** or catkins, these possessing a central **stalk** through its **length** to which the fls are attached..**C**
Individual tiny fls or florets, grouped into dense, branched, often irregular, globular or rounded **clusters**, but **without** any collar or ruff of lf-like bracts around the base of the fl-cluster (though sometimes with one or two lf-like bracts)..**D**

5 Individual fls **small** (5 mm or less across), grouped into **umbels**, in which the several fl-stalks all radiate from **one** point, like the ribs of an umbrella, on the tip of a stem – the umbels may be simple, or compound (an umbel of umbels) ....**E**
Individual fls of varied size, either solitary on the tip of a stem, or grouped into branched **infls** of various forms and shapes, but **never** with all fl-stalks radiating from **one** point as in an umbel..**6**

6 Herbs without any green chlorophyll in lvs or stem – lvs reduced to **scales** ....**F**
Plants submerged or floating in water with inconspicuous greenish or brownish fls and lvs with green chlorophyll ....**A**
Green-lvd plants, **either** on dry **or** marshy ground, **or** water plants with **conspicuous** yellow or white fls **7**

B   B   B   B   C   C

D   D   E   E

7 Fls **regular**, with the perianth segments or petals spreading out equally from the centre in a star, cup, or bell, and all of equal (or ± equal) length **8**
Fls **irregular**, with obvious 'upper' and 'lower' sides to the fl, usually with an obvious lower lip and sometimes an upper lip as well – perianth segments or petals not all of equal length **16**
Apparent fl like an Arum lily, with one v large 'petal' and a club-shaped 'style', the whole 15–30 cm long, in fact a bract (spathe) enclosing a spike-like infl within its base – lvs triangular to arrow-shaped, stalked, net-veined, 10–20 cm long.......***ARACEAE***, p 495 (*Arum*)

8 Lvs simple, in successive whorls of 4 or more along the stems – fls with 4–5-lobed corolla, **inferior** two-celled ovary, and calyx inconspicuous..***RUBIACEAE***, p 418
Lvs simple, in successive whorls of 3–6 – fls with a single six-lobed perianth and superior ovary ....***LILIACEAE***, p 503 (*Polygonatum*)
Lvs alt, **or** in opp pairs (**or**, rarely, in whorls of 3), **or** all in a basal rosette, or in **one** whorl only some distance up stem...**9**

9 Fls with a perianth comprising a distinct **calyx** and **corolla**, these differing markedly from each other in shape, colour and/or size **10**
Fls with perianth of **one whorl** only, **or** of 2 or more whorls **similar** in shape, colour and size, **or** perianth **abs**...**14**

10 Fls with the petals **joined**, at least their bases often forming a **corolla-tube**, which is usually lobed above **13**
Fls with the petals **separate** from one another to their bases **11**

11 Fls with the ovary **superior** **12**
Fls with the ovary **inferior** ....**G**

12 Carpels and styles both **free** and **separate** from one another, or almost so ....**H**
Carpels or styles, or both, all **joined together** into a single ovary, or ovary apparently of one carpel only ....**I**

8   8   16   16

Rubiaceae

10   14   14   13   13

11   12   G   H   I

13 Fls with the ovary **superior** ....**J**
    Fls with the ovary **inferior** ....**K**

14 Perianth coloured or white, of petal-like segments or lobes ....**L**
    Perianth either green, of sepal-like segments, or chaffy, or membranous, or scale-like, or abs..**15**

15 Trees and **shrubs** ....**M**
    Herbs ....**N**
16 Petals **free** ....**O**
    Petals joined below into a corolla-tube, with several lobes at its mouth ....**P**
    Perianth of one oval pointed lobe, forming a tube below – ovary inferior ....
                                                        ARISTOLOCHIACEAE, p 98 (*Aristolochia*)

J   K   O   P

A. Submerged or floating aquatic plants with inconspicuous, greenish to brownish fls
(in case of difficulty use also keys to aquatic plants not in flower, pp 79–82. If fls
conspicuous, 5 mm or more wide, and coloured or white, return to 7 in Master Key above.)

1 Plants consisting **either** of small, round or oval green discs, 3–10 mm across; **or** of pointed
    elliptical translucent green scales, 5–12 mm long, joined together by stalks, without
    distinction into stems and lvs, with one or more unbranched **roots** hanging from their
    undersides – fls in each case minute, in tiny pockets on edge of plant, hard to see and
    often abs – the whole plant free-floating ....***LEMNACEAE***, p 497 (*Lemna* & *Spirodela*)
    Plants consisting of tiny (0.5–1.0 mm wide) green globular granules, **without** any roots or
    fls – free-floating ....***LEMNACEAE***, p 497 (*Wolffia*)
    Plants with distinct **stems** and **lvs** present **2**

2 Plants composed of **rosettes** or **crowns** of narrow, linear lvs, 5–10 cm long, attached to a **v
    short stout stem**, rooted on bottom of ponds or lakes **3**
    Plants with long **stems**, bearing lvs along their length, floating or submerged **4**

3 Fls solitary – male fls erect and emerging from water on a stalk 2–10 cm long, each with
    4 tiny green sepals, and 4 stamens on stalks 1–2 cm long – female fls ± stalkless, 4–5 mm
    long, each with 3–4 tiny green sepals 4–5 mm long and an ovary with a style 1 cm long –
    in shallow water or on exposed sandy pond edges ....***PLANTAGINACEAE***, p 387 (*Littorella*)
    Fls grouped in a stalked infl, or abs ....see Vegetative Key V section D, p 82

4 Lvs **undivided**, linear to oval, toothed or not...**5**
    Lvs much **divided** into narrow, linear segments, in whorls along the stems **9**

5 Lvs (at least some of them) in **whorls** of 3 or more **6**
    Lvs **never** in whorls, but either in **opp** pairs or **alt** and borne singly along the stems **7**

6 Lvs strap-like, in whorls of **8 or more**, along stout stems (0.5–1.0 cm across) that are
    spongy inside – some stems erect and emergent from water, others submerged and
    trailing below water surface – fls in lf-axils, each fl with one reddish stamen and tiny
    green ovary with one long style only on emergent stems ....***HIPPURIDACEAE*** (p 382)
    Lvs short and scale-like, in whorls of 4–30, on erect ridged rough-pointed stems arising
    from water – sometimes with branches in whorls, sometimes with oval **cones**, 1–2 cm long
    and containing powdery spores on tips of stems ....***EQUISETACEAE***
                                                        (**Horsetails**, not flowering plants)

Lvs linear, translucent, strap-shaped, in **whorls of 3–5,** along slender submerged flaccid stems – female fls on long (to 20 cm) stalks, 5–10 mm across, green-purple or whitish, with 3 sepals, 3 petals and inferior ovary ....
    ***HYDROCHARITACEAE**, p 485 (Elodea and relatives) – see also Vegetative Key IV, p 76*
Lvs ± linear, fine-toothed, along slender submerged flaccid stems, mostly in **whorls of 3** (some only in **opp** pairs) – fls 1–3 together in lf-axils – male fls enclosed each in a little sheath, with minute two-lipped perianth and one stamen – female fls without sheath or perianth, with one carpel bearing 2–4 stigmas – fr ovoid, 3 mm long ....***NAJADACEAE**, p 494 (Najas)*

7 Infls of dense, stalked leafless **spikes** of fls, borne in lf-axils or on ends of stems – spikes 2–8 cm long, their stalks 2–25 cm long, sometimes stout – fls in these spikes in opp pairs, small (to 4 mm wide), each with 4 green to brown perianth segments, 4 stamens and 4 (or fewer) separate carpels that form nutlets 2–5 mm long – lvs either narrow-linear and stalkless, or ± oval and stalked and up to 20 cm long x 7 cm wide, floating or submerged, translucent or opaque, with translucent narrow stipules in lf-axils – in fresh or brackish water ....***POTAMOGETONACEAE**, p 488*
Infl of narrow **flattened spikes enclosed** within basal sheaths of lvs – lvs linear, alt – in sea or estuaries ....***ZOSTERACEAE**, p 494*
Infl of globose clusters – lower lvs long (to 60 cm) ribbon-like, floating, with a keel beneath – infl-stalk emerging from water, bearing usually some erect grass-like lvs and fls in globular clusters alt up the infl-stem, which may be branched or not – **upper** heads of **male** fls only, each with 3 or more stamens, **lower** ones of **female** fls only, each with a one-celled ovary, all fls with a perianth of 3–6 spoon-shaped membranous scales – females fl-heads form **spiky** globes to 3 cm across in fr ....***SPARGANIACEAE**, p 500*
Infls **not** of spikes or globose clusters of fls, but either of v **short racemes,** ± **stalkless clusters,** or of **solitary** fls in lf-axils **8**

8 Lvs linear to oval, in most spp **opaque,** green and partly floating in terminal rosettes, in some all submerged and translucent, **always opp,** ± stalkless 0.6–2.0 cm long – fls ± stalkless in lf-axils in **pairs,** one pair male with one yellow stamen, the other female with a four-lobed **oval** to globular ovary with 2 styles – in fresh water or mud ....
    ***CALLITRICHACEAE**, p 384*
Lvs **linear,** alt **or** opp, **always** translucent and submerged, **3–5 cm** long – infls short and **umbel-like** on **long** stalks (up to 10 cm in fr), in lf-axils – all fls with perianth abs, ± short stamens, and an ovary producing a single short-beaked nut in fr – in brackish water near sea ....***RUPPIACEAE**, p 493*
Lvs linear, nearly all **opp, always** translucent and submerged, 2–3 cm long – infls in lf-axils, each of 2–6 tiny fls in **stalkless or v short** clusters in a tiny cup – male fls of one stamen, female of one carpel with **beak**-like style, forming a nutlet 2–3 mm long with ± wavy edge and a **beak** 0.5–1.5 mm long – in fresh or brackish water ....***ZANNICHELLIACEAE**, p 494*
Lvs linear to spoon-shaped, **opp,** 3–10 mm long x 2 mm wide, thin, untoothed, ± submerged, short-stalked – fls 2–3 mm across, stalked or not, **solitary** in lf-axils, each fl with 3 or 4 sepals, 3 or 4 pinkish-white petals, 6 or 8 stamens, and a tiny 3–4-celled globular superior ovary forming a capsule in fr – in fresh water or mud ....
    ***ELATINACEAE**, p 174*
Lvs linear and opp, rather like the last, but slightly **fleshy, joined** at the base **in pairs round the stem,** 3–20 mm long – fls solitary and short-stalked in lf-axils, 1–2 mm across, with 4 free sepals and 4 larger white petals, 4 stamens and 4 separate carpels – in fresh water ....
    ***CRASSULACEAE**, p 238 (Crassula)*

9 Lvs repeatedly **forked,** 1–3 times – fls **solitary** in lf-axils, with tiny perianth of many narrow lobes – male fls with 10–20 stamens – female with one carpel – fr an ovoid nut with long 4 mm beak – in fresh water...***CERATOPHYLLACEAE**, p 99*
Lvs **pinnately** divided into narrow segments – infl a ± lfy **terminal whorled spike** – fls with 4 tiny sepals, 4 tiny petals, 8 stamens in male fl, four-celled ovary in female fl forming 4 separate nutlets in fr – in fresh water ......***HALORAGACEAE**, p 294 (Myriophyllum)*

B. Plants with individual fls, or florets, massed together into close heads, which may be either rounded, button-, disc-, or shaving-brush shaped – all florets stalkless and seated on a common disc or receptacle, which usually has a collar or ruff of lf-like bracts around it at its base – fl-heads stalked, often superficially resembling a single fl

1 Fls mixed with, or ± replaced by, ovoid bulbils – plant garlic-scented when bruised – lvs narrow-linear ....*LILIACEAE*, p 503 (*Allium*, Key p 513)
Fls neither mixed with, nor replaced by, bulbils – no garlic scent **2**

2 Ovaries of florets **inferior** **3**
Ovaries of florets **superior** **9**

3 Petals 5, **free**, not joined into a corolla-tube – stamens 5, free – lvs lobed, spiny-edged ....
APIACEAE, p 325 (*Eryngium*)
Petals **joined** into a corolla-tube **4**

4 Florets only 5 per head, each 6 mm wide, yellow-green, 4 facing **outwards** like the four faces of a public clock, one facing **upwards** – lvs yellow-green, slightly fleshy, twice-trifoliate ....*ADOXACEAE*, p 426
Florets few to many, all **side by side** in a flat or rounded head **5**

5 Climbing shrubs with **opp** lvs – **no** distinct whorl of bracts at base of head – florets trumpet-like, 4 cm or more long, two-lipped, with 5 long free stamens ....
CAPRIFOLIACEAE, p 424 (*Lonicera*)
Herbs, with distinct **whorl** of bracts at base of head **6**

6 Stamens 5, **joined** into a tube round the style, scarcely or not projecting from corolla-tubes..**7**
Stamens **free**, **projecting** from corolla-tubes on obvious stalks **8**

7 **Calyx abs** (a ruff of bracts is not a calyx because it is not below an *individual* fl), or represented by **hairs** (forming a **pappus** in fr) or by papery scales mixed in between florets – florets **all** tubular; or **all** strap-shaped; **or** tubular in centre and strap-shaped around margin of head (as in a Daisy) ....*ASTERACEAE*, p 433
Calyx of 5 **conspicuous**, narrow **green** teeth – fls sky-blue – lvs alt, strap-shaped.....................
CAMPANULACEAE, p 412 (*Jasione*)

8 Lvs **opp** – stamens 3 – infl-stem repeatedly **forked** – plant to 20 cm – fls v small (2 mm wide), lilac-mauve ....*VALERIANACEAE*, p 428 (*Valerianella*)
Lvs **opp** – stamens 5, long-projecting – corolla-lobes shorter than tube – long bristles between the florets – stem **not** forked – plants erect, usually 30 cm or more tall – calyx cup-like...*DIPSACACEAE*, p 430
Lvs **alt** – stamens 5 – corolla-lobes much **longer** than tube, their tips at first cohering into a tube over stamens – calyx of 5 long narrow green teeth ....
CAMPANULACEAE, p 412 (*Phyteuma*)

9 Lvs linear, all in basal rosette – infl-stem lfless – fl-heads pink or white – florets with **stalks equalling calyces** – petals free almost to base – stamens 5 – ovary superior, styles 5
PLUMBAGINACEAE, p 172 (*Armeria*)
Lvs linear, all in a basal rosette, **deeply submerged in fresh** water – infl-stem leafless – fls in a greyish-white head 0.5–2.0 cm across – r – Scot, Ire ....*ERIOCAULACEAE*, p 499

C. Plants with individual tiny fls massed into dense, unbranched, cylindrical or ovoid spikes or catkins, these possessing a central stalk through their length to which fls are attached

1 Trees or shrubs...**2**
Herbs **12**

2 Male catkins (with stamens) and female catkins (with styles) or frs on **separate** plants......**3**
Male and female catkins separate, but on the **same** plant – male catkins always ± pendulous, female often shorter and erect **6**
Tiny pink fls in catkin-like spikes, but fls have 5 petals, 5 sepals, 5 stamens, and 3 styles **all** together in one fl – fls 3 mm wide – lvs tiny, scale- or needle-like – shrub or tree nearly always by the sea...***TAMARICACEAE***, p 189 (*Tamarix*)

3 Lvs narrow lanceolate, covered in silvery-grey scales – stems thorny – fls tiny in short catkins, male with 2 sepals and 4 stamens, female with 2 sepals and one style – frs orange globular berries 6–8 mm wide – on dunes etc by sea ......***ELAEAGNACEAE***, p 294 (*Hippophae*)
Lvs narrow-linear, evergreen, arranged pinnately along thornless stems – male catkins tiny (3–5 mm long), yellow – female fls solitary, green – fr a naked oval green seed seated in a fleshy, pinkish-red cup, to 1 cm long ....***CONIFERALES***
(*Taxus*, **Yew**; no text entry, see illustration p 50, in Vegetative Key)
Lvs **not** silvery-scaly, **deciduous** – no thorns on stems – frs dry **4**

4 Lvs strongly **resinous-aromatic**, no stipules – male catkins 7–15 mm long, fls each with about 4 red stamens – female catkins 5–10 mm long, oval, green, fls each with 2 red stigmas – fr a dry nutlet ....***MYRICACEAE***, p 123
Lvs not aromatic, with stipules – stamens 2 to many, stigmas 2 – fr a tiny capsule opening into 2 halves to release seeds plumed with long silky hairs.....................**5** (***SALICACEAE***, p 190)

5 Catkin scales **toothed** into narrow segments – catkins long, **pendulous**, always **before** the lvs – fl-bases **cup-like** – stamens 4 to many, red – no nectaries – wind-pollinated ....
Populus, p 194
Catkin scales untoothed – catkins shorter (5 cm or less) erect, before or with lvs – no cuplike base to each fl – stamens 2, rarely 3 or 5, golden nectaries present – insect-pollinated ....
Salix, p 191

6 Lvs narrow-linear, needle-like, mostly evergreen – male fls in oval yellow catkins – female infls form woody cones, which when ripe bear **naked** seeds on their scales.................................
CONIFERALES (not described in text nor illustrated)
Lvs broad-bladed, mostly deciduous – male fls in oval to cylindrical, pendulous yellow catkins – female infls erect catkins or few-fld clusters, sometimes superficially cone-like (*Alnus*) but seeds **always enclosed** inside large (or tiny) nut-like frs **7**

7 Frs large and nutlike (beechnuts, acorns, chestnuts), seated in a **hard woody**, often deeply lobed cup – female fls with 3 (or more) styles each ....***FAGACEAE***, p 123
Frs either large and nutlike (hazelnuts or walnuts) or else tiny, often winged nutlets, seated either in a **lfy** cup (*Corylus*), a papery cup (*Carpinus*) or naked – female fls with 2 styles each **8**

8 Lvs simple **9**
Lvs pinnate – frs large drupe (walnut) ....***JUGLANDACEAE***, p 123 (*Juglans*)

9 Female fls in **erect**, ovoid or cylindrical, **catkins** – frs **tiny flattened**, **winged nutlets**, **2 or 3** on each **scale** of the fruiting catkin **10**
Female fls either in short erect **bud**-like spikes, or in **drooping**, lfy catkins – frs ± **rounded nuts**, each with a **lfy lobed cup** or **bract** surrounding, or to one side of each.........................**11**
10 Catkins appearing **with** the lvs – stamens **2** per male fl, but each two-lobed – female catkin **cylindrical**, its scales three-lobed, **falling** as frs fall..............***BETULACEAE***, p 128 (*Betula*)

Catkins appearing **before** the lvs – stamens **4** per male fl – female catkin **ovoid** and cone-like, its scales five-lobed, **persistent** and **woody** (like a tiny pine cone) after frs have fallen from it ....*BETULACEAE*, p 126 (*Alnus*)

11 Female fls **many** in each 2-cm-long **drooping** catkin, styles **greenish** – ripe fr with a three-lobed papery **bract** 2.5–4.0 cm long, to **one side** of the 5–10-mm long ± flattened oval nut ....*BETULACEAE*, p 126 (*Carpinus*)
Female fls **few** in each 5-mm-long **erect bud-like** catkin, styles **crimson** and protruding from tip of **bud** – ripe fr with a much-lobed lfy **cup**, surrounding base of the globose 1.5–2.0-cm-long woody hazel nut ....*BETULACEAE*, p 126 (*Corylus*)

12 Plants with broad, ribbon-like parallel-veined lvs and stout terminal or lateral fl-spikes....**13**
Plants grass-like, with narrow parallel-veined lvs with sheathing bases, and slender, often branched infls ....**Grasses, Sedges** and **Rushes**
Plants with net-veined lvs, not parallel-sided but ovate, lanceolate or lobed **14**
Plants v fleshy, with no obvious lvs, and with cylindrical jointed branches in opp pairs – fls minute (with 1 stamen only), protruding from pores in branch segments ....
CHENOPODIACEAE, p 132 (*Salicornia*)

13 Lvs ribbon-like, **linear**, to 1 m long, ± erect, **parallel-veined**, **grey**-green – spike or catkin of fls **terminal** on an erect **unbranched** stem 1–2 m tall – **lower** part of spike **brown, sausage-shaped**, velvety, of many tiny female fls – **upper** part of spike golden, conical, furry, of many tiny male fls – growing in water or swamps ....*TYPHACEAE*, p 502
Lvs ribbon-like, **linear**, to 60 cm long, ± erect, parallel-veined, **fresh** green with **wavy edges** – dense spike or catkin of fls to 8 cm long, borne **on the side** of flattened lf-like stem **below** its tip – fls **each** with 6 perianth segments, 6 stamens, and three-celled ovary – plant with odour of dried oranges when bruised – growing in shallow water...........................
ARACEAE, p 495 (*Acorus*)

14 **Climbing** or erect herbs with **deeply palmately lobed lvs** – male and female fls on separate plants – female catkins cone-like, 15–20 mm long in fl, 5 cm long in fr (**hop cones**) – male infls much branched panicles of small (5 mm wide) fls with five-lobed perianth and 5 stamens – female fls with 2 stigmas ....
CANNABACEAE, p 121 (*Humulus* or *Cannabis*)
Lvs not palmately lobed **15**

15 Lvs opp – male and female fls on separate plants – stipules small, green, spreading **16**
Lvs alt – tubular, ± translucent, brown or silvery, membranous stipules (ochrea) surrounding stems above lf-bases...**17**
Lvs **all** in a basal rosette – infl an erect long-stalked leafless spike – calyx and corolla each four-lobed, 2–3 mm wide ....*PLANTAGINACEAE*, p 387

16 Herbs 10–130 cm tall, with coarse-toothed oval to triangular lvs bristly with **stinging hairs** – male and female infls similar, spike-like, fls with 4 sepals, male with 4 stamens, female with one-celled ovary and one style ....*URTICACEAE*, p 121 (*Urtica*)
Herbs 10–40 cm tall, with fine-toothed oval to lanceolate downy or hairless lvs **without** stinging hairs – male catkins stalked, 3–5 cm long, erect in lf-axils, fls with 3 green sepals and 8–15 stamens – female fls either solitary in lf-axils or 1–3 on long stalks, with 3 sepals and a two-celled ovary with 2 styles ....*EUPHORBIACEAE*, p 306 (*Mercurialis*)

17 Infls of dense unbranched cylindrical spikes, **each** with 5 pink or white ± equal sepals, 4–8 stamens and 2–3 styles – fr a hard three-sided or two-faced nutlet, enclosed in the sepals – lvs oval, lanceolate or arrow-shaped – stems smooth ....
POLYGONACEAE, p 162 (*Polygonum* or *Persicaria*)
Infls of **interrupted** ± whorled, branched spikes, fls each with 6 perianth segments, the 3 outer small, narrow, greenish, the 3 inner larger, triangular, enlarging round the fr – fr a hard three-sided nutlet – lvs oval, lanceolate or arrow-shaped – stems with ridges lengthwise ....*POLYGONACEAE*, p 161 (*Rumex*)

D. Plants with individual tiny fls or florets grouped into dense rounded clusters, which are often branched or irregular in shape, but without any collar or whorl of narrow bracts at base of cluster as in key B (though sometimes with one or a pair of lf-like bracts at the base) – ovary always superior

1 Shrubs or trees..**2**
Climbing herbs..**3**
Non-climbing herbs **4**

2 Parasitic shrub on branches of trees with strap-shaped lvs 3–5 cm long – branches forked – fls in tiny yellow clusters – frs white berries...***VISCACEAE***, p 304
Shrub or tree with evergreen, leathery, oval-oblong, untoothed lvs – male fls with 4 sepals and 4 stamens – female fls with three-celled ovary producing a dry capsule ....
BUXACEAE, p 305
Low shrubs of salt-marshes with fleshy, cylindrical lvs – 5 sepals, 5 stamens, 2–3 stigmas....
CHENOPODIACEAE, p 129 (*Suaeda*)
Low shrub of salt-marshes with fleshy, flat, grey-mealy lvs – 5 sepals, 5 stamens (in male fls), stigmas 2 ....***CHENOPODIACEAE***, p 129 (*Atriplex portulacoides*)
Trees with deciduous, toothed, flat lvs, oval to diamond-shaped – lf-blade extending further down stalk on one side than on other – fls in tiny stalkless clusters appearing before lvs – fls with bell-shaped perianth 4–5-lobed and 2 mm long with 4–5 reddish stamens and 2 styles producing an oval winged fr to 2 cm long ....***ULMACEAE***, p 120

3 Lvs palmately three-lobed, bristly – male and female fls on separate plants – female infls like cones (hops), male fls in branched panicles........................***CANNABACEAE***, p 121 (*Humulus*)
Lvs cordate-triangular, hairless but ± mealy below – fls in elongated clusters with both stamens and ovary – frs three-sided nuts enclosed by 3 enlarged sepals 3–4 mm long ....
POLYGONACEAE, p 162 (*Fallopia*)

4 Plants with narrow-linear grass-like, or narrow-cylindrical lvs ..**Grasses, Sedges** and **Rushes**
Plants with broad (1 cm wide or more) linear lvs, alt on stems arising from water....................
SPARGANIACEAE, p 500
Plants with fleshy, jointed stems but apparently without lvs – tiny fls 1 or 3 together in pits in stem-joints – in salt-marshes ....***CHENOPODIACEAE***, p 132 (*Salicornia*)
Plants not grass-like – lvs, if narrow-linear, v short (otherwise lvs broad).................................**5**

5 Lvs palmately lobed, with lobed lfy stipules at their bases – fls in clusters, with 4 green sepals, epicalyx of 4 extra outer sepals, but no petal ....
ROSACEAE, p 262 (*Aphanes* or *Alchemilla*)
Lvs digitate, 3-9 lobed (almost to base) – stipules present – male and female fls on separate plants – male fl with 5 petals and 5 stamens, female fl no petals and sepals reduced to sheath and 2 stigmas ....***CANNABACEAE***, p 121 (*Cannabis*)
Lvs **pinnate**-compound, with toothed **lfy** stipules at their bases – plants 15–40 cm tall – fls in terminal globular heads on long stalks – perianth four-lobed – male fls (lower on head) with many long stamens – female fls with purple feathery stigmas..................................
ROSACEAE, p 264 (*Sanguisorba*)
Lvs not compound, and **without lfy green** stipules...**6**

6 Lvs alt, with tubular, ± translucent, brown or silvery, membranous stipules (ochrea), often fringed, enclosing stem above each lf-base..***POLYGONACEAE***, p 160
Lvs without tubular membranous stipules, though sometimes with small free ones **7**

7 Plants with milky **latex**– lvs untoothed – stipules abs – infl with **forking** branches, ending in fl-like clusters of tiny fls with a pair of lf-like bracts at base of each cluster – each cluster with one stalked female fl with a three-celled ovary, surrounded by several male fls each of one stamen ....***EUPHORBIACEAE***, p 306 (*Euphorbia*)
Plants **without** milky latex, infl-branches not forked **8**

8 Lvs **alt**, at least in upper part of stem, toothed or not **9**
Lvs all **opp**, untoothed **12**

9 Fls with **2–3** styles each, stigmas **short**, **undivided** – fls with **5** tiny sepals and **5** stamens
(sometimes female fls separate, then with 2 triangular sepal-like bracts only)......................**10**
Fls with **1 style** each, stigma **tufted**, sometimes **pinkish**, **feathery** – fls with tiny **four-lobed**
calyx and **4 stamens** – lvs either tiny (< 4mm) almost round, untoothed or oval-
lanceolate, untoothed..**URTICACEAE**, p 122 (*Parietaria* or *Soleirolia*)

10 Small prostrate plant of shingly lake shores (vr) – sepals 5, green or red in centre, with
broad white margins – petals 5, white or red-tipped, equalling petals – lvs strap-shaped,
1–2 cm long x 2–3 mm wide – fls in dense terminal clusters ....
....**CARYOPHYLLACEAE**, key p 140 (*Corrigiola*)
Taller or larger plants, with lvs oblong to triangular, or fleshy and cylindrical **11**

11 Sepals green, normally 5 – lvs and stems hairless, but often also mealy or fleshy ....
....**CHENOPODIACEAE**, p 129
Sepals chaffy, 3 or 5 – lvs and stems ± downy, never mealy or fleshy ..**AMARANTHACEAE**, p 138

12 Calyx six-lobed, reddish – style 1 – small creeping plant of wet muddy or sandy bare
ground with obovate lvs ....**LYTHRACEAE**, p 295 (*Lythrum portula*)
Calyx 4–5-lobed, whitish, green or greyish – styles 2 or more..**13**
(small members of **CARYOPHYLLACEAE**, p 139)

13 Lvs grey-green, awl-shaped, joined in pairs at base – stipules abs – fls in terminal clusters –
sepals 5, grey-green with **white** edges, no petals – tiny prostrate plants of dry, open ground
Scleranthus, p 158
Lvs green, oval – fls in whorled clusters in lf-axils **14**

14 Fls with 5 fleshy hooded white sepals 2 mm long with long pointed tips – in moist places,
r ....*Illecebrum*, p 158
Fls with 5 thin flattish, greenish sepals 2 mm long, with blunt tips – in dry places, r ....
Herniaria, p 158

**E. Plants with individual fls small (5 mm or less across), grouped into umbels,
with individual fl-stalks all radiating from one point on tip of infl-stem like the ribs
of an umbrella – umbels either simple, or compound (umbels of umbels)**

1 Shrubs or woody climbers – ovary inferior – fr berry-like **2**
Herbs **3**

2 Shrub ± erect, with **opp**, oval, untoothed lvs, their side-veins turning up ± parallel to
midrib – 4 sepals, 4 petals, 4 stamens, ovary two-celled, with one style – fls
cream ....**CORNACEAE**, p 302 (*Cornus*)
Woody climber with **alt**, oval or palmately lobed leathery evergreen lvs – 5 sepals, 5 petals,
5 stamens, ovary five-celled, with one style – fls yellow-green ..**ARALIACEAE**, p 324 (*Hedera*)

3 Lvs **parallel-veined**, linear, ± cylindrical, or elliptical – ovary superior – perianth segments
6, all similar and petal-like – stamens 6 – one or two large, often papery, bracts at base of
umbels ....**LILIACEAE**, p 503 (*Allium* and *Gagea*)
Lvs **net-veined, oval, oblong**, or ± **pinnately divided** – petals 5 (or 4) **4**

4 Lvs **all** in a **basal rosette**, elliptical to spoon-shaped – sepals and petals respectively **fused**
to form a calyx-tube and a corolla-tube below – ovary superior – 5 stamens attached to
inside of corolla-tube – 1 style – fls pink or yellow (usually **more** than 5 mm wide).................
PRIMULACEAE, p 231 (*Primula*)
Lvs **not** all in a basal rosette, but at least **some** lvs **up the stem** **5**

5 Lvs **alt** (or, rarely, opp), undivided, ± unstalked – umbels simple, but bearing **forked clusters** of tiny green-yellow fls on tips of umbel-rays – each simple fl-cluster **looking like a little fl**, with one stalked central female fl with superior three-celled ovary and 3 styles, surrounded by several one-stamened male fls – the fl-cluster seated in a **calyx-like** toothed cup which bears fleshy yellowish petal-like bracts (**glands**) alternating with its teeth – **milky latex present** in plants ....***EUPHORBIACEAE***, p 306 (*Euphorbia*)
Lvs **alt**, stalked, often pinnate or twice-pinnate – umbels often compound, often with tiny bracts at base of main umbel and of secondary umbels – fls with 5 tiny sepals (or sepals abs), 5 petals (often notched), 5 stamens, and inferior ovary of 2 joined carpels and 2 styles – fr of 2 dry separating nutlets – no milky latex ....***APIACEAE***, p 325
Lvs **opp**, undivided, ± unstalked, untoothed – no milky latex...
CORNACEAE, p 303 (*Cornus suecica*)

F. Herbs without green colour (chlorophyll) in lvs and stem – lvs reduced to small scales

1 Climbing parasitic plants, with ± reddish leafless thread-like stems attaching themselves to other plants by means of suckers – fls small (3–5 mm long) in globular clusters, with .... five-lobed calyx and five-lobed corolla ....***CUSCUTACEAE***, p 356 (*Cuscuta*)
Erect (not climbing) herbs, with fls in spikes or racemes **2**

2 Ovary **inferior** – perianth segments 6, **free**, with one petal different from others and .... enlarged into a lip (either at bottom or top of fl), and sometimes spurred...................................
ORCHIDACEAE, p 521
Ovary **superior** **3**

3 Fls **regular**, pale yellow or cream – calyx of 5 sepals, corolla of 4 or 5 **separate** and **equal** petals converging into a narrow bell or tube – stamens 8 or 10, free ....
MONOTROPACEAE, p230
Fls **irregular**, two-lipped, of various colours – stamens 4, attached to five-lobed corolla-tube – parasitic on various plants...***OROBANCHACEAE***, p 407

G. Plants with regular fls with ovary inferior (or apparently so), petals free from one another, calyx distinct from corolla (or sometimes calyx abs) – lvs not in whorls of 4 or more along stems

1 Shrubs or trees...**2**
Herbs **5**

2 Lvs **evergreen**, leathery, palmately lobed or oval, **alt – woody climber** – fls each with 5 sepals, 5 petals, 5 stamens – ovary five-celled with 1 style – fls yellow-green, in umbels ....
ARALIACEAE, p 324 (*Hedera*)
Lvs usually **deciduous**, sometimes + evergreen but plants not climbing.....................................**3**

3 Lvs **opp**, oval, **untoothed**, with side-veins curving round and ± parallel to midrib – fls each with 4 sepals, 4 petals, 4 stamens, ovary two-celled with 1 style – fls cream, in umbel-like heads ....***CORNACEAE***, p 302 (*Cornus*)
Lvs **opp, minutely toothed** – fls pendant, 4 pink sepals and 4 purple petals both appearing petal-like, stamens > 8, much longer than petals ....***ONAGRACEAE***, p 302 (*Fuchsia*)
Lvs **alt**, ± toothed...**4**

4 Petals 5, **shorter** than the 5 sepals – stamens 5 – styles 2 – fr a berry, **soft** inside – lvs ± palmately lobed ....***GROSSULARIACEAE***, p 236
Petals 5, as **long** as or much **longer** than the 5 sepals – stamens 10 or v many – styles 1 to many – fr fleshy and coloured, but **firm** (when opened, carpels may sometimes be found to be **free** of fr-wall, but ovary appears to be inferior) ....***ROSACEAE***, p 245

5 Lvs **fleshy, opp**, narrow-linear, triangular in cross-section, 7–10 cm long, joined in pairs at their bases – calyx-teeth lf-like, 5 – petals **numerous**, v narrow – stamens many – fr fleshy – fls 5 cm or more across, yellow or mauve, ± resembling large daisy fl-heads – on sea cliffs ....***AIZOACEAE***, p 128 (*Carpobrotus*)
Lvs **not** fleshy, not triangular in cross-section – petals few, only 2, 3, 4 or 5 per fl **6**

6 Petals either 2 and notched, or 4 (rarely abs) – sepals 2 or 4 – stamens 2, 4 or 8 – ovary four-celled – stigma simple or four-lobed – lvs opp (or alt), unlobed ....***ONAGRACEAE***, p 296
Petals 3 – sepals 3 ....**7**
Petals 5 – sepals 5 (or abs) **10**

7 Plants **floating** in water – sepals green, distinct from the white petals **8**
Plants **not** floating in water (though sometimes growing erect out of water) – sepals never wholly green – sepals and petals either **both** coloured or **both** white, often v alike – lvs linear, grass- or sword-like, untoothed, with veins all parallel...**9**

8 Lvs **kidney-shaped**, **untoothed**, 3 cm wide, floating on water on long stalks (like tiny Water-lily lvs) ....***HYDROCHARITACEAE***, p 485 (*Hydrocharis*)
Lvs **sword-shaped, sharp-toothed**, to 20 cm long, in a floating crown or rosette ....
HYDROCHARITACEAE, p 485 (*Stratiotes*)

9 Stamens 6 – petals and sepals alike – (fl sometimes with a central trumpet as in Daffodil)
LILIACEAE, p 503
Stamens 3 – petals and sepals alike, or of different shape...................................***IRIDACEAE*** p 517

10 Creeping marsh or bog plants with circular, **parasol**-shaped lvs 1–5 cm across, held aloft and horizontal on 2–10 cm stalks attached to their centres – infls 3–5 cm tall, with fls in a head or a whorled spike – 5 sepals, 5 petals, 5 stamens – ovary two-celled, with 2 styles ......
APIACEAE, p 325 (*Hydrocotyle*)
Plants with lvs **not** circular, not parasol-shaped **11**

11 Fls in long **spikes** or **racemes** or **panicles** **12**
Fls in **heads** or **umbels** (simple or compound) – sepals small, or abs – stamens 5 – ovary of 2 completely fused carpels, with 2 styles – fr of 2 dry, separating, one-seeded nutlets – lvs alt, often compound ....***APIACEAE***, p 325

12 Fls in spikes, **yellow** – stamens 10–20 – receptacle-cup below sepals with many **hooked spines** – carpels 2, **free** within oval or bell-shaped receptacle-cup (so technically ovary is not inferior though superficially so) – lvs alt, pinnate, with tiny lfts alternating with larger ones ....***ROSACEAE***, p 262 (*Agrimonia*)
Fls in a spike, **greenish**, later red-tinged, with **petals fringed** by long whisker-like lobes – stamens 10 – lvs palmately lobed ....***SAXIFRAGACEAE***, p 244 (*Tellima*)
Fls in loose panicles, or solitary – stamens 10 – ovary of 2 carpels, **fused** in their lower parts to calyx-tube and to each other, but **free** above, each carpel with a style – fr a **many**-seeded capsule – lvs varied in shape, from wedge-, kidney- or strap-shaped or palmately lobed ....***SAXIFRAGACEAE***, p 240

H. Plants with regular fls with free petals and sepals, and superior ovary of 2 to several carpels that are free and separate one from another (or nearly so) and not joined into an ovary – lvs not in whorls or 4 or more along stems

1 Plants with **v fleshy**, oval, ovoid or cylindrical **undivided** lvs – petals 3, 4, 5 or 6 or more – stamens as many as (or twice as many as) petals – carpels as many as petals, each carpel forming a 2- to many-seeded **follicle** in fr, opening to release seeds...
CRASSULACEAE, p 238
Plants with lvs **not** fleshy **2**

2 Lvs **parallel**-veined – 3 sepals, 3 petals, stamens and carpels 6 to many – plant always in wet places or in water **3**

Lvs with **net-veining**, often lobed or compound – 4–5 sepals, 4–5 petals, stamens and carpels v many...**4**

Lvs with net-veining, rounded-triangular – 3 green sepals, 8–9 yellow petals, stamens and carpels v many, carpels one-seeded achenes....***RANUNCULACEAE***, p 102 (*Ranunculus ficaria*)

3 Fr-head of numerous one-seeded **achenes** – sepals **green**, petals white or pink – lvs with blade and stalk usually **distinct** ....***ALISMATACEAE***, p 482

Fr-head of 6–9 several-seeded **follicles**, each opening to release seeds – sepals and petals **both** pink, but sepals shorter and narrower than petals – fls in an umbel – lvs **linear**, 1 m or more long, triangular in cross-section ....***BUTOMACEAE***, p 482

4 **Epicalyx** of extra green sepals **present** below calyx – fr-head of many **dry** achenes (in Strawberries embedded in a red fleshy mass) ....

ROSACEAE, p 258 (*Potentilla, Fragaria, Geum, Sibbaldia*)

Epicalyx **abs** **5**

5 Fr-head of many separate **fleshy globules** (like tiny berries cohering together, as in a Blackberry)..***ROSACEAE***, p 256 (*Rubus*)

Fr-head of many **dry achenes** (or of **follicles**, each of which opens to release several seeds)......**6**

6 Lfy **stipules** present at base of lf-stalks – receptacle concave ....

ROSACEAE, p 262 (*Filipendula,Dryas*)

Stipules **abs** – receptacle convex ....***RANUNCULACEAE***, p 100

I. Plants with regular fls with calyx and corolla distinct, the petals separate from one another to their bases and not forming a corolla-tube – the carpels (or their styles, or both) either several but all joined together into a single, superior ovary, or else ovary apparently of one superior carpel only

1 Aquatic plants with **large**, **rounded**, ± cordate-based **floating** lvs, and **solitary**, long-stalked, usually floating fls 5–20 cm across – sepals 4–6, green – petals 3 to about 25, yellow or white, the innermost often grading into stamens – stamens many – carpels 8 to many, forming a rounded or flask-shaped capsule 1.5–4.0 cm wide in fr

NYMPHAEACEAE, p 99

Plants **without** large, rounded floating lvs, not truly aquatic **2**

2 Trees, shrubs, and woody climbing plants..**3**

Herbs, sometimes woody at the (usually creeping) base **14**

3 Stamens **many** (**more** than twice number of petals)...**4**

Stamens **equalling**, or **up to** twice number of petals only **7**

4 Calyx-tube forming a globular **cup** round the single carpel (but **not** attached to it) with 5 calyx-teeth – the 5 petals and many stamens attached to top of calyx-cup – fr a cherry, or plum-like, with a 'stone' inside the fleshy part containing a single seed – petals white ....

ROSACEAE, p 249 (*Prunus*)

Calyx of **free** sepals, or sepals ± joined into a tube, but petals attached to base of fl **below** ovary, and not to calyx-tube or to sepals **5**

5 Trees – lvs **alt**, rounded, cordate-based, hairless above, with long-pointed tips and **long** (**2–5 cm**) **stalks** – infl with a conspicuous **oblong bract partly fused** to its stalk (lime wing) – fls with 5 sepals, 5 petals, many stamens, and five-celled ovary with five-lobed stigma, fragrant – fr a small nut ....***TILIACEAE***, p 179

Low shrubs or herbs, 10–40 cm tall – lvs **opp** (or rarely, alt), oblong to oval, short-stalked or stalkless – **no** bract fused to infl-stalk...**6**

6 Stalks of stamens **joined** into **5 bundles** at their bases, free above – styles **3 or 5** – sepals 5, ± **equal** – lvs oval, **stalkless, no stipules** – fls yellow ....***CLUSIACEAE***, p 175 (*Hypericum*)
Stamens fused into a tube, free above, styles many – sepals 5, **epicalyx** of 3 lobes present – lvs **palmately-lobed** ....***MALVACEAE***, p 180 (*Lavatera*)
Stamens **all free** to base – **2 of the sepals far smaller** than the other 3 – lvs narrow-oblong, **short-stalked**, often with stipules ....***CISTACEAE***, p 183 (*Helianthemum*)

7 **Evergreen** shrubs – lvs leathery, spiny, or heath-like...**8**
Deciduous shrubs – lvs broad, oval or lobed **11**
Deciduous shrub or small tree (2–6 m) with **tiny** (to 2 mm) **scale-like**, **overlapping**, grey-or fresh-green lvs – fls in **catkin-like spikes**, **pink** or **white**, each fl 3 mm across, with 5 sepals, 5 petals, 5 stamens and 3 styles – fr a capsule – plant near sea ....
TAMARICACEAE, p 189

8 Low (to 80 cm) much-branched shrub with oval, dark green, spine-tipped apparent lvs 1–4 cm long, that are in reality flattened stems (**cladodes**) and bear on their surfaces 1 or 2 greenish fls 3 mm across in the axils of the true lvs, which are chaffy scales under 5 mm long – fls with 3 sepals, 3 greenish petals and **either** 3 stamens, **or** a three-celled ovary, forming a red round berry 1 cm wide in fr..***LILIACEAE***, p 503 (*Ruscus*)
Shrubs with fls **not** apparently growing from spine-tipped lvs, but on **stalks** from normal stems, singly or in clusters **9**

9 Low (to 40 cm) **heath-like** shrubs with **linear, alt** lvs 4–6 mm long x 2 mm wide, their edges rolled under – fls 1–2 mm across, with 3 sepals, 3 pinkish petals, and **either** 3 stamens, **or** an ovary forming a black berry 5 mm wide in fr (rarely fls with stamens and ovary together) ....***EMPETRACEAE***, p 220
Taller (to 100 cm) Rhododendron-like shrubs with linear to oblong **alt** lvs 1–5 cm long x 0.5–1.0 cm wide, dark green above, rusty-hairy below – fls 1.0–1.5 cm across in umbel-like heads with 5 green sepals, 5 cream petals, 10 stamens and five-celled ovary – fr a capsule – in bogs, r ....***ERICACEAE***, p 220 (*Ledum*)
Lvs holly-like, with spines round the margins **10**

10 Lvs pinnate, lfts like thin holly lvs – fls yellow, 6–8 mm across, in clusters, each fl with usually 5 whorls of perianth segments of 3 each, yellow, not all of same size – frs blue-black, like clusters of tiny grapes...***BERBERIDACEAE***, p 113 (*Mahonia*)
Lvs undivided, oval – fls white, in clusters, with 4 sepals, 4 petals (joined together at extreme base) and either 4 stamens or a four-celled ovary – fr red berry ....
AQUIFOLIACEAE, p 305 (*Ilex*)

11 Lvs opp **12**
Lvs alt **13**

12 Lvs **oval-elliptical**, fine-toothed – fls in clusters – 4 sepals, 4 greenish petals, 4 stamens, four-celled ovary, forming a salmon-coloured four-lobed capsule in fr, with orange seeds inside ....***CELASTRACEAE***, p 304 (*Euonymus*)
Lvs **palmately lobed** – fls in pendulous racemes – sepals and petals 5, stamens 8, ovary two-celled – fr dry, with 2 propeller-blade-like wings ....***ACERACEAE***, p 316

13 Lvs oval, toothed around edges (or not) – fls in loose clusters – sepals 4–5 – petals 4–5 – stamens 4–5 – ovary 2–4-celled – fr an oval dark red or black berry – stems sometimes thorny ....***RHAMNACEAE***, p 312
Lvs obovate – spines present in threes on shoots – fls clustered, perianth segments yellow, in up to 5 whorls of 3, of different sizes – stamens 6 – style 1, v short – fr an oblong red berry, 8–12 mm long ....***BERBERIDACEAE***, p 113 (*Berberis*)

14 Sepals **fused** into a **long calyx-tube**, with toothed tip – lvs opp **15**
Sepals **separate** or **abs** (or fused, either forming a hood or only at base into a v short cup)......**17**

15 Calyx-tube six-toothed – petals 6 (or sometimes fewer), pink or purple – fls in spike-like racemes, or solitary in lf-axils ....***LYTHRACEAE***, p 295
Calyx-tube five-toothed – petals 5, long-stalked, arising from within base of calyx-tube ..**16**

16 Prostrate salt-marsh plant – stems ± **woody** – lvs heath-like, linear, 2–4 mm long, margins rolled under – fls stalkless, 5 mm wide, **solitary** on stems and in forks – petals **pink**, **crinkly** – stamens 6 – style 1, long, with 3 stigmas at tip – fr a capsule ....***FRANKENIACEAE***, p 189
Plants ± erect, not heath-like, **not** woody – petals **flat** – stamens (when present) twice number of petals – styles 2–5, free to base – ovary one-celled – fr a capsule (rarely a berry) with many seeds – infl a dichasium, with a fl in each fork ....***CARYOPHYLLACEAE***, key p 140

17 Lvs oval-elliptical, in a **single whorl** of 4 (3, 5 or 6) at top of stem just below solitary terminal fl – fl with 4 narrow green sepals and 4 similar petals, 8 stamens, four-celled ovary – fr a black berry ....***LILIACEAE***, (*Paris quadrifolia*) p 503
Lvs **opp**, undivided **18**
Lvs **alt** **21**

18 Infl a dichasium – 4–5 petals, 4–5 sepals **19**
Fls solitary – 5 petals, 5 sepals **20**
Fls in a loose raceme – lvs in a basal rosette and also in pairs up stem, stalkless, elliptical-lanceolate, downy – fls pale yellow with a **red spot** on each of the 5 petals – sepals 5, **v unequal** – stamens many – stigma ± stalkless, on globular ovary ***CISTACEAE***, p 184 (*Tuberaria*)

19 Sepals, petals and stamens 5 – ovary five-celled – fr a five-celled capsule – petals white – lvs stalkless ....***LINACEAE***, p 313 (*Linum*)
Sepals, petals and stamens 4 – ovary four-celled, fr a four-celled capsule – tiny bushy plant to 5 cm tall ....***LINACEAE***, p 313 (*Radiola*)
Sepals 2 only, petals 5 – ovary one-celled, capsule 1–3-seeded...............***PORTULACACEAE***, p 138
Sepals, petals 4–5 – stamens usually **twice number** of petals – white or pink – ovary one-celled, capsule many-seeded ....***CARYOPHYLLACEAE***, key p 140
Sepals, petals 5 – stamens many, in 3 or 5 **bundles** – petals yellow – ovary 3- or 5-celled – lvs **stalkless** – fr a capsule..***CLUSIACEAE***, p 174

20 Lvs all in basal rosette, oval, long-stalked, untoothed, hairless, 1–2 cm long, glossy – fl on a long erect stalk, drooping, 15 mm wide, cream – 10 stamens with pores at tips – ovary five-celled, with one style ....***PYROLACEAE***, p 228 (*Moneses*)
Lvs stalkless, close-set on cushion-forming and creeping stems – lvs 2–6 mm long, oval, with chalky white pit on blunt tip of each – fls 12–20 mm across, rose-purple – stamens 10 – carpels 2, free above, fused to each other and to calyx-cup below – in mts
SAXIFRAGACEAE, p 243 (*Saxifraga oppositifolia*)

21 Stamens as many as, or up to twice number of petals **22**
Stamens far more numerous than petals...**28**

22 Sepals 3, narrow – petals 3, 2–8 mm long, oval-triangular, erect, greenish or reddish, sometimes with a large wart on each – stamens 6 – fr a three-angled nut 0.8–4.0 mm long – infl a panicle of erect, whorled, dense spikes or racemes – lvs simple, oval, lanceolate or arrow-shaped – tubular, ± translucent, brown or silvery, membranous stipules (ochrea) round stem ....***POLYGONACEAE***, p 160 (*Rumex*)
Sepals 4 – petals 4, white or coloured (rarely abs, or with 2 petals rather longer than the other 2) – stamens 6 or 4 – ovary of 2 joined carpels – fr a capsule dividing into 2 parts vertically, **or** crosswise, or else a nutlet – no stipules present ....***BRASSICACEAE***, p 196
Sepals 4 – petals abs, but **epicalyx** of 4 bracts below sepals (so sepals could be mistaken for petals) – stamens 1–4 – carpel 1, in a little receptacle-cup – lvs palmately lobed or veined – stipules lfy ....***ROSACEAE***, p 262 (*Alchemilla, Aphanes*)
Sepals and petals 5 of each, rarely more **23**

23 Lvs **trifoliate**, long-stalked, **clover-like** – stamens 10 – ovary five-celled – fr a capsule
without a beak...***OXALIDACEAE***, p 317
Lvs **not** clover-like **24**

24 Lvs **all** in a basal rosette, infl-stalk leafless, erect **25**
Lvs **both** basal **and** alt up stem...**27**
Lvs narrow, linear, stalkless, **all** alt up stem – no basal lf-rosette – infl a dichasium –
stamens 5 – ovary five-celled – fr a globose capsule ....***LINACEAE***, p 313 (*Linum*)

25 Lvs with many **long-stalked**, **sticky**, **red-tipped hairs**, insect-trapping – fls white, 5 mm
wide – stamens same number as petals (5 or 6) – infl erect, forked ......***DROSERACEAE***, p 182
Lvs **without** sticky red-tipped hairs **26**

26 Lvs linear, 2 mm wide – infl a rounded head of pink or white fls, button-shaped, 2–3 cm
across ....***PLUMBAGINACEAE***, p 172 (*Armeria*)
Lvs oval or round, long-stalked, hairless, ± glossy, 2–4 cm long x 2–3 cm wide – infl a
raceme of white or greenish fls, each with 10 stamens (with pores at their tips) and a five-
celled ovary with a single stout style ....***PYROLACEAE***, p 228 (*Pyrola*)

27 Lvs palmately or pinnately lobed or divided – lower lvs long-stalked – stamens 10, 5 of
them sometimes without anthers – ovary five-celled, with a long (1–3 cm) central beak
bearing 5 stigmas at its tip – fr dry ....***GERANIACEAE***, p 318
Lvs varied in shape, from wedge-, kidney- or strap-shaped to deeply palmately lobed into
narrow segments – stamens 10 – carpels 2, joined below, separate above – fr a capsule
dividing into 2 parts ....***SAXIFRAGACEAE***, p 240 (*Saxifraga* or *Tellima*)
Lvs undivided – basal lvs cordate-oval, long-stalked – stem-lf one only, cordate-oval,
stalkless, clasping – fl solitary, erect on stem 10–30 cm tall, 2 cm across, white – 5 normal
stamens alt with 5 larger, sterile palmately-branched ones – ovary with 4 stigmas ....
SAXIFRAGACEAE, (*Parnassia palustris*) p 244

28 Lvs twice-pinnate, with glossy, oval, sharp-toothed lfts – fls in **spikes** – sepals 5 – 5 tiny
cream petals – 1 carpel – fr an ovoid black berry.....................***RANUNCULACEAE***, p 113 (*Actaea*)
Lvs toothed, or pinnate, with narrow segments – sepals 2 (usually falling as fl opens) –
petals 4 – stamens many, free – carpels 2 to many, joined.........................***PAPAVERACEAE***, p 113
Lvs palmately veined or lobed – sepals 5 – petals 5 – stamens all joined into a tube below
– carpels many, joined into a flattened disc-like ovary separating into many one-seeded
nutlets in fruit ....***MALVACEAE***, p 179

**J. Plants with regular fls with calyx and corolla distinct, petals joined at least at
base, sometimes forming a corolla-tube and ovary superior**

1 Petals joined only at extreme base, so at first sight petals appear free.......................................**2**
Petals joined into an obvious corolla-tube, 1/3 or more length of free corolla-lobes **3**

2 Shrub or tree with evergreen, spiny lvs – fls white – petals 4 ....***AQUIFOLIACEAE***, p 305
Herbs with **either** 2 ± fleshy round lvs, joined and encircling stem, and white fls, **or** with a
pair of lvs, rounded but not encircling stem, and pink fls – either variant with sepals 2,
petals 5 ....***PORTULACACEAE***, p 138
Herbs with lvs in a basal rosette only, fls with a funnel-shaped chaffy calyx persisting in fr
PLUMBAGINACEAE, p 172

3 Stamens 8 or 10, twice number of corolla-lobes...**4**
Stamens as many as, or fewer than, corolla-lobes **5**

4 Shrubs or trees with linear, elliptical or oval lvs – carpels 4 or 5, joined together into an
ovary ....***ERICACEAE***, p 220

Succulent herb with circular **fleshy** lvs attached by a central stalk – fls greenish, in a spike – carpels separate from one another – on rocks or walls..***CRASSULACEAE***, p 240 (*Umbilicus*)

5 Ovary of 4 almost separate lobes from between which arises a style, sometimes forked at tip – fr of 4 nutlets – lvs and stems usually bristly, lvs alt ....***BORAGINACEAE***, p 358
Ovary not divided into 4 lobes **6**

6 Stamens 2 – corolla-lobes 4 – lvs opp ....***OLEACEAE***, p 390 (*Ligustrum* or *Syringa*)
Stamens 4 – corolla-lobes 5 – lvs alt, usually with a basal rosette ....
SCROPHULARIACEAE, p 391 (*Erinus*)
Stamens **same** number as corolla-lobes..**7**

7 Stamens **opp** the corolla-lobes..**8**
Stamens **alternating** with the corolla-lobes **9**

8 Style 1 – lvs all in a basal rosette, **or** opp on stem **or** on stem in a single whorl below fl – fr a globose capsule with many seeds – calyx green ....***PRIMULACEAE***, p 231
Styles 5 – fr a one-seeded capsule – lvs all basal – calyx chaffy ....***PLUMBAGINACEAE***, p 172

9 Shrub to 2.5 m tall, with **alt** lanceolate grey-green lvs and ± **spiny stems** – fls 1-2 cm across, rose-purple, five-lobed – stamens 5 ....***SOLANACEAE***, p 352 (*Lycium*)
Shrubs of various sizes without spines – **opp** usually toothed lvs, white-downy below – infl in dense spike-like panicle, fl mauve to purple, four-lobed – stamens 4..***BUDDLEJACEAE***, p 389
Low creeping or cushion-forming **alpine** shrubs, with small narrow leathery opp or alt lvs – calyx and corolla each five-lobed – stamens 5...**10**
Herbs...**11**

10 Creeping shrub with tiny **oval** opp lvs, and pink fls 4–5 mm wide – ovary 2–3-celled – stigma undivided ....***ERICACEAE***, p 224 (*Loiseleuria*)
Cushion-forming shrub with **spoon-shaped** alt lvs in rosettes, and white fls 8–15 mm wide – ovary three-celled – stigma three-lobed..***DIAPENSIACEAE***, p 230

11 Bog or swamp plant with erect, long-stalked **trifoliate** lvs – lfts 3–7 cm long, untoothed – fls in erect racemes – each fl 15 mm wide, with 5 white corolla-lobes fringed with long white hairs ....***MENYANTHACEAE***, p 356 (*Menyanthes*)
Floating aquatic plant with rounded, cordate-based, long-stalked floating lvs 3–10 cm across (like small Water-lily lvs) – fls floating, in few-fld long-stalked clusters – each fl 3 cm wide, with 5 yellow corolla-lobes fringed with long yellow hairs ....
MENYANTHACEAE, p 356 (*Nymphoides*)
Plants of dry (or merely damp) ground, with stalkless or short-stalked mostly simple lvs (sometimes with 2 small lobes at lf-base) **12**

12 Lvs **opp**, always **simple**, without stipules – carpels 2 ....**13**
Lvs **alt** – calyx and corolla each five-lobed **14**
Lvs **all** in a **basal rosette** – fls v small, in **spikes** – calyx and corolla each four-lobed, 2–3 mm wide...***PLANTAGINACEAE***, p 387

13 Plants **creeping** – lvs glossy, short-stalked, oval-elliptical – fls solitary in lf-axils, violet – corolla five-lobed, **twisted** in bud, saucer-shaped when open – carpels **free** below, **joined** above into a common style surrounded by a **ring** above ....***APOCYNACEAE***, p 352
Plants **erect** (rarely with stem creeping **at base only**) – carpels **fused** together below – **no** ring round style – lvs **stalkless** – infl a **dichasium**, or else of a solitary terminal fl – stamens **not** covered each by a fleshy scale – style **slender**, single or forked into two – calyx and corolla four – or five-lobed – seeds tiny, **not** plumed..............***GENTIANACEAE***, p 348

14 Ovary **three**-celled, with **3** stigmas – lvs pinnate ....***POLEMONIACEAE***, p 356
Ovary two-celled, with 1 or 2 stigmas – lvs **not** pinnate, but simple (or with 2 lobes at base only) **15**

15 **Twining** herbs with cordate-triangular lvs – corolla funnel-shaped, scarcely lobed
CONVOLVULACEAE, p 354
Herbs **not** twining their stems around other plants – lvs oval-elliptical – corolla **deeply** five-lobed **16**

16 Fls in terminal spikes or racemes, one corolla-lobe slightly larger than others – corolla-tube short – stamens spreading ....*SCROPHULARIACEAE*, p 391 (*Verbascum*)
Fls solitary, or in curved cymes – **either** anthers converging into a column, **or else** corolla-tube v long ....*SOLANACEAE*, p 352

K. Plants with regular fls with calyx and corolla distinct – petals joined into a corolla-tube – ovary inferior

1 Small (to 10 cm) herb with twice-trifoliate, yellow-green, slightly fleshy lvs, from base and also in a pair on stem – fls yellow-green, to 6 mm wide, in a **close**, **four-angled head of 5 fls only** – 4 of the fls facing **outwards** like the **four faces of a clock tower**, each with 3 calyx-teeth, a five-lobed corolla, and 10 stamens – 1 fl terminal, facing **upwards**, with 2 calyx-teeth, a four-lobed corolla, and 8 stamens – ovary 3–5-celled – fr a green berry – woods in spring...*ADOXACEAE*, p 426
Fls **not** in close four-angled heads of 5 fls only, **but** in (a) many-fld heads, or (b) loose infls, or (c) solitary **2**

2 Stamens 8 or 10 – low shrubs (under 1 m) either erect in drier ground, or prostrate and creeping over *Sphagnum* in bogs – lvs simple, **alt** – corolla 4–5-lobed, either bell-shaped or with corolla-lobes arched back – frs berries ....*ERICACEAE*, p 222 (*Vaccinium*)
Stamens 1 to 5 only **3**

3 Fls in **close heads**, with **many** unstalked florets in each – heads each surrounded below by a collar or ruff (involucre) of **many** bracts – herbs...**4**
Fls **neither** in close heads, **nor** with an involucre of many bracts – herbs or shrubs **7**

4 Anthers of each floret **joined into a tube** round the style – lvs alt **5**
Anthers **free**, **long-stalked** **6**

5 Calyx of 5 conspicuous **long green teeth** – fr a capsule with **several** seeds
CAMPANULACEAE, p 412 (*Jasione*)
Calyx **abs** (a ruff of bracts (involucre) is not a calyx because it is not below an *individual* fl) or represented by a **pappus** of hairs or papery scales mixed in between florets – frs tiny **nutlets** ....*ASTERACEAE*, p 433

6 Corolla-lobes **long**, **narrow**, **joined** at tips in bud – calyx of 5 **long teeth** – lvs alt – fr a **capsule** with **several** seeds ....*CAMPANULACEAE*, p 412 (*Phyteuma*)
Corolla-lobes **shorter** than tube – calyx cup-like – lvs **opp** – frs tiny **nutlets** *DIPSACACEAE*, p 430

7 Lvs simple, in successive **whorls** of 4 or more up stems – corolla 4–5-lobed, wheel- or funnel-shaped..*RUBIACEAE*, p 418
Lvs **not** in whorls up stems – corolla 4–5-lobed **8**

8 Herbs climbing by means of **coiled tendrils** – lvs palmately-lobed – frs red berries – plants dioecious ....*CUCURBITACEAE*, p 190
Tendrils **abs** **9**

9 Lvs **opp** – herbs or shrubs **10**
Lvs **alt** – all herbs **11**

10 **Stamens** 4 or 5 – calyx of 5 **teeth** – **either** erect, creeping or twining shrubs **without** stipules to lvs, **or** herbs **with** stipules and pinnate lvs ....*CAPRIFOLIACEAE*, p 424

Stamens 1–3 – calyx **either** a ± **untoothed** tiny ring **or else** forming a **hairy pappus** – herbs **without** stipules to lvs ....***VALERIANACEAE***, p 428

11 Stamens 5, **opp** corolla-lobes – stigma **1**, pinhead-shaped – fls small, white, cup-like, 3 mm wide x 3 mm long ....***PRIMULACEAE***, p 232 (*Samolus*)
Stamens 5, **alternating** with corolla-lobes – stigmas 2–5 – fls blue, purple or white, bell- or wheel-shaped, from 8 mm wide and from 6 mm long..........................***CAMPANULACEAE***, p 412

L. Plants with regular fls with perianth (a) of one whorl only, or (b) of 2 or more whorls similar in shape, colour and size, coloured or white, of petal-like segments or lobes

1 Ovary inferior **2**
Ovary superior **9**

2 Perianth tubular below **3**
Perianth segments 6, free, ascending **8**
Perianth segments 5, free, spreading – fls in simple or compound umbels – lvs often pinnate, alt – stamens 5, styles 2 – fr of 2 segments, separating when ripe..***APIACEAE***, p 325

3 Perianth unlobed, forming a single erect oval dull yellow flap, pointed at tip, 1–2 cm long – perianth-tube with a swelling at base above ovary – stamens 6, styles 6 – erect herbs with oval cordate, blunt, alt, stalked lvs – fls 4–8 in lf-axils ....
....***ARISTOLOCHIACEAE***, p 98 (*Aristolochia*)
Perianth-tube symmetrically lobed into 3–6 segments or teeth in upper part **4**

4 Fls with 6 ± equal petal-like perianth-lobes – ovary three-celled **5**
Fls with 3 v short triangular perianth segments – fls bell-shaped, solitary, brownish, downy, 15 mm long – plant creeping, with long-stalked, kidney-shaped, thick, glossy dark green lvs – r ....***ARISTOLOCHIACEAE***, p 98 (*Asarum*)
Fls with 4 or 5 petal-like perianth-lobes **7**

5 Stamens 3 ....***IRIDACEAE***, p 517
Stamens 6 ....**6**

6 **Twining, climbing** herb with glossy, **cordate, stalked** lvs and small (4–5 mm wide) fls in racemes in lf-axils – plants dioecious – fr a red berry ....***DIOSCOREACEAE***, p 520 (*Tamus*)
Non-climbing erect herbs with linear, parallel-veined lvs – fls solitary or in loose umbels, all with stamens and ovary ....***LILIACEAE***, (*Narcissus*, *Leucojum* or *Galanthus*) p 503

7 Lvs simple, in **whorls** of 4 or more up the stem – stamens 4–5, ovary two-celled ....
....***RUBIACEAE***, p 418
Lvs in **opp** pairs, simple or pinnate – stamens 1–3 ....***VALERIANACEAE***, p 428
Lvs **alt**, strap-shaped to lanceolate, or linear, slightly fleshy, yellow-green, 1–2 cm long – plant prostrate – fls star-like, 3 mm across, perianth lobes 4 or 5, white, star-like – stamens 5 ....***SANTALACEAE***, p 303

8 Stamens 3 ....***IRIDACEAE***, p 517
Stamens 6 ....***LILIACEAE***, (*Narcissus*, *Leucojum* or *Galanthus*), p 503

9 Stamens v many in each fl – carpels many, in a head, all free from one another **10**
Stamens equal in number to, or up to twice as many as, perianth segments **11**

10 Woody climber with opp trifoliate or pinnate lvs – perianth segments 4, cream-coloured, free ....***RANUNCULACEAE***, p 110 (*Clematis*)
Herbs with alt, simple, or ± trifoliate or pinnate lvs – perianth segments 5 or more, white or coloured, free – sepal-like bracts sometimes present below fl, but separated from it by a shorter or longer length of fl-stalk.......................................***RANUNCULACEAE***, p 100

11 Shrubs with red or black berries...**12**
Herbs or climbing plants **13**

12 Heath-like shrubs to 30 cm tall with narrow, alt lvs 4–6 mm long – fls with 6 free pinkish perianth segments, each fl 1–2 mm across, in lf-axils – fr a black berry 5 mm..................................
EMPETRACEAE, p 220
Shrubs to 2 m tall, with alt, obovate, or pinnate lvs – **either** stems (or lf-margins) with **spines** – perianth 6–8 mm across, yellow, of **several** whorls, each of 3 **free** petal-like segments – stamens 6 ....***BERBERIDACEAE***, p 113
Shrubs to 1 m tall, with alt elliptical to obovate lvs 5 to 10 cm long – **no** spines on stem – perianth **tubular**, of one whorl, four-lobed at tip, pink or greenish-yellow, 8–12 mm long – stamens 8 ....***THYMELAEACEAE***, p 296

13 Perianth segments 6, petal-like, **either** forming a tube below, **or** free to base, when usually in 2 similar whorls of 3 each – stamens 6 – lvs **parallel-veined**, **linear** or **elliptical**, either all from base of plant or alt up the stem as well **14**
Perianth segments 4 – fls in oblong terminal heads or spikes – stamens 4 – lvs ± stalked, either pinnate or cordate-oval..**15**
Perianth segments 5, joined into a tube at base, pink or white...................................**16**

14 Carpels **free from one another** at least above, follicles pink-red – lvs linear, to 1 m, keeled – fls pink, in umbels – erect plant growing in water.........................***BUTOMACEAE***, p 482
Carpels joined into a three-celled ovary..***LILIACEAE***, p 503

15 Lvs **pinnate**, with oval toothed lfts – plant to 1 m tall – fls **red**, in oblong heads 1–2 cm long x 1 cm wide – perianth four-lobed – fr a one-seeded nutlet ***ROSACEAE***, p 264 (*Sanguisorba*)
Lvs cordate-oval, 1 long-stalked lf from base, stem-lf often stalkless – plant 10–15 cm tall – fls white, in oblong racemes to 5 cm long – 4 free perianth segments – fr a two-celled red berry ....***LILIACEAE***, p 503 (*Maianthemum*)

16 Creeping plant of salt-marshes – lvs **opp**, strap-shaped to obovate, no stipules – fls in lf-axils, pink or white, 5 mm across – stamens 5 – fr a **capsule** ....***PRIMULACEAE***, p 236 (*Glaux*)
Erect or procumbent or climbing plants – lvs **alt**, linear to elliptical-lanceolate, with tubular, ± translucent, brown or silvery, membranous stipules (ochrea) sheathing stem above each lf – fls pink, in lf-axils or in spikes, whitish or greenish – stamens 5–8 – fr a three-angled or lens-shaped brown **nutlet** ....***POLYGONACEAE***, p 162 (*Polygonum, Persicaria, Fagopyrum* or *Fallopia*)

M. Trees or shrubs with fls not two-lipped, with perianth (a) of one whorl of sepal-like green segments, or (b) of chaffy, membranous or scale-like segments, or (c) perianth abs

1 Fls in ovoid or cylindrical **catkins** ....see Key C, p 29
Fls not in catkins, but in small clusters, panicles, or solitary in lf axils ....**2** (or see Key D, p 31)

2 Small bushy shrub, parasitic on branches of trees, with forked green branches and opp, untoothed, pale green, evergreen oblong or obovate lvs – fls in tiny clusters – berries white ....***VISCACEAE***, p 304
Trees or shrubs **not** parasitic on branches of trees **3**

3 Erect much-branched evergreen shrub to 1 m tall, with dark green, oval, spine-tipped lf-like blades 1–4 cm long (**cladodes**), some of which bear 1 or 2 tiny (3–5 mm wide) greenish six-petalled fls **on their surfaces**, in the axil of a tiny chaffy bract – stamens 6 – fr a red berry 1 cm wide ....***LILIACEAE***, p 503 (*Ruscus*)

4 Fls **not** borne on the surfaces of lf-like blades...**4**
Lvs alternate **5**
Lvs opposite **6**

5 Shrub 1–2 m tall, with strap-shaped, silvery-grey deciduous lvs and orange berries – mostly on sand dunes ....***ELAEAGNACEAE***, p 294
Heath-like shrub to 30 cm tall, with linear-oblong, blunt evergreen lvs 4–6 mm long x 2 mm wide with rolled-under edges – fls 1–2 mm across, pinkish or purplish, with 6 free perianth segments – male fls with 3 stamens, female with a green superior ovary – fr a black berry 5 mm wide..***EMPETRACEAE***, p 220
Salt-marsh shrub to 1 m tall, with fleshy, **cylindrical**, grey-green lvs and tiny green fls, 2–3 together in lf-axils – each fl 2 mm wide with 5 sepals and 5 stamens....................................
....***CHENOPODIACEAE***, p 134 (*Suaeda*)
Little-branched shrub, to 1 m tall, with obovate or elliptical evergreen glossy lvs – fls borne in short racemes in lf-axils, green, 8–12 mm long – long perianth-tube, with 4 spreading teeth at mouth, 8 stamens and an ovary – fr an ovoid black berry 12 mm ....
....***THYMELAEACEAE***, p 296

6 Shrub or small tree 1–5 m tall, with oval, glossy, untoothed **evergreen** simple lvs 1.0–2.5 cm long – clusters of tiny 2 mm-wide yellow-green fls in lf-axils – perianth segments 4, male fls with 4 stamens, female with three-celled ovary – fr a three-horned capsule 8 mm long...***BUXACEAE***, p 305
Tree with **deciduous pinnate** lvs, black buds, and tiny purplish fls in branched clusters, each with 2 stamens, no perianth, and a two-celled ovary – frs oblong, winged, 3 cm long..
....***OLEACEAE***, p 390 (*Fraxinus*)
Salt-marsh shrub to 50 cm tall, with fleshy, **flat** elliptical **grey-mealy** lvs, and tiny yellow fls in spikes – male fls with 5 sepals and 5 stamens, female with 2 broad sepal-like bracts and a tiny ovary ....***CHENOPODIACEAE***, p 136 (*Atriplex portulacoides*)

N. Herbs with fls not two-lipped, but with (a) perianth of green, sepal-like segments, or (b) of chaffy, membranous or scale-like segments, or (c) perianth abs

1 Plants with individual fls v small (3 mm wide or less), and in either dense, catkin-like spikes or dense irregular clusters ....return to stage 4 in Master Key
Plants with grass-like lvs (lvs linear, non-fleshy, long and narrow, flat, folded or bristle-like) with blade and sheathing base – florets, perianth abs, in axils of scales ....
....**Sedges** and **Grasses**
Plants **not** as either of above...**2**

2 Lvs **long** and **narrow**, **linear**, **cylindrical** (or **half-cylindrical**) – fls under 6 mm wide..............**3**
Lvs **neither** long and narrow, **nor** cylindrical (nor half so)...**5**

3 Lvs ± fleshy – infl **an erect spike** or **raceme** – each fl with 6 tiny (to 3 mm long) **green** perianth segments...**4**
Lvs **not** fleshy, but **firm**, and **either** spongy, **or** hollow with cross-partitions, within – infl a branched panicle or cluster of fls, either terminal or from the **side** of a stem – each fl with 6 tiny brown or purplish perianth segments, 6 or 3 stamens, and 3 stigmas ...***JUNCACEAE***, p 486

4 Lvs each with a **pore** at the tip, alt along stem – infl loose, 3–10-fld, 5 cm long, overtopped by lvs – fls 5 mm wide with 3 ovoid spreading carpels, only joined at the base – in bog pools, vr ....***SCHEUCHZERIACEAE***, p 486 (*Scheuchzeria*)
Lvs with **no** pore at tip, mostly from base – infl dense, many-fld, 10–20 cm long, longer than lvs – fls 3 mm wide, with 6 carpels all **fused** together – in salt and fresh water marshes..***JUNCAGINACEAE***, p 486 (*Triglochin*)

5 Stamens **numerous** and **conspicuous** in each fl, **many more** than the number of perianth segments – carpels **free** from each other, either **achenes** or **follicles** – lvs without stipules, compound, pinnate or palmate......................***RANUNCULACEAE***, p 100 (*Helleborus, Thalictrum*)
Stamens **not** numerous in each fl, at most twice number of perianth segments **6**

6 Fls **solitary, terminal**, 6–8 mm wide, erect on stalks of 5–10cm – petals and sepals 5 each, greenish – stamens 5 to 10 – fr of many free nutlets in a long (to 7 cm) **cylindrical spike** rather like a Plantain infl – lvs linear, to 5 cm, all basal ....***RANUNCULACEAE***, p 111 (*Myosurus*)
Fls **grouped** in umbels, in spikes, or in axils of stem-lvs...**7**

7 Fls in **loose, umbel-like heads**, with spreading, sepal-like green bracts round edges of infl or fl-clusters – lvs simple **8**
Fls in tall, often branched **spikes** or **racemes** – lvs simple, mostly alt **9**
Fls in erect racemes – lvs pinnate, alt, with pointed clasping bases – sepals 4, stamens 6 ....
BRASSICACEAE, p 210 (*Cardamine impatiens*)
Fls in loose or dense **panicle-like cymes** – lvs **palmately-veined** or lobed with conspicuous lfy stipules at base of each lf-stalk – perianth of 4 triangular green sepals and epicalyx of 4 similar green bracts outside the sepals, all attached to a calyx-cup – stamens 4, carpel 1
ROSACEAE, p 262 (*Alchemilla*)
Fls in loose **forked** cymes (dichasia) – lvs **simple**, **opp**, **untoothed**, no stipules –sepals 4–5 – **no** epicalyx – fr a one-celled capsule ....***CARYOPHYLLACEAE***, pp 148, 160 (*Stellaria, Sagina*)
Fls in **small** close clusters, in **axils** of **alt**, **simple** lvs ....***POLYGONACEAE***, p 170 (*Koenigia*)
Fls **solitary**, or in **small** close clusters, in **axils** of **opp**, **simple** lvs along **prostrate**, rooting stems – in damp places **10**

8 Plants with **milky** latex – infl a true umbel with several branches radiating from **one point** – fl-clusters, with opp bracts beneath each, fl-like, no perianths, but with several tiny yellow stamens surrounding a three-celled ovary in each fl-cluster – lvs alt (v rarely opp) ....***EUPHORBIACEAE***, p 306 (*Euphorbia*)
Plants **without latex** – infl a repeatedly forked cyme with a fl in each fork – fls with 4 green sepals, 8 stamens, 2 styles – lvs opp or alt ....***SAXIFRAGACEAE***, p 244 (*Chrysosplenium*)

9 Fls in ± whorled spikes or racemes – tubular, ± translucent, brown or silvery, membranous stipules (ochrea) surrounding stem above each lf – lvs not fleshy – each fl with perianth of 3 tiny spreading outer segments, 3 larger (2–8 mm long) erect, ± triangular inner segments, often with a wart on outside, surrounding fr of a three-sided nutlet ....
POLYGONACEAE, p 161 (*Rumex*)
Fls in ± irregular spikes, racemes or axillary clusters – **no** stipules – lvs often fleshy – perianth of 5 small greenish ± fleshy segments.....................................***CHENOPODIACEAE***, p 129

10 Ovary **inferior** – perianth 3 mm wide, of 4 spreading triangular teeth – stamens 8 – lvs **oval**, **pointed**, 2–5 cm long – whole plant ± red-flushed – vr................***ONAGRACEAE***, p 302 (*Ludwigia*)
Ovary **superior**, in perianth-cup – perianth 2 mm wide, with 6 erect short pointed teeth – stamens 6 or 12 – lvs **obovate**, green, 1–2 cm long – stem and fls ± reddish ....
LYTHRACEAE, p 295 (*Lythrum portula*)

O. Plants with irregular fls, with petals ± free from one another, not forming a corolla tube

1 **Large trees** – fls showy, in **erect** spike-like panicles – lvs op, palmate into 5–7 obovate lfts 10–20 cm long – sepals joined into a five-toothed calyx-tube – petals 4, white or pink, spotted – fr a large prickly green globe (to 6 cm) with shiny nut-like seeds inside ....
HIPPOCASTANACEAE, p 315
Herbs, shrubs, or trees – if trees, infls **pendulous** racemes..**2**

2 Ovary inferior – lvs **parallel-veined** – perianth segments 6, in 2 whorls, the 3 outer ± alike, 2 of inner ones similar to each other, the third forming a lip, often lobed or spurred, either at top or at bottom of fl – anther 1 (or 2), stalkless – fls in spikes or racemes with pollinia ..
ORCHIDACEAE, p 521
Ovary **superior** – lvs **not** parallel-veined or if apparently parallel-veined, without pollinia ..**3**

3 Fls shaped like those of **Peas** (see illustration at step 16 of Master Key, p 26), with a five-lobed calyx – 5 petals, one ± ascending at **rear** of fl (the **standard**), 2 at sides (the **wings**), and 2 ± joined together below in a boat-shape (the **keel**) – stamens 10, ± joined into a tube round the single carpel, which usually ripens into a fr like a little pea-pod with several seeds (but may be jointed, one-seeded, or not opening).....................***FABACEAE***, p 266

Fls **not** pea-flower like **4**

4 Fls with either a **spur**, or a swollen **sac**, below each one, formed from a petal or a sepal **5**

Fls **without** a spur or sac below them **7**

5 Lvs much divided ± pinnately into narrow lfts – fls elongated and narrow with ± parallel petals, **flattened** on left and right of centre line in a vertical plane – sepals 2, small, coloured, lateral – petals 4, long and narrow, the uppermost one with a spur, or a blunt sac, at its rear end – stamens 2, each three-lobed ....***FUMARIACEAE***, p 116

Lvs simple – fls not flattened to left and right, but with **spreading** petals – stamens 5, in a close ring round the style..**6**

6 Plants with erect lfy translucent stems – lvs alt or opp, without stipules – infls short racemes or clusters in axils of upper, oval-oblong lvs – fls with 3 **petal-like** sepals (the lowest one with a curved pointed spur, the other 2 much smaller) and 5 petals (the upper largest, the lower 4 joined in pairs on either side, so total number apparently only 3) – frs oblong capsules that explode when ripe to release seeds.........................***BALSAMINACEAE***, p 323

Plants with decumbent, or v short erect, opaque, stems – lvs alt, long-stalked, cordate-triangular to elliptical, with a pair of long narrow ± toothed stipules at base of each – fls solitary, long-stalked, in lf-axils – sepals 5, short, green, ± equal, each with a **backward-pointing** as well as a forward-pointing lobe – petals 5, spreading outwards, lowest petal with a back-pointing **spur** – fr an ovoid three-sided capsule with a single style ....***VIOLACEAE***, p 184

7 Small low herbs, sometimes ± woody at base, with narrow untoothed alt or opp **lanceolate** or **obovate** lvs and no stipules – fls in racemes – 3 tiny sepals, plus 2 **large** 3–7 mm oval white or coloured lateral **petal-like sepals** – **true** petals 3, tiny, fringed, joined to tube of 8 stamens ± hidden inside fl – fr a flat oval, two-celled capsule ....

POLYGALACEAE, p 314

Erect herbs with lfy stems terminating in long erect racemes of fls **8**

8 Lvs **deeply palmately** lobed – fls with 5 petal-like blue-purple sepals, the rear sepal forming a **hood** 18–20 mm high over rest of fl – petals small and hidden in hood – stamens many – carpels 3–5, free, forming follicles in fr...***RANUNCULACEAE***, p 110 (*Aconitum*)

Lvs **undivided**, or **pinnately** lobed – no hood at rear of fl – fls ± flat, under 10 mm across – fr a capsule **9**

9 Sepals 4, equal – petals 4, undivided and similar in shape, but the 2 lower **longer** than the 2 upper – stamens 6 – fr a two-celled capsule.........................***BRASSICACEAE***, p 216 (*Iberis, Teesdalia*)

Sepals 4 or 6 – petals 4 or 6, spreading, repeatedly **lobed** or **divided** into narrow segments – stamens **free**, **many**, crowded to lower side of fl – ovary of one cell but with 3 or 4 stigma-bearing **lobes** at tip, **open** between the lobes even before fr is ripe ....***RESEDACEAE***, p 220

P. Plants with irregular fls, with the petals joined into a corolla-tube below, either two-lipped, one-lipped, or with several unequal corolla-lobes

1 Ovary inferior **2**

Ovary superior **3**

2 Lvs long, linear to sword-shaped, ± erect, **parallel-veined** – fls in long spikes – perianth segments 6, unequal, the 3 upper ones each broader than the 3 lower ones which form a lower lip – all joined at base into a short tube – stamens 3 ....***IRIDACEAE***, p 517 (*Gladiolus*)

Lvs **either** elliptical and net-veined, alt up stem (on dry ground) **or** all in a basal rosette and strap-shaped (in shallow water) – fls in a loose raceme, sepals 5, corolla-tube five-lobed, the 3 lower lobes forming a lower lip, the 2 upper an upper lip – stamens 5 .... ***CAMPANULACEAE***, (*Lobelia*) p 412

Erect or climbing shrubs with oval to elliptical opp lvs – corolla trumpet-shaped, two-lipped, five-lobed ....***CAPRIFOLIACEAE***, p 424 (*Lonicera*)

3 Stamens numerous >10 – petals and sepals looking ± identical but with upper sepal forming long spur ....***RANUNCULACEAE***, p 110 (*Consolida*)

Stamens ffi 10 ....**4**

4 Lvs all in a **basal rosette**, oval to oblong, simple, covered with **sticky**, **insect-trapping** glands – fls **solitary**, on long leafless erect stalks, each with 5 sepals and five-lobed and two-lipped corolla bearing a spur beneath the lower lip – in wet places ....
LENTIBULARIACEAE, p 410 (*Pinguicula*)

Lvs floating or submerged in water, much divided into narrow linear segments, some of which bear tiny (1–4 mm long) insect-trapping **bladders** – fls in an erect, few-fld, leafless **raceme** emerging from water, with two-lipped calyx and five-lobed, two-lipped corolla with a spur below lower lip, yellow like small Snapdragon fls ....
LENTIBULARIACEAE, p 410 (*Utricularia*)

Lvs alt or opp on **erect**, or sometimes creeping, **lfy stems** – corollas 4–5-lobed **5**

5 Ovary of 4 small, **separate**, **rounded** lobes, with a long style, forked at tip, arising from **between** the 4 lobes – fr of 4, one-seeded nutlets (open corolla-tube to see this) **6**

Ovary globular, ovoid, conical, or lens-shaped and flattened on each side, with a long style, often forked at tip, arising from the **top of the ovary**..**7**

6 Lvs in **opp** pairs – fls in densely branched clusters or whorls in the lf-axils – fls usually strongly two-lipped, with 4 or 2 stamens – stems four-angled – plants ± hairy (but not usually with stiff bristles all over stem and lvs)...***LAMIACEAE***, p 366

Lvs **alt** – infls one-sided clusters (cymes) **drooping** from lf-axils – fls with lowest of the 5 corolla lobes rather larger than the others, forming an indistinct lip – stamens 5, of **unequal** lengths – plants with ± **stiff bristles** on **round** stems and lvs..
BORAGINACEAE, p 358 (*Echium*)

7 Lvs **always** opp – ovary four-celled – fr of 4 nutlets – fls in long spikes in lf-axils, their bracts v short (half length of calyx only)..***VERBENACEAE***, p 366

Lvs alt, or sometimes opp – ovary two-celled – fr a globular or ovoid **capsule**, producing **many** seeds – infl varied, but if fls in spikes, bracts **longer** than calyx ....
SCROPHULARIACEAE, p 391

THE VEGETATIVE KEYS TO PLANTS NOT IN FLOWER

Introduction
These keys are designed to enable you to identify common plants without flowers. There are still no other vegetative keys available for the majority of British flowering plants, over twenty years since the first publication of this book. Vegetative keys are available for trees, shrubs, sedges, grasses and water plants, and simple vegetative keys to the very common species of a limited number of habitats are available from the Field Studies Council. *Plant Crib* also has some vegetative keys. The relevant texts are listed in the **Further Reading** section on p 23.

These vegetative keys are very useful for ecology students and professional ecologists, who often need to identify non-flowering plants. However, the best way to learn vegetative plant identification is to botanise with an experienced botanist, as non-flowering plant parts can be extremely variable. There are no short cuts to vegetative identification; you must simply practise regularly.

The plants are grouped by **habitat** rather than by families because it is in their habitats that they are naturally grouped in the countryside. Some of the keys, particularly those to **woodlands** (**I**), **chalk grasslands** (**II**) and **heaths** (**III**), will be useful even in winter, as many plants of these habitats remain visible above ground at that time. Other keys will be more useful from spring to autumn, at times when, although some species are in flower, many others are only in leaf. Because some aquatic plants are shy flowering, the flowers of many others inconspicuous and their structure difficult to interpret, the key to **aquatic plants** (**IV**) should be useful throughout summer and early autumn (many water plants are not visible in winter and spring).

Limitations
It proved impossible to include in these keys all the species described in the main, illustrated text (pp 98 – 546) of this book. Exclusions comprise some annuals that normally have flowers present nearly all the time they are above ground and may not be visible at all for the rest of the year. Critical (ie difficult), rare, or garden plants have also been omitted. In some cases firm identification beyond genus to distinct species cannot be achieved without flowers; in this case, the keys take you as far as possible, with an indication of further likely separation. However, many plants of grazed grassland, where animals have eaten all the flowers, should be identifiable with these keys. More detailed tips on the vegetative differences between plants in flower that look much alike are given in the main text, on the pages to which the key will have directed you.

How to use the vegetative keys

1 First identify the habitat – if necessary read the descriptive introduction at the beginning of the key.

2 Always use a hand lens for studying fine details.

3 It is important to find the best mature leaves, shoots or fruits of each plant studied – immature, tattered or broken leaves may be misleading.

4 Work carefully through the master key to find the sub-key that clearly describes the salient features of the plant concerned.

5 Work through the sub-key until a reasonable correspondence of characters denotes a particular species. Some species 'key out' in more than one place; this has been done to allow for plant variation, or for possible different interpretations of the features.

6 In many places in the keys there are small drawings of the plant concerned, or of a particular feature of it; these should help in identification. But if there is no illustration in the key, look at the description and illustration in the main text, using the page references usually given. The absence of a page reference means that the species concerned is not mentioned elsewhere in this book; in these cases, the English plant names are given as well as the scientific, if the former exist.

7 If nothing in a key agrees with the plant being studied, try the key to an adjacent and possibly overlapping habitat.

8 If there is some, but not complete, agreement between plant and key, weigh up the balance of characters for and against the plant being the species described, on the majority principle referred to above. If a decision cannot be reached, go through the key again, to ensure that a wrong choice was not made at an earlier stage, if need be using a fresh section of the plant.

Tips for using the vegetative keys

• In many steps in the keys more than one character is given, in order to make identification more certain.

• Plants are very variable (it is humans, not nature, that categorizes them), and sometimes two out of three (or three out of four) characters suggest one genus or species, but a fourth (or fifth) character another. In such cases, follow the 'majority verdict'.

• You have to accept that identification will often be tentative rather than conclusive and remember that these keys omit many less common species.

• The keys will become much easier to use with experience!

I. PLANTS OF WOODLANDS, SHADY HEDGEBANKS AND SUNKEN LANES

Woodland of some kind is the natural 'climax' vegetation of most of Western Europe below the tree line on mountains, so very many of our native plants are woodland species. Woodlands vary in flora according to the type of geological parent rock and hence of soil, according to the degree of drainage, and, of course, according to the type of human management exercised over the centuries. Today, much woodland is composed of planted evergreen conifers; such woodlands, unless of some age and partially thinned out, are of very little botanical interest except along the open rides or paths where more light penetrates. Deciduous woodlands, formerly often managed by coppicing (cutting the shrub layer every 8–15 years), more closely reproduce the light and shade conditions of primeval forests, and tend to have retained a continuity of flora from prehistoric times. Coppicing, when practised, lets in the light, and the plants respond by a burst of flowering one or two years later. But many old coppices have now not been cut for years, this form of management being no longer considered economic, and in such woods many plants do not flower at all. In less dense woodlands, with a less closed shrub layer and tall mature trees (and hence woodlands of a more natural structure) there is a 'light phase' in spring when most plants tend to flower before the leaves of the trees unfold and reduce the light, but when it is nevertheless warm and sunny enough for herb growth. In many such woods comparatively few plants are still in flower during mid to late summer, but many woodland herbs, most of them perennials, remain growing and visible as leaf-rosettes, leafy runners or non-flowering shoots, in the autumn and even throughout milder winters. Most of the commoner of these species can thus be identified with this key throughout much of the year.

Woodland soil variations greatly affect the flora. On heavy clay soils water logging is common in winter, and many marsh species may occur that can tolerate some shade (so see also Key IV). Light sandy or gravelly soils, or those on gritstones, granites or other ancient rocks, are often very poor in nutrients, particularly in lime, and are consequently very acid (with a pH often below 5). On such soils, where oak, birch or pine predominate, the woodland ground flora is often very limited and composed of heathland-type plants, especially where there is enough light (see Key III). On calcareous (lime-rich) soils, the pH is high (often above 7) and the soils are thus alkaline. Here ash, elm or hazel are often dominant, with beech in SE England or France, in drier areas; the ground flora is rich and may contain many species also found on chalk scrub or grassland, especially by rides and in glades where the light is good. It is not usual to find plants of acid and of alkaline soils side by side, but if there is a geological change they may occur in the same wood; in such cases it should be possible to see a change in the soil, eg, from sand to loam, or where pieces of chalk become noticeable.

Shady hedgebanks and hollow sunken lanes have been included in this key because they tend to have shade-loving (or shade-tolerant) woodland-type floras; indeed, ancient hollow lanes and hedges may preserve actual relics of former woodlands from the time when old forests bordered them.

MASTER KEY

A. Trees and taller shrubs
(see also Key II F)

1 Lvs (or buds) in opp pairs............**2**

Lvs (or buds) alt along stems........**5**

2 Lvs pinnate **3**

Lvs palmately-lobed **4**

Lvs palmately compound – lfts large (10–20 cm long), obovate
Aesculus hippocastanum (p 315)

3 Lfts oval to elliptical, 5 (or, rarely, 7), hairy on veins beneath – buds small (3–5 mm long) – twigs fluted, brown – frs black berries ..*Sambucus nigra* (p 424 **A**)

Lfts oval, 9–13, hairless when mature – buds large (5–10 mm long), black – twigs smooth, round, grey – frs bunches of dry 'keys', with flat oblong tips..............
Fraxinus excelsior (p 390 **A**)

4 Lvs 3–5-lobed, 3–5 cm wide, usually broader than long, lobes blunt, with few blunt teeth ....
Acer campestre (p 316 **A**)

Lvs five-lobed, 7–15 cm wide, as long as broad, lobes pointed, with many pointed teeth ....
Acer pseudoplatanus (p 316 **B**)

5 Lvs evergreen, tough or thick ..**6**

Lvs ± thin, deciduous.........................**7**

6 Lvs glossy dark green, oval, ± edged with spinous teeth ....
....................*Ilex aquifolium* (p 305)

Lvs dull dark green, palmate to elliptical, untoothed – plant either climbing or creeping on the ground *Hedera helix* (p **A** 468)

Lvs linear, needle-like, 3–8 cm long x 1–2 mm wide, borne in pairs on short side-shoots
Pinus sylvestris (**Scots Pine**)

Lvs linear, tapered to sharp points, 1.0–1.5 cm long, grey-green, borne in whorls of 3 on twigs ....
Juniperus communis (**Juniper**)

Lvs linear, strap-shaped, 1.0–2.5 cm long x 2–3 mm wide, in two spreading rows along twigs............
Taxus baccata (**Yew**)

7 Lvs pinnate, downy below – no prickles on stem – frs red berries *Sorbus aucuparia* (p 248 **A**)

Lvs pinnate – stems with prickles – frs rose-hips........*Rosa spp* (p 253)

Lvs simple **8**

8 Lvs unlobed, toothed or not **9**

Lvs ± deeply lobed.............................**21**

9 Lvs rounded-obovate, and as if cut across at tip.....................................
Alnus glutinosa (p 126 **C**)

Lvs ± pointed at tip **10**

10 Lvs rounded in outline **11**

Lvs oval to elliptical-lanceolate in outline **13**

11 Lvs with broad shallow blunt wavy teeth, tip not drawn-out ......
(*Populus*) **12**

Lvs with small fine teeth and drawn-out tip, base cordate – lvs v **downy** both sides, also twigs......
Corylus avellana (p 126 **B**)

Lvs with fine teeth and pointed tip, base cordate – lvs **hairless** above, but with hair tufts in vein axils below – mature twigs hairless..........
Tilia spp, especially *T. cordata* (p 179)

12 Lvs grey-felted below, dark glossy green above – stalk 1.5–4.0 cm ....*Populus* x *canescens* (p 194 **F**)

Lvs green below, but downy when young – stalk v long, flexible and flattened (2.5–6.0 cm long)...............
Populus tremula (p 194 **E**)

13 Lvs with blade extending further down stalk on one side than on other ....(*Ulmus*) **14**

Lvs symmetrical, blade extending ± equally down each side of stalk... **15**

14 Lvs large, obovate or elliptical to diamond-shaped, 8–16 cm long on short shoots ....
Ulmus glabra (p 120 **B**)

Lvs small, oval-elliptical, 4–6 cm long on short shoots ....
Ulmus procera (p 120 **A**)

15 Lvs oval, – untoothed **16**

Lvs oval, toothed **17**

Lvs elliptical-lanceolate **18**

16 Lvs smooth, hairless except on edges and on veins below – buds pointed, cigar-shaped, brown – twigs grey – tree to 30 m tall ....
Fagus sylvatica (p 124 **G**)

Lvs ± wrinkled, downy (especially below) – buds blunt ..*Salix caprea* (p 193 **A**)

Lvs smooth, hairless, glossy green – buds without scales – twigs green and downy when young – shrub to 4–5 m – frs berries, red then black....
Frangula alnus (p 312 **B**)

17 Lvs smooth and glossy above, elliptical, hairless below, with 2 small red warts on stalk just below lf-blade – fr a cherry..............
Prunus avium (p 249 **A**)

Lvs ± wrinkled and deep green above, oval, **white-felted** below, no red warts on lf-stalk – fr a hard red berry......................................
Sorbus aria (p 249 **C**)

Lvs oval, smooth, green both sides, with blunt teeth, some borne on short spur-shoots in clusters – thorns often present on twigs ....
Malus sylvestris (p 250 **G**)

Lvs triangular-oval, to 5 cm long, flat ....(*Betula*) **20**

Lvs oval, to 10 cm long, with pleated side-veins and strongly doubly-toothed ....
Carpinus betulus (p 126 **A**)

18 Twigs ending in strong straight thorns – lvs elliptical, 1–4 cm long, toothed – fr black with blue bloom like a small round plum ....*Prunus spinosa* (p 249 **C**)

Twigs without thorns....................**19**

19 Lvs elliptical-lanceolate, sharply-toothed, 10–20 cm long, hairless when mature – twigs hairless when mature........................
Castanea sativa (p 124 **F**)

Lvs elliptical-obovate, scarcely toothed, 3–7 cm long, v downy below (and also above when young) – twigs downy ....
Salix cinerea (p 193 **B**)

20 Lvs with ± truncate base, teeth themselves toothed, hairless – twigs hairless, warted ....
........................*Betula pendula* (p 128 **E**)

Lvs with rounded base, simple teeth, downy – twigs downy, no warts ..*Betula pubescens* (p 128 **D**)

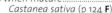

21 Lvs 5–12 cm long, margin with deep, wavy but rounded lobes – no thorns – frs acorns ....(*Quercus*) **22**

Lvs 1.5–5.0 cm long, margins deeply lobed – thorns present on twigs – frs hard red berries .... (*Crataegus*) **23**

Lvs 7–10 cm long, margin pinnatifid or almost palmately-lobed, with few, deep, pointed, toothed lobes – no thorns present – fr a brown, spotted hard berry.............................. *Sorbus torminalis* (p 248 **B**)

22 Lvs hairless below, with **rounded basal lobes**, and stalks 1 cm long or less.................................... *Quercus robur* (p 123 **A**)

Lvs, with tufts of hairs in vein-axils underneath, **tapered** into stalks 2 cm or more long .... *Quercus petraea* (p 124 **B**)

23 Lvs on short shoots **oval** in outline, with **deep**, **sharp-pointed** triangular lobes longer than broad – style one per fr .... *Crataegus monogyna* (p 250 **H**)

Lvs on short shoots **obovate** in outline, with **shallow rounded** lobes broader than long – styles 2–3 per fr .... *Crataegus laevigata* (p 250 **I**)

B. Low erect shrubs (under 1 m tall), or climbing or creeping plants with woody stems

1 Lvs **compound**, **stems** sometimes with prickles **2**

Lvs **simple**, at most lobed only, **stems** not prickly **3**

Much-branched lfless shrub (Nov–April), with younger stems green, four-angled .... *Vaccinium myrtillus* (p 222 **A**)

2 Lvs alt, **pinnate**, with stipules at bases of stalks – stems with prickles along them – frs red hips ....(*Rosa*) **4**

Lvs alt, **palmate**, with **3–7 lfts**, with stipules at bases – stems and lvs prickly – frs blackberries.... *Rubus fruticosus* (p 256 **A**)

Lvs **opp**, **pinnate**, with 3–5 lfts – woody climber – frs with white feathery styles .... *Clematis vitalba* (p 110 **A**)

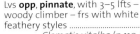

3 Lvs **palmately-lobed**, leathery, evergreen – no stipules – plant creeping on (and carpeting) ground in shade,or climbing trees by means of sucker roots – frs purple-black berries .... *Hedera helix* (p 324 **A**)

Lvs **oval**, with **spines** on edges, leathery, evergreen, glossy – erect shrub (or tree) – frs red berries...... *Ilex aquifolium* (p 305)

Lvs unlobed, spineless........................**5**

4 Lvs with blunt, feebly-toothed, grey-green lfts – prickles on stem weak, without broad bases, hooked at tips...............*Rosa arvensis* (p 254 **A**)

Lvs with pointed, strongly-toothed, bright green lfts – prickles on stem strong, broad-based, arched .... *Rosa canina* agg (p 256 **D**)

5 Lvs **opp**, dull grey-green, ± downy, oval-elliptical, untoothed, 3–6 cm long – plant carpeting ground in dry wds or climbing trees or hedges – frs red berries .. *Lonicera periclymenum* (p 424 **F**)

Lvs **alt**, glossy dark green, elliptical-lanceolate, hairless, 5–12 cm long, untoothed – evergreen, erect shrub, little-branched, rarely over 1 m tall – frs flattened oval black berries .... *Daphne laureola* (p 296 **A**)

Lvs **alt**, fresh green, oval, pointed, 1–3 cm long, finely-toothed all round, with net veins conspicuous – stems much-branched, plant bushy, younger stems green, four-angled – frs globular black berries with grey-blue bloom and flattened tips – in hthy wds on acid soils.................. *Vaccinium myrtillus* (p 222 **A**)

C. Climbing herbs, twining or scrambling up the stems of other plants

1 Lvs in **opp** pairs, trilobed or palmately-lobed, rough – stems twining clockwise (as viewed from above) – in moist wds, hbs
Humulus lupulus (p 121 **A**)

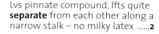

Lvs **alt** along stems **2**

2 Stems climbing by **coiled tendrils** in axils of **palmately 3–5-lobed** lvs – stems bristly – frs clusters of red berries ....
Bryonia dioica (p 190)

Stems **without** coiled tendrils, but **twining** up stems of other plants – lvs cordate to arrow-shaped, un-toothed **3**

Stems **scrambling**, **not** twining up stems of other plants – **lvs oval**, sometimes with 2 pinnate basal lobes, not glossy – frs clusters of red berries ....
Solanum dulcamara (p 352 **F**)

3 Lvs **v glossy**, **oval**, with pointed tips and deeply **cordate** bases, twining anti-clockwise – frs clusters of **red berries** ....
Tamus communis (p 520)

Lvs **not** glossy, arrow-shaped, bases cordate to square-cut – frs **dry** **4**

4 Lvs mealy below, smooth above, with fringed chaffy stipules at bases – stem twining clockwise – frs three-angled ....*Fallopia convolvulus* or *F. dumetorum* (p 170 **B**, **C**)

Lvs **not** mealy on either side, no stipules present – stem twining anticlockwise – frs dry globular capsules ....
Calystegia sepium (p 354 **D**)

D. Herbs with once-pinnate lvs

1 Lvs pinnately-lobed but with **at least** a narrow flange of lf-blade along stalk connecting lf-lobes – milky latex present........................
Mycelis muralis (p 472 **F**)

Lvs pinnate compound, lfts quite **separate** from each other along a narrow stalk – no milky latex **2**

2 Lvs with a **terminal** lft as well as side lfts – lfts all ± **toothed**......**3**

Lvs with no terminal lft, but sometimes with a branched **tendril** instead – stipules present at lf-base – lfts **never** toothed **8**

3 Side lfts of **different** sizes, **small** ones alt with **larger** ones down lf-stalk – stipules **always** present at lf-base **4**

Side lfts all ± of **same size** – stipules **abs**........................**6**

4 Terminal lft much broader and longer than side ones........................**5**

Terminal lft not, or hardly, broader or larger than side ones – in damp or marshy places............
Filipendula ulmaria (p 264 **B**)

5 Terminal lft of (at least) basal lvs broadly orbicular to kidney-shaped, much broader than rest of lf, unlobed in wetter places ....*Geum rivale* (p 264 **E**)

Terminal lft, even of basal lvs, ± triangular and ± trilobed, little or no broader than rest of lf – in drier places ....
Geum urbanum (p 264 **D**)

6 Terminal lft of basal lvs (**a**) as broad as long, orbicular, **much** broader than side lfts – **upper** lvs (**b**) with all lfts **narrow**-oblong to linear ....
Cardamine pratensis (p 210 **A**)

Terminal lft of basal lvs longer than broad, little or no broader than side lfts – **upper** lvs with all lfts **oval** **7**

7 Terminal lft scarcely-toothed – **all** lfts of similar size and shape *Cardamine amara* (p 210 **D**)

Terminal lft with large coarse blunt teeth, side lfts narrower but also coarsely-toothed................
Cardamine flexuosa (p 210 **B**)

8 Lfts 5–12 pairs per lf **9**
Lfts 1–3 pairs per lf **11**

9 Lfts **oval**, widest near **base**,
5–6 pairs per lf, ± hairless ....
.......................*Vicia sepium* (p 282 **C**)

Lfts **elliptical**, widest near
middle, 6–12 pairs per lf **10**

10 Lfts thinly **downy** – stipules
half-arrow-shaped ....
Vicia cracca (p 282 **A**)

Lfts **hairless** – stipules long-
toothed and **rounded** at
base............*Vicia sylvatica* (p 284 **B**)

11 Lfts **one** pair only, narrow-
lanceolate, 7–15 cm long – lf
ending in branched **tendril** –
climbing plant ....
Lathyrus sylvestris (p 286 **B**)

Lfts **2–3** pairs, elliptical-lanceolate,
1–4 cm long – **no** tendrils – erect
plant......*Lathyrus linifolius* (p 284 **E**)

E. Herbs with lvs twice, or more times pinnate

1 Lvs **fern-like**, with lfts deeply
lobed **2**

Lvs **not** fern-like, lfts rounded,
oval or coarsely-toothed **5**

2 Lf-stalk with **papery brown
scales** – **no** erect stem present ......
Dryopteris (**Buckler-ferns**)

Lf-stalk **without** papery brown
scales – **erect stem** (or last year's
dead one) present **3**

3 Lfts **hairless**, 1–3 cm long, with
bases **wedge**-shaped, tips **bluntly**
2–4 lobed – in wet places ....
Oenanthe crocata (p 341 **A**)

Lfts **hairy**, bases **rounded**,
tips finely **cut** **4**

4 Lf-stalks **purple**-spotted – lfts v
downy, **grey**-green, **blunt**..................
Chaerophyllum temulum
(p 330 **F**)

Lf-stalks **not** purple-spotted – lfts
v sparsely and shortly downy,
fresh green, **sharp**-toothed
Anthriscus sylvestris (p 329 **A**)

5 Lfts irregularly lobed, **coarsely**-
toothed, **roughly hairy** –
secondary lfts **webbed** together,
not oval ....
Heracleum sphondylium (p 336 **D**)

Lfts **rounded**, **finely** and regularly
toothed, **smooth**, scarcely hairy –
secondary lfts mostly **separate**,
oval-elliptical **6**

6 Lfts **elliptical**, the side veins
4–5 mm apart – lvs and stalks
uniformly fresh green – lf-stalk
triangular, hardly inflated at base
– lvs twice-**trifoliate** rather than
twice-pinnate ....
Aegopodium podagraria (p 334 **F**)

Lfts **oval**, side veins 2–3 mm
apart..**7**

7 Lfts v **sharply**-toothed,
sometimes lobed, **dark glossy**
green above – stalks **all uniform**
green, **not** inflated below – vl in
calc wds ....*Actaea spicata* (p 113 **E**)

Lfts v finely-toothed, fresh green,
not glossy above – stalks **purple**-
stained below, **purple**-spotted
where lfts
are attached – lf-stalk **inflated** at
base – in moist wds
Angelica sylvestris (p 336 **A**)

F. Herbs with trifoliate lvs

1 Lfts simple, toothed or not........**2**
Lfts deeply palmately-lobed or
divided **4**

2 Lfts untoothed, obovate,
notched at tips, green on
uppersides, purplish on lower
sides, drooping from tips of long
stalks – only a few scattered
adpressed hairs present....................
Oxalis acetosella (p 317 **A**)

Lfts shallow-toothed, v hairy,
oval-obovate, 1–4 cm long, lvs all
from base...**3**

Lfts deeply-toothed, diamond-shaped, hairy, 5–7 cm long, their bases wedge-shaped – lvs alt along stem – **lower** lvs pinnate – stipules lf-like ....
Geum urbanum (p 264 **D**)

3 Lfts pointed, bright green, end tooth of each lft longer than its neighbours on each side ....
Fragaria vesca (p 260 **J**)

Lfts blunt, blue-green, end tooth of each lft not longer (often shorter) than its neighbours on each side ....
Potentilla sterilis (p 260 **I**)

4 Lvs v hairy – lfts oval, deeply and sharply palmately-lobed – no stipules – lfy rooting runners extend from lf-rosettes ....
............*Ranunculus repens* (p 102 **A**)

Lvs hairless – lfts ± fleshy, yellowish-green, deeply palmately-lobed – lobes blunt, but tipped with tiny spines ....
Adoxa moschatellina (p 426 **E**)

Lvs hairless – lfts thin, dark green, lobed ± to base into narrow-oblong sharp-toothed segments which are not spine-tipped ....
Anemone nemorosa (p 109 **B**)

G. Herbs with lvs palmately divided or lobed, ie, with 5 or more lfts or lobes radiating from tip of lf-stalk (on larger lvs at least)

1 Lvs **hairless**..**2**

Lvs ± **hairy** **5**

2 Lvs **divided to base** of blade into **separate**, toothed lfts **3**

Lvs **lobed**, but **not** divided to base **4**

3 Lfts **dark blue**-green, narrow-lanceolate (to 1.5 cm wide), never forked, arising from **2** '**arches**' at tip of lf-stalk, and **not** radiating from one central point – in open calc wds ....
Helleborus foetidus (p 104 **F**)

Lfts **bright** green, broad-lanceolate (to 3 cm wide), lobed or forked, radiating from **one** central point at tip of lf-stalk – in ash and hazel wds ....
Helleborus viridis (p 104 **E**)

4 Lf-lobes **rounded**, untoothed, veins **pale** – lvs arising from **creeping stems** – older stems **woody** ....*Hedera helix* (p 324 **A**)

Lf-lobes **pointed**, separated _ way to lf-base, themselves pinnately-lobed and toothed, veins not v pale – lvs in basal **rosette**, and alt up **erect non-woody** stems ....
Sanicula europaea (p 336 **E**)

5 Plants with ± creeping stems, lvs **alt** along them **6**

Plants with lvs in **basal rosette** on long stalks, and also **alt** along erect stems **7**

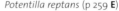

6 Lvs **long-stalked** (5–8 cm), lvs **4–6** cm wide, divided to **base** into 5 (or 7) obovate, toothed lfts – **stipules** present where stalk joins stem...
Potentilla reptans (p 259 **E**)

Lvs **short-stalked** (0.5–1.0 cm), lvs 1.0–1.5 cm wide, **shallow**-lobed (like ivy lvs) – **no** stipules present
Veronica hederifolia (p 402 **G**)

7 Basal lvs (**a**) with palmate veins, but outline varying from round, through kidney-shaped to palmately-lobed – hairs sparse – stem-lvs (**b**) ± stalkless, divided to base into narrow segments – lvs all pale green ....
Ranunculus auricomus (p 102 **G**)

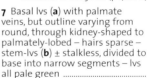

Basal (and stem) lvs palmately **divided to base** into triangular shiny lfts, themselves deeply twice-pinnate, stalks and lf margins ± reddish-flushed ....
Geranium robertianum (p 319 **A**)

Basal lvs deeply **palmately-lobed**, **nearly** to base, into elliptical-obovate lobes, themselves deeply pinnately-lobed – lvs and stalks green, not shiny – in N of Br Isles
Geranium sylvaticum (p 322 **D**)

Basal lvs (**a**) long-stalked, divided to **base** into 3 toothed lfts – stem-lvs (**b**) stalkless, with apparently 5 obovate toothed lfts (actually the 2 outer shorter lfts are stipules resembling lfts) – in hthy wds ....
Potentilla erecta (p 259 **A**)

H. Herbs with linear, grass-like lvs (Note: on calc soils other grasses, sedges or orchids may occur in woods; if plants fail to key out here, consult Key II)

1 Lvs brittle, fleshy-textured, bright glossy green, without distinction into blade and basal sheath – lvs in an erect rosette ..**2**

Lvs firm, thin-textured, clearly divided into blade and ± cylindrical or angled sheathing base – lvs usually many together in a dense tuft **3**

2 Lvs oblong-lanceolate, 1.5–4.0 cm wide x 12–30 cm long, with obovoid **seed-pod** 3–5 cm long arising from centre of a group of lvs on a short stalk in April–May
Colchicum autumnale (p 508 **E**)

Lvs linear, ± hooded at tip, 7–15 mm wide x 20–45 cm long, appearing Feb–March – fl-spike arising from centre of lvs in April–May – only dry spike of fr capsules (20–40 cm tall) present Aug–Feb ....
Hyacinthoides non-scripta (p 510 **A**)

3 Lvs alt, but in **2 opp** vertical ranks on **round, hollow,** pointed stem ....(**Grasses**) **4**

Lvs in **3 vertical** ranks on stem – stem **solid,** ± **triangular** in cross-section ....(**Sedges**) **12**

4 Lvs **bristle-like,** dark green, not more than 1.5 mm wide, rough edged – lf-sheath pinkish – in hthy wds ....
Deschampsia flexuosa
(**Wavy Hair-grass**)

Lvs **flat** or **furrowed,** over 1.5 mm wide **5**

5 Lvs to 3 mm wide **6**

Lvs over 3 mm wide **7**

6 Lvs dark green, smooth, keeled, 1–2 mm wide, ligule v **short** or abs – sheaths **smooth** – infl drooping – in dry wds ....
Poa nemoralis
(**Wood Meadow-grass**)

Lvs yellow-green, 2–3 mm wide, flat, ligule **long,** pointed –sheaths **rough** – infl erect – in damp wds ....*Poa trivialis*
(**Rough Meadow-grass**)

7 Lvs flat, rigid, 3–5 mm wide, hairless, grey-green, with deep parallel **furrows** on upperside, **v rough** when rubbed downwards – plant forms dense tussocks in damper wds ....
Deschampsia caespitosa
(**Tufted Hair-grass**)

Lvs without deep parallel furrows on upperside **8**

8 Lvs **bright deep** green, 10–20 cm long x 3–5 mm wide, soft, drooping at tips, with only a few hairs – sheaths downy with pointed flap at tip on side of stem opp to lf-blade – infl loose, fls purple-brown – in dry wds ....
Melica uniflora (**Wood Melick**)

Lvs **yellow**-green, **light** green, or **grey**-green **9**

9 Lvs v **hairy** **10**
Lvs ± **hairless** **11**

10 Lvs 6–13 mm wide, yellow-green, **soft, drooping,** downy – sheaths downy..................................
Brachypodium sylvaticum
(**False Brome**)

Lvs 5–10 mm wide, light green, **rough**-hairy, **not** drooping – sheaths with many long downward-pointing bristles ....
Bromopsis ramosa (**Hairy-brome**)

11 Lvs to 18 mm wide, rough, **shining bright** green, with a pair of **pincer**-shaped purplish lobes (**auricles**) where lf-blade joins sheath..
Festuca gigantea (**Giant Fescue**)

Lvs 5–10 mm wide, rough, **whitish-grey**-green, with **no** auricles – infl (50–100 cm), persisting in winter, is loose, of whorled, egg-shaped spikelets, 2–3 mm long ....*Milium effusum* (**Wood Millet**)

12 Plants forming large (50–100 cm wide) **tussocks** of lvs, often **raised** on a shaggy column 30–100 cm off ground – lvs 3–7 mm wide, deeply grooved above, with v **rough** edges – in v wet swampy wds ....
Carex paniculata
(**Greater Tussock-sedge**)

Plants forming **smaller** tufts or tussocks, **not** raised off ground – lvs ± **smooth**-edged **13**

13 Lvs 2 mm wide, **grooved** above, bright green – ligules 1–2 mm long, triangular – in wet places ....
Carex remota (**Remote Sedge**)

Lvs 3–6 mm wide, ± **flat**, bright green – ligules 2 mm long, triangular – in drier places ....
Carex sylvatica (**Wood-sedge**)

Lvs 6–10 mm wide, ± flat, yellow-green, tips turning **orange-brown** in winter – ligules 10 mm long, triangular – in damp places ....
Carex strigosa
(**Thin-spiked Wood-sedge**)

Lvs 15–20 mm wide, ± flat, **dark green** on upperside, **paler** beneath – ligules 3–6 mm long, lanceolate – on heavy soils ....
Carex pendula (**Pendulous Sedge**)

I. Herbs with lvs untoothed, with many parallel-veins, and no cross-veined network, but not grass-like – lvs in a basal rosette, or at or near base of stem, or along stem

1 Lvs ± linear, grass-shaped but brittle and rather fleshy, arising at an angle of 50°–80° from ground, appearing in spring only **2**

Lvs not linear or grass-shaped **3**

2 Lvs oblong-lanceolate, 1.5–4.0 cm wide x 12–30 cm long, with obovoid fleshy seed pod 3–5 cm long arising in April–May from centre of lf-group on a short stalk – local only...
Colchicum autumnale (p 508 **E**)

Lvs linear, ± hooded at tips, 7–15 mm wide x 20–45 cm long, arise Feb–March – fl-spike arises April–May from centre of lf-group – only fr-spike visible July–Feb – capsules dry papery brown in autumn, winter ....
Hyacinthoides non-scripta (p 510 **A**)

3 Lvs elliptical, stalked, 1 or several, arising ± **vertically** from base of plant or on a short erect stem **4**

Lvs in an **opp pair** only, blunt, oval-elliptical, **spreading** at a **wide angle**, arising either directly from ground, or from a short stem a few cm above it **5**

Lvs elliptical, **alt** along a stem 20–60 cm tall that is erect below, arching above.........................
(*Polygonatum*) **6**

Lvs several, in a basal **rosette** arising from ground, some lvs lying flat, some ascending at an angle **7**

4 Lvs ± fleshy, **garlic**-scented when bruised, **bright** green, arising direct from bulb – lf-stalks twisted through 180° – in moist wds ....*Allium ursinum* (p 515 **C**)

Lvs **not** fleshy, **not** garlic-scented, **grey**-green, arising in a pair on a short erect stem – in dry wds ....
Convallaria majalis (p 505 **E**)

5 Lvs elliptical to elliptical-lanceolate, in **one opp** pair arising **directly** from ground, and **spreading** ....
Platanthera chlorantha (p 538 **B**)

Lvs oval, in one opp pair a few cm **above** ground, on a fleshy **stem** – fl-spike (or bud of one) may be present above lvs ....
Listera ovata (p 532 **B**)

6 Stem **round**...
Polygonatum multiflorum (p 504 **A**)

Stem **angled** – local, on
limestone ....
Polygonatum odoratum (p 505 **C**)

7 Lvs **unspotted**....................................
Orchis spp (see Key p 534)

Lvs **purple**-spotted **8**

8 Rosette-lvs ± in 2 vertical **ranks**,
grey-green – spots **transverse**
across lvs ....
Dactylorhiza fuchsii (p 542 **A**)

Rosette-lvs radiating ± **equally**
from base, glossy **deep** green –
spots elongated **lengthwise**
along lvs ....
Orchis mascula (p 535 **A**)

**J. Herbs with lvs in successive
whorls of 4 or more along the
stems**

Stems erect smooth – lvs 6–8 per
whorl, dark green, firm, hairless,
with tiny **forward**-pointing
prickles, lanceolate or
elliptical, pointed ....
Galium odoratum (p 422 **D**)

Stems erect at first, then
scrambling, **v rough** with down-
pointing prickles – lvs 6–8 per
whorl, light green, hairy, with **v
strong** backward-pointing
prickles, obovate to lanceolate ......
Galium aparine (p 420 **H**)

Stems ascending or decumbent,
weak, smooth or rough, lvs
bluntly lanceolate with no point
at tip, margins with **tiny** back-
pointing teeth, 4–6 per whorl –
in wet places..
Galium palustre (p 420 **E**)

**K. Herbs with lvs in opp pairs
along the stems – lvs simple,
not with many parallel veins**

1 Lvs quite untoothed......................**2**

Lvs with (at least) some shallow
blunt teeth, or deeply-toothed **7**

2 Lvs linear-lanceolate, 4–6 cm
long x 0.5 mm wide, fine-pointed,
rough-edged, hairless, stalkless –
stems square in cross-section ....
Stellaria holostea (p 150 **B**)

Lvs elliptical-lanceolate and
stalkless on stem, obovate and
broadly-stalked at base of stem,
all **hairy**, smooth-edged, 5–20 cm
long x 2–4 cm wide – stem
rounded........*Silene dioica* (p 144 **B**)

Lvs all elliptical-lanceolate,
stalkless, hairless, soft, smooth-
edged – stem rounded,
branched below....................................
Melampyrum pratense (p 404 **A**)

Lvs ± oval..**3**

3 Lvs with **2 strong side-veins** ±
parallel to midrib, lvs 1–2 cm
long, v sparsely hairy – stem
round ....
Moehringia trinervia (p 154 **A**)

Lvs with pinnate veins **4**

4 Lvs 1–3 cm long, thin and
flaccid, hairless, on stalks 2–5 cm
long – stem weak, straggling, ±
hairy, round ....*Stellaria neglecta*
(or large form of *S. media*)
(p 150 **E**, **A**)

Lvs with stalks under 1 cm long –
stem round – stem and lvs ±
hairless – creeping plants **5**

5 Lvs oval-elliptical, 3–7 cm long,
grey-green, often slightly downy
– stem somewhat woody below
and far-creeping on ground............
Lonicera periclymenum (p 424 **F**)

Lvs oval-elliptical, 0.5–2.0 cm
long, hairless, glossy green –
stem not woody but tough ....
Veronica serpyllifolia (p 400 **C**)

Lvs oval-orbicular, glossy bright
green – stem weak and brittle,
never woody, shortly **creeping** **6**

6 Lvs blunt, orbicular, 1.5–3.0 cm
long..
Lysimachia nummularia (p 234 **C**)

Lvs pointed, oval, 2–4 cm long
Lysimachia nemorum (p 234 **B**)

7 Stems round **8**

Stems square (or four-winged) in
section..**14**

8 Stems erect **9**
Stems creeping or ascending **10**

9 Lvs oval-triangular, 3–6 cm
long, feebly wavy-toothed,
hairless above, stalk 2–10 cm
long – stem green, swollen at
nodes, ± hairless ....
Circaea lutetiana (p 300 **A**)

Lvs oval-elliptical, 3–8 cm long,
finely and sharply-toothed,
downy, stalk 3–10 mm long –
stem green, not swollen at
nodes, hairy ....
Mercurialis perennis (p 310 **E**)

Lvs oval-lanceolate, fine-toothed,
hairless, stalk 3–6 mm only –
stem reddish, ± hairless ....
Epilobium spp, especially *E.
montanum* (p 298 **A**)

10 Lvs oval-elliptical **11**

Lvs ivy-shaped, 1–2 cm wide,
long-stalked – lower lvs opp, but
upper lvs alt..
Veronica hederifolia (p 402 **G**)

11 Lvs with v shallow teeth, stalks
2 mm long or less **12**

Lvs with coarse blunt teeth **13**

12 Lvs 1–2 cm long, ± untoothed,
hairless – stem downy ....
Veronica serpyllifolia (p 400 **C**)

Lvs 2–3 cm long, shallow-toothed
all round, hairy (as is stem) ....
Veronica officinalis (p 400 **D**)

13 Lvs with stalks 5–15 mm long
– stem hairy all round.......................
Veronica montana (p 400 **B**)

Lvs with stalks to 5 mm, or none
– stem hairy only in 2 lines on
opp sides...
Veronica chamaedrys (p 400 **A**)

14 Lvs 1–2 cm long, pale green,
fleshy, brittle, stalked, orbicular,
with wedge-shaped base and
blunt rounded tips, sparsely
bristly – in creeping mats in
boggy ground...
Chrysosplenium oppositifolium
(p 244 **J**)

Lvs various, over 2 cm,
not brittle...**15**

15 Lvs hairless **16**

Lvs hairy **18**

16 Lvs obovate-oblong, blunt,
glossy dark green – teeth few,
blunt wavy – rosette lvs long-
stalked – plant with creeping lfy
runners ....*Ajuga reptans* (p 382 **A**)

Lvs oval-triangular – stems all
erect **17**

17 Lvs with blunt rounded teeth
and tips – stem four-winged ....
Scrophularia auriculata (p 394 **B**)

Lvs with pointed teeth and tips
–stem square, unwinged..................
Scrophularia nodosa (p 394 **A**)

18 Lvs kidney-shaped, glossy,
with large blunt teeth all round
edges – long lfy runners present,
with long-stalked lvs held erect ....
Glechoma hederacea (p 378 **J**)

Lvs oval, oblong, or oval-
triangular, (not kidney-shaped) **19**

19 Lvs bluntly oval, with sparse
short bristles, teeth v few and
shallow – runners with spreading
lvs present ....
Prunella vulgaris (p 380 **F**)

Lvs oblong, long-stalked in basal
rosette, shiny dark green with
coarse rounded teeth and blunt
tip, ± hairless above, hairy
underneath ....
Stachys officinalis (p 380 **A**)

Lvs oval-triangular, tips pointed,
teeth pointed, ± hairy above......**20**

20 Lvs with sharp teeth all
round, and covered with stinging
hairs – stipules present ....
Urtica dioica (p 121 **A**)

Lvs without stinging hairs – no
stipules **21**

21 Lvs cordate-based, v hairy,
long-pointed, **sharp**-toothed,
unpleasant smell – **no lfy
runners** *Stachys sylvatica* (p 379 **A**)

Lvs cordate-based, downy,
wrinkled, grey-green, point rather
blunt, **blunt**-toothed, no smell –
no lfy runners.......................................
Teucrium scorodonia (p 382 **C**)

Lvs **rounded** at base, ± wrinkled,
shiny dark green, sparsely hairy,
teeth and **tip pointed** – no smell
– lfy runners present ....
Lamiastrum galeobdolon (p 374 **A**)

L. Herbs with lvs in a rosette raised 10–20 cm or more above ground, at tip of stem

Lvs **obovate**, hairless, 5–10 cm long, with 3–5 main veins and a network of veins between these – lvs are 3–8 (most commonly 4) in a whorl on tip of a **fleshy** stem – in wds on calc or richer soils ......
Paris quadrifolia (p 506 **D**)

Lvs oblong-**lanceolate**, **downy**, 3–8 cm long – lvs **many** in a rosette-like cluster on tip of a ± **reddish**, ± **woody** stem 20–30 cm tall, persisting through winter (infl appears from rosette in spring) – latex present ....
Euphorbia amygdaloides (p307 **A**)

Lvs obovate-lanceolate, **hairless**, ± untoothed, shiny grey-green, 2–8 cm long, with one main vein and pinnate side-veins – lvs in a whorl of 5–6 on top of v slender stem – mostly in pinewoods in N of GB......................................
Trientalis europaea (p 236 **I**)

M. Herbs with lvs alt along, or at base of, the stems – lvs simple, not with many parallel veins

1 Lvs elliptical to oblong-lanceolate, broadest near or above middle........**2**

Lvs kidney-shaped to oval-triangular in outline, broadest near base **7**

2 Lvs hairless (or nearly so) **3**

Lvs obviously downy, woolly, or bristly **4**

3 Lvs all oval-lanceolate, 10–15 cm long, scarcely toothed, **stalked**, with bases rounded, tips pointed, and with chaffy tubular **stipules** around lf-bases ....
Rumex sanguineus (p 163 **A**)

Lvs obovate in basal rosette, elliptical-lanceolate up stem, upper lvs pointed – all lvs 2–10 cm long, sparsely-toothed, narrowed to base – no stipules ....
Solidago virgaurea (p 442 **E**)

4 Lvs 10–40 cm long, **shallow**-toothed, mainly in a dense **spreading** basal winter rosette – in **open** glades etc **5**

Lvs 5–20 cm long, **untoothed**, in a sparse **ascending** winter rosette, or ± erect **6**

5 Lvs densely **whitish**-woolly, obovate-lanceolate to oblong, 15–40 cm long, short-stalked, sparsely-toothed – on ± calc soils
Verbascum thapsus (p 392 **A**)

Lvs **softly** downy and **green** above, grey-downy below, oval to obovate-lanceolate, 15–30 cm long, narrowed into short stalks, with rounded teeth – stems grey-green – on acid soils ....
Digitalis purpurea (p 394 **D**)

As the last, but lvs **harsh** to touch, teeth sharper, stems **purplish** – on calc soils ....
Inula conyzae (p 444 **899**)

6 Lvs roundish-oval and stalked in rosette, oblong-lanceolate, pointed and stalkless on stem – all lvs with dense spreading hairs on both sides, unspotted, all 5–10 cm long *Myosotis sylvatica* (per or ann) and *M. arvensis* (ann; spp only distinguishable by fls) ....
(p 360 **D**, **E**)

Lvs lanceolate, bristly-hairy, with pale **spots**, all 5–10 cm long (basal lvs may be up to 60 cm in autumn) ....
Pulmonaria longifolia (p 364 **F**)

7 Lvs ± hairless (except lf-stalks sometimes hairy) **8**

Lvs obviously hairy or bristly **14**

8 Lvs **garlic**-scented when bruised, 3–8 cm long and wide – lower lvs kidney-shaped, upper rounded-triangular, all cordate-based, with wavy margins ....
Alliaria petiolata (p 212 **C**)

Lvs **not** garlic-scented......................**9**

9 Lvs cordate-based, thin, with **fringed stipules** at base of each lf-stalk ....(**Violets**) **10**

Lvs **without stipules** at base of lf-stalk..**11**

10 Lf-stalks downy – lvs oval-rounded, blunt or ± pointed, dark glossy green – stipules oval, fringes short – plants with thick rhizomes and rooting runners ......
Viola odorata (p 188 **D**)

Lf-stalks ± hairless – lvs oval-triangular, pointed, not glossy dark green – stipules narrow-lanceolate, fringes long – plants with central lf-rosette, without central stem or rooting runners, but with fl-bearing lfy side branches........*Viola riviniana* and *V. reichenbachiana* (indistinguishable except by fls or in fr; p 186 **A**, **B**)

11 Basal lvs thin, long-stalked (8–12 cm), 3–4 cm long, pale green, varying from rounded **kidney-shaped** to **trifoliate** with deep-toothed lobes – stem-lvs stalkless, divided to base into narrow, **linear** lobes ....
Ranunculus auricomus (p 102 **G**)

Lvs **all** similar, dark green, glossy **12**

12 All lvs on long (10–20 cm) stout stalks, triangular **arrow**-shaped, untoothed, blades 10–20 cm long, often purple-spotted ......
Arum maculatum (p 496 **A**)

All lvs rather fleshy, cordate-based, **rounded** to **oval**-triangular, bluntly-angled or -toothed, lower long-stalked, upper ± stalkless **13**

13 Lvs 2–4 cm long and wide, rounded-triangular – stems **slender** (2–3 mm wide) – Jan–April..
Ranunculus ficaria (p 102 **J**)

Lvs 5–10 cm long and wide, kidney-shaped to blunt-triangular, lower lvs broader than long – stems **stout** (6–12 mm wide) hollow – in wet places – all year ....*Caltha palustris* (p 104 **C**)

14 Low creeping plant of **wet** springy ground under alders – lvs long-stalked (4–8 cm), kidney-shaped, round, v bluntly-toothed, 2–3 cm across, ± fleshy, with scattered bristles on upper surface – lvs arise from creeping

stems all year (and are alt on erect fl-stems in spring)....................
Chrysosplenium alternifolium (p 244 **K**)

Erect plants of **dry** ground with cordate-triangular lvs, long-stalked at base of stems, short-stalked along stems........................**15**

15 Lvs v **large** (15–30 cm long and 10–20 cm wide), rough, **downy** and green above, ± grey-downy beneath, on stout (1 cm or more wide) stalks ....
Arctium minus (lf-stalk hollow) and *A. lappa*
(If-stalk solid; p 468 **A**, **A.2**)

Lvs **smaller** (5–10 cm long x 3–5 cm wide), bristly, on slender stalks – on ± calc soils ....
Campanula trachelium (p 414 **A**)

N Herbs with lvs entirely from base of stem in a spreading or ± erect rosette (not in opp pairs)

1 Basal rosette with central lfy bud, producing erect lfy stem in following year...
(see key M, section 5)

Basal rosette producing **only** fl-buds, and not a lfy stem – lvs blunt, fine-toothed, oval to obovate ....*(Primula)* **2**

2 Lvs narrowed **gradually** to base, **dark green** and **hairless** above, grey-downy underneath, **v wrinkled**, fine-toothed all round, stalk short or **abs** – all side-veins in lf at acute angle to midrib..........
Primula vulgaris (p 231 **A**)

Lvs narrowed ± **suddenly** into a winged **stalk** as long as the blade – side-veins in lower part of lf at 90° to midrib **3**

3 Lvs **wrinkled**, downy with sticky hairs on **both** sides, 5–15 cm long, narrowed v suddenly into stalk, **bright** green ....
Primula veris (p 232 **C**)

Lvs almost **smooth**, crisped-downy beneath, almost **hairless** above, 10–20 cm long, narrowed more **gradually** into stalk, **yellow**-green – plant v local ....
Primula elatior (p 232 **B**)

II. PLANTS OF CHALK AND LIMESTONE GRASSLAND AND SCRUB

Most chalk and limestone grasslands are semi-natural areas of vegetation, originally forested, cleared long ago by man, and eventually used as pasture, normally for sheep. They occur especially on steeper slopes difficult to plough, wherever chalk or limestone rocks are close to the surface. If ungrazed or unmown, as is often the case nowadays, they soon revert to scrub, and eventually to a more or less stable woodland 'climax' of beech or ash; in these areas will also be found many shrubs and the more light-demanding woodland herbs. The shallow, well-drained soils, alkaline from excess of lime (calcium carbonate), and with a pH often as high as 8 or more, favour the spread of many low-growing species that can well tolerate these conditions, while excluding the competition of other, often taller plants, that require either a lower pH or a better balance of nutrients, or cannot survive grazing, when it still occurs. But most chalk grassland plants are highly adapted to being constantly grazed, many having either well-developed basal leaf-rosettes or being grasses that readily produce new shoots. This key concentrates very much on the leaf-rosettes.

The flora of chalk grassland is not only rich in species, but most (like those of woodlands) are perennials and usually have leaves above ground through much of the year, and not only when in flower or fruit. The flowering season is a long one, beginning in April, with peaks in late May to early June for some plants and in late July to August for others. Thus it is rare for all the species to be in flower concurrently, and this key should be useful in summer as well as at other times of year.

MASTER KEY

Herbs with grass-like lvs..**A**

Herbs with lvs not grass-like, but with strong parallel main veins – all lvs untoothed, and in a basal rosette ....**B**

Herbs with compound, pinnate or trifoliate lvs...**C**

Herbs with simple, oval, oblong or cordate lvs, toothed or not, with one main vein only not several parallel main veins) ....**D**

Herbs with lvs in whorls of 4–8 along stems ....**E**

Trees or shrubs, erect, usually 60 cm tall or much more ....**F**

Dwarf creeping shrubs, woody below, not over 20 cm tall, with lvs in opp pairs, untoothed ....**G**

Herbs with lvs deeply pinnatifid, but with their side-lobes joined by at least a narrow flange of lf-blade and themselves ± pinnatifid ....**H**

A. Herbs with grass-like lvs

1 Lvs in 2 opp ranks – stems hollow, round ....(**Grasses**) **2**

Lvs in 3 vertical ranks – stems solid ....(**Sedges**) **8**

2 Basal lvs bristle-like, not easily flattened...**3**

Basal lvs flat or folded lengthwise, not bristle-like (except perhaps when dead) **4**

3 Stem-lvs (a) flat (when fresh) – lf-sheaths forming a closed tube round stem – basal lvs (b) 10–30 cm long – plant with runners .... *Festuca rubra* (**Red Fescue**)

Stem-lvs all bristle-like – lf sheaths forming a tube round stem, split at least half way down – basal lvs 3–8 cm long – no runners .... *Festuca ovina* (**Sheep's-fescue**)

4 Basal lvs short (to 5 cm), lanceolate, ± flat on gd, hairless.... *Briza media* (**Quaking-grass**)

Basal lvs longer, not flat on gd, ± erect **5**

5 Basal lvs ± inrolled at edges, upper lvs flat c. 5 mm wide, hairless except for stiff bristles held at 45° from lf-blade along edges .... *Bromopsis erecta* (**Upright Brome**)

Lvs all with inrolled edges, narrow, **many-grooved** above – lvs and stems **all v downy**............... *Koeleria macrantha* (**Crested Hair-grass**)

Lvs keeled, pointed, with folded blunt ligule and flattened sheath .... *Dactylis glomerata* (**Cock's-foot**)

Lvs strongly keeled, blunt-tipped, grey-green, ligule flat, not folded **6**

Lvs flat, ribbon-like, broad, yellow-green **7**

6 Lvs waxy, hairless, with white dots on underside.............................. *Helictotrichon pratense* (**Meadow Oat-grass**)

Lvs hairy, especially on sheaths, no white dots .... *Helictotrichon pubescens* (**Downy Oat-grass**)

Lvs hairless, no white dots .... *Sesleria caerulea* (**Blue Moor-grass**)

7 Lvs downy above, hairless below, to 10 cm long x 2–4 mm wide........... *Trisetum flavescens* (**Yellow Oat-grass**)

Lvs stiff, edges slightly inrolled, rough, hairless, to 8 mm wide........ *Brachypodium pinnatum agg* (**Tor-grass**)

Lvs drooping, quite flat, downy, soft, to 13 mm wide .... *Brachypodium sylvaticum* (**False-brome**)

8 Lvs waxy grey, carnation-like .... *Carex flacca* (**Glaucous Sedge**)

Lvs bright green, flat, 2 mm wide, spreading .... *Carex caryophyllea* (**Spring-sedge**)

Lvs yellow-green, channelled, rough, 1.0–1.5 mm wide, erect – vlc.. *Carex humilis* (**Dwarf Sedge**)

B. Herbs with lvs not grass-like, but with strong parallel main veins; all lvs untoothed, and in a basal rosette

1 Lvs strongly corrugated lengthwise, stalked, ± hairy .... (*Plantago*) **2**

Lvs smooth, stalkless, hairless .... (Orchids and *Centaurium*) **3**

(Note: Orchids and *Centaurium* have lvs up the stem in summer, as well as a basal rosette; this key is for identification when infl-stem is abs)

2 Lvs lanceolate (length 4–5 times breadth), scarcely hairy ....
Plantago lanceolata (p 387 **A**)

Lvs oval (length 2–3 times breadth), v downy ....
Plantago media (p 387 **C**)

3 Lvs spotted **4**

Lvs unspotted **5**

4 Lvs grey-green, spots transverse ....
Dactylorhiza fuchsii (p 542 **A**)

Lvs bright glossy green, spots lengthwise ....
Orchis mascula (p 535 **A**)

5 Lvs **alt**, but in two opp ranks, keeled, narrow-lanceolate, dark glossy green ....
Gymnadenia conopsea (p 538 **I**)

Lvs in **opp pairs**, oblong-obovate, blunt, unkeeled, all in a flat rosette ....
Centaurium erythraea (p 350 **F**)

Lvs in a ± flat spreading rosette, oblong, not in opp pairs or opp ranks **6**

6 Lvs grey-green, unwrinkled ....
Ophrys apifera (p 545 **A**)

Lvs bright green, unwrinkled ....
Orchis spp, probably *O. morio* (p 536 **F**)

Lvs bright green, with transverse wrinkles near base ....
Aceras anthropophorum (p 536 **G**)

C. Herbs with compound, pinnate or trifoliate lvs

1 Lvs trifoliate **2**
Lvs pinnate **8**

2 The 3 lfts of each lf deeply lobed, lobes toothed, hairy – no stipules at lf-base ....
Ranunculus bulbosus (p 102 **D**)

The 3 lfts ± toothed, but not lobed – stipules present at lf-base **3**

3 Lvs all from base of plant, lfts sharply-toothed, hairy................
Fragaria vesca (p 260 **J**)

Lvs borne along erect or creeping stems **4**

4 Stipules **lfy** – lfts pointed, toothed, sticky-hairy ....
Ononis repens (p 270 **A**)

Stipules small, membranous – lfts blunt, not sticky-hairy **5**

5 Lfts with whitish band or spot across each **6**

Lfts without whitish band **7**

6 Plant erect – lf-stalks to 6 cm long at most – lfts untoothed, hairs on margins only ....
Trifolium pratense (p 274 **A**)

Plant with creeping and rooting runners – lf-stalks to 10 cm long – lfts fine-toothed, hairless ....
Trifolium repens (p 274 **D**)

7 Stem and lfts v downy – lfts mucronate, toothed ....
Medicago lupulina (p 289 **E**)

Stem and lfts usually hairless – lfts blunt, with an extra pair of lfts at lf-base (resembling stipules) untoothed ....
Lotus corniculatus (p 290 **E**)

8 Lvs once-pinnate **9**

Lvs twice or more pinnate or with lfts cut more than halfway to their midribs, hairy or downy **15**

9 Lfts quite untoothed **10**

Lfts toothed **12**

10 Lfts blunt – side lfts in 3 or more pairs, end lft of similar size to side ones **11**

Lfts pointed – end lft of lower lvs much larger than side ones ......
Anthyllis vulneraria (p 292 **B**)

Lfts ± blunt – side lfts in 2 pairs only, lower pair arising from junction of lf-stalk and stem – upper pair same size as end lft......
Lotus corniculatus (p 290 **E**)

11 Lfts quite hairless ....
Hippocrepis comosa (p 292 **A**)

Lfts downy ....
Astragalus danicus (p 272 **G**)

12 Lvs with pairs of tiny lfts alt with larger ones along midrib ..**13**

Lvs with all side-lfts of similar size, no smaller intermediate ones......**14**

13 Lvs hairy – lfts oval-lanceolate, coarse-toothed.....................................
Agrimonia eupatoria (p 262 **H**)

Lvs hairless – lfts oblong-lanceolate, v deeply pinnatifid – broken stem smells of germaline ..
Filipendula vulgaris (p 262 **C**)

14 Lfts of **all** lvs **hairless**, round to oval, deeply toothed, 0.3–2.0 cm long – **lf**-stalks **purplish** – lvs **cucumber**-scented when crushed..
Sanguisorba minor (p 264 **J**)

Lfts of **basal** lvs **downy**, oval, ± deeply toothed, 1.0–2.5 cm long – lfts of **stem-lvs narrow-linear**, themselves **pinnate** again – lf-stalks green – no cucumber scent ....
Pimpinella saxifraga (p 334 **C**)

Lfts of **all** lvs **rough-hairy**, oval, coarse-toothed to **pinnatifid**, **3–6 cm long** ..*Pastinaca sativa* (p 358 **A**)

15 Lvs triangular in outline, lobes oval to lanceolate, 5–10 mm long ....*Daucus carota* (p 346 **C**)

Lvs triangular in outline, lfts oval, 3–4 x 1–2 cm ....
Pastinaca sativa (p 358 **A**)

Lvs oblong in outline, to 4 cm wide, lobes of twice-pinnate stem-lvs linear, to 1 cm long x 2 mm wide ....
Pimpinella saxifraga (p 334 **C**)

Lvs narrow-oblong in outline, to 2 cm wide, **repeatedly** lobed into **narrow linear** segments 2–3 mm long x **1 mm wide** ....
Achillea millefolium (p 452 **A**)

> **D. Herbs with simple, oval, oblong or cordate lvs, toothed or not, with one main vein only (not several parallel main veins)**

1 Lvs in **opp** pairs along stems **2**

Lvs **alt** along stems **11**

Lvs **only** in a basal rosette (so that not evident if alt or opp) ..**20**

Lvs in a **distinct** basal **rosette** and **also alt** along stems (in spring, no stem may be present so see also 20) **16**

2 Plant quite **hairless** – lvs untoothed, unstalked **3**

Plant **hairy**, at least on stem and lf-stalks **6**

3 Lvs **joined** in pairs, forming rings round stem, waxy grey-green, oval-triangular, pointed ......
Blackstonia perfoliata (p 350 **J**)

Lvs **not** joined in pairs round stem **4**

4 Lvs ± oval, **blunt** **5**
Lvs oval-lanceolate, 1–2 cm long, tapered to sharp **points** ....
Gentianella amarella or *G. germanica* (p 349 **C**, **D**)

5 Lvs **under** 1 cm long – stems thread-like ....
Linum catharticum (p 313 **A**)

Lvs **1–2 cm long, with** translucent **dots**, **grey**-green, oval-oblong ....
Hypericum perforatum (p 176 **A**)

Lvs 1–3 cm long, **without** translucent dots, **glossy** green with 3–7 strong veins, oval-obovate (often with basal rosette)..
Centaurium erythraea (p 350 **F**)

6 Lvs oblong, 2–5 cm long, blunt, hairy, ± **stalkless**, **untoothed**...........
Hypericum hirsutum (p 176 **D**)

Lvs oblong, 3–7 cm long, blunt, **long-stalked**, with few hairs and coarse blunt **teeth** (with basal rosette) *Stachys officinalis* (p 380 **A**)

Lvs with **cordate** base, oblong, pointed, wrinkled, **sage**-scented, 7–15 cm long, blunt-toothed ....
Salvia pratensis (p 374 **G**)

Lvs oval-triangular, to 1 cm long, with sharp **bristle**-pointed **teeth**..............*Euphrasia spp* (p 405)

Lvs oval, 2–4 cm long, base rounded, teeth v blunt or abs **7**

Lvs elliptical-lanceolate, 5–10 cm long, tapered into stalk **9**

7 Lvs **long** (2–3 cm)-stalked, **blunt**, oval, downy, **unscented** ....
Prunella vulgaris (p 380 **F**)

Lvs **short** (to 1 cm)-stalked, **pointed**, oval, **scented**, hairy **8**

8 Lvs ± smooth, side-veins **faint**, sweetly aromatic, ± purple underneath, teeth almost abs ......
Origanum vulgare (p 376 **G**)

Lvs ± wrinkled (with side-veins **impressed**), weakly aromatic, green underneath, teeth small but distinct ....
Clinopodium vulgare (p 376 **A**)

9 Lvs quite **untoothed**, sparsely hairy, blunt or pointed ....
Succisa pratensis (p 432 **E**)

Lvs with shallow blunt **teeth**, sometimes with side-lobes at base **10**

10 Basal lvs short-stalked, **hairy**, **pointed**, rough, to 12 cm long ....
Knautia arvensis (p 432 **D**)

Basal lvs long-stalked, ± **downy**, **blunt**, smooth, to 8 cm long............
Scabiosa columbaria (p 432 **C**)

11 Lower lvs (and perhaps others) with distinct narrow stalks over 2 cm long and with broad (1 cm or more) expanded blades **12**

All lvs elliptical-lanceolate, 5 cm long or more, usually untoothed, 1–2 cm wide, tapering to base to form indistinct stalk, green, rough, with short bristly hairs ......
Centaurea nigra (p 466 **A**)

Lvs narrow-linear (3 mm wide or less) one-veined, ± fleshy, yellow-green, with stalks 1–2 mm long or abs ....
Thesium humifusum (p 303)

12 Lvs cordate-based, triangular-rounded, with toothed stipules at base of stalk..................................**13**

Lvs rounded-oblong, cordate-based or not, without stipules at base..**14**

13 Lvs with ± **hairless** stalks and blades v sparsely hairy or not, **oval-triangular** – stipules narrow, their teeth long and hair-like ....
Viola riviniana (p 186 **A**)

Lvs with **v hairy** stalks and blades, blades **narrow-triangular**, pale green – stipules oval, their teeth not much longer than width of stipule ....*Viola hirta* (p 186 **C**)

Lvs with **sparsely hairy** stalks and blades, blades oval to **rounded**, **dark green** – stipules and their teeth as in *Viola hirta* ....
Viola odorata (p 188 **D**)

14 Lvs **downy**, cordate-based, oblong-triangular, basal lvs and stem-lvs similar ....
Campanula glomerata (p 414 **D**)

Lvs hairless – basal lvs and stem-lvs different – basal lvs long-stalked **15**

15 Basal lvs spoon-shaped, wedge-shaped at base, with small pointed teeth – stem-lvs oblong, deeply-toothed, ± stalkless..
Leucanthemum vulgare (p 448 **A**)

Basal lvs almost round, rounded at base, with few blunt teeth – stem-lvs narrow-linear ....
Campanula rotundifolia (p 414 **C**)

Basal lvs oval-oblong, with shallow rounded teeth, rounded or cordate at base – stem-lvs linear-lanceolate ....
Phyteuma tenerum (p 417 **C**)

16 Lvs hairless, untoothed **17**

Lvs hairy or woolly **19**

17 Rosette-lvs 3–5 in number, under 20 mm long, elliptical or obovate, stalked, smooth **18**

Rosette-lvs many, to 15 cm long x 1.0–1.5 cm wide, linear-oblong, stalkless, wavy-edged and crinkled transversely, with distinct pale midrib ......*Reseda luteola* (p 220 **B**)

18 Rosette-lvs **pointed**, **oval**, **shorter** than the pointed, elliptical-lanceolate stem-lvs..........
Polygala vulgaris (p 314 **A**)

Rosette-lvs v **blunt**, **obovate**, **longer** than the blunt, obovate stem-lvs...
Polygala calcarea (p 314 **B**)

19 Rosette-lvs v **bristly**, oblong, untoothed, short-stalked – stem-lvs stalkless, to 15 cm long x to 2 cm wide....*Echium vulgare* (p 362 **B**)

Rosette-lvs v **woolly**, elliptical-lanceolate, to 20 cm long x 3–6 cm wide, pointed, bluntly-toothed, some ascending ....
Verbascum thapsus (p 392 **A**)

Rosette-lvs **cottony** below and above, especially when young, 3–5 cm long x 2 cm wide, broad-**oval**, blunt, **untoothed**, **flat** on gd ....
Tephroseris integrifolius (p 442 **B**)

20 Lvs spoon-shaped, blunt, smooth, sparsely hairy, shallow-toothed, short-stalked, 2–4 cm long, mostly flat on gd ....
Bellis perennis (p 450 **D**)

Lvs oval, blunt, wrinkled, with impressed veins, downy, 5–15 cm long, blunt-toothed, narrowed into long (5–6 cm) winged stalk ..
Primula veris (p 232 **C**)

Lvs untoothed, elliptical-lanceolate, 3–8 cm long, white-woolly on underside, green on upperside with long scattered white hairs – short-lvd runners present ....
Pilosella officinarum (p 480 **A**)

Lvs blunt-toothed, oblong-lanceolate, 10–20 cm long, green and v hairy both sides – no runners present ....
Leontodon hispidus (p 474 **D**)

Lvs untoothed, hairless, shiny, 1–3 cm long, glossy green, oval-obovate, with 3–7 strong veins ......
Centaurium erythraea (p 350 **F**)

E. Herbs with lvs in whorls of 4–8 along the stems

1 Lvs oval-elliptical, three-veined, hairy both sides, 4 per whorl, to 25 mm long x 6–10 mm wide ....
Cruciata laevipes (p 422 **A**)

Lvs linear-obovate, one-veined, hairless or downy **2**

2 Lvs linear, 8–12 per whorl, hairless, 0.5–2.0 mm wide – stems erect ....*Galium verum* (p 422 **B**)

Lvs obovate below and 4 per whorl, linear above and 4–6 per whorl, 1–2 mm wide, all mucronate, hairless – stems ± prostrate...............
Asperula cynanchica (p 423 **F**)

Lvs obovate, 6–8 per whorl, mucronate, 2–4 mm wide, with forward-pointing prickles on edges, hairless or downy ....
Galium mollugo (p 419 **A**)

F. Trees or shrubs, erect, usually 60 cm tall or much more

1 Lvs narrow-**linear**, tough, evergreen **2**

Lvs broad, lvs (or their lfts) oval or lanceolate **3**

2 Lvs parallel-sided, blunt-tipped, dark glossy green, spreading on either side of younger branches (like barbs of a feather) – frs red, fleshy – male fls like tiny cones ....
Taxus baccata (**Yew**)

Lvs tapered to v sharp points, grey-green, in whorls of 3 along stem......
Juniperus communis (**Juniper**)

3 Lvs (or buds in winter) in **opp** pairs **4**

Lvs (or buds in winter) **alt** along twigs **11**

4 Lvs evergreen, leathery, oval-elliptical, edges curved back, untoothed, blunt, dark shiny green ....
Buxus sempervirens (p 305)

Lvs semi-evergreen, elliptical, pointed, untoothed, not leathery – edges flat – frs black berries........
Ligustrum vulgare (p 390 **C**)

Lvs deciduous **5**

5 Lvs pinnate (or, in winter, with stout black pointed buds) **6**

Lvs palmately-lobed...........................**7**

Lvs simple, unlobed, toothed or not **8**

6 Shrub climbing (by means of coiling lf-stalks or scrambling stems) – lfts tapered to pointed tips – bark fibrous, peeling in strips – frs clusters of nutlets with long white feathery-plumed styles ......*Clematis vitalba* (p 110 **A**)

Erect, non-climbing shrub or tree with smooth grey bark and large (5–10 mm long) pointed black buds – lfts oval, fine-toothed – frs like a single propeller blade ....
Fraxinus excelsior (p 390 **A**)

7 Lvs usually five-lobed, broader than long, 3–5 cm wide, lobes blunt – twigs downy – frs like a complete, two-bladed propeller ....
Acer campestre (p 316 **A**)

Lvs usually five-lobed, as long as broad, 7–16 cm wide, lobes pointed, toothed – twigs hairless – fr as above ....
Acer pseudoplatanus (p 316 **B**)

Lvs three-lobed, as long as or longer than broad – lobes sharp-toothed – frs red berries in umbels..
Viburnum opulus (p 424 **D**)

8 Lvs elliptical to oval-lanceolate, fine-toothed, 5–10 cm long – twigs hairless – frs four-lobed, salmon-coloured, seeds orange ....
Euonymus europaeus (p 304)

Lvs narrowly oval-lanceolate, untoothed, 3–5 cm long – twigs downy when young – frs black berries ....
Ligustrum vulgare (p 390 **C**)

Lvs oval ....**9**

9 Lvs wrinkled, v downy, fine-toothed, 5–10 cm long – side-veins spreading at wide angle to midrib – buds without protective scales in winter – frs red (then black) berries, flattened oval, in umbels..
Viburnum lantana (p 424 **E**)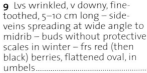

Lvs smooth, hairless – side-veins turning up towards apex – frs globular black berries....................**10**

10 Lvs untoothed – twigs purple in winter
Cornus sanguinea (p 302 **A**)

Lvs fine-toothed all round – twigs grey-brown in winter ....
Rhamnus cathartica (p 312 **A**)

11 Lvs compound – stems prickly...**12**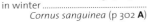

Lvs simple – stems smooth or with thorns **14**

12 Lvs compound palmate, with 3–7 oval lfts – prickles on lfts as well as on stems ....
Rubus fruticosus (p 256 **A**)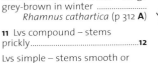

Lvs pinnate – thorns or prickles on stems ....(*Rosa*) **13**

13 Lvs **sticky**-hairy, apple-scented – arched thorns along stems – fr **red**, oval
Rosa rubiginosa agg (p 256 **C**)

Lvs hairy or not, but **not** sticky-hairy – arched thorns along stems – fr **red**, oval ....
Rosa canina agg (p 256 **D**)

Lvs hairless above – **numerous** stiff but **slender** bristles and prickles clothing stems – fr **black**, globose ..
Rosa spinosissima (p 255 **B**)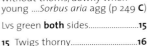

14 Lvs oval, toothed, green above, **white-felted** below – twigs without thorns, downy when young ....*Sorbus aria* agg (p 249 **C**)

Lvs green **both** sides........................**15**

15 Twigs thorny..............................**16**

Twigs without thorns **17**

16 Lvs elliptical, toothed but not lobed – frs black with bluish bloom, globular ....
Prunus spinosa (p 249 **C**)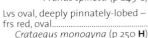

Lvs oval, deeply pinnately-lobed – frs red, oval..
Crataegus monogyna (p 250 **H**)

17 Lvs oval, smooth, quite untoothed, hairless except for silky hairs on margins and on veins below – twigs hairless – buds cigar-shaped ....
Fagus sylvatica (p 124 **G**)

Lvs toothed **18**

18 Lvs rounded, but with toothed edges and long point at tip – twigs and lvs downy .... *Corylus avellana* (p 126 **B**)

Lvs oval in outline **19**

19 Lvs oval-triangular, hairless, their teeth themselves toothed – twigs hairless, with warts .... *Betula pendula* (p 128 **E**)

Lvs rounded-oval, ± downy, teeth untoothed – twigs downy, without warts .... *Betula pubescens* (p 128 **D**)

G. Dwarf creeping shrubs, woody below, not over 20 cm tall, with lvs in opp pairs, untoothed

1 Young stems hairy – lvs under 1 cm long, oval-elliptical, green underneath, edges flat....................**2**

Young stems hairless – lvs to 2 cm long, oblong, white-downy beneath, green and ± hairless above, edges ± turned down .... *Helianthemum nummularium* (p 183 **A**)

2 Erect infl-stems v hairy on the 4 corners, hairless on the 2 wider faces .... *Thymus pulegioides* (p 371 **B**)

Erect infl-stems hairy on two opp faces, hairless on other two .... *Thymus polytrichus* (p 371 **C**)

Erect infl-stems ± round, equally hairy all round .... *Thymus serpyllum* (p 372 **D**)

H. Herbs with lvs deeply pinnatifid, but with their side-lobes joined by at least a narrow flange of lf-blade and themselves ± pinnatifid

1 Lvs with **spines** along margins ..**2**

Lvs not spiny **3**

2 Lvs hairless and shiny above, all in a close spreading rosette .... *Cirsium acaule* (p 463 **E**)

Lvs rough, grey-hairy above, in a basal rosette and also up erect stem ......*Cirsium vulgare* (p 463 **A**)

Lvs whitish-cottony both sides ...... *Carlina vulgaris* (p 470 **F**)

3 Lvs (at least those of basal rosette) with small pinnately arranged lobes at base only – upper part of lf formed by one large undivided terminal lobe **4**

Lvs with terminal lobe not obviously larger than side-lobes ..**7**

4 Lvs alt, smooth, hairless – terminal lobe of lf lanceolate, all lobes v finely toothed .... *Serratula tinctoria* (p 468 **D**)

Lvs downy, or rough-hairy, or cottony, at least below **5**

5 Lvs **alt** – side-lobes of winter rosette-lvs oblong, pointed, 3–8 mm long, shiny bright green above, cottony below – end-lobes of lvs large (to 3 cm long), oval, blunt .... *Senecio jacobaea* (p 440 **A**)

Lvs **opp** – side-lobes of rosette-lvs v short, broad – end-lobes elliptical-lanceolate to obovate, 3–6 cm long, hairy **6**

6 Lvs rough, hairy – upper lvs with broad side-lobes .... *Knautia arvensis* (p 432 **D**)

Lvs smooth, ± downy – upper lvs with v narrow side-lobes .... *Scabiosa columbaria* (p 432 **C**)

7 Side-lobes of lvs oblong, blunt, sparsely hairy, shiny dark green, often themselves pinnatifid into blunt lobes – rosette-lvs 10–25 cm long .... *Centaurea scabiosa* (p 466 **B**)

Side-lobes of lvs oblong, 1–2 cm long x 2–3 mm wide, blunt, hairless, themselves ± pinnatifid, – rosette-lvs 5–10 cm long .... *Reseda lutea* (p 220 **A**)

Side-lobes of lvs repeatedly cut into narrow linear segments 2–3 mm wide x 1 mm wide, downy ...... *Achillea millefolium* (p 452 **A**)

Side-lobes of lvs cut again into short triangular pointed segments 2–10 mm long x 3–4 mm wide, ± hairless above, cottony below .... *Senecio jacobaea* (p 440 **A**)

III. PLANTS OF HEATHLANDS (DRY AND WET), MOORS, ACID *SPHAGNUM* BOGS AND HEATHY GLADES IN WOODLANDS

These plant habitats are combined in one key because their vegetation and flora have much in common. All these habitats are acidic, and very poor in plant nutrients, so the flora is restricted to species able to tolerate these inhospitable environments.

Dry **lowland heaths** occur on sandy and gravelly geological strata in several parts of England, especially on the Lower Greensand and Bagshot Sands of the south and on Glacial Sands, washed out from former ice-sheets, further north and east, and in western Europe. Most are semi-natural habitats, resulting from clearance of former natural forests, mostly by prehistoric man. Although subsequently often cultivated, their very porous soils allowed most of the plant nutrients to be quickly washed down into lower levels, beyond the reach of plant roots, once the forest oaks were no longer present to recycle these with their deep root systems. Indeed, apart from the surface black humus, the upper layers of heathland soils are often bleached a whitish grey by the acid percolating rain (though the **deeper** subsoil is usually red-brown from the presence of washed-down iron and other minerals). The pH of the surface soil is often very low (3.5 or less), and hence there is great acidity.

Heathland vegetation is very limited in species; mostly low evergreen shrubs of the Ericaceae and some low evergreen spiny shrubs of the Leguminosae predominate. Most of these can be identified without flowers at any season. More open areas beside paths often have a greater variety of herbs.

Wetter areas of heathland have some of the same plants, but the better water supply brings in a greater variety of species. In areas permanently wet, even in summer, such as some valleys and hollows on lowland heathland, the water logging prevents decay of vegetable litter and peat is formed, leading to the development of acid bog. This is built up and becomes dominated by *Sphagnum* mosses of various species; the yellow or green *Sphagna* tend to colonize the water-filled hollows first, followed by the red species. The latter build up hummocks, on which various heath-type shrubs tolerant of wet conditions, such as Cross-leaved Heath and Cranberry, will grow. Various sedges and insect-catching plants also occur, but these are not easy to identify in winter.

In wetter upland areas **heather moors** occur with a similar flora. In wet lowland basins extensive **raised** or **domed** bogs (so-called from their convex form) or concave **valley** bogs develop, and blanket peat bogs on the flat tops of the wetter hill and mountain areas, and at low levels in parts of western Ireland. Where, however, nutrient-rich water enters a bog or moorland, or there is a more rapid movement of water as on a slope, a **flush** area may develop, often much greener (because less acid) and more species-rich than the surrounding areas, that are dull brown from the prevailing heath-type vegetation. For some flush species see Key IV.

MASTER KEY (Plants confined to bogs, or wetter heaths, are indicated in key)

1 Trees, and large shrubs over 2 m tall ....**(see Key IA, p 50)**
Low shrubs, under 2 m tall, with **woody stems**, erect or creeping along the ground ....**A**
Herbs **without** woody stems **2**

2 Lvs ± linear, narrow, **grass-like**, in tussocks, tufts, rosettes, or along stems ....**B**
Lvs **not** grass-like, but with broader blades ....**C**
Ferns with pinnately-divided fronds..**D**

A. Low shrubs, under 2 m tall, with woody stems, erect or creeping along the ground

1 Evergreen shrubs with **spiny shoots** **2**

Shrubs (evergreen or not) **without** spines...**5**

2 Without obvious lvs – stems bearing **pinnately-branched green spines** along their length – frs like tiny **hairy** pea-pods ....(**Gorses**) **3**

With oval, pointed, hairless lvs 2–8 mm long – stems bearing **unbranched, yellowish-brownish** spines, 1–2 cm long, along their length – frs like tiny **hairless** pea-pods ....*Genista anglica* (p 268 **A**)

3 Spines **deeply furrowed**, v rigid, 1.5–3.0 cm long – main branches with few hairs – shrub 60–200 cm tall – fls mostly in **spring** ....
Ulex europaeus (p 268 **D**)

Spines **faintly furrowed**, 1.0–2.5 cm long – main branches **v hairy** – shrubs 10–100 cm tall – fls in **late summer** **4**

4 **Spines stout**, **v rigid**, 1.0–2.5 cm long – main branches ± erect, to 1 m tall.................*Ulex gallii* (p 268 **F**)

Spines **weak**, **flexible**, **thin**, to 1 cm long – main branches creeping or straggling, usually under 50 cm tall ....*Ulex minor* (p 268 **D**)

(Note: certain identification of *Ulex gallii* and *U. minor* is impossible without fls; but the former is c in W of GB and Ire, **whilst** the latter ± replaces it in SE Eng)

5 Erect shrub to 2 m, with masses of long, green, **five-ridged**, hairless little-branched twigs – lvs on young shoots trifoliate or undivided, lfts elliptical – frs like tiny (4 cm long) black, ± hairy pea-pods......*Cytisus scoparius* (p 268 **G**)

Tiny prostrate shrub, creeping over mosses in **sphagnum** bogs, with leathery, oval-oblong lvs 4–8 mm long, dark green above, waxy-grey green below, edges rolled under, alt along thin woody stems – frs red berries 6–8 mm long, on 2–3 cm slender stalks – local....................
Vaccinium oxycoccos (p 223 **D**)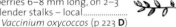

Bushy shrubs, ± **erect**, but with creeping rhizomes at base, 10–60 cm tall, with oval or linear, **unlobed** lvs 1–30 mm long **6**

6 Lvs narrow-linear to oval, evergreen, **stalked**, edges sharply rolled under, lvs **2–7 mm long**, **spreading**, in **whorls of 3–4** (or 5) – frs dry ....(*Erica*) **7**

Lvs **narrow-linear**, evergreen, **stalked**, edges strongly rolled under, **4–6 mm long**, **spreading**, dense-set, and **alt** along stems, **not** whorled – frs black berries – N of Br Isles ....
Empetrum nigrum (p 220 **C**)

Lvs v narrow **triangular**, evergreen, ± **stalkless**, edges strongly rolled under, 1–2 mm long, **adpressed** to stem in **opp pairs**, resembling tiny hat-pegs – frs dry – in dry hths and bogs ....
Calluna vulgaris (p 226 **A**)

Lvs **oval** to linear-**lanceolate**, **alt**, **3–30 mm long**, edges flat or slightly rolled under, evergreen or not..**8**

7 Lvs in **whorls of 4, 2–4 mm long**, **linear**, **grey-downy**, with long white sticky-tipped hairs – dead fls in terminal **heads** – in wet hths and bogs ....
Erica tetralix (p 226 **B**)

Lvs in **whorls of 3, 5–7 mm long** linear, **hairless**, **dark green** or **purplish**, **shiny** – dead fls in **short racemes** – on dry hths only ....
Erica cinerea (p 226 **C**)

Lvs in **whorls of 3, 2–3 mm long**, **oval**, ± **hairless** above, with long sticky-tipped hairs on **edges**, **white** beneath, margins only slightly rolled under – dead fls in **long** racemes – in damp hths in SW Eng ....
Erica ciliaris (p 226 **D**)

8 Lvs **deciduous**, **thin**-textured, toothed or not...................................**9**

Lvs **evergreen**, **leathery**, untoothed, hairless..........................**11**

9 Young twigs and buds silky-downy – lvs oval to elliptical-lanceolate, greyish and ± silky-hairy below, ± hairless on upper-sides, 1–4 cm long – grey-green (female) or golden (male) catkins in spring – frs **dry capsules** with white plumed seeds in summer ..
Salix repens (p 194 **D**)

Young twigs and buds downy – lvs oval, blunt, **downy** below, hairless on uppersides, 3–5 mm long – frs like tiny **downy** pea-pods – GB vr ....
Genista pilosa (p 268 **B**)

Young twigs and buds hairless or ± so – lvs hairless, 10–30 mm long – fls globular – frs **blue-black berries**.....................................**10**

10 Lvs **oval**, **pointed**, **toothed all round**, **light green** – younger twigs conspicuous when lfless in winter, **bright green**, with **4 angles** or **flanges** along their length – c in dry wds, hths, moors
Vaccinium myrtillus (p 222 **A**)

Lvs oval-**obovate**, blunt, untoothed, blue-green – twigs **brownish**, **cylindrical** on moors, bogs – N of GB lc...................................
Vaccinium uliginosum (p 222 **C**)

11 Lvs **flat**, **obovate**, blunt, fresh green, 1–2 cm long, with net-veining conspicuously pale – frs red berries – on hths, moors – N of Br Isles ....
Arctostaphylos uva-ursi (p 228 **A**)

Lvs with **rolled-under edges**, 1–3 cm long, obovate, blunt, **dark glossy** green above, **paler** green and dotted below, net-veining **inconspicuous** frs red berries – on moors – N of Br Isles ....
Vaccinium vitis-idaea (p 222 **B**)

Lvs with **strongly** rolled-under edges, linear-lanceolate to elliptical, ± pointed, ± glossy green above, **waxy-white** below, 1.5–3.5 cm long – frs dry capsules – If in *sphagnum* bogs ....
Andromeda polifolia (p 228 **B**)

B. Herbs with lvs ± linear, narrow, grass-like, in tussocks, tufts, rosettes or along stems

1 Lvs (or narrow lf-like stems, if no lvs visible) all v narrow (1–2 mm wide), and **bristle-like**, not hollow, but cylindrical or three-edged, in a ± erect dense basal **tussock** or **tuft** **2**

Lvs **not** bristle-like, each with an obvious **blade**, either flat or folded along its length...................**5**

Lvs cylindrical, 1–4 mm wide, hollow, with **cross-partitions** inside, borne from base or **alt** along stems – in bogs...................**12**

Lf-like stems cylindrical, 1.5–3.0 mm wide, **spongy** inside, in tuft from base of plant, bearing lateral infls near tip – in bogs **13**

2 Tussock composed of **dark green**, **three-angled lvs**, **1** mm wide and to 20 cm long, with **pale inflated** sheathing bases – stems (if present) bear shorter lvs with v inflated sheathing bases – in bogs and wet hths – r to S, c to N and W of Br Isles..............................
Eriophorum vaginatum
(**Hare's-tail Cottongrass**)

Tussock composed of **cylindrical, ridged stems**, 1–2 mm wide and to 30 cm long, each stem with tiny narrow basal **lf-sheaths**, with blades only 2–3 mm long, the whole tussock conspicuously **orange-brown** from Oct to March – remains of infls (ovoid, 3–5 mm long) sometimes visible on stem-tips – in bogs, wet hths ....
Trichophorum cespitosum
(**Deergrass**)

Tussock composed of shorter (10–15 cm long) ± **cylindrical lvs** 0.2–0.4 mm wide, grooved or not on upper side – in dry hths............**3**

3 Plant with a dense crowded basal tuft of erect, **pale**, tough, **stout shining lf-sheaths**, from which spread the grey-green ascending **bristle-like** lf-blades ± at 90°, in a rosette – infls one-sided spikes – on hths, moors ....
Nardus stricta (**Mat-grass**)

Plant with the bristle-like lf-blades rising at an **acute** angle to their **slender** sheaths, which are **neither** tough nor shiny – in dry hths **4**

4 Lvs grooved above, **grey**-green and rough beneath – ligule (**a**) at tip of sheath (**b**) **narrow, pointed, 2–4 mm** long – florets of infl with projecting **bristles** – in dry hths in SW of GB ....
Agrostis curtisii (**Bristle Bent**)

Lvs grooved above, bright green and smooth – ligule (**a**) **blunt, less than 0.5 mm** long, scarcely visible – florets **without** bristles
Festuca filiformis
(**Fine-leaved Sheep's-fescue**)

Lvs **dark** green, shiny, rough-edged, **not** grooved above – ligule c. **1 mm long**, with **square-cut** tip – florets **silvery**, with bristles ....*Deschampsia flexuosa*
(**Wavy Hair-grass**)

5 Lvs like those of a **tiny iris, flattened in one plane**, all **edge-on** to stem, orange in autumn – erect dead straw coloured fr-spikes persistent in winter, with frs conical – in *Sphagnum* bogs ....
Narthecium ossifragum (p 511 **F**)

Lvs **not** iris-like, **not** flattened in one plane..**6**

6 Lvs **tough**, glossy dark green, wiry, grooved above, 3–5 mm wide at base, **spreading at 90°** to their erect basal sheaths in a **flat** rosette like a sweep's brush – on bare or trampled places on hths ..
Juncus squarrosus (**Heath Rush**)

Lvs **not** spreading at 90° to their sheaths, not in a **flat** rosette **7**

7 Lvs in an ascending rosette, with channelled blades, to 6 mm wide, and to 12 cm long, bright or grey-green, with **long white hairs** thinly clothing the blades – no ligule to lvs ....*Luzula multiflora*
(**Heath Wood-rush**)

Lvs with ± **hairless** blades..............**8**

8 Lvs arranged in 3 vertical ranks, on solid stem that is triangular in section below **9**

Lvs in 2 opp vertical ranks on hollow round stem **10**

9 Lvs with purplish-red three-angled tips – fr-heads 2–5 per infl, cottony – in bogs ....
Eriophorum angustifolium
(**Common Cottongrass**)

Lvs with tips neither purplish-red nor three-angled ....
(**Sedges**, *Carex* spp) **14**

10 Lvs flat, **ribbon**-like, to 10 mm wide x 50 cm long, grey-green, almost hairless, in **large, dense tussocks** to 80 cm across – ligules of lvs composed of a fringe of hairs only – dead lvs in winter (Nov–May) **straw**-coloured, curled – in wet hths, bogs........................*Molinia caerulea*
(**Purple Moor-grass**)

Lvs **not** ribbon-like, **not** in dense tussocks – lf-blades **linear-lanceolate**, to 15 cm long x 1–5 mm wide – **ligules membranous** – creeping lfy stems present............**11**

11 Ligules 2–5 mm long, pointed – infl-branches ascending................
Agrostis canina (**Velvet Bent**)

Ligules 0.5–2.0 mm long, blunt –
infl-branches spreading ....
Agrostis capillaris (**Common Bent**)

12 Erect plant, 30–100 cm tall –
lvs 2–4 mm wide, stiff, ascending,
each of one tube, with internal
cross-partitions v distinct – in
bogs.......................*Juncus acutiflorus*
(**Sharp-flowered Rush**)

Creeping plant with ± erect
tufted stems, 10–20 cm tall – lvs
1–2 mm wide, flaccid, each of
two tubes, with many indistinct
cross-partitions – in bog pools,
wet hollows.........*Juncus bulbosus*
(**Bulbous Rush**)

13 Stems ± smooth............................
Juncus effusus (**Soft-rush**)

Stems with many lengthwise
ridges.........*Juncus conglomeratus*
(**Compact Rush**)

14 Lfy shoots many, in a dense
tussock – lvs 2–3 mm wide,
deeply channelled, bright green,
arched outwards – on dry hths ....
Carex pilulifera (**Pill Sedge**)

Lfy shoots few, in a loose tussock
– lvs 5–10 mm wide, slightly
channelled, glossy green but
orange-brown tipped in winter –
on dry hths................*Carex binervis*
(**Green-ribbed Sedge**)

Lfy shoots ± separate, hardly in
tussocks – lvs 3–5 mm wide,
waxy grey-green – on wet hths
and bogs....................*Carex panicea*
(**Carnation Sedge**)

Not as any of above 3
alternatives ....
Carex spp, see vegetative key in
BSBI Sedges Handbook.

**C. Herbs with lvs not grass-
like, but with broader blades**

1 Lvs ± circular, green, hairless,
2–3 cm wide, held aloft on
stalks, like tiny, flat umbrellas –
in **bogs** ....
Hydrocotyle vulgaris (p 344 **E**)

Lvs **all** in **a basal rosette**..................2

Lvs along the stems in **whorls** of
4 or more **6**

Lvs **alt** along the stems, **not** in
whorls, **not** all in a basal rosette –
on dry hths **8**

Lvs in opp pairs **along** stems – lvs
orbicular-oval, 3–8 mm long –
plant creeping in **bogs** ....
Anagallis tenella (p 234 **F**)

2 Lvs **without** stalks, tapering
from broad base to narrow tip,
untoothed **3**

Lvs with **distinct** stalks – lf-blades
undivided, untoothed, broader
than their stalks, clothed with
bright **red sticky hairs** like
tentacles – in wet hths or **bogs** ....
(*Drosera* spp) **5**

Lvs **pinnatifid** or **pinnately-
compound**, without red sticky
hairs – on open dry sandy hths ..**7**

3 Lvs **greasy-looking**, spreading
star-wise on ground, unspotted –
in bogs.........................(*Pinguicula*) **4**

Lvs glossy, but not greasy-looking,
in 2 ± opp vertical ranks, with
transverse purple spots ....
Dactylorhiza maculata (p 542 **B**)

4 Lvs yellow-green, **pointed**, in a
starfish-like rosette – in **basic**
bogs *Pinguicula vulgaris* (p 410 **A**)

Lvs **olive**-green, **blunt**, purplish-
tinged – in **acid** bogs ....
Pinguicula lusitanica (p 410 **C**)

5 Lf-blades almost **circular**, 1–2
cm wide, on stalks 2–4 cm long –
lc *Drosera rotundifolia* (p 182 **A**)

Lf-blades **obovate** to spoon-
shaped, **0.5–1.0 cm long, quickly
narrowed** into stalk – 1 ....
Drosera intermedia (p 182 **B**)

Lf-blades **linear-obovate** to
**linear-oblong, 2–5 cm long,
gradually narrowed** into stalk – vl
Drosera anglica (p 183 **C**)

6 Lvs 6–8 per whorl, obovate,
tips **mucronate**, prickles on edges
pointing **forwards** – on dry grassy
hths ....*Galium saxatile* (p 419 **B**)

Lvs 4–6 per whorl, linear-oblong to obovate, tips **blunt**, prickles on edges pointing **backwards** – in **bogs** ......*Galium palustre* (p 420 **E**)

7 Lvs pinnatifid, 2–6 cm long, with lobes **linear**, downy, often lobed themselves, in outline like deers' antlers – on dry sandy gd ......
Plantago coronopus (p 388 **E**)

Lvs **pinnate**, 2–10 cm long, with lfts **oval**, **pinnatifid**, **hairy** – on dry sandy gd...
Erodium cicutarium (p 322 **E**)

Lvs **pinnate**, 1.5–4.0 cm long, with lfts **elliptical**, **untoothed**, **downy** – on dry sandy gd ....
Ornithopus perpusillus (p 272 **H**)

Lvs **pinnatifid**, 2–5 cm long, with lobes **themselves deeply pinnatifid**, ± **hairless** – on damp hths. ......
Pedicularis sylvatica (p 406 **B**)

8 Lvs **stalkless** on stem, **palmately divided** into 5 obovate, sharply-toothed lfts (actually 3 lfts + 2 lf-like stipules) – basal lvs long-stalked, trifoliate ....
Potentilla erecta (p 259 **A**)

Lvs **stalked** in **rosette and on stem**, **hastate-shaped**, with 2 narrow side-lobes spreading at 90° and turning forward at tips ....
Rumex acetosella (p 166 **C**)

D. Ferns with pinnately-divided fronds

Tall (1–2 m) fern with v broad 3-times pinnate fronds – fronds arising separately from ground – plant forming brownish litter on ground Nov–May – in dry places.
Pteridium aquilinum (**Bracken**)

Shorter (0.5–1.0 m) ferns with oblong-lanceolate twice-pinnate fronds – fronds arising in crowns or rosettes together – in damp places...........................*Dryopteris* etc
(**Buckler-ferns**)

Short (20–50 cm) fern with narrow lanceolate once-pinnate fronds arising in crowns – in damper places, ditches.....................
Blechnum spicant (**Hard-fern**)

(For further details consult *The Fern Guide*, details in the **Further Reading** section, p 23.)

IV. PLANTS OF FENS AND MARSHY MEADOWS

This key covers the vegetation of nutrient-rich, often alkaline, waterlogged areas. Where such water-logging has been prolonged, preventing the decay of plant remains, peat will develop (though less rapidly than in an acid bog), and a fen results.

Fens tend to have very different floras from those of acid bogs. Instead of *Sphagnum* moss with various heath-type shrubs and a few sedges, many different kinds of sedges, grasses and broad-leaved herbs occur, so that the flora may be almost as rich as in a chalk grassland. Unfortunately, however, most of the herbs die right back in autumn, while the grasses and sedges leave only very similar-looking leaves, so that identification without flowers in winter (and, indeed, until early summer) is difficult. But this key should enable the user to name at least some of the commoner or more distinctive species on leaf characters.

Fens are unstable, and if not mown regularly are soon invaded by shrubs and become wet woodland. Meadows are often partly drained areas of former fen, mown for hay, or else used for summer grazing; they (like mountain flush areas, see Key III) have some of the same plants as fens. For acid bogs, see also Key III.

1 Lvs **circular**, 2–4 cm across, held ± horizontal on erect stalks, margin bluntly-toothed – stem creeping ....
Hydrocotyle vulgaris (p 344 **E**)

Lvs trifoliate – lfts 5–10 cm long, oval-elliptical, untoothed, borne alt and ± erect on **long** fleshy stalks from base of plant..................
Menyanthes trifoliata (p 356 **B**)

Lvs trifoliate – lfts 5–10 cm long, elliptical-lanceolate, toothed, borne opp on **short** stalks on erect stem to 1 m tall ....
Eupatorium cannabinum (p 453 **C**)

Lvs **linear**, 30–60 cm long x 3–4 cm wide, erect, **folded** lengthwise, and all **flattened** in one plane ....
Iris pseudacorus (p 518 **A**)

Lvs **once-pinnate**....................................**2**

Lvs **twice-pinnate** **7**

Lvs **palmately-lobed** **8**

Lvs in **whorls** of **4 or more** lvs, up stems..**9**

Lvs simple, oval-lanceolate, **alt** up stems **10**

Lvs **simple** oval-lanceolate in outline, either lobed, toothed or untoothed, in opp pairs (or **whorls** of 3) up stems **11**

Lvs **linear**, with **grass-like blades 14**

Lvs either **abs**, plant with cylindrical stems only, or with the lvs **linear**, **cylindrical** **17**

2 Lfy **stipules** present at base of lf-stalks – lfts 5–7 in number, 3–6 cm long, obovate, deeply-toothed, dark green above, grey below ....
Potentilla palustris (p 262 **A**)

Lfy stipules **abs** **3**

3 All lvs pinnate **4**

Upper lvs pinnate or pinnatifid, with terminal lft wider and longer than side ones, all lfts untoothed – **basal lvs** oval-elliptical, simple, untoothed, long-stalked ....
Valeriana dioica (p 428 **B**)

4 Lfts blunt, oval-elliptical, **scarcely-toothed** – terminal lft larger than side ones ....
Rorippa nasturtium-aquaticum (p 210 **E**)

Lfts pointed, oval-elliptical, teeth prominent – terminal lft not obviously larger than side-lfts **5**

Lfts linear, untoothed, stems and lf-stalks hollow, inflated, ridged ....
Oenanthe fistulosa (p 342 **D**)

5 Lfts smooth, veins **pinnate**, **not** impressed – stem hairless **6**

Lfts with veins prominent but **pinnate**, side veins **not** parallel – lfts ± **white downy** below – terminal lft three-lobed – pairs of tiny lfts alt with pairs of larger ones along lf-stalk – stem ± hairless – broken stem smells of germaline ....
Filipendula ulmaria (p 264 **B**)

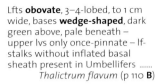

No pairs of tiny lfts between larger ones – lfts with midrib and side veins prominent and **deeply impressed**, side veins running almost **parallel** with midrib – stem **hairy** below ....
Valeriana officinalis (p 428 **A**)

6 Dark line visible (against the light) across lf-stalk below lowest pair of lfts – lfts v sharp-toothed, **dull** green, in 7–10 pairs ....
Berula erecta (p 344 **B**)

No dark line visible across lf-stalk – lfts with blunt teeth, **shiny** bright green, in 4–6 pairs ....
Apium nodiflorum (p 344 **C**)

7 Lfts **oval**, **pointed**, fine-toothed, 2–8 cm long x 2–3 cm wide ....
Angelica sylvestris (p 336 **A**)

Lfts **linear-lanceolate**, sharp-toothed, 2–10 cm long x 1 cm wide..............*Cicuta virosa* (p 336 **C**)

Lfts **pinnatifid**, their lobes lanceolate, bluntish, 1–2 cm long x 0.2–0.5 cm wide, mucronate, with **rough** edges ....
Peucedanum palustre (p 346 **D**)

Lfts linear, untoothed, ± blunt, with **smooth** edges ....
Oenanthe lachenalii and other *Oenanthe* spp (p 342 **G**)

Lfts **obovate**, 3–4-lobed, to 1 cm wide, bases **wedge-shaped**, dark green above, pale beneath – upper lvs only once-pinnate – lf-stalks without inflated basal sheath present in Umbellifers ......
Thalictrum flavum (p 110 **B**)

8 Lvs fleshy, hairless, glossy, lobes blunt ....
Ranunculus sceleratus (p 102 **H**)

Lvs firm, downy, dull green, lobes v pointed and narrow............................
Ranunculus acris (p 102 **B**)

9 Lvs **obovate**-lanceolate, **mucronate** – backward-pointing prickles (v tiny) on lf-edges...............
Galium uliginosum (p 420 **D**)

Lvs **elliptical**-lanceolate, **blunt** – prickles as above ....
Galium palustre (p 420 **E**)

10 Stem-lvs to 25 cm long x 2.5 cm wide, lanceolate – **basal** lvs oval, blunt, cordate-based, long-stalked, to 20 cm long x 8 cm wide (autumn to spring only) ....
Ranunculus lingua (p 104 **B**)

Stem-lvs to 10 cm long x 1.5 cm wide, lanceolate – basal lvs as above in shape, but to 4 cm long x 3 cm wide only, persisting in summer...
Ranunculus flammula (p 104 **A**)

11 Lvs oval, with **blunt teeth**, **stalked**, **mint-scented** – stems four-angled ....*Mentha aquatica*, and other Mints (p 370 **A**)

Lvs elliptical in outline, **deeply-toothed** in upper part, to **pinnatifid** especially in lower part of lf, **stalked**, **unscented** – stems four-angled ....
Lycopus europaeus (p 371 **A**)

Lvs **untoothed**, **stalkless**................**12**

12 Tiny creeping plant, rooting at intervals – lvs oval, under 1 cm long, in opp pairs ....
Anagallis tenella (p 234 **F**)

Tall erect (30–100 cm) plants – lvs in opp pairs or whorls of 3, pointed and downy **13**

13 Lvs 5–12 cm long, **oval**-lanceolate, light green, sticky-hairy, with orange or black glands present – side-veins pinnate – fr-head **pyramidal** – stem **round** ....
Lysimachia vulgaris (p 234 **A**)

Lvs 4–7 cm long, lanceolate, not **sticky**-hairy, dark green – main side-veins turning up **parallel with midrib** – fr-head **cylindrical** – stem **four-angled**, at least when young ....
Lythrum salicaria (p 295 **E**)

14 Lvs with only shortly sheathing bases, flat, **grey-green**, 1 cm wide or more, x 60–130 cm long, not keeled, all from base of plant..........*Typha latifolia* (p 502 **A**)

Lvs alt with sheathing bases, up stem in two vertical rows – stems hollow ....**Grasses**

Lvs with sheathing bases, in 3 vertical ranks on stem – stems solid ....(**Sedges**) **15**

15 Shoots forming large **tussocks** raised 30–60 cm above ground, on a cylindrical fibrous pillar......**16**

Shoots not in large tussocks ....
other Sedges

16 Lvs 5–7 mm wide, fresh green
Carex paniculata
(**Greater Tussock-sedge**)

Lvs 3–6 mm wide, grey- or blue-green ..*Carex elata* (**Tufted-sedge**)

17 Stems tall (1–3 m), stout (over 1 cm thick), interior spongy..............
Schoenoplectus lacustris
(**Common Club-rush**)

Stems less than 0.5 cm thick......**18**

18 Lvs abs – stems spongy inside – infl on side of stem below tip **19**

Cylindrical hollow lvs present, with cross-partitions inside **20**

19 Stems ribbed lengthwise, brightish green, 3–5 mm thick ......
Juncus conglomeratus
(**Compact Rush**)

Stems not noticeably ribbed lengthwise, bright green, 3–5 mm thick ....
Juncus effusus (**Soft-rush**)

Stems ribbed, grey-green, 2–3 mm thick ....*Juncus inflexus*
(**Hard Rush**)

20 Vertical as well as cross-partitions inside hollow lvs ....
Juncus subnodulosus
(**Blunt-flowered Rush**)

Cross partitions only, inside hollow lvs **21**

21 Lvs flattened strongly – frs oval, pointed abruptly, usually about 30 seeds....................................
Juncus articulatus (**Jointed Rush**)

Lvs scarcely flattened – frs conical, tapered to a long point, usually about 12 seeds ....
Juncus acutiflorus
(**Sharp-flowered Rush**)

V. FLOATING AND SUBMERGED AQUATIC PLANTS

Open water is a very interesting and important plant habitat, but it is a specialized one, and gives its flora a very different appearance from that of land habitats. The vegetation varies according to a number of factors: (1) the depth of the water (there is more light near the surface, and it is warmer); (2) the amount of current (plants adapted to grow in fast-running rivers, where there is much more oxygen, are often very different species from those of stagnant ponds); (3) the nature of the bottom (the species growing on the bottoms of stony or sandy-bedded lakes or rivers are different from those growing on muddy ones); (4) the turbidity or clarity of the water (upland streams or lakes over hard rocks are much clearer and hence let more light through to plants than lowland ones over clays or loams – though lowland chalk streams are often very clear); (5) the nutrient content, or the pH of the water. Some of these factors are also affected by environmental pollution; where this is severe, almost no plants except some algae (and a few Water-lilies) will grow, as in parts of the Norfolk Broads, and in some midland and northern English rivers.

Aquatic plants tend to be either submerged, with narrow translucent leaves, or floating, when their leaves may be broad and opaque; both types of leaf may occur on one plant. Most aquatic plants pass the winter as buds at the bottom of the water, so this key is unlikely to be very useful then. However, many water plants have such tiny inconspicuous flowers or even inflorescences, extremely difficult to interpret or even to dissect, that vegetative identification is often easier than that based on flower structure. Thus, this key will be most useful in high summer.

In general, the user of this book is asked not to collect plants at all, but to bring the book (and the hand lens) to the plant *in situ*. With water plants, however, this rule must be modified. In most cases it is necessary to collect bits of plants in order to see their details; a grapnel (three hooks fastened together on the end of a long string) is most effective for this purpose. The plant should be thrown back into the water after study. It will almost certainly grow again if it has not been allowed to dry out.

A very few flowering plants (species of Eelgrass, *Zostera*) are confined to salt water, either on the shallow offshore sandbanks or mudbanks that exist on more sheltered coastlines, or else in shallow estuaries. Quite a number of aquatics, however, occur in slightly salt (brackish) water in estuarine areas behind sea walls out of direct reach of the tides.

MASTER KEY

Rounded, flat, long-stalked floating lvs like those of Water-lilies...**A**

Lvs grass-like to oval, opp or alt, undivided and unlobed, normally **over** 5 cm long,
borne along **branched** stems, but **not** in close whorls ....**B**

Lvs translucent, **strap**-shaped, **under** 3 cm long and less than 10 times as long as
wide, in **whorls** of 3 or more along the stems..**C**

Lvs linear, strap- or quill-like, undivided, in ± erect or spreading crowns, tufts or
rosettes, growing on the bottom of lakes or ponds...**D**

Lvs **much divided** into narrow segments, normally submerged (flat, shallow-lobed
lvs with broad blunt divisions may also be present) ....**E**

Lvs, oval or strap-shaped under 3 cm long, less than 10 times as long as wide,
in **opp** pairs along the stems ....*Callitriche* (p 384)

Lvs like long (50–100 cm) green ribbons to 2 cm wide, floating on surface, but
attached to submerged stem ....*Sparganium* spp (p 500)

Plants **either** small, round or oval discs 3–10 mm across, **or** of pointed elliptical
translucent green scales 5–12 mm long, jointed together by stalks, **without**
distinction into stem and lvs – one or more unbranched slender roots
hanging from undersides – free-floating ....*Lemna* spp or *Spirodela* (p 497)

Plants like tiny (0.5–1.0 mm) green grains of sand, without roots, floating
on water ....*Wolffia* (p 497)

A. Plants with lvs rounded, flat, floating, like those of Water-lilies

Lvs 3–10 cm across, orbicular, cordate-based – lvs of floating stems opp ....
Nymphoides peltata (p 356 **C**)

Plants without floating stems – lvs normally over 10 cm across, all from a rhizome deep in water or mud ....**Water-lily family**, *Nymphaeaceae* (p 99)

Lvs 2–3 cm across, round kidney-shaped, borne in tufts from floating stems, with large chaffy stipules at their bases ....
Hydrocharis morsus-ranae (p 485 **A**)

B. Plants with lvs grass-like to oval, opp or alt, undivided and unlobed, over 1.5 cm long, borne along branched stems, but not in close whorls

1 In the sea – lvs **ribbon**-like, 6–50 cm long, 1–10 mm wide – fls in **flattened spikes** ± **enclosed** within **lf-sheaths** ..*Zostera* (p 494)

In fresh or **brackish water** – lvs **not** ribbon-like, mostly (if narrow) under 10 cm long – fls **not** in flattened spikes enclosed within lf-sheaths **2**

2 Frs and fls in stalked **spikes** in axils of upper lvs, each fl with 4 perianth segments or, if **without** fls or frs, lvs elliptical to oval ....
Potamogetonaceae (see key p488)

Frs and fls in **stalked umbels** – lvs narrow, grass-like ..*Ruppia* (p 493)

Frs and fls in small **clusters** in lf axils – lvs narrow **3**

Frs and fls **abs** **4**

3 Lvs in **opp** pairs, **untoothed** – frs stalked, with beaks 0.5–2.0 mm long..........*Zannichellia* (p 494)

Lvs **opp**, or in **whorls** of 2–3, **toothed** – frs unstalked, ovoid, with 3 styles................*Najas* (p 494)

4 Lvs **opp** **5**

Lvs **alt** (but may be in an opp pair when infls present in their axils), linear to oval, with membranous **stipules** ....
Potamogeton (see key p 488)

Lvs 3 per whorl, or in opp pairs, toothed, linear to oblong, with no stipules but with sheathing bases ....*Najas* (p 494)

5 Lvs **oval-triangular**, **without** stipules, **finely-toothed**, 1.5–4.0 cm long x **5–15 mm wide**.................
Groenlandia densa (p 493)

Lvs **linear**, **flat**, with **sheathing stipules**, **untoothed**, 1.5–6.0 cm long x **0.5–2.0 mm wide** ....
Zannichellia (p 494)

Lvs **linear**, or **thread-like**, with **sheathing bases** but **no** stipules, untoothed, 5–10 cm long x **0.5–1.0 mm wide**....*Ruppia* (p 493)

C. Submerged waterweeds with lvs translucent, ± strap-shaped, under 5 cm long and less than 10 times as long as wide, in whorls along the stems or densely spirally arranged, resembling Canadian Waterweed (p 486 C)

1 Stems **stout** (to 6 mm wide), **spongy** inside: leaves light green, **strap**-shaped, to 5 cm long x to 3 mm wide, in **whorls** of 6–12............
Hippuris vulgaris (p 382 **43**)

Stems **slender**, **weak**, **translucent**, **not more** than 3 cm long...............**2**

2 Lvs **spirally** arranged along stem, blunt-toothed, usually crowded and arched back................
Lagarosiphon major
(**Curly Waterweed**)

Lvs in **whorls** of 3 (to 5) the whorls crowded or loose **3**

3 Lvs **linear-oblong** to oblong-ovate, mostly **blunt**, 0.7–2.3 mm wide *c.* 0.5 mm below apex ....
Elodea canadensis (p 486 **C**)

Lvs linear or linear lanceolate, **tapered to narrow points**, 0.2–0.7 mm wide *c.* 0.5 mm below apex **4**

4 Lvs linear, sharp-pointed, to 4 cm long ....*Egeria densa*
(**Large-flowered Waterweed**)

Lvs oblong-linear, under 2 cm long...**5**

5 At least some lvs strongly **arched back**; lf-blades often strongly **twisted**; side-root tips white or greyish...................................
Elodea nuttallii (p 486 **D**)

Lvs **never** strongly arched back; lf-blades rarely much twisted; side-root tips red...
Elodea callitrichoides
(**South American Waterweed**)

D Aquatic plants with lvs linear, strap- or quill-like, undivided, most in ± erect crowns, tufts or rosettes, growing on the bottom of lakes or ponds

1   Plants with tufts of lvs **connected** by creeping runners or rhizomes **2**
Plants with tufts of lvs in **separate** crowns or rosettes, **not** connected to other tufts by runners or rhizomes **5**

2   Lvs firm, opaque, v narrow (1.5 mm wide or less) either borne singly or in tufts....................**3**

Lvs either fleshy, or translucent, 2.0 mm wide or more in erect crowns or rosettes **4**

3   Lvs **solid**, arising **singly** from creeping rhizome, tips of young lvs coiled in a spiral – small fern, with hard pill-like swellings on rhizome – in ponds on hths ....
WO (NI) FPO (Eire) BAP * *Pilularia globulifera* (**Pillwort**)

Lvs **hollow**, each composed of **two tubes** within, arising in **tufts**, tips **not** coiled..
Juncus bulbosus (**Bulbous Rush**)

4 Lvs fleshy, **opaque**, **cylindrical**, **parallel**-sided to **blunt** tip, **spongy** inside ....
Littorella uniflora (p 389 **F**)

Lvs **translucent**, flattened, **tapered** to pointed tip, with the **hollow** inside divided by **cross-partitions** – W Ire, NW Scot only, vl ....
Eriocaulon aquaticum (p 499 **G**)

5 Lvs ± **cylindrical** above, **tapered to pointed** tips, composed of **4 tubes** (**a**) internally, **with** cross-partitions – spore-capsules (**b**) usually present sunken in expanded lf-bases – (mostly in mt lakes, lf – non-flowering plants) ....
Isoetes spp (**Quillworts**)

Lvs **flattened**, with **blunt**, rounded tips, composed of **2 tubes** internally, **without** cross-partitions – no spore-capsules but sometimes with tall (30–40 cm) fl- or fr-spike – N of Br Isles ....
Lobelia dortmanna (p 417 **F**)

Lvs **flattened**, **tapered** to pointed tips, **triangular** in cross-section, **solid** – short erect racemes of white four-petalled fls 2.5 mm wide, producing egg-shaped flattened frs 2.5 mm long, sometimes present – a Crucifer of mt lakes, lf ....
Subularia aquatica (**Awlwort**)

E Aquatic plants with lvs much divided into narrow segments, normally submerged (flat, shallow-lobed lvs with broad blunt divisions may also be present)

1 Lvs alt along stems **2**
Lvs opp or whorled **3**

2 Tiny bladders present on some or all lf-segments ....
Utricularia (p 410)

Bladders abs, lvs palmately divided........*Ranunculus* subgenus *Batrachium* (p 106)

Bladders abs, lvs pinnately lobed or divided ....
Oenanthe (p 341), or *Apium inundatum* (p 344 **D**)

3 Lvs pinnate **4**

Lvs forked 1–3 times...........................
Ceratophyllum (p 99)

4 Lvs in rosettes, with flat linear segments ..*Hottonia* (p 232)

Lvs in whorls spaced out along stems, with hairlike segments ......
Myriophyllum (p 294)

VI. HERBS OF ROADSIDES, WASTE GROUND, WELL LIT HEDGEBANKS, ARABLE AND DRY MEADOWS

In all these habitats a great many species of plants occur. In some very cultivated or suburbanized areas these may be almost the only habitats available for study. Well lit hedgebanks are best grouped here (for shady hedgebanks see Key I).

It was difficult to decide which species of the roadside and waste ground habitats to include and which to omit from this key, because their incidence is so varied, many plants being present in one place and many others in another but similar habitat. The choice ultimately made is of plants most likely to be found and, of these, those easiest to identify. All sorts of introduced or garden plants can turn up in such places, but these are not included, for reasons of space.

With arable weeds the problem is that most of them are annuals; when present they may quite likely be at least partially in flower, while in winter many of them completely disappear. Hence the key contains few plants that are strictly weeds of arable land. The key concentrates very much on leaf-rosette form; but it is worth remembering that many plants whose leaves are confined to a basal rosette occur in trampled places, and that some other plants which only show a rosette in winter may in summer bear a tall leafy stem.

MASTER KEY (For roadsides in calc areas use also Key II; for shady lanes or dense tall roadside hedges use Key I)

1 Lvs grass-like, linear, with sheathing bases ....**A**
Lvs not grass-like..**2**

2 Lvs compound with separate lfts, or deeply divided into lobes more than halfway to the midrib or base of lf **3**
Lvs simple, and, if toothed or lobed, not divided into lobes more than 1/3 way to midrib, or to base, of lf...**4**

3 Lvs **palmately** divided or lobed, into **5 or more** lfts or main lobes all radiating from tip of lf-stalk..**B**
Lvs **trifoliately** divided or lobed, with 3 main lfts or lobes only ....**C**
Lvs **pinnately** divided or lobed, with the lfts or lobes in one to many **pairs** along central midrib of lf, with sometimes a terminal lft at lf-tip...............................**D**

4 Lvs simple, in opp pairs along stems ....**E**
Lvs in **whorls** of 4 or more along stems ....**F**
Lvs simple, alt along stems ....**G**
Lvs all in a spreading **basal rosette** (so impossible to see whether alt or opp) ....**H**

A. Herbs with lvs grass-like, linear, with sheathing bases

1 Lvs in 3 vertical ranks, stems solid, ± triangular............(**Sedges**) **2**

Lvs in 2 vertical ranks, stems hollow, round ....(**Grasses**) **3**

2 Lvs hairy ....
Carex hirta (**Hairy Sedge**)

Lvs hairless, with grey waxy bloom on upperside ....
Carex flacca (**Glaucous Sedge**)

Lvs hairless, bright green on upperside ....
other *Carex* spp (see vegetative key in *BSBI Sedges Handbook*)

3 Basal lvs bristle-like, 1mm wide or less...**4**

Root-lvs not bristle-like, over 2 mm wide, either flat, or folded lengthwise..**5**

4 Basal lvs (**b**) grooved on upperside, stem-lvs (**a**) flat – ligule v short, blunt ....
Festuca rubra (**Red Fescue**)

Basal lvs not grooved on upperside, stem-lvs bristle-like – ligule pointed – on sandy or hthy rdside banks ....
Deschampsia flexuosa (**Wavy Hair-grass**)

5 Basal lvs flat....................................**6**

Basal lvs strongly folded and keeled, rough – sheath flattened – ligule blunt, folded ....
Dactylis glomerata (**Cock's-foot**)

6 Lvs linear-lanceolate – ligules long, pointed – stems creeping ....
Agrostis stolonifera (**Creeping Bent**)

Lvs parallel-sided most of their length – stems tufted.....................**7**

7 Lvs with several deep parallel grooves above, v rough when rubbed downwards – lvs 20–30 cm long in large tussocks – ligules long, pointed ....
Deschampsia caespitosa (**Tufted Hair-grass**)

Lvs without deep grooves above ..
8

8 Lvs grey-green, v downy, sheaths pinkish – ligules long, blunt..
Holcus lanatus (**Yorkshire-fog**)

Lvs fresh green, flaccid, not downy, sheaths green – ligules long, blunt ....
Arrhenatherum elatius (**False Oat-grass**)

Lvs as last, but rigid, rough, hairless – long underground runners ....
Elytrigia repens (**Common Couch**)

B. Herbs with lvs palmately divided or lobed into 5 or more lfts or main lobes, all radiating from tip of lf-stalk

1 Lvs palmately **divided** into 5 (or 7) **separate lfts** – lfts obovate, deeply toothed – lvs long-stalked – stem creeping................................
Potentilla reptans (p 259 **E**)

Lvs ± deeply palmately **lobed**, but **not** divided to base into separate lfts **2**

2 Lvs hairless, shiny, ± red-tinged – lobes v blunt – local, in hbs..........
Geranium lucidum (p 320 **D**)

Lvs hairy, main lobes divided again into pointed narrow segments **3**

3 Basal lvs **five-angled** in outline – lf-stalks **green** – lvs with 3 larger twice-pinnately cut lobes, and 2 **smaller** basal lobes cut into few sharp teeth ....:...............
Ranunculus acris (p 102 **B**)

Basal lvs seven-angled in outline – lf-stalks red-flushed below – lvs with 5–7 main lobes, all twice-pinnately cut – calc soils ....
Geranium pratense (also other *Geraniums*, pp 319–22)

C. Herbs with lvs trifoliately divided or lobed, with 3 main lfts or lobes only, which may (or may not) themselves be lobed again

1 Lfts simple, unlobed, with small teeth or none **2**

Lfts deeply divided into lobed segments **4**

2 Stems **creeping** – lfts oval-obovate, **finely-toothed**, ± hairless, each often with a whitish crescent across it – lf-stalks 5–14 cm long ....
Trifolium repens (p 274 **D**)

Stems tufted, ± erect – lfts obovate-elliptical, **scarcely-toothed**, margins downy – stalks of upper stem-lvs short (to 5 cm) **3**

3 Lfts usually with a whitish crescent-shaped band across each, oval-obovate, blunt, dull green – upper parts of stipules **broad**-triangular with fine brown **erect bristle**-points.............................
Trifolium pratense (p 274 **A**)

Lfts usually without a whitish band across each, narrow-elliptical, pointed, **bright** green – upper parts of stipules **narrow-oblong**, with green **awl-shaped spreading** points ....
Trifolium medium (p 274 **B**)

4 Plant with **creeping** and **rooting runners** – the 3 lfts v hairy, mid lft **stalked**, **three-lobed**, the lobes divided into deeply but **bluntly** lobed segments ....
Ranunculus repens (p 102 **A**)

Plant **erect**, **without** runners – the 3 lfts ± hairy, mid lft **unstalked**, each lft **twice-pinnately** divided into **narrow-**lanceolate, **pointed** segments........
Ranunculus acris (p 102 **B**)

Plant **erect**, **without** runners, but with corm at stem-base – the 3 lfts ± hairy, mid lft **stalked**, three-lobed, the lobes divided into rather narrow, **pointed** segments
Ranunculus bulbosus (p 102 **D**)

D Herbs with lvs pinnately divided or lobed, with the lfts or lobes in two to many pairs along central midrib of lf, with sometimes a terminal lft at lf-tip

(Note: if lobes only shallow and divided less than halfway to midrib of lf, like coarse teeth, return to 4 in Master Key VI above)

1 Lvs **pinnatifid** (ie, with the lobes on each side of main lf midrib joined together by a **continuous**, though sometimes **v narrow**, **flange** of lf-blade)..............................**2**

Lvs **pinnately compound** (ie, with **separate lfts**, the pairs of lfts along the main lf midrib **separated** from each other, at least in **lower** part of lf, by a length of stalk **without** any flange of lf-blade)...........................**14**

2 Lvs **spine**-edged and/or -tipped
....**11**

Lvs **without** spines (though sometimes bristly-hairy) **3**

3 Lvs with **terminal lobe much larger** than side-lobes **4**

Lvs with terminal lobe **not** conspicuously larger than side-lobes...**5**

4 Lvs soft, not glossy, with v large (4–6 cm long x 3–5 cm wide), **oval**, often cordate, wavy-toothed terminal lobe – and few, much smaller, undivided side-lobes along the flanged lf midrib ....
Lapsana communis (p 473 **D**)

Lvs of **basal rosette** (a) with terminal lobe **oval**, firm, glossy, 2–3 cm long – side-lobes themselves ± pinnately lobed – lvs (**b**) **above** rosette have terminal lobe no larger than side-lobes ..*Senecio jacobaea* (p 440 **A**)

Lvs of basal rosette with terminal lobe **triangular**, **pointed**, 1–2 cm long – side-lobes undivided ....
Crepis spp (p 476)

5 Lvs **white-felted** below, green and ± hairless on upperside, aromatic when bruised, twice-pinnatifid into **flat**, pointed lobes
Artemisia vulgaris (p 456 **929**)

Lvs **not** white-felted below **6**

6 Lvs **twice-pinnatifid, not** aromatic when bruised – lf outline oval-oblong, 3 cm or more wide – segments **blunt**, furrowed, dull green, ± sparsely downy above and below, not cottony above.....................................
Senecio jacobaea (p 440 **A**)

Lvs **twice-pinnatifid**, but **aromatic**, with **pointed**, **fine-toothed**, ± hairless segments ....
Tanacetum vulgare (p 448 **D**)

Lvs once-pinnatifid only.................**7**

7 Lvs **deeply** pinnatifid into **narrow**, ± **linear** lobes **8**

Lvs **shallowly** pinnatifid into **broad**, ± triangular lobes – lvs all in a basal rosette **10**

Lvs **deeply** pinnatifid (**a**) into **broad**, oblong, pointed side-lobes – lvs **waxy grey**-green – stem-lvs (**b**) with pointed basal lobes – latex **milky**...
Sonchus oleraceus (p 480 **E**)

8 Lvs **aromatic** when bruised – segments flat, linear, finely-toothed, ± hairless – lvs 3 cm wide or more, **broad**-oblong in outline ....
Tanacetum vulgare (p 448 **D**)

Lvs **not** aromatic when bruised – lvs to 2 cm, wide only, **narrow**-oblong to lanceolate in outline ..**9**

9 Lvs ± **cottony** above, especially near their base – lf-lobes all blunt, **crisped**, **wavy**, coarse-toothed – end lobe **no longer than** side-lobes – lvs 3–6 cm long x 1–2 cm wide, mostly clasping and alt up stem – basal rosette usually abs – ann plant, often with erect infl in winter....................
Senecio vulgaris (p 440 **F**)

Lvs **not** cottony, but with ± **sparse stiff** hairs above – lf-lobes all narrow, pointed, flat, ± untoothed – end lobe **much longer than** side-lobes – lvs 10 cm or more long x 1–2 cm wide, **all** in a basal rosette – per plant...
Leontodon autumnalis (p 474 **E**)

Lvs ± **hairless** above, v **variable** from broad-lanceolate and shallow-toothed, to deeply pinnatifid into narrow, toothed lobes – terminal lf-lobe **oval**, toothed – alt stem-lvs like basal lvs but with clasping bases – ann plant ....
Capsella bursa-pastoris (p 216 **E**)

10 Lvs ± **hairless**, **glossy**, **bright** green – side-lobes broad-triangular, sharp-toothed, teeth of upper lobes **backward-pointing** – much **milky latex** in lvs.....................
Taraxacum officinale (p 470 **A**)

Lvs hairless, **waxy grey-green** – side-lobes irregularly-toothed, endlobes triangular and often wider than side-lobes – much **milky latex** in lvs.................................
Sonchus oleraceus (p 480 **E**)

Lvs roughly **hairy**, **dull** green – side-lobes blunt, rounded, wavy-edged, shallow, untoothed, side-ways-pointing – little or no milky latex in lvs ....
Hypochaeris radicata (p 474 **A**)

11 Lvs **waxy grey-green** with **milky latex** – all in basal rosette in winter ....*Sonchus asper* (p 480 **F**)

Lvs **not** waxy grey-green, juice **watery** **12**

12 Lvs **prickly** and **rough**-hairy on upperside, hence dull grey-green above ....
Cirsium vulgare (p 463 **A**)

Lvs **not** prickly on upperside, only sparsely (or not at all) hairy there, hence ± glossy above **13**

13 Lvs **dark green** and ± **purple-flushed**, all in a basal rosette in winter – stem, when present, has continuous spiny wings up it and alt lvs – bi plant ....
Cirsium palustre (p 463 **C**)

Lvs **grey-green**, **never** purplish – stem **unwinged** – per plant ....
Cirsium arvense (p 463 **D**)

14 Lvs **once**-pinnate – lfts lobed or toothed or not so, but not divided again into lfts **15**

Lvs **twice or more times pinnate** into separate lfts **25**

15 Lfts quite untoothed – tip of lf with a branched tendril instead of a lft **16**

Lfts toothed or lobed – tip of lf with a lft...**18**

16 Lvs hairless, with 1 pair of lfts only – stem often **square**, **angled** – large triangular lfy stipules at lf-base ......*Lathyrus* spp, especially *L. pratensis* (p 284 **D**)

Lvs ± hairy, with 2–12 pairs of lfts – stem **round**...................................**17**

17 Lfts 2–8 pairs – ann plant ....
Vicia sativa (p 282 **D**)

Lfts 6–12 pairs – per plant ....
Vicia spp (p 280)

18 Side-lfts all ± **equal** in size **19**

Tiny side-lfts alt with **larger** ones along lf-stalk **24**

19 Lfts strongly lobed and toothed, 2 cm wide (or much more)...**20**

Lfts not lobed or only weakly so, teeth blunt or shallow – side-lfts under 2 cm wide **21**

20 Lvs hairless, glossy dark green
Pimpinella major (p 334 **C**)

Lvs roughly hairy, dull greyish-green ....
Heracleum sphondylium (p 336 **D**)

21 Terminal lft of basal lvs **pointed**-oval, blunt-toothed, similar in shape to side-lfts – lvs grey-green – crushed lf-stalks produce **orange** liquid ....
Chelidonium majus (p 114 **H**)

Terminal lft of basal lvs **rounded** to kidney-shaped, **much** wider than side-lfts – lvs fresh to dark green – crushed lf-stalk produces watery liquid **22**

22 Stem-lvs (**B**) with all lfts v **narrow**-lanceolate – basal lvs (**A**) with terminal lft kidney-shaped, dark glossy green, to 2 cm wide – per plant of damp places ....
Cardamine pratensis (p 210 **A**)

Stem-lvs with all lfts **oval**-elliptical, bluntly-toothed – basal lvs with terminal lft kidney-shaped, but pale green, to 1 cm wide **23**

23 Basal lvs many in a rosette – ann plant of dry open ground........
Cardamine hirsuta (p 210 **B.1**)

Basal lvs few, in sparse rosette – per plant of damp ground ....
Cardamine flexuosa (p 210 **B**)

24 Lfts **silvery**, silky-hairy – stems **creeping** ....
Potentilla anserina (p 260 **H**)

Lfts **green**- or purple-flushed, downy – stems **erect** ....
Agrimonia eupatoria (p 262 **H**)

25 Lvs with lfts narrow-linear, almost bristle-like – lf-outline oblong **26**

Lvs with lfts broad and flat – lf-outline **triangular**............................**27**

26 Lvs narrow oblong in outline, end-lfts grey-green, downy, dense-set, only 1–2 mm long ....
Achillea millefolium (p 452 920)

Lvs broad-oblong in outline, end-lfts dark green, ± fleshy, hairless, loosely-set, 5–15 mm long ....
Tripleurospermum maritimum (p 455 **B**)

27 Secondary lfts **oval**, 3–6 long x 3–4 cm wide, **finely**-toothed..**28**

Secondary lfts **deeply** cut, 0.5–2.0 cm long x 0.5 mm wide – whole lf fern-like **29**

28 Lfts oval-elliptical, 4–8 cm long, borne mostly in **threes** on side-branches of lf – lfts and stalks **wholly** fresh green – sheathing lf-bases **not** inflated ....
Aegopodium podagraria (p 334 **F**)

Lfts **oval**, 2–8 cm long, **pinnately** arranged on side branches of lf – **purple** blotches where side branches of lf join main stalk – sheathing lf-bases **inflated** ....
Angelica sylvestris (p 336 **A**)

29 Lf-stalks with **papery brown scales** numerous below ....
(**Ferns**) *Dryopteris* spp

Lf-stalks **without** papery brown scales below......................................**30**

30 Lfts and lf-stalks hairy **31**
Lfts and lf-stalks hairless..............**32**

31 Lfts grey-green, **downy**, blunt – lf-stalks and stems **purple**-spotted, softly downy ....
Chaerophyllum temulum (p 330 **F**)

Lfts grey-green, **rough**, v pointed – lf-stalks and stems **unspotted**, rough ......*Torilis japonica* (p 330 **C**)

Lfts fresh green, smooth, ± hairless on uppersides, with pointed teeth – lf stalks and stems unspotted...............................
Anthriscus sylvestris (p 329 **A**)

32 Lfts with narrow, flat lobes – lf-stalks and stems **unspotted** – ann plant ....
Aethusa cynapium (p 330 **H**)

Lfts with many fine-cut lobes – lf-stalks and stems **purple**-spotted – bi plant ....
Conium maculatum (p 330 **G**)

E. Herbs with lvs simple, in opp pairs along stems

1 Lvs fleshy, linear, 2 mm wide, or less..**2**

Lvs not fleshy, but flat, over 4 mm wide **3**

2 Lvs blunt, 1–3 cm long, half-cylindrical – papery stipules at lf-bases – clusters of lvs in lf-axils (thus lvs appearing ± whorled) – plant to 20 cm tall – on dry open sandy wa and ar ....
Spergula arvensis (p 156 **E**)

Lvs pointed, under 1 cm long – no stipules – tiny creeping plant (to 5 cm tall) of paths etc ....
Sagina procumbens (p 156 **F**)

3 Basal rosette-lvs 10–20 cm long, oblong-lanceolate, untoothed, with strong swollen-based **prickles** on **uppersides** – stem-lvs sometimes toothed, with prickles only on **underside** of **midrib** – stem angled, prickly....
Dipsacus fullonum (p 430 **A**)

All lvs **without** spines or prickles**4**

4 Lvs v **hairy**, **downy**, or **rough** on upper and on lower surfaces........**5**

All lvs ± **hairless** and **smooth** on upper and lower surfaces (but **edges** may be **rough**) **11**

5 Rosette-lvs untoothed – stem-lvs usually ± **pinnatifid**, with elliptical end-lobes – all lvs oblong-lanceolate to elliptical, 4–12 cm long x 1–3 cm wide............
Knautia arvensis (p 432 **D**)

All lvs **untoothed** – **stems** round **6**

All lvs toothed – **stems round** – lvs **not** cordate-based **7**

All lvs toothed – **stems four-angled** – lvs **cordate**-based **9**

6 Lvs narrow-oblong, 1.0–2.5 cm long x 0.3–0.5 cm wide, short-stalked at base, stalkless above, all ± blunt – stems creeping and ascending...
Cerastium fontanum (p 152 **A**)

Lvs oval-oblong, 2–5 cm long x 1–2 cm wide, all **blunt** and unstalked – stems all erect, **not** swollen at joints ....
Hypericum hirsutum (p 176 **D**)

Lvs oblong-lanceolate to elliptical, 5–10 cm long x 2–3 cm wide, lower lvs narrowed to short stalks, upper stalkless – stems **swollen** at joints, erect ....
Silene latifolia (p 144 **A**)

7 Rosette-lvs with **coarse**, blunt, often **backward**-pointing **teeth** – upper lvs ± coarsely pinnatifid – **all** lvs opp ....
Knautia arvensis (p 432 **D**)

All lvs with sparse, tiny, pointed, forward-pointing teeth – upper lvs often alt, lower always **opp**....**8**

8 Lvs 6–12 cm long, v hairy (except when young often ± hairless), their bases **clasping** stem – plant to 150 cm tall..............
Epilobium hirsutum (p 298 **E**)

Lvs 3–7 cm long, **softly** hairy, **not** clasping stem – plant to 60 cm tall ....
Epilobium parviflorum (p 298 **F**)

9 Lvs bluntly oval, with coarse, blunt teeth – ann plants ....
Lamium purpureum (p 372 **F**)

Lvs oval-triangular, pointed, teeth sharp – per plants **10**

10 Lvs and stem with long rigid stinging hairs – stem ± purplish, with more than 4 ridges ....
Urtica dioica (p 121 **A**)

Lvs and stem with soft, non-stinging hairs – stem green, with 4 angles only..................................
Lamium album (p 372 **J**)

11 Lvs untoothed **12**

Lvs finely-toothed, oval-lanceolate to oblong – stems ± red-flushed below **15**

12 Lvs **oblong**, **blunt**, with **transparent** pellucid **dots** (hold up to light), all stalkless, 1–2 cm long x 0.5–1.0 cm wide – stems round with 2 raised ridges, often **reddish** below......................................
Hypericum perforatum (p 176 **A**)

Lvs **oval**, **pointed**, the lower **stalked**, the upper **stalkless** – **no** pellucid dots – stems round, green **13**

Lvs **linear-lanceolate**, **pointed**, **all** stalkless – no pellucid dots – stems **square** in cross-section, green **14**

13 Lvs yellowish to bright green, **flat**, **soft**, 0.3–2.5 cm long x 0.5–1.5 cm wide – plant **spreading**, much-branched, floppy ......*Stellaria media* (p 150 **A**)

Lvs greyish-green, lower ± **folded** lengthwise, **firm**, 4–5 cm long x 1.0–1.5 cm wide – plant **erect**..........
Silene vulgaris (p 146 **A**)

14 Lvs rigid, sharp-pointed, edges rough, broadest at base ....
Stellaria holostea (p 150 **B**)

Lvs less rigid, short-pointed, edges smooth, lvs broadest near middle ....
Stellaria graminea (p 150 **C**)

15 Stems **cylindrical**, ± **hairless** – lvs **oval**-lanceolate, short-stalked
Epilobium montanum (p 298 **A**)

Stems with **4 raised**, **lengthwise ridges** above, with **many curly** and **sticky hairs** above – lvs **oblong**-lanceolate, short-stalked
Epilobium spp, probably *E. ciliatum* (p 298 **C**)

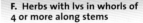

F. Herbs with lvs in whorls of 4 or more along stems

1 Lvs **fleshy**, **cylindrical**, **linear**, **blunt** – **clusters** of lvs in **lf-axils** can **appear** like whorls of lvs ....
Spergula arvensis (p 156 **E**)

Lvs **not** fleshy or cylindrical **2**

2 Lvs with 3 main veins, **oval**-elliptical, 4 per whorl, hairy ....
Cruciata laevipes (p 422 **A**)

Lvs one-veined, **lanceolate** to **linear**, 6 or more per whorl, never oval **3**

3 Lvs **linear**, 0.5–2.0 mm wide, **8–12** per whorl, hairless
Galium verum (p 422 **B**)

Lvs obovate to linear-lanceolate, **mucronate**, **6–8** per whorl, ± hairy or prickly **4**

4 Lvs obovate – lvs and stem not prickly, ± hairy but scarcely rough
Galium mollugo (p 419 **A**)

Lvs linear-lanceolate – lvs and stem with large downward-pointed prickles ....
Galium aparine (p 420 **H**)

G. Lvs simple, alt along the stems – if toothed or lobed, not divided into lobes more than 1/3 of way to midrib, or to base, of lf

1 Lvs with veins **palmately** arranged, lower always with obvious stalks....................................**2**

Lvs with veins **pinnately** arranged, or with only one vein visible **4**

2 Stipules at base of lvs tiny, 2–3 mm long x 1–2 mm wide, oval-lanceolate – lvs **not** pleated along veins, ± glossy **3**

Stipules at base of lvs conspicuous, 5–10 mm long x 5–10 mm wide, lf-like, toothed – upper lvs short-stalked, lower long-stalked – lvs pleated along veins, not glossy ....
Alchemilla vulgaris agg (p 262 **C**)

3 Basal lvs 5–10 cm across, sparsely hairy, outlines ± angled, lobed 1/3 way to base........................
Malva sylvestris (p 180 **A**)

Basal lvs 4–7 across, ± downy, rounded, kidney-shaped, lobed 1/4 way to base or less ....
Malva spp, especially *M. neglecta* (p 180 **B**)

Basal lvs 2–3 cm across, ± deeply lobed...
Geranium spp (pp 319–322)

4 Stems erect – lvs over 5 cm long and over 1 cm wide **5**

Stems ± **prostrate**, or **climbing** ..**14**

5 Lvs **hairy**...**6**
Lvs **hairless** **10**

6 Lvs **cordate**-based, oval-triangular, 10–30 cm long x 10–20 cm wide, the lower lvs on long ± erect stalks, upper lvs short-stalked – all lvs green on uppersides, grey-cottony on undersides – stout erect plants to 1 m tall ....
Arctium spp (p 468 **A**, **A.1**, **A.2**)

Lvs not cordate-based but

elliptical-lanceolate, or oblong, or linear **7**

7 Basal lvs ± erect, **not** in a spreading rosette (or no basal lvs present) **8**

Basal lvs in a **large spreading rosette** – lvs 10–30 cm long, elliptical-lanceolate to oblong, narrowed into a v short stalk **9**

8 Basal lvs 5–15 cm long, elliptical-lanceolate to oblong, narrowed into long stalks – stem-lvs stalkless – all lvs ± roughly hairy, untoothed to ± pinnatifid, dark green ....
Centaurea nigra (p 466 **A**)

Basal lvs 5–8 cm long, obovate-lanceolate, soon withering – stem-lvs **numerous**, **linear**-lanceolate, pointed – all lvs pale green, short-stalked, ± hairy, with few or no teeth ....
Conyza (p 446)

9 Lvs clothed with dense soft white wool..
Verbascum thapsus (p 392 **A**)

Lvs dark green and thinly downy on upperside, hairier underneath
Verbascum spp, probably *V. nigrum* (p 392 **B**)

10 Lvs numerous, in spiral up **unbranched** stem, **without** stipules – lvs 5–15 cm long, oblong-lanceolate, ± untoothed, narrowed to each end, uppersides dark green, lower grey-green ....
Chamerion angustifolium (p 300 **J**)

Lvs well spaced-out up **branched** stem, with papery sheathing **stipules** at base of each **11**

11 Lvs **arrowhead-shaped**, with **down**-pointing basal lobes – stipules fringed with bristles – infls branched – stems weakly ribbed lengthwise ....
Rumex acetosa (p 166 **B**)

As the last, but basal lobes of lvs arching **forwards** or **outwards** ......
Rumex acetosella (p 166 **C**)

Lvs **not** arrowhead-shaped **12**

12 Lvs **lanceolate** – stipules **fringed** with bristles – lvs **not over 2 cm** wide x 10 cm long, v **short**-stalked, often **black-blotched** – stems slender (2–4 mm wide at base), **smooth** and round.....................................
    Persicaria or *Fallopia* spp, especially *P. maculosa* (p 168 **C**)

Lvs oval-oblong to lanceolate, the lowest **over 3 cm wide** x **15 cm long**, **long**-stalked – lvs **never** blotched – stipules **without** fringes – stems **stout** (5 mm or more wide at base), **ridged** length-wise **13**

13 Lower lvs **oval-oblong**, **cordate**-based, **blunt**-tipped, to 25 cm long – edges wavy but not crisped – stems and infls **erect**......
    Rumex obtusifolius (p 164 **D**)

Lower lvs **oblong-lanceolate**, with **crisped wavy** edges, blunt-tipped, to 30 cm long, **narrowed** to or **rounded** at base – stems and infls **erect** ....
    Rumex crispus (p 164 **F**)

Lower lvs **oblong**, with wavy edges and **pointed** tip, 15–20 cm long, rounded at base – stems ± **wavy** – infls ± **spreading** from vertical...
    Rumex conglomeratus (p 163 **B**)

14 Lvs elliptical-lanceolate, not over 3 cm long x 1 cm wide, untoothed, v short-stalked, with papery sheathing stipules at lf-bases ....
    ·*Polygonum aviculare* (p 166 **E**)

Lvs arrow-shaped to cordate-triangular, 2–6 cm long x 2–3 cm wide, long-stalked, with pointed basal lobes – plants both creeping and climbing **15**

15 Papery sheathing stipules at lf-**bases** – lvs ± mealy on undersides, pointed
    Fallopia convolvulus (p 170 **B**)

No stipules at lf-bases – lvs not mealy on either side **16**

16 Basal lobes pointed, spreading, lvs 2–3 cm long ....
    Convolvulus arvensis (p 354 **C**)

Basal lobes blunt, backward-pointing, lvs 5–8 cm long ....
    Calystegia sepium (p 354 **D**)

H. Herbs with lvs simple, all in a spreading basal rosette – no lvs on stems (or no stems present, so impossible to see whether lvs alt or opp)

Note: this section of the key covers some plants that **may** have lfy stems in summer, but in winter have lvs only in a basal rosette.
    Some of the spp appearing below occur also in other sections of the Roadsides Key (VI), but are repeated here because (a) some spp included in section D may occur with scarcely pinnatifid lvs, and (b) because some spp in sections G and E may be without stems **in winter**. Species with **compound** or **deeply-lobed** lvs should be sought in sections B, C and D.

1 Lvs with **broad** (8 cm or more wide), **rounded** blades, and **cordate** bases, borne **erect** on **long stout** stalks that are **grooved** above **2**

Lvs with several ± **parallel main veins**, with inconspicuous cross-veins – lvs 10–15 cm long **4**

Lvs with **spiny** margins...................**5**

Lvs with **swollen-based** prickles on upperside **8**

Lvs ± **hairless** **9**

Lvs **hairy** **10**

2 Lvs up to 90 cm wide in summer (only about 10–12 cm wide at fl-time in spring), rounded-triangular, grey-downy below, green above (but not glossy), with few, sharp, distant teeth and stout grooved stalks......
    Petasites hybridus (p 450 **F**)

Lvs 10–20 cm wide in summer **3**

3 Lvs kidney-shaped, equally fine-toothed, **blunt**, **glossy** green above, persistent through winter
Petasites fragrans (p 452 **G**)

Lvs rounded-**polygonal**, with 5–12 shallow, **pointed** lobes, white-felted on both sides in spring, then light green above, grey beneath..
Tussilago farfara (p 450 **E**)

4 Lvs elliptical-lanceolate to **linear**, shiny and scarcely hairy above, scarcely toothed, **gradually** narrowed to a **short** stalk ....
Plantago lanceolata (p 387 **A**)

Lvs **oval** or elliptical, not shiny, downy or not downy, feebly-toothed, **abruptly** narrowed to a stalk **equalling** lf ....
Plantago major (p 387 **B**)

5 Lvs waxy grey-green, with milky latex ....
Sonchus asper (p 480 **F**)

Lvs not waxy grey-green, with watery juice **6**

6 Lvs **prickly-hairy** on upper surface, **not shiny** ....
Cirsium vulgare (p 463 **A**)

Lvs **not** prickly above, though ± hairy, ± glossy **7**

7 Lvs cottony above, ± **purple-flushed** – bi plant ....
Cirsium palustre (p 463 **C**)

Lvs scarcely cottony above, wholly **grey**-green, **without** purple flush – per plant with **creeping** rhizome ....
Cirsium arvense (p 463 **D**)

8 Lvs hairless and shiny, ± pointed, with **few** scattered swollen-based prickles – lvs **15–30** cm long ....
Dipsacus fullonum (p 430 **A**)

Lvs **v bristly**, with each bristle arising from a whitish wart – **blunt** lvs 5–10 cm long ....
Picris echioides (p 478 **H**)

9 Lvs tapered to base, stalkless, fresh glossy green, with triangular teeth, the upper ones ± arched backwards – milky latex in lvs ....
Taraxacum officinale agg (p 470 **A**)

Basal lvs with **long** narrow stalks, roundish **spoon-shaped**, few-toothed, 1–2 cm long x 1.5 cm wide, green, ± ascending – no milky latex in lvs ....
Leucanthemum vulgare (p 448 **A**)

10 Lvs **spoon**-shaped, tapered below into a **broad stalk**, with **sparse** hairs and few blunt teeth – lf-blade 1–2 cm long x 1.0–1.5 cm wide ..*Bellis perennis* (p 450 **D**)

Lvs ± **unstalked**, **oblong** to **lanceolate** **11**

11 Lvs elliptical-lanceolate, 10–30 cm long, only **feebly**-toothed ....
(*Verbascum*) **12**

Lvs oblong, 5–10 cm long, ± **deeply**-toothed **13**

12 Lvs clothed with **soft white** woolly fur ....
Verbascum thapsus (p 392 **A**)

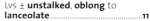

Lvs **dark** green, **thinly** downy on upperside, hairier underneath ......
Verbascum nigrum (p 392 **B**)

13 Lvs with **scattered** white bristles – side and terminal lobes **narrow-linear**, flat, untoothed ......
Leontodon autumnalis (p 474 **E**)

Lvs **rough-hairy** – side-lobes **blunt**, rounded, wavy-edged, shallow, untoothed ....
Hypochaeris radicata (p 474 **A**)

VII. PLANTS OF SALT-MARSHES

Salt-marsh vegetation is that of the zone between low and high tide marks on flat, muddy or sandy substrates in sheltered places, along estuaries or behind beaches that protect it from the wave action of the sea. In fact, most of the flowering plants of salt-marshes only occur in the uppermost part of the intertidal zone; lower down, where the mud or sand flats are covered for more than about a fifth of the time, only seaweeds (algae, non-flowering plants) occur. There is also, within the salt-marsh community, a clear zonation of plant species, according to the length of time per day (or per month) each zone is inundated with salt water. Many more species only occur at the very uppermost edge of the marshes, by sea walls, or on banks which are dry nearly all the time but are percolated by salt water from below (thus excluding from the habitat most ordinary land or fresh water plants).

Salt-marsh plants have several devices that enable them to exist in such a habitat. Some are very fleshy and swollen, often with reduced or modified leaves; others have very mealy leaves. Some have creeping rhizomes that help to bind the sand, or mud, thereby gradually raising the level of the marsh until it becomes dry land. Most have in their cell-sap very high concentrations of salt (much higher than that of the sea water). This enables them, by the process known as osmosis, to absorb water through their roots, even when the external salt concentration is high.

Most salt-marsh plants flower in late summer, or even autumn. This key will therefore be useful in spring or early summer, at least for those plants with vegetative parts above ground at that time.

1 Low bushy shrubs, to ± 1 m tall, evergreen **2**

Herbs **3**

2 Erect much-branched shrub to 1 m, with alt, sausage-shaped to **cylindrical** fleshy, hairless, green lvs 5–18 mm long x 2 mm wide, **blunt** at **both** ends (if tips **pointed**, see stage **10** below) ....
Suaeda vera (p 134 **E**)

More spreading shrub to 40 cm, with opp, **oval-elliptical**, **mealy-grey**, **flat**, short-stalked lvs 3–5 cm long x 1–2 cm wide ....
Atriplex portulacoides (p 136 **F**)

3 Plant **without** obvious lvs, composed only of opp-branched, fleshy stems, jointed into cylindrical or ovoid, swollen segments – (in spring only **shrivelled** last year's plants of ann spp will be visible)......................
Salicornia spp (pp 132–3)

Plants with **obvious lvs** **4**

4 Lvs **grass-like** (flat, grooved, folded or rolled, narrow-linear, **non-fleshy** and **thin-bladed** **5**

Lvs narrow-linear but **fleshy**, either flattish or half cylindrical, all in a **basal** tuft or rosette, 10 cm long or more **8**

Lvs ± **cylindrical** or strap-shaped, stalks v short or abs, **fleshy**, **1–5 cm long only**, borne either alt, or in opp pairs, or in clusters, **along** the stems..**10**

Lvs with definite **blade** and **stalk**, blade oval, triangular, or elliptical-lanceolate **11**

Lvs and stems **cylindrical**, **not** fleshy but tough and rigid, sharp-tipped, in erect tufts 20–60 cm tall – lvs both basal and alt up stem ....
Juncus maritimus (**Sea Rush**)

Lvs twice-**pinnately** divided into linear 1 mm-wide **white-woolly** segments ....
Seriphidium maritimum (p 458 **E**)

5 Plant with **creeping lfy** rooting runners forming a compact **turf** – lvs grey-green, hairless, ± folded or inrolled, 1–3 mm wide x to 20 cm long, with hooded tips ....
Puccinellia maritima
(**Common Saltmarsh-grass**)

Plant **erect**, with underground rhizomes **6**

6 Lfy stems arising separately from mud, **stout** (1 cm or more thick), 40–80 cm tall – lvs tough, **ascending**, 7–15 mm wide, with sharp rigid points, flat or inrolled, smooth, ribbed above – ligule a fringe of hairs – infl of several spreading spikes like fingers of a spread hand ....
Spartina angelica
(**Common Cord-grass**)

Lvs and lfy stems densely **tufted**, erect, stems thin (3 mm or less thick) **7**

7 Lvs **rigid**, **grey**-green, rough-edged, **ribbed** on upperside, sharp-pointed, erect, 2–6 mm wide ....
Elytrigia atherica (**Sea Couch**)

Lvs **flexible**, **bristle**-like, 0.5 to 1.0 mm wide, dark green, mostly from base ....
Festuca rubra (**Red Fescue**)

Lvs narrow-linear, 1–2 mm wide, flattened above, dark green, mostly on infl-stem which is three-angled above and bears branched cluster of brown capsules each 1.5–2.0 mm long
Juncus gerardii (**Saltmarsh Rush**)

8 Lvs **spreading** in a dense basal rosette or cushion, ± fleshy, ± hairless, 2–10 cm long x 1–2 mm wide – infls button-shaped heads 2–3 cm wide on stems 5–20 cm tall ......*Armeria maritima* (p 172 **A**)

Lvs ± **erect** in a rosette or tuft, hairless, 5–20 cm long **9**

9 Lvs **flattened**, v fleshy, 3–5-veined, 5–15 cm long x 5–10 mm wide ....
Plantago maritima (p 388 **D**)

Lvs **half-cylindrical**, fleshy, grooved on upperside, vein not obvious, lvs 10–20 cm long x 2–3 mm wide
Triglochin maritimum (p 487 **C**)

10 Bushy, **much**-branched plant often red-tinged, with stout stems and **alt**, half-cylindrical **pointed** lvs 5–25 mm long x 1–2 mm wide (if lvs **blunt**, see **2** above) – stipules abs ....
Suaeda maritima (p 134 **D**)

Sprawling or creeping **little**-branched plant with weak stems and **opp** lvs 1.0–2.5 cm long x 1–2 mm wide, ± half-cylindrical, pointed, with membranous translucent **stipules** at bases ....
Spergularia spp (p 156)

Erect ± bushy plant, 20–80 cm tall, branched above only, with lvs **alt**, hairless, fleshy, strap-shaped, to 4 mm wide and 2.5–5.0 cm long, sometimes three-lobed at tip, cross-section elliptical – fls in yellow daisy-like heads in umbel-like clusters at stem-tips in July–Sept...
Inula crithmoides (p 444 **E**)

11 Lvs **alt**, oval or obovate to oblong or elliptical-lanceolate – stems erect **12**

Lvs **mostly alt** (some lower lvs may be opp), diamond-shaped to triangular or strap-shaped **13**

Lvs in **opp** pairs on prostrate stems, **small**, oval-oblong, 0.5–2.0 cm long, flat, fleshy ....
Glaux maritima (p 236 **G**)

12 Lvs firm, but **not** fleshy, elliptical-lanceolate, with **mucronate** tips, 4–12 cm long x 2–4 cm wide, stalked, **all in a basal rosette** ....
Limonium spp, mostly *L. vulgare* (p 172 **B**)

Lvs **fleshy**, hairless, basal lvs stalked, obovate, to 2 cm wide, **not mucronate**, upper **stem**-lvs narrower, oblong – fls or frs in daisy-like heads ....
Aster tripolium (p 446 **A**)

Basal lvs in rosette, oval to oblong-ovate, shiny dark green, **fleshy**, bases **tapered** into long stalks – stem-lvs oval-elliptical, stalkless, clasping – white four-petalled fls in racemes in spring ..
Cochlearia anglica (p 218 **C**)

As above, but basal lvs **cordate**-based ....
Cochlearia officinalis (p 218 **A**)

13 Lvs alt, **glossy dark** green, to 10 cm long ....
Beta vulgaris (p 136 **G**)

Lvs alt or opp, **grey-mealy** ....
Atriplex spp (pp 135–6)

VIII. PLANTS OF SHINGLE BEACHES AND SAND DUNES

These maritime habitats have a specialized flora. The action of the sea often makes them unstable, and their plants have to be able to recolonize disturbed ground. Because of the instability, plant communities are often open, with much ground totally bare; hence competition between those plants that can grow there is often less intense than in closed, inland plant communities. Modifications to withstand drought (a common feature of such maritime, well-drained habitats), exposure to wind, and to salt spray, are frequent. Leaves are often fleshy, wax-covered (hence greyish or bluish), tough, or fleshy, many plants present a 'low profile' to the wind, and have either thick protective layers of hairs or wax to check water loss, or very deep roots to tap underground fresh-water supplies.

Further from the sea, these habitats become with age more stable (and more closed), and the vegetation may become more heath- or scrub-like. Or, if there is much lime in the dunes from incorporated shells of shellfish, the vegetation may resemble that of chalk grassland (so for sand dunes see also Key II). Many plants of open waste ground, including many introduced species, sometimes colonize sand-dunes, particularly near ports, or in areas where there are gardens nearby.

Successive dune ridges may be built up, and fixed over time by the agency of Marram (*Ammophila*), with its creeping rhizomes and fibrous roots. This plant can grow up through new layers of sand that may cover it when onshore winds bring in sand from a beach dried out at low tide, forming a new layer of rhizomes and roots on top of the old ones. Thus Marram-dominated dunes may grow to a considerable height.

In between dune ridges there are often fresh water marshes with fen vegetation, because the dunes and ridges act as natural spongy reservoirs of rain water, which may seep out into the intervening hollows. For such places, Key IV may be useful.

In winter, most dune plants, except the grasses and shrubs such as Sea-buckthorn, tend to die back; but the key should be useful especially in autumn or late summer, when many dune plants are visible but not in flower.

For this habitat also consult Key VI.

1 Thorny Shrub with linear lanceolate lvs, silver-grey, with scales – orange berries......................
Hippophae rhamnoides (p 294 **A**)

Creeping tufted herb with small **opp** fleshy oval pointed lvs, 6–20 mm long – on dunes.........................
Honckenya peploides (p 154 **B**)

Herbs with lvs over 2 cm long, not in opp pairs.................................**2**

2 Lvs pinnately lobed or divided ..**3**

Lvs palmately lobed or veined, waxy bluish-grey, edged with **spines** – on dunes...............................
Eryngium maritimum (p 344 **F**)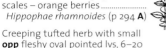

Lvs oval-oblong, fleshy, waxy grey-green, not spine-edged, 1–2 cm long, alt and closely set up reddish stems – on dunes...............
Euphorbia paralias (p 308 **B**)

Lvs elliptical-oblong, bright green, edges wavy and crisped, 10–15 cm long, in erect crowns from root and up stems – on shingle ......*Rumex crispus* (p 164 **F**)

Lvs grass-like.......................................**4**

3 Basal lvs v thick and fleshy, 20–40 cm long, oval, **hairless**, waxy grey-green, often purplish-flushed when young, with shallow pinnate lobes – each fr size and shape of garden pea, borne in large open much-branched rounded heads – on shingle ....
Crambe maritima (p 218 **F**)

Basal lvs **v hairy**, 10–20 cm long, waxy-grey, deeply pinnately-lobed – frs 15–30 cm long, rough, hairless, sickle-shaped – on shingle *Glaucium flavum* (p 114 **F**)

Basal lvs hairless, 3–10 cm long, **shiny dark** green, twice-pinnate into **hair**-like segments – on dunes, beaches.....................................
Tripleurospermum maritimum (p 455 **B**)

Basal lvs hairless, 10–15 cm long, **grey**-green, twice-pinnate into v **fleshy** blunt segments – on shingle and rocks ....
Crithmum maritimum (p 338 **D**)

Basal lvs hairy, to 10 cm long, once-pinnate into thin, flat, rounded, toothed segments – on grassy dunes.......................................
Erodium cicutarium (p 322 **E**)

Basal lvs hairy, with sharp spines on edges and upper surface ....
Cirsium vulgare (p 463 **A**)

4 Lvs in 3 vertical ranks, on shoots in rows along rhizomes – on dunes..
Carex arenaria (**Sand Sedge**)

Lvs in 2 vertical ranks on stem **5**

5 Lvs in dense tufts 50–100 cm tall, grey-green, strongly inrolled, cylindrical, rigid, sharp-tipped – on dunes..
Ammophila arenaria (**Marram**)

Lvs in small tufts on shoots arising from rhizomes creeping in sand, flat when moist, to **1 cm wide**, **grey-green**, strongly ribbed above – on open dunes ....
Elytrigia juncea (**Sand Couch**)

Lvs **blue-grey**, to 2 cm wide, flat, rigid, ribbed, pointed – on dunes..
Leymus arenarius (**Lyme-grass**)

Lvs grey-green, bristle-like, to 2 mm wide – on dunes ....
Festuca arenaria (**Rush-leaved Fescue**)

PLANT DESCRIPTIONS – DICOTYLEDONS

BIRTHWORT FAMILY *Aristolochiaceae*

A family of per herbs, regarded as one of the most primitive in evolutionary terms of those plants included in this book. Easily distinguished by their lvs with cordate bases and tubular fls.

A Asarabacca, *Asarum europaeum*, creeping evergreen herb with fleshy downy stems; lvs glossy, dark green, kidney-shaped, untoothed, 4–8 cm across, on stalks to 10 cm long, in habit reminiscent of Ivy. Fls solitary, bell-shaped, 15 mm long, on short stalks with a three-lobed downy brownish single perianth, 12 short stamens, and six-celled ovary inferior to calyx. Eng only, vr, probably introd (C Eur) in a few wds, hbs. Fl 5–8.

B Birthwort, *Aristolochia clematitis*, v different from A, is a hairless erect per herb, 20–80 cm, with simple stems bearing alt bluntly cordate, stalked lvs; yellow fls in clusters of 4–8 in lf-axils; trumpet-shaped calyx-tubes, 2–3 cm long, with an oval blunt lip on top, and a basal swelling. Eng, CI vr, introd (E and C Eur); on wa, hbs. Fl 6–9.

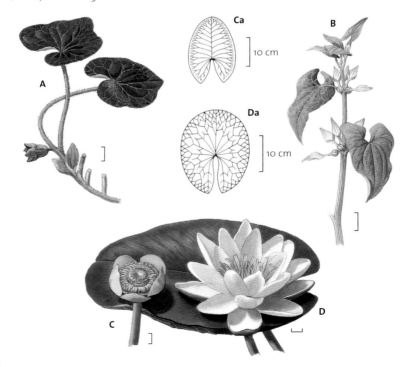

Ca

10 cm

B

Da

10 cm

A

C

D

WATER-LILY FAMILY   *Nymphaeaceae*

Aquatic per herbs with fls floating on or near water surface. **Do not confuse** yellow-flowered plants with **Fringed Water-lily** (p 356), which is similar but in a different family, having fringed petals and different lf-venation.

C Yellow Water-lily, *Nuphar lutea*, has a massive fleshy rootstock which creeps in underwater mud, and bears leathery cordate floating lvs (**Ca**) on long elastic stalks, triangular in cross-section and some thinner submerged lvs. Lvs with **23 or more** main veins, each vein dividing in two to form a **parallel**, 'tuning fork' pattern. Fls (**C**) (4–6 cm across) have 5–6 yellow concave sepals, about 20 smaller, narrower yellow petals, and many large stamens in a dense whorl; flask-shaped ovary has a flat cap with 10–25 **radiating stigmas** with margin of stigma-disk **entire**. Br Isles, c in ponds, canals, slow rivers. Fl 6–8. **C.1** *Least Water-lily, N. pumila*, is smaller in all parts than C, with **key difference** in lvs with **11-18** main veins in a similar pattern to C; 7–12 stigma-rays and fls 1.5–3.5 cm across; stigma-disc deeply-**scalloped**. N GB only, r in mostly acid lakes. Fl 7–8. Hybridizes with C (*N. x spenneriana*) which is often confused with C.1, having stigma-disk margin **undulate** but not uniformly deeply divided and pollen 75% sterile.

D White Water-lily, *Nymphaea alba*, has almost circular lvs (**Da**), with lf-veins joining up laterally to form a **network** (not like C or C.1); much larger fls (10–20 cm across), with 4 green sepals and 20–25 petals, white in the usual wild form. Br Isles, c in ponds and lakes. Fl 7–8.

HORNWORT FAMILY   *Ceratophyllaceae*

Aquatic per herbs growing wholly submerged; **key difference** from all other submerged water plants in their lvs are divided in twos (**dichotomous**) at each branch.

A Rigid Hornwort, *Ceratophyllum demersum*, resembles a Water-milfoil (p 294), but the whorled lvs (**Aa**) are **1-2 forked**, and **toothed** (not pinnately-divided); fls solitary in lf-axils; female fls produce rarely **variable** frs (**Ab**) with spines at base as shown or without. Br Isles: Eng, Ire lf; Wales, Scot r; in still fresh and brackish water. Fl 7–9. **Do not confuse** young, shaded or winter shoots of A, which may appear v soft and flaccid, with B.

B Soft Hornwort, *C. submersum*, close to A, **key differences** are softer, **3-4 forked** lvs (**B**) and **frs** without spines. Br Isles: S Eng, Wales r-o; Ire vr; Scot abs; in still fresh and brackish water. Fl 7–9.

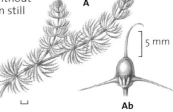

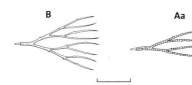

BUTTERCUP FAMILY	*Ranunculaceae*

A family of herbs (except the woody climber *Clematis*) with alt lvs (except *Clematis* with opp lvs), and with no stipules. Lvs mostly palmately-lobed or divided (linear in *Myosurus*). Fls usually have 5 sepals (4 in *Clematis*) and often 5 petals. The features common to ± all are the many stamens and the many carpels, free from each other and superior ovary. **Do not confuse** with some herbs in the Rose family which have similar-looking fls, but these have an epicalyx, or at least a slight cup to the fl below the sepals, and lfy stipules, all features abs in the Buttercup family.

KEY TO THE BUTTERCUP FAMILY

1 Woody climber with opp lvs and 4 petal-like sepals..*Clematis* (p 110)
   Herbs with spirally-arranged alt lvs – not climbers **2**

2 Fls with 1 carpel only..**3**
   Fls with 2 or many carpels...**4**

3 Fr a black oval berry – fls small, white, without spurs ....*Actaea* (p 113)
   Fr a dry follicle – fls blue (pink or white), with long spurs ....*Consolida* (p 110)

4 Carpels ripening into one-seeded achenes **5**
   Carpels ripening into many-seeded follicles that open at top when ripe **10**

5 Fls with both petals and sepals..**6**
   Fls with one perianth-whorl only **8**

6 Fls solitary and terminal on a lfless stem – lvs all basal and linear – fr-head a long slender spike covered with achenes ....*Myosurus* (p 110)
   Fls in branched lfy infls – lvs variously lobed or lanceolate...**7**

7 Lvs bi- or tri-pinnate, with linear, ± hair-like segments – petals red with no nectary – achenes wrinkled ....*Adonis* (p 104)
   Lvs palmately-lobed or divided, or undivided – petals yellow or white with a nectary at base..*Ranunculus* (p 102–109)

8 Fls small (1 cm or less across) – stamens conspicuous – fls many together in a loose infl – lvs 2–4 times pinnate, with small toothed lfts..*Thalictrum* (p 110)
   Fls large (2 cm or more across), solitary, or 2 together only – sepals white or coloured, resembling petals – a whorl of 3 lvs below the fl (or pair of fls)..**9**

9 Stem-lvs some way below fl, each lf with 3 broad spreading lobes – achenes unplumed
                                               Anemone (p 109)
   3 stem-lvs close below fl, erect, deeply-cut into narrow lobes – achenes with long feathery styles ....*Pulsatilla* (p 109)

10 Fls two-lipped, with a hood and a lip (but petals not joined) ....*Aconitum* (p 110)
   Fls radially symmetrical **11**

11 All 5 petals with long spurs – sepals coloured like petals ....*Aquilegia* (p 110)
   No spurs on perianth segments **12**

12 One whorl of (yellow) perianth segments only ....*Caltha* (p 104)
   Sepals petal-like – petals represented by narrow nectaries within fl **13**

13 Nectaries narrow, strap-shaped, yellow – sepals many (about 10), yellow, forming a globe-shaped fl ....*Trollius* (p 104)
Nectaries tubular, two-lipped...**14**

14 Fls solitary, with a ruff of deeply-cut lvs below fl – sepals and nectaries yellow ....
Eranthis (p 110)
Fls in branched infls, no ruff of lvs below fls – lvs lobed like fingers of a hand – sepals and nectaries greenish..*Helleborus* (p 104)

ID Tips   Yellow Buttercups

- **What to look at**: **(1)** presence or not of runners **(2)** lf-shape and hairiness **(3)** whether sepals are spreading or reflexed **(4)** shape of, and presence of warts or spines on achenes (which as a group form Buttercup fr-heads). Achene characters are v important for correct identification of several spp (**E, F, H** and **I** on p 102)

- **Creeping Buttercup** is the only common yellow buttercup with runners (actually stolons).

- Always look at a **fully open** flower to see if the sepals are reflexed or spreading.

KEY TO YELLOW BUTTERCUPS

1 Stem-lvs lanceolate ....Spearworts (p 104 **A, B**)
Stem-lvs lobed or round **2**

2 Sepals 3 (at least on some plants), petals 7–12 – lvs cordate ....*Ranunculus ficaria* (p 102 **J**)
Sepals 5, petals 5 (sometimes fewer in *R. auricomus*) **3**

3 Achenes with spines or hooks **4**
Achenes smooth, hairy or warted only...**5**

4 Sepals reflexed – achenes with short hooks on faces but not edges ..*R. parviflorus* (p 102 **F**)
Sepals spreading – achenes with long spines, especially on edges ....*R. arvensis* (p 102 **E**)

5 Sepals reflexed **6**
Sepals not reflexed **7**

6 Base of stem swollen – achenes without warts ....*R. bulbosus* (p 102 **D**)
Base of stem not swollen – achenes warted.......................................*R. sardous* (p 102 **I**)

7 Lvs (at least basal) hairless **8**
Lvs hairy **9**

8 Fls less than 1 cm across – fr-head elongated ....*R. sceleratus* (p 102 **H**)
Fls more than 1 cm across – fr-head globose ....*R. auricomus* (p 102 **G**)

9 Achenes in a v pronounced elongated head – Jersey only, vr ....*R. paludosus* (p 102 **C**)
Achenes in a globose head – vc and widespread **10**

10 Fl-stalks smooth – plant without runners – lf-lobes 3–5, sharp-toothed..........*R. acris* (p 102 **B**)
Fl-stalks grooved – plant with runners – lf-lobes 3, stalked, shallow-toothed *R. repens* (p 102 **A**)

Buttercups (*Ranunculus*) are herbs with alt lvs and yellow fls (white in **Water-crowfoots**, subgenus *Batrachium*, p 106). They have 5 (or 3) free green sepals, 5 free petals (except **Lesser Celandine** with different numbers of sepals and petals), many free stamens, and many tiny carpels in a head, which each develop into one-seeded achenes.

A Creeping Buttercup, *Ranunculus repens*, per herb to 60 cm tall, with creeping **runners**, **rooting** at the **nodes**; lvs (**Aa**) hairy, three-lobed, middle lobe long-stalked; fl-stalk **grooved**; sepals (**Ab**) **spreading**; fls deep yellow. Achenes (**Ac**) hairless with curved beak. Br Isles, vc in most habitats. Fl 5–8.

B Meadow Buttercup, *R. acris*, erect hairy per herb, 20–100 cm tall; stems branched, no runners; lvs polygonal in outline, **deeply** three-five lobed, lobes deeply-toothed; fl-stalks **not** grooved; sepals **spreading**; fls bright yellow. Achenes (**Ba**) 2–3.5 mm long, smooth, hairless with a short, hooked beak in a **globose** head. Br Isles, vc in mds, gsld. Fl 5–8.

C Jersey Buttercup, *R. paludosus*, similar to B but with most lvs being **basal lvs**, three-lobed and 1-2 small stem lvs; **key difference** from B is **achenes** (**Ca**) usually smaller and borne in a much **elongated** head (**Cb**). **Jersey only**, vr on gsld dry in summer and damp in winter. Fl 5.

D Bulbous Buttercup, *R. bulbosus*, erect hairy per, 15–40 cm tall, with basal tuber (**Db**), no runners; lvs three-lobed; fl-stalk **grooved**; sepals **reflexed**; fls bright yellow. Achenes (**Da**) 2–4 mm long, hairless, finely pitted. Br Isles, vc on dry gsld, especially on calc soils. Fl 4–6.

E CR Corn Buttercup, *R. arvensis*, **ann**, erect; stem-lvs deeply-cut; fls 4–12 mm across, yellow; sepals spreading; achenes (**Ea**) with strong **border** and **long spines**. Br Isles, r on ar, especially on calc soil. Fl 6–7. **Do not confuse** with F, as fl-size overlaps; instead look at the sepals and achenes.

F Small-flowered Buttercup, *R. parviflorus*, spreading hairy **ann**, to 20 cm; lvs lobed; sepals reflexed; fls 3–6 mm across, pale yellow; achenes (**Fa**) 2.5–3.0 mm, narrowly-bordered, with **hooked** spines, but not on their edges. Br Isles: Eng, Wales, o; Ire r; on ar, dry banks, open gsld on calc soil. Fl 6–7.

G AWI Goldilocks Buttercup, *R. auricomus*, erect ± hairless per; **key difference** from all other Buttercups is basal lvs **kidney-shaped**. Lower stem-lvs three-lobed; upper stem-lvs with narrow divisions. Fls few, 1.5–2.5 cm across, with smooth stalks, the golden-yellow petals often partly missing; achenes (**Ga**) smooth, downy. Br Isles, f-lc in wds on basic soils. Fl 4–6.

H Celery-leaved Buttercup, *R. sceleratus*, **ann**, with stout short stem (20–60 cm); basal lvs palmately three-lobed, long-stalked, shiny; stem-lvs short-stalked, more divided. Fls in much-branched infl; sepals reflexed; fls 5–10 mm across, petals shiny pale yellow; **achenes** (**Ha**) many in an **elongated head** (**Hb**). Br Isles: Eng c; Wales, Ire, S Scot, o-lc; N Scot r; on mud by ditches, ponds, streams. Fl 5–9.

I Hairy Buttercup, *R. sardous*, is like D, with sepals reflexed and fl-stalks grooved, but **key differences** are that I is ann plant, has no tubers, has **pale** yellow fls, and **warts** in a ring on achenes (**Ia**) inside the **thick border**. Plant usually more hairy than D but **not** always, so its name may be misleading, referring in fact to the plant's hairy **receptacle**. Br Isles: GB N to mid-Scot c on damp gsld near sea, inland on ar on clay. Fl 6–9.

J Lesser Celandine, *R. ficaria*, per herb to 20 cm, with stems creeping and rooting, then

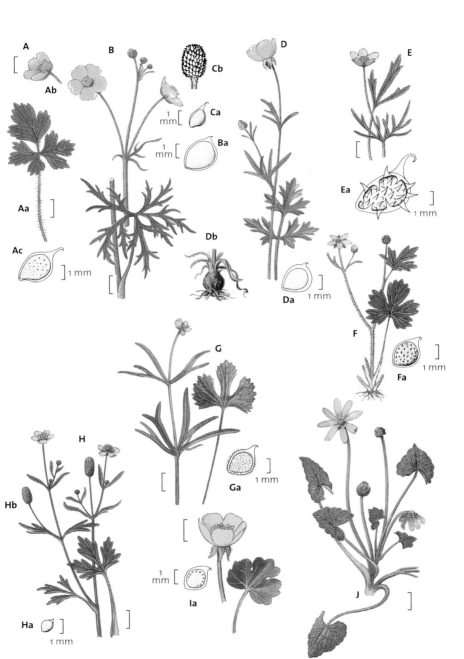

103

ascending; lvs long-stalked, cordate, fleshy, blunt, hairless, glossy green, 1–4 cm long and wide, in rosettes. Fls 2–3 cm across, solitary, long-stalked; usually **3** oval sepals; 8–12 **narrow** oval petals, glossy yellow, 1.0–1.5 cm long; achenes smooth, round. Sometimes with **bulbils** in lf-axils **after** flg (ssp *bulbilifer*). Br Isles, vc prefers damp, loamy or clay soils in wds, mds, hbs, rds. Fl 3–5.

A   Lesser Spearwort, *Ranunculus flammula*, variable hairless erect or creeping per, with hollow stem, sometimes rooting at **some** nodes; lanceolate stem-lvs, and oblong, stalked basal lvs (**Aa**). Fls few, 8–20 mm across, petals overlapping, with **furrowed** stalks; sepals yellowish; achenes **smooth**. Br Isles, c in marshes, wet places. Fl 5–9. **A.1** Sch 8 VU ****Adder's-tongue Spearwort**, *R. ophioglossifolius*, of low habit, but can be tall in sheltered situations; has broadly oval, cordate-based lvs on long stalks; petals not over-lapping and paler yellow than A. **Key difference** from A is achenes covered with **minute bumps.** Br Isles: **Glos only**, vvr; wet mud by ponds with seasonally fluctuating water levels. Fl 6–7.   **A.2**   VU ****Creeping Spearwort**, *R. reptans*, has runners and generally roots at **every** node but requires expert id (see *Plant Crib*); vvr by lakes in N GB. Fl 6–8.

B   Greater Spearwort, *R. lingua*, resembles a taller (50–120 cm) and stouter A, with lanceolate stem-lvs to 25 cm long; fls brighter yellow, 2–5 cm across; **unfurrowed** fl-stalks. Br Isles, o-lc native in fens but f planted in ponds. Fl 6–9.

C   Marsh-marigold, *Caltha palustris*, hairless stout buttercup-like per herb, with hollow stems and cordate or kidney-shaped lvs to 10 cm across, long-stalked below; fls, 15–50 mm across, have 5 yellow petal-like sepals, **no** true petals and many yellow stamens. Carpels form into green sac-like **follicles** to 1 cm long, each opening to release several seeds. Br Isles, c in wet mds, fens, wet wds, by rivers. Fl 3–5.

D   FPO (Eire) WO (NI) ******Ire AWI **Globeflower**, *Trollius europaeus*, resembles C, but is more erect, with lvs palmately-lobed, ± deeply-toothed. The numerous (5–15) petal-like sepals form a **globe**, enclosing the small strap-shaped nectar-secreting **petals** and the many stamens; frs as in C. Br Isles: Wales to N Scot lc; NW Ire lc; on upland wet mds, mt cliffs and mt woods. Fl 6–8.

E   AWI **Green Hellebore**, *Helleborus viridis*, bushy, ± hairless per herb; lvs long-stalked, with 7–11 digitate lfts, arising from a central point; lvs are 15 cm across, deciduous, deep green; fls like those of C in structure, but with 5 green spreading sepals, 9–12 nectaries. Br Isles: S Eng lf (but often as gdn escape); Wales, N Ire r; **S Ire abs**; in wds on basic soils. Fl 3–4.

F   AWI***Stinking Hellebore**, *H. foetidus*, is more erect than E, with long-stalked **dark blue-green** evergreen lower lvs, each with 3–9 **narrower** (0.5–2.0 cm) segments that rise from two 'arches' on tip of lf-stalks and **not** from a single point at stalk-tip. Fls more **cup-like** than in E and with **purple** edges to green sepals. Br Isles: S Eng, Wales vlf; Scot, Ire vr; in open wds on shallow chk and limestone soils. Fl 1–4.

G   EN ****Pheasant's-eye**, *Adonis annua*, erect hairless ann herb, with 2–3-times pinnate lvs with hair-like segments; scarlet fls with dark stamens and wrinkled achenes (**Ga**). Br Isles: S Eng, r on ar on chk. Fl 7.

A

Aa

B

C

D

E

F

G

Ga

2 mm [

Water-crowfoots   (*Ranunculus* subgenus *Batrachium*)

All Water-crowfoots have **white** fls, with **yellow base** to petals, and grow in water or on mud. Some only have **capillary** (ie finely divided) lvs, some only have **laminar** lvs (ie with a broad blade) and some have both sorts of lvs. All are f in Br Isles unless stated, and fl 4–9.

Many Water-crowfoots are difficult to identify because of their great variability due to factors such as the level or speed of water and possibly due to hybridisation. With experience, most populations can be identified when they have abundant fls; however some populations will always be difficult to identify with confidence. Some plants, or even entire populations, may lack diagnostic features and it may not be possible to arrive at a confident identification. In these cases, an identification may be made based on judgement of a combination of all available features. The best source of guidance for difficult material is *Plant Crib*. For species with capillary lvs, **only** aquatic plants with fls and frs can be identified.

ID Tips   Water-crowfoots

• Capillary lvs of Water-crowfoots are divided in threes (into **three segments**) at initial divisions but may be divided in twos (**dichotomous**) toward shoot tips - no other aquatic plants do this.

• **What to look at**: **(1)** for floating laminar lvs: the outline and the degree to which they are divided **(2)** for capillary lvs: their length relative to adjacent internodes and the number of times that they divide **(3)** petal length **(4)** the length of the fl-stalk in fr relative to that of the opposed lf-stalk **(5)** receptacle shape and hairiness **(6)** presence of short bristly hairs on frs, at least when young.

• **Brackish Water-crowfoot** (*R. baudotii*) is the only Water-crowfoot in which the receptacle **elongates** in fruit (**E**). The fruits are hairless and the capillary leaves are shorter than adjacent internodes, even when young.

KEY TO WATER-CROWFOOTS
by R. V. Lansdown
Use this key for well-developed plants with abundant fls

1 Only laminar lvs present...**2**
Only capillary lvs present **3**
Both capillary and laminar lvs present ....**table 1**

2 Most petals >4.5mm, 2-3 x as long as sepals – lvs (**Ga**) relatively uniform, rounded in
outline with short incisions between lobes – fl receptacle hairless – on mud and in
shallow water in S and W of GB, S Ire ....
Round-leaved Crowfoot *Ranunculus omiophyllus* (p 107 **G**, **Ga**)
Most petals ≤4.5mm, slightly longer than sepals – lvs (**Da**) relatively uniform, shallowly-
divided, angular to rounded-triangular in outline without incisions between lobes – fl
receptacle hairless – on mud and in shallow water in GB, Ire I in SE ....
Ivy-leaved Crowfoot *R. hederaceus* (p 107 **D**, **Da**)
Most petals ≤4.5mm, slightly longer than sepals – all lvs (**F**) deeply divided with wide
divisions, variable in outline, mainly three to five-lobed but some smaller lvs three-lobed
– fl receptacle hairy – r in S Eng and Wales on mud and in shallow water on rutted tracks,
ditches, edges of shallow pools often on hths ....
EN BAP **Ire***Three-lobed Crowfoot** *R. tripartitus* (p 107 **F**)

3 Lvs (**A**) much divided, held in a fan, with rigid segments in a single plane (like spokes of a
bicycle wheel with no rim) – in ditches, lakes and gravel pits (not N Scot), usually in
calcareous water ....**Fan-leaved Water-crowfoot** *R. circinatus* (p 107 **A**)
Lvs soft or slightly stiff and spreading in 3 dimensions **4**

4 Lvs on the middle part of stems shorter than and never >2x adjacent internodes ....**table 3**
Lvs on the middle part of stems as long as or up to >2x adjacent internodes ....**table 2**

Table 1 Capillary and laminar leaves present				
	petals (mm)	fl-stalk in fruit relative to stalk of opposite laminar lf	capillary lvs relative to adjacent internode	young fruits with short bristles
Common Water-crowfoot *R. aquatilis* (**H**, **Ha**)	5–10	shorter	shorter	yes
Pond Water-crowfoot *R. peltatus* (**C**, **Ca**)	11–22	longer	shorter	yes
Brackish Water-crowfoot *R. baudotii* (**E**, **Ea**)	5.5–10	longer	shorter	no
EN BAP **Ire***Three-lobed Crowfoot* *R. tripartitus* (**F**)	1.25–4.5	variable	shorter	no
***Stream Water-crowfoot** R. penicillatus* ssp *penicillatus* var *penicillatus* (**I**, **Ia**)	10–22	longer	longer	yes

Table 2 Laminar leaves lacking; lvs on the middle part of stems longer than adjacent internodes

	capillary lvs	receptacle
WO (NI) **River Water-crowfoot** R. fluitans	forked <4 times	hairless to sparsely hairy
*Stream Water-crowfoot** R. penicillatus ssp pseudofluitans var pseudofluitans	forked >4 times	densely hairy
*Stream Water-crowfoot** R. penicillatus ssp penicillatus (**I, Ia**)	forked >4 times	densely hairy

Table 3 Laminar leaves lacking; lvs on the middle part of stems shorter than adjacent internodes

	petals (mm)	young fruits with short bristles
Common Water-crowfoot R. aquatilis (**H, Ha**)	5–10	yes
Pond Water-crowfoot R. peltatus (**C, Ca**)	11–22	yes
Brackish Water-crowfoot R. baudotii (**E, Ea**)	5.5–10	no
Thread-leaved Water-crowfoot R. trichophyllus (**B, Ba**)	3.5–5.5	no
*Stream Water-crowfoot** R. penicillatus ssp pseudofluitans var vertumnus	10–22	yes

There appears to be no absolutely reliable means of separating *R. penicillatus* ssp *pseudofluitans* var *vertumnus* from Pond Water-crowfoot when Pond Water-crowfoot does not have laminar lvs.

Anemones (A, B below) have coloured petal-like sepals, but no petals, and many stamens; also a whorl of 3 lvs on fl-stalk, sometimes resembling a calyx.

▼ **A** VU *Pasqueflower**, *Pulsatilla vulgaris*, hairy per herb with feathery lvs twice-pinnate, from base; large, solitary, rich purple fls, 5–8 cm across, first erect, then drooping and bell-like; 6 sepals, silky outside; no petals, many golden anthers; a whorl of bracts with many narrow hairy lobes, 2–5 cm, below fl. The head of achenes develops silky plumes 3.5–5.0 cm long. Br Isles: S, mid, E Eng vlf; in chk gslds especially on S slopes. Fl 4–5.

▼ **B** AWI **Wood Anemone**, *Anemone nemorosa*, per hairless herb with creeping rhizome and star-like white fls with 6 (or more) spreading sepals. Halfway up stem is a whorl of 3 palmately-lobed lvs; other long-stalked similar lvs arise from root. Fls often pink-purple-flushed, especially outside. Br Isles, c in dry deciduous wds, old hbs, upland mds. Fl 3–4.

C **Winter Aconite**, *Eranthis hyemalis*, has a ruff of 3 deeply-cut lvs below fls; fls 20–30 mm across, with yellow sepals only (no petals); frs are many-seeded follicles. Introd (S Eur); Br Isles, o in wds, parks, open gslds. Fl 1–3.

D AWI ***Monk's-hood**, *Aconitum napellus*, hairless per to 1 m or more, with deeply-cut palmate lvs, **divided to the base**; hooded-tipped, blue or violet fls, 3–4 cm across; upper 'petal' (actually the sepal) c. **as high as wide**; fl-stalks **v densely hairy**. Br Isles: restricted to SW Eng and Wales, r by streams and in wds as a native; plants elsewhere are generally gdn escapes. Fl 5–9. **Do not confuse** with natd gdn *Aconitum* spp which either do not have deeply divided lvs or fl-stalks hairless (or sparsely hairy) and upper 'petals' higher than wide.

E AWI **Columbine**, *Aquilegia vulgaris*, tall hairless per with large twice-trifoliate lvs; rounded, stalked, lobed lfts; large blue fls with 5 spurred sepals and 5 petals, all blue or violet-blue, sometimes pink or white. Br Isles SW, S Eng, Wales, f; S Scot lf; Ire o; in calc scrub, wds, fens. Fl 5–6.

F **Larkspur**, *Consolida ajacis*, ann to 1m, palmate lvs with **finely divided** segments; fls blue, pink or white with 4 petals and 5 petal-like sepals, numerous stamens; upper sepal with a long **spur** 12-18mm; fr dry many-seeded follicle. Introd (Mediterranean and SW Asia) formerly with grain, now o gdn escape in S Eng on wa, rds and vr as an ar weed in E Eng. Fl 6–8.

▼ **A** **Traveller's-joy**, *Clematis vitalba*, woody climber with peeling fibrous bark, and opp, pinnate compound lvs with narrow oval pointed, usually toothed lfts. Fls 2 cm across, fragrant, terminal and in cymes, in lf-axils, with 4 greenish-creamy sepals, hairy outside and inside, and many stamens. Achenes (**Aa**) each develop long white plumed styles in fr. Br Isles: Eng, Wales, N to Morecambe Bay, c; Ire f but introd; on chk and limestone. Fl 7–8.

▼ **B** **Common Meadow-rue**, *Thalictrum flavum*, erect, nearly hairless per herb, 50–100 cm tall; lvs 2–3-times pinnately compound with wedge-shaped lfts longer than broad above, 1–2 cm long; dense panicles of erect small fls (**Ba**) with 4 narrow whitish sepals and many conspicuous **erect** yellow stamens; achenes (**Bb**) 1.5–2.5 mm long with **6 ribs**. Br Isles: GB N to Inverness lf; SW and SE Eng r; Ire o; in fens, streamsides, mds. Fl 7–8.

▼ **C** **Lesser Meadow-rue**, *T. minus*, more branched, spreading plant than B, with more divided (3–4-times pinnate) lvs, and smaller lfts as broad as long. Fls (**Ca**) in more diffuse spreading panicles, with drooping stamens; achenes (**Cb**) larger (3–6 mm) with **8–10 ribs**. Br Isles, especially in Breckland and in N and W vlf; on dry gsld on chk or dunes, rocky river banks. Fl 6–8.   **C.1** **Alpine Meadow-rue**, *T. alpinum*, small hairy per herb, 8–15 cm tall, with lvs 2–5 cm long and twice-trifoliate; **key difference** from B and C is fls in simple racemes. Mts of GB, vl on rocks and in flushes. Fl 6–7.

▼ **D** VU **Mousetail**, *Myosurus minimus*, small hairless ann herb, with basal rosette of linear fleshy lvs and erect stems, 5–12 cm long, bearing solitary erect greenish-yellow fls (**Da**); 5 sepals, 5 petals, 5–10 stamens and a cylindrical, mouse-tail-like receptacle which elongates to 3–7 cm in fr, bearing many tiny achenes. Br Isles: Eng, Wales, r; Ire, Scot abs; on damp sandy ar, open ground. Fl 5–7.

A

◀ B

C

D

E

F

111

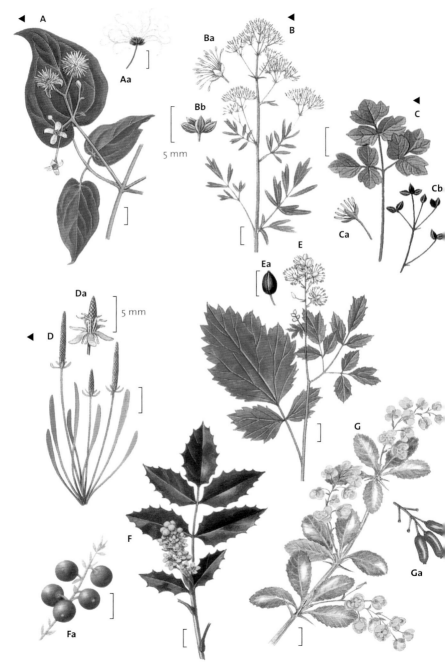

A

Aa

B

Ba

Bb

5 mm

C

Ca

Cb

Da

5 mm

D

E

Ea

F

Fa

G

Ga

E **Baneberry*, *Actaea spicata*, per, almost hairless herb, 30–60 cm tall; glossy, long-stalked basal lvs, twice-trifoliate or twice-pinnate, with toothed lfts; stem-lvs smaller. Fls creamy-white in dense stalked spikes, with 3–6 sepals, 4–6 petals, many stamens, and a single carpel which ripens to a black berry (**Ea**) 1 cm long. Br Isles: N Eng, Scot, vl on limestone pavements and calc wds. Fl 5–6.

BARBERRY FAMILY   *Berberidaceae*

F Oregon-grape, *Mahonia aquifolium*, shrub with holly-like lvs, yellow fls in dense spikes, and frs (**Fa**) in racemes like tiny blue-black grapes. Introd (NW of N America); Br Isles: Eng, Wales, lowland Scot c; Ire r; in hbs, scrub, wds, hths. Fl 1–5.

G Barberry, *Berberis vulgaris*, shrub with oval toothed lvs, and three-pointed spines; fls in drooping spikes, usually with 5 whorls of yellow perianth segments, 6–8 mm across, and 10 mm-long red berries (**Ga**). Introd (Eur) possibly during the Neolithic period; GB much destroyed in 19th century as it is a secondary host of wheat rust, but still f in hds, scrub. Fl 5–6.

POPPY FAMILY   *Papaveraceae*

This family has fls with 2 sepals that soon fall, four petals, many stamens, and milky, orange or yellow juice (**latex**) in canals in stem and lvs. **Poppies** (*Papaver*) have crinkly petals to the large bright fls, and a fr of 4–20 carpels forming a capsule that opens by pores below the stigma-disc; and white or coloured latex.

KEY TO POPPIES (*PAPAVER*)

1 Plant with waxy-grey bloom, ± hairless – lvs simple, toothed or lobed – fr globular...............
Papaver somniferum (p 114 **E**)
Plant green, no waxy bloom, hairy – lvs 1–2-times pinnate **2**

2 Capsule hairless **3**
Capsule bristly..**5**

3 Capsule globular ....*P. rhoeas* (p 114 **A**)
Capsule oblong, at least twice as long as wide **4**

4 Latex white – petal-bases overlapping ....*P. dubium* ssp *dubium* (p 114 **D**)
Latex turning yellow – petal-bases not overlapping ....*P. dubium* ssp *lecoqii* (p 114 **D.1**)

5 Capsule globose – bristles many, spreading ....*P. hybridum* (p 114 **B**)
Capsule narrow, oblong, at least some capsules twice as long as wide or more – bristles few, erect...*P. argemone* (p 114 **C**)

A Common Poppy, *Papaver rhoeas*, bristly erect ann, 20–60 cm tall, with pinnate lvs and large fls, 7–10 cm across; scarlet petals, often with a dark basal blotch; capsule globular, hairless. Br Isles c on ar, wa, rds. Fl 6–8.

B FPO (Eire) **Ire Rough Poppy**, *P. hybridum*, similar, but has smaller crimson fls, 2–5 cm across; globose capsules (**B**) covered in stiff bristles. Br Isles: S Eng r, Scot, Ire almost abs; ± restricted to ar on chk. Fl 6–7.

C VU **Prickly Poppy**, *P. argemone*, resembles B, but has smaller red fls with narrower petals; **key difference** in **long** narrow bristly capsules (**C**), at least some twice as long as wide or more. Br Isles: S & C Eng, SE Scot, r-o; C Scot, Ire vr; on chk or sandy soils on ar, vr on other disturbed gd. Fl 6–7.

D Long-headed Poppy, *P. dubium* ssp *dubium*, has capsules smooth as in A, but elongated (**D**) as in C. Petals usually paler than A; sometimes with dark basal blotch; anthers bluish-black. Br Isles c, more c than A in Scot; on ar, wa on disturbed light soils but also c in and near gdns. Fl 6–7. **D.1 Yellow-juiced Poppy**, *P. dubium* ssp *lecoqii*, has **yellow latex** (on exposure to air) and anthers often yellow. Br Isles: S Eng, Ire f, but rarer than D; usually on chalk.

E Opium Poppy, *P. somniferum*, stouter, taller (to 80 cm), almost **hairless** per herb with oblong wavy **waxy-grey** coarsely-toothed lvs (not pinnate and bristly as in A and D). Fls 8–18 cm wide and white, pale lilac (or pink in cultivated forms), with purple-black centres. Capsule (**E**) oval, smooth, with disc at top 2 cm long. Opium and its derivatives are produced from its latex. Introd (Asia); Br Isles c on ar, wa. Fl 6–8.

F Yellow Horned-poppy, *Glaucium flavum*, tall (30–90 cm) branched per herb, with a waxy bloom. Basal lvs pinnately-lobed, waxy, hairy, in a rosette, over-wintering; stem-lvs less divided; fls yellow, 6–9 cm; sepals hairy. Capsule (**Fa**) is two-celled, sickle-shaped, to 30 cm long, rough; latex yellow. Br Isles (not N Scot), f-la on coastal shingle and o inland on wa. Fl 6–9.

G *Welsh Poppy*, *Meconopsis cambrica*, almost hairless per herb, 30–60 cm tall, with pinnate lvs, the lower long-stalked; **yellow** latex; **yellow** poppy fls; capsules rather like those of D, 2.5–3.0 cm long, elliptical, with 4–6 stigmas, but with a distinct style and opening by **slits** about 1/4 of fr's length. SW, N Eng, Wales, W Ire lc in damp shady rocky places, also f as gdn escape throughout Br Isles. Fl 6–8.
G.1 Californian Poppy, *Eschscholzia californica*, hairless ann or per to 60 cm tall, lvs grey-green with more **finely-divided** segments than G and **watery sap**; fl **yellow** to **orange**, poppy-like; sepals 2, forming a **pointed cap**, soon falling off, with a ring round the base; fr linear capsule, splitting lengthwise into 2 parts. Introd (N America); o gdn escape in S Eng, CI, in wa, quarries, dunes but not usually persisting. Fl 7-9.

H Greater Celandine, *Chelidonium majus*, branched slightly hairy per herb, to 90 cm tall, with grey-green pinnately-lobed lvs and **orange latex**. Fls are like tiny Poppies, four-petalled, yellow and 20–25 mm across, in groups; fr (**Ha**) a narrow capsule, 3–5 cm long, opening into 2 parts from **below**. Br Isles (especially S), c on hbs, pavements, walls, wa. Fl 5–8. **Poisonous**.

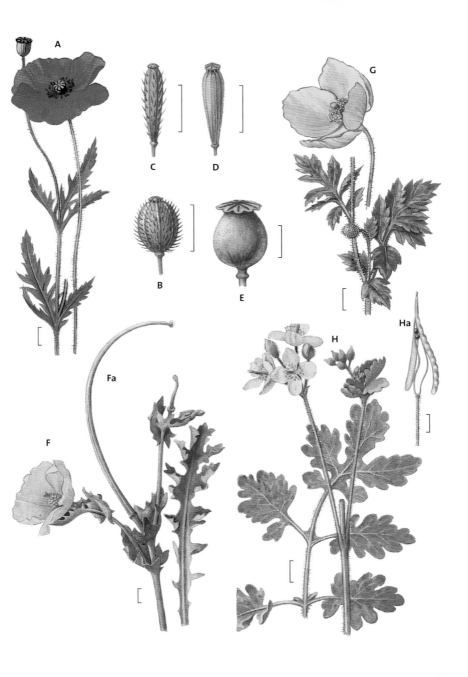

A

C   D

B   E

G

Fa

F

H

Ha

FUMITORY FAMILY by Dr Tim Rich   *Fumariaceae*

This family consists of delicate, usually sprawling, hairless, often waxy-lvd plants with repeatedly pinnate lvs; racemes of tubular two-lipped fls with 2 tiny sepals, 2 outer petals (with 1 or both spurred, or with basal sacs), 2 narrower inner petals (often partly joined), and 2 stamens, each three-lobed. Fr of Fumitory (*Fumaria*) is an 1-seeded achene; that of *Corydalis*, *Ceratocapnos* and *Pseudofumaria* is a two-valved many-seeded oblong capsule. *Ceratocapnos* has lvs with tendrils; *Corydalis* has simple stems and *Pseudofumaria* branched stems. Related to Poppies, but with watery latex and v different fls.

ID Tips   Fumitories

• **What to look at**: **(1)** sepals are particularly useful for id and tend to maintain their shape irrespective of other factors (but they fall off readily), measure maximum length and width including teeth **(2)** fl shape and size (measure largest fresh fls) **(3)** fr shape and size (measure mature frs).

• Properly **developed fls** on healthy plants in open situations are **crucial** – ignore shaded plants and the small whitish fls on old straggly plants.

1 **Fl-stalk**
2 **Spur**
3 **Keel of upper petal**
4 **Wing of upper petal**
5 **Lateral (inner) petals**
6 **Wing of lower petal**
7 **Keel of lower petal**
8 **Sepal**
9 **Bract**

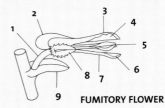

FUMITORY FLOWER

KEY TO FUMITORIES (*FUMARIA*)

1 Largest fls 5–8(–9) mm long...**2**
   Largest fls 9–14 mm long **5**

2 Largest sepals 2–3.5 x 1–3 mm **3**
   Largest sepals 0.5–1.2 x 0.5 mm **4**

3 Fr truncate or notched at tip, usually wider than long – fls c.3 x as long as sepals ....
                                      Fumaria officinalis (p 117 **A**)
   Fr rounded at tip, about as wide as long – fls c.2 x as long as sepals.... *F. densiflora* (p 118 **B**)

4 Most bracts as long as or longer than fr-stalks – fls white ....*F. parviflora* (p 118 **C**)
   Most bracts c.1/2 – 3/4 as long as fr-stalks – fls pink......................................*F. vaillantii* (p 118 **D**)

5 At least some infls (especially older ones) with most fl- and/or fr- stalks strongly recurved (can be variable) – fresh fruits with distinct or indistinct fleshy neck.......................................**6**
   Fl- or fr- stalks erect to spreading (rarely a few fls recurved) – fleshy neck indistinct...........**7**

6 Fls whitish flushing pink with age – sepals ovate, toothed – wing of upper petal folded upwards but only half way up keel ....*F. capreolata* (p 118 **E**)

Fls purple in looser spikes – sepals oblong, toothed – wing of upper petal folded upwards to cover most of keel ....*F. purpurea* (p 118 **F**)

Fls rosy-pink – sepals (**Fig 1**) ovate ± untoothed – wing of upper petal folded upwards to cover most of keel – fr acute at apex – infl usually much longer than its stalk – vr in IoW, probably gone from Cornwall (probably introd)...

Sch 8 **Martin's Ramping-fumitory** *F. reuteri* (**L**, **Fig. 1**)

7 Largest sepals 2–3 x 1.5 mm, strongly toothed ....*F. bastardii* (p 118 **G**)

Largest sepals 3–5.5 x 1.5–4 mm, weakly toothed **8**

8 Fls 9–12 mm, usually pink to purplish – fr 2-2.5 x 2 mm usually ± smooth *F. muralis* (p 118 **H**)

Fls 12–15 mm, usually rosy-white, wing of upper petal pink with white margin in young fls, later pink with purple blotch at base – fr 3 x 3 mm, wrinkled when dry – r, Cornwall and IoS only........ ....End BAP***Western Ramping-fumitory** *Fumaria occidentalis* (**M**, **Fig. 1**)

Fig. 1 *Fumaria* sepals

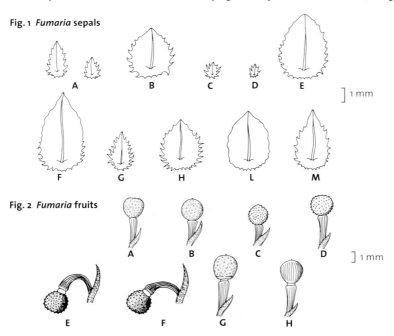

] 1 mm

Fig. 2 *Fumaria* fruits

] 1 mm

▼ **A** **Common Fumitory**, *Fumaria officinalis*, a weak scrambling or ± erect ann, to 20 cm tall; 2–3 pinnate lvs, with flat lanceolate narrow grey-green lfts; fls 6–8 mm long; sepals (**Fig 1**) 2–3.5 x 1–2.5 mm (ssp *officinalis*), **narrow**; **bracts** mostly a little **shorter** than the straight, erect-spreading fr-stalks; fr (**Fig 2**) 2 x 2–3 mm, **truncate to weakly notched** at tip. Br Isles, vc (the commonest sp in S, rarer to N & W); on ar, especially on chk and sand, wa. ssp *wirtgenii* has smaller sepals, to 2 x 1 mm. Fl 5–10.

B Dense-flowered Fumitory, *Fumaria densiflora* (*F. micrantha*) resembles A, with 5–7 mm pink fls (**B**), but in initially dense, later widely-spaced racemes with **broader rounded sepals (Fig 1)** 2.0–3.5 x 1.5–3 mm, **bracts longer than fr-stalks**, fr (**Fig 2**) rounded at tip. SE Eng, E Scot, lc, elsewhere r, Ire abs to vr; on ar usually on chk. Fl 6–10.

C VU *****Fine-leaved Fumitory**, *F. parviflora*, differs from A in its smaller (5–6 mm long) white or merely pink-tinged fls (**Ca**), sepals (**Fig 1**) tiny, usually under 1mm long and under 1mm wide; narrow, **very fine** channelled lf-segments (flat when shaded), bracts about as long or longer than fr-stalks. S Eng o; restricted to ar on chk. Fl 6–9.

D VU *****Few-flowered Fumitory**, *F. vaillantii*, has small fls (**Da**) like C, but pink, and lfts flat; sepals (**Fig 1**) tiny, to 1.2 x 0.5mm; **bracts c. ¹/2 to ³/4 as long as fr-stalks**. S Eng r; almost only on ar on chk. Fl 6–9.

E White Ramping-fumitory, *F. capreolata*, sprawls to about 1 m tall; infls dense with fls 10–12(-14) mm long (**Ea**), **whitish** often flushing pink with age; sepals (**Fig 1**) 4–6 x 2.5–3 mm ovate, and (in older infls at least) strongly **recurved** or **arched** fl- and fr-stalks. Fresh fr (**Fig 2**) blunt at tip, with distinct **fleshy neck** where they join the stalks. Br Isles (mostly S and W) lf; on wa, ar, hbs, old walls, most frequent nr coast. Fl 4–6(-9).

F BAP*****Purple Ramping-fumitory**, *F. purpurea*, is rather like E but differs in having laxer but still recurved infl, purplish-pink flowers (**Fa**) **± oblong sepals (Fig 1)** 5–6.5 x 2–3 mm, and the wing at the tip of the upper petal upturned so as to almost hide the keel in side view (upturned to c. ¹/2 way in E). Br Isles (mostly in W, abs from SE), r on wa, ar, hbs. Fl 4–6(-9).

G Tall Ramping-fumitory, *F. bastardii*, fls large 9–11(-12) mm long (often smaller at end of flowering), salmon-pink, spreading (rarely flexuous) (**Ga**), and **smaller** (usually < 3 x 1.5 mm) **more strongly toothed sepals (Fig 1)** than the other Ramping-fumitories; **bracts up to ¹/2** as long as fr-stalk; fl and fr-stalks usually spreading; fr (**Fig 2**) c. 2–2.5 x 2–2.5 mm, wrinkled, fleshy neck obscure. W of Br Isles widespread and lc (one of the commonest spp in W); on ar, wa, hbs. Fl 4–10.

H Common Ramping-fumitory, *F. muralis*, resembles G, but the **sepals (Fig 1)** are larger usually 3–5 x 1.5–3 mm, the fls (**H**) 9–12 mm, purplish-pink, the **bracts c. 2/3** as long as fr (**Fig 2**) stalk; fl- and fr-stalks spreading; fr (**Fig 2**) c. 2–2.5 x 2 mm, smooth, fleshy neck obscure. Br Isles (one of the commonest spp in N), lf on ar, hbs, wa on open soils. Fl 5–10.

I AWI **Climbing Corydalis**, *Ceratocapnos claviculata* (*Corydalis claviculata*), delicate climbing ann with bipinnate lvs ending in **tendrils**, and short spikes of **creamy fls (Ia)** each 5–6 mm; frs (**Ib**) 6 mm and three-seeded. Br Isles: Eng, Scot c (more c in Wales and the W), Ire o-r; in dry wds, hths, scrub, on shady rocks on more acid soils. Fl 5–9.

J Yellow Corydalis, *Pseudofumaria lutea* (*Corydalis lutea*), stout per plant with no tendrils and bright **yellow fls (J)** 12–18 mm long. Introd (S Eur); Eng, Wales vc, Scot f and Ire o; on walls, stony wa. Fl 5–8.

K Bird-in-a-bush, *Corydalis solida*, is like L, but with purple fls (**K**) 15–22 mm long, with **long** straight spurs, toothed bracts to fls, fleshy scales on stem below lvs. Introd (E Eur), possibly native in Kent; Eng r; in dry wds, hds, rds. Fl 4–5. Other introd *Corydalis* spp occur but are rare.

A

E

B

Ea

H

C

Ca

D

G

Ga

I

Ib

Ia

Da

5 mm

F

Fa

J

K

ELM FAMILY

Ulmaceae

Elms (*Ulmus*) are deciduous trees with alt, unlobed, but toothed, ± hairy, rough lvs, with blade often extending lower down stalk on one side than the other (**asymmetrical**). Tiny fls in globular heads, with calyx 4–5-lobed and bell-shaped, no petals, and many purplish-red stamens in early spring; fr a tiny achene surrounded by a thin, oval, notched wing. A difficult group with hybrids and many planted cultivars (consult Coleman (2002) in *British Wildlife Magazine* 13 (6) p 390 for a beginner-friendly guide).

A English Elm, *Ulmus procera*, tall tree (to 30 m) with erect trunk, rough bark, and oblong outline. Twigs short, hairy; lvs 4–9 cm, rounded-oval, rough-hairy on upperside, with v unequal bases. Suckers produced freely from base of trunk. Catkins (**Aa**) purplish, tassel-like; fr (**Ab**) rarely produced; seeds above middle of fr-wing. Br Isles: Eng, Wales, f in lowlands; Scot, Ire r; full-sized trees now r due to Dutch Elm Disease in 1960s and 70s, but suckers survive; in hds, parks, riverside wds. Fl 2–3. **A.1 Small-leaved Elm**, *U. minor* agg (*U. carpinifolia*), is a complex group, including several forms described as separate spp or ssp (eg *U. plotii*), now known to be clones within *U. minor* agg. Often difficult to distinguish from A. Has hairless twigs; narrower elliptical shiny lvs, hairless below, usually **smooth** above. Br Isles: SE, mid, SW Eng, lf; o introd elsewhere; on hths, wds on acidic, nutrient-poor soils. Fl 2–3.

B ᴀᴡɪ **Wych Elm**, *U. glabra*, tall tree, with broad, spreading outline; suckers abs or few. Twigs only hairy when young; lvs **large** (7–16 cm long), oval to diamond-shaped or elliptical, v rough, bases unequal, stalks v short, <3 mm. **Many** rust-coloured hairs on buds (not none or few, as in A). Catkins (**Ba**) larger than in A; fr (**Bb**) with seed in centre of fr-wing. Br Isles: c in hilly or rocky wds on calc soils, especially in N and W of GB, also planted. Fl 2–3.

HEMP FAMILY   *Cannabaceae*

A Hop, *Humulus lupulus*, roughly hairy, square-stemmed, herbaceous climbing per, twining clockwise; lvs deeply palmately-lobed, long-stalked, toothed, 10–15 cm; male fls in branched yellow-green clusters (**Aa**), female catkins (**Ab**) resembling yellow-green pine cones, 15–20 mm long in fl, enlarging to 5 cm long in fr. Br Isles: Eng c; Wales, Ire o; Scot r; in hbs, wds, fen carr (probably native in S Eng but often originally an escape from cultivation). Fl 7–8.

B Hemp, *Cannabis sativa*, erect strong-smelling ann herb to 2.5 m; lvs digitate, 3–9 lobed almost to base. Older lvs have a greater number of lobes and lf edges become more toothed with age. Plants dioecious. Fls green, male (**Ba**) in branching cluster with 5 petals and 5 stamens, female (**Bb**) with no perianth (like A) in spikes. Introd (SW & C Asia) an escape from cultivation or introd with bird-seed; S Eng, Wales o-r, Scot, Ire r; on wa, ar, rds. Fl 7–9.

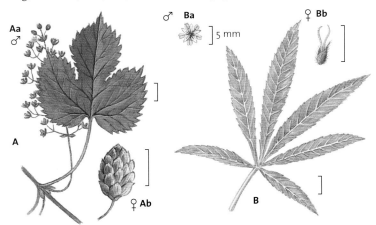

NETTLE FAMILY   *Urticaceae*

Herbs, often with stinging hairs and simple lvs. Spp looking v different vegetatively but with the same fl structure. Fls small, of separate sexes, usually on the same plant (**monoecious**); calyx four-lobed, green; petals none; stamens 4–5; ovary one-celled; fr a tiny nut.

▼ **A Common Nettle**, *Urtica dioica*, coarse, roughly hairy erect **per** herb with yellow fleshy far-creeping rhizomes. Lvs in opp pairs, oval-cordate, pointed, short-stalked, coarsely-toothed, covered (like stem) with stinging hairs that release irritant histamine-containing juice when broken. Plants **dioecious**; fls in drooping catkin-like infls, have 4 sepals; male have 4 or 5 stamens, female one carpel, enclosed by a larger sepal in fr. Br Isles, vc in wa, wds, rds, hbs, fens especially on phosphate-enriched soils. Fl 6–8. **A.1 Fen Nettle**, *Urtica dioica* ssp *galeopsifolia*, is a nettle sometimes (but not always) without any stinging hairs. **Key differences** from A are much **narrower lvs**; **lowest** infl is at stem **node (12–)13–22** from base, ie higher than A's lowest infl at nodes 7–14. Br Isles r, but probably over-looked in fens and damp places, not wa. Fl later than A, 7–8.

B   Small Nettle, *U. urens*, smaller (10–50 cm) **ann** plant; longer-stalked, more deeply-toothed lvs (**B**), 1.5–4.0 cm long; infls (**Ba**) short (1 cm), borne on leafy side-branches; plants **monoecious**. Br Isles: Eng, E Scot c; Wales, Ire o-lf; W Scot r; in wa, ar, gdns, especially on light soils. Fl 6–9.

C   Pellitory-of-the-wall, *Parietaria judaica* (*P. diffusa*), softly hairy per herb, 30–60 cm, with cylindrical, much-branched, reddish stems not rooting at nodes and stalked, alt, oval-lanceolate **untoothed lvs** to 7 cm long. Fls in clusters, male (**Ca**) and female separate, with greenish-red-tinged, four-toothed calyx; female fls terminal, male to sides of infl. Br Isles: Eng, coastal Wales f-lc; Scot o-r, **abs to N**; Ire f in S; on rocks, old walls, hbs. Fl 6–10.

D   Mind-your-own-business, *Soleirolia soleirolii*, evergreen mat-forming per herb; lvs alt, hairy, **circular** 4–5 mm long, untoothed; thread-like stems, **rooting at the nodes**; fl tiny and solitary, pink; plants monoecious. Introd (W Mediterranean); S Eng and Wales c, Ire f, Scot r; shady damp places on rds, pavements, walls and churchyards. Fl 5–8.

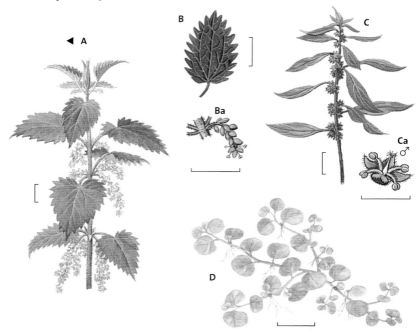

WALNUT FAMILY

Juglandaceae

Walnut, *Juglans regia*, tree with alt **pinnate** lvs without stipules, 3-9 oval leaflets, entire, aromatic; chambered pith inside twigs. Plant monoecious, fls yellow-green, male in catkins with 3 stamens and female in short erect clusters with 2 styles and branched stigmas; frs large drupe (**a**) (walnut) with green husk. Introd (SW & C Asia) probably by the Romans, sometimes self-sown; S Eng f; Scot, Ire, Wales o in wds, hds, rds, wa. Fl 4–5.

BOG-MYRTLE FAMILY

Myricaceae

Bog-myrtle, *Myrica gale*, deciduous **shrub** to 2 m tall; red-brown twigs; buds oval, red-brown, blunt. Lvs 2–6 cm, narrow-oblong, tapered to base, wider above middle, scarcely stalked, ± toothed near tip, grey-green, hairless above, downy below, with many stalkless yellow glands producing resinous **fragrance** if lvs bruised or rubbed. Usually dioecious; male catkins (**a**) red-brown, 7–15 mm, with red stamens; female catkins (**b**) 4–5 mm in fl, dark green, oval with red styles, 5–10 mm in fr; nutlets small, winged. Br Isles, lc in W, but vr-abs most of central Eng and vr and declining in SE Eng; in bogs, wet hths, fens. Fl 4–5.

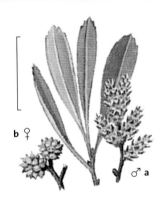

BEECH FAMILY

Fagaceae

Trees with alt undivided lvs, and male and female fls in separate catkins on the same plant (**monoecious**). Male catkins many-fld, mostly long; perianth 4–6-lobed, stamens twice number of lobes. Female fls in groups of 1 or 3, with involucre of **bracts**, that grows later to form a woody **cup** round fr; fr a one-seeded nut. Wind pollinated – hence fls have no petals or nectar to attract insects.

▼ **A Pedunculate Oak**, *Quercus robur*, large, spreading deciduous tree, to 30 m or more, with rugged bark; hairless twigs; lvs **hairless**, oblong, pinnate-lobed with rounded side-lobes, basal lobes (**auricles**) rounded and exceeding the short lf-stalk; acorns (**Ab**) in cups, 1, 2 or 3 together on **long** common stalk; male catkins (**Aa**) pendulous. Br Isles (except outer **Scot islands**), vc in wds, hbs, etc, except on v poor soils. Fl 4–5; fr 9–10.

B AWI **Sessile Oak**, *Quercus petraea*, differs from A in more ascending branches and narrower crown; lvs (**B**) long-stalked, lf-base tapering into long stalks (**without** basal auricles), less deeply-lobed, broadest above middle, with **star-shaped hairs** in axils of side veins beneath (**Ba**); acorns (**Bb**) almost **stalkless** on twigs. GB, lc (dominant Oak of the poorer soils of N, W Britain, and of some sandy areas of SE Eng). Fl 4–5; fr 9–10. The hybrid between A and B (*Q.* x *rosacea*) is vc and overlooked but difficult to identify; generally having some degree of rounded auricles but with a long lf-stalk (see *Plant Crib*).

C **Turkey Oak**, *Q. cerris*, tree to 40 m, has similar lvs to B but **v variable**, often with deeper-cut lobes, to $^1/_2$ way or more, with **short points** at the tips of the lobes; end twigs with persistent stipules, looking like long whiskers; acorns (**Ca**) with bristly cups. Introd (S Eur), often planted and c natd in S Eng, Wales; Scot, Ire f; on well-drained soils on rds, wa, hths, railways. Fl 4–5; fr 9.

D **Red Oak**, *Q. rubra*, tree to 34 m; large ± hairless lvs 10– 23 cm long with pointed lobes, that often turn red in autumn; dark red twigs, acorns (**Da**). Introd (N America) much planted and self-sowing; Eng, Wales f; Scot, Ire 0; on sandy soils. Fl 4–5; fr 9.

E **Evergreen Oak**, *Q. ilex*, **evergreen** tree with dark brown-black scaly bark; lvs **v variable**, elliptical, toothed (juvenile lvs) or not, dark glossy green on upperside, woolly-downy below (**E** shows underside); acorn (**Ea**) with woolly cup. Introd (Mediterranean); Br Isles: Eng c; Wales, Scot, Ire 0-r; in dry wds, especially on calc soils. Fl 5; fr 9. **Do not confuse** with **Holly** (p 305) which sometimes has v similar, almost entire lvs (especially on tall trees) but **bark** is grey-brown often with circular warts.

F **Sweet Chestnut**, *Castanea sativa*, large spreading deciduous tree with bark longitudinally fissured, fissures often in spiral curves round trunk. Lvs 10–25 cm, oblong-lanceolate, pointed, with large pointed teeth; catkins 12–20 cm long, erect; male fls (**Fb**) above, with many conspicuous yellowish stamens and a sickly scent, female fls (**Fc**) at base of catkins, 3 together in scaly cups; frs (**Fa**) large, shiny-brown, 1–3 together in a green cup 5–7 cm wide, with many long spines. Introd (S Eur); Br Isles, c (probably introd by the Romans). Fl 7; fr 10.

G **Beech**, *Fagus sylvatica*, large deciduous tree to 30 m or more, usually with smooth, grey bark. Buds red-brown, cigar-shaped. Lvs 4–9 cm, stalked, oval, pointed, scarcely-toothed, veins prominent at edges; edges and veins beneath silky-hairy, rest of **mature** lf hairless; male catkins (**Gb**) tassel-like on long stalks; female fls (**Ga**) in **pairs** on a long stalk, surrounded by a scaly cup, which develops into the deeply-lobed woody cup that encloses the beech nut; frs 1–2 together, 12–18 mm long, with one flat and one curved face. Native S Eng, S Wales, widely introd rest of Br Isles; in wds on chk, limestone, also on sands, light loams. Fl 4–5; fr 9–10.

A
Aa ♂
Ab
B
Ba
5 mm
Bb
C
Ca
♂
D
Da
E
Ea
F
Fa
Fb ♂
Fc
G
Ga ♀
Gb ♂

BIRCH FAMILY   *Betulaceae*

All trees or shrubs in the Birch Family have separate male and female catkins on the same plant (**monoecious**). **Hornbeam** (*Carpinus*) and **Hazel** (*Corylus*) differ from the Beech family in lacking perianth to male fls, and in their **leafy**, lobed cup (not woody, scaly case) around nut. Fls of **Alders** (*Alnus*) and **Birches** (*Betula*) are v like those of Hazels, but with 3 (not 1) male fls in each catkin-scale; no perianth; fr a **flattened, tiny nutlet** on surface of a fleshy or woody scale, so female catkins cone-like in fr; frs with rounded wing (for wind dispersal). **Birches** (*Betula*) are deciduous trees or shrubs with alt simple toothed lvs; male catkins drooping, female erect, catkin-scales three-lobed; frs with rounded papery wings.

A   AWI **Hornbeam**, *Carpinus betulus*, deciduous tree to 30 m (usually much less); smooth grey bark, as in G p 124, but trunk fluted or angled (not rounded). Buds shorter, broader, less pointed than in G p 124, pale brown, hairless; twigs **downy**. Lvs 3–10 cm long, oval, rather like those of G p 124 in shape, but more pointed and strongly, sharply **doubly-toothed**, pleated along veins, hairless above, hairy on veins below. Male catkins (**Ab**) 2.5–5.0 cm, with oval greenish bracts; female catkins (**Aa**) 2 cm in fl, 5–10 cm in fr; each fl with green three-lobed involucre, mid-lobe much the longest and enlarging in fr to a 4 cm trident-shaped lfy bract around the 5–10-mm, green, oval, ribbed nut. SE Eng, N to Norfolk, W to Hants (possibly further W) native and lc, widely introd elsewhere in Br Isles; in wds, hbs, especially on loams, sandy clays. Fl 4–5.

B   **Hazel**, *Corylus avellana*, many-stemmed deciduous shrub to 8 m; coppery-brown smooth bark tends to peel. Twigs v hairy, with reddish glandular hairs; buds oval, blunt, hairy; lvs 5–12 cm, rounded-oval with drawn-out point, cordate at base, sharply doubly-toothed, downy; stalk hairy. Male catkins (**Ba**) appear before lvs, in groups 2–8 cm long, pendulous, with yellow stamens and oval yellowish bracts; female catkins (**Bb**), 5 mm long, resembling lf-buds, erect, with red styles protruding from tip. Frs (**Bc**) 1.5–2.0 cm, in clusters of 1–4, globular to oval, with brown woody shell, surrounded by involucre deeply divided into toothed lfy lobes. Br Isles, vc in wds, scrub, hbs, on less acid soils. Fl 1–3; fr 9–10.

C   **Alder**, *Alnus glutinosa*, deciduous tree to 20 m; v dark fissured bark, twigs hairless; buds blunt, purplish. Lvs 3–9 cm, **rounded**, tip usually truncate with no point, tapered into stalk, doubly-toothed, dark green, **4–8** pairs of veins, hairless except in vein-axils below, sticky when young; stalk to 3 cm. Male catkins (**Ca**), 2–6 cm, pendulous, yellow with dark tips to bracts, appear before lvs; female catkins (**Cb**) green, egg-shaped, 1 cm long, 3–8 on each stalk, become in fr (**Cc**) woody (like small pine-cones) and to 28 mm long. Br Isles, c in wet wds, fen carr, streamsides etc. Fl 2–3.   **C.1 Grey Alder**, *A. incana*, has smooth grey bark; lvs with more **pointed tips** than C, **paler** on lower-side with **7–15** pairs of veins and not sticky when young. Introd (Eur); f throughout Br Isles, natd most often by suckers, sometimes self-sown; tolerating poor, damp soils. Fl 2–3.
C.2 Italian Alder, *A. cordata*, differs from C in having greyish bark and heart-shaped lvs with **cordate bases**; large female cones 15–30 mm, **1–3 together** on a stalk. Introd (Italy); o natd in S Eng, especially urban areas on dry soils and unlike other *Alnus* spp, even on chk. Fl 2–3.

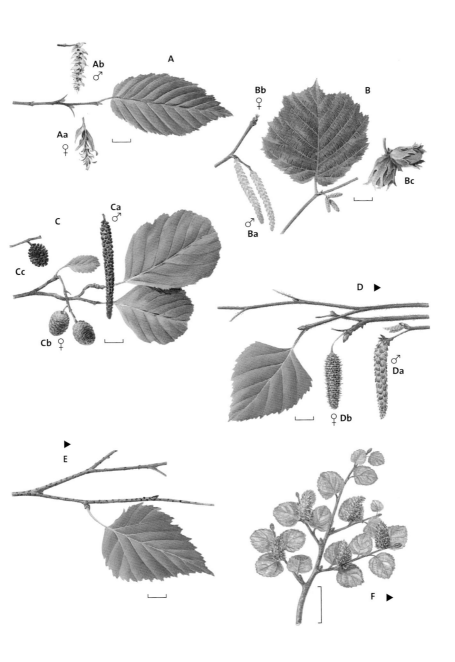

A

Ab ♂
Aa ♀

Bb ♀
B
Ba ♂
Bc

C
Ca ♂
Cc
Cb ♀

D ▶
Da ♂
Db ♀

▶
E

F ▶

127

▲ **D Downy Birch**, *Betula pubescens*, tree to 25 m, with brownish peeling papery bark; **downy** twigs; oval-triangular, **simply-toothed** lvs, downy below; male catkins (**Da**) 3–6 cm long, female (**Db**) shorter. Br Isles, c especially to N; in woods, hths, bogs. Fl 4–5.

▲ **E Silver Birch**, *B. pendula*, tree to 25 m, differs from D in white papery bark when young (dark and fissured when old); twigs **hairless**, but **warted**; lvs more pointed, triangular, **doubly-toothed**, hairless. Br Isles, c especially to S; in dry wds, downs, hths. Fl 4–5. Hybridizes with D (*B.* x *aurata*), c and overlooked, first year twigs with **both** hairs and warts may indicate a hybrid. Catkins and mature lvs needed for accurate id (see *Plant Crib*).

▲ **F *Dwarf Birch**, *B. nana*, tiny, ± prostrate shrub (to 30 cm tall) with downy twigs; almost **circular**, deeply but bluntly-toothed lvs, 5–15 mm long, hairless when mature; tiny male catkins 6–8 mm long. Br Isles: Scot Highlands o-vlf, N Eng vvr; on mt moors and in *Sphagnum* bogs. Fl 5–6. Hybridizes with D (*B.* x *intermedia*), f in Scot (see *Plant Crib*).

DEW PLANT (MESEMBRYANTHEMUM) FAMILY   *Aizoaceae*

Hottentot-fig, *Carpobrotus edulis*, **creeping** per, with angled ± woody stems below, and narrow-linear, opp, **fleshy curved lvs**, 7–10 cm long, **triangular** in cross-section, toothed on the back. Fl-stalks 3 cm, v swollen below fls; fls to 10 cm wide, solitary, superficially like large Daisies (but **single** fls, not **heads** of fls); calyx-tube with 5 lfy lobes; petals **v many**, linear, **purplish-pink or yellow**; stamens many; fr fleshy, rather fig-like, edible. Introd (S Africa) and often invasive. Sea cliffs of SW Eng, IoS, Ire, vla. Fl 4–7. Other members of the family also introd on IoS and SW Eng coast.

GOOSEFOOT FAMILY   *Chenopodiaceae*

Goosefoot (*Chenopodium*) is a genus of ann herbs (**Good-King-Henry** is per), with grooved, often striped stems, often mealy lvs, and clusters of tiny fls in branched cymes; 5 tepals, sometimes partly fused; no petals; stamens usually 2–5; fr a tiny nutlet with one seed. **Oraches** (*Atriplex* p 135) are v close, but have **separate** male and female fls, the latter enclosed by conspicuous **triangular bracteoles**.

KEY TO THE GOOSEFOOT FAMILY

1 Plant without obvious lvs, composed of opp-branched, cylindrical, jointed, fleshy stems with swollen segments – lvs in fact adpressed to and fused with stem, only their short tips free – fls minute, not obvious – stamens or styles emerging from pores in segments in Sept ....*Salicornia* and *Sarcocornia* (p 132)
Plant with obvious lvs **2**

2 Lvs cylindrical or half-cylindrical, alt..**3**
Lvs flat, though often thick and fleshy **4**

3 Lvs spine-tipped – on sandy shores ....*Salsola kali* (p 134 **F**)
Lvs not spine-tipped – in salt-marshes ....*Suaeda* (p 134)

4 Fls with both stamens and style – sepals 5, ± equal..**5**
Fls either male or female – male fls with 5 sepals and 5 stamens, female with 2 triangular sepal-like bracts only and a tiny ovary **6**

5 Calyx-lobes much thickened at base and adhering in clusters in fr – lower lvs untoothed
Beta vulgaris (p 136 **G**)
Calyx-lobes not thickened at base, not adhering in clusters in fr – all lvs ± lobed or toothed ....*Chenopodium* (key p 130)

6 Small shrubs with untoothed elliptical, mealy-grey lvs, lower lvs opp – 2 bracts of the female fls joined into a cup well over half way up – in salt-marshes (or a vr herb)
Atriplex portulacoides (p 136 **F**)
Herbs with lvs triangular, diamond- or strap-shaped, ± toothed and alt, sometimes opp below – 2 bracts of female fls either free or only joined less than half way up ....
Atriplex (key p 135)

ID Tips   Goosefoots

- **What to look at**: **(1)** lf-shape **(2)** colour of the stem **(3)** whether plant mealy or not **(4)** various characters (eg colour, seed-coat features, etc) of ripe seeds are important for correct id of some spp (although these need to be studied with a microscope).

- **Fat-hen** (*Chenopodium album*) is v variable and v common; familiarise yourself with this sp so that you can distinguish it from the other, much less c spp included here.

- **Red, Oak-leaved and Saltmarsh Goosefoot** (*C. rubrum*, *C. glaucum* and *C. chenopodioides*) are the only 3 spp included here which have **fruiting heads** mostly **longer** than wide.

KEY TO GOOSEFOOTS (*CHENOPODIUM*)

1 Tufted per, 30–50 cm – long (> 0.8 mm) stigmas projecting from fls...
Chenopodium bonus-henricus (p 132 **F**)
Ann – stigmas short **2**

2 Lvs cordate-based, with few large wavy teeth ....*C. hybridum* (p 132 **G**)
Lvs never cordate, tapered to base **3**

3 Stem of infl and fls hairless **4**
Stem of infl and fls mealy **8**

4 Lvs oval, untoothed, to 5 cm long, green or purple both sides – stems four-angled – seeds
black..*C. polyspermum* (p 132 **E**)
Lvs toothed (except uppermost) – stems ridged but not four-angled **5**

5 All fls with 5 tepals, 5 stamens – seeds black (essential to confirm ID) – plant tall (40–70
cm) – r plant of disturbed wa gd in GB (Ire abs)................CR ** **Upright Goosefoot** *C. urbicum*
All fls (except at stem-tip) with 2–4 tepals and stamens – seeds red-brown **6**

6 Lvs mealy-white below, green above, oval, wavy-edged, 1–5 cm long – plant ± creeping
C. glaucum (p 130 **B1**)
Lvs not mealy-white, ± red on underside...**7**

7 Lvs diamond-shaped, much-toothed, thick, fleshy, ± reddish – tepals green with a red
flush, not joined above middle ....*C. rubrum* (p 132 **D**)
Lvs triangular, scarcely-toothed, thin, purple below – tepals bright red, joined almost to
tips, forming a little bag round fr..*C. chenopodioides* (p 132 **H**)

8 Lvs oval, grey – tepals rounded on back – plant stinking of rotten fish, ± prostrate
C. vulvaria (p 132 **E.1**)
Lvs narrow to triangular – tepals keeled on back – plant not stinking, ± erect **9**

9 Lvs toothed, but not clearly three-lobed **10**
Some lvs clearly three-lobed..*C. ficifolium* (p 130 **B**)

10 Infl lfy to top, with short spreading branches – tepals with many small teeth (x 20) ....
C. murale (p 132 **C**)
Infl lfless above, branches long, erect – tepals entire, without teeth..........*C. album* (p 130 **A**)

A   Fat-hen, *Chenopodium album*, v common and **v variable** ann weed, 30–100
cm; lvs grey-green, ± covered with whitish-grey meal, lanceolate to diamond-
shaped, lf-edges entire, lobed **or** bluntly-toothed; infls erect, dense, greyish, not
spreading, lfless above. Fls 2 mm wide, five-tepalled (**Aa**), tepals **not** toothed. Br
Isles, vc (except N Scot); in ar, wa etc. Fl 7–10. **Do not confuse** with **Common
Orache** (p 136 B) which has similar lvs but with lobes at base of lf larger and
forward-pointing and fls with conspicuous bracteoles.

B   Fig-leaved Goosefoot, *C. ficifolium*, **key difference** from A in its lower lvs (**B**)
deeply **three-lobed**, mid-lobes largest. Br Isles: S Eng, CI c (but less f than A); o
scattered elsewhere in ar, wa etc. Fl 7–9. **B.1** VU * **Oak-leaved Goosefoot**, *C.
glaucum*, smaller, often prostrate plant, with wavy-edged narrow oblong lvs 1–5 cm
long, mealy below, green above. Infl dense, spike-like, to 3 cm long. Probably introd
(Eur); Br Isles, r on wa, docks, seashores. Fl 6–9.

A

Aa] 3 mm

B

C ▶

D ▶

Da 3 mm

Db

E ▶

F ▶

G ▶

H ▶

▲ **C** vu **Nettle-leaved Goosefoot**, *Chenopodium murale*, has slightly mealy, diamond-shaped, sharp-toothed lvs (**A**); infl more **spreading** than in A, and lfy above; **key differences** from A in tepals with minute teeth and in seed shape. Br Isles: Eng, o; CI, IoS f; Scot, Wales, Ire r; on nutrient-rich ar, wa, dunes. Fl 7–10.

▲ **D** **Red Goosefoot**, *C. rubrum*, glossy (**not mealy**) ann, often red-tinged (especially on stem and underside of lvs); lvs (**Db**) diamond-shaped, irregular-sharp-toothed; infl reddish, dense; fls (**Da**) have 2–4 tepals, not joined above middle. GB c to S, r to N; Ire o; on wa, manure heaps, especially near sea. Fl 7–9.

▲ **E** **Many-seeded Goosefoot**, *C. polyspermum*, spreading or erect ann, usually not at all mealy, except perhaps below, with square, ± red stems; lvs undivided, **oval, ± untoothed**, 3–5 cm long; loose, hairless, infls of tiny green fls. Br Isles: Eng, CI, IoS c, r elsewhere; on wa, ar. Fl 7–10. **E.1** EN Sch 8 ****Stinking Goosefoot**, *C. vulvaria*, similar plant, but with no red colouring, often prostrate, smelling of rotten fish; oval mealy-grey lvs, ± untoothed. S Eng vr; drier salt-marshes and shingly or sandy beaches; inland in nutrient-rich ruderal places. Fl 7–9.

▲ **F** vu **Good-King-Henry**, *C. bonus-henricus*, **per** herb, 30–50 cm, with triangular, wavy-edged, scarcely-toothed lvs to 10 cm long and 8 cm wide at base; lvs mealy at first, later becoming smooth and green. Fls in erect terminal pyramidal lfless spikes, all with 5 sepals, 5 stamens, and 2–3 **projecting** stigmas **>0.8 mm** long. Introd (mts C Eur); Br Isles, o-lf (but declining) in disturbed places on ar, rds, wa. Fl 5–7.

▲ **G** **Maple-leaved Goosefoot**, *C. hybridum*, tall ann like C and D, not mealy; lvs (**G**) hairless, **cordate at base**, with few but large acute triangular teeth; infls spreading, usually lfless. Introd (Eur); Br Isles, S and E Eng o-lf on nutrient-rich ar and wa. Fl 8–10.

▲ **H** ***Saltmarsh Goosefoot**, *C. chenopodioides* (*C. botryodes*), is related to D, but is a smaller, usually prostrate plant, ± always **red** in fr on stem and **lf-undersides**; lvs triangular, hardly-toothed, 2–4 cm long; dense **red clusters** of v numerous fls, **smaller** than in D; the tepals joined, persisting, and forming a cup around the frs; each tepal keeled on back. SE Eng, CI vlf, especially on Thames Estuary; in brackish and saline hollows and creeks near the sea. Fl 8–10.

Glassworts (*Salicornia* and *Sarcocornia*) are strange-looking herbs, erect or prostrate, usually v branched and bushy. The main stem and the branches (in opp pairs) are succulent and jointed into swollen segments. Each segment has on its upper margins a pair of opp triangular lobes that represent the lvs. Fls are borne on the upper stems and are in opp groups of 1–3 (a **triad**) in the axils of the lobes representing the lvs. Each fl consists of a small fleshy disc or segment with a central pore, from which emerges in Aug or Sept a single tiny anther; the bifid stigma usually remains hidden. The frs are small nutlets, released as the stems break up in winter. There are many spp of *Salicornia*, and they are not easy to distinguish because they are v variable and at least 20-30 'sorts' can be identified in SE Eng. Only the broad groups are included here.

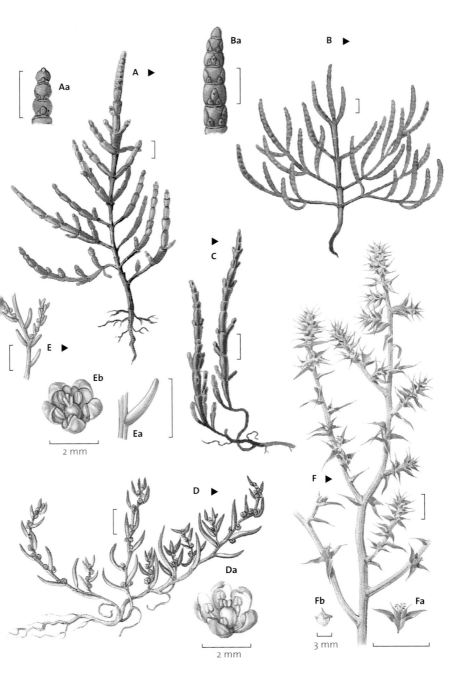

Aa

Ba

A ▶

B ▶

C ▶

E ▶

Eb

Ea

2 mm

D ▶

Da

2 mm

F ▶

Fb

Fa

3 mm

133

▲ **A** **Purple Glasswort**, *Salicornia ramosissima*, is a **v variable** ann, erect to prostrate dark shining green, or ± red- or purple tinged, with each segment v convex; contractions between segments give the stems a beaded effect (**Aa**). Fls always in threes, **unequal** in size, the disc of the **central** fl of the triad **much enlarged**. Br Isles: S and E Eng, S Wales, f-lc in the middle consolidated parts of salt-marshes. Fl 8–9.

▲ **B** **Yellow Glasswort**, *S. procumbens* agg (includes *S. dolichostachya*, **S. nitens* and **S. fragilis*) is a group of closely related spp, sprawling, bushy, pale grey-green to yellow anns, with many, hardly beaded, long, cylindrical branches (**Ba**) from the base; the three fls in each triad **equal** in size; **anthers 0.5–1 mm** long, always projecting. Br Isles: S coasts of GB, Ire lc; on lower unconsolidated salt-marshes, on mud. Fl 8–9. **B.1** **Common Glasswort**, *S. europaea* agg (includes A, *S. europaea* and ***S. obscura*) is similar to B in branch shape, but erect, with short opp branches, not bushy; three fls **not equal** in size, **central fl larger**; **anthers 0.2–0.5 mm** long, often not projecting. Br Isles, f–lc on lower muddy parts of salt-marshes. **B.2** ***One-flowered Glasswort**, *S. pusilla* is like a small A in its beaded branches, but its fls are borne **singly** in lf-axils (not in threes). Br Isles: SE Eng, S Ire, lf on drier uppermost parts of salt-marshes.

▲ **C** ***Perennial Glasswort**, *Sarcocornia perennis* (*Salicornia perennis*) is a **per** with creeping woody stems to 1 m long that produces low bushy tussocks of ascending green to orange-yellow shoots to 20 cm tall. Some **non-flg shoots** present. Fls in threes in each lf-axil. Br Isles: SE Eng o-la; Wales, N Eng, SE Ire vr; on upper, firmer often gravelly parts of salt-marshes. Fl 8–9.

▲ **D** **Annual Sea-blite**, *Suaeda maritima*, another fleshy ann salt-marsh plant, but its lvs and fls are distinct, not fused into a cylindrical mass with the stem. The branched stems, prostrate to erect, glaucous-green to red-flushed, are up to 60 cm tall, and bear **half**-cylindrical fleshy alt lvs, 3–25 mm, **pointed** at tip and **tapered** to base. Fls (**Da**), 1–2 mm across, are in small clusters; 5 unkeeled succulent sepals; 5 stamens; 2 stigmas. Br Isles, c in salt-marshes generally, also on beaches. Fl 7–10.

▲ **E** ***Shrubby Sea-blite**, *S. vera* (*S. fruticosa*), similar but larger per shrub, 40–120 cm, with ± **cylindrical** evergreen lvs (**Ea**), 5–18 mm long, **rounded** at tip and at base; fls (**Eb**) with 3 stigmas. Br Isles: S Eng from Dorset to S Lincs (as a native; vr introd elsewhere) o but vlc on shingle above high-water mark. Fl 7–10.

▲ **F** vu **Prickly Saltwort**, *Salsola kali*, distinctive, ± prostrate prickly ann; stems 20–40 cm long, striped pale green or reddish, branched; lvs 1–4 cm long, succulent, rounded, tapered into a **spine** at the tip; tiny fls (**Fa**) in lf-axils, each with a pair of bracts and 5 sepals; fr (**Fb**) 3–5 mm across, winged on back, top-shaped. ssp *kali* is the coastal plant; **Spineless Saltwort**, ssp *iberica*, with lvs usually **not** spine-tipped, is a vr introd inland. Coasts of Br Isles, f-lc on sandy shores along drift line. Fl 7–9.

Oraches (*Atriplex*) are ann (**Sea Purslane** is per), rather fleshy, often mealy herbs, prostrate to erect, 10–60 cm, resembling Goosefoots, with alt lvs and spikes of inconspicuous fls, but with male and female fls separate; female fls sepal-less, within a pair of triangular **bracteoles** (eg Aa p 137) that enlarge in fr. They are plants of wa or ar, usually on nutrient-rich soil, or on the coastline.

ID Tips   Oraches

- **What to look at: (1)** the **bracteoles** (eg Aa p 137) on the female fls **in fr** are crucial for correct id; look at their size, shape, how far fused together, whether thickened or not at base **(2)** shape of lower lvs.

- Look at several plants in any group to find **both** plants with fr **and** younger plants with lower lvs intact.

- If a plant does not key out, it may be one of several hybrids or other vr causal introd spp.

KEY TO ORACHES (*ATRIPLEX*)

1   Shrub with opp lower lvs .... *Atriplex portulacoides* (p 136 **F**)
    Ann herb usually with alt lvs (sometimes opp).. **2**

2   Bracteoles on female frs fused ± to tip with 2 long side lobes and 1 short end lobe ....
    A. *pedunculata* (p 136 **F.1**)
    Bracteoles without 3 lobes and rarely fused > half way **3**

3   Lower lvs linear-lanceolate, sometimes toothed but without basal lobes ....*A. littoralis* (p 136 **C**)
    Lower lvs triangular or diamond-shaped and toothed with large basal lobes **4**

4   Bracteoles fused at base only or < 1/4 way **5**
    Bracteoles fused > 1/3 of their length, often c. 1/2 way **7**

5   Some bracteoles >10 mm long, united only at base and with stalks ≥ 5 mm long ....
    A. *longipes* (p 136 **D.1**)
    Bracteoles all <10 mm long, united only at base with stalks ≤ 5 mm long **6**

6   Lower lvs diamond-shaped, rounded at base – Scot & Northumberland only, just above
    seaweed zone ....*A. praecox* (p 136 **A.1**)
    Lower lvs triangular, ± truncate at base – widespread on coasts and inland ....
    A. *prostrata* (p 135 **A**)

7   Plant white or silvery – bracteoles in fr hardened below, 6–7 mm long *A. laciniata* (p 136 **E**)
    Plant greenish, sometimes mealy – bracteoles in fr not hardened below, 3–10 mm or
    more long...**8**

8   Lower lvs lanceolate to diamond-shaped – bracteoles thin, lf-like at base ....*A. patula* (p 136 **B**)
    Lower lvs triangular to diamond-shaped – bracteoles thickened and spongy at base ....
    A. *glabriuscula* (p 136 **D**)

▼ **A   Spear-leaved Orache**, *Atriplex prostrata* (*A. hastata*), is usually an **erect** plant to 1m; has **triangular** lower lvs, 3–6 cm long, the base of the lf making a right-angle with its stalk; bracteoles (with fr, **Aa**) fused **only at the base**. Br Isles, vc on moist fertile wa, ar and salted rds inland, salt-marshes, sea walls, beaches. Fl 7–10.

A.1 ***Early Orache**, *Atriplex praecox*, close to A but lvs more **diamond-shaped** and lf-base more **rounded,** plant reddening in summer**.** Br Isles, r in **Scot** and **Northumberland** only; on beaches just above seaweed zone. Fl 6–7.

B Common Orache, *A. patula*, similar to A, but lower lvs (**B**) **diamond-shaped**, with the lf-base tapering **gradually** into the stalk, lowest lobes **forward-pointing,** tips sharply **pointed** and more toothed than in A; upper lvs oblong-linear, narrower than in A. Fr-bracteoles (**Ba**) fused to c. **1/2 way** (unlike A). Br Isles, c in similar places to A. Fl 7–10. **Do not confuse** young lvs of B with **Fig-leaved Goosefoot** (p 130 B) which has similar 3-lobed lvs but with **blunt** lobe-tips or with **Fat-hen** (p 130 A) which has lvs with less pronounced or no side lobes.

C Grass-leaved Orache, *A. littoralis*, differs from both A and B in that the lvs (**C**) are **all linear** to lanceolate, and the plants are erect, 50–100 cm (whereas in A and B they tend to sprawl). Br Isles, c on coasts (shores and sea walls, upper salt-marsh edges) but also spreading along inland salted rds. Fl 7–10.

D Babington's Orache, *A. glabriuscula*, is similar to A but is **procumbent** to rarely erect; its bracteoles (**D**) are fused to c. **1/2 way** up, and are more diamond-shaped than triangular. Female bracteoles **stalkless**, **thickened** and **spongy** at base in fr. Br Isles, f on coastal beaches and wa near sea. Fl 7–10.   **D.1** ***Long-stalked Orache**, *A. longipes*, similar to D, but bracteoles fused only **at base** with **stalks** at least on the larger bracteoles. Br Isles: o-r and vl on coasts from N Eng to C Scot in tall saltmarsh vegetation. Fl 7–9.   D.1 hybridizes with D and A, the fertile hybrids often ± **procumbent**, growing with D or A and may be similar to but much **more c** than D.1.

E Frosted Orache, *A. laciniata* (*A. sabulosa*), is distinct in its thick, **silvery**, frosted-looking, whitish-grey diamond-shaped lvs with wavy edges, and its usually reddish stems. The plants are sprawling, and up to 30 cm; bracteoles silvery, fused to c. **1/2 way** up. Male fls (**Ea**) with 5 stamens. Br Isles, lc on shingle or sandy beaches. Fl 7–10.

F Sea-purslane, *A.* (*Halimione*) *portulacoides*, low **shrub** to 80 cm, with elliptical, untoothed, stalked, **opp**, whitish-mealy lower lvs; upper lvs narrower. Infls of dense spikes of tiny greyish fls, each with 5 sepals like an Orache; fls stalkless; stamens (**Fa**) in separate fls; paired bracteoles enclosing frs (**Fb**), **fused nearly to top**, are ± stalkless and reversed triangular (broadest at **3-lobed** tip). Br Isles: GB to S Scot f-la; Ire o-lc; on drier salt-marshes, especially forming fringes to channels and pools. Fl 7–9.   **F.1** Sch 8 CR ****Pedunculate Sea-purslane**, *A.* (*Halimione*) *pedunculata* is a small erect or spreading **ann**; similar to F but **alt** lvs; 3-lobed bracteoles but **long-stalked** in fr. **Essex only**, vr on sandy salt-marshes. Fl 8–9.

G   Sea Beet, *Beta vulgaris*, tap-rooted sprawling hairless per herb with lfy shoots to 100 cm; fleshy stems usually red-striped; lvs from cordate (below) to diamond-shaped or oblong (above), **untoothed**, dark **glossy** green; fls (**Ga**) resemble those of Goosefoots (p 130) in having 5 equal sepals, but fleshy, conspicuous, and thickened in fr; frs adhere in groups by their swollen sepal-bases. The seaside plant is usually ssp *maritima*; the form cultivated for its root, ssp *vulgaris*, also occurs in similar habitats and wa, ar. Br Isles, c on seashores, sea walls, etc. Fl 7–9.

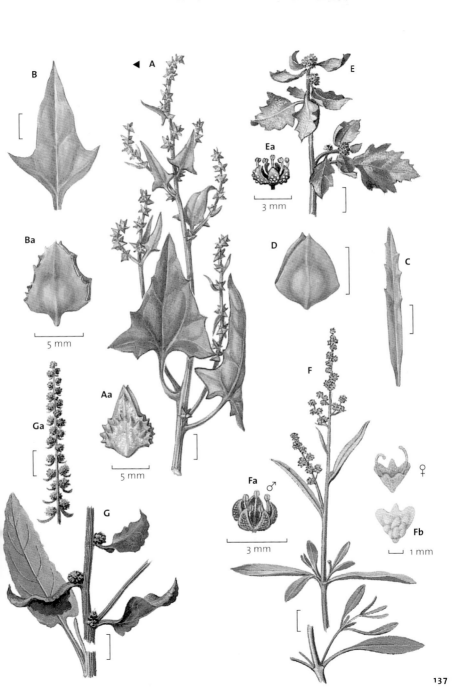

B

A ◀

E

Ea
3 mm

Ba
5 mm

D

C

Ga

Aa
5 mm

F

♀

Fa ♂

Fb
1 mm

3 mm

G

137

AMARANTH FAMILY *Amaranthaceae*

Members of this family somewhat resemble a downy Goosefoot (p 130) but Amaranths have brown, papery bracts (**a**) around the fls.

Common Amaranth, *Amaranthus retroflexus*, erect downy branched grey-green ann, 20–80 cm tall; rough stem; lvs oval, blunt but mucronate, long-stalked. Fls small, green, with calyx of spoon-shaped sepals 2 mm long; 5 stamens; fr (**b**) a rounded flattened nut with 2–3 persistent stigmas.
Introd (N America); Br Isles: S Eng, f; o-r elsewhere; in wa, ar. Fl 7–9.

Other *Amaranthus* spp sometimes occur on ar or wa.

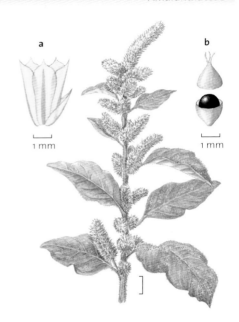

a

b

1 mm

1 mm

PURSLANE FAMILY *Portulacaceae*

Members of this family resemble the Campion Family (*Caryophyllaceae*) except for their calyx of 2 sepals and their three-seeded (not many-seeded) capsules.

A Pink Purslane, *Claytonia sibirica* (*Montia sibirica*), erect hairless fleshy ann to per herb, 15–40 cm; long-stalked oval basal lvs, tapered at both ends, stem-lvs in a single, opp, stalkless, but **separate** pair. Fls in raceme-like cymes, 15–20 mm across; 5 notched petals, **pink** with darker veins; 2 sepals, 5 stamens to each fl. Introd (N America); Br Isles: GB lf in S and Wales, vc in SW, CI and to N; Ire o-r; in damp wds, streamsides, rds. Fl 4–7.

B Springbeauty, *Claytonia perfoliata* (*M. perfoliata*), though of similar floral structure, looks v little like A; it is a hairless ann, 10–20 cm, with long-stalked basal lvs rather like those of A, but with its fleshy stem-lvs (**B**) **joined in a pair**, forming a ring round the stem; tiny (5–8 mm) white fls have scarcely-notched petals. Introd (western N America); Br Isles, f (except SW Eng, Wales, Ire, r); on light sandy open gd, ar, wa. Fl 5–7.

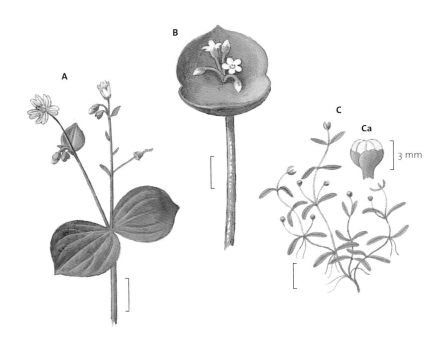

A

B

C

Ca

3 mm

C Blinks, *M. fontana*, looks rather unlike A and B, but more like a Water-starwort (p 384). Straggling plant, 2–50 cm long, with **round** stem and narrow spoon-shaped lvs in opp pairs, 2–20 mm long; tiny fls, 2–3 mm across, 2 or 3 together on **stalks**; each fl (**Ca**) has 2 sepals and **5** white **petals** joined below; 3 seeds to each fr, dark brown or black. Br Isles: Scot, Wales, N Ire vc; elsewhere f-lc; on non-chalky springs, bare, usually damp gd, wds or moist ar. Fl 5–10.

CAMPION FAMILY   *Caryophyllaceae*

Herbs with lvs nearly always in opp equal pairs, unlobed and untoothed, mostly without stipules. Fls in forked cymes, rarely solitary or in lf-axils. Petals and sepals 5 (sometimes 4), petals sometimes abs; stamens normally twice number of sepals. Ovary superior, styles 2, 3 or 5; fr a one-celled capsule (berry in *Cucubalus*), opening by teeth at top, equal (or twice) number of styles. Larger members of the family are easily recognized; v small ones are more difficult, because fl-details are so hard to see.

KEY TO CAMPION FAMILY

1 Lvs alt – small prostrate plant of lake shores ....*Corrigiola litoralis* (p 159 **F**)
  Lvs opp **2**

2 Lvs without stipules **3**
  Lvs with stipules **20**

3 Fls with a receptacle-cup – fr a one-seeded nut, enclosed in the cup – small prostrate
  herbs with linear lvs, joined in pairs ....*Scleranthus* (p 158)
  Fls with no receptacle-cup – ovary free from sepals **4**

4 Fr a berry – climbing herb with joined sepals, 3 styles ....*Cucubalus* (p 154)
  Fr a capsule **5**

5 Sepals joined into a calyx-tube **6**
  Sepals all free **11**

6 Styles 2 ....**7**
  Styles 3 to 5 (or male plants with stamens only)...**9**

7 Calyx-tube with white seams alternating with the teeth – fl-head surrounded by a
  sheath of scales ....*Petrorhagia* (p 142)
  Calyx-tube without white seams – no sheath of scales to fl-heads...................................**8**

8 Base of calyx enclosed tightly by epicalyx of 1–3 bracts....................................*Dianthus* (p 141)
  No epicalyx around base of calyx ....*Saponaria officinalis* (p 142 **D**)

9 Calyx with long narrow teeth longer than petals or ± as long as petals – fls large, 2–5 cm
  across, purple ....*Agrostemma githago* (p 142 **E**)
  Calyx-teeth shorter than petals – fls pink, cream or white **10**

10 Styles 3 or 5 (normally) – capsule opening has twice as many teeth as there are styles ....
  ....*Silene* (key p 143)
  Styles 5 – capsule opening with 5 teeth ....*Lychnis* (p 143)

11 Petals present..**12**
  No petals present..**19**

12 Petals deeply bifid **13**
  Petals not bifid, sometimes toothed **15**

13 Styles 3, lvs almost hairless ....*Stellaria* (p 148)
  Styles 5 ....**14**

14 Lvs almost hairless, 2–5 cm, cordate at base – petals bifid for more than 1/2 their length
  ....*Myosoton aquaticum* (p 150 **G**)
  Lvs v hairy, 0.5–2.5 cm, rounded at base – petals bifid to less than 1/2 their length.................
  ....*Cerastium* (key p 149)

15 4–5 sepals – 4–5 styles **16**
  5 sepals – but 2–3 styles only **17**

16 Small erect (3–10-cm tall) herb – narrow strap-shaped grey-green lvs – sepals with white
  margins – fls erect, not often opening – capsule with 8 teeth ......*Moenchia erecta* (p 152 **G**)
  Low (2–5 cm tall) creeping to ascending herb – linear lvs less than 2 mm wide – capsule
  splitting to base into 4 or 5 parts..*Sagina* (key p 160)

17 Lvs broad fleshy close-set – greenish petals – spherical frs – seashore plant ....
  ....*Honckenya peploides* (p 154 **B**)
  Lvs not fleshy **18**

18 Lvs oval, to 2.5 cm, with 3 strong veins in each – in shady places ....
Moehringia trinervia (p 154 **A**)
Lvs oval, less than 1 cm, one strong vein in each – in open places......................*Arenaria* (p 155)
Lvs linear ....*Minuartia* (p 155)

19 Lvs linear, rounded on back ....*Sagina* (key p 160)
Lvs oval, 5–7 mm long x 4–5 mm wide..*Stellaria pallida* (p 150 **D**)

20 Tiny obovate lvs in whorls of 4 (2 lf-pairs often smaller than other 2) – fls 3 mm across ......
Polycarpon tetraphyllum (p 158 **C**)
At least the lower lvs in opp pairs **21**

21 Fls in open forked, or one-sided cymes – conspicuous petals, at least as long as sepals –
lvs almost cylindrical, fleshy **22**
Fls minute in dense, clusters in lf-axils – petals and sepals, 5, minute – lvs oval – low
creeping plants **23**

22 Styles 5..*Spergula arvensis* (p 156 **E**)
Styles 3 ....*Spergularia* (p 156)

23 Fls white, in apparent whorls in lf-axils – sepals hooded with a little white point on tip of
each – upper lvs in equal pairs ....*Illecebrum verticillatum* (p 158 **A**)

Fls green, in axillary clusters, not whorled – sepals not white, not hooded – upper lvs alt....
Herniaria (p 158)

Pinks (*Dianthus*) have stiff, linear opp lvs, often waxy-grey; fls pink to red, or white,
sometimes fragrant, with petals toothed or fringed, often spotted. Sepals fused
into a cylindrical tube; epicalyx of scales below the calyx. **Campions** and **Catchflies**
(*Silene* and *Lychnis*) are usually taller than Pinks, with 5 petals, cleft ± deeply into 2
lobes; ribbed, often inflated calyx-tubes with **no** epicalyx; styles 5 or 3.

KEY TO PINKS (*DIANTHUS*)

1 Fls in heads with bracts around head...**2**
Fls solitary or few, not in heads **3**

2 Bracts and epicalyx hairy – fls 8–13 mm across ....*Dianthus armeria* (p 142 **A**)
Bracts and epicalyx hairless – fls 20 mm or more across..........................*D. barbatus* (p 142 **B.1**)

3 Petals cut 1/3 – 1/2 way down..**4**
Petals undivided or only toothed to 1/3 length or less **5**

4 Stem downy below – petals cut 1/3 way down.................................*D. gallicus* (p 142 **C.2**)
Stem hairless below – petals cut to middle ....*D. plumarius* (p 142 **C.1**)

5 Stem downy below – each petal with a dark bar across base, and pale dots ....
D. deltoides (p 142 **B**)
Stem hairless – petals without bars or dots....................................*D. gratianopolitanus* (p 142 **C**)

A Sch 8 (ENG & WALES ONLY) EN *****Deptford Pink**, *Dianthus armeria*, rigidly erect ann, 30–60 cm, with linear, pointed, keeled stem-lvs, and a basal rosette of lanceolate lvs, all dark **green** (**not grey**). Fls stalkless, rose-red, 8–13 mm, in close terminal heads with erect green hairy calyx, bracts and epicalyx scales. Br Isles: Eng, Wales, CI r; Scot, Ire vvr (first recognised 1993); on open, disturbed places in hbs, rds, dry sandy gslds. Fl 7–8. **A.1** Sch 8 VU ****Childing Pink**, *Petrorhagia nanteuilii* (*Kohlrauschia nanteuilii*), resembles a slender ± hairless A; linear lvs fused into short sheath below, and the small (6–9 mm) pink fls in dense oval heads (opening one at a time), with white **seams** on the calyx-tube, surrounded by a group of broad, papery, shiny-brown bracts; seed surface covered with minute **bumps**. S Eng, CI vr on coastal shingle or sandy gd; vvr inland as introd. Fl 7–8. **A.2 Proliferous Pink**, *P. prolifera* (*K. prolifera*) **key difference** from A.1 in seed surface covered with a **net-like** pattern. Br Isles: Norfolk only as a native, introd elsewhere, vr on dry sandy gsld. Fl 7–8.

B NT *****Maiden Pink**, *Dianthus deltoides*, per herb with both creeping stems and erect fl-shoots 10–20 cm. Lvs 10–25 mm, linear, pointed, grey-green, edges **hairy**. Fls larger than in A, 18 mm, **solitary** or **2–3** in terminal groups; rose-red, toothed petals (**Ba**) with a dark stripe across base and pale dots above. Epicalyx of 2–4 oval hairy scales 1/2 length of calyx-tube. Br Isles: (not Ire or IoS) r – vlf, scattered throughout; on dry sandy fields, hbs, dunes or on dry chk or limestone gsld. Fl 6–9.
B.1 Sweet-William, *D. barbatus*, sometimes introd (S Eur), has dense flat-topped fl-heads and (unlike B) epicalyx-scales longer than calyx. Br Isles: 0 gdn escape, scattered throughout (except Ire, vvr) on dunes, rds, wa. Fl 7–8.

C Sch 8 VU ****Cheddar Pink**, *D. gratianopolitanus*, **hairless tufted** per herb, 10–20 cm; creeping runners and erect fl-stems with **rough**-edged grey linear lvs, 2–6 cm. Fls usually **solitary**, about 25 mm across, fragrant; petals toothed to 1/6 of length, rose-pink, bearded; epicalyx of 4–6 blunt grey-green scales, 1/4 length of calyx. S Eng (**Cheddar only** as native) vr; on limestone rocks or gsld. Fl 6–7.
C.1 Pink, *D. plumarius*, resembles C in its rough-edged lvs, but has petals cut **1/2 way** into slender teeth; fls 25–35 mm; epicalyx of 2–4 pointed lobes to 1/3 length of calyx. Introd (E & C Eur); GB, r on old walls. Fl 7–8. **C.2 Jersey Pink**, *D. gallicus*, also resembles C, but has deeply cut petals (to 1/3 length), stiff short lvs and downy stems. **Jersey only**, vr on dunes. Fl 6–8.

D Soapwort, *Saponaria officinalis*, robust **hairless** erect per herb (30–90 cm) with runners. Branched fl-stems bear **pink** fls, about 2.5 cm across, in forked cymes. Sepals joined into a cylindrical tube with 5 teeth; no epicalyx. Petals long-stalked, unnotched. Stamens 10. Br Isles (probably an ancient introd), f on hbs, rds, wa. Fl 7–9.

E ******Ire ****Corncockle**, *Agrostemma githago*, tall hairy ann with narrow lanceolate lvs; pale reddish-purple fls, 2–5 cm across; calyx-tube v hairy, ribbed; sepals hairy, linear, much longer than the broad, notched petals. Forms with sepals ± same length as petals are introd cultivars and are probably more widespread than sown native forms. GB vvr, almost extinct as a native, probably naturally occurring at one or two sites only, elsewhere introd from 'wildflower' seed; on ar. Fl 6–8.

KEY TO CAMPIONS AND CATCHFLIES (*SILENE* AND *LYCHNIS*)

1 Styles 5 ....**2**
  Styles 3 ....**7**

2 Plants dioecious **3**
  Plants bisexual (fls with both stamens and styles) **4**

3 Fls white – capsule-teeth upright ....*Silene latifolia* (p 144 **A**)
  Fls red (sometimes white) – capsule-teeth strongly recurved...........................*S. dioica* (p 144 **B**)

4 Stems less than 15 cm tall – fls pink, 6–12 mm across, in a dense head – on mts, vr ....
  ....*Lychnis alpina* (p 144 **E.1**)
  Stems over 30 cm – fls in a loose infl **5**

5 Lvs and stems with dense white-woolly hairs ....*L. coronaria* (p 144 **B.1**)
  Lvs and stems hairless or hairy but not white-woolly **6**

6 Stems v sticky below fl-pairs – fls in a spike-like panicle – petals slightly notched ....
  ....*L. viscaria* (p 144 **E**)
  Stems not sticky below lf-pairs – fls in loose forking cymes – petals deeply four-times-
  cleft...*L. flos-cuculi* (p 144 **D**)

7 Calyx with 20–30 veins, inflated..**8**
Calyx with 10 veins, not inflated **10**

8 Calyx downy, swollen below, conical, strongly ribbed, not net veined, teeth narrow...............
....*S. conica* (p 146 **F**)
Calyx hairless, bladder-like, net-veined, teeth broad, triangular.................................**9**

9 Plants forming patches of prostrate non-flowering shoots – fl-stems erect, to 20 cm, one-
four-fld – calyx cylindrical ....*S. uniflora* (p 146 **B**)
Plants without prostrate non-flowering shoots – erect fl-stems 25–80 cm, many-fld –
calyx egg-shaped..*S. vulgaris* (p 146 **A**)

10 Calyx hairless **11**
Calyx hairy or sticky...**12**

11 Mt plant, forming cushions with linear lvs 6–12 mm long – fls solitary, pink, 9–12 mm
across, on short (2 cm) stalks ....*S. acaulis* (p 144 **C**)
Lowland plant of sandy soils, r – basal lvs narrow, spoon-shaped – stem sticky below – fls
yellow-green, 3–4 mm across, in an erect panicle 20–30 cm tall ....*S. otites* (p 146 **C**)

12 Per plant – non-flowering shoots – infl an erect loose panicle of pairs of forked cymes,
25–60 cm tall – fls drooping, cream-coloured, 18 mm across ....*S. nutans* (p 146 **E**)
Ann plant – no non-flowering shoots – erect infl, few-fld, 15–30 cm tall – petals rosy
above, rolled up during day, spreading at night..*S. noctiflora* (p 146 **D**)

A White Campion, *Silene latifolia* (*S. alba*), sticky-hairy ann to per herb, 30–100 cm, with elliptical – lanceolate opp lvs, 3–10 cm. Infl forks repeatedly, with a fl in each fork. Fls white, 25–30 mm; petals deeply-forked with a scale on each petal where it narrows to its stalk. Plants dioecious, male fls with 10 stamens, female with 5 styles. Capsules (**Aa**) with 10 **erect** teeth at top. Br Isles, c in ar, wa, hbs. Fl 5–10.

B AWI **Red Campion**, *S. dioica*, v like A, but usually per, with creeping rhizomes; narrower blunt calyx-teeth; bright rose-pink petals (rarely but sometimes white); capsule-teeth (**Ba**) on female plants are **strongly recurved**. Br Isles, vc (but locally vr, eg Cambridgeshire fens); in hbs, wds, on base-rich soils. Fl 3–10. Hybrids (pinkish-white petals) between A and B (*S. x hampeana*) are c where the parents meet. **B.1 Rose Campion**, *Lychnis coronaria*, per with stems and lvs covered with dense **white-woolly** hairs; deep pink-purple fl on long stalks. Br Isles, o gdn escape in wa, hths, rds, dunes. Fl 4–7.

C WO (NI) ****Ire Moss Campion**, *S. acaulis*, forms dense green cushions, dotted with solitary pink 9–12 mm fls on v short stalks. N Wales to N Scot, f-la NW Ire, r; on mt cliffs and screes. Fl 7–8.

D Ragged-Robin, *Lychnis flos-cuculi*, resembles B but fls 30–40 mm, petals **deep** pink or white, usually divided each into **4 v narrow lobes**. Lvs narrow-lanceolate, rough but **hairless**, broader from root. Both stamens and styles are present in each fl (unlike A and B). Br Isles, f in moist mds, fens, wet wds. Fl 5–8.

E NT ***Sticky Catchfly**, *L. viscaria*, differs from D in having stems **sticky** below the lf-pairs; infls more whorled, denser; petals **red**, only slightly notched into 2 lobes; fls 18–20 mm. Br Isles: Wales, Scot r (introd C Eng, vr); on dry volcanic rocks. Fl 6–8.
E.1 Sch 8 VU ****Alpine Catchfly**, *L. alpina*, smaller (5–15 cm) than E, fls 6–12 mm in

A

Aa

D

E

B

Ba

C

rounded heads; stems not sticky. Br Isles: vr on mts Cumbria, E Scot on metalliferous rocks. Fl 6–7.

A Bladder Campion, *Silene vulgaris*, either waxy ± hairless or sometimes hairy per herb, 25–90 cm, with erect shoots bearing branched forking infls. Lvs oval, pointed, greyish; fls white, 18–20 mm across, with bifid petals; calyx-tubes strongly ribbed and inflated, 10–12 mm long, like **oval** bladders. **Key difference** from White Campion (p 144 A) is **3** (not 5) **styles**. Br Isles, c in S, rarer to N; on ar, rds, hbs, open gd, especially on chk and sand. Fl 6–8.

B Sea Campion, *S. uniflora* (*S. maritima*), resembles a lower-growing (8–25 cm) version of A, with cushions or mats of non-flowering shoots of smaller, fleshier, more **waxy** lvs. Infls few (1–4); fls larger (20–25 mm), **erect**, (not drooping); calyx-tubes **cylindrical**-oblong, not **oval**, 17–20 mm long. Br Isles, c on coasts, shingle, cliffs; also on mts inland. Fl 6–8.

C EN ****Spanish Catchfly**, *S. otites*, erect per herb, 20–90 cm, with **much smaller** fls than in A or B, and **sticky** stems below; basal rosette-lvs **narrow spoon-shaped, stalked**; upper lvs linear-lanceolate, stalkless, with close short hairs. Infl branched, with open apparent whorls of fls 3–4 mm across, greenish-yellow; calyx shortly bell-shaped and ten-veined; petals undivided, lack scales above; plants usually dioecious, but sometimes bisexual fls occur on otherwise male plant. Eng vl and r (**Breckland only** as a native) on dry, disturbed places on sandy chk gsld. Fl 6–8.

D VU **Night-flowering Catchfly**, *S. noctiflora*, hairy erect ann, 15–60 cm, rather like a much smaller less branched sticky White Campion (p 144 A), but fls open at night, with petals rolled inwards by day to show yellow undersides; 10 stamens, 3 styles in each fl. Br Isles: lf in S, vr to N; on ar on sand and chk. Fl 7–9.
D.1 EN BAP *Small-flowered Catchfly, *S. gallica*, stickily-hairy ann, 15–40 cm; narrow lanceolate sessile upper lvs, spoon-shaped lower ones; elongated spike-like infl of white fls 10–12 mm across, often red-spotted, with shallow notched petals. Br Isles, r (except CI, IoS, c); on sandy ar, wa, old walls. Fl 6–10.

E NT *Nottingham Catchfly, *S. nutans*, erect per herb ± downy below and sticky in upper parts, 25–80 cm; basal lvs (**Eb**) spoon-shaped, stalked; upper lvs narrow-lanceolate, sessile, pointed, all hairy. Fls (**Ea**) 18 mm, creamy-white, drooping, opening fragrantly at night with rolled back deeply-cleft narrow petals in loose apparently whorled infls; some populations have yellow or pink-tipped fls. Fr **short-stalked** within calyx, mature capsules <1 cm long. Br Isles: SE and C GB, N to Aberdeen r-vla; CI la; on shingle beaches; inland on chk gsld or limestone cliffs. Fl 5–8. All British sp are usually regarded as ssp *smithiana*; however the author notes that the plants of Hampshire and Sussex on coastal dunes and shingle are distinct, with mature capsules >1 cm long and referable to ssp *salmoniana*.

F VU *Sand Catchfly, *S. conica*, short erect few-fld sticky-hairy grey-green ann; narrow lanceolate lvs; fls 4–5 mm across; petals rose-coloured, bifid, with scales; long, stickily-hairy, oval-conical calyx-tube with about 30 green ribs on a grey background, enlarged in fr. Styles 3. S Eng, Wales, CI only o-lf; open gd, trackway edges on dunes, sandy gsld, hths, especially near coast. Fl 5–6.

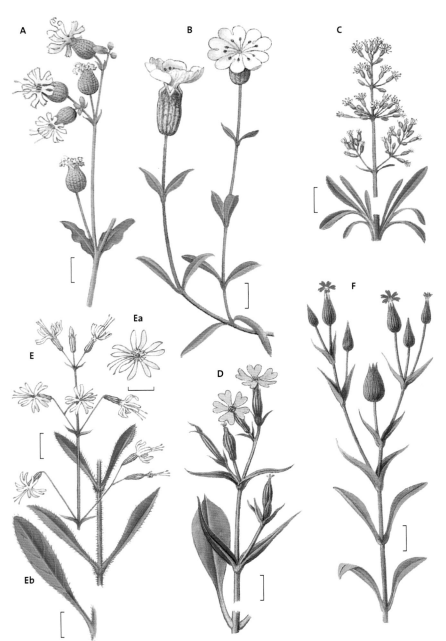

A

B

C

Ea

E

Eb

D

F

Chickweeds, Stitchworts, etc (*Stellaria* and *Myosoton*)
Mostly ± hairless slender brittle-stemmed herbs; small white fls in forked cymes, with a fl in each fork (**dichasia**); petals forked; sepals not forming a calyx-tube, but spreading; capsules oval. Styles 3, capsule-teeth 6 (*Stellaria*); styles 5, capsule with 5 split teeth (*Myosoton*).

ID Tips   Chickweeds

- **What to look at: (1)** number of stamens **(2)** ratio sepals:petal length **(3)** length of sepals **(4)** whether bracts green or with white margins.

- When counting stamens, don't rely on the number of anthers visible as these may fall off; instead count the filaments.

- **Do not confuse** early plants of **Common Chickweed** (*S. media*) which have shrunken or abs petals with **Lesser Chickweed** (*S. pallida*).

- **Common, Greater** and **Lesser Chickweed** all may have a **single line** of hairs running down the stem.

KEY TO CHICKWEEDS AND STITCHWORTS (*STELLARIA*)

1 Lvs oval or cordate, lower stalked – stems round **2**
   Lvs narrow, linear to narrow-oblong, stalkless – stems four-angled **5**

2 Erect plant – petals twice length of sepals ....*Stellaria nemorum* (p 150 **H**)
   Sprawling plant – petals none, or equalling, or scarcely longer than sepals.............................**3**

3 Stamens 10 – fls 12 mm – sepals 5.0–6.5 mm – seeds mostly >1.3 mm ....
                                                                                                            S. neglecta (p 150 **E**)
   Stamens 1–8 – fls 4–8 mm – sepals 2–5 mm – seeds mostly <1.3 mm **4**

4 Lvs green – petals present – sepals 3–5 mm – stamens 3–8 – anthers reddish ....
                                                                                                            S. media (p 150 **A**)
   Lvs yellow-green – no petals – sepals 2.0–3.5 mm – stamens 1–3 – anthers grey-violet ....
                                                                                                            S. pallida (p 150 **D**)

5 Petals shorter than sepals – calyx funnel-shaped below ....*S. alsine* (p 150 **I**)
   Petals equal to or longer than sepals – calyx rounded below **6**

6 Bracts wholly green, without whitish papery margins – lvs rough-edged – fls 20–30 mm, petals bifid less than 1/2 their length ....*S. holostea* (p 150 **B**)
   Bracts with whitish papery margins – lvs smooth-edged – fls 5–18 mm – petals bifid more than 1/2 their length..**7**

7 Bracts with hairs on edges – lvs green – fls 5–12 mm ....*S. graminea* (p 150 **C**)
   Bracts hairless – lvs ± waxy grey-green – fls 12–18 mm ....*S. palustris* (p 150 **F**)

Mouse-ears (*Cerastium*) resemble **Chickweeds** and **Stitchworts** (*Stellaria*), but mostly have roughly hairy blunt, oblong, stalkless lvs, 5 styles (except for *C. cerastoides*, which has 3) and **cylindrical** (**not oval**) capsules; 5 free sepals, 5 free bifid petals. **Sandworts** (*Arenaria, Minuartia*) are similar, but have undivided petals, see key, p 155.

KEY TO MOUSE-EARS (*CERASTIUM*)

1 Styles usually 3, capsule-teeth 6 – r in high mts in Scot ....
Starwort Mouse-ear *Cerastium cerastoides*
Styles 5, rarely 4 – capsule-teeth twice number of styles **2**

2 Fls 12 mm across or more – petals usually twice length of sepals – pers with many creeping, non-flowering shoots...**3**
Fls less than 12 mm across – petals not more than 1¹/₂ times length of sepals – mostly ann plants **5**

3 Lvs linear-lanceolate to narrow-oblong, with close grey down, hairs short, <1 mm long
....*C. arvense* (p 152 **F**)
Lvs linear-lanceolate to narrow-oblong, with dense white-woolly hairs, some hairs >1mm long ....*C. tomentosum* (p 152 **F.1**)
Lvs oblong to almost circular, with spreading hairs, some hairs >1mm long – r in mts of N GB and N Wales **4**

4 Stem and young lvs woolly, with long soft white hairs – lvs oval to elliptical – sepals narrow – petal-tips flared – fr curved (at least in upper half) – restricted to soft calc rocks or flushes...*C. alpinum* (p 152 **F.2**)
Stem and young lvs with short yellowish hairs – sepals oval-lanceolate – petal-tips reflexed – fr straight – prefers acidic or hard rocks ....NT ***Arctic Mouse-ear** C. nigrescens* (ssp *arcticum* (*C. arcticum*) with lvs narrow or wide-elliptical – few sticky hairs – on high mts of N GB; End BAP EN **Shetland Mouse-ear** ssp *nigrescens* (*C. nigrescens*) with lvs almost round with many short sticky hairs occurs on **Shetland only**)

5 Per or ann – no sticky, glandular hairs anywhere **6**
Ann, all shoots have fls – sticky, glandular hairs on lvs and sepals **7**

6 Per – upper bracts with silvery margins – widespread ....*C. fontanum* (p 152 **A**)
Ann – upper bracts wholly green – vr in S Eng only ....*C. brachypetalum* (p 152 **A.1**)

7 Fls in compact heads – sepals with long, non-glandular hairs projecting beyond sepal-tips – frs 7–10 mm, curved, with stalks mostly shorter than sepals ....*C. glomeratum* (p 152 **D**)
Fls in loose infls – sepals without long, non-glandular hairs projecting beyond sepal-tips – frs 4.5–7.0 mm, ± straight, with stalks mostly longer than sepals **8**

8 Bracts of infls wholly green – fl-stalks erect – fls usually four-petalled (rarely five-petalled) ....*C. diffusum* (p 152 **E**)
Bracts of infls with silvery margins – fl-stalks bent down – 5 petals **9**

9 Upper third-half of bracts silvery – petals 2/3 length of sepals – fr-stalk bent down, but not curved – lvs all green ....*C. semidecandrum* (p 152 **B**)
Bracts with only a narrow silvery edge (at most upper quarter) – petals equalling sepals – fr-stalk curved down – lower lvs often reddish ....*C. pumilum* (p 152 **C**)

A **Common Chickweed**, *Stellaria media*, v variable ann of sprawling habit, with a single line (rarely 2 lines) of hairs running down round stems; cordate lvs with one strong vein; fls with 5 deeply bifid white petals (rarely abs), 1 – 3 mm long, no longer than the spreading, oval, usually hairy **3–5 mm sepals** without strong veins; **3–8 red-violet** anthers; 3 styles. Seeds round, 0.8–1.4 mm across, red-brown, with rounded warts on surface. Br Isles, vc in ar, wa everywhere. Fl 1–12. **Do not confuse** with **Three-nerved Sandwort** (p 154 A) which is v similar but has **undivided** petals and usually has lvs and sepals with 3 strong veins.

B AWI **Greater Stitchwort**, *S. holostea*, per herb with weak brittle ± erect stems 15–60 cm long, square, rough-angled; lvs arranged in pairs, each pair at right-angles to the next, so appearing as 2 ranks from above; lvs narrow-lanceolate, rigid, 1.5–5.0 cm long, stalkless, ± hairless, but rough-edged (**Ba**); bracts all green. Fls 20–30 mm across, petals divided **halfway or less**. Br Isles, vc in hbs, wds, except on v acid soils. Fl 4–6.

C **Lesser Stitchwort**, *S. graminea*, is similar to I below, but has lvs 1.5–4.0 cm long, smooth-edged; bracts hairy with whitish papery margins; fls only 5–18 mm across, petals split **more than** halfway, **equalling** sepals. Br Isles, c on acid gslds, hths, open wds. Fl 5–8.

D **Lesser Chickweed**, *S. pallida*, is like small version of A, but yellow-green lvs; no petals (**D**), **1–3 grey-violet** anthers; **sepals 2–3mm**; seeds 0.6–0.9 mm across, pale yellow brown, with blunt warts. Br Isles: Eng, Wales (mainly S and E, CI, IoS) lf-c; Scot, Ire o; on dunes, sandy wa. Fl 3–4, then dries up so often over-looked.

E AWI **Greater Chickweed**, *S. neglecta*, is like large version of A; fls (**E**) 10–12 mm across; **10 reddish stamens**; seeds mostly >1.3 mm across, dark red-brown, with narrow conical warts. Br Isles: Eng, Wales lf; Scot vr; Ire, CI abs; on hbs, damp wds, shady riversides. Fl 4–7.

F VU **Marsh Stitchwort**, *S. palustris*, resembles B in general habit, with square weak stems and narrow lanceolate lvs; but differs in the wide, whitish, papery, smooth, **hairless** bract-margins (**Fa**); lvs smooth-edged, usually **waxy grey-green**. Fls 12–20 mm across, in few-fld cymes; petals split to base. Br Isles: GB (not N Scot, IoS) r-vlf; Ire o; (decreasing due to drainage) on fens, basic marshes. Fl 5–8.

G **Water Chickweed**, *Myosoton aquaticum*, resembles a large (20–100 cm), more erect version of E; larger, ± hairy, oval, cordate lvs; fls 12–15 mm across, with sticky-hairy stalks, petals far longer than sepals; 10 stamens, **5 styles**; oval capsules open by **5 split teeth**. Br Isles: Eng, Wales, lc (Ire, much of Scot, abs); in marshes, river and streamsides, wet wds on basic soils. Fl 6–10.

H AWI **Wood Stitchwort**, *Stellaria nemorum*, per of 15–60 cm, resembling tall erect version of E, but with stems hairy all round; lvs (**H**) long-stalked, **cordate**; fls 13–18 mm across, petals twice length of sepals. Br Isles: W and N of GB lf (abs SE Eng); abs Ire, CI, IoS; in damp wds, especially by streams. Fl 5–6.

I **Bog Stitchwort**, *S. alsine* (*S. uliginosa*), can be confused with C, but lvs shorter (5–20 mm), **hairless**, ± **grey-green**; lvs on **non-flg shoots stalked** (C stalkless); petals (**Ia**) **shorter than** sepals; calyx-base funnel-shaped. Stem usually square (**Ib**) (although C may also have square stem and I may have a round stem). Br Isles, lc in marshes, wet wds, especially on acid soils. Fl 5–6.

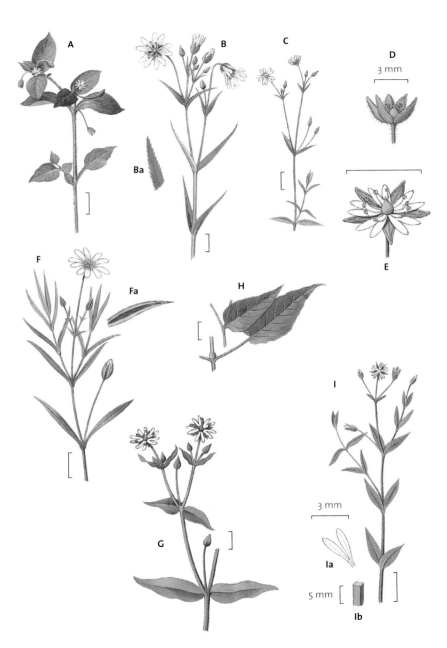

A Common Mouse-ear, *Cerastium fontanum* (*C. holosteoides*), is the commonest *Cerastium* sp; a low **per** herb, with runners and erect fl-shoots; stems and lvs all densely **hairy** (no glandular hairs). Fls in loose cymes, 3–12 mm across; petals white, bifid, equal to or slightly longer than sepals; upper bracts (**Aa**) and sepals hairy, with narrow **silvery margins**. Stamens usually 10; capsule 9–12 mm long, curved. Br Isles, vc in gsld, beaches, dunes, rds, wa, mds, etc (grows much taller in wet mds). Fl 4–9. End AWI ***ssp *scoticum*, differs from A in generally larger seeds and petals but it is so v close to A that the *New Flora* considers it would be best grouped with A. Scot (Angus) only, vvr. **A.1 Grey Mouse-ear**, *C. brachypetalum*, similar to A but a silvery-grey hairy **ann** herb (no glandular hairs); **key differences** from A are bracts wholly **green**; petals **1/2 as long as** sepals. Br Isles discovered 1947 and arguably native; S Eng vr (possibly over-looked elsewhere); on chk. Fl 4–6.

B Little Mouse-ear, *C. semidecandrum*, similar to A but **ann** and **glandular-hairy**. Fls 5–7 mm across; petals only slightly notched, petals **2/3** length of sepals; 5 stamens; upper bracts (**Ba**) and sepals with **broad** silvery margins, central green part v narrow; capsule (**Bb**) 4.5–6.5 mm long, scarcely curved; fr-stalks shortly bent down. Br Isles, f-lc on dry open sandy gd. Fl 4–6.

C NT ***Dwarf Mouse-ear**, *C. pumilum* is close to B in size and habit, but **key differences** are upper bracts with only a **narrow** silvery margin (**C**); petals **equal** sepals; fr-stalk curved. S Eng, N Wales r-vlc; in bare places on calc gslds. Fl 4–5.

D Sticky Mouse-ear, *C. glomeratum*, similar size to A, but like B erect, **ann, sticky, glandular-hairy**. Fls 8–10 mm, in rather **dense** terminal heads, lvs yellowish; **10 stamens**; bracts (**Da**) **without** silvery margins; capsule (**Db**) 7–10 mm long, curved. Br Isles, vc on ar, wa, walls, hbs, dunes. Fl 4–9.

E Sea Mouse-ear, *C. diffusum* (*C. atrovirens*), another low glandular-hairy **ann** resembling B, but of much more open branching, and with **all bracts** (**Ea**) wholly **green** (no silvery margins); fls 3–6 mm across; capsule (**Eb**) 5.0–7.5 mm long, straight. Br Isles, lc on open sandy gd, usually by sea; scattered inland on rds, paths, dry gsld. Fl 4–7.

F Field Mouse-ear, *C. arvense*, low creeping **per**, large fls (12–20 mm across), petals twice length of sepals, and narrow lvs downy with **no long hairs**. Br Isles: E Eng lf-lc, o elsewhere; Ire r; on dry gslds, hbs, rds, especially on chk and sand. Fl 4–8. **F.1 Snow-in-summer**, *C. tomentosum*, also has large fls, a mat-forming per; lvs with dense, matted **woolly-white hairs**, some **long** >1 mm. Introd (Italy and Sicily), much more c than F and increasing; Br Isles, c gdn escape on wa, rds, dunes, shingle. Fl 5–8. **F.2** VU ***Alpine Mouse-ear**, *C. alpinum* also has large fls (18–25 mm across) and v hairy, **rounded** lvs. On basic rock ledges of mts of Scot, N Wales, and Cumbria only. Fl 6–8.

G Upright Chickweed, *Moenchia erecta*, is not much like a Chickweed; it is a short stiffly erect hairless ann, 3–12 cm; lvs strap-shaped below, acute above, v rigid and waxy. Fls about 8 mm long, usually with 4 petals, 4 sepals, erect on long stalks, only opening briefly in fine weather; sepals pointed with conspicuous white edges. Stamens and styles usually 4. Br Isles: S Eng, S Wales o-lf ; CI c; Scot, Ire abs; on dry gsld, cliffs, dunes. Fl 4–6.

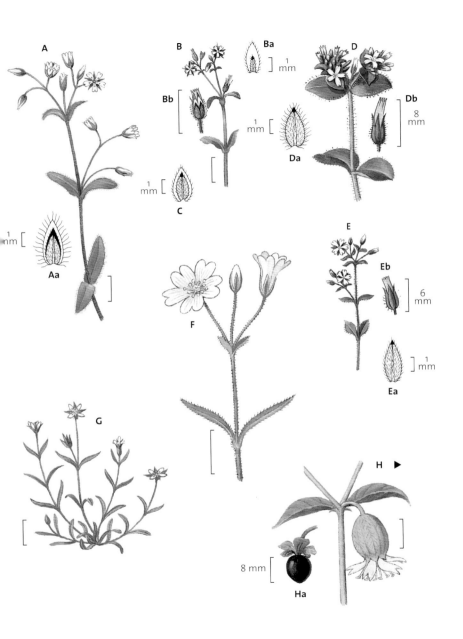

A

Aa
$\frac{1}{mm}$ [

B

Ba
$\frac{1}{mm}$]

Bb

$\frac{1}{mm}$ [

C
$\frac{1}{mm}$ [

Da
$\frac{1}{mm}$ [

D

Db
$\frac{8}{mm}$]

E

Eb
$\frac{6}{mm}$]

Ea
$\frac{1}{mm}$]

F

G

H ▶

Ha
8 mm [

▲ **H Berry Catchfly**, *Cucubalus baccifer*, straggling hairy per herb, 60–100 cm, scrambling over other plants; it resembles a large **Water Chickweed** (p 150 G) with hairy oval lvs, but fls are 18–20 mm across, sepals joined into a short calyx-tube as in the Campions, and the greenish-cream petals are deeply bifid with scales above their spreading narrow lobes. Frs (**Ha**) – unique in European *Caryophyllaceae* – are round black berries, 6–8 mm. Br Isles: Norfolk only, vr but arguably native; in damp wds, riversides. Fl 6–8.

A AWI **Three-nerved Sandwort**, *Moehringia trinervia*, a small, slightly downy ann, 10–30 cm; rather like **Chickweed** (p 150 A), but usually with **3 strong veins** in each lf, narrow 3-veined sepals longer than petals, and petals **undivided**. Br Isles, c (Hebrides, Mull and Orkney & Shetland abs) in wds on richer soils, shady hbs. Fl 5–6.

B Sea Sandwort, *Honckenya peploides*, **creeping** per, with opp stalkless **fleshy** lvs (oval-pointed) and stems; fls, among lvs, 6–10 mm across; 5 petals, not notched, greenish-cream, blunt, undivided, not longer than sepals. Br Isles, lc on beaches and open sand dunes by the sea. Fl 5–8.

C NT *****Spring Sandwort**, *Minuartia verna*, low cushion-forming per herb, either hairless or sticky-hairy, with many dark green rigid 3-veined linear lvs, 6–15 mm long; fls (**Ca**) with petals white, oval, not notched, a little longer than narrow whitish-edged 3-veined sepals; 3 styles. Br Isles: W and N of GB, scattered and vl; Ire r; on basic open rocks and screes, especially on old lead workings. Fl 5–9. **Do not confuse** with **Knotted Pearlwort** (p 158 I) which has similar lvs but petals **much longer** than sepals and 5 styles.   **C.1** VU ****Mountain Sandwort**, *M. rubella* is a low per, forming cushions 2–8 cm across, with few-fld stems, 2–6 cm tall; lvs only 4–8 mm; petals 2/3 length of sepals; vr on mts of Scot. Fl 6–8.   **C.2** VU *****Cyphel**, *M. sedoides* (*Cherleria sedoides*), is a cushion-forming alpine herb rather like C, but with tiny green almost stalkless fls of sepals only, 4–5 mm across. Mts of Scot lf. Fl 6–8.   **C.3** FPO (Eire) ******Ire **Recurved Sandwort**, *M. recurva*, close to C but sepals 5–7 veined and more densely tufted. **Ire only** vr, W Cork and S Kerry on mt rocks. Fl 6–10.

D EN *****Fine-leaved Sandwort**, *M. hybrida*, slender erect hairless branched ann, 5–20 cm tall, with acute linear lvs and no sterile shoots; fls 6 mm across, white, petals much shorter than white-bordered sepals. Br Isles: S Eng, Wales, CI r-lf; Ire r; on walls, dry stony gsld, sandy ar etc. Fl 5–6.   **D.1** Sch 8 VU ****Teesdale Sandwort**, *M. stricta*, also erect tufted per herb, 5–10 cm, with v narrow one-veined lvs 6–12 mm, in remote pairs on fl-stems; fls 5–8 mm across, white, in few-fld cymes, petals equalling sepals. **Teesdale only**, in calc flushes. Fl 6–7.

E Thyme-leaved Sandwort, *Arenaria serpyllifolia*, low, spreading to bushy-branched hairy ann, 2.5–20 cm tall, with many tiny hairy oval pointed lvs. White fls 5–6 mm across, petals undivided, shorter than sepals; fr flask-shaped, thick-walled. Seeds 0.6 mm wide, black. Br Isles, c on bare gd, chk gsld, ar, walls, wa. Fl 6–8.   **E.1** End FPO (Eire) ******Ire **Fringed Sandwort**, *A. ciliata* is a hairy creeping per with lvs oblong to 5 cm tall; fls 12–16 mm across; sepals hairy. **W Ire only**; vvr on limestone cliffs near Sligo. Fl 6–7.
E.2 Sch 8 ******Ire VU ****Arctic Sandwort**, *A. norvegica* ssp *norvegica* is similar to E.1, per with **many** non-flg shoots; but with sepals **hairless**, lvs almost so; fls **9–10 mm** across. Br Isles: Scot; vr on basic rocks. Fl 6–7.   **E.3** Sch 8 End NT ****English Sandwort**, *A. norvegica* ssp *anglica*, differs from E.2 in **ann** or bi with **few** non-flg shoots and fls **11-23 mm** across; occurs in **Yorks only**, vr on limestone.

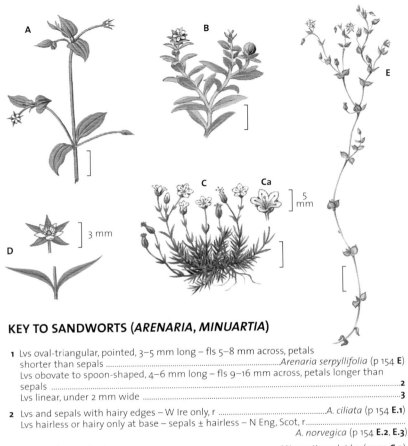

KEY TO SANDWORTS (*ARENARIA, MINUARTIA*)

1 Lvs oval-triangular, pointed, 3–5 mm long – fls 5–8 mm across, petals shorter than sepals ....*Arenaria serpyllifolia* (p 154 **E**)

Lvs obovate to spoon-shaped, 4–6 mm long – fls 9–16 mm across, petals longer than sepals **2**

Lvs linear, under 2 mm wide **3**

2 Lvs and sepals with hairy edges – W Ire only, r ....*A. ciliata* (p 154 **E.1**)

Lvs hairless or hairy only at base – sepals ± hairless – N Eng, Scot, r................................
A. norvegica (p 154 **E.2**, **E.3**)

3 Petals abs or minute ....*Minuartia sedoides* (p 154 **C.2**)

Petals present **4**

4 All stems erect and with fls, not tufted – fls 6 mm across – petals much shorter than sepals – ann plant ....*M. hybrida* (p 154 **D**)

Plant with tufts or cushions of non-flowering shoots as well as erect fl-shoots – petals 2/3 length of sepals, or more – per plant..**5**

5 Lvs 4–8 mm long – petals only 2/3 length of sepals – fls 5–9 mm across – Scot mts, vr ....
M. rubella (p 154 **C.1**)

Lvs 6–15 mm long – petals as long or longer than sepals **6**

6 Lvs three-veined – fl-stalks sticky-hairy – fls 8–12 mm across, infls ascending from dense cushion of lfy shoots – lf-pairs ± own length apart on fl-stems ....*M. verna* (p 154 **C**)

Lvs one-veined – fl-stalks hairless – fls 5–8 mm across – infls strictly erect from small tuft of lfy shoots – lf-pairs much more than own length apart – Teesdale, vr....................................
M. stricta (p 154 **D.1**)

Spurreys (*Spergula* and *Spergularia*) are low herbs with cylindrical, ± fleshy lvs, **stipules** and small fls, with 5 white or pink uncleft petals and 5 spreading sepals; either 3 styles (*Spergularia*) or 5 (*Spergula*). **Pearlworts** (*Sagina*, see key p 160), v small per or ann plants, tufted; linear fleshy lvs fused at base in pairs, and **no stipules**; creeping or ascending fl-stems. Fls globular in bud, with 4–5 sepals; petals white, not notched, 4, 5, or none; capsules v tiny.

A Greater Sea-spurrey, *Spergularia media*, almost hairless, creeping to ascending per herb, to 30 cm long, with v fleshy pointed lvs; fls (**Aa**) white or pink, 8–12 mm across; capsule 7–10 mm long, equalling or longer than stalk; seeds (**Ab**) with a clear **winged** border. Br Isles, c in salt-marshes, coastal ditches, dune-slacks; vr inland on salted rds. Fl 6–9.

B Lesser Sea-spurrey, *S. marina*, only to 20 cm long, ann; fls (**B**) always **deep** pink, 6–8 mm across; capsule 4–6 mm long, shorter than stalk; seeds (**Ba**) usually (but not always) **without winged** border. Br Isles, c salt-marshes, f and increasing inland along salted rds. Fl 6–9.

C Sand Spurrey, *S. rubra*, is ann, 5–25 cm long, sticky-hairy above, with grey-green, non-fleshy stiff-pointed lvs; stipules silvery (not pale brown as in A and B), narrow; fls (**Ca**) 3–5 mm across, pink; seeds **unwinged**. Br Isles: GB f-lc; Ire o; on open sandy or stony acid gslds, hths, wa. Fl 5–9. **C.1 Greek Sea-spurrey**, *S. bocconei*, like C in lvs and seeds, but stipules broad-triangular, dull, fl-stalks shorter than sepals, and fls 2 mm across in curved, one-sided, unforked racemes. Introd (Mediterranean); Cornwall, E Kent, CI; vr on sandy rocky gd by sea. Fl 5–9.

D Rock Sea-spurrey, *S. rupicola*, per resembling A, but is 5–15 cm long, with **densely stickily-hairy** stems and lvs (**D**), and **unwinged** seeds. Br Isles: c on W coasts, vc IoS, CI; vr to S and E; on sea cliffs, also short coastal gsld and walls. Fl 6–9.

E VU **Corn Spurrey**, *Spergula arvensis*, sticky-hairy ann, 10–30 cm, looking taller than A and C, with weak ascending stems; lvs in conspicuous whorled clusters, **furrowed** below, blunt, with **tiny** stipules that soon fall; fls (**Ea**) 4–7 mm across, white, in forked cymes; seeds with **v narrow wing** or none. Br Isles, c on sandy ar and wa, also coastal. Fl 6–8.

F Procumbent Pearlwort, *Sagina procumbens*, low **per** tufted herb, with a **central non-flowering rosette**, from which arise rooting **runners** that turn up and bear the infls; lvs linear, with bristle –like points at tips; fls on long stalks, **4** sepals, usually **no petals** (but **Fa** with petals). Frs (**Fb**) 2–3 mm long. Br Isles, vc everywhere on bare or trampled gd, especially on paths, lawns, wall tops, wa, hbs. Fl 5–9. *S. boydii* was considered endemic to Scot but genetic studies show that it is best regarded as a cultivar of *S. procumbens*. **F.1** EN *****Alpine Pearlwort**, *S. saginoides*, is like F, but larger, mat-forming, with lfy flowering shoots from basal rosette as well as from runners; fls about 4 mm across, borne singly on stalks 1–2.5 cm; **5** sepals, usually **5** petals; 10 stamens. **Frs larger** than F's, 3.5–4 mm long. Scot on high mt ledges. Fl 6–8.

G Annual Pearlwort, *S. apetala* ssp *erecta*, is close to F, but is **ann**, 3–10 cm tall, and the main stem from the central rosette (which withers early) produces fls (**Ga**); the side branches do not root; lvs have small hairs on edges (**Gb**) and bristle-like points; 4 blunt sepals, **no petals**. Br Isles, c (except N Scot, Ire o); on walls, paths, bare gd. Fl 5–8. **G.1** *S. apetala* ssp *apetala* (*S. ciliata*) is ann v similar to G, but with outer sepals **pointed** (not all blunt as in G) and **adpressed**, not spreading in fr (as in F and G). Br Isles, o, lc to S; dry sandy gd. Fl 5–7.

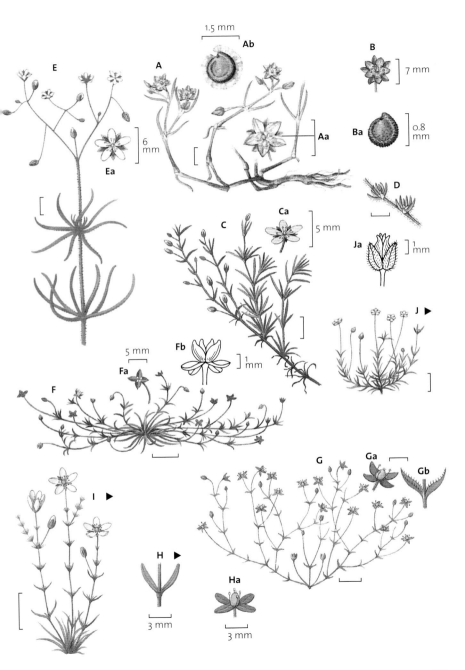

▲ **H Sea Pearlwort**, *Sagina maritima*, resembles G p 156, **ann**, but has dark green **fleshy, blunt** lvs (**H**) **without** (or v minute) bristle-like point; fls (**Ha**) with 4 sepals usually adpressed in fr as in G.1 p 156. Br Isles, widespread but only lc on bare saline gd on sea coasts, tidal river-banks, salted inland rds and mts inland in Scot. Fl 5–9.

▲ **I Knotted Pearlwort**, *S. nodosa*, low tufted **per**, 5–15 cm, with narrow cylindrical pointed lvs in stiff lateral groups or knots along the stems; fls few, 0.5–1.0 cm across, with 5 white unnotched **petals twice length** of sepals; 10 stamens, 5 styles. Br Isles: l mostly to W and N; on bare damp sandy or peaty gd on dunes, hths, mts. Fl 7–9.

▲ **J Heath Pearlwort**, *S. subulata*, **per** forming low cushions of dark green sticky-hairy lvs with bristle-like points, from which arise long, 2–4 cm, erect, ± leafless fl-stalks, bearing erect white fls 4–5 mm across; 5 petals and **glandular hairy sepals** (**Ja**) of equal length. A variety with **hairless** sepals occurs in S Eng. Br Isles: N of GB, Wales and NW Ire, f-lc; CI, c; on dry sandy hths. Fl 6–8.

A   VU ****Coral-necklace**, *Illecebrum verticillatum*, prostrate hairless ann of v distinctive form: tiny stipulate oval grey-green lvs (2–6 mm long) in pairs subtend conspicuous shining whorl-like clusters of tiny fls, 8–12 per whorl. Each fl, about 2–3 mm long, has 5 thick corky white sepals, hood-shaped with bristle-tips; 5 v tiny petals; 5 stamens. S Eng only (vr introd N Eng); vlf on damp acid sandy gd by pounds, or in trackways on hths or wds. Fl 7–9.

B   FPO (Eire) EN **Annual Knawel**, *Scleranthus annuus*, low spreading **ann**, with linear-pointed lvs (no stipules), joined in pairs at base around stem. Clusters of minute fls, 4 mm across, with 5 sepals and no petals, occur in lf-axils and terminally. Sepals, straight, grey-green, pointed, with narrow white margins, arise from a short hairless calyx-tube (**Ba**) that encloses the 10 or fewer stamens and tiny ovary. GB o-r, Ire vr and declining throughout; on dry sandy and gravelly wa, ar, gsld. Fl 6–8. **B.1** Sch 8 CR ****Perennial Knawel**, *S. perennis* ssp *perennis*, erect to ascending **per**, woody below; sepals **blunt** with broad white margins, incurved in fr; basal cup hairy; **achene** (including sepals) (3-)**3.5–4.5 mm** long. End BAP EN ****ssp** *prostratus* is usually procumbent and **key difference** in **achene** (including sepals) **2–3**(-3.5) **mm** long. Br Isles: E Anglia (ssp *prostratus*), central Wales (ssp *perennis*) vr; on basic rocks or chk gsld. Fl 6–8.

C ****Four-leaved Allseed**, *Polycarpon tetraphyllum*, is another small ann plant, with branched erect stem, 5–15 cm tall, hairless, with blunt oval lvs apparently in whorls of 4 (really **2 pairs**, sometimes one pair much smaller than the other, close-set), 8–13 mm long. Fls tiny, 2–3 mm across, in much-forked cymes (dichasia); 5 white-margined, hooded sepals, 5 narrow tiny petals. SW Eng, r and l; IoS, CI, c; in open sandy wa near sea. Fl 6–7.

D ****Smooth Rupturewort**, *Herniaria glabra*, prostrate mat-forming ann or bi herb, ± hairless, with clusters of tiny (2 mm) green fls in axils of oval lvs 3–7 mm long. 5 sepals, green, hairless; 5 petals, white, minute; ovary in a cup; frs acute, longer than sepals. E Eng r and l (scattered elsewhere as r introd, apparently increasing); on bare chalky gd. Fl 7.

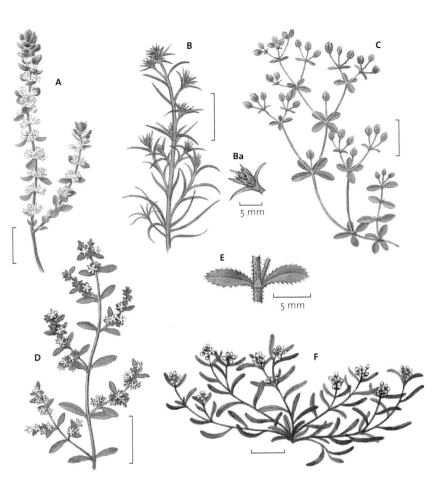

A

B

C

Ba

5 mm

E

5 mm

D

F

E vu ****Fringed Rupturewort**, *H. ciliolata*, differs from D in being a per, woody at base, with **hairy** lvs (**E**), and **blunt** frs no longer than calyx. Br Isles: Cornwall and CI only, vr and I on sandy gd and rocks. Fl 6–8.

F Sch 8 CR ****Strapwort**, *Corrigiola litoralis*, is another small, branched prostrate ann. Stems often reddish; lvs linear-lanceolate, all alt, fleshy glaucous-grey, with blunt apices, widest near tips, tiny stipules. Fls minute, 1–2 mm across, in crowded terminal and axillary clusters; 5 blunt green-red sepals with **white** edges; 5 white or **red**-tipped petals about equal length. S Devon only (as a native; vr introd in N Eng), vr but sometimes la; in damp sandy or gravelly open places, usually on lakeshores flooded in winter. Fl 7–8.

KEY TO PEARLWORTS (*SAGINA*)

1 Fls with 4 (v rarely 5) sepals, 4 (5) stamens – petals usually abs (or else 4, v tiny) **2**
Fls with 5 sepals, 8 or 10 stamens – 5 obvious petals...**5**

2 Per plant – creeping, rooting runners bearing fl-stems – basal central lf-rosette without
fl-stems from it ....*Sagina procumbens* (p 156 **F**)
Ann plant – erect or ascending branches – fls on central stem from basal rosette................**3**

3 Lvs blunt-tipped (or with minute bristle point < 0.1 mm) ....*S. maritima* (p 158 **H**)
Lvs bristle-pointed ....(*S. apetala*) **4**

4 Sepals spreading in fr, all blunt ....*S. apetala* ssp *erecta* (p 156 **G**)
Sepals adpressed to fr, 2 outer sepals pointed ....*S. apetala* ssp *apetala* (p 156 **G.1**)

5 Petals twice length of sepals – fls short-stalked – upper lvs in tight clusters on stems
S. nodosa (p 158 **I**)
Petals not or slightly longer than sepals...**6**

6 Calyx sticky-hairy – fls long-stalked, erect – on dry hths ....*S. subulata* (p 158 **J**)
Calyx neither sticky nor hairy – in mts only **7**

7 Tiny tufted plant, 1–2 cm tall– petals not notched – ripe fr 2.5–3 mm long, greenish, dull –
on Ben Lawers (Scot) only – vr ....VU ****Snow Pearlwort** *S. nivalis* (*S. intermedia*)
Spreading plants, over 2 cm tall– petals often notched – ripe fr 3–4 mm long,
yellowish, shiny...**8**

8 Stems ascending, not creeping – petals notched – fr 3.5–4 mm long (with seeds) ....
S. saginoides (p 156 **F.1**)
Stems creeping, usually rooting – petals sometimes shallowly notched – fr 3–3.5 mm long,
usually seedless – v close to *S. saginoides*; Scot mts, r ......**Scottish Pearlwort** *S.* x *normaniana*

DOCK FAMILY   *Polygonaceae*

Herbs with alt, simple lvs with tubular, ± translucent, brown or silvery,
membranous stipules (**ochrea**). Tepals 3–6, free or joined, small; stamens 6–9;
superior ovary ripens into a hard three-angled, one-seeded nut enclosed in the
tepals. **Docks** (*Rumex*) have erect, much-branched infls, with the 6 greenish
tepals in 2 whorls, the 3 outer tiny and narrow, the 3 inner broad, enlarging to
enclose the ripe fr. **Iceland–purslane** (*Koenigia*) has only 3 tepals and **Mountain
sorrel** (*Oxyria*) has 4. **Knotweeds** and **Bindweeds,** (*Fallopia*) have 3 outer tepals
winged (or a strong keel) in fr, sometimes with branched infls whereas
Knotgrasses and **Persicarias** (*Polygonum* or *Persicaria*) have no tepal wings in fr,
usually unbranched infls, with 5 pink or white tepals, apparently in one whorl,
not swelling up around ripe fr.

ID Tips   Docks

- **What to look at:** Dock fls have inner tepals, which enlarge to enclose the fr. The shape of these inner tepals **in fr**, and whether or not they bear warts (also called tubercles), are vital for correct identification of most spp.

- Young lvs of **Broad-leaved Dock** (*R. obtusifolius*) can appear fiddle-shaped so may be confused with the much less c **Fiddle Dock** (*R. pulcher*).

- If your plant does not key out, it may be one of several large introd Docks (as the key only includes the two most widespread introd spp) or a hybrid, which are widespread and usually sterile.

- The *BSBI Docks & Knotweeds Handbook* has keys to, and line illustrations of, all spp and many hybrids found in the Br Isles.

KEY TO DOCKS (*RUMEX*)

1 Lvs hastate-shaped or arrow-shaped – slender plants – dioecious...**2**
Lvs oblong, lanceolate or round (not arrow- or hastate-shaped) – stout plants – not dioecious **3**

2 Lf-lobes spreading out sideways or even forwards – upper lvs not clasping stem ....
Rumex acetosella (p 166 **C**)
Lf-lobes backward-pointing (so lf ± arrow-shaped) – upper lvs clasping stem............................
R. acetosa (p 166 **B**)

3 All 3 inner tepals without warts **4**
One, at least, of 3 inner tepals with a distinct, sometimes coloured wart **6**

4 Fr-stalk without visible joint – inner tepals in fr longer than wide ......*R. aquaticus* (p 164 **K**)
Fr-stalk with an obvious joint – inner tepals in fr ± rounded, not longer than wide **5**

5 Lvs round, 20–40 cm wide and long, with cordate base – inner tepals with truncate bases
R. alpinus (p 164 **E**)
Lvs oval-lanceolate – inner tepals with cordate bases ....*R. longifolius* (p 164 **G**)

6 Inner tepals long-toothed (teeth 1 mm or more)..**7**
Inner tepals untoothed (or with tiny teeth under 1 mm) **11**

7 Branches spreading nearly at right angles – lvs violin-shaped...................*R. pulcher* (p 166 **A**)
Branches erect or at acute angle to stem – lvs usually not violin-shaped...................................**8**

8 Inner tepals 4–7 mm long **9**
Inner tepals 2.5–4 mm long, yellow or golden when ripe, with bristle-like teeth – lvs narrow, oblong, tapered into stalk **10**

9 Inner tepals $\geq$ 5 mm wide (excluding teeth) – lf-tip pointed.......................*R. cristatus* (p 164 **M**)
Inner tepals < 5 mm wide (excluding teeth) – lf-tip blunt ....*R. obtusifolius* (p 164 **D**)

10 Inner tepals greenish-yellow, with blunt tips and rigid bristle-like teeth shorter than outer tepals – wart 1.5–2 mm long ....*R. palustris* (p 164 **H**)
Inner tepals golden, with pointed tips and fine hairlike teeth, some longer than outer tepals – wart 0.75–1 mm long ....*R. maritimus* (p 164 **I**)

11 Infl of loose distant whorls – inner tepals 3–4 mm long – all 3 with warts 2/3 length of tepals – lvs grey-green – on rocks by sea in SW of GB....................................*R. rupestris* (p 163 **C**)
Infl of dense crowded whorls – inner tepals 4–8 mm long...**12**
Infl of loose distant whorls – inner tepals to 3 mm only **16**

12 Inner tepals with only one large wart **13**
Inner tepals all 3 with large warts **15**

13 Lvs wavy edged, crisped – inner tepals <5 mm wide – largest wart > half as long as inner tepal ....*R. crispus* (p 164 **F**)
Lf-edges flat or only slightly wavy – inner tepals > 5mm wide – largest wart < half as long as inner tepal **14**

14 Lvs with side veins at >60° to midrib – lower lf-bases cordate – margins of inner tepals with teeth ....*R. cristatus* (p 164 **M**)
Lvs with side veins at acute angle, <60° to midrib – lower lf-bases tapering to truncate – margins of inner tepals entire...*R. patientia* (p 164 **L**)

15 Inner tepals triangular – wart >3 mm – lvs lanceolate, not v wavy-edged – v large plant of wet places ....*R. hydrolapathum* (p 164 **J**)
Inner tepals oval or round – wart <3 mm – lvs oblong-lanceolate, wavy-edged, crisped ....
R. crispus (p 164 **F**)

16 Stems straight, erect, branching at acute angle <30° – infls not lfy except at base – one tepal only having large round wart, others with smaller or no warts ....
R. sanguineus (p 163 **A**)
Stems wavy, branching widely 30 –90° – infl lfy nearly to top – all 3 inner tepals with large oblong warts...*R. conglomeratus* (p 163 **B**)

KEY TO KNOTGRASSES AND PERSICARIAS etc
(*POLYGONUM, PERSICARIA, FAGOPYRUM & FALLOPIA*)

1 Outer 3 tepals (looking like sepals) winged or strongly keeled ....(*Fallopia*) **2**
Outer 3 tepals not winged..**5**

2 Plants ± erect, free-standing, tall (usually over 1m) **3**
Plants scrambling or creeping, if >1m high then supported by other plants **4**

3 Lvs normally <12 cm long – lf-base truncate.......................................*F. japonica* (p 170 **D**)
Lvs normally >12 cm long – lf-base cordate ....*F. sachalinensis* (p 170 **E**)

4 Fr 4 mm long, dull brown ....*F. baldschuanica* (p 170 **C.1**)
Fr 4–5 mm long, dull black...*F. convolvulus* (p 170 **B**)
Fr 2.5–3 mm long, shiny black ....*F. dumetorum* (p 170 **C**)

5 Plants ± prostrate – lvs narrow elliptical-lanceolate, 1–5 cm long – fls in clusters of 1–6 in lf-axils...**6**
Plants ± erect – lvs 5–15 cm long – fls in many-fld spikes, some or all at tops of stems **10**

6 Fr smooth, shiny, as long as or longer than calyx **9**
Fr dull, rough, ± enclosed in calyx **7**

7 Lvs all similar in size – perianth fused for $\geq \frac{1}{3}$ – achene 1.5–2.5 mm long with 2 concave and 1 convex sides .... *Polygonum arenastrum* (p 166 **E.2**)
Lvs of side stems much smaller than those on main stem – perianth fused for $\leq \frac{1}{4}$ – achene 2.5–4.5 mm long with 3 concave sides...**8**

8 Lvs ovate-elliptic – achene < 3.5 mm long – widespread ....*P. aviculare* (p 166 **E**)
  Larger lvs ± obovate – achene >3 mm long – Scot only..................................*P. boreale* (p 166 **E.1**)

9 Stipules with 4–6 unbranched veins, short – fr 4–5 mm, longer than calyx – plant not
  woody – lvs 1.0–3.5 cm, elliptical-lanceolate ....*P. oxyspermum* (p 166 **F**)
  Stipules with 8–12 branched veins, extending to base of next lf above – fr 5–6 mm, longer
  than calyx – plant woody – lvs 2–5 cm, oval-elliptical, edges rolled under..................................
  P. maritimum (p 166 **G**)

10 Infl usually unbranched, erect – stamens always projecting **11**
  Infl branched – stamens not projecting...**13**

11 2 stigmas ....*Persicaria amphibia* (p 168 **F**)
  3 stigmas **12**

12 Lf-stalks unwinged – lower part of infl bearing purple bulbils ....*P. vivipara* (p 168 **E**)
  Lf-stalks of lower stem lvs and basal lvs winged – no bulbils in the infl *P. bistorta* (p 168 **D**)

13 3 stigmas **14**
  2 stigmas **15**

14 Fr >2x as long as perianth – lvs arrow-shaped ....*Fagopyrum esculentum* (p 170 **F**)
  Fr no longer than perianth – lvs ovate to lanceolate ....*Persicaria wallichii* (p 170 **A**)

15 Infl dense with fls crowded and overlapping **16**
  Infl open with fls not or hardly overlapping..**17**

16 Fl-stalks with numerous yellow glands – ochrea hairless or with short fringe ....
  P. lapathifolia (p 168 **B**)
  Fl-stalks with few or no glands – ochrea with long fine hairs ....*P. maculosa*(p 168 **A**)

17 Fr matt, 2.5–3.8 mm, brown or occasionally blackish – pink in infl pale or greenish ....
  P. hydropiper (p 168 **H**)
  Fr shiny, 1.8–2.5 mm, black – pink in infl bright ....*P. minor* (p 168 **C**)
  Fr very shiny, 2.8–4 mm, black – pink in infl bright ....*P. mitis* (p 168 **I**)

▼ **A** **Wood Dock**, *Rumex sanguineus*, erect per to 100 cm, with branches at an acute
  angle (<30°) to stem; lvs oval-lanceolate, their bases rounded, rarely with red
  veins. Infl much branched, lfy **only at base**; whorls of fls distant. Inner tepals (**Aa**)
  3 mm, oblong, **untoothed**, blunt, **one** only having a round wart 1.5 mm across (the
  others with none, or smaller ones); fr 1.2–1.8 mm long. GB N to mid Scot, Ire, c; in
  wds, hbs, wa. Fl 6–8.

▼ **B**   **Clustered Dock**, *R. conglomeratus*, is close to A, except that branches make a
  wider (30°–90°) angle with the wavy stem; infl **lfy nearly to top**; and **all three**
  untoothed, blunt, inner tepals (**Ba**) have **oblong** warts 1.2–1.8 mm long. Fr 1.8–2.0
  mm long. Br Isles c (except Scot r-lf); on rds, hbs, wa, mds. Fl 7–9.

▼ **C**   EU Sch 8 BAP EN ***Shore Dock**, *R. rupestris*, has grey-green, oblong blunt lvs, inner
  tepals (**C**) 3–4 mm long, oblong, untoothed, blunt, **all three** with **oblong** warts
  2.5–3.0 mm long covering **2/3 length** of the inner tepal; branches strictly erect.
  SW Eng, Wales, IoS, CI r to lf; on rocky shores, dunes. Fl 6–8.

D Broad-leaved Dock, *Rumex obtusifolius*, stouter and more robust than A and B, up to 120 cm tall; lower lvs (**Db**) oval-oblong, blunt, only slightly wavy-edged, to 25 cm long, with **cordate** bases; upper lvs narrower. Infl branched, lfy below, whorls of fls separate. Inner tepals (**Da**) 5–6 mm long, triangular, at least **one** of them with a **large wart**, edges with 3–5 long, **coarse** teeth (rarely no or short teeth). Fr 3 mm. Br Isles, vc in wa, hbs, rds, field edges. Fl 6–10.

E NT *****Monk's-rhubarb**, *R. alpinus* (*R. pseudoalpinus*), has v large round, cordate-based lvs, 20–40 cm long and wide; inner tepals (**E**) untoothed, without warts. N Eng, E Scot, o near houses. Fl 7.

F Curled Dock, *R. crispus*, resembles D in build, but has oblong-lanceolate lvs (**F**), **rounded** or **tapered** to bases, with strongly **crisped** edges; inner tepals (**Fa**) 4–5 mm long, **without** teeth and usually with **warts on all three** (sometimes only one wart). Br Isles, vc on wa, rds, hbs, ar, shingle beaches by sea. Fl 6–10.

G Northern Dock, *R. longifolius*, stout erect per to 120 cm, with wavy **crisped** lvs as in F, though cordate-based; inner tepals (**G**) **6 mm long, with cordate bases, without warts**; fr-stalk jointed. N half GB lc; on riversides, wa. Fl 6–7.

H Marsh Dock, *R. palustris*, erect branched ann or per to 60 cm, with linear-lanceolate pointed lvs; lfy spreading infl-branches. Inner tepals (**Ha**) lanceolate, blunt, 3–4 mm long, all warted, **yellowish** in fr, with each edge bearing 2 or 3 **bristle-like teeth shorter** than tepals; outer tepals curved inwards. SE, E Eng, r-lf; on bare muddy gd by lakes and rivers. Fl 6–9.

I **Ire Golden Dock**, *R. maritimus*, v like H but whorls of fls much more crowded, fl-stalks **longer, v slender**. Inner tepals (**I**) **triangular**, 2.5–3 mm long, all warted, golden-brown in fr, with each edge bearing 2 almost hair-like teeth **longer** than tepals; outer tepals curved outwards. Br Isles: S, mid, E Eng, o-lf; Ire vr; habitat as for I, but sometimes coastal. Fl 6–9.

J Water Dock, *R. hydrolapathum*, v large erect per, to 200 cm; lanceolate-oval pointed lvs (**Jb**) to 100 cm long; much-branched erect infl has crowded whorls of fls, with the inner tepals (**Ja**) 6–8 mm long, triangular, truncate-based, **each** with a long wart, and a few **short** teeth at base. Fr 4 mm long. Eng f-lc, rarer N and W in Br Isles; in shallow water, ditches, riversides, fens. Fl 7–9.

K VU *****Scottish Dock**, *R. aquaticus*, is like J, but tepals (**K**) have no warts, and fr-stalks v slender, without joint. Scot; Loch Lomond area only; by lake and stream shores. Fl 7–8.

L Patience Dock, *R. patientia*, an erect per to 2m, with large oval-lanceolate pointed lvs, with side veins at acute angle, <60° to midrib; lower lf-bases tapering to truncate; large fls with large inner tepal (**L**) in fr > 5 mm wide, oval or rounded, cordate-based, one with a large wart and **entire** margins without teeth. Introd (E Eur); about London and scattered in N Eng on wa. Fl 5–6.

M Greek Dock, *R. cristatus*, close to L but lvs with side veins at >60° to midrib; lower lf-bases **cordate**; large inner tepal in fr (**M**) with one large wart and **teeth**; most f of the large introd (SE Eur) Docks and probably overlooked; S Eng vla around London, also f in Bristol area, S Wales; on wa, riverbanks, dunes. Fl 6–7.

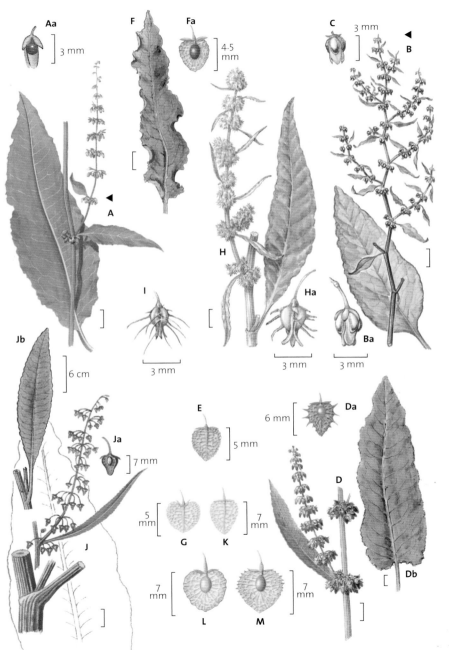

Aa

3 mm

F

Fa

4·5 mm

C

3 mm

B

A

Jb

6 cm

H

Ha

I

3 mm

3 mm

3 mm

Ba

Ja

7 mm

J

E

5 mm

Da

6 mm

G

5 mm

K

7 mm

D

Db

L

7 mm

M

7 mm

A Fiddle Dock, *Rumex pulcher*, per, distinct in its branches spreading at up to 90° from the main stem, and in its violin-shaped lvs (**Ab**), up to 10 cm long, with a 'waist'; whorls of fls distant, inner tepals (**Aa**) all with warts and each tepal-edge bearing 3 or 4 short broad teeth. S, E Eng o-lf; on dry gsld, rds on sand or limestone, especially near coast. Fl 6–7.

B Common Sorrel, *R. acetosa*, erect hairless per, 30–80 cm tall, more slender than other Dock spp; lvs shiny, **arrow-shaped**, to 10 cm long, 2–6 times as long as wide, stalked below, stalkless and clasping stem above. Tubular brown fringed stipules above lvs are obvious. Branched infl has loose whorls of small reddish fls (female **Ba**, male **Bb**) with cordate inner tepals without warts; plant dioecious. Br Isles, vc in gslds, open wds. Fl 5–6.

C Sheep's Sorrel, *R. acetosella*, appears as much smaller version of B, 10–30 cm tall, with narrower lvs, **hastate-shaped**, to 4 cm long, their side-lobes spreading or pointing forwards (not backwards like B), all lvs stalked. Fls (female **Ca**, male **Cb**) reddish; inner tepals **not enlarging in fr**. Br Isles, vc on dry acid gsld, hths, shingle beaches, poor ar. Fl 5–8.

D Mountain Sorrel, *Oxyria digyna*, resembles B in habit, but has fleshy, **kidney-shaped** lvs, 1–3 cm long, infl leafless, and 4 wartless tepals, only the 2 inner enlarging around **flat**, broadly-winged fr (**Da**). Br Isles: N Wales, Cumbria, Ire r; Scot lc; in damp rocky places on mts. Fl 7–8.

E Knotgrass, *Polygonum aviculare*, low, prostrate (sometimes erect), hairless ann, forming patches up to 1 m across, usually much less. Lvs ovate, elliptical to linear or lanceolate, **shorter** on flowering than on main stems. Stipules silvery, with fringed edges; veins not clearly seen; infl of 1–6 fls in lf-axils; calyx of 5 tepals, with green bases and **white** (**striped green**) petal-like lobes above, fused for $\leq$ 1/4 of their length. Fr (**Ea**) an achene, 1.5–3.5 mm long, three-angled, pointed, brown, not shiny, enclosed in persistent calyx. Seeds with **3 equal** concave sides (use hand lens). Br Isles, vc on wa, ar, seashores, etc. Fl 7–10. **E.1** *****Northern Knotgrass**, *P. boreale*, is v close to E but larger lvs obovate and achenes >3 mm long. Much more c than E in and confined to **Scot only**. Habitat and flg as E. **E.2 Equal-leaved Knotgrass**, *P. arenastrum*, as E but **key differences** are: all lvs $\pm$ **equal** in length; petal-like lobes fused $\geq$ 1/3 of their length and in shape of seeds (use hand lens) having **2 convex** and **1 concave** sides. Habitat and flg as E. (**Cornfield Knotgrass**, *P. rurivagum* is v close to and best amalgamated with *P. aviculare*).

F Ray's Knotgrass, *P. oxyspermum* (*P. raii*), prostrate like E, has grey-green elliptical-lanceolate lvs (**F**), 1.0–3.5 cm long with slightly turned-back edges; stipules silvery, with 4–6 **unbranched** veins; frs (**Fa**) shiny, three-angled (use hand lens), chestnut brown, 4–5 mm long, **longer** than persistent calyx which has broad white margins. S, W coasts of GB; coasts of Ire o-lf; on sand or fine shingle beaches. Fl 7–10.

G Sch 8 **Ire VU ****Sea Knotgrass**, *P. maritimum*, prostrate per, with stem woody at base; lvs 2–5 cm long, waxy, grey-green, oval to elliptical-lanceolate, with edges rolled-back underneath. Frs (**Ga**) chestnut-brown, 5–6 mm long, **longer** than white-edged calyx; silvery stipules (**Gb**) have 8–12 **branched** veins. S Eng, CI, S Ire vr but apparently expanding its range; on coastal sand and shingle beaches. Fl 4–10.

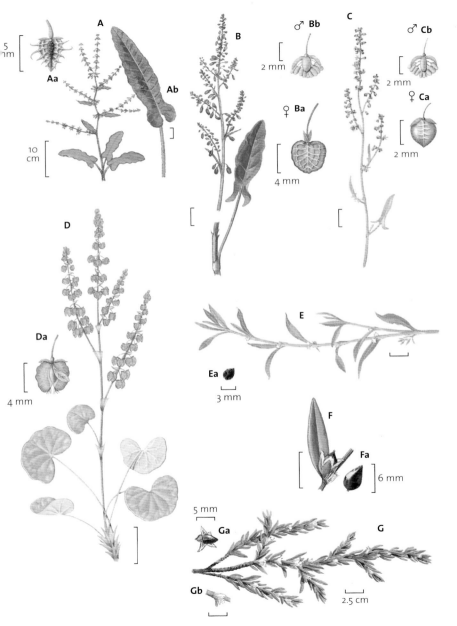

A **Water-pepper**, *Persicaria* (*Polygonum*) *hydropiper*, ± erect, ± hairless ann, with branched stem to 70 cm; lvs 5–10 cm, thin, **lanceolate**, ± stalkless; stipules truncate, hardly fringed; infl slender, pointed, often nodding, lfy below; fls (**Aa**) pinkish or greenish-white; **2 stigmas**; tepals in fr (**Ab**) covered with many dense, yellow **raised gland-dots** (use hand lens). Fr 2.5–3.8 mm long, oval, dark brown, matt, biconvex. Br Isles (except N Scot r), c; in damp gd, mds, paths, shallow water. Fl 7–9.

B VU *Tasteless Water-pepper, *P. mitis* (*Polygonum mite*), is close to A, but infl is slenderer, fewer-fld, more pointed and drooping; lvs 12–30 mm wide; fls (**B**) pink, 3–4 mm long; sometimes **no** gland-dots, if present then few sparse, **flat**; lvs **never** black-blotched. Fr 2.8–4 mm, biconvex, shiny. Br Isles (except Scot), r in ditches, pond edges. Fl 6–9.

C **Redshank**, *P. maculosa* (*Polygonum persicaria*), branched, ± erect, ± hairless ann to 70 cm, with reddish stems swollen above the lf-bases; ochrea fringed **with long, fine hairs**; lanceolate, often **black-blotched** lvs, 5–10 cm long. Infls (**Cb**) stout, blunt, fls pink with **2 stigmas**; usually (but not always) **without** yellow gland-dots on calyx. Fr (**Ca**) 2–3 mm long, shiny, **biconvex** or **triangular**. Br Isles, c in wa, ar, wet places. Fl 6–10. **Do not confuse** with other *Persicaria* spp which may also sometimes have black-blotched lvs.

D **Pale Persicaria**, *P. lapathifolia* (*Polygonum lapathifolium*), larger than C, ann, slightly hairy, with a few gland-dots on calyx; lvs often **black-blotched** as in C, but infl stouter; fls (**Da**) normally greenish-white. **Key difference** from C in ochrea **hairless** or with **short** fringe and fr 2–3 mm, **biconcave**, never triangular, 2–3 mm. Br Isles: S of GB c, Scot o-f; Ire f; in wa, ar, wet places. Fl 6–10.

E VU **Small Water-pepper**, *P. minor* (*Polygonum minus*), resembles C, but lvs only 2–15 mm wide, **never** black-blotched; infl narrow; fls pink; calyx with **no** gland-dots; fr (**E**) 1.8–2.5 mm, oval, shiny. Br Isles, r on bare wet gd, pond edges. Fl 8–9.

F **Common Bistort**, *P.* (*Polygonum*) *bistorta*, erect, **± hairless unbranched** per, 30–100 cm tall; at least lower stem and basal lvs with **winged** lf-stalks; basal lvs broad-oval, blunt, with ± cordate bases; upper lvs triangular, pointed; stipules fringed. Infl terminal, a stout, dense blunt spike (10–15 mm wide) of **pink** fls, 4–5 mm across, with **3 stigmas**. Br Isles: GB widespread, but only c in N Eng, Wales, r-o elsewhere; Ire r; in damp places on neutral or acid soils in mds, rds, often wds. Fl 6–8.

G FPO (Eire) **Ire **Alpine Bistort**, *P. vivipara* (*Polygonum viviparum*), shorter than F, only 6–30 cm tall; lower lvs linear-lanceolate, **tapering** into **unwinged** stalks, upper lvs stalkless; infl as in F, but more slender (4–8 mm wide), of white 3–4-mm fls above, purple bulbils replacing fls in lower part. Br Isles: mt areas of N Eng, Scot, lf-lc; N & W Ire, N Wales vr; on mt gslds, moist cliffs. Fl 6–8.

H **Amphibious Bistort**, *P. amphibia* (*Polygonum amphibium*), hairless or bristly hairy branched **per**, either floating in water or ± erect on dry gd, 30–70 cm; lvs 5–15 cm, oval to oblong-lanceolate, blunt or pointed, base cordate and lvs **± hairless** in floating form, rounded and **hairy** in land form; may have **black-blotched** lvs. Infl 2–4 cm long, dense, many-fld, terminal; infl-stalk lfless above; fls pink, 2–3 mm long, no gland-dots; stamens 5, protruding; fr 2 mm, biconvex, brown, shiny. Br Isles, f in fresh water, wa, banks near water and on ar. Fl 7–9. **Key differences** from other c spp are **2 stigmas** in most fls, **per** plant and hairy lvs (except when growing in water).

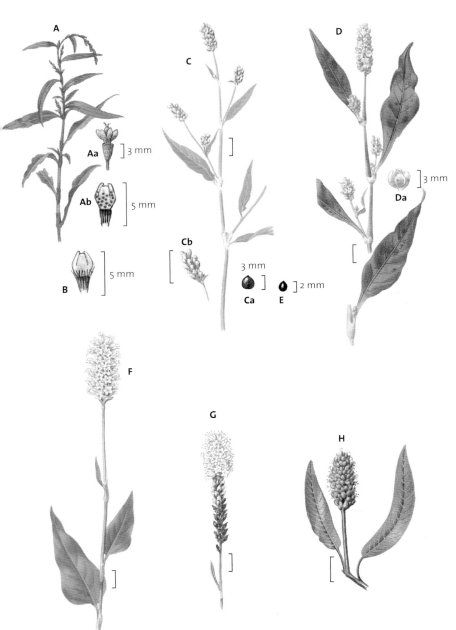

A Himalayan Knotweed, *Persicaria wallichii*, a tall per herb to 1.8 m with rhizomes; lvs **ovate to lanceolate** and **cordate or truncate** to base, variably hairless to densely hairy on lower side; fls in loose panicle, white or pinkish, 5-lobed, the outer 2 perianth segments much narrower than inner; fr a smooth three-angled nut. Introd (Himalayas); Br Isles, o-vla gdn escape natd in gslds, rds, wa. Fl 8–9.

B Black-bindweed, *Fallopia convolvulus* (*Polygonum convolvulus*), twining, scrambling or climbing ann, with stems to 1 m or more; stem angular, mealy; lvs 2–6 cm long, oval-triangular with cordate or truncate bases, v like those of Hedge Bindweed (p 354 D), but mealy below. Fls in long racemes, each fl on 1–2 mm stalk with joint above middle; the 3 outer sepals 5–6 mm long, greenish-grey, with white keels or narrow wings in fr; fr (**Ba**) 4–5 mm, **dull** black, enclosed in sepals. Br Isles (except NW Scot), c ar ar, wa, hbs, wd borders. Fl 7–10.

C vu ***Copse-bindweed**, *F. dumetorum* (*P. dumetorum*), closely resembles B, but may be taller, ann; lvs (**Cb**) as in B; fr-stalks up to 8 mm long, jointed below middle; 3 outer sepals have broad, wrinkled, membranous pinky-white wings (**Ca**) (although a variety of B also has wings). **Key difference** from B is **fr**, 2.5–3 mm, **shiny**. SE and C Eng r; climbing in hbs, open wds. Fl 7–10. **C.1 Russian-vine**, *F. baldschuanica*, similar to C but more rampant, **per** climber to 10 m with stems **woody** below; lvs oval-triangular with heart-shaped bases; showy **white** fls in **branched** infls; fr dull brown nut. Introd (C Asia); Br Isles: S Eng, Wales f, Ire, Scot o-r; natd in hds, cliffs, scrub. Fl 7–10.

D Japanese Knotweed, *F. japonica* (*Reynoutria japonica*), tall erect hairless per to 2 m, spreading by rhizomes to often form thickets; broad **oval-triangular** pointed lvs, with lf-base **truncate** and zigzag reddish stems; fls white, in branched spikes; fr encloses a three-angled dark-brown glossy nut. Introd (Japan), originally as a gdn plant, highly invasive; Br Isles, c (except for N Scot) on wa, railways, rds, streamsides. Fl 8–10.

E Giant Knotweed, *F. sachalinensis*, a taller version of D to 3 m or more with lf-bases (**E**) strongly **cordate**; fls greenish-white. Introd (Japan); Br Isles, o-vla in similar places to D. Fl 8–9. Hybridizes with D, (*F. x bohemica*) plants usually having both truncate and more rounded (but not strongly cordate) lf-bases; overlooked, especially in Ire and probably widespread.

F Buckwheat, *Fagopyrum esculentum*, erect ann to 60 cm; lvs **arrow-shaped**, upper clasping stem, lower stalked; fls in clusters, 5-lobed, cream edged pink, 8 stamens and 3 stigmas; fr a three-angled matt brown nut. Introd (C Asia) and formerly much cultivated; Eng o, Scot, Ire r; casual on wa, wdland rides, field margins. Fl 7–9.

G ****Iceland-purslane**, *Koenigia islandica*, superficially resembles **Blinks** (p 139 C). Prostrate ann, 1–6 cm long; stems red; lvs opp, rounded-obovate, elliptical, 3–5 mm long, with membranous stipules 1 mm long. Fls in tiny clusters at stem-tips; calyx of 3 pale green, rounded sepals 1 mm long; 3 tiny red-anthered stamens alt with 3 yellow glands; fr three-angled with 2 stigmas. Scot: **Skye, Mull only**, vr on stony moist mt tops (arctic plant). Fl 7–8.

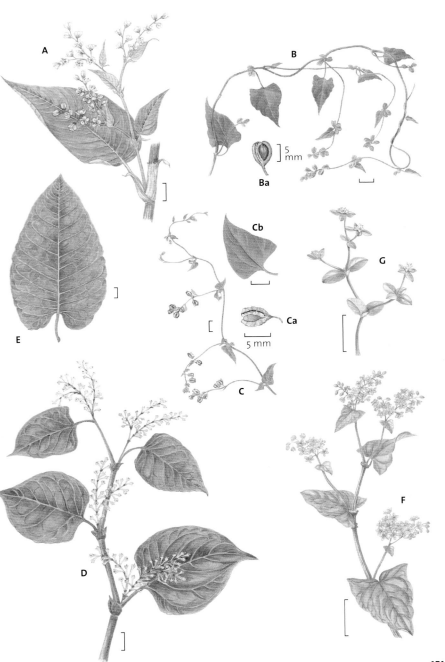

A

B

Ba

]5 mm

Cb

Ca

5 mm

G

E

D

C

F

SEA-LAVENDER FAMILY *Plumbaginaceae*

Per herbs with alt, simple lvs, in basal rosettes only; infl a dense round head in **Thrift** (*Armeria*) or a branched panicle in **Sea-lavenders** (*Limonium*). Fls with funnel-shaped, 5–10-ribbed, papery, often coloured calyx, persisting in fr; corolla of 5 petals joined only at base, 5 stamens opp petals, 5 separate styles. Fr a one-seeded capsule. **Sea-lavenders** are a v difficult and controversial group and many spp are vr, so only the more widespread broad groups are included here.

A   Thrift, *Armeria maritima* ssp *maritima*, cushion-forming per; woody rootstock bears rosettes of narrow-linear, ± hairless, one-veined lvs, 2–10 cm long. Infl stalks, 5–30 cm tall, ± downy, bear no lvs but terminal rounded heads (1.5–2.5 cm across) of pink or white fls; a brown, sleeve-like sheath, toothed at lower end, extends 2–3 cm down stem from fl-head. Fls five-petalled, 8 mm across; calyx five-ribbed, hairy. Br Isles, c all round coasts on cliffs and salt marshes, inland on mt ledges. Fl 4–10. A taller (to 50 cm) form (CR**ssp *elongata*) occurs inland on lowland dry gsld in Lincs, vr.   **A.1 Jersey Thrift**, *A. arenaria* (*A. plantaginea*), has narrow-**lanceolate** 3–5-veined lvs and spine-like calyx-teeth. **Jersey only**, vl on dunes. Fl 6–9.

B   Common Sea-lavender, *Limonium vulgare*, per with wdy rootstock, bearing rosettes of ascending, long-stalked, elliptical-lanceolate, 4–12 cm-long lvs, with **pinnate** veins and a tiny spine on tips. Infl **rounded** or **flat-topped**, lfless, 8–40 cm tall, branching only above middle, bearing **dense**, arching spikes of fls, gathered in small clusters; each cluster with green bracts, outer one **rounded** on back. Papery, pale lilac, ribbed calyx bears 5 sharp teeth and 5 smaller teeth between. Corolla 8 mm across, with 5 pale purplish-blue petals; anthers **yellow**. Br Isles: coasts of Eng, Wales f, lc in S and E; Scot vr; **Ire abs**; in salt-marshes. Fl 7–8.

C   *Lax-flowered Sea-lavender, *L. humile*, close to B, also has **pinnate** lf-veins, tall infl, and calyx with small intermediate teeth; but lvs **narrower**, infl divides **below middle** into long erect branches on which fl-clusters are **widely spaced-out**; outer green bracts of clusters have **keeled** backs. Anthers **reddish-brown**. GB, more local than B, but o-lc N to S Scot; Ire f; in salt-marshes. Fl 7–8. Both B and C are v variable and hybridize (*R.* x *neumanii*); so identifying the sp for certain can be v difficult and requires microscopic examination of the pollen and stigmas.

D   WO (NI) BAP **Rock Sea-lavender**, *L. binervosum* agg, a group of many v similar spp, (or arguably sspp) smaller, usually shorter (8–20 cm) than B and C; lvs shorter (2–5 cm) obovate, with 3 veins, **winged** stalks and **no** pinnate lf-veins. Erect infls branch low down, and fl-clusters more spaced out along branches than in B. Coasts of Eng, Wales lc; S Scot vr; Ire o-lf; on sea cliffs, shingle drier salt-marshes. Fl 7–9.   **D.1 Matted Sea-lavender**, *L. bellidifolium*, low per resembling D in its short (1.5–4.0 cm) obovate lvs with three-veined base and no pinnate veins; but **key differences** from D are stalks slender, lvs blunt, and almost prostrate infls fork **repeatedly** into many, **mostly fl-less**, branches; fl-clusters only on **uppermost** branches, their **outer** bracts **white, papery**. Corollas 5 mm across, pale pink. Eng: **Norfolk, Lincs only**, vl but lf; on dry sandy upper parts of salt-marshes. Fl 7–8.   **D.2   Broad-leaved Sea-lavender**, *L. auriculae-ursifolium*, robust plant like B but, like D, has no pinnate veins in lvs; lvs longer (to 10 cm) than in D, broader, spoon-shaped, with **broadly-winged** 5–9-veined stalks. Tall (to 45 cm) infls crowded with fl-clusters as in B but inner fl-cluster bracts have bright **red** papery margins. Jersey only, vl rocks by sea. Fl 7–8.

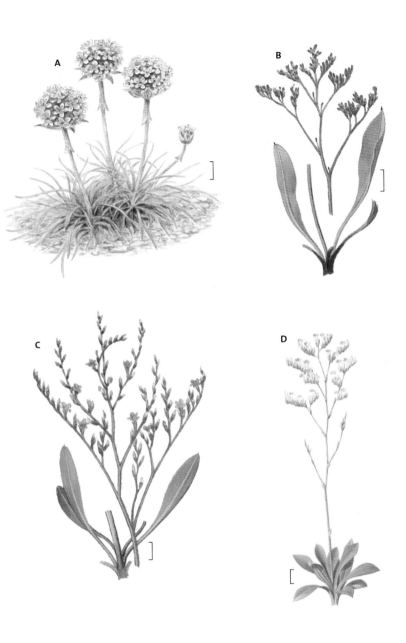

173

WATERWORT FAMILY

Elatinaceae

A Six-stamened Waterwort, *Elatine hexandra*, an ann with short-stalked undivided strap-shaped lvs, widened above, in whorls of 4 or in opp pairs; tiny **stalked three**-petalled and **three**-sepalled pinkish fls (**Aa**) in lf-axils. Br Isles, v scattered and o -vla; in ponds and exposed mud pond-margins with acid peaty water. Fl 7–9.

A.1 WO (NI) *****Eight-stamened Waterwort**, *E. hydropiper*, is v like A, but **key difference** is fls are **stalkless** with **four** sepals and petals. Similar habitat to A, but vr in Scot, N Ire, Wales in ponds and lakes. Fl 7–8. *Elatine* spp often grow partly submerged in silt on the bed of water bodies, with only paired lvs showing.

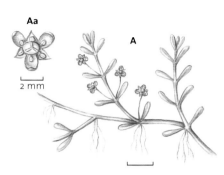

Aa

A

2 mm

ST JOHN'S-WORT FAMILY

Clusiaceae

Also known as the *Guttiferae*, per herbs (or shrubs, F p 176, G p 178); lvs opp, untoothed, stalkless, simple, with translucent veins; fls in branched cymes (or solitary, G p 178); fls yellow, 5 sepals and 5 petals, many stamens; frs capsules (except F, F.1 and F.2 p 176, berry-like) with 3 or 5 cells.

ID Tips   St John's-worts

• **What to look at**: **(1)** stem shape, whether round, square or with 2 ridges **(2)** whether lvs hairy or hairless **(3)** whether lvs have white, translucent dots when held up to the light **(4)** length of sepals in relation to petals **(5)** presence or absence of black dots on lvs, sepals and /or petals.

• **Irish St John's-wort** is the only *Hypericum* with stamens in a **continuous ring**, not grouped in bundles.

• **Marsh St John's-wort** looks unlike any other *Hypericum* spp and is the only **hairy** *Hypericum* with **rooting** nodes.

KEY TO ST JOHN'S-WORTS (*HYPERICUM*)

1 Low shrubby plants – stamens in 5 bundles **2**
Herbs, rarely woody and at base only – stamens in 3 bundles **5**

2 Creeping woody rhizomes – evergreen lvs – styles 5................*Hypericum calycinum* (p 178 **G**)
Erect, no rhizomes – deciduous lvs – styles 3 ....**3**

3 Petals >15mm – sepals falling soon after flowering – lvs usually smelling unpleasant, goat-like when crushed ....*H. hircinum* (p 176 **F.1**)
Petals <15 mm – sepals persistent – lvs often scented but no strong unpleasant goat-like smell **4**

4 Petals shorter than or equal to sepals – fr succulent when ripe ......*H. androsaemum* (p 176 **F**)
Petals longer than sepals – fr dry when ripe ....*H.* x *inodorum* (p 176 **F.2**)

5 Plant downy, at least on undersides of lvs...**6**
Plant hairless...**8**

6 Lvs hairless above, ± downy below...*H. montanum* (p 176 **E**)
Lvs hairy on both sides...**7**

7 Bog plant with runners, rooting at lower stem nodes – lvs round to oval, woolly – red dots on sepal margin ....*H. elodes* (p 178 **H**)
Dry-gd plant with erect stems – lvs oblong to elliptical, only downy – black dots on sepal margin ....*H. hirsutum* (p 176 **D**)

8 Stems with 4 angles (often 2 weak and 2 strong) or 4 wings **9**
Stems round, or with 2 raised ridges only..**12**

9 Petals shorter than, or equalling sepals – stem with four broad wings >0.25 mm wide **10**
Petals 2–3 times length of sepals – stem narrow four-winged <0.25 mm wide or square in section **11**

10 Stems to 20 cm, v slender – lvs narrow-elliptical to 0.5 mm wide, without black dots – petals golden yellow ....*H. canadense* (p 176 **C.2**)
Stems 30–60 cm, stout – lvs broad oval >0.5 mm wide, usually with black dots – petals pale yellow ....*H. tetrapterum* (p 176 **C**)

11 Sepals unequal, blunt – stem square, not winged – lvs narrowed to base – translucent lf-dots few or none..*H. maculatum* (p 176 **B**)
Sepals ± equal, pointed – stem with four narrow wings <0.25 mm wide – lvs wavy-edged, clasping stem – translucent lf-dots many...*H. undulatum* (p 176 **C.1**)

12 Stems creeping to ascending – petals not much longer than sepals – sepals unequal, 3 longer and wider than other 2 ....*H. humifusum* (p 178 **J**)
Stems erect – petals at least twice length of sepals – sepals ± equal **13**

13 Lvs of main stem widest at base, almost clasping stem – sepals oval, blunt ....
H. pulchrum (p 178 **I**)
Lvs of main stem narrowed to base – sepals narrow, pointed...**14**

14 Sepals without fringe of stalked black glands – many translucent lf-dots – oval lvs ....
H. perforatum (p 176 **A**)
Sepals with fringe of stalked black glands, no or v few translucent lf-dots – linear lvs ....
H. linarifolium (p 178 **K**)

A Perforate St John's-wort, *Hypericum perforatum*, erect per herb, 30–90 cm; round stems with **two** opp raised ridges; lvs (**Aa**) hairless, elliptic to oblong, blunt, 1–2 cm, with many **translucent dots** (hold up to light). Golden-yellow fls, 20 mm across, with black dots on petal edges; **pointed sepals**. Br Isles, vc (except N Scot, W Ire o) in scrub, hbs, rds, gsld, especially on chk and sand. Fl 6–9.

B AWI **Imperforate St John's-wort**, *H. maculatum*, is like A, but has **square, unwinged** stems (**Bb**), **no** translucent leaf dots (**B**), **blunt sepals** (**Bc**) and fls (**Ba**) with black dots on petal surfaces. Br Isles: Wales vc; elsewhere o-lf; in wd-borders, hbs, rds, on damp or heavier soils. Fl 6–8. Hybridizes with A (*H.* x *desetangsii*).

C Square-stalked St John's-wort, *H. tetrapterum*, resembles B, but has **square, four-winged,** red stems (**Ca**), oval lvs with **translucent dots**, paler fls (**C**), 10 mm across, narrow **pointed sepals** (**Cb**) without black dots. Br Isles, c (except N Scot, r) in damp mds, marshes, by water. Fl 6–9. **C.1** *Wavy St John's-wort**, *H. undulatum*, has four-winged stems like C, but differs in red-flushed fls, **wavy-edged** lvs, pointed oval sepals **with black dots**. Br Isles: SW Eng, Wales; r and vl in marshes. **C.2** FPO (Eire) **Ire **Irish St John's-wort**, *H. canadense*, has four-angled stems like C, but has slender erect habit of I p 178; narrow, linear, three-veined lvs; petals, 3–4 mm across, have crimson lines on backs, and spread like stars. **Ire only**; vr on wet peaty gd. Fl 7–9.

D FPO (Eire) **Ire AWI **Hairy St John's-wort**, *H. hirsutum*, resembles A in build and round stem, but is **downy** on lvs (**Db**) and stem; lvs oblong, much longer (2–5 cm), and **strongly veined**, with translucent dots; petals (**Da**) **pale yellow**; pointed sepals have **stalked black dots** on edges. Br Isles: GB c, vc to S (but IoS, CI abs); N & E Ire r; on gsld on chk and clay, scrub, open wds, rds. Fl 7–8.

E NT AWI **Pale St John's-wort**, *H. montanum*, has the stiff erect habit, pale yellow fls, longer lvs and round stem of D, but its oval-oblong lvs are **hairless** on upper sides, hairy on lower-sides and **lack translucent dots**; black dots on lf-edges beneath; sepals with stalked black dots on edges as in D. Br Isles: Eng, Wales, r-lf; **not Scot or Ire**; on scrub, wds, hbs, on calc soil, especially where sand or gravel overlies thinly. Fl 6–8.

F AWI **Tutsan**, *H. androsaemum*, hairless low deciduous shrub, 40–100 cm, with **two-edged**, usually red, stems; oval lvs, 5–10 cm long, **± stalkless**, aromatic when bruised; fls few together, **petals < 15 mm** long, **equal to or shorter** than sepals; stamens c. **as long as** petals; **3** styles; sepals blunt, unequal, oval, **persistent** on frs. Fr (**Fa**) a red berry turning black, 5–8 mm wide. Br Isles: c in S, W, CI and Ire as a native; f elsewhere as gdn escape; in wds, hbs, cliffs (avoids v acid soils only). Fl 6–8. **F.1 Stinking Tutsan**, *H. hircinum*, similar to F but with **square** stems; lvs usually (but not always) smelling unpleasant, goat-like, when crushed; **petals > 15 mm** long; stamens **slightly longer** than petals; sepals soon **falling off** fl or before fr ripens. Introd (Mediterranean); Br Isles: r gdn escape, lf around London; in shady places and wa. Fl 6–8. **F.2 Tall Tutsan**, *H.* x *inodorum*, fertile hybrid of F and F.1, with slightly to strongly aromatic lvs when crushed (despite its name). Closest to F in petal size and sepals persistent but petals **longer** than sepals; stamens c. **1.25 x** as long as petals; fr red berry turning brown. Br Isles: S, SW Eng o-f, scattered elsewhere; popular gdn plant and probably overlooked; in damp, shady places on hbs, streamsides. Fl 6–8.

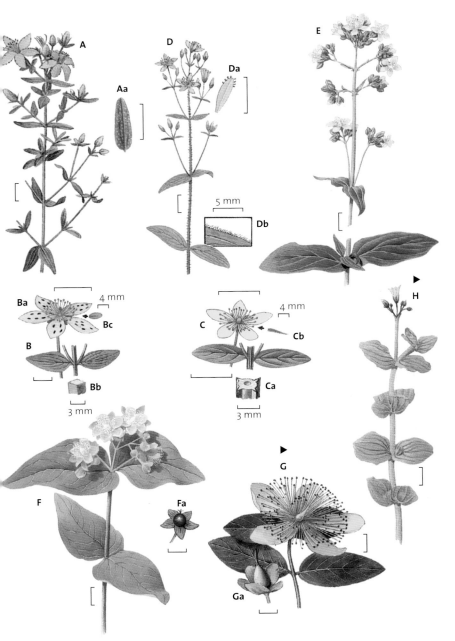

▲ **G   Rose-of-Sharon**, *Hypericum calycinum*, creeping low **evergreen** shrub up to 60 cm, with four-angled stems and elliptical lvs 5–10 cm on **short** lf-stalks; fls **solitary**, v large, 7–8 cm across; **5** styles; fr (**Ga**). Introd (SE Eur); Br Isles, f gdn escape in wds, railway banks, rds, by gdns. Fl 7–9.

▲ **H   Marsh St John's-wort**, *H. elodes*, creeping **grey, woolly**-hairy per herb, **rooting** at lower stem nodes; almost circular to oval lvs, and upright fl-stems bearing groups of yellow fls, 15 mm across, **less open** than in other spp; sepals have **red dots** on margin. Br Isles: lc in W (not C Scot), scattered elsewhere; by and in acid bog pools and hollows, wet hths. Fl 6–9.

I   AWI **Slender St John's-wort**, *H. pulchrum*, hairless, strictly erect, slender, 20–40 cm; round reddish stems and blunt oval lvs with **cordate** bases and translucent dots; fls 15 mm; petals, orange-yellow, red-dotted, reddish below, have black dots on edges, as do sepals. Br Isles f-lvc; on gsld, scrub, dry wds, usually on non-calc soils. Fl 6–8.

J   Trailing St John's-wort, *H. humifusum*, **procumbent, creeping** hairless per, with **two-ridged** stems (**Ja**) and elliptic-oval, gland-dotted lvs 0.5–1.5 cm long. Fls few, 10 mm across, petals slightly longer than sepals; sepal margins with no or few stalked black dots (**Jb**). Br Isles, c on hths, open acid wds. Fl 6–9.

K   NT ****Toadflax-leaved St John's-wort**, *H. linariifolium*, resembles J, but at least central flowering stem **erect**, with **round** stems (**K**); lvs longer (1.0–2.5 cm), linear, almost without translucent dots; petals **more than twice** length of sepals, which have fringes of many stalked black dots (**Ka**). Br Isles: SW Eng, Wales, CI vr; on acid rocks. Fl 6–7.

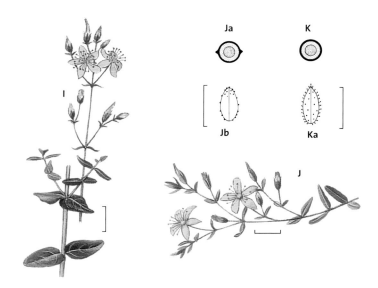

LIME (LINDEN) FAMILY   *Tiliaceae*

Limes (*Tilia*) are trees with alt, long-stalked, cordate lvs; blunt buds; fragrant fls in small clusters whose common stalk is fused halfway to an oblong bract; globular nut-like frs 6–10 mm long.

A Lime, *Tilia* x *europaea* (*T.* x *vulgaris*) is the hybrid of B and C. Large irregular bosses on trunk. Lvs 6–10 cm long, usually (but not always) with lopsided lf-base, **hairless** below, except for tufts of white **hairs in vein-axils**. Fls 4–10 per infl; fr (**Aa**) oval with a pointed tip. Br Isles, planted and vc, but rarely self-sown; also vr as a natural hybrid. Fl 7.

B *Large-leaved Lime, T. platyphyllos*, has smooth dark bark; lvs 6–12 cm long, **grey-downy** all over below, and sometimes above, sometimes with lopsided lf-base; side veins of lvs strong; infl drooping, with 2–5 fls; frs (**Ba**) strongly ribbed. Br Isles: Eng, SE Wales r but lf, N to Yorks, as native but also widely introd; in old wds on base-rich soils, and on wooded cliffs. Fl 6.

C ᴬᵂᴵ **Small-leaved Lime**, *T. cordata*, has a smooth bark usually without swollen bosses, lvs only 3–6 cm long, rounded, cordate, with abrupt points, grey-green, **hairless** below (except for **tufts** of **rusty hairs** in vein-axils), side veins **not** prominent. Infl 4–10-fld, not drooping; frs (**Ca**) smooth, with pointed tips. Br Isles: Eng, Wales o, only lc; Scot, Ire r; in old wds on fertile soils, wooded limestone cliffs. Fl 7.

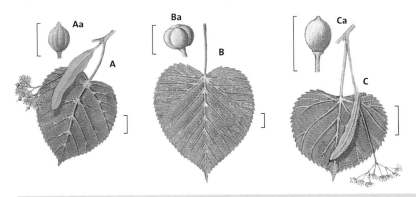

MALLOW FAMILY   *Malvaceae*

Softly hairy or downy plants with palmate, lobed, stalked, alt lvs, with stipules at bases. Fls 1–6 together in lf-axils, usually showy and open, with 5 sepals, 5 notched petals, and many stamens joined into a tube below; a superior ovary of many carpels joined into a flattened ring ripens into a fr that divides into many one-seeded segments. An epicalyx is present below the true calyx. In *Malva* the epicalyx is of 3 free segments; in *Lavatera* the 3 epicalyx segments are joined together; in *Althaea* and *Alcea* there are 6–9 epicalyx segments.

KEY TO MALLOW FAMILY

1 Epicalyx of 3 lobes **2**
Epicalyx of 6–9 lobes **3**

2 Epicalyx of free parts **5**
Epicalyx a cup of lobes joined below..**7**

3 Tall per herb – lvs velvety-downy **4**
Shorter (to 40 cm) ann or bi herb – stems bristly ....*Althaea hirsuta* (p 182 **E.1**)

4 Stamen–tube (pull off petals and sepals to view column formed by fused stamens) hairy
– petals 15–20 mm long ....*A. officinalis* (p 182 **E**)
Stamen–tube hairless – petals 25–50 mm long ....*Alcea rosea* (p 182 **F**)

5 Stem-lvs deeply-cut, with pinnately-cut lobes – fls rose-pink........*Malva moschata* (p 180 **C**)
Stem-lvs bluntly-lobed **6**

6 Per – nutlets with strong ridges, netted ....*M. sylvestris* (p 180 **A**)
Ann – nutlets smooth ....*M. neglecta* (p 180 **B**)

7 Fl 1 in each lf-axil – plant usually (but not always) sterile......*Lavatera* x *clementii* (p 182 **D.2**)
Fl 2 or more in each lf-axil – plant fertile **8**

8 Bushy plant, 60–300 cm, woody below – epicalyx enlarged in fr – lvs velvety – nutlets
wrinkled ....*L. arborea* (p 180 **D**)
Plant 50–150 cm, bristly not woody below – epicalyx not enlarged in fr – nutlets smooth
L. cretica (p 180 **D.1**)

A Common Mallow, *Malva sylvestris*, robust **per** herb, 45–90 cm, erect or spreading; lvs sparsely hairy, palmately-lobed, the lobes shallowly-toothed. Fls stalked, in axillary clusters up the stem; petals, (12–)15–30 mm, rose-purple with darker veins, are 2–4 times length of sepals; fr (**Aa**) has segments sharp-angled, **netted**. Br Isles: c in S, rarer in N (especially r N Scot); on rds, wa, hbs. Fl 6–9.

B Dwarf Mallow, *M. neglecta*, prostrate downy **ann**; kidney-shaped to roundish lvs, v shallowly-lobed, 5–7 times. Fls (**Ba**) in clusters along stems; petals 8–13 mm, whitish with lilac veins, are only 2–3 times length of calyx; fr (**Bb**) with segments blunt-angled, **smooth**. Br Isles: SE Eng c, rarer N and W; on dry soils on wa, rds, hbs, beaches. Fl 6–9.

C Musk-mallow, *M. moschata*, erect per herb with sparse simple hairs; stem-lvs deeply palmately-cut into narrow pinnately-cut segments; basal lvs kidney-shaped with 3 shallow lobes, long-stalked. Fls, 3–6 cm, rose-pink, solitary in lf-axils and in a terminal cluster; epicalyx-segments v narrow. Br Isles: GB f-c (NW Scot r); Ire o-f; on rds, hbs, gslds, especially on sand and clay. Fl 7–8.

D Tree-mallow, *Lavatera arborea*, erect shrub-like bi, to 3 m, woody below, downy with star-shaped hairs above. Lvs rounded, 5–7-lobed, velvety; fls 3–5 cm, purplish-pink and purple-veined, in terminal racemes, 2 or more per lf-axil; epicalyx a three-lobed cup of broadly oval segments wider than calyx, especially in fr; nutlets wrinkled. Br Isles: S and W coasts of GB, N to Ayr, f; Ire o-f; on nutrient-rich soils on rocks and wa by sea. Fl 7–9. **D.1 Smaller Tree-mallow**, *L. cretica*, ann or bi herb to 150 cm, resembles A in appearance, but has the cup-like epicalyx of D, though with lobes shorter than the calyx and not enlarged in

A

Aa

B

Ba

Bb

C

D

E

F

181

fr; nutlets smooth. Br Isles: coasts of IoS, CI, o and W Cornwall (possibly introd), r; vr introd inland in Eng. Fl 6–7. **D.2 Garden Tree-mallow**, *L.* x *clementii* (*L. thuringiaca* x *L. olbia* in the *New Flora*), is probably the most widespread of escaped gdn tree mallows, a robust per shrub to 2 m (soft-wooded as is a hybrid of a herb and a woody shrub), usually **sterile**, with only **1 fl per lf-axil** rather than 2 or more in D and D.1, the flowers mostly snapping off leaving a persistent fl-stalk. Much grown in gdns and o escaping but usually short-lived on rds, wa. Fl 7–10.

▲ **E** *Marsh-mallow*, *Althaea officinalis*, erect velvety-downy per herb, 60–120 cm, with a slightly-branched stem bearing stalked, roundish to oval, slightly 3–5-lobed velvety-downy lvs, folded like fans above. Fls 2.5–4.0 cm across, 1–3 together in lf-axils; epicalyx a cup of 8–9 narrow triangular lobes; sepals velvety; petals pale pink; fr downy. Br Isles: coasts of S half of Eng and Wales o-lf; S Ire r; upper parts of salt- and brackish-marshes, and ditches, banks near salt water. Fl 8–9. **E.1** Sch 8 **Rough Marsh-mallow**, *A. hirsuta*, ascending **bristly** ann or bi herb, 8–60 cm, bristles have swollen bases; lower lvs stalked, kidney-shaped, bluntly five-lobed; upper lvs deeply palmately-cut. Fls cup-shaped, 2.5 cm across, their stalks long; sepals and epicalyx-lobes lanceolate; petals longer than sepals, broadest at tips, rose-pink. Possibly introd (Mediterranean); S Eng vr; on chky ar, wa, gsld. Fl 6–7.

▲ **F Hollyhock**, *Alcea rosea*, tall bi to per herb to 3m with 3-9 lobed palmate lvs, hairy beneath; fls in a terminal racemose infl, or clustered in axils of upper lvs, red, pink, yellow or white, **large petals** 25–50 mm; **hairless stamen-tube**, characteristic of all *Alcea* spp. Introd and f gdn escape in S Eng on wa, rds, railway banks. Fl 7–9.

SUNDEW FAMILY   *Droseraceae*

Sundews (*Drosera*) are small per herbs with undivided, long-stalked lvs, in basal rosettes, and covered with sticky, red-tipped, gland-bearing hairs. They trap, then digest the proteins in, small insects, probably to supplement their nitrate nutrition because their normal acid bog habitats are nutrient-deficient. Fls, 5 mm across, in erect sometimes forked infls, with 5–8 white petals, 5–8 sepals, 5–8 stamens, only open briefly in sunshine; frs tiny many-seeded capsules.

A Round-leaved Sundew, *Drosera rotundifolia*, has almost **circular** lf-blades, to 1 cm across, on long horizontally-spreading hairy stalks; infl lfless, 10–15 cm tall, sometimes forked; fls usually six-petalled. Br Isles, lc in wet acid peat on hths or among *Sphagnum* in bogs, decreasing through habitat destruction. Fl 6–8.

B Oblong-leaved Sundew, *D. intermedia*, has narrow **spoon-shaped lvs** (**Ba**) with blades to 1 cm long x 0.5 cm wide, narrowing **quickly** into long (1–3 cm) **erect**, **hairless** stalks; infl **barely longer** than lvs, arising from **below** rosette, upward-curving. Br Isles widespread but vl: W Scot, W Ire and hths of S Eng lc; on bare acid damp peat on hths or in bog pools also on *Sphagnum*. Fl 6–8. Hybridizes with A (vu *D.* x *belezeana*), hybrid with lf-shape closer to A but infl like B; sterile.

C ɴᴛ **Great Sundew**, *D. anglica*, has v narrow obovate lvs (**C**) widest below tips, **gradually** narrowing into v long stalks; blades to 4 cm long, stalks up to 12 cm; infl, to **twice** length of lvs, springs from **centre** of rosette. Br Isles: W, N Scot, W Ire f-la; Wales, Eng r (but lf in Norfolk, New Forest–Dorset); decreasing, but vlf to W and N; in *Sphagnum* in bog pools and open fen pools. Fl 6–8.

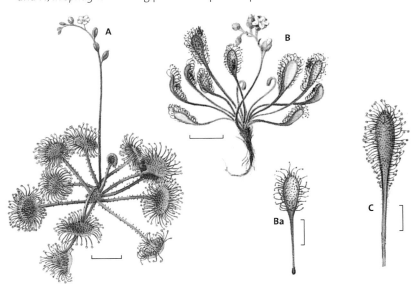

ROCK-ROSE FAMILY

Cistaceae

The **Rock-rose Family** (*Cistaceae*), comprises low downy shrubs (but *Tuberaria* is an ann herb) with undivided elliptical lvs in opp pairs, and fls with 5 strong-veined sepals (3 large, 2 tiny) and 5 spreading petals; many stamens, and egg-shaped capsules splitting into 3 lobes.

▼ **A** ꜰᴘᴏ (Eire) **Ire **Common Rock-rose**, *Helianthemum nummularium* (*H. chamaecistus*), a ± **prostrate** shrub, 5–30 cm, with many branches from thick woody stem. Lvs 0.5–2.0 cm, oblong (or oval), green, scarcely hairy above, **white-woolly** below, with untoothed, slightly inrolled margins, and long **stipules** at bases. Fls (in 1–12-fld lax infls) 2.0–2.5 cm across, bright yellow, with crinkly petals in bud, stalks downy. Br Isles: GB f-lc, abs some parts; Ire vvr; on gsld, scrub on chk, and on acid soils in Scot. Fl 5–9.

▼ **B** ᴠᴜ **White Rock-rose, *H. apenninum*, resembles A, but has lvs **grey-downy above** with margins strongly inrolled; stipules no longer than lf-stalks; petals **white**. Eng vr, but vla Somerset, Devon; on limestone rocks. Fl 4–7.

5 mm

C **Ire ***Hoary Rock-rose**, *Helianthemum oelandicum* (*H. canum*), has yellow fls as in A, but only 1.0–1.5 cm across; strongly bent (instead of straight) styles (**Ca**); and narrower lvs (less than 2 mm wide), without stipules. Br Isles: Wales, W Ire vl on limestone rocks, gsld. Fl 5–7. End VU **ssp *levigatum* has lvs ± hairless on upper side and occurs in Yorks and Cumbria only.

D **Ire NT ****Spotted Rock-rose**, *Tuberaria guttata*, ann, 6–20 cm tall, with three-veined, elliptical, hairy lvs; yellow fls 10–20 mm across, often with **red spot** at petal bases. Cl r, N Wales, W Ire, Scot (Coll only) vr; on rocky cliffs and dry gd near sea. Fl 4–8.

VIOLET FAMILY

Violaceae

There is only one genus, *Viola*, of this family in the Br Isles. Per or ann herbs, with alt, stalked lvs with stipules in pairs at lf-bases. Fls solitary, irregular; 5 sepals, all with back-pointing appendages; 5 petals, the lower lip-petal with a spur; frs three-valved capsules. Many Violets form hybrids which are hard to name. Pansies have conspicuous divided lf-like stipules, side petals directed upwards, often forming a flat 'face'; Violets have narrow unlobed stipules, and side petals horizontally spreading.

ID Tips   Violets & Pansies

- **What to look at**: **(1)** whether runners (stolons or rhizomes) present or not **(2)** whether lf-stalk has hairs or not and if hairs are adpressed to lf-stalk or spreading **(3)** on violets, stipule-teeth hair-like or triangular **(4)** sepals blunt or pointed (Aa, Ca p 187).

- Violets often produce petal-less, small, rounded fls later in the year, which are self-fertilizing and do not open (**cleistogamous** fls).

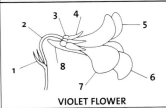

1	**Bracteoles**
2	**Fl-stalk**
3	**Sepal appendage**
4	**Sepal**
5	**Dorsal petals**
6	**Lip (lower petal)**
7	**Side petals**
8	**Spur**

VIOLET FLOWER

STIPULES

teeth
hair-like

teeth
triangular

KEY TO VIOLETS AND PANSIES (*VIOLA*)

1 Stipules undivided to fringed, with hair-like or triangular teeth, not pinnately- or palmately-lobed – side petals spreading outwards ....(**Violets**) **2**
Stipules lfy, pinnately- or palmately-lobed – side petals arching up to top of fl, often forming a 'face' to fl ....(**Pansies**) **10**

2 Bog or marsh plant with no stems above gd – undergd rhizome from which arise long-stalked, blunt, kidney-shaped lvs and lilac, dark-veined fls ....*Viola palustris* (p 188 **G**)
Stems above gd – plants with tufts of lvs, or lfy shoots – all lvs ± heart-shaped **3**

3 Plant with lvs and fl-stems arising from the base – sepals blunt **4**
Plant with leafy flowering stems (although small plants may appear to have no fl-stem) – sepals pointed **5**

4 Stolons present, often as merely procumbent green stems – lf-stalks adpressed hairy – mature lf-tips ± blunt to rounded...*V. odorata* (p 188 **D**)
Stolons abs – lf-stalk with long, spreading hairs – mature lf-tips distinctly pointed ....
V. hirta (p 186 **C**)

5 Lvs and lf-stalks hairy – capsule usually hairy, globose (vr, Teesdale)....*V. rupestris* (p 186 **A.1**)
Lvs and lf-stalks with only v sparse hairs – capsule hairless, ovoid **6**

6 Basal rosette usually present – lvs little longer than wide, oval-cordate – stipule-teeth hair-like...**7**
Basal rosette abs – lvs at least twice as long as wide, oval-lanceolate – stipule-teeth triangular **8**

7 Fl -spur paler than petals, blunt, curved, notched at tip – sepals with back-appendages >1.5 mm long...*V. riviniana* (p 186 **A**)
Fl -spur darker than petals, pointed, straight, unnotched – sepals with back-appendages tiny, < 1.5 mm long...*V. reichenbachiana* (p 186 **B**)

8 Rhizomes – lvs thin, ± translucent – spur greenish, not or a little longer than sepal appendages ....*V. persicifolia* (p 188 **F.1**)
No rhizomes – lvs thick – spur yellowish or greenish, c. twice as long as sepal appendages **9**

9 Lvs oval-lanceolate, base truncate or cordate – stipules to only $^1/2$ length of lf-stalk – petals broad, 1.5-2 x as long as broad ....*V. canina* (p 188 **E**)
Lvs lanceolate, base tapered into stalk or truncate – upper stipules equalling or longer than length of lf-stalk – petals narrow, 3–4 x as long as broad ....*V. lactea* (p 188 **F**)

10 Spur 10-15 mm – NE Scot f, r elsewhere; introd (Pyrenees) gdn escape; lf NE Scot; vr scattered elsewhere; Ire abs; on rds, wd edges, gsld ....**Horned Pansy** *V. cornuta*
Spur < 7mm **11**

11 Fl usually >35mm from top to bottom – petals overlapping ....*V. x wittrockiana* (p 188 **I.1**)
Fl < 35mm from top to bottom – petals not or scarcely overlapping **12**

12 Per plants with multiple stems arising from soil surface and some stems without fls **13**
Ann plants with single stems and no non-flg stems...**14**

13 End-lobe of stipule ± same width as other stipule lobes ....*V. lutea* (p 188 **J**)
End-lobe of stipule much broader than other stipule lobes ....*V. tricolor* ssp *curtisii* (p 188 **I**)

14 Fl 4–8 mm from top to bottom – IoS and CI only ....*V. kitaibeliana* (p 188 **H.1**)
Fl > 8 mm from top to bottom – widespread **15**

15 Petals longer than sepals ....*V. tricolor* ssp *tricolor* (p 188 **I**)
Petals shorter than sepals ....*V. arvensis* (p 188 **H**)

A AWI **Common Dog-violet**, *Viola riviniana*, our commonest Violet, is an unscented per herb, 2–20 cm, with lfy shoots bearing fls, arising round a central flless lf-rosette. Lvs often with few sparse hairs, long-stalked with cordate blades, 0.5–4.0 cm long, and 2 narrow stipules with stipule-teeth **hair-like**. Fls 14–25 mm across, on stalks 5–10 cm. Sepals (**Aa**) pointed, with large square-cut **appendages** (**Aa**) **>1.5 mm** long. Petals blue-violet, but variable. Spur **curved upward, blunt, notched** at tip, **paler** than petals. Strong purple **veins** (nectar-guides) especially on lower petal, **much branched**. Capsule **hairless**, ovoid, three-angled, pointed. Br Isles, vc in wds, hbs, gslds, rocks. Fl 3–5; often again 8–10, often without petals.
A.1 **Teesdale Violet**, *V. rupestris*, resembles A in having fls on lfy side-shoots (not direct from flless central rosette), and C below in closely **downy lvs, lf-stalks** and **capsule**. Lvs 5–10 mm, cordate to **kidney-shaped**; spurs thick, pale, furrowed. Capsule globose, hairy. N Eng; vr on limestone rocks in Teesdale. Fl 5. **Do not confuse** A.1 with small plants of A, which has much less hairy lvs and lf-stalks (usually only a few hairs on veins and marginal area of upper lf) and a hairless, ovoid capsule.

B AWI **Early Dog-violet**, *V. reichenbachiana*, is close to A but lvs and stipules narrower; **sepal appendages** v short (**<1.5 mm** long); fls with spur **straight, pointed, unnotched, violet, darker** than petals; upper petals narrower than in A, erect, (looking rather like rabbit ears) with purple veins on lower petal generally **not** or slightly branched. Br Isles: S Eng, Wales lc; N Eng, Ire vl; Scot vr; CI abs; in dry, especially chky wds and hbs. Fl 3–5.

C FPO (Eire) **Ire **Hairy Violet**, *V. hirta*, differs from A and B in its **spreading** hairs on lf-stalks, fl-stalks and capsules; **blunt** sepals (**Ca**); all lvs and fls arise directly from **rootstock** of plant, no lfy stems or runners. Petals pale-violet blue, unscented. Br Isles; GB S Eng to mid Scot f-la; Ire vr; CI abs; on calc gsld, scrub, rocks. Fl 3–5.

A

B

Aa

Ca

C

D ▶

E ▶

F ▶

G ▶

H ▶

I ▶

J ▶

187

▲ **D** AWI **Sweet Violet**, *Viola odorata*, our only fragrant Violet, like C is hairy, has blunt sepals, and lvs and fls all arise from rootstock; but it is **closely downy** (**not spreading**) hairy, has broader, glossier lvs, (becoming bigger in summer) and rooting runners; fls (**D**) dark violet or white (sometimes other tints). Br Isles: Eng, Wales, CI vc; Ire f; Scot vr-o in N, f-lc in S; in hbs, wds, scrub, usually on calc soils (often a gdn escape). Fl 3–5.

▲ **E** NT **Heath Dog-violet**, *V. canina*, is ± hairless, with creeping stems, but without a basal lf-rosette. Lvs narrowed, **oval-lance-shaped**, with cordate bases; stipules 1/3 to 1/2 length of lf-stalk, teeth **triangular** not hair-like. Fls deep slate-blue (not violet); spur straight, blunt, yellowish (sometimes greenish). Br Isles, lf only on hths, acid gsld, fens. Fl 4–6.

▲ **F** FPO (Eire) ******Ire VU ***Pale Dog-violet**, *V. lactea*, resembles E, but has still narrower lanceolate lvs with rounded to wedge-shaped or truncate bases, **not cordate** below. Fls all pale milky-violet to **greyish-pink**, with **short greenish** spurs, upper stipules triangular **equalling or longer than** lf-stalks. Br Isles: Eng, Wales vl on hths, c in SW only; S & W Ire r in damp gsld. Fl 5–6. **F.1** Sch 8 WO (NI) ******Ire EN ****Fen Violet**, *V. persicifolia* (*V. stagnina*), has narrow lanceolate lvs, long stipules, and pale fls like F, but lvs v **thin**, ± translucent; petals round, blunt, bluish-white; spurs v short, **not or only a little longer than** sepal-appendages (twice sepal-appendages in F). E & C Eng, vr (only 2 recent sites); Ire, r; in fens and turloughs. Fl 5–6.

▲ **G** AWI **Marsh Violet**, *V. palustris*, low per herb; long creeping runners bear long-stalked hairless **kidney-shaped** lvs 1–4 cm wide. Fls 10–15 mm wide, with **pale lilac, blunt** petals with **dark** purple veins, and blunt spurs. Br Isles, f-lvc in N and W; o-r in S and E; in acid marshes and bogs (decreased through drainage). Fl 4–7.

▲ **H** **Field Pansy**, *V. arvensis*, **ann** with oval, deeply- but bluntly-toothed lvs; pinnate stipules with large, toothed, lf-like end-lobes. Fls creamy-yellow, 8–20 mm from top to bottom of fl, petals **shorter** than sepals. Br Isles, c (except W Ire, NW Scot, r) on ar, wa. Fl 4–10. **H.1** NT ****Dwarf Pansy**, *V. kitaibeliana*, tiny downy ann with concave fls 4–8 mm top to bottom. IoS, CI, vr but often vla on dunes and short coastal gsld. Fl 4–7.

▲ **I** **Wild Pansy**, *V. tricolor*, is variable: either **ann** weed of wa and ar (NT ssp *tricolor*) or tufted **per** (ssp *curtisii*) of dry gsld and dunes. Pinnate stipules have end-lobes narrow, untoothed in ssp *tricolor* but in ssp *curtisii* end-lobe much broader than other stipule lobes; fl 1.5–2.5 cm from top to bottom, yellow, blue-violet or partly both; petals **longer** than sepals, flat. Br Isles, f-lc but o-r in SE, SW Eng and S Ire. Fl 4–9. **I.1 Garden Pansy**, *V*. x *wittrockiana*, is much more c than I, especially in urban areas; per plant comprising a group of gdn hybrids that are fertile; often close to I but usually having larger fl (>35 mm top to bottom; although fls decrease in size with subsequent generations) but always with **overlapping** petals. Br Isles, o gdn escape on wa, rds and pavement cracks. Fl 4–9.

▲ **J** **Mountain Pansy**, *V. lutea*, **low creeping** per with fls large (2.0–3.5 cm from top to bottom), on **long** stalks (5–9 cm), bright yellow or sometimes blue-violet; stipules with palmate lobes, end lobe **± same width** as other stipule lobes. N and W Br Isles, f on upland and mt gslds, moors on less acid soils. Fl 5–8.

TAMARISK FAMILY

Tamarisk, *Tamarix gallica*, hairless evergreen shrub, to 3 m; reddish twigs; lvs (**a**) minute, scale-like, grey-green, overlap closely on the twigs; fls (**b**) pink, 3 mm, in dense spike-like racemes; 5 sepals, petals and stamens, 3 styles; fr a capsule. Introd (W Mediterranean); Br Isles: S, E coasts of Eng f-lc; IoS, CI c; Ire r; by the sea. Fl 7–9.

1 mm

3 mm

a   b

SEA-HEATH FAMILY

*****Sea-heath**, *Frankenia laevis*, prostrate mat-forming shrub resembling a Heather; opp linear lvs, 2–4 mm long, with inrolled edges; crinkly five-petalled open pink fls (**a**), 5 mm across; 6 stamens; three-celled capsules. Br Isles: coasts of S Eng (Cornwall to Norfolk), Wales, CI r but lf on dry stony or sandy salt-marshes. Fl 7–8.

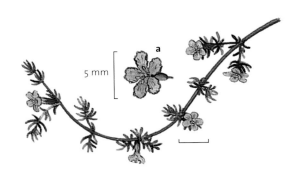

5 mm

a

GOURD FAMILY

Cucurbitaceae

White Bryony, *Bryonia dioica*, climbing per herb up to 4 m, with angled bristly stems, climbing by unbranched **spirally-coiled** tendrils arising from side of lf-stalk; lvs v variably **palmately-lobed**, dull green. Plant dioecious; fls in axillary cymes, pale green; male fls 12–18 mm across, with 5 sepals, 5 net-veined hairy petals, 5 stamens (one free, 2 joined in pairs); female fls smaller (10–12 mm across) with 3 downy bifid stigmas, ovary inferior, oval; fr (**a**) a red berry 5–8 mm across. Br Isles: Eng N to Northumberland, mid, E, SE Eng vc; Wales r; SW Eng, Scot, Ire vr-abs; on hbs, scrub, wd edges, especially on calc soils. Fl 5–9.

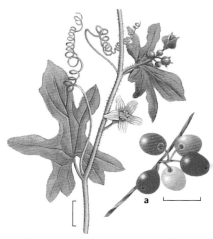

a

WILLOW FAMILY

Salicaceae

A family of deciduous, dioecious trees and shrubs with (usually) alt simple lvs with stipules. Fls in catkins; no perianth; carpels 2; fr a one-celled **capsule**, which bursts to release the **many** silky-plumed seeds. **Willows** (*Salix*) have narrow, lanceolate-oval lvs; catkins usually erect, a single outer scale enclosing the bud; catkin-scales untoothed; 2–5 stamens; nectar glands present (insect pollination). **Poplars** (*Populus*) have broad, triangular, cordate or rounded lvs; catkins hanging, several toothed outer scales to the bud; fls cup-like, with many red stamens but no nectar (wind pollination).

ID Tips   Willows
• **What to look at: (1) mature lvs** are the more reliable character and are best looked at from July – Sept. Always use unshaded lvs from full-grown trees/shrubs (not suckers), especially when looking at lf hairiness as young lvs are atypical. **(2) catkins** are also important and you may have to check a tree again when flg the following spring to confirm your id **(3)** 2nd year twigs (not the youngest twigs growing at the tip of the branches) **(4)** stamen number.
• Willows can not be identified by using catkins alone.
• Many willows form hybrids with one another. These are not included in these keys. Plants found with a mixture of characters of two spp, or intermediate, are probably hybrids; only specialists can normally name them accurately.
• See *BSBI Willows Handbook* for illustrations and keys to all the spp and many hybrids in Br Isles. There are useful scans of willow lf-silhouettes in the *New Flora*.

KEYS TO WILLOWS (*SALIX*)

A. Key to lowland spp of Br Isles

1 Shrubs ± prostrate (under 50 cm) with some creeping stems – lvs to only 25 mm long, silvery or silky below – on hths, moors ....*Salix repens* (p 194 **D**)
Taller shrubs or trees – no creeping shoots **2**

2 Lvs mostly opp, oblong to linear-lanceolate, hairless, dull bluish-green – young twigs purplish ....*S. purpurea* (p 192 **C**)
Lvs all alt...**3**

3 Lvs ± hairless **4**
Lvs clearly hairy, at least below – stamens always 2 per male fl **7**

4 Lvs 2–6 cm, oval, pointed but not long-tapering, waxy-grey below, 2–3 times as long as wide – stamens 2 – lf in N Eng, N Ire & Scot ....FPO (Eire) ******Ire **Tea-leaved Willow** *S. phylicifolia*
Lvs 5–15 cm long, elliptical-lanceolate, long-tapering **5**

5 Bark peeling off main stems in patches – 3 stamens per male fl ....*S. triandra* (p 193 **E**)
Bark not peeling – male fls with 2 or 5 stamens **6**

6 Twigs v brittle, especially where they join stems – lvs lanceolate, grey to pale green below – lf-tip not symmetrical – 2 stamens ....*S. fragilis* (p 192 **B**)
Twigs pliant, not brittle – lvs green below, dark, polished, leathery green above – lf-tip symmetrical – 5 stamens...*S. pentandra* (p 194 **C.1**)

7 Lvs lanceolate or linear, over 5 times as long as wide **8**
Lvs oval or oblong, 4 times as long as wide (or less) **9**

8 Lvs 5–10 cm long, 5–8 times as long as wide, with silky adpressed hairs on both sides ....
S. alba (p 192 **A**)
Lvs 10–18 cm long, 7–18 times as long as wide, dark green and hairless above, silky hairs below only ....*S. viminalis* (p 192 **D**)

9 Shrub under 150 cm – lvs 3.5 cm long or less, pointed, elliptical, with silky adpressed hairs both sides – on dunes, fens..*S. repens* (p 194 **Da**)
Larger shrub or tree – lvs with spreading hairs, ± woolly below, tips ± rounded...................**10**

10 Lvs round to oval, wrinkled, 2–6 cm long, with large round stipules at lf-bases – shrub to 2.5 m with spreading branches...*S. aurita* (p 193 **C**)
Lvs oval to oblong, not wrinkled, 2–12 cm long, stipules less conspicuous – tall shrub (3–10 m) or tree, with erect branches...**11**

11 Lvs oval or oval-oblong, 1–2 times as long as wide – 2nd year twigs without raised ridges under bark..*S. caprea* (p 193 **A**)
Lvs obovate to oblong-lanceolate, 2–4 times as long as wide – 2nd year twigs with raised ridges under bark (peel off bark to see)..*S. cinerea* agg (p 193 **B**)

B. Key to Willows of mt rocks and moors over 600 m – r-o in N Eng & Scot – all small shrubs

1 Low prostrate creeping shrub – lvs rounded at tips **2**
Shrub not creeping – stems ± erect – lvs ± pointed at tips...**4**

2 Lvs oval to elliptical, densely silky, with adpressed hairs at least below – catkins on sides of shoots, appear before lvs ....*S. repens* (p 194 **D**)
Lvs almost round, ± hairless – catkins on tips of shoots, appearing with lvs............................**3**

3 Lvs shiny green both sides, 3–20 mm long, less than 1$^{1}/_{2}$ times as long as wide, veins not strongly imprinted above – lf-stalks 4 mm or less – branches 2–3 cm long – N Eng, Wa, Ire & Scot only, r but sometimes vla on mts ....**Dwarf Willow** *S. herbacea*
Lvs green above, waxy-grey below, 12–40 mm long, less than twice as long as wide, veins strongly imprinted above, forming a raised network below – lf-stalks 10–40 mm long – branches to 10 cm long – r in Scot only, on mts ....****Net-leaved Willow** *S. reticulata*

4 Mature lvs ± hairless **5**
Mature lvs hairy, at least below on midrib **7**

5 Shrub 1 m or more, ± erect – lvs 2–6 cm, pointed, leathery texture, lf-underside waxy grey-green – in N Eng, N Ire & Scot ....FPO (Eire) **Ire **Tea-leaved Willow** *S. phylicifolia*
Low shrub, 60 cm or less – branches spreading, bushy **6**

6 Lvs equally green both sides, 1.5–7.0 cm long – stipules mostly large and persistent – r, Scot only...EN ***Whortle-leaved Willow** *S. myrsinites*
Lvs pale or grey-green below, 1.5–3.0 cm long, sometimes with a few silky hairs below – stipules mostly small and soon dropping – vla on mts in Scot only ....
Mountain Willow *S. arbuscula*

7 Lvs hairless above, hairy below (at least on mid rib) **8**
Lvs densely silky or woolly with white hairs both sides **9**

8 Lvs thinly downy below, especially – on veins, not wrinkled – catkins appear with lvs – lf in N Eng & Scot only ....**Dark-leaved Willow** *S. myrsinifolia*
Lvs wrinkled, densely downy below – catkins appear before lvs – f-lc throughout Br Isles (except C Eng) ....*S. aurita* (p 193 **C**)

9 Lvs oval, 1$^{1}/_{2}$–2 times as long as wide – branches stout – buds stout, woolly – vr, Scot only ....VU BAP ****Woolly Willow** *S. lanata*
Lvs elliptical-oblong, 2–4 times as long as wide – branches slender – buds small, ± downy only - r, Scot only ....VU ***Downy Willow** *S. lapponum*

A **White Willow**, *Salix alba*, tree to 25 m with rugged bark; twigs not fragile. Lvs hairy, silver-grey, toothed, narrow, lanceolate-acute; male catkins (**Aa**) slender, cylindrical, 2.5–5.0 cm; 2 stamens per male fl. Fls appear with the lvs. Br Isles, c to S, r to N (probably native SE Eng, but much planted); on streamsides, wet wds, fens on richer soils in lowlands. Fl 4–5 (planted **Weeping Willows** are usually *S.* x *sepulcralis* and are r natd).

B **Crack-willow**, *S. fragilis*, more spreading tree, with **fragile** twigs; lvs **hairless** at maturity; female catkin (**Bb**), male (**Ba**); 2 stamens per male fl. Br Isles, distribution, habitat as A (but more c in Eng as native than A). Fl 4.

C **Purple Willow**, *S. purpurea*, shrub with long slender purplish twigs; upper lvs ± **opp**, narrow-oblong, obovate, tapered to base, finely-toothed, hairless, blue-green above, waxy below. Male catkins with purple-tipped scales and stamens, female red-purple. Fls appear before lvs. Br Isles, f-la in fens and by fresh water. Fl 3–4.

D **Osier**, *S. viminalis*, shrub with long straight flexible branches, downy young twigs; lvs 10–18 cm long (7–18 times as long as wide), untoothed, linear-lanceolate, with inrolled edges, dark green, hairless above, silvery-silky below. Oblong catkins appear before lvs. Br Isles, lowlands c, hills r; by ponds, streams, in fens, marshes. Fl 4–5.

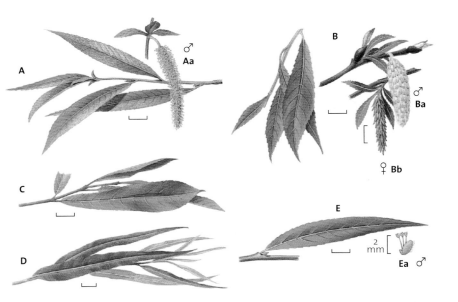

E Almond Willow, *S. triandra*, shrub or small tree with brown smooth peeling bark, hairless olive twigs; lvs, 5–10 cm long, oblong-obovate, shiny green above, waxy green below. Catkins among lvs; male fls (**Ea**) with **3** stamens each. Br Isles: S, E Eng f-lc; Scot, Ire o; by water and in marshes. Fl 3–5.

▼ **A Goat Willow**, *S. caprea*, shrub or (especially in W Scot, Ire) tree, to 10 m; 2nd year twigs **without** raised, lengthways ridges under bark. Lvs **oval** with pointed tips, hairless above, densely grey-downy below, 5–12 cm long, 1.2–2 times as long as wide. Catkins appear before lvs, erect, stout, male 2–4 cm, female 3–7 cm when mature, unstalked, oval to oblong; scales dark brown with many silvery-white hairs, male with 2 golden stamens per fl, female with downy green-grey ovaries and 2 green stigmas. Br Isles, vc (although in some areas, eg Glos, vr and replaced by hybrids) in wds, hds, scrub, wa. Fl 3–4.

▼ **B Grey Willow**, *S. cinerea* agg, is close to A, but 2nd year twigs with **raised ridges** on the wood under bark (peel to see this). Lvs usually **obovate** (but v variable), narrower than in A, tapered to stalk, 2–9 cm long, 2–4 times as long as wide, downy above when young, persistently downy below. Catkins as in A, but smaller, male (**Ba**), female (**Bb**). There are 2 ssp, one o and the other c; ssp *cinerea*, with lvs softly woolly-grey below, no rusty hairs, is c **only in E Eng**; ssp *oleifolia*, with lvs only downy on veins below when mature, some **hairs rusty** (sometimes only developing July onwards), is c in most of Br Isles (not E Eng); in wet wds, fen carr, fens, by fresh water. Fl 3–4.

▼ **C Eared Willow**, *S. aurita*, shrub to only 2.5 m, smaller than A or B; twigs more slender, **angled, spreading**, with ridges as B. Lvs 1.5 to 2.5 times as long as wide, obovate, rounded, v wrinkled, woolly-grey below, with conspicuous large rounded stipules (the 'ears'). Catkins smaller (1–2 cm) versions of B. Br Isles, f but much more c to N and in uplands; on hths, moors, damp wds, especially on acid soils. Fl 4. **C.1 Bay Willow**, *Salix pentandra*, shrub or small tree with grey rugged bark, shiny hairless twigs; lvs 5–12 cm long, broadly elliptical, dark shiny green (polished-looking) above; lvs and buds sticky and fragrant when young. Catkins appear after lvs; male fls with 5 stamens each. Br Isles: **S Eng vr-abs**; N Eng to mid Scot f-lc; N Ire c. Fl 5–6.

D Creeping Willow, *S. repens*, low shrub with creeping and ascending stems to 1.5 m tall (variable; only 50 cm in hth form, **D**). Lvs 1.0–3.5 cm long, oval, usually with white silky hairs below (dune-slack forms (**Da**) with larger lvs, silky both sides, ± toothed). Catkins short, 5–25 mm, appear before lvs; male catkins (**D**) oblong with golden stamens, female globose; scales dark-tipped; seeds (**Db**) 1.5 mm. Br Isles, lc especially to N, on hths, fens, dune-slacks. Fl 4–5.

E AWI **Aspen**, *Populus tremula*, tree to 20 m with suckers, sticky hairless buds, **hairless** mature twigs. Lvs on v long **flattened** stalks (2.5–6.0 cm), v flexible, trembling in any breeze. Lf-blades almost round, with large rounded teeth, hairless except when v young. Sucker-lvs downy below. Catkins 4–8 cm, pendulous with deeply jagged scales; red stamens in male, purple stigmas in female (**Ea**). Br Isles c in N and W, f-lc in S and E; in damp wds on heavier poorer soils. Fl 2–3.

F Grey Poplar, *P.* x *canescens*, (hybrid between E and G) large tree (to 30 m) with smooth yellowish-grey bark with horizontal lines (like Birches) when young, roughly fissured below when older, and with suckers. Twigs and buds thinly downy; lvs of short shoots (**F**) 3–6 cm, wide, oval, rounded, coarsely-toothed, grey above, **grey-downy** below, then later ± hairless; lvs of suckers and summer lvs larger (to 10 cm long), oval-triangular, more coarsely toothed, remaining grey-woolly below. Lf-stalks to 4 cm, not flattened. Catkins to 10 cm, stout, stigmas yellow, scales jagged. Native in S and E Eng, planted in rest of Br Isles, f-lc in damp wds. Fl 2–3.

G White Poplar, *P. alba*, has **thickly downy** buds and twigs, wider lvs to short shoots, **white** below, then ± hairless; lvs of suckers and summer lvs (**G**) **palmately-lobed** and maple-like, remaining white-woolly below. Catkin scales only toothed, not jagged. Br Isles, c and often planted, especially in S Eng. Fl 3.

H Black-poplar, *P. nigra*, tall tree with rugged black bark and **trunk-bosses**; **no** suckers; branches arching down; lvs (**H**) green, triangular, **longer than wide**, **hairy when young**, lf-margins with teeth but **not** hooked. Native in E, mid Eng o; planted elsewhere; by streamsides in wet wds. Fl 4. **Do not confuse** with I, which is much more widespread.

I Hybrid Black-poplar, *P.* x *canadensis* agg (*P. nigra* x *P. deltoides*), is a group of many cultivars, v like H, but has **no trunk-bosses**; ascending branches curve upwards; lvs (**I**) v variable, usually **wider than long**, **hairless**, lf-margins usually with **hooked** teeth; glands at base of lf-blade; always male. Introd (N America) and widely planted. Fl 3–4. **I.1 Lombardy Poplar**, *P. nigra* 'Italica', has a narrow column-like form with stiffly erect branches; often planted.

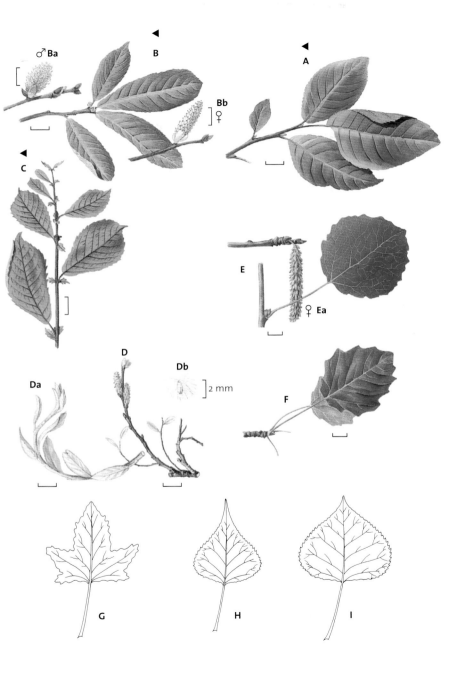

CABBAGE (CRUCIFER) FAMILY   *Brassicaceae*

A large and distinctive family of herbs, also known as the *Cruciferae*, or Crucifers; none is poisonous, and many (eg, Cabbage, Turnip, Mustard, Cress, Water-cress etc) are useful vegetables. Lvs alt, without stipules; fls in a racemose infl. Fls have 4 free sepals, **4** free (usually equal) **petals**, arranged in a **cross**, usually 6 (sometimes fewer) stamens, and 2 carpels joined along their inner sides to form a superior ovary. Ovary normally ripens to a two-celled fr, which opens from the bottom upwards by 2 valves (strips of tissue on outside of carpels). Fr either long and narrow (a **siliqua**), or less than 3 times as long as wide (a **silicula**); seeds remain attached to central framework of fr for a time. Some Crucifers have a specialized **beak** on top of the two-valved part of the fr; beak may contain 1 or more seeds. In a few (eg, Radish), seeds are confined to the beak, which either breaks transversely into one-seeded joints or remains in one piece (as in Sea-kale or Bastard Cabbage). A Crucifer of mt lakes, **Awlwort** (*Subularia aquatica*), is not illustrated here but included in the vegetative key part V section D.

ID Tips   Crucifers

- **What to look at: (1)** petal length and colour (white, pink/purple, orange, yellow) or if petals abs **(2)** sepal length in relation to petals **(3)** lf-characters such as whether lvs are entire, lobed, pinnate or pinnatifid; fleshy or thin; clasp stem or not **(4)** types of hairs (whether simple, forked or star-shaped etc) on stems and lvs (look at both) **(5)** ripe frs.

1 Fr-stalk
2 Stipe
3 Persistent style or beak
4 Valves
5 Septum

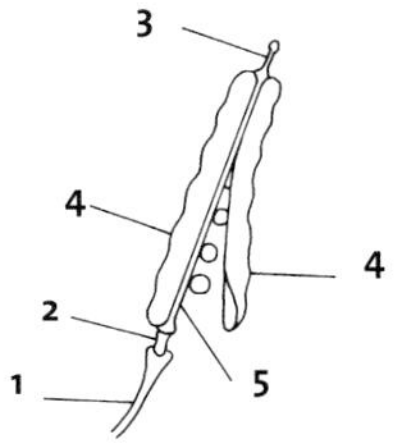

CRUCIFER FRUIT

- Frs are often essential for correct id of many Crucifer spp. The most important characters are: size and shape of fr (measure several frs and use the average); whether fr has **valves (4)**; and whether fr has a persistent **style** or **beak (3)**.

- The *BSBI Crucifers Handbook* has a key to non-fruiting plants (although this is less reliable for id than keys using frs).

- **Charlock etc** (*Sinapis*) is the only member of the Crucifer family often having **sepals reflexed** to about 90°.

- **Candytuft** (*Iberis*) and **Shepherd's Cress** (*Teesdalia*) are the only white fld Crucifers with unequal petals, outer petals larger than the inner.

- **Common Whitlowgrass** (*Erophila verna* agg) is the only white fld Crucifer with **notched** petals and a **lfless** fl-stem.

KEYS TO CABBAGE (OR CRUCIFER) FAMILY
(Consult the main headings A, B, C, D, E and F first)

A. Crucifers with white or pink fls, and either (a) with frs lacking 2 obvious, parallel, valves or (b) not developing frs at all

1 Plants not producing ripe frs – lvs once-pinnate, with violet-brown bulbils in lf-axils – erect wdland herb – rose-pink fls in spring ....*Cardamine bulbifera* (p 208 **G**)
Plant not producing ripe frs – lvs v large (to 60 cm long), glossy, oblong, wavy, sometimes pinnatifid – fls white, in large panicles in May-June ....*Armoracia rusticana* (p 220 **G**)
Plant producing frs – not in wdlands **2**

2 Large cabbage-like seashore plant – broad rounded heads of white fls – frs shape and size of garden peas, 6–15 mm long, on stalk-like lower joints................*Crambe maritima* (p 218 **F**)
Frs not pea-like **3**

3 Fr with upper part 8–20 mm long, shaped like bishop's mitre, on a stalk-like stout lower joint – lvs fleshy, pinnatifid – fls pink (or white) in racemes ....*Cakile maritima* (p 208 **D**)
Fr cylindrical, tapered to a long point..**4**

4 Fr constricted into bead-like joints that separate when ripe ....
Raphanus raphanistrum (p 202 **E** & p 204 **A**)
Fr sausage-shaped (except for tapered point), not constricted ....*R. sativus* (p 204 **F**)

B. Crucifers with white, pale yellow, pink or purple fls, fr of 2 obvious parallel valves, and less than 3 times as long as wide

1 Petals pale yellow ....*Alyssum alyssoides* (p 218 **D.1**)
Petals white, pink or purple...**2**

2 Plants small, prostrate – fls tiny (2.0–2.5 mm across) – frs of 2, almost distinct, round carpels, covered with network of pits or ridges – carpels finally separate into 2 one-seeded achenes ....*Coronopus* (p 216)
Plants ± erect – 2 carpels joined along their length **3**

3 Petals deeply notched – stems lfless – rosettes of undivided root-lvs – tiny (2–10 cm) plant of early spring ....*Erophila verna* agg (p 216 **G**)
Petals not or v slightly notched and stem with lvs..**4**

4 Petals unequal-sized (2 outer of each fl much larger than inner) **5**
Petals equal-sized (or abs)..**6**

5 Lvs up stem – shorter petals twice length of sepals – on chk...................*Iberis amara* (p 216 **F**)
Lvs ± confined to basal rosette, pinnatifid – shorter petals ± same length as sepals – on sandy or stony gd ....*Teesdalia nudicaulis* (p 216 **D**)

6 Petals > 10 mm long, usually 12mm or more **7**
Petals < 10 mm long **8**

7 Fruit large, 25 mm x 15mm (or more), very flat, ± circular, membranous with 2 rows of flattened seeds..*Lunaria annua* (p 208 **F**)
Fruit ellipsoid, not or slightly flattened, < 20mm long with a long, persistent style > 4mm
Aubrieta deltoidea (p 208 **E**)

8 Fr egg-shaped, convex in cross-section, ± swollen – lvs fleshy.......................*Cochlearia* (p 218)
Fr distinctly flattened in cross-section..**9**

9 Fr two-celled, each cell with only 1 seed **10**
Fr two-celled, each cell with 2 or more seeds **12**

10 Lvs all unlobed, untoothed, linear – partition of fr-cells across widest diameter
Lobularia maritima (p 218 **D**)
Lvs (some at least) lobed or toothed – partition of fr-cells across narrowest diameter **11**

11 Fls in a much-branched umbel-like head – lvs grey – fr without wings or flanges, cordate
at base, tapering into style above ....*Lepidium draba* (p 214 **H**)
Fls in spike-like infls – fr without cordate base, but with wings or flanges at edges, or
downy..*Lepidium* (p 214)

12 Fr with broad winged edges..*Thlaspi* (p 214)
Fr without winged edges (or v narrow wings) **13**

13 Fr shaped like triangle standing on its point, notched above *Capsella bursa-pastoris* (p 216 **E**)
Fr ± as long as wide with sepals persistent on ripe fr ....*Alyssum alyssoides* (p 218 **D.1**)
Fr oval to elliptical without sepals persistent on ripe fr **14**

14 Lvs unlobed – fr elliptical, flat or twisted – partition of fr-cells parallel with flat sides (ie
across widest diameter) ....*Draba* (p 212)
Lvs pinnately-lobed – fr oval, short – partition of fr-cells at right angles to broader sides
(ie. across narrowest diameter)..*Hornungia petraea* (p 218 **E**)

C. Crucifers with white, pink or purple fls, fr of 2 obvious valves, long and narrow, at least 3 times as long as wide

1 Style of fl deeply lobed into 2 stigmas **2**
Style of fl with a single disc-shaped stigma **3**

2 Stigmas with a horn-like swelling on back of each – seaside plant ....*Matthiola* (p 208 **A**)
Stigmas with no swelling on backs – tall inland plant of wa and riversides................................
Hesperis matronalis (p 208 **C**)

3 Fr strongly flattened, cross-section elliptical..**4**
Fr either four-angled, or cross-section cylindrical **6**

4 Fr's 2 valves with no obvious veins, rolling up on opening – lvs compound, pinnate................
Cardamine (pp 208–10)
Fr's 2 valves with a strong central vein – lvs simple, untoothed to pinnatifid only................**5**

5 Basal lvs long-stalked, pinnatifid – pods held at an angle....................*Arabis petraea* (p 21 **A.1**)
Basal lvs elliptical, short-stalked, only slightly toothed – pods adpressed to stem....................
Arabis (pp 206, 210–12)

6 Lvs pinnate ....*Rorippa nasturtium-aquaticum*, R. microphylla or hybrid (p 210)
Lvs all undivided, toothed only..**7**

7 Frs with long persistent style > 4mm – plant mat-forming, prostrate ....
Aubrieta deltoidea (p 208 **E**)
Frs with short persistent style < 2mm or none – plant erect..**8**

8 Lvs heart- to kidney-shaped, ± hairless, long-stalked – tall (50–100 cm) plant with garlic
smell when bruised..*Alliaria petiolata* (p 212 **C**)
Lvs elliptical, ± hairy, short-stalked, mostly in basal rosette – shorter (5–30 cm) plant with
no garlic smell ....*Arabidopsis thaliana* (p 212 **B**)

D. Crucifers with yellow fls, and fr not obviously composed of 2 parallel valves

1 Fr elongated, composed of bead-like rounded joints with constrictions between, when ripe breaking into one-seeded segments ....*Raphanus* (p 202) **2**
Fr with 2 joints: the upper globular, 3 mm wide, with a long beak like an old Chianti bottle; lower joint like a stout stalk..*Rapistrum rugosum* (p 208 **H**)

E. Crucifers with yellow fls, and frs in 2 parts: lower part of 2 obvious parallel valves; upper part (the beak), between top of valves and stigma, does not split open vertically, but may contain 0–6 seeds

1 Fr-valves one-veined (may be difficult to see; easier on dried pods) – plant usually ± hairless ....*(Brassica)* **2**
Fr-valves 3–7-veined – plant hairy...**5**

2 Upper stem lvs stalked, not clasping stem – frs stiffly erect, adpressed to stem (0–20°)
Brassica nigra (p 200 **B**)
Upper stem lvs stalkless with auricles usually clasping stem – frs either erect but not adpressed (>30°) or spreading **3**

3 Petals 18–30 mm long – fr-valves rounded (so fr cylindrical), with faint veins – beak only 1/10 of whole fr's length, usually with 1 seed – lower lvs fleshy................*B. oleracea* (p 200 **A**)
Petals 6–18 mm long – beak usually seedless – lower lvs not fleshy..............................**4**

4 Most petals 6–13 mm long – open fls over-topping or equalling buds – first year rosettes green ....*B. rapa* (p 200 **C**)
Most petals 13–18 mm long – buds over-topping or equalling open fls – first year rosettes grey-green ....*B. napus* (p 200 **D**)

5 Beak short, swollen below, to 1/2 length of fr – fr erect, adpressed to stem – plant v hairy below, ± hairless above..*Hirschfeldia incana* (p 202 **D**)
Beak long, not swollen below – fr not adpressed to stem **6**

6 Sepals erect ....*Coincya* (p 202)
Sepals spreading ....*(Sinapis)* **7**

7 Upper stem lvs simple (or with 2 shallow lobes) – beak conical to ellipsoid ....
Sinapis arvensis (p 202 **G**)
Upper stem lvs pinnately-lobed – beak flat, sabre-shaped ....*S. alba* (p 202 **F**)

F. Crucifers with yellow fls, and fr of 2 obvious valves, without an upper joint or beak between top of valves and stigma

1 Fr only 4 times as long as wide (or less) **2**
Fr 6 or more times as long as wide **4**

2 Fr flattened **3**
Fr rounded in cross-section, ± veinless ....*Rorippa* (p 206 **A–D**)

3 Fr elliptical, erect – small tufted plant – lvs in basal rosette ....*Draba aizoides* (p 212 **D.2**)

Fr oblong, winged, pendulous, like an Ash 'key' – tall robust plant........*Isatis tinctoria* (p 206 **G**)

4 Style deeply divided into 2 spreading stigmas – lvs oblong, untoothed – bushy plant of old walls ....*Erysimum cheiri* (p 206 **F**)
Style with a single, ± disc-shaped (or v slightly divided) stigma..**5**

5 Fr strongly flattened, elliptical in cross-section – fls v pale yellow – on rocks near Bristol – only ....*Arabis scabra* (p 212 **A**)

Fr ± cylindrical or four-angled in cross-section...**6**

6 Lvs unlobed, ± toothed – plant hairy throughout ....*Erysimum cheiranthoides* (p 204 **D**)

At least basal lvs pinnatifid **7**

7 Lvs twice pinnate into fine, almost hairlike, hairy segments ....*Descurainia sophia* (p 204 **E**)

Lvs no more than once pinnate..**8**

8 Valves of fr one-veined **9**

Valves of fr 3–7 veined – lvs all ± pinnatifid to pinnate ....*Sisymbrium* (pp 204–5)

9 Seeds in 2 vertical rows in each cell of fr...**10**

Seeds in 1 vertical row in each cell of fr...**11**

10 Upper stem-lvs clasping stem, hairless, waxy grey-green, with half-arrow-shaped bases – basal lvs hairy, pinnatifid – frs adpressed to stem – fls v pale yellow*Arabis glabra* (p 206 **E**)

Upper stem-lvs not clasping stem – frs spreading – fls bright yellow ....*Diplotaxis* (p 202)

11 Fr with convex rounded valves ....*Brassica* (go to key E, couplet 3)

Fr four-angled in cross-section..**12**

12 Stem and lvs all ± hairless – all lvs glossy deep green – rosette-lvs stalked – upper lvs stalkless and clasping ....*Barbarea* (p 204)

Lower stem and lvs bristly, grass-green – upper stem and lvs hairless, grey-green – all lvs stalked..*Brassica nigra* (p 200 **B**)

Yellow fld (except F p 204) Crucifers with jointed frs

A **Wild Cabbage*, *Brassica oleracea*, the ancestor of cabbage, kale, etc, is a robust hairless herb, with stout stem bearing lf-scars on its **woody** base; lvs waxy-grey green, oblong, thick and fleshy, pinnately-lobed in lower part; upper lvs **clasping** stem. Fls in long v **spaced-out** spikes, lemon-yellow, **large** compared to other *Brassica* spp, 20–30 mm across and petals **>18** mm long. Frs 5–10 cm, cylindrical, lower part comprising two cells, each with several seeds, and a seedless beak 5–10 mm. Br Isles: S Eng, S Wales; vla on calc sea cliffs, o inland as casual gdn escape. Fl 5–8.

B **Black Mustard**, *B. nigra*, tall ann (to 1 m); lyre-shaped, pinnately-lobed, bristly bright green lvs; upper lvs undivided, elliptical, narrowed below, hairless, glaucous and **not clasping** stem. Racemes of yellow fls, 10–15 mm across. Frs (**B, Ba**) erect, **adpressed**, **four-angled**, hairless, 12–20 mm long, beaded, with a **seedless slender beak** (although may occasionally be swollen by insect larvae). Br Isles: Eng, Wales f-lc; Scot, Ire r; on riverbanks, cliffs, damper wa. Fl 6–9. **Do not confuse** with **Hoary Mustard** (p 202 D) which has ripe frs with a **seed** in the beak and prefers drier places.

C **Turnip**, *B. rapa*, tall ann or bi like B, but with upper stems-lvs **clasping**, with cordate bases; infl-spike flat-topped, **fls overtopping** the buds above them and most petals **6–13** mm long. Br Isles, c on riverbanks, cliffs, wa. Fl 6–8.

D **Rape**, *B. napus*, is v similar to C, with all lvs waxy-grey-green but **key differences** are **buds overtopping** the pale yellow fls; most petals **13–18 mm** long. Check that the fls are developing normally and that no aborted buds (small, dead clusters of buds) are present. Plants of D with damaged infl may have fls over topping the buds and appear like C. Br Isles, c escape from cultivation on wa, rds, ar and disturbed gd. Fl 5-8.

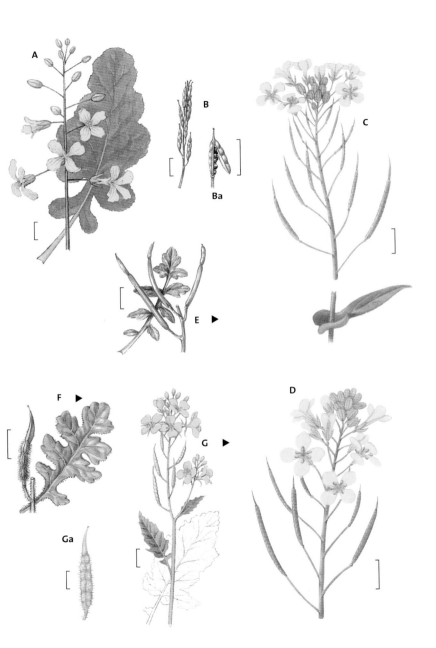

▲ **E** End **Isle-of-Man Cabbage*, *Coincya monensis* ssp *monensis* (*Rhynchosinapis monensis*), **hairless** bi, 15–30 cm; lvs nearly confined to a basal rosette, glaucous, deeply pinnately-lobed; petals pale yellow; frs spreading, 4–7 cm, on stalks 6–10 mm; **beak** (1/3 of fr) **flat and sword-like**, with up to 5 seeds. Br Isles: W coasts of GB (Devon–Argyll, including IoM); lc on cliffs, beaches (Scot vr introd inland). Introd plants (**Wallflower Cabbage**, ssp *cheiranthos*) occur as r casuals (o-lf in S Wales, CI). Fl 6–8. **E.1** Sch 8 End BAP VU ***Lundy Cabbage*, *C. wrightii*, confined to Lundy Island, Devon, has lvs v hairy on both sides, and **hairy** fr.

▲ **F** **White Mustard**, *Sinapis alba*, is often confused with G, but **key differences** are **all** lvs pinnately-lobed; beak of fr is **flat** and sabre-like, and as **long** or **longer** than lower two-valved part. Br Isles, f on ar, wa and apparently declining. Fl 6–8.

▲ **G** **Charlock**, *S. arvensis*, roughly hairy ann, 30–60 cm, with stalked lyre-shaped lower lvs, **bristly** and strongly-toothed; upper lvs **simple** (or 2 shallow lobes at base). Fls yellow, 15–20 mm across; sepals **spreading to 90°**; frs (**Ga**) hairy, 25–40 mm long, spreading; beak of fr ellipsoid, not much more than 1/3 as long as lower two-valved part. Br Isles, vc on ar, dry wa, rds and often near rd works. Fl 5–8.

A **Annual Wall-rocket**, *Diplotaxis muralis*, erect to spreading branched **ann or bi**; basal lvs to 10 cm, elliptical, stalked, deeply pinnate. Fls (**Ab**) lemon-yellow, petals 4–8 mm long; frs (**Aa**) narrowed at both ends, ascending at an angle to stalk; **not stalked** above sepal-scars. Br Isles: S Eng c; Ire, Scot o-r; introd (C and S Eur) in dry places on ar, wa, walls, pavements, rocks. Fl 6–9.

B **Perennial Wall-rocket**, *D. tenuifolia*, similar, but taller more erect **per**, quite hairless, with larger fleshier lvs. Petals to 15 mm long; **key difference** from A is fr (**Ba**) **stalked** above sepal-scars, stalk ± same length as fr. GB (Ire vr), lc in dry places on old walls, wa. Fl 5–9.

C **Hedge Mustard**, *Sisymbrium officinale*, v variable stiffly erect ann to bi herb, 30–90 cm, branched above, bristly (sometimes hairless). Basal lvs deeply pinnately-lobed, with round terminal lobe; stem lvs with arrow-shaped bases clasping stem; fls 3 mm across, short-stalked, in dense long racemes; frs (**Ca**) pale yellow, cylindrical, with v short beak, two-valved, stiffly erect, adpressed to stems. Br Isles, vc on hbs, wa, rds, etc. Fl 6–7. **Do not confuse** with D, which often has superficially similar clusters of fls on long, branching stems, with frs all adpressed to stem below fl-heads.

D **Hoary Mustard**, *Hirschfeldia incana*, ann or per herb to 1.3m, similar to **Black Mustard** (p 200 B), hairy below and usually hairless above; grey-green lvs variable but lower lvs (**Da**) usually deeply pinnate with up to 5 lobes, with larger terminal lobe; fls having larger petals than Black Mustard, 5–10 mm long; thin frs that are **adpressed** to the stem, especially pronounced when unripe; when frs ripe (July onwards) looking like an old-fashioned clothes peg, with a **seed** in the **beak** of the **ripe** fr (**Db**). Introd (Mediterranean); Eng, Wales c but o-r in Scot and Ire, increasing in dry places on wa, rds, urban areas. Fl 5–10.

E **Wild Radish**, *Raphanus raphanistrum* ssp *raphanistrum*, bristly ann with lyre-shaped pinnatifid lvs, rather like **Charlock** (p 202 G), but has long-beaked frs (**Ea**), with 3–8 **weakly-ribbed, bead-like** joints which easily break apart when ripe,

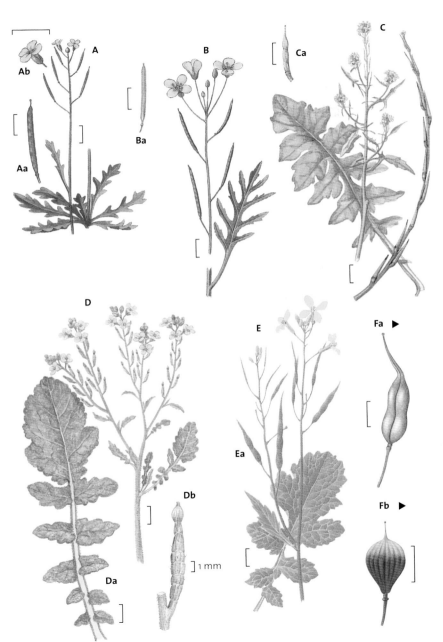

beak to 5 times length of top fr-joints. Fls may be yellow, white or lilac, usually veined with deeper lilac. Br Isles, possibly originally introd; vc on ar, wa. Fl 5–9.

▲ **F   Garden Radish**, *Raphanus sativus*, is close to E (p 202) but fls white, pink or violet; tuberous radish root present; **fr v variable**, either **inflated**, like small sausages (**Fa**) tapered to a long beak, hardly constricted, not easily breaking into separate joints, or sometimes ± globose (**Fb**) with long persistent style. Br Isles, r but widespread escape from cultivation, growing on wa, rds, disturbed gd. Fl 5–10.

A   Sea Radish, *R. raphanistrum* ssp *maritimus*, differs from other Radishes in having fls (**A**) normally always yellow; frs (**Aa**) with **fewer** (1–5) bead-like joints, which are **strongly ribbed**, do not readily break apart when ripe; beak of fr v short, only to twice length of top joint. Br Isles, f on beaches, cliffs. Fl 6–8.

Yellow or orange fld Crucifers with elongated frs

B   Winter-cress, *Barbarea vulgaris*, erect bi or per herb, with strong tap-root, upright parallel branches and hairless, with pinnately-lobed, **deep green, shiny**, lower lvs, terminal lobe oval and larger than others, but shorter than rest of lf; **upper lvs ± simple**, oval. Dense racemes of small (7–9 mm) yellow fls with ± hairless buds, petals 2x as long as sepals; erect, four-angled narrow frs (**Ba**), 15–25 mm long, on stalks of 4–5 mm with **thin** persistent style **>1.8mm**. Br Isles, vc on streamsides, hbs, damp places but also sometimes drier wa. Fl 5–8.   **B.1 Small-flowered Winter-cress**, *B. stricta*, close to B but having **hairy fl buds**, shorter petals only about 1.5x as long as sepals; and a **short**, **stout** persistent style **<1.8mm** in fr. Introd (E Eur); Eng, vl Scot, Ire, Wales r; canal and riverbanks, ditches and as r casual on wa. Fl 5–9.

C   American Winter-cress, *B. verna*, v variable, either small ann or large bi; has **upper stem lvs pinnately-lobed**; basal lvs with terminal lobes much **smaller** and narrower, 6–10 pairs of side lobes; frs (**Ca**) **curved upwards**, at least some **frs >40 mm** long. Introd (SW Eur); Eng, Wales o-lf; Scot, Ire r; escape from cultivation on ar, wa. Fl 3–7. **C.1 Medium-flowered Winter-cress**, *B. intermedia*, close to C, but smaller **frs <40 mm** long with a short, stout persistent style <1.8 mm long. Introd (W Eur); Br Isles, c in a range of wa and disturbed gd and more c than C. Fl 3–7. **Do not confuse** with B, which has simple upper stem lvs and a longer persistent style.

D   Treacle-mustard, *Erysimum cheiranthoides*, resembles B in its erect spikes of small yellow fls and narrow erect pods (**Da**), but has a **square** stem with adpressed hairs, and all lvs lanceolate and **undivided**, though shallowly-toothed. Lowland and S in Br Isles, f-lc in ar and wa. Fl 6–8.

E   Flixweed, *Descurainia sophia*, erect ann to 60 cm, with elongated frs; lvs and lower stem **grey** with star-shaped adpressed hairs; lvs 2 or 3 times **pinnately** cut, with **linear** segments; fls (**Ea**) **pale** yellow, frs (**Eb**) rounded, 15–25 mm by 1 mm, curving upwards, erect, from long slender stalks. Br Isles (not N Scot), lf on sandy wa, rds. Fl 6–8.

F   Tall Rocket, *Sisymbrium altissimum*, resembles E in its finely-cut lvs and long frs; but taller (to 100 cm), stem-lvs ± stalked, only **once pinnate**, and **hairless**, with longer narrower segments; hairless frs (**Fa**) far longer, 50–100 mm (basal lvs pinnately-lobed and hairy, but soon wither). Fl yellow (rarely cream). Introd (E Eur); Br Isles, o in wa, especially sandy gd. Fl 6–8.

▲ **G Eastern Rocket**, *Sisymbrium orientale*, resembles F p 204, but **stalked**, pinnate stem-lvs have broader, **hastate**-shaped terminal lobes and short side-lfts; uppermost stem-lvs merely lanceolate, undivided; frs (**Ga**) like those of F, 40–100 mm long, but hairy at first. Introd (Mediterranean); Br Isles, o on wa. Fl 6–8.

Yellow-cresses (*Rorippa*) are yellow-fld per Crucifers with fls in erect racemes, petals < 6mm long (v small compared to other yellow Crucifers).

A Creeping Yellow-cress, *Rorippa sylvestris*, 20–30 cm, straggling and hairless, with pinnately-lobed lvs, the lobes toothed (**Aa**); **petals longer than sepals**; frs (**Ab**) v variable, cylindrical-oblong to linear, 9–22 mm long. Br Isles, f-lc on wa, gdns, wet places. Fl 6–10.

B Marsh Yellow-cress, *R. palustris* (*R. islandica*), is taller (to 60 cm), stronger and more erect than A; lvs (**B**) often auricled and less deeply divided; fls 3 mm across **petals ± equal sepals,** sepals > 1.6 mm; frs (**Ba**) **shorter** (5–10 x 1.5–3.0 mm), ellipsoid, swollen and **curved**, contracted into short style above. Frs <2x as long as fr-stalks. Br Isles, f in wet places. Fl 6–10.

C **Ire ***Northern Yellow-cress**, *R. islandica*, v close to B (previously not separated from B) but generally with lvs (**C**) without auricles, smaller petals and sepals <1.6 mm; frs usually held over to one side, 2-3x as long as fr-stalk. Microscopic examination of ripe seeds needed to confirm id (see *BSBI Crucifers Handbook*). Ire, Scot and Wales r, but recently found in Eng; damp places, reservoir margins. Fl 7–10.

D Great Yellow-cress, *R. amphibia*, erect hairless per with stout stem 40–120 cm; lvs (**D**) **lanceolate** erect, mostly toothed only, sometimes pinnately-lobed below. Fls usually larger than in A or B, petals about 3–6 mm long and longer than sepals; fr (**Da**) 3–6 x 1.5–3 mm, oval, straight, with style (1–2 mm) longer than in B; fr-stalks to 16 mm long. Br Isles, lc by streams, ponds etc. Fl 6–8.

E BAP EN****Tower Mustard**, *Arabis glabra* (*Turritis glabra*), stiffly erect bi with a basal rosette of hairy, deeply-toothed basal lvs, and **glaucous-waxy**, hairless, grey-green, arrow-shaped stem-lvs that **clasp** the hairless stem. Fls in a long raceme, creamy yellow, about 6 mm; frs (**Ea**) stiffly erect, stalked, 30–60 mm x 1.0 mm, cylindrical. Br Isles: S Eng only, especially E, vl and decreasing; in dry sandy hbs, rds, dry wds. Fl 5–7.

F Wallflower, *Erysimum cheiri* (*Cheiranthus cheiri*), per herb 20–60 cm tall, with stem woody below, lfy and angled above; lvs 5–10 cm, oblong lance-shaped, untoothed, sessile above; adpressed forked hairs on lvs and upper stems; fls in short racemes, bright **orange-yellow**, 2.5 cm across; sepals 1/2 length of petals. Frs (**Fa**) 2.5–7.0 cm long, narrow, flattened, hairy, valves one-veined. Introd (SE Eur); Br Isles f-lc on old walls, especially castles. Fl 4–6.

G ***Woad**, *Isatis tinctoria*, source of the blue dye, is a tall, branched, ± hairless, usually bi herb, with basal rosettes of stalked, wavy-edged, lanceolate downy lvs; upper lvs (**Ga**), arrow-shaped, hairless, grey-green, clasp the stem. Branched infl (**G**) bears yellow fls, 4 mm across; only Crucifer with frs (**Gb**) **pendulous**, obovate, (8–20 x 3.6 mm) and purplish-brown. Introd (S and C Eur) Eng r, on cliffs (Tewkesbury, Guildford) and scattered elsewhere in S Eng on wa. Fl 7–8.

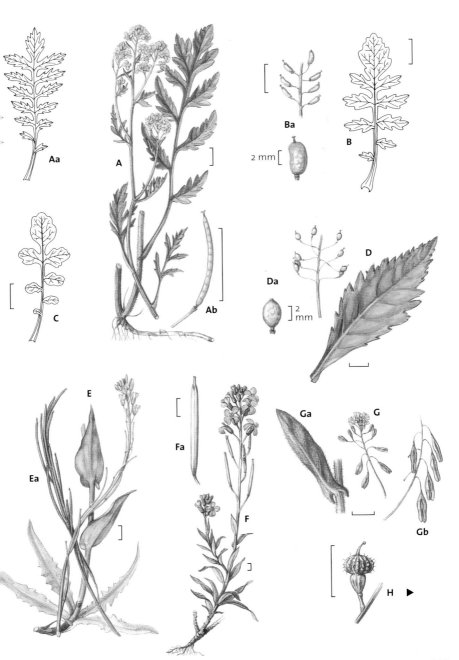

Aa

A

Ba

2 mm

B

C

Ab

Da

2
mm

D

E

Ea

Fa

F

Ga

G

Gb

H ▶

▲ **H Bastard Cabbage**, *Rapistrum rugosum*, is a hairy branched ann with pinnately-lobed lvs and pale yellow fls, 5 mm across. The erect frs (**H**) are distinctive (shaped like old Chianti bottles), each with an upper **globular** (about 3 mm wide) segment having strong wavy vertical ribs, abruptly narrowed to the shorter style above; and a narrow cylindrical lower segment $1/2 - 1/3$ length of stalk. Introd (Mediterranean); S Eng, Dublin area in Ire, o-vla on wa, gsld. Fl 5–9.

Pink and/or white fld Crucifers with elongated frs

A Hoary Stock, *Matthiola incana*, showy ann or per of erect, bushy habit, woody and lfless below like **Wallflower** (p 206, F); lvs narrow, lanceolate, the lower in rosettes, the upper sessile, all ± **untoothed**, grey-downy. Fls 2.5–5.0 cm across, in loose racemes, with hairy sepals and purple, red or white petals; fragrant. Frs (**Aa**) 4.5–13.0 cm long, ± erect, narrow, compressed, downy, **not glandular**. Br Isles: probably native on S coast Eng and S Wales; scattered elsewhere, Ire, Scot vr; on sea cliffs. Fl 4–7.

B **Ire VU ****Sea Stock**, *M. sinuata*, bi with non-woody stem; lvs **wavy-edged** to **lobed**, grey-woolly, not bushy; looser infl of purple-pink fls (**Ba**); frs 7–12 cm, downy, and with **sticky glands**. SW Eng, S Wales, CI vr; on sea cliffs and dunes. Fl 6–8.

C Dame's-violet, *Hesperis matronalis*, tall bi or per herb, with lfy stems, 40–90 cm tall, branched, ± hairy. Lvs lanceolate, toothed, pointed, stalked, and hairy. Fls 18–20 mm across, stalked, v fragrant, violet or white, in racemes. Frs (**Ca**) 9 cm long, curving upwards, hairless. Introd (S Eur); Br Isles, vc in shady, damp places on riverbanks and wa. Fl 5–7.

D Sea Rocket, *Cakile maritima*, ann herb with stems that sprawl, then rise to bear succulent, shiny, hairless, simple or pinnate-toothed, linear to oblong lvs; and lilac-pink or white fls, 6–12 mm across, in racemes. Frs (**Da**) 10–25 mm, on thick stalks, lower joint shorter, top-shaped, upper joint longer and mitre-shaped. Br Isles, f-lc on sandy seashores. Fl 6–9.

E Aubretia, *Aubrieta deltoidea*, a prostrate, mat-forming hairy per to 30 cm, with **star-shaped** hairs; lvs kite-shaped, apparently stalkless; fl violet (occasionally white), petals 12–28 mm long; fr ellipsoid with a long **persistent style** >4 mm. Introd (Sicily, E Eur); Eng, Wales lf (c in Devon, Somerset); Ire, Scot r; gdn escape natd on paths, wa, walls, quarries. Fl 3–7.

F Honesty, *Lunaria annua*, erect hairy bi to 1.5 m; basal lvs heart-shaped and long-stalked, lower stem lvs opposite; fls purple (or white), petals 15–24 mm long; fr (**Fa**) **distinctive**, ± flat and circular, membranous, looking like a silver coin, with large flattened seeds in 2 rows. A well-known gdn escape, the frs used for decoration. GB c, Ire o-f; increasingly natd on rds, wa, wds. Fl 4–6.

G ***Coralroot**, *Cardamine bulbifera* (*Dentaria bulbifera*), per herb with slender erect hairless unbranched stems, 30–70 cm, rising from a fleshy rhizome (**Gb**). Lower lvs stalked, pinnate, upper three-lobed or simple; lfts **dark** green, stiffly spreading, lanceolate, acute, ± toothed, v sparsely hairy, with strong veins. Axils of upper lvs bear purple-brown **bulbils** (**Ga**) by which plant normally reproduces. Fls in a short raceme, almost like an umbel, are rose-pink, 12–18 mm across. Frs elongated, but rarely ripening in GB. SE Eng r to vla; in dry sandy or chky wds. Fl 4–5 (v briefly only). **Do not confuse** with a garden variety (forma *ptarmicifolia*) which occurs as an o gdn escape,

distinguished from G by lf-shape, having broader, strongly toothed lvs (see illustration in *Plant Crib*).

A Cuckooflower (Lady's-smock), *Cardamine pratensis*, v variable per herb to 60 cm, sometimes with runners. Lower lvs (**Aa**) have large round lfts, terminal lft much larger than side ones, and kidney-shaped; upper lvs with **narrow** lfts, including terminal lft; all sparsely hairy. Fls rose-pink to white, 12–18 mm across; fr 24–40 mm, long narrow and straight. Br Isles, vc on mds, streamsides, moist wds. Fl 4–6.

B Wavy Bitter-cress, *C. flexuosa*, slender hairy bi to per herb to 50 cm, with erect but **v wavy stem**. Lvs pinnately compound; lfts oval or rounded below, narrower on upper lvs, all sparsely hairy, especially terminal lft. Raceme short at first, then elongates; fls white; small narrow petals, 2–4 mm long, are twice length of sepals. **6 stamens** (do not overlook 2 stamens that are shorter than the rest); frs narrow, 12–25 mm, ascending at an angle. Br Isles, vc usually in damp places, streamsides, wa, gdns. Fl 3–9. **B.1 Hairy Bitter-cress**, *C. hirsuta*, is smaller, ann, straight or wavy stemmed, v like B, but **key difference** is fl with **4 stamens**. Br Isles, vc in **drier** places than B, wa, gdns, walls, rocks. Fl 3–9.

C FPO (Eire) **Ire AWI NT *Narrow-leaved Bitter-cress**, *C. impatiens*, is like a tall version of B, with a stouter, ridged stem to 60 cm, but lvs have ± **strongly-toothed** lfts, in up to 6–9 pairs per lf; conspicuous **stipule-like** pointed **auricles** (**Cc**) clasp stem at lf-base. Fls (**Ca**) like B, but usually **without** petals (**Cb**); 6 yellow stamens. Frs, 18–30 mm x 1 mm, ascend stiffly at 45°. Pods burst explosively to scatter seeds. GB to Mid Scot lf (abs SW Eng); C Ire vr (probably only native in Westmeath); in rocky (limestone) wds, especially on scree, and on riverbanks in SE Eng. Fl 5–8.

D **Ire AWI **Large Bitter-cress**, *C. amara*, also resembles B but is per, ± hairless, taller, and all its pale green lvs (**Db**) have **oval** side and terminal lfts. Fls (**Da**) **large**, 12 mm across, white, with conspicuous **violet** anthers; frs 20–40 mm long, spreading at 45°. Br Isles: Eng, Wales, NE Ire; la in springy wds, fens, streamsides, often with Golden-saxifrages (*Chrysosplenium* spp p 244). Fl 4–6.

E Water-cress, *Rorippa nasturtium-aquaticum*, hairless per with creeping stems and erect flowering shoots; lvs, rather like those of D, but darker green, and with broader lfts below, persisting in winter. White fls (**Ea**) much smaller (4–6 mm across) than D's; stamens **pale**. Frs (**Eb**) only 13–18 cm, stouter (2–4 mm wide), curved, and with seeds visible in **2 rows** inside pod. Br Isles, vc in streams, ditches etc, in running water. Fl 5–10. **Do not confuse** with **Fool's Water-cress** (p 344 C) or **Lesser Water-parsnip** (p 344 B) which both have similar lvs but with **sheathing**, inflated lf-bases typical of the Umbellifer family.

F Narrow-fruited Water-cress, *R. microphylla*, differs from E in having longer pods (**F**) with only **one row** of seeds, and lvs turning purple-brown in autumn. Hybridizes with E (*R.* x *sterilis*) which is c, much planted and overlooked, infls often elongating to >30 cm and sterile, having only 0–3 seeds per rather distorted pod. However, sterility is **not** necessarily the **key difference** as the parents E and F are sometimes sterile (see *BSBI Crucifers Handbook*).

G Hairy Rock-cress, *Arabis hirsuta*, short to medium hairy bi or per herb, 10–60 cm; basal rosette of hairy oval-oblong lvs, scarcely toothed, narrowed to base; narrower oblong **stalkless**, **clasping** lvs up stem. Fls many, white, small 3–4 mm across, in a long erect raceme; frs (**Ga**) narrow, stiffly erect, ± flattened, 15–50 mm x 1.3 mm. Br

4 mm

4 mm

4 mm

Isles, lc on calc gslds, rocks, dunes (dune forms look different, with non-clasping stem-lvs). Fl 5–8. **G.1** Sch 8 EN ****Alpine Rock-cress**, *Arabis alpina*, has oval, **toothed**, basal lvs with **lf-stalk** much **shorter than** leaf; fls 6–10 mm across. Scot: mt rocks in Skye, introd in single localities in Somerset and Yorkshire, vr. Fl 6–8. **G.2 Garden Arabis**, *A. caucasica*, a per mat-forming herb close to G.1 but with **only 1-2 teeth** on lf-margins and basal lvs with lf-stalk about as long as lf; densely **star-shaped** hairy; fls white, **large**, each petal > 10mm long. Introd (S Eur); Br Isles, o but vla in places (eg Derbyshire limestone); natd on walls, rocks and cliffs. Fl 3–6.

A Sch 8 VU ****Bristol Rock-cress**, *A. scabra* (*A. stricta*), lower (8–25 cm) per herb, with one or **more** stems less erect than in G (p 210), stem hairless above, hairy like the lvs below. Rosette-lvs oblong-lanceolate and wavily-lobed, darker green than in G (p 210); stem-lvs hairy, few, clasping as in G (p 210). Infl of only 3–6 fls, each 5–6 mm across; petals creamy-yellow with a **claw without teeth**; frs (**Aa**) 25–40 mm, spreading. Eng: **about Bristol only**, vr; on open limestone rocks. Fl 3–5.
A.1 FPO (Eire) ******Ire VU ***Northern Rock-cress**, *A. petraea* (*Cardaminopsis petraea*), another similar-looking per, v variable in habit and much more f than A; lvs shallowly-lobed with larger end-lobe; fls white (sometimes lilac), **claw** with a **pair of teeth**; frs spreading. Br Isles: N Wales, Scot, W Ire; lf on mt rocks. Fl 6–8.

B **Thale Cress**, *Arabidopsis thaliana*, resembles a smaller ann to bi slender version of G (p 210), 5–50 cm tall, with a similar lf-rosette of stalked elliptical hairy lvs, but is hairless on upper stems and lvs. Fls 3 mm across, white; frs 10–18 x 0.8 mm, narrow, **curved**, borne **obliquely** on slender **spreading** stalks, **cylindrical** with **convex strong-veined** valves. Br Isles, c on walls, pavements, dry sandy open gd, not on chk. Fl 4–7. A plant much used in laboratory experiments due to its short life-cycle, and the first living thing to have its genome mapped.

C **Garlic Mustard**, *Alliaria petiolata*, distinctive ± hairless, erect per herb, 20–120 cm; cordate, toothed, glossy, thin lvs, long-stalked below, **smell of garlic** when bruised. Fls in a raceme, white, 6 mm across; frs (**Ca**) diverging from stem, curving up erect, 35–60 mm, and angled. Br Isles, vc in hbs, open wds (especially on chk). Fl 4–7.

D ******Ire **Hoary Whitlowgrass**, *Draba incana*, bi to per herb with short creeping stock, bearing erect hairy stems 7–50 cm tall. Basal lvs in rosettes, v hairy (hairs star-shaped), oblong, narrowed to a stalk below, entire or with a few teeth. Stems with **narrow**-elliptical, hairy, erect lvs, stalkless and slightly clasping at base, usually toothed. Infl at first dense, elongating as frs ripen, with white fls, 3–5 mm, with slightly notched petals (**Da**); frs (**Db**) 7–9 x 2.0–2.5 mm, erect, elliptical, **twisted** when ripe. Br Isles: N Ire, N Wales and N Eng to Scot; vl on screes, cliffs, rocks, dunes, especially on limestone. Fl 6–7. **D.1** ***Rock Whitlowgrass**, *D. norvegica*, smaller plant with ± leafless fl-stems and pods **not twisted**, 5–6 mm long. Scot, vr on a few high mt rocks. Fl 7–8. **D.2** NT ****Yellow Whitlowgrass**, *D. aizoides*, has narrow lvs, all basal, heads of **yellow** fls, 8 mm wide, and flat elliptical frs 6–12 mm. S Wales only (introd Dorset); vr on limestone. Fl 3–5.

E ***Wall Whitlowgrass**, *D. muralis*, is ann, shorter (8–20 cm) than D, much less hairy above; **broadly oval**, clasping, **toothed stem-lvs**; petals (**Ea**) unnotched; frs (**Eb**) 3–6 x 1.5–2.0 mm, flat, not twisted. Eng, vl on limestone rocks; also introd on walls scattered throughout Br Isles. Fl 4–5.

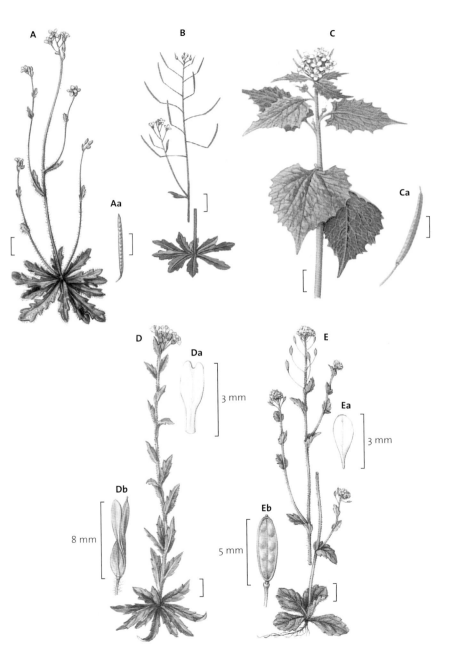

A

Aa

B

C

Ca

D

Da

3 mm

Db

8 mm

E

Ea

3 mm

Eb

5 mm

213

White fld Crucifers with short frs

Pepperworts (*Lepidium*) and **Penny-cresses** (*Thlaspi*) both have winged frs, but *Lepidium* has **one** seed per cell (2–5 in *Thlaspi*).

A Sch 8 BAP VU ****Cotswold (Perfoliate) Penny-cress**, *Thlaspi perfoliatum*, ann, 5–25 cm, smaller than C below; fl (**Ab**); lvs **waxy** grey-green, basal lvs obovate, stalked, stem-lvs with **deeply clasping rounded bases**; cordate winged frs (**Aa**) only 4–6 mm long x 3.5 mm wide, narrowed to stalk; style shorter than fr-notch. Gloucs and Oxfordshire, vr (r introd elsewhere); on limestone scree. Fl 4–5.

B ***Alpine Penny-cress**, *T. caerulescens* (*T. alpestre*), per of varied height (10–40 cm) with fls and fr (**Ba**) similar to those in A, except that the style is as long as or longer than the notch, anthers are usually violet, not yellow, and the basal rosette lvs **spoon-shaped** and **long-stalked**, while auricles of the green stem-lvs are pointed. GB, vl on limestone and other basic rocks from Somerset, Wales to Scot. Fl 4–8.

C **Field Penny-cress**, *T. arvense*, branched erect hairless ann, 10–60 cm, with broad-lanceolate, toothed clasping stem-lvs; no basal rosette; fls white, 4–6 mm, with yellow anthers; frs (**Ca**) almost circular and coin-shaped, flattened, with v **broad wings**, 12–22 mm across, deeply-notched above, with style shorter than notch, on stalks curved upwards. Br Isles, c in ar, disturbed rds and wa. Fl 5–7.

D **Field Pepperwort**, *Lepidium campestre*, ann or bi herb 20–60 cm; basal lvs entire or lyre-shaped, soon withering; stem-lvs triangular, stalkless, clasping with pointed basal lobes; all softly hairy and toothed. Inconspicuous white fls 2.0–2.5 mm, in long racemes; 6 stamens, anthers yellow. Frs (**Da**) 5 x 4 mm, covered with minute bumps (use hand lens), **style short**, not projecting beyond notch of fr. Br Isles, generally f in lowlands, except to N and IoS, CI; on dry sandy or gravelly soils on gslds, ar, wa, hbs, rds. Fl 5–8.

E **Smith's Pepperwort**, *L. heterophyllum*, differs from D in being per (although dying back after fruiting so not obviously per), less erect; many-stemmed; **key differences** from D are the frs (**Ea**) with **few or no** minute bumps and **style projecting** well beyond notch of fr (**E**). Br Isles, including IoS and CI, f in similar habitats to D. Fl 5–8.

F **Garden Cress**, *L. sativum*, is the salad plant; v variable, looking nothing like the seedlings in egg sandwiches; a single erect-stemmed **hairless** ann with Cress smell; upper stem-lvs pinnate, not clasping; fr (**F**) style not projecting on capsule. Br Isles, r gdn escape in ruderal habitats, wa. Fl 5–11.

G **Narrow-leaved Pepperwort**, *L. ruderale*, is hairless, strongly cress-smelling, more slender than D and E but often forming stiff, dense tufts; lower lvs pinnate, upper lvs simple, not clasping. Usually 2 or 4 stamens (not 6); **key difference** from other *Lepidium* spp in frs (**Ga**), shorter than their stalks, **less than 2 mm wide**; style v short, **within** fr-notch. Eng, Wales f in SE, r elsewhere; Scot vr; in ruderal habitats, on grassy banks near sea, wa, rds, pavements. Fl 5–7.

H **Hoary Cress**, *L. draba* (*Cardaria draba*), bushy per herb, 30–90 cm, with runners; branched stems; oblong auricled, clasping wavy-toothed and usually hairy (but not always), **grey-green** lvs (**Hb**); fls 5–6 mm, white; fr (**Ha**) 4 x 4 mm, cordate, tapering into the style above. Racemes in dense umbel-like heads. Introd (S Eur) with bedding of sick troops from 1809 Walcheren expedition; Br Isles: f throughout (except Ire, r-o) and vc in S Eng; on ar, wa, rds, salt-marshes. Fl 5–6.

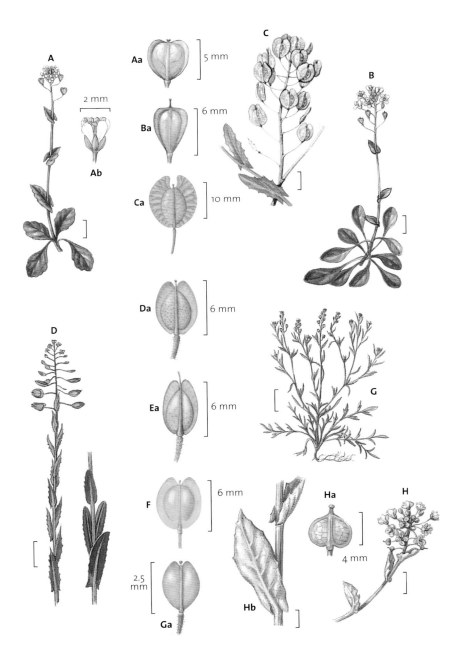

A ***Dittander**, *Lepidium latifolium*, much the tallest, stoutest *Lepidium*; a per, 50–130 cm, with long-stalked oval simple basal lvs to 30 cm x 5 cm, sometimes with side-lobes below, grey-green, but **hairless**. Infl a large branched pyramidal panicle; fls (**Aa**) 2.5 mm across, white; sepals **white-margined**; frs (**Ab**) elliptical to rounded with no notch, downy, hardly winged, style v short. Br Isles: Eng, Wales, CI o-lf; S Ire vl; Scot vr; on saltmarsh edges, wa inland and near sea. Fl 6–7.

B **Swine-cress**, *Coronopus squamatus*, prostrate ann or bi, with spreading stems 5–30 cm long; stalked, deeply pinnate, scarcely hairy lvs with ± toothed segments; stalkless heads of fls on stem-tip and **in lf-axils**; fls 2–2.5 mm across, white; stamens 6; frs (**Ba**) 2.5 x 4 mm, notched below, tapered into short style above, with **strong ridges and pits** on the carpels, not opening when ripe, instead breaking off whole to be pressed into the gd by hoofs, soles and treads. Br Isles, c (except Ire, f; Scot, o-r); on wa, ar, especially on trampled places near gates. Fl 6–9.

C **Lesser Swine-cress**, *C. didymus*, a smaller version of B, with a strong cress smell when bruised; stamens usually 2, petals often none; **key difference** from B is the frs (**C**) only 1.5 x 2.5 mm, shorter than stalks, notched above as well as below; carpels rounded, pitted but **not ridged**. Introd (S America); Br Isles (except Scot, o-r); c in similar habitats as B. Fl 7–9.

D WO (NI) **Ire NT **Shepherd's Cress**, *Teesdalia nudicaulis*, neat ann or bi, hairless to ± downy, 8–20 cm tall; basal rosette of pinnate-lobed lvs with broader end-lobes and few stem-lvs (**none** on erect flowering central stem). Fls 2 mm, white, in short racemes, outer petals **longer** than inner; frs (**Da**) widely cordate, concave above, narrowed to base, 3–4 mm long; style v short. Br Isles: lc to S; Scot, Ire r; on open sandy gd, dry gsld, shingle, avoids chk. Fl 4–6.

E **Shepherd's-purse**, *Capsella bursa-pastoris*, variable ann or bi herb with erect main stem, 3–40 cm, hairy or not; rosette of pinnately-lobed to undivided lvs (**Ea**); **clasping stem-lvs** with basal arrow-shaped **auricles**. Fls in erect racemes; fls white, 2.5 mm across; frs (**Eb**) erect, like a triangle standing on its point, narrowed to base, notched above, **flat**, 6–9 mm long, on long stalks. Br Isles, vc weed of wa, rds, ar. Fl ± 1–12.

F VU ***Wild Candytuft**, *Iberis amara*, ann, with toothed glossy green lvs on ± erect, usually hairy (at least below) stems. Fls (**Fa**), 6–8 mm across, white or lilac, with two outer petals **much longer** than two inner; in flat-topped heads elongating in fr. Frs 3–6 mm, almost round, with wings and a deep notch above, in which is set the short style, convex when ripe. S and mid Eng vlc; on bare chky gd on downs and ar, especially where rabbits are c. Fl 7–9. **Do not confuse** F with **Sweet Alison** (p 218) or with **F.1** **Garden Candytuft**, *I. umbellata*, also an ann but is usually hairless, often with coloured petals and has larger frs 7–10 mm long with more pronounced, pointed lobes. Br Isles f gdn escape but casual on wa, rds, near gdns. Fl 5–10.

G **Common Whitlowgrass**, *Erophila verna* agg, is a group of v similar spp. Ann, 2–10 cm (rarely more) with a basal rosette of broad lanceolate lvs narrowed into wide stalks, minutely hairy. Fl-stems lfless; fls (**Ga**) in short racemes, petals white, 2.5 mm, deeply notched; frs (**Gb**) oval to elliptical, flattened, 1.3–3.8 x 1.5–9 mm long, long-stalked. Only white Crucifer with **notched petals** and **lfless fl-stem**. Br Isles, c-la on walls, rocks, open dry sandy gd. Fl 3–5, then soon disappearing.

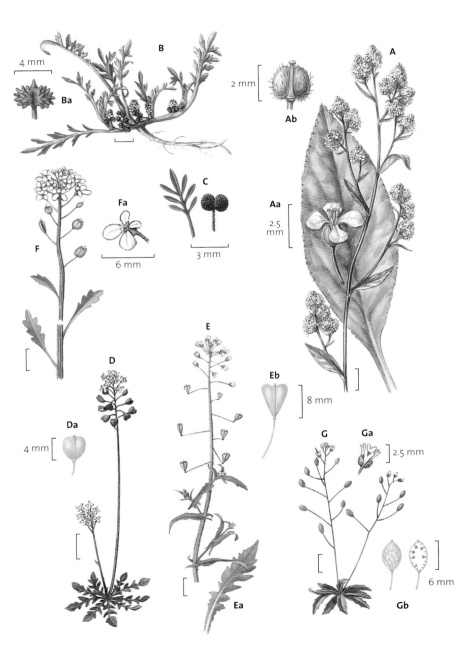

4 mm

B

Ba

A

2 mm

Ab

Aa

2.5 mm

F

Fa

C

6 mm

3 mm

E

Eb

8 mm

D

Da

4 mm

G   Ga

2.5 mm

Ea

Gb

6 mm

Scurvygrasses (*Cochlearia*) are seaside, inland salted rds or mt rock plants with **fleshy lvs**, white fls, and ± oval **swollen** frs. The spp are sometimes difficult to distinguish.

A   Common Scurvygrass, *Cochlearia officinalis* agg (including End BAP **Scottish Scurvygrass,** spp *scotica* and End BAP *****Mountain Scurvygrass**, *C. micacea*), is a group of v similar bi or per hairless herbs, with 1 or more ascending stems, 5–50 cm. Basal lvs in a loose rosette, long-stalked, kidney- to heart-shaped; stem-lvs triangular-ovate, fleshy, clasping, the upper lvs stalkless. Fls (**Aa**) 8–10 mm, white, rarely lilac; frs 3–7 x 2.5–6.0 mm, globose, rounded below, **swollen**, narrowed into style. Br Isles, lc on salt-marshes and cliffs (not SE Eng), also inland on basic rocks on hills. Fl 5–8.

B   Danish Scurvygrass, *C. danica*, low-growing plant, with stems 2–20 cm long; long-stalked cordate basal lvs (**B**), stem-lvs **stalked**, the lower ivy-shaped. Fls 4–5 mm. Frs (**Ba**) oval, **swollen, narrowed at both ends**. Br Isles, lc on sandy, shingly and rocky shores, banks near sea; also vc along salted main rds and in railway ballast inland. Fl 2–6.

C   English Scurvygrass, *C. anglica*, like A, but basal lvs (**C**) have bases **wedge-shaped, tapering** into stalks; fls 5–7 mm across; frs less swollen, blunt-tipped. Br Isles (including SE Eng), lc on coasts, salt-marshes. Fl 4–7.

D   Sweet Alison, *Lobularia maritima*, short tufted grey-hairy per with narrow-lanceolate, **untoothed** lvs. Fls (**Da**) 6 mm across, white, fragrant, in dense racemes which elongate (**Db**); frs (**Dc**) obovate, 2.5 mm, hairy, with convex sides, on spreading stalks. Introd (Mediterranean); Br Isles, f on coasts of Eng (especially S), also c gdn escape on wa, pavement cracks, rds in urban areas. Fl 4–9. **Do not confuse** with **Candytufts** (p 216) which have **unequal** petals and frs with **deep notch** above.   **D.1** Sch 8 **Small Alison**, *Alyssum alyssoides*, densely hairy ann or bi with erect branching stems and narrow-lanceolate, untoothed lvs similar to D but **key differences** are: fls **pale yellow** fading to white; frs ± as long as wide with narrow wing and **persistent sepals** that remain on the mature fr, dropping only when fr turns brown. Introd (W Eur); previously widespread in S and E Eng, now vvr in Suffolk on disturbed grassy and ar fields. Fl 5–9.

E   *****Hutchinsia, *Hornungia petraea*, small ann herb rather like **Shepherd's Cress** (p 216 D); stalked rosette lvs pinnately cut into small elliptical segments, but with similar stalkless lvs **up stem**. Fls 1.3 mm across, greenish-white; frs (**Ea**) 2–4 mm, narrow, **elliptical** to oval, hardly notched, compressed, held spreading on the stem. Br Isles: W, N Eng, Wales, CI r-lf; on limestone rocks and calc dunes. Fl 3–5.

White fld Crucifers with v large lvs

F   **Ire **Sea-kale, *Crambe maritima*, large, cabbage-like, squat per herb, 40–60 cm tall and to 100 cm across; basal lvs oval, long-stalked, waxy-grey, hairless, **v fleshy**, up to 30 cm long, with lobed, wavy margins; upper lvs smaller and narrower. Fls (**Fa**), white, 10–16 mm across, borne in large, rather flat-topped, much-branched heads to 30 cm across. Frs (**Fb**) are 6-15 mm long, with a stalk-like lower joint, and a spherical one-seeded non-opening **pea-shaped** upper joint. Br Isles (not N Scot) o but vla (Ire r-o); on shingle beaches and sand by sea. Fl 5–6 (rarely to 8).

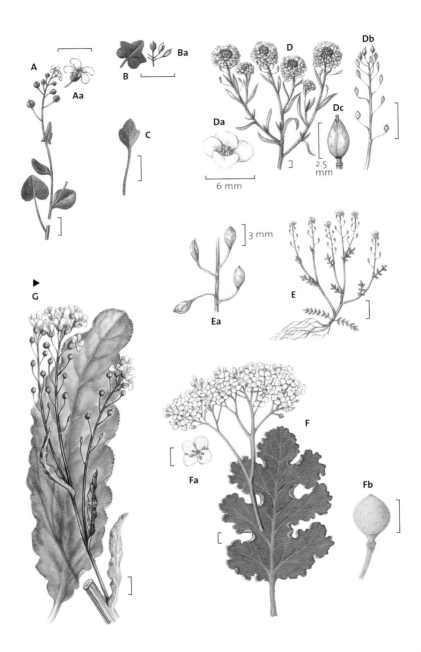

A

Aa

B

Ba

C

D

Da

Db

Dc

2.5 mm

6 mm

G

3 mm

Ea

E

F

Fa

Fb

▲ **G Horse-radish**, *Armoracia rusticana*, tall hairless per to 1.50 m with tap-root; erect lfy branched stems; erect **large** basal lvs 30–100 cm long, oval or oblong, shiny dark green, long-stalked, toothed or sometimes pinnately-cut. Panicles of white fls 8–9 mm across; ovoid frs that rarely ripen in Br Isles. Introd (SE Eur); Br Isles f-lc on wa, rds. Fl 5–8.

MIGNONETTE FAMILY   *Resedaceae*

A Wild Mignonette, *Reseda lutea*, a bi to per hairless herb, with ± erect or sprawling branched **ribbed** stems, 30–75 cm. Basal lvs in a **rosette**; stem-lvs many; all lvs deeply once or twice **pinnately** cut; fls 6 mm across, **greenish-yellow**, in conical racemes, not scented; **6** sepals; **6** petals, upper petals two- or three-lobed, the 2 lower undivided; stamens 12–20, bent down. Fr (**Aa**) an oblong warty capsule open at top even in fl; 3 carpels. Br Isles: Eng c; Wales, Ire, E Scot o-lf; on wa, gsld, disturbed gd, especially on chk. Fl 6–9. **A.1 White Mignonette**, *R. alba*, close to A but has **white** fls and 4 carpels. Introd (Mediterranean); Eng, Wales o; Scot, Ire vr; on rds, wa, cultivated gd. Fl 6–8.

B Weld, *R. luteola*, a bi rather similar to A, but 50–150 cm, stiffly erect, hardly branched, with **key difference** in all lvs lanceolate, **entire,** wavy-edged with pale midrib. Fls (**Ba**) in **long narrow** racemes, shorter-stalked, with **4 sepals, 4 petals**; fr (**Bb**) globular. Br Isles, c (r Scot, Wales in uplands); on disturbed gd, especially on chk. Fl 6–9.

CROWBERRY FAMILY   *Empetraceae*

C Crowberry, *Empetrum nigrum*, much resembles a member of the Heath family, with its linear alt untoothed glossy green lvs, 4–6 mm long, with rolled-back edges; its fls (male, **Ca**), however, have **6** tiny separate **pink sepals**, and are only 1–2 mm across. Plants dioecious, female plant bears black berries 5 mm wide. Br Isles, c in upland and mt areas, **abs S and E Eng – E of Exmoor**; on hths, moors, bogs. Fl 5–6.

HEATH FAMILY   *Ericaceae*

Shrubs (and, rarely, trees) with simple, usually narrow leathery, mostly evergreen lvs, without stipules. Fls normally radially symmetrical, with 4 or 5 of each part. Petals joined into a tube; stamens usually twice number of corolla-lobes, joined into an ovary of as many cells as carpels; ovary normally superior, but inferior in **Bilberries** (*Vaccinium*). Style undivided, with cap-like stigma. Anthers open by pores at tip, not by slits. Fr a berry or a capsule. Attractive, distinctive family, mainly low shrubs of acid soils.

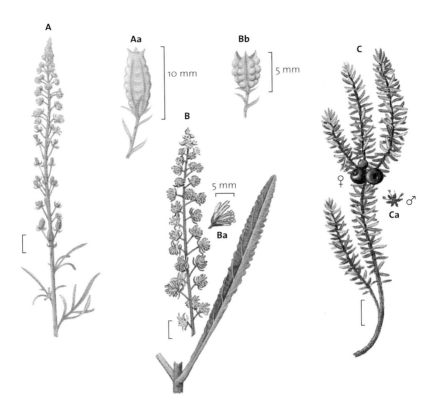

A

Aa

10 mm

Bb

5 mm

C

B

5 mm

Ba

♀

Ca ♂

KEY TO HEATH FAMILY

1 Ovary inferior ....*Vaccinium* (p 222)
   Ovary superior **2**

2 Lvs opp or in whorls **3**
   Lvs alt..**5**

3 Creeping tiny shrub of mts – lvs in opp pairs – funnel-shaped pink fls...
   Loiseleuria procumbens (p 224 **B**)
   Erect shrubs, 10–60 cm tall – fls not widely open – dead corolla persists at fr stage............**4**

4 Calyx coloured, four-lobed, longer than corolla – lvs in opp pairs, only 1–2 mm long,
   adpressed to stems...*Calluna vulgaris* (p 226 **A**)
   Calyx tiny, green or dark, shorter than corolla – lvs in whorls, spreading..............*Erica* (p 224)

5 Corolla widely bell-shaped, 5 lobes not all equal – shrubs with large (6 cm or more long)
   elliptical lvs ....*Rhododendron ponticum* (p 224 **A**)
   Corolla urn-shaped, lobes all equal **6**

6 Corolla four-lobed at tip of tube – fls in loose racemes – lvs elliptical, 5–10 mm long, white-woolly below ....*Daboecia cantabrica* (p 226 **F**)
Corolla five-lobed at tip of tube – lvs not white-woolly (but sometimes white-waxy) below...**7**

7 Fr berry-like, fleshy – lvs oval or elliptical **8**
Fr a dry capsule – lvs linear or linear-elliptical **11**

8 Trees or erect shrubs over 50 cm tall – lvs 4 cm or more long, veins not conspicuous **9**
Creeping shrubs less than 30 cm tall at most – lvs at most 2.5 cm long, conspicuously net-veined...**10**

9 Tree or tall shrub, 3–12 m tall – lvs tapering to base – fr red, warted, 1.5–2.0 cm wide, globular ....*Arbutus unedo* (p 224 **C**)
Low erect shrub to 1 m tall – lvs rounded to cordate at base – fr black, hairy, to 1 cm wide, globular...*Gaultheria shallon* (p 224 **D**)

10 Lvs evergreen, leathery, untoothed – fr red ....*Arctostaphylos uva-ursi* (p 228 **A**)
Lvs deciduous, thin, toothed – fr black ....*A. alpinus* (p 228 **A.1**)

11 Lvs linear, densely-set on stems, fine-toothed, blunt, green below, 5–9 mm long – calyx and fl-stalks sticky – corolla oval, purple – on mt moors in Scot – vr ....
Phyllodoce caerulea (p 226 **F.1**)

Lvs elliptical-linear, spaced out on stems, untoothed, pointed, edges rolled back, waxy-white below, 15–35 mm long – calyx and fl-stalks neither sticky nor hairy – corolla ± globular, rosy-pink or white – in bogs – vlf....................................*Andromeda polifolia* (p 228 **B**)

A   AWI **Bilberry** (**Whortleberry**), *Vaccinium myrtillus*, low hairless deciduous shrub, 20–60 cm; erect **four-angled** green twigs; lvs alt, oval, **finely-toothed**, flat, bright green, 1–3 cm. Fls 1–2 together in lf-axils; corolla globular-urn-shaped, greenish pink, 4–6 mm long, with 5 tiny teeth; **no** calyx-lobes; ovary inferior, ripening to an edible black berry (**Aa**) with a violet bloom. Br Isles (except mid, E Eng) c; on dry acid wds, upland hths, moors. Fl 4–6; fr 7–8.

B   **Cowberry**, *V. vitis-idaea*, has creeping stems from which ± erect branched shoots arise, to 30 cm tall. Twigs **round, downy** at first. Lvs bluntly oval-obovate, **evergreen** and like those of **Box** (p 305); dark green and glossy above, leathery, with edges **turned down**; pale green and dotted below, ± **untoothed**, 1–3 cm. Fls in short drooping racemes; calyx of 4 short rounded **lobes**; corolla bell-shaped, white or pink-flushed, the 4 lobes bent back half length of tube. Ovary inferior, ripening to an edible **red** berry (**Ba**). Br Isles: GB, SW Eng (vr) through Wales to Scot, c in hills, **abs in lowland S, mid and E Eng** (except vr introd, Berks); Ire r and l; on upland moors, wds, especially pine on acid soils. Fl 6–8. Hybridizes with A (*V. x intermedium*) (see *Plant Crib*).

C   **Bog Bilberry**, *V. uliginosum*, is deciduous and has urn-shaped corollas like A, but twigs **round, brown**; lvs **oval, blunt, untoothed**, blue-green with conspicuous netted veins. Fls 1–4 together in lf-axils, with **short** calyx-lobes; corolla 4 mm, **oval**-urn-shaped, pale pink. Berry (**Ca**) 6 mm, with a blue bloom. Br Isles: mts of N half of Scot f-la; N Eng vr; **abs in S Eng** (except vr Exmoor); **Ire abs**; high moors, hths. Fl 5–6.

D Cranberry, *V. oxycoccos*, tiny prostrate creeping shrub with **threadlike stems**, bearing oblong to narrow-elliptical, alt pointed lvs, 4–10 mm long, widely spaced along them; lvs dark green above, waxy grey below, with rolled-under edges. Fls 1–2, drooping, on erect **hairy** stalks 1.5–3.0 cm; 4 strongly **arched-back rosy-pink** petals 5–6 mm long; dark **stamens project** in a column (rather like miniature Cyclamen fls). Fr (**Da**) 6–8 mm, spherical or pear-shaped, red- or brown-spotted, edible. Widespread and c in N GB, Wales, Ire (but vr and l in S, E Eng, N Scot, SW Ire); in acid *Sphagnum* bogs. Fl 6–8. **D.1** *Small Cranberry*, *V. microcarpum*, close to D, **key differences** being fl-stalk ± hairless and lvs smaller, 2-6 mm x 1-2.5 mm wide. Scot only, o-lf on *Sphagnum* bogs. Fl 7. Hybridizes with D (see *Plant Crib*).

A **Rhododendron**, *Rhododendron ponticum*, tall, ± hairless evergreen shrub; lvs elliptical, leathery, pointed, dark green, entire, 6–12 cm long; fls conspicuous in rounded heads; corolla widely bell- or funnel-shaped, 5 cm across, dull violet purple, brown-spotted, with 5 long, unequal lobes, and 10 stamens. Ovary superior; fr a dry oblong capsule. Br Isles, c, widely introd and invasive (was native at least in Ire in Great Interglacial Period); in wds, hths, on acid, peaty or sandy soils. Fl 5–6. **Do not confuse** lvs with **Cherry Laurel** (p 250 E) which has tiny **teeth** on lf-margin.

B **Trailing Azalea**, *Loiseleuria procumbens*, tiny, prostrate mat-forming shrub; lvs **opp**, oblong, 3–8 mm long, with rolled-under edges; fls funnel-shaped, five-lobed, pink, 4–5 mm across, with 5 stamens. **Scot only**; f-lc on high mts. Fl 5–7.

C **Strawberry-tree**, *Arbutus unedo*, shrub or small tree to 12 m tall, with reddish-brown bark; lvs 4–10 cm, elliptical, leathery, shiny dark green above, paler below, toothed, pointed at both ends, short-stalked. Fls in panicles, with corolla oval urn-shaped, creamy, five-toothed, 7 mm long; 10 stamens. Fr 1.5–2.0 cm across, globular, dull red, warted. SW Ire, also by Loch Gill near Sligo, r but vla as a native; r introd elsewhere in GB; rocky oak wds. Fl 9–12; frs next 9–12.

D **Shallon**, *Gaultheria shallon*, evergreen shrub to 1.5m; alt lvs 5–10 cm, oval, leathery, rounded to cordate at base and edged with minute teeth; infl solitary in lf-axils or in terminal racemes; fls pink-white, 5 petals fused below into a tube, 10 stamens; fr (**Da**) a purple-black berry. Introd (N America), often planted for game cover, sometimes invasive. Br Isles o, except SE Eng f; on hths, wds, scrub usually on sandy or peaty soils. Fl 5–6.

KEY TO HEATHS (*ERICA*) AND LING (*CALLUNA*)

1 Lvs in opp pairs, stalkless adpressed upwards to stem, 1–2 mm long – calyx coloured like corolla, petal-like, deeply four-lobed, longer than corolla in fl and fr ....
Calluna vulgaris (p 226 **A**)
Lvs in whorls, stalked spreading, 2 mm long or more – calyx v small, shorter than corolla – corolla urn- or bell-shaped, conspicuous ....(*Erica*) **2**

2 Stamens enclosed wholly inside corolla-tube...**3**
Stamens projecting from corolla-tube...**6**

3 Lvs and sepals hairless – lf-edges rolled under completely obscuring underside ....
E. cinerea (p 226 **C**)
Lvs and sepals edged with bristles, usually with glands on tips – lf-edges partly rolled under, at least basal half of lf-underside visible **4**

4 Fls in racemes – lvs 3 in a whorl, oval ....*E. ciliaris* (p 226 **D**)
Fls in umbel-like heads – lvs 4 in a whorl, linear or oblong..**5**

5 Lvs grey-downy above, with long gland-tipped bristles – lf-edges curved under nearly to midrib – widespread...*E. tetralix* (p 226 **B**)
Lvs dark green, hairless above – lf-edges only slightly recurved, leaving white underside showing – W Ire only ....*E. mackaiana* (p 226 **B.1**)

6 Lvs 4 in a whorl – corolla tubular, ± parallel-sided – anthers only projecting halfway – fl-stalks shorter than fls – W Ire only ....*E. erigena* (p 226 **E.1**)
Lvs 4–5 in a whorl – corolla widely bell-shaped – anthers fully projecting– fl-stalks 3–4 times as long as fls – GB, N Ire ....*E. vagans* (p 226 **E**)

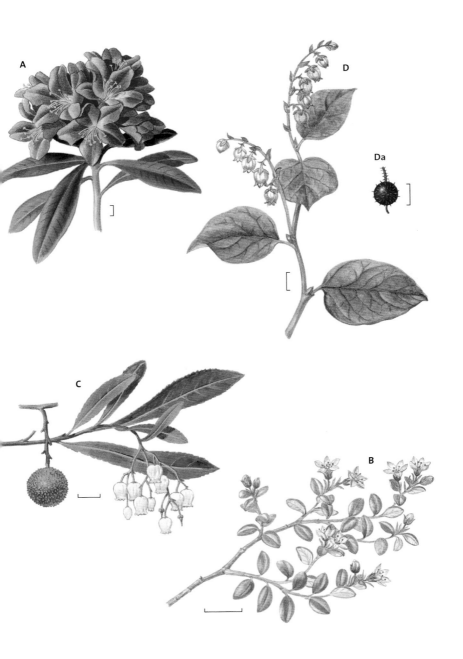

A Heather (Ling), *Calluna vulgaris*, bushy evergreen shrub to 60 cm, with many often tortuous stems. Lvs opp, 1–2 mm, **stalkless**, adpressed, triangular in shape, hairless or ± downy, close-set in 4 vertical rows. Fls (**Aa**) in dense spikes, each with 4 bracts; calyx 4 mm long, pinky-purple, deeply four lobed, longer than pink corolla, both persisting in fr; fr a capsule 2.0–2.5 mm long. Br Isles, vc, locally dominant; in hths, bogs, upland moors, open wds on acid soils. Fl 7–9.

B Cross-leaved Heath, *Erica tetralix*, shrub shorter than A (to 30 cm), with ± erect less-branched stems. Twigs downy; lvs **stalked, 4 in a whorl, greyish**, 2–5 mm long, linear, edges rolled under but **not** meeting below (at least basal half of lf-underside visible), downy above and also with glandular bristles. Fls in **umbel-like heads**; corolla 6–7 mm long, rose-pink, oval to urn-shaped; sepals 2 mm, both downy and with spreading glandular bristles like those on lvs. Stamens hidden in corolla. Br Isles, f-lc (r in highly-farmed areas); in wet hths, bogs, always on acid soils. Fl 6–9.
B.1 **Ire **Mackay's Heath**, *E. mackaiana*, like a denser-branched B, but lvs oblong-lanceolate to elliptical, edges only **slightly** rolled under, showing white underside (as in D below), hairless except for glandular bristles on their edges; heads of fls as in B, but corolla deeper pink; and sepals reddish, not downy but with short straight gland-tipped bristles above. **GB abs**; Ire: Connemara and Donegal r and vl; in blanket-bogs. Fl 8–9.

C Bell Heather, *E. cinerea*, is like B in habit, but quite **hairless**; lvs stalked, **3 in a whorl**, 3–6 mm long, linear, dark green, with edges rolled under, completely **obscuring** lf-underside; fls in short racemes; corolla 5–6 mm long, crimson-purple, oval to urn-shaped; sepals hairless, dark green, stamens hidden in corolla. Br Isles, vc except in highly-farmed areas of central Eng, central Ire; on dry acid hths. Fl 7–9.

D **Ire ****Dorset Heath**, *E. ciliaris*, often to 60 cm; lvs 3 in a whorl, 2–3 mm long, oval, not downy, but with long gland-tipped bristles on edges, (edges slightly turned under, showing white underside). Fls in racemes 5–12 cm long; corolla 8–10 mm long, oblong, urn-shaped, deep pink; sepals downy, with bristles. Eng: New Forest, Dorset, S Devon, Cornwall, vl but vla; Ire: Connemara vr; on dampish hths. Fl 6–9.

E WO (NI) **Ire ****Cornish Heath**, *E. vagans*, hairless shrub to 80 cm, with long erect branches; lvs 4–5 in a whorl, 7–10 mm long, linear, edges strongly bent back; fls in long dense racemes, lfy at tips; fl-stalks 3–4 times as long as fls; corolla 3–4 mm, widely bell-shaped, pale lilac, pink or white, with deep purple-brown anthers fully protruding. Eng: S Cornwall, vla; Ire: Fermanagh vr but la; vr introd elsewhere in GB; on dry hths. Fl 7–8. **E.1 Irish Heath**, *E. erigena*, is taller than other Heaths (to 2 m); lvs in **whorls of 4**, linear, 5–8 mm long, hairless; fls in long dense racemes, lfy at tips as in E; corolla **5–7 mm**, tubular, dull pink-purple, with reddish anthers protruding halfway. **GB abs**; Ire: Galway and Mayo vla on bogs and wet hths. Fl 3–5.

F St Dabeoc's Heath, *Daboecia cantabrica*, heath-like shrub to 50 cm; lvs 5–10 mm, elliptical, pointed, alt, green above, with scattered gland-tipped bristles, white-downy below, edges slightly rolled under. Fls in loose racemes; corolla 8–12 mm, rosy-purple or white, oval, urn-shaped, with 4 teeth; calyx-teeth 4, capsule hairy. Br Isles: Ire: Galway and Mayo lc as a native; S Eng, CI vr introd; on dry hths, rocks. Fl 7–9.
F.1 Sch 8 VU ****Blue Heath**, *Phyllodoce coerulea*, low erect shrub to 15 cm; lvs blunt, fine-toothed, **linear**, heath-like, green **both** sides; fls five-lobed, drooping, purple, urn-shaped, oval, long-stalked, 7–8 mm long, in terminal heads of 2–6; fl-stalks and calyx reddish, sticky. Central Scot, vr on rocky moors in mts. Fl 6–7.

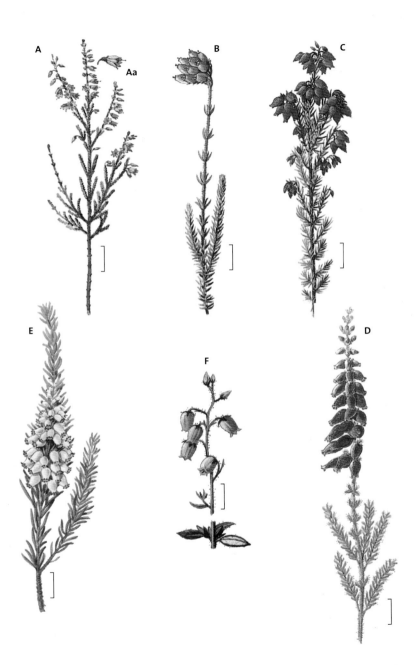

A

Aa

B

C

E

F

D

A Bearberry, *Arctostaphylos uva-ursi*, prostrate, mat-forming, ± hairless evergreen shrub; lvs (**Aa**) obovate, alt, blunt, leathery, dark green above, with conspicuous network of paler veins, paler below, **untoothed**, flat. Fls in short dense clusters; corolla 4–6 mm, globular to urn-shaped, white with pink flush; fr (**Ab**) a red shiny berry. GB: N Eng r, Scot vc; Ire to NW r-lf; on dry acid moors, especially in mts. Fl 5–7.
A.1 *Alpine Bearberry**, *A. alpinus* (*Arctous alpinus*), has similar habit to A, but lvs net-veined, **thin, toothed, deciduous**; fls white; berry black. NW Scot, vl on mt moors. Fl 8–10.

B wo (NI) **Bog-rosemary**, *Andromeda polifolia*, low hairless shrub to 20 cm, little branched; lvs 1.5–3.5 cm, elliptical-lanceolate, pointed, glossy grey-green above, edges rolled under, white below. Fls in small umbel-like clusters, long-stalked; corolla rosy-pink or white, globular-urn-shaped, five-toothed; fr a dry capsule. Br Isles: central Wales to mid Scot vl, decreasing to S; central Ire c; among *Sphagnum* in bogs in mt areas. Fl 4–9.

WINTERGREEN FAMILY   *Pyrolaceae*

A small family, related to the Heaths, the **Wintergreens** are creeping hairless per herbs with stalked **evergreen** simple lvs all in rosettes; infl erect, ± lfless, bearing racemes of symmetrical fls with 5 tiny sepals, 5 **free**, usually white, petals, 10 free stamens opening by **pores** at ends of the anthers as in Heaths; a superior ovary of 5 joined carpels with a single stout style on top; fr a round capsule. **Do not confuse** with **Lily-of-the-valley** (p 505 E) which has lvs with **parallel** veins.

C AWI **Common Wintergreen**, *Pyrola minor*, has lvs light green, oval, alt, ± blunt, 2.5–4.0 cm long, on stalks shorter than blades; infl 10–30 cm tall, rather dense; globular, pinkish-white fls (**Ca**) each 6 mm long; styles (**1–2 mm**) **shorter** than petals, straight; stigma with 5 spreading lobes. Br Isles, widespread but o: Scot, N Eng, f-lc; S Eng vl-lf; Ire r; in pinewoods, moors, damp rocks, acid dunes. Fl 6–8.

D ******Ire VU *Intermediate Wintergreen**, *P. media*, taller (15–30 cm) than C; lvs **larger** (3–5 cm-long), almost **round**, shiny **dark green**; fls in looser racemes, and larger (to 10 mm long), but also globular; petals whiter, with **longer** styles (5 mm) **projecting** from fls; stigma with 5 erect lobes. Br Isles: Scot Highlands o-lf; N Eng, vr; N Ire r; in pinewoods, moors. Fl 6–8. **D.1** WO (NI) **Serrated Wintergreen**, *Orthilia secunda*, is shorter (3–10 cm) than C and D, differs also in that lvs oval, **pointed, toothed**; fls, in **one-sided racemes**, are **greenish-white**, 5 mm long, cup-shaped rather than globular, with straight styles **longer** than petals. Br Isles: Scot Highlands o-lf; Ire vr; in pinewoods, and on damp rock ledges. Fl 7–8.

E FPO (Eire) ******Ire *Round-leaved Wintergreen**, *P. rotundifolia*, has dark green, shiny rounded lvs as in D, but longer-stalked; infl (10–30 cm) loose, fls (**E**) white, wide-open, almost **flat**, 12 mm across; **v long style** (8–10 mm) **curves down, then up** at the end. Br Isles: GB scattered and r; mainly Norfolk Broads, Kent, W Coast, Scot Highlands; Ire vr; on mt rock ledges, fens, open wds, dune-slacks, quarries, usually on **calc** soils. Fl 7–9.

F VU ****One-flowered Wintergreen**, *Moneses uniflora*, with **solitary**, drooping, large (15 mm), wide-open fls with straight, long styles, occurs in pinewoods in **Scot** vr. Fl 5–7.

229

BIRD'S-NEST FAMILY

Monotropaceae

WO (NI) EN AWI **Yellow Bird's-nest**, *Monotropa hypopitys*, erect per herb with **yellow** stem and **scale-like lvs** (8–20 cm tall) without any green chlorophyll, lives on decaying organic matter (a saprophyte). Infl **drooping** in fl, erect in fr, a raceme. Fls 10–15 mm long on short stalks; 4 or 5 strap-shaped sepals; 4 or 5 equal free petals with ± out-curved tips, all waxy pale yellow and of ± equal length, sometimes hairy within fl; stamens 10 in side fls, 8 in terminal one. Eng, o-lf in S and E, r to N and in Wales; Scot, Ire, vr; in wds, especially of beech or pine, dune hollows. Fl 6–8. **Do not confuse** with **Bird's-nest Orchid**, p 530 D, which is honey-brown, not yellow, and has obvious hanging two-lobed **lips** to the fls.

DIAPENSIA FAMILY

Diapensiaceae

Sch 8 VU ****Diapensia**, *Diapensia lapponica*, dwarf cushion-forming evergreen shrub, a few cm tall; lvs oval, blunt, leathery, toothless, 5–10 mm long, tapered to base, in dense rosettes; fls solitary, on stalks 1–3 cm; calyx five-lobed, leathery; corolla white, 1.0–1.5 cm wide, with 5 oval spreading lobes; stamens 5, short, yellow, alt with corolla-lobes; stigma three-lobed; fr a capsule. **NW Scot only**, vr on open rocky mt tops (arctic plant). Fl 5–6. **Do not confuse** with members of Heath family. Differs in 5 stamens only, and **three-lobed** stigma.

PRIMROSE FAMILY   *Primulaceae*

Per or ann herbs with lvs either in basal rosettes only, or along the stems; lvs without stipules; fls regular, usually with parts in fives, rarely more; corolla of petals joined into a tube below, wheel-, bell- or funnel-shaped, rarely abs. Stamens attached to corolla-tube and **opp** its lobes. Ovary superior (except in **Brookweed**, p 232 G) with 1 style; fr a one-celled capsule with many seeds. **Sea Milkwort** (p 236 G) is the only member of the Primrose Family with a perianth of **one whorl only** (ie. no petals, only sepals present).

KEY TO PRIMROSE FAMILY

1 Lvs all basal – fl-stems lfless **2**
Stem-lvs present..**3**

2 Corolla-lobes spreading or turned inwards ....*Primula* (pp 231–2)
Corolla-lobes strongly bent back....................................*Cyclamen hederifolium* (p 232 **H**)

3 Lvs pinnately-divided, with linear divisions ....*Hottonia palustris* (p 232 **F**)
Lvs simple **4**

4 Lvs alt (but often appear all basal) – calyx-tube adhering to ovary ....
....*Samolus valerandi* (p 232 **G**)
Lvs opp or whorled – calyx-tube free from ovary **5**

5 Lvs in one whorl on unbranched stem – fls white – corolla and calyx usually seven-lobed ..
....*Trientalis europaea* (p 236 **I**)
Lvs in several opp pairs (or whorls of 3) along stems..**6**

6 Fls yellow **7**
Fls pink, red or blue...**8**

7 Fls singly in lf-axils or in terminal panicles ....*Lysimachia* (p 234)
Fls in dense racemes in lf-axils ....*L. thyrsiflora* (p 234 **A.2**)

8 Calyx and corolla both present – capsule splitting across, its top falling off like a little hat
....*Anagallis* (p 234)
Corolla abs – calyx petal-like, pink – capsule bursting by 5 slits down sides ....
....*Glaux maritima* (p 236 **G**)

Most **Primulas** have two types of fl, on separate plants: one kind (**'pin-eyed'**) has pinhead-like styles visible at corolla-mouth, with stamens hidden below in tube; the other (**'thrum-eyed'**) has long stamens visible at corolla-mouth, with short style hidden below.

▼ **A** WO (NI) AWI **Primrose**, *Primula vulgaris*, per herb with rosette of v wrinkled obovate to spoon-shaped lvs, 8–15 cm long, **± unstalked** (rarely on a lfless common stalk), but narrowed gradually to base, downy beneath, shiny and ± hairless above. Fls upright, wheel- or saucer-shaped, are borne **singly** on woolly lfless stalks 5–12 cm long, arising from lf-rosette. Corolla (**Aa**) 30–40 mm across, pale yellow, with greenish veins, five-lobed, each lobe with shallow notch; thick folds at mouth of corolla-tube nearly close it. Calyx-tube woolly, nearly cylindrical, has 5 triangular narrow teeth. Scent violet-like in hot sunshine. Br Isles f-lc, r in mts; in wds, hbs, gslds in W. Fl 3–6.

B NT AWI *Oxlip, *Primula elatior*, per herb with lvs ± **abruptly** narrowing into **long winged stalk**, ± **downy** both sides, **paler green, less wrinkled** than in A and C. Fls 10–20 in **umbel**, turned to **one side**, ± **drooping**, on lfless common stalk (or **scape**) 10–30 cm tall; corolla (**Ba**) clear pale yellow, 10–15 mm across, **funnel-shaped**, throat of tube **open, no folds**; fls peach-scented in hot sun. Calyx and infl-stalk downy, not woolly. Eng: area from W of Bucks E into Suffolk and N Essex on chky boulder clay, replacing Primrose there, vla (outliers vr in nearby counties); in wds on damp base-rich soils. Fl 3–5. **Do not confuse** with hybrid of A and C (*P.* x *polyantha*), f with parents, resembles B with fls in umbel on a scape but throat of corolla-tube has folds and corolla-lobes have orange streaks within. Fertile hybrids also f between B and A (*P.* x *digenea*) in E Eng, which are intermediate in lf and fl shape and hairiness, and may be easily confused with C.

C WO (NI) **Cowslip**, *P. veris*, per herb v like B, but lvs more wrinkled, lf-stalks more gradually tapered to base, 5–15 cm long, downy both sides. Fls apricot-scented, 10–30 in an umbel, spreading, drooping, **not turned** to one side; corolla (**Ca**) apricot-yellow with orange streaks inside, 8–10 mm across, **cup-shaped** with **folds** in throat of tube; calyx downy, teeth oval, **blunt**. Br Isles: Eng, S Ire f-vla; Wales, Scot o-lf (W Scot vr); N Ire r; in mds, gslds, open wds, on calc or basic soils. Fl 4–5.

D VU *Bird's-eye Primrose**, *P. farinosa*, per herb; lvs only 1–5 cm long, obovate, shallow-toothed, **mealy-white** below; infl up to 15 cm tall, stalk mealy; fls 1 cm across, **rosy-violet**, with yellow eye, gaps between petals; calyx (**Da**) mealy, teeth pointed. N Eng, from near Skipton N to Durham and N Cumbria, o-la; **Scot, Ire abs**; on moist mds, open boggy gd on calc soils. Fl 5–6.

E End *Scottish Primrose**, *P. scotica*, v close to D, but lvs **untoothed**, oval, **widest in middle**, and equally mealy below; infl only to 10 cm; fls (**Ea**) on shorter stouter stalks, **purple** with yellow eye; stamens and style of same length on all fls, petals broader, oval, **touching** each other; calyx-teeth (**Eb**) **blunt**. Scot: N coast of Sutherland, Caithness, Orkney, vla; in damp mds, cliff tops, dunes. Fl 6–9.

F WO (NI) **Water-violet**, *Hottonia palustris*, has whorls of 2–10 cm long pinnate lvs (**Fa**) with linear lfts, like a **Water-milfoil** (p 294), but lfts **flattened**, and floating stems produce erect ± hairless, lfless infls 20–40 cm tall, with long-stalked, lilac-pink fls in whorls of 3–8 up stems. Corolla 20–25 mm across, with 5 lobes and yellow eye; calyx divided into narrow teeth nearly to base. Fr a globular capsule. Br Isles: E, SE Eng o-lf; Wales vr; **Scot, Ire abs** except as introd; in ponds, ditches with basic water. Fl 5–6.

G **Brookweed**, *Samolus valerandi*, unlike other members of Primrose family, erect hairless per with obovate untoothed lvs short-stalked at base of stem, stalkless and **alt** up infl-stalk (or appearing all basal). Fls in raceme, long-stalked, with cup-like calyx-tube **fused** to lower half of ovary and bearing triangular teeth; bell-shaped, five-lobed white corolla, 2–3 mm across. When not flg, plant forms a basal rosette resembling that of a **Daisy** (p 450 D). Br Isles: Eng, Wales, Ire, SW Scot o-lc in E and near coasts; on wet open grassy gd, fens. Fl 6–8.

H **Sowbread (Cyclamen)**, *Cyclamen hederifolium*, ± hairless per; lvs 4–8 cm, oval, cordate, five-angled, dark green above with whitish band near edge, purplish below, on long stalks arising from a large corm, (appearing in 9 after fls). Fls pink (or white), solitary on v long stalks, coiling spirally in fr; 5 reflexed petals 2.5 cm

A ◀ Aa

B Ba

C Ca

D

Eb Ea ⌉7 mm

E

F Fa

G

H

233

long, arising from short, five-angled corolla-tube. Gdn cultivar, long introd mainly in S Eng, CI in wds and shady places. Fl 8–9 (Spring flg gdn spp are vr natd).

A   AWI **Yellow Loosestrife**, *Lysimachia vulgaris*, erect downy semi-evergreen per, 60–150 cm tall; lvs 5–12 cm, opp or in whorls of 3–4, oval-lanceolate, pointed, **±
stalkless**, black-dotted; fls in terminal panicles, short-stalked; corolla 15 mm across, bright yellow; calyx-teeth narrow-triangular, with long hairs, **orange-
edged**. Br Isles f-lc (except N Scot vr); in fens, riversides, lakesides. Fl 7–8.
A.1 Dotted Loosestrife, *L. punctata*, v like A, but evergreen, lvs clearly **stalked**, oval,
not black-dotted, but hairy-edged; fls to 35 mm across, yellow, with a **purple-
brown eye**; calyx-teeth narrower than in A, sticky-hairy, **wholly green**. Introd (SE
Eur) and increasing; GB, f-lc in mds, hbs. Fl 7–10.   **A.2** *****Tufted Loosestrife**, *L.
thyrsiflora* (*Naumburgia thyrsiflora*), is like A in habit, but yellow fls are only 5 mm
across, in dense, rounded stalked **racemes**, 5–10 cm long, **in axils** of mid stem-lvs
only. Lvs blunt, stalkless, hairless, clasping, with many black glands; stamens
protrude slightly from fls. Br Isles: N Eng r; central Scot o-r; vr introd elsewhere; in
fens, lakesides, ditches. Fl 6–7.

B   AWI **Yellow Pimpernel**, *L. nemorum*, ± prostrate hairless per, to 40 cm long; lvs
oval, **pointed**, 2–4 cm long, short-stalked, in opp pairs along stems. Fls solitary in
lf-axils, on fine stalks as long as lvs; corolla yellow, wheel-shaped, with 5 oval
spreading **hairless** lobes; calyx-teeth 5 mm, v narrow. Br Isles, vc in wds, hbs on
less acid soils. Fl 5–9. **Do not confuse** with C, which has gland-dots (usually black)
on lvs.

C   **Creeping-Jenny**, *L. nummularia*, prostrate, creeping, **rooting** hairless per, rather
like B, but lvs **wide-oval** to **almost round, blunt**, with **gland-dots**, 1.5–3.0 cm long;
fls (**Ca**) 15–25 mm across, on **stout** stalks shorter than lvs; calyx-teeth 8–10 mm
long, **widely oval**, pointed; corolla yellow, **cup-shaped**, lobes **fringed** with tiny
hairs. Br Isles: GB N to central Scot, c to S and E, r to N and in SW Eng; Ire o; in
moist wds, damp mds, hbs. Fl 6–8.

D   **Scarlet Pimpernel**, *Anagallis arvensis*, prostrate to ascending hairless ann;
stems, to 30 cm long, four-angled, bear opp pairs of lvs, oval, pointed, stalkless,
with black dots below. Fls **solitary** on slender stalks in lf-axils; narrow, pointed
calyx-teeth nearly as long as corolla; corolla 10–15 mm across, flat, wheel-shaped;
usually **scarlet**, sometimes **blue** or pink; edges of oval lobes with **dense** fringe of
tiny hairs; capsule five-veined, opening transversely by a lid. Br Isles, c (except N
Scot r); on dunes, open gslds. Fl 6–8.

E   *****Blue Pimpernel**, *A. arvensis* ssp *foemina*, as D, but fls (**E**) always blue, up to 12
mm across, with v few tiny hairs on petal edges; calyx conceals corolla in bud.
Key difference from blue-fld forms of D is in number of cells in petal hairs (see
New Flora). S Eng, CI r; on ar on chk. Fl 6–8.

F   **Bog Pimpernel**, *A. tenella*, tiny hairless, **creeping** per, rooting at nodes; lvs short-
stalked, oval or ± circular, 5 mm long, in pairs along the delicate stems; fls solitary in
lf-axils on stalks 10 mm long; calyx-teeth v narrow; corolla to 14 mm across, funnel-
shaped, five-lobed, 2–3 times length of calyx, **white** with fine **crimson veins** (so
appearing pink); fr 3 mm across. Br Isles f-lc, more c in W; in wet mds, less acid bogs,
fens. Fl 6–8.

▲ **G Sea-milkwort**, *Glaux maritima*, creeping and ascending per, 10–30 cm long; stalkless, **fleshy**, opp, strap-shaped, blunt lvs, 4–12 mm long. Fls (**Ga**) solitary in lf-axils, 5 mm across, with **pink** five-lobed calyx, no petals, 5 stamens. Br Isles, vc on coasts, vr inland near salt springs etc; on salt-marshes, damp sand and dunes, shingle, seaside rocks, tidal riverbank mud. Fl 6–8. **Do not confuse** with **Sea Sandwort** (p 154 B) which has v similar opp fleshy lvs but 5 greenish-white petals, preferring drier habitats than G.

▲ **H** NT AWI **Chaffweed**, *Anagallis minima*, amongst smallest flowering plants of Eur, 2–7 cm tall; ± erect hairless ann; usually alt lvs 3–5 mm long, stalkless, oval, un-toothed, with black border beneath. Fls solitary in lf-axils, 1 mm across, with 5 tiny narrow pointed sepals and a shorter pinkish five-lobed corolla. Frs 1.5 mm across, globular, ± pink-flushed, like tiny apples. Br Isles, widespread but r; on open damp sandy gd on hths, wdland paths etc, often with **Four-Leaved Allseed** (p 158 C). Fl 6–7.

▲ **I Chickweed-wintergreen**, *Trientalis europaea*, slender erect per, 10–20 cm tall; single whorl of hairless obovate lvs, 2–8 cm long, at top of stem, from which arise 1 or 2 long-stalked white star like fls, 12–18 mm across; corolla usually with 7 oval lobes; fr a globular capsule. Br Isles: Suffolk vr, N Eng r, Scot o-lc, **Ire abs**; in wds, mostly of pine in Scot. Fl 6–7.

CURRANT FAMILY   *Grossulariaceae*

Small shrubs with alt, palmately-lobed lvs; fls in racemes; 5 petals, 5 sepals, ovary **inferior**; 2 styles, joined below; fr a **berry**.

A AWI **Red Currant**, *Ribes rubrum*, (*R. sylvestre*), deciduous shrub 1–2 m tall; lvs 3–5-lobed, **downy, unscented**, cordate at base; spreading racemes of 6–20 greenish fls; receptacle-cup (on which sepals sit) **five-angled** (**Aa**), **saucer-shaped**, with a raised rim round base of styles (section **Ab**); berries (**Ac**) **red**, 6–10 mm across, globular. GB, vc; Ire o; in moist wds, probably native but often a gdn escape. Fl 4–5.

B AWI **Black Currant**, *R. nigrum*, looks like A, but lvs hairless above, with glands beneath that have a **slightly aromatic to catty smell** when crushed; glands are usually **sessile**; berries (**B**) **black**, 12–15 mm across. Br Isles: GB, N Ire vc, S Ire o; in moist wds, fen carrs (native or introd as A). Fl 4–5.

C *Downy Currant**, *R. spicatum*, close to A, with red fr, but lvs (**C**) usually **without** a cordate base (check *several* lvs); the **key difference** from A is the shape of the fl-receptacle: **circular**, with **no raised rim** and **not saucer-shaped** above (section **Ca**). Br Isles: N Eng, Scot lf and native but elsewhere a gdn escape; Ire abs; wds on limestone. Fl 4–5.

D *Mountain Currant**, *R. alpinum*, has red frs like A, but plants **dioecious**; racemes always **erect**, bracts of fls **exceed** fl-stalks. Br Isles: N Wales, N Eng o-lf; native on limestone rocks (gdn escape elsewhere in GB), o-r. Fl 4–5.

E Gooseberry, *R. uva-crispa*, shrub with **spines** at lf-bases, fls only 1–3 together, reflexed petals; frs (**Ea**) oval, 10–20 mm long, bristly, yellow-green. GB, N Ire c; S Ire, NW Scot o-f; in wds, hbs (probably sometimes native in wet wds, also a gdn escape). Fl 3–5.

F Flowering Currant, *R. sanguineum*, lvs ± hairless above, with glands beneath that have an unpleasant, **tom-cat smell** when crushed; lvs close to B but glands are usually **stalked;** fls pink-red (rarely white) with a **corolla-tube** (unlike A–E) forming a drooping raceme, often fragrant; fr (**Fa**) blue-black, with a whitish bloom. Br Isles, f gdn escape in wds, hbs, rds. Fl 3–4.

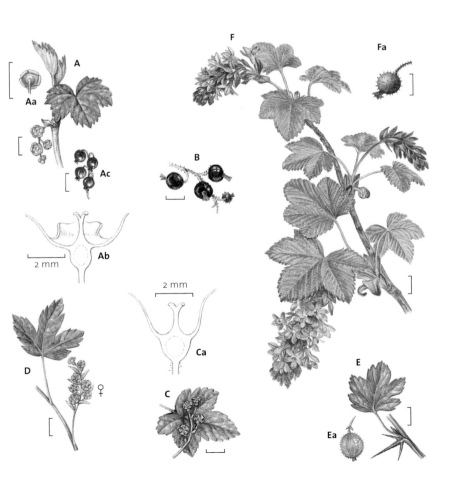

STONECROP FAMILY   *Crassulaceae*

A family of **succulents**, with undivided **fleshy** lvs, no stipules; fls mostly small, star-like, five-petalled, with 5 sepals, 5 or 10 stamens, and 5 separate carpels which develop into follicles. In **Stonecrops** and **Pigmyweeds** (*Sedum* and *Crassula*) petals are free; in **Wall Pennywort** (*Umbilicus*) petals are joined into a cylindrical tube, and lvs circular, attached by a central stalk. Other introd spp of *Sedum* occasionally occur but are not included here.

KEY TO STONECROPS & PIGMYWEEDS
(*SEDUM, SEMPERVIVUM, CRASSULA*)

1 Stamens as many as petals – lvs opp ....*Crassula*) **2**
Stamens 2x as many as petals – lvs alt or spiralled ....(*Sedum* or *Sempervivum*) **4**

2 Fl-stalks $\geq$ 2 mm ....*Crassula helmsii* (p 240 **H**)
Fl-stalks abs or < 1 mm **3**

3 Oval lvs 1– 2 mm long – petals shorter than sepals............................*Crassula tillaea* (p 240 **G**)
Linear lvs 3–5 mm long – petals longer than sepals................................*C. aquatica* (p 240 **H.1**)

4 Lvs broad, obovate – procumbent plant, stems rooting at nodes – 5 petals ....
Sedum spurium (p 239 **A.1**)
Lvs various but stems not rooting at nodes, erect plants **5**

5 Lvs narrow, much wider than thick, untoothed – 6 petals..
Sempervivum tectorum (p 239 **A.3**)
Lvs broad, flat, ± toothed – 4 or 5 petals **6**
Lvs narrow, either round in cross-section or flat above and rounded below, about as thick as wide, untoothed – 5 petals **7**

6 Fls yellow – petals 4 – obvious fleshy rhizome – on mts...................................*S. rosea* (p 239 **A.2**)
Fls pink-purple – petals 5 – rhizome neither obvious nor fleshy............*S. telephium* (p 239 **A**)

7 Petals pink – lvs alt, v sticky-hairy, reddish-flushed – on mts ....*S. villosum* (p 239 **F.1**)
Petals yellow **8**
Petals white..**11**

8 Small plants (to 10 cm tall) – lvs blunt, ± oval or cylindrical, 7 mm long or less **9**
Larger plants (15–30 cm tall) – lvs pointed, linear, parallel-sided, 8 mm long or more **10**

9 Lvs oval to blunt-triangular, adpressed to stem, taste peppery – widespread ....
S. acre (p 239 **D**)
Lvs cylindrical, spreading, taste mild – r ....*S. sexangulare* (p 239 **D.1**)

10 Lvs round in cross-section, spread evenly along fl-less shoots....................*S. rupestre* (p 239 **C**)
Lvs flattened on upper side, round below, clustered at tips of fl-less shoots
S. forsterianum (p 239 **C.1**)

11 Lvs opp, ± downy, oval, round in cross-section ....*S. dasyphyllum* (p 239 **F**)
Lvs alt, hairless..**12**

12 Lvs 6–12 mm long, green or reddish, elliptical – infl stalked, umbel-like, 7–15 cm tall ....
S. album (p 239 **B**)
Lvs 3–5 mm long, glaucous, often red-tinged, ± oval – infl little-branched, 2–5 cm tall ....
S. anglicum (p 239 **E**)

▼ **A** AWI **Orpine**, *Sedum telephium*, hairless, succulent, erect, rather waxy-lvd, per herb, 20–60 cm tall, often reddish below, with flat, toothed, alt oval-oblong lvs 2–8 cm long, up the stems; fls in dense umbel-like heads, red-purple, petals 3–5 mm long; stamens 10, 5 purple erect carpels. GB widespread but only lf (N Scot vr); Ire, IoM probably introd; on dry wd borders, hbs, especially on sand or gravel, also among rocks. Fl 7–9. **A.1 Caucasian-stonecrop**, *S. spurium*, creeping and **mat-forming** per herb; 5 petals 5–12 mm long, pink-purple (rarely white); f gdn escape on walls, rocks, rds. Fl 6–8. **A.2 Roseroot**, *S. rosea*, similar to A, but with obvious fleshy rhizome above gd level; dioecious **orange-yellow four**-petalled fls. Br Isles: Wales, N Eng, S Scot, W and N Ire, vlf; Scot Highlands f-lc; introd elsewhere; on mt rocks, sea cliffs. Fl 5–8. **A.3 House-leek**, *Sempervivum tectorum*, erect per with large basal rosette, lvs narrow but wider than thick, flat on upperside, untoothed; fl pink-purple, **6 petals**; Introd (mts C Eur) c. 1200, planted on roofs as a protection against fire and lightening; o on roofs, walls, quarry cliffs, sometimes dunes. Fl 6–7.

▼ **B** **White Stonecrop**, *S. album*, hairless evergreen per, 7–15 cm tall, with shiny, green to red-tinged, cylindrical-oblong, blunt lvs 6–12 mm long, alt along both creeping and erect sterns. Fls 6–9 mm across, in branched umbel-like cymes; 5 petals, **white** or pink-tinged, spreading. Br Isles widely introd, c (probably native SW Eng); on rocks, walls, sea cliffs. Fl 6–8.

▼ **C** **Reflexed Stonecrop**, *S. rupestre* (*S. reflexum*), taller than B (10–30 cm); lvs, on creeping stems, deeply-set, longer (8–20 mm), alt, cylindrical, pointed; spur-like projections on the lf-bases; dense umbel-like heads of **yellow** fls 15 mm across, usually with 7 petals. Br Isles, widely introd (S Eur) in Eng, Wales, Ire; on walls, rocks, dry banks. Fl 6–8. **C.1** *****Rock Stonecrop**, *S. forsterianum*, similar, but native plant with lvs **flat** above, forming dense heads on tips of sterile shoots. Br Isles: SW Eng and Wales on rocks in well-drained soils; o introd elsewhere on wa, railways. Fl 6–7.

▼ **D** **Biting Stonecrop**, *S. acre*, low (2–10 cm) mat-forming plant; lvs adpressed, **egg-shaped**, fleshy, yellow-green, only 3–5 mm long, with **peppery** taste; infls branched, not forming dense heads, fls 12 mm across, yellow. Br Isles, c on dry gsld (often where chky), dunes, beaches, walls. Fl 6–7. **D.1 Tasteless Stonecrop**, *S. sexangulare*, has **no** peppery taste; lvs spreading, **cylindrical**; fls smaller (9 mm across), yellow. Introd (C Eur); GB r; S Ire vr; on old walls, hbs, cliffs. Fl 7–8.

▼ **E** **English Stonecrop**, *S. anglicum*, low (2–5 cm) mat-forming evergreen per; lvs alt, egg-shaped, glaucous, fleshy, usually red-tinged; little-branched infls of few white fls, 12 mm across, pink-tinged on back of petals. Br Isles, vc W coast, lc S, E coasts, o scattered inland; Ire f-lvc; on acid rocks, coastal shingles, dry gsld. Fl 6–9.

▼ **F** **Thick-leaved Stonecrop**, *S. dasyphyllum*, is close to E, but has **sticky-downy** grey-green lvs in **opp** pairs on **erect** sterile shoots (**F**); fls white, 6 mm across. Introd (S Eur); Br Isles: S Eng, Wales, S Ire o introd; on old walls, quarries, limestone rocks. Fl 6–7. **F.1** NT *****Hairy Stonecrop**, *S. villosum*, has erect stems 5–15 cm; lvs alt, **linear**, blunt, **reddish, hairy** with many **sticky** glands, **flat** above; fls in open cymes, **pink**. N Eng to central Scot lowland o-vlf; in wet stony places on basic soils in mts. Fl 6–7.

G ***Mossy Stonecrop**, *Crassula tillaea*, resembles a minute ann *Sedum*, with creeping and ascending lfy stems 1–5 cm long, that become **bright red in fr**; dense-set fleshy lvs, 1–2 mm long, joined in opp pairs; fls (**Ga**) white, 1–2 mm across, **stalkless**, in axils of most lvs; usually 3 sepals, 3 (or 4) petals, **shorter than** sepals. S Eng, E Anglia, vlf; CI c; introd N Scot, N Ire vr; on bare sandy, gravelly gd on hths. Fl 6–7.

H New Zealand Pigmyweed (**Australian Swamp Stonecrop**), *C. helmsii*, per aquatic plant with stems either erect or trailing in water or on mud; lvs linear, 4-15 mm long, ± fleshy with **dark rings** below nodes; fls (**Ha**) whitish**,** on **long stalks** (> 2 mm) with 4 petals longer than sepals. Introd from discarded aquarium plants. Highly invasive and increasing; S Eng c, Ire, Scot, Wales r, but vla; aquatic or on mud around pond or lake margins. Fl 6–8. **Do not confuse** with **Water Starworts** (p 384) which also have opp linear lvs but are **not** fleshy and lf-tip often notched; or with **Water Purslane** (p 295 F) which **lacks** dark rings below nodes.   **H.1**   Sch 8 VU
****Pigmyweed**, *C. aquatica*, is an ann of pond margins; similar to H, but narrower, green, linear lvs 3-5 mm long; 4 petalled fls, **stalkless**, petals **longer than** sepals. Probably native; vr, one place in Scot, but vla. Fl 6–8.

I **Navelwort** or **Wall Pennywort**, *Umbilicus rupestris*, erect hairless per herb, 10–40 cm tall; lvs, mostly from the base, circular, attached by a central long fleshy stalk (resembling hollow-centred navels), fleshy, bright green with rounded teeth, 1–7 cm across. Infl a long spike of many greenish-white fls (**Ia**) with cylindrical or bell-shaped five-toothed corollas, 8–10 mm long. Br Isles: GB vc to W, N to Argyll, rarer E to Kent; Ire f-c; in crevices of acid rocks and walls. Fl 6–8.

SAXIFRAGE FAMILY   *Saxifragaceae*

A family of mostly per herbs, with simple but often lobed lvs, 2 stigmas, and capsular frs formed of 2 carpels joined only below; sepals joined below to form a shallow cup. Stamens twice number of sepals. **Saxifrages** proper (*Saxifraga*) have 5 sepals, 5 petals, 10 stamens; **Golden-saxifrages** (*Chrysosplenium*) have 4 sepals, no petals, 8 stamens. **Grass of Parnassus**, sometimes placed in its own family *Parnassiaceae*, has white fls with green veins, 5 fan-shaped sterile stamens alternating with normal ones, and 4 stigmas. Several other members of the family occur as r introd spp but only the most widespread are included here. Except for A (p 242) and B (p 243) *Saxifraga* are mostly mt plants only.

KEY TO SAXIFRAGE FAMILY

1 Petals abs – sepals 4 – stamens 8 – fls green ....*Chrysosplenium* (p 244 **J**, **K**)
  Petals normally present – irregular red/brown fl – sepals 5 – stamens 3 ....
                                                                              Tolmiea menziesii (p 244 **M.1**)
  Petals normally present – regular fl – sepals 5 – stamens 10 ....**2**

2 Fl greenish, petals with long, whisker-like lobes ....*Tellima grandiflora* (p 244 **M**)
  Fl white, yellow or purple, no whisker-like lobes on petals **3**

3 Lvs alt or basal only – fls white or yellow **4**
  Lvs opp – fls purple – stem creeping – mts only, r....................*Saxifraga oppositifolia* (p 243 **C**)

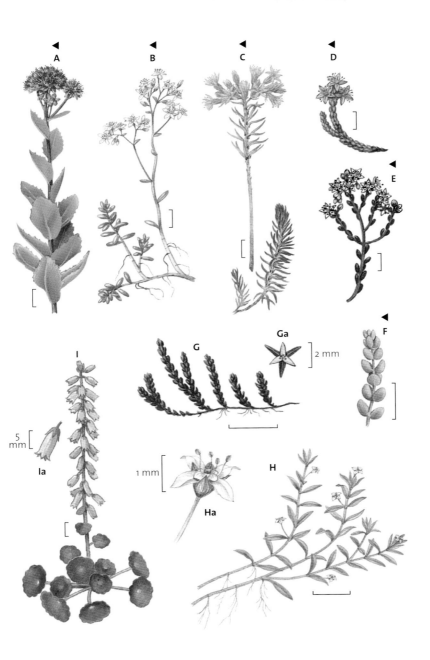

A

B

C

D

E

Ga

G

2 mm

F

I

5
mm

Ia

1 mm

Ha

H

4 Fls yellow **5**
Fls white, sometimes spotted red or yellow **6**

5 Ovary free from calyx – petals 10–15 mm long – lower lvs stalked – stem with red hairs ....
S. hirculus(p 244 **I**)
Ovary fused to calyx – petals 3–6 mm long – all lvs ± stalkless – stem with no red hairs ....
S. aizoides (p 244 **H**)

6 Fl-stem naked **7**
Fl-stem lfy (or one lf only) **10**

7 Lvs ± stalkless, teeth few – infl loose ....*S. stellaris* (p **G.1**)
Lvs stalked, closely toothed **9**

8 Fls unspotted, in a dense head – mts of Scot, Wales, N Eng, NW Ire vr ....
FPO (Eire) **Ire *Alpine Saxifrage* *S. nivalis*
Fls spotted, in an open branched infl **8**

9 Lvs kidney-shaped, bases cordate – lf-stalk hairy, to 1 mm wide ....*S. hirsuta* (p 244 **F**)
Lvs wedge-shaped at base – lf-stalk hairless, 2 mm broad or more..
S. spathularis, and hybrids (p 243 **D**, **D.1**, p244 **E**)

10 Erect plants – no barren, creeping or rosette-forming shoots **11**
Creeping plants of mossy habit..**14**

11 5 large staminodes present, alternating with 5 stamens – fl-stem with 1 fl and 1 lf only
Parnassia palustris (p 244 **L**)
No staminodes present – fl-stems with > 1 fl and/or > 1 lf...**12**

12 Lvs 3 or more fingered, tapered to stalk – ann ....*S. tridactylites* (p 243 **B**)
Lower lvs cordate at base, rounded at top – per..**13**

13 Tall (10–50 cm) lowland plant – lvs kidney-shaped, bluntly-lobed – fls many – widespread
S. granulata (p 242 **A**)
Shorter (3–15 cm) alpine plant – lvs kidney-shaped, the lower sharp-toothed – fls one or
none – red bulbils in upper lf-axils – Scot only, vr ....*S. cernua* (p 243 **A.1**)
Short (2–8 cm) alpine plant – lvs palmate, lobes blunt – fls 1–3 – mts of Scot, r ....
Highland Saxifrage *S. rivularis*

14 Lf-lobes linear, pointed, rarely >1 mm wide, lf-tip with long bristle-like point – never any
glandular hairs on lvs – fl-buds drooping...*S. hypnoides* (p 244 **G**)
Lf-lobes oblong, blunt, often >1 mm wide, lf-tip without or v short bristle-like point –
sometimes glandular hairs on lvs – fl-buds erect **15**

15 Lvs with many short glandular hairs – petals dull cream or greenish white – mts of Scot, vr
Sch 8 EN **Tufted Saxifrage** *S. cespitosa*
Lvs with long glandular and non-glandular hairs many >0.5 mm long – petals pure white
– S&W Ire, r; introd N Scot (previously N Wales), vr ....
EW**Irish Saxifrage** *S. rosacea* ssp *rosacea*
Lvs with all hairs glandular and <0.5 mm long – petals pure white – W Donegal only, vr ....
End FPO (Eire) **Ire *S. rosacea* ssp *hartii*

A FPO (Eire) **Ire **Meadow Saxifrage**, *Saxifraga granulata*, per herb 10–50 cm tall,
with bulbils in axils of lower lvs, and no runners. Basal rosette-lvs (**Ab**) kidney-
shaped, long-stalked, bluntly-toothed, 2–3 cm across, with scattered hairs;
smaller, shorter-stalked, narrower lvs up the erect stems. Fl-stems loosely-
branched, sticky-hairy, with 2–12 white fls, 1–2 cm across. Receptacle-cup fused to
ovary, which it encloses more than halfway up; two carpels (**Aa**), conspicuous
both in fl and in the two-celled capsular fr. Br Isles: Eng f-lc; Scot o-lf in E and S,

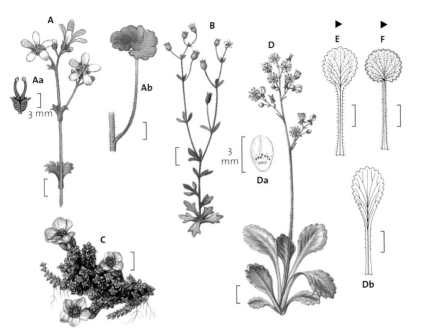

NW and mts abs; Ire vr (E coast only as native); on dry, sandy gsld, mds by rivers, avoids v acid soils. Fl 4–6. **A.1** Sch 8 VU ****Drooping Saxifrage**, *S. cernua*, is close to A, but 3–15 cm tall, with lower lvs sharp-lobed; red **bulbils** in axils of upper lvs; 1 or no fls. Scot alpine, vr. Fl 7 (stems not always flg).

B Rue-leaved Saxifrage, *S. tridactylites*, small erect sticky-hairy ann, 2–15 cm tall. Red-tinged lvs, lower with broad, stalked, **3–5-lobed** or **fingered** blades 1 cm long or less, upper lvs simple, smaller. Fls solitary or in branched open cymes, 3–5 mm across, with bell-shaped reddish calyx; 5 white petals only 2–3 mm long. Br Isles: Eng, Wales, CI f-lc; Scot r, mostly E; Ire f; on dry open sandy gd, wall-tops, pavements and railway lines. Fl 4–5.

C WO (NI) ****Ire (NI) Purple Saxifrage**, *S. oppositifolia*, creeping, mat-forming per herb with densely-set, tiny, oval lvs (2–6 mm long) in opp pairs along stems, with **chalky pits** on tips and bristly margins; fls solitary, almost stalkless, **rosy-purple**, 10–20 mm across, five-petalled. Br Isles: from Brecon and Yorks N on basic mt rocks and cliffs, at sea level in N Scot, r-lf (Scot Highlands f-lc); NW Ire vr. Fl 3–5.

D St Patrick's-cabbage, *S. spathularis*, has basal rosettes of ± erect, hairless, spoon-shaped lvs (**Db**), sharp-toothed above, tapered into broad, long stalks very sparsely hairy, 3–5 cm long (with stalks); fl-stems lfless, branched above, sticky-hairy, with **many** fls 8–10 mm across; petals (**Da**) white, with 1–3 yellow spots at base, many red spots above. GB vr introd; W Ire lf-la; among acid rocks, usually where damp, in mt areas. Fl 6–8. **D.1 Londonpride**, *S.* x *urbium*, v close to D but has larger, more erect lvs than D, with rounded teeth on each side and lf-stalk margins v hairy. Gdn hybrid and escape, f-lc throughout Br Isles (except S Ire); in damp, shady places in wds, streamsides, rocks, hbs. Fl 5–7.

▲ **E   False Londonpride**, *Saxifraga* x *polita*, is D x F and has lvs (**E**) intermediate between these. Native in W Ire only, r introd elsewhere in GB. Fl 6–7.

▲ **F   Kidney Saxifrage**, *S. hirsuta*, differs from D in its hairy **kidney-shaped** lvs (**F**), cordate (not wedge-shaped) at base; lf-stalks narrow, 1 mm wide. As a native, lc in SW Ire only; r introd elsewhere. Fl 5–7.

G   VU **Mossy Saxifrage**, *S. hypnoides*, mat-forming per herb, with both flowering rosettes and long trailing sterile lfy shoots; rosettes bear mostly three-lobed lvs, 1 cm long; lobes linear, pointed, hairy only on their long stalks; sterile-shoot-lvs ± undivided. Infls erect, **few**; fls white, five-petalled, 10–15 mm across. Br Isles: Wales, Derby, N Eng to N Scot lf-la; Mendips r; Ire vl; o introd S Eng as gdn escape; on rock ledges and scree, mostly basic, in hills and mts. Fl 5–7. **G.1   Starry Saxifrage**, *S. stellaris*, has similar fls to G but oval, toothed lvs and fls with **2 yellow spots** at base of white petal. Br Isles: N Wales, N Eng, Scot, Ire o-vla; on mts in flushes, streamsides, rock ledges. Fl 6–8.

H   WO (NI) **Ire **Yellow Saxifrage**, *S. aizoides*, has many creeping sterile shoots and erect lfy fl-shoots, stems without red hairs; lvs stalkless, pointed, undivided, narrow-oblong, 10–20 mm long; fls 3 or more in loose heads, petals **3–6 mm** long, **yellow** sometimes with **red** spots. Mts N Eng to N Scot lc; NW Ire r. Fl 6–8.

I   EU Sch 8 WO (NI) FPO (Eire) **Ire BAP VU ****Marsh Saxifrage**, *S. hirculus*, has yellow fls like H but **key differences** are: stems with red hairs; lower lvs (**I**) **stalked**; fls usually solitary, petals **10-15 mm** long. N GB, N & W Ire, vr; on mt bogs. Fl 7–9 (rarely flg).

J   AWI **Opposite-leaved Golden-saxifrage**, *Chrysosplenium oppositifolium*, low per herb with mats of creeping shoots; **opp pairs** of stalked, rounded lvs with wedge-shaped bases; erect 5–15 cm-tall fl-stems bear repeatedly forked umbel-like cymes of tiny 3–5 mm-wide golden fls (**Ja**), surrounded by large green-yellow bracts; calyx four-toothed, no petals, 8 stamens, 2 carpels. Br Isles (except in E mid Eng) c-vla; in springy areas and by streams in wds, mt rocks. Fl 4–5.

K   AWI **Alternate-leaved Golden-saxifrage**, *C. alternifolium*, larger version of J, with creeping **lfless** stolons (no creeping lfy stems); **long-stalked**, kidney-shaped **alt** basal lvs, 1.0–2.5 cm across, with scattered bristles; stem-lvs usually only 1, not opp; fl-heads slightly larger, with a more golden sheen on bracts. Br Isles: GB o-lf to E; **Ire abs**; in boggy wds and mt rocks on more base-rich soils. Fl 4–5.

L   Grass-of-Parnassus, *Parnassia palustris*, erect hairless per herb, 10–30 cm tall; basal lvs oval-cordate, long-stalked; one stalkless cordate stem-lf low down; fl (**La**) **solitary**, 15–30 mm across; 5 white, oval, green-veined petals; 5 fertile stamens; 5 alt **fan-branched sterile stamens** (**Lb**) with **yellow glands on tips**; **4 stigmas**; fr a capsule. Br Isles, N Eng, Wales f, Scot c, S Eng vr; Ire lf; in marshes, fens, especially on basic soils. Fl 7–10.

M   Fringecups, *Tellima grandiflora*, erect hairy per with basal rosette of palmately-lobed lvs; infl a spike-like raceme, fls greenish, later reddish-tinged, bell-shaped with 5 petals, each fringed by long, **whisker-like lobes** with **10 stamens**. Introd (N America); Br Isles, o gdn escape in wds, hds and damp, shady places. Fl 4–7.
M.1   Pick-a-back-plant, *Tolmiea menziesii*, is similar to M but has more heart-shaped lvs, an irregular red-brown fl with 4 petals (rarely 5), **3 stamens** and **no** long whisker-like lobes. Introd (N America); Br Isles, o but increasing gdn escape (S Scot f) in similar habitats to M**.** Fl 4–8.

ROSE FAMILY

Rosaceae

The Rose family members are trees, shrubs or herbs, always with alt lvs, often much lobed or divided (though sometimes not so), always with **stipules**. Fls regular, mostly with their parts in fives (but Ladies-mantles, Burnets, and some Cinquefoils have them in fours). **Epicalyx present** in many of the herb members. Petals not joined into a tube below. Stamens usually 2, 3 or 4 times number of the sepals, rarely fewer. A **receptacle-cup** surrounds stamens and carpels and bears the calyx-teeth, petals and stamens; it may be fused to the carpels, making the fl either **perigynous** (ie semi-inferior ovary) when unfused, or **epigynous** (ie inferior ovary) when fused to them. Frs either **achenes**, which sit on a fleshy receptacle (Strawberry) or on a non-fleshy one (Cinquefoils) or may be drupes (Plums and Cherries) or pomes (Apples, Pears, *Sorbus*, etc) or in other forms.

KEY TO TREES AND SHRUBS OF THE ROSE FAMILY

1 Trees or upright shrubs **2**
Herbs, sometimes with a slightly woody creeping stem ....see key to herbs (p 247)

2 Fls bright yellow, with epicalyx – lvs pinnate – shrub without thorns – less than 1 m
Potentilla fruticosa (p 259 **C**)
Fls white or pink – no epicalyx **3**

3 No cup below sepals, carpels on a conical receptacle – fr blackberry- or raspberry-like, red
or black – prickles on stem and lvs – lvs with 3–7 lfts ....*Rubus* (p 256)
Receptacle forming a deep cup below sepals **4**

4 Many styles and carpels within a fleshy cup – pinnately-lvd shrubs with broad-based
prickles on sides of shoots ....*Rosa* (key p 253)
Only 1–5 styles and carpels – trees or shrubs with no prickles along stems or lvs, but
sometimes with woody thorns at tips of the short shoots – lvs usually undivided (if
pinnate then not thorny) **5**

5 Only one carpel, forming a cherry- or plum-like fleshy fr with receptacle-cup not fused to
it – ovary superior..*Prunus* (p 249)
Carpels 5 – ovary superior – fr a dry follicle ....*Spiraea* agg (p 252 **L**)
Carpels 1–5, fused to receptacle cup (so that ovary inferior or semi-inferior) **6**

6 Infls in erect racemes...*Amelanchier lamarckii* (p 252 **K**)
Infls umbel-like, or fls only 1 or 2–3 together..**7**

7 Fls 1–3 together, or in simple umbels **8**
Fls in dense compound umbel-like heads..

8 Fls less than 1.5 cm across – frs 1 cm or less, red, purple or black, berry-like **11**
Cotoneaster (p 252)
Fls 3 cm across or more – frs 2 cm or more, green or brown, firm **9**

9 Fls borne singly – fr like small apple, but with the 5 carpels exposed on the top in a broad
disc, which is surrounded by long (10–15 mm) sepals – twigs thorny – lvs downy elliptical
– S Eng r – in wds, hbs ....*****Medlar** *Mespilus germanica*
Fls several together in an umbel – fr with the small (3–7 mm) sepals surrounding a tiny
pore **10**

10 Petals pink-tinged – stamens yellow – fr like a small apple, fleshy not gritty..*Malus* (p 250)
Petals white – stamens purple – fr pear-shaped with gritty flesh.........................*Pyrus* (p 250)

11 Woody thorns at ends of short shoots – lvs lobed or pinnatifid – stamens purple – carpels
inside the fr stony (haw)..*Crataegus* (p 250)
No thorns – lvs oval, lobed or pinnate – stamens cream to pink – carpels gristly (not
stony)..**12**

12 Lvs pinnate, green both sides, lfts toothed – frs scarlet ....*Sorbus aucuparia* (p 248 **A**)
Lvs oval, white woolly-hairy underneath (although wearing off with age) – frs scarlet.........
S. aria agg (p 249 **C**)
Lvs shallowly lobed, grey to yellowish woolly-hairy underneath (although wearing off
with age) – frs red ....*S. intermedia* agg or other *Sorbus* agg (p 249)
Lvs deeply-lobed, some $\geq$ 1/3 to midrib, green both sides (although woolly-white
underneath when young) – frs brown ....*S. torminalis* (p 248 **B**)

KEY TO HERBS OF THE ROSE FAMILY

1 Epicalyx present below calyx of fls **2**
No epicalyx present **8**

2 Styles long, zigzag in middle in fl, persisting in fr as hooked bristles – basal lvs pinnate – end lft much larger than side lfts...*Geum* (p 264)
Styles short, not zigzag, not persisting in fr – if lvs pinnate, end lft hardly larger than side lfts**3**

3 Petals conspicuous, yellow or white **4**
Petals abs or tiny – fls v small **6**

4 Lvs palmate or pinnate – fls yellow or white.......................................*Potentilla* (key p 258)
Lvs trifoliate – fls white **5**

5 Terminal tooth of lft longer than adjacent teeth (Ja p 260); fr fleshy (like little Strawberry) – fr-receptacle hairless..*Fragaria* (key p 258)
Terminal tooth of lft shorter, so end of lft appears slightly indented (Ia p 260); fr dry – fr-receptacle hairy..*Potentilla sterilis* (p 260 **I**)

6 Lvs trifoliate – petals none or tiny – carpels 5–12 – on mts only ....
Sibbaldia procumbens (p 260 **I.1**)
Lvs palmately-lobed – petals none – fls with 4 green sepals only **7**

7 Lvs normally over 2 cm wide, clearly stalked – fls in loose terminal heads – per plant often (but not always) over 10 cm tall ....*Alchemilla* (p 262)
Lvs less than 1 cm wide, scarcely stalked – fls in dense clusters in lf-axils – ann plant under 10 cm tall ....*Aphanes* (p 262)

8 Lvs pinnate – plant ± erect...**9**
Lvs trifoliate or undivided – plant low, creeping – fls white **13**

9 Petals present – lvs with pairs of smaller lfts between larger lfts **10**
Petals abs – lvs without smaller lfts between larger lfts – fls in dense heads or spikes **11**

10 Fls white (or pink-tinged), in loose umbel-like panicles – receptacle-cup of fl and fr, spineless...*Filipendula* (p 264)
Fls yellow, in long loose spikes – receptacle-cup of fl and fr with a crown of conspicuous hooked spines ....*Agrimonia* (p 262)

11 Plant creeping – erect fl-stems – fls in round heads – conspicuous hooked spines on fr-head – (introd from Australia) o-vla...............................**Pirri-pirri-bur** *Acaena novae-zelandiae*
Plant erect – no spines on fr-head **12**

12 Fls in oblong spikes – each fl bisexual with 4 crimson sepals, 4 stamens, undivided stigma *Sanguisorba officinalis* (p 264 **B**)
Fls in globular heads – each fl with 4 green sepals – lower fls male or bisexual with many stamens – upper fls female with a red-purple feathery stigma only ....*S. minor* (p 264 **A**)

13 Petals usually 8 – stems creeping, ± woody below – lvs oval, lobed like tiny oak-lvs, green above, white below – fr-head a group of achenes with long white feathery styles..................
Dryas octopetala (p 262 **B**)
Petals 4 or 5 – stems not woody – fr a head of red or orange fleshy berry-like drupes (rather resembling a raspberry)..**14**

14 Lvs simple, palmately-lobed, rounded (like mulberry lvs) from underground rhizomes – fls solitary, not always produced – fr red, then orange ....*Rubus chamaemorus* (p 257 **D**)
Lvs trifoliate – fls 2–8 in a loose head – plant with creeping runners above ground – fr scarlet ....*R. saxatilis* (p 256 **B**)

Whitebeams, Service-trees, Rowan (*Sorbus*) is a genus of trees with white five-petalled fls in branched, umbel-like heads (technically compound corymbs) and pome-type fruits in which the gristly-walled carpels are enclosed in the fleshy receptacle-cup, as with apples and pears. The twigs have no spines and the lvs are toothed, lobed or pinnate.

A ᴬᵂᴵ **Rowan**, *Sorbus aucuparia*, smooth-barked tree, ± 15 m; lvs **pinnate**, compound, 10–20 cm long, bearing strongly-toothed lfts 3–6 cm long, terminal lft **no longer** than side ones, hairless above, **whitish-grey** and **downy** below when young. Dense corymbs of fls resemble branched umbels; petals 5, creamy-white, 3–5 mm long; fl-stalks woolly; styles 3–4, stamens many. Frs (**Aa**) scarlet, globose, 6–9 mm long, like tiny apples in structure, a fleshy outer case enclosing gristle-like carpel walls, each containing 1 or 2 seeds. Br Isles, c especially to N, in dry wds usually on acid soils, and on mt rocks. Fl 5–6; fr 9.

B ᴬᵂᴵ **Wild Service-tree**, *S. torminalis*, has fls similar to A but is larger tree (to 25 m) with rough bark of a pear tree (not smooth as in A), and lvs v different, 7–10 cm long, oval-oblong, not divided into lfts, but with triangular-pointed, toothed lobes that diverge at ± 45°, resembling those of Maples (p 316), but narrower, with lowest 2 lobes not (or scarcely) wider than rest of lf; lvs white downy **below** when young, ± hairless, dark green, leathery when mature. Frs (**Ba**) similar in form to A but 12–16 mm long, oblong, not globose, **brown** with **dark spots**. Eng, Wales, scattered and lf only; in old wds on either clay or calc soils, rarely many trees together. Fl 5–6.

C Whitebeams, *S. aria* agg, a complex of closely related micro-spp, with simple, but always toothed, often ± pinnately-cut lvs, dark green, ± hairless above when mature, woolly-white to yellow-grey below. White fls and reddish to orange frs (**Ca**) as in A. Br Isles, f-lc in dry wds and usually on limestone rocks. Fl 5–6; fr 9. **C.1 Common Whitebeam**, *S. aria* in narrow sense, has lvs egg-shaped, small-toothed, but not lobed, pure white below; oval red frs. S Eng, c on chk, limestone, o on sandy soils, in wds and scrub. (Illustrated, C)   **C.2** *Rock Whitebeam**, *S. rupicola*, lvs untoothed at base with ± **spoon-shaped** lvs but can be easily mistaken for other *Sorbus* spp (significant records should be expertly confirmed) o in N Eng, Wales only. **C.3 Swedish Whitebeam**, *S. intermedia* agg is more readily recognised with v variable lvs but usually at least the **lowest 3** lf-teeth **distinctly lobe-like**, divided < 1/6 way to midrib and grey–yellowish woolly-hairy on underside. Introd (Sweden and Baltic area), planted as an ornamental tree and f-c self-sown in wds, wa.

Plums (*Prunus*) have five-petalled fls with concave or cup-like receptacles; **superior** ovary with 1 carpel which in fr forms a **drupe** (with fleshy outer part and a stony inner part, containing the seed or kernel).

▼ **A** AWI **Wild Cherry (Gean)**, *Prunus avium*, tree, to 25 m, with suckers and peeling bark with **horizontal lines** on it when young, rough when old. Lvs oval-elliptical, doubly-toothed, pointed, hairless above, downy below, **not shiny**, 6–15 cm, with two red knobs (**Ab**), or glands, on stalk just below lf-blade. Fls 2–6, in umbels without a common stalk, white, cup-shaped; petals 8–15 mm, narrowed to base, arising from reddish cup which bears bent-back sepals and also the stamens, and is 'waisted' above (**Ac**). Fr (**Aa**) 1 cm, globular, red, hairless, like a tiny garden cherry. Br Isles: Eng, Wales, Ire f-lc; Scot r to N; in wds, hbs on better soils. Fl 4–5.

▼ **B Dwarf Cherry**, *P. cerasus*, is similar to A, but usually only a shrub; lvs **broader, shorter** (5–8 cm), **thicker**, dark green, **shiny**, ± hairless below; flat fls, receptacle with **no 'waist'** (**B**), petals rounded below. Introd (SW Asia); Br Isles: S Eng, Wales, Ire o-f; N Eng, Scot r; in hbs, wds. Fl 4–5.

▼ **C Blackthorn (Sloe)**, *P. spinosa*, spiny rigid shrub, 1–4 m, forming thickets; twigs downy when young, dark grey-brown, with many straight side-shoots that become thorns; lvs elliptical-lanceolate, dull, tapered below, toothed; **fls appear before lvs**, 1 or 2 together, petals white, 5–8 mm; fr (**Ca**) 10–15 mm, globular, blue-black with a grey bloom, v astringent. Br Isles, c and la in scrub, wds, hbs. Fl 3–5. **C.1 Wild Plum**, *P. domestica*, is similar to C, but v variable, usually more of a tree and **not** thorny; **grey-brown** twigs soon hairless as are fl-stalks; fls 1–3, appear with lvs; petals 7 mm or more; fr 4–8 cm, oval-oblong, colour varied. Br Isles, escape from cultivation, c in hbs, wds. Fl 4–5. **C.2 Cherry Plum**, *P. cerasifera*, has **hairless** glossy **green** (**usually spineless**) twigs and lf upper sides (only hairy on lf-stalks and lower midribs); lvs oval, glands small or abs; petals 7–11 mm, oval; fr 2.0–2.5 cm, globose, yellow or red. Introd (E Eur, Asia); Br Isles, c in hbs. Fl 3–4, **earliest** Plum to fl, **fls appear w lvs.**

▼ **D** **Ire AWI **Bird Cherry**, *P. padus*, tree, 3–15 m, with odorous peeling bark; lvs 5–10 cm, elliptical, toothed, thin, ± hairless (or hairs on lf-underside in vein-axils), with **red glands on stalk** as in B. Fls in **long ± erect or drooping racemes**, 10–40 together; petals white, **toothed**, 6–9 mm on fl-stalks 8–15 mm; fr (**Da**) black.

Native in N half of GB, Wales f-lc, E. Anglia; elsewhere introd gdn escape (in S Eng easily confused with introd *Prunus* spp) in wds (especially moist) in mt areas. Fl 5.   **D.1   Rum Cherry**, *Prunus serotina*, often mistaken for D, is a shrub to small, sprawling tree with similar drooping racemes of white fls and large, thin lvs, usually hairless (or hairs on lf-underside) but red glands often **not** present (as in D); lvs oblong-obovate toothed, with wedge-shaped to truncate base; **fl-stalks 2–7** mm and **small petals** 3–4.5 mm. Fr dark purple. Introd (N America); bird-sown, increasing and becoming invasive; S Eng, o in wds, hds, rds, hths usually on clay or sandy soils. Fl 4–5 (often not flg if shaded).

E   Cherry Laurel, *P. laurocerasus*, evergreen, with large shiny leathery lvs, 5–8 cm with small, spaced-out **teeth** on lf-margin; 1st year twigs **green**; fls in **erect** racemes, 5–12 cm long; petals white, 4 mm long. Introd (SE Eur), often invasive where natd. Br Isles, c in wds, wa, rds. Fl 3–5.   **E.1   Portugal Laurel**, *P. lusitanica*, has evergreen, leathery lvs but with **deep red** lf-stalks and bark; fls in long **drooping** raceme. Introd (Spain and Portugal). Less c than E but can be equally invasive; Br Isles, f and increasing in wds, wa, scrub, especially on chk. Fl 4–5.

F   Pear, *Pyrus communis*, deciduous tree, to 20 m; rough-fissured bark; usually spineless twigs; oval, scarcely-toothed lvs. Fls in umbel-like heads; 5 petals, white, 12–15 mm, styles free; stamens purple; sepals persisting; fr familiar pear-shape. Br Isles, widely introd (S and E Eur); in hbs. Fl 4.   **F.1** Sch 8 VU ****Plymouth Pear**, *P. cordata*, v close to F but has spines, petals 7.8–12.7 mm, sepals usually falling off and smaller frs < 5 cm which are not pear-shaped. Br Isles, vr Devon and Cornwall; in hds. Fl 4.   **F.2 Wild Pear**, *P. pyraster*, also v difficult to distinguish from F, but usually spiny and has inedible, rounded (sometimes pear-shaped) frs. Fl 4. As it is difficult to distinguish *Pyrus* spp, you can record finding '*Pyrus communis sensu lato*', ie Pear in the broad sense.

G   AWI **Crab Apple**, *Malus sylvestris*, **thorny** small rounded tree, to 10 m; irregularly **scaly** bark; toothed, oval pointed deciduous lvs, **hairless when mature**. Five-petalled fls in umbel-like heads as in F, but petals pink-flushed on white, 1.3–3.0 cm long, stamens **yellow**, styles joined below. Outside of **calyx and fl-stalks hairless**. Fr globular, 2 cm, apple-shaped, yellow to red-flushed, acrid. Eng, Wales f; Ire, S Scot r; native in old oakwds, scrub. Fl 5.   **G.1   Apple**, *M. pumila* (*M. domestica*), v variable (often close to G), generally has mature lvs **downy** on lower side, **no thorns**, **outside** of **calyx** and **fl-stalks hairy** and larger frs. Often introd from cultivation or discarded cores in hbs, scrub, rds throughout Br Isles, **much more c than** G. Fl 4–5.

H   Hawthorn, *Crataegus monogyna*, hairless, v thorny deciduous shrub or tree to 10 m; **deeply** 3–5-lobed lvs; lobes ± pointed; fls in flat corymbs, 10–16 together; 5 petals 4–6 mm long, white (rarely pink); **one style**. Fr (**Ha**) of one stony carpel in a hollow, 8–10 mm long crimson oval cup, forming the familiar 'haw' (**Ha**). Br Isles (except N Scot r), vc in hbs, scrub, wds; r on v acid or wet soils. Fl 5–6.

I   AWI **Midland Hawthorn**, *C. laevigata*, similar to H, but lvs (**I**) **rounded**, lf tapering to the base; 5 petals 5–8 mm, **styles 2 or 3**; carpels, 2 or 3, joined below, free above. SE, mid Eng f-lc; mostly in old wds on heavy soils. Fl 4–5. Hybridizes with H (*C. x media*) f and easily confused with I; hybrids often have fls with mixture of 1 and 2 styles, and have an intermediate lf shape (see *Plant Crib*).

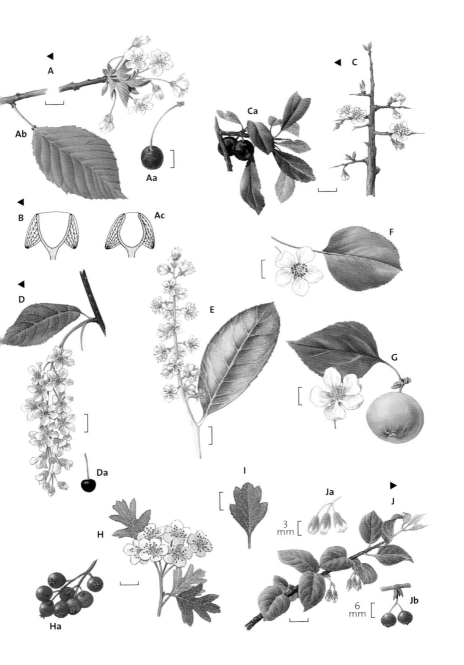

▲ **J** End Sch 8 BAP CR ****Wild Cotoneaster**, *Cotoneaster cambricus*, (*C. integerrimus*), shrub rarely over 1 m, without spines; hairless twigs except when young; lvs **alt**, deciduous, shortly oval, 15–40 mm, **untoothed**, matt above, grey-woolly below. Fls (**Ja**) 2–4 together in clusters, 5 petals, 3 mm, pink; fr (**Jb**) 6 mm, globular, red, like a small 'haw', but the 2–5 carpels are not (as in I p 250) joined on their inner sides. **Wales only**, vr on limestone rocks near Llandudno. Fl 4–6. **J.1 Wall Cotoneaster**, *C. horizontalis*, is the most widely recorded introd sp of *Cotoneaster*, shrub having stems spreading **horizontally**, regularly branched in herring-bone pattern, flattened in **one plane**; lvs shiny, almost hairless below, 0.6–1.2 cm long. Fls cream tinged pink and fr similar to J. Introd (W China); Br Isles, c on rocky gsld, quarries, pavements, walls, wa. Fl 4–6. Many gdn *Cotoneaster* spp are widespread introd throughout much of Br Isles (Consult the *New Flora*, although the key requires **both** flowers and ripe frs).

K   Juneberry, *Amelanchier lamarckii*, shrub or tree to 10m high; lvs alt, **oblong to oval**, ± hairless, toothed. Narrow, hair-like stipules (which often fall off); infl a raceme, 5 ± **linear-oblong white petals**, 35-40mm, yellow anthers; 5 sepals ± triangular; **inferior** ovary; fr (**Ka**) a pome, green turning red and becoming purplish-black. Lvs turn yellow, orange or crimson in autumn. Introd (N America) and o natd in SE Eng, r elsewhere, abs-vr Ire; mainly on sandy soils in wds, scrub, hths and rds. Fl 4–5.

L   Bridewort, *Spiraea salicifolia* agg, is a confusing group of **v variable** spp and hybrids. *S* x *pseudosalicifolia* is the most widespread (illustrated here). In general, suckering shrubs to 2m; lvs alt, ovate–oblong, toothed, without stipules; infl a dense panicle of pink fls (some spp white to pink-purple in more lax panicles or corymbs); carpels 5, free; fr many-seeded follicle but usually sterile, without seeds. Introd gdn escape; Br Isles, c on rds, hbs, river-banks, wa. Fl 5–7.

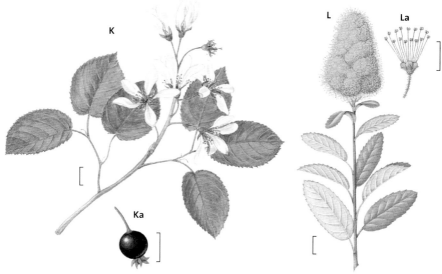

Wild Roses (*Rosa*) are shrubs with prickly stems, pinnate lvs with obvious stipules; fls solitary or in clusters, 5 sepals and petals, many stamens, styles and carpels, enclosed in, but not fused to, a deep receptacle-cup, ripening to a coloured 'hip'. Petals large, soon falling.

ID Tips   Roses

- **What to look at:** It is difficult, if not impossible, to identify Roses in fl; you need fully developed (but not necessarily ripe) hips. Also look at **(1)** whether lvs hairy or not **(2)** the arrangement of the stigmas on hips **(3)** presence of glands on lvs or fr-stalks.

- Roses hybridize frequently and this key cannot cover these. If a plant will not key out, it may be a hybrid or a rare introd sp.

- The *BSBI Roses Handbook* has illustrations and keys to all the spp and many hybrids in the Br Isles.

KEY TO ROSES (*ROSA*)
by Rev Dr Anthony Primavesi

1   Lvs hairless (occasionally with few sparse hairs on lf-midrib) **2**
    Lvs hairy, at least on veins on lf-underside **7**

2   Styles on hip united, forming a long, projecting column from tip of hip **3**
    Styles on hip free, not projecting **4**

3   Fls and hips numerous, in clusters of 10 or more – tall bush climbing with arching stems – gdn escape f in GB (S Ire vr) ....**Many-flowered Rose** *Rosa multiflora*
    Fls and hips <10 – low trailing shrub, or if climbing, with ends of stems hanging vertically downwards ....*R. arvensis* (p 254 **A**)

4   Sepals entire – erect, free standing shrub..**5**
    Sepals lobed – climbing plant with arching stems **6**

5   Low-growing shrub up to usually 50 cm high, suckering to form dense masses – lfts small, usually 9–11 in number – prickles straight and slender – ripe hips purplish-black ...... *R. spinosissima* (p 255 **B**)
    Taller shrub not spreading widely by suckering (young stems and lvs tinged reddish) – lfts 5–7 – prickles curved – ripe hips red – gdn escape o throughout Br Isles .... **Red-leaved Rose** *R. ferruginea*

6   Fr-stalks short and partly hidden by large bracts – sepals erect or spreading, persisting at least until hips ripen – stigmas in a domed woolly head, covering the disc at the top of the hip ....**Glaucous Dog-rose** *R. caesia* ssp *vosagiaca*
    Fr-stalks not concealed by bracts – sepals reflexed, falling early – stigmas in small globose head, not covering the disc at top of hip ....*R. canina* agg (p 256 **D**)

7   Disc at top of hip strongly conical – styles projecting from hip, forming a column at first but soon separating – f in S GB, midlands r, N abs; Ire r ....**Short-styled Field-Rose** *R. stylosa*
    Disc not conical – styles not projecting or if projecting, permanently united in a column **8**

8   Lft underside with glands with fruity odour **9**
    Lft underside without glands or if present, odourless or with an unpleasant, resinous scent when crushed **11**

9 Fr-stalks short c. 1 cm, with stalked glands – sepals erect or spreading, persisting until hips ripen – free-standing shrub with erect straight stems ....*R. rubiginosa* (p 256 **C**)
Fr-stalks >1 cm long, glandular hairy or not – sepals reflexed, falling early – stems arching or winding **10**

10 Climbing plant with arching stems – fr-stalks with stalked glands – lfts rounded at base –
Br Isles f in S, r to N; Scot abs; on calc soils ....**Small-flowered Sweet-briar** *R. micrantha*
Free-standing shrub with winding stems – fr-stalks smooth without glands – lfts wedge-shaped at base – Br Isles, o S Eng to Cheshire and N Wales; C Ire f; on calc gsld ....
NT **Small-leaved Sweet-briar** *R. agrestis*

11 Fls and hips numerous, in clusters of ≥10 – styles united into long projecting column – f
gdn escape ....**Many-flowered Rose** *R. multiflora*
Fls and hips <10 – styles free and not projecting **12**

12 Stems densely hairy with numerous ± straight prickles of various sizes, the larger prickles
hairy at base **13**
Prickles of same size, straight or curved but not hairy at base **14**

13 Hips large to 2 cm or more wide – fr-stalks curved when hips ripe ....*R. rugosa* (p 256 **E**)
Hips <2 cm long – fr-stalks straight when hips ripe – r-o gdn escape ....
Dutch-rose *R.* 'Hollandica'

14 Fr-stalks smooth, without glands – lvs moderately hairy, at least on veins on underside ..**15**
Fr-stalks with stalked glands – lvs densely hairy, at least on underside **17**

15 Fr-stalks partly concealed by large bracts – sepals erect or spreading, persisting until hips
redden – stigmas in a large domed woolly head, covering the disc at the top of the hip ....
Hairy Dog-rose *R. caesia* ssp *caesia*
Fr-stalks not concealed by bracts – sepals reflexed, falling early – stigmas in small globose head not covering the disc **16**

16 Fr-stalks short – lfts with reddish brown glands on teeth – GB f in S; r in N; Scot abs; Ire vr
Round-leaved Dog-rose *R. obtusifolia*
Fr-stalks longer to 2 cm – lfts without glands – distribution unclear, probably widespread preferring shady places ....*R. canina* (p 256 **D**)

17 Climbing plant with arching stems – sepals spreading but falling early – GB f in S and
Midlands, r N; Scot vvr-abs ....**Harsh Downy-rose** *R. tomentosa*
Free standing shrub with erect or winding stems – sepals erect to spreading persistent at least until hips ripe **18**

18 Stems straight and erect, suckering – prickles straight – sepals ± entire, erect, red and
fleshy at base when hips ripe, persisting until hips rot – Br Isles: Scot, N Eng, N Wales c; o
N Ire; S GB abs...**Soft Downy-rose** *R. mollis*
Stems zig-zag and winding, not suckering – prickles curved – sepals lobed, erect or spreading, thin and papery at base persisting until hips are ripe – N Eng, Scot, Wales c ; o elsewhere..**Sherard's Downy-rose** *R. sherardii*

A AWI **Field-rose**, *Rosa arvensis*, low scrambling or trailing shrub to 1 m; weak green stems; narrow-based arching prickles; lvs hairless with 2–3 pairs 1.0–3.5cm long, oval lfts. Fls white, 30–50 mm, 1–6 together, styles joined in a column above (**Aa**). Sepals usually hairless, less than 1 cm, almost unlobed, falling before oval fr ripens red. Br Isles: S Eng, Wales, Ire, vc; N Eng r; **Scot abs**; in wds, scrub, hbs especially on heavy soils. Fl 6–7. Rarely hybridizes, therefore A is the most easily recognised *Rosa* sp.

B Burnet Rose, *R. spinosissima* (*R. pimpinellifolia*), low, suckering patch-forming shrub, to 50 cm, sometimes ± creeping; stems with many long straight narrow prickles and stiff bristles (looking like prickles of different lengths); lvs with 3–5 pairs small 0.5–1.5-cm long oval lfts; often flushed purple (as may be stems). Fls solitary, cream, 20–40 mm across, styles free, woolly; sepals unlobed, persisting on globose **purple-black** fr (**Ba**). Br Isles, c near coasts, o inland; on dunes, hths, shingle, calc gsld, scrub. Fl 5–7.

▲ **C   Sweet-briar**, *Rosa rubiginosa*, not unlike D below, but less tall and vigorous (to 1.5 m); lfts and fl-stalks covered with brownish sticky **gland-bearing** apple-scented hairs. Prickles unequal, ± hooked; sepals pinnately-lobed, sticky-hairy, persistent in fr (**Ca**); styles hairy, free. Br Isles: Eng, Wales f-lc; Scot r; Ire o-lf; on gsld, scrub, usually on calc soils. Fl 6–7.

▲ **D   Dog-rose**, *R. canina* agg, v variable, has strong arching stems to 3 m, broad-based strongly-hooked prickles; lvs with 2–3 pairs of toothed lfts, hairless or ± hairy; fls 1–4 together, 4–5 cm, pink (or white), styles free; sepals (**Da**) spreading, falling before fr. Most c rose in S Br Isles (r N Scot). Fl 6–7.

▲ **E   Japanese Rose**, *R. rugosa*, a suckering shrub to 2m, stems with numerous mixture of hairy, broad-based and small thin-based prickles; lfts wrinkled, hairy and rather shiny; large, fuchsia-pink fls (sometimes white) 6–9 cm across; fr (**Ea**) **large**, 1.5–2.5 cm, broader than long, usually **pumpkin-shaped**. Introd (E Asia), often planted but natd; c and increasing in hds, rds, wa and especially near sea on sandy soils, dunes, shingle. Fl 6–7.

Brambles (*Rubus*) are a genus of scrambling, erect, or creeping shrubs and herbs, mostly spiny; lvs undivided, or usually with 3–5 (or more) pinnately or palmately arranged lfts, with stipules; fls with no epicalyx, 5 sepals joined below in a cup, 5 separate petals, many stamens, many separate carpels on a conical receptacle; frs aggregates of many tiny fleshy drupes, each with a tiny stone or pip containing a seed. Cultivated *Rubus* spp are sometimes natd, the most widespread being **Salmonberry** (*R. spectabilis*) with trifoliate lvs and large pink fls 20–30mm across (consult the *New Flora* for illustration).

A   Bramble (Blackberry), *Rubus fruticosus*, is really an agg of v numerous microspecies, differing in types of stem armament, hair and gland distribution, lf shape, fl colour, and fr shape, colour and flavour. In general, a scrambling shrub, 1–3 m, with usually arching and angled stems bearing hooked spines, prickles, hairs (sometimes gland-tipped, sticky), and lvs with 3–5 (rarely 7) oval or oblong palmately arranged lfts, which also vary in hairiness and prickliness etc. Fls white or pink (**Ab**), in panicles on ends of last year's stems. Petals often crinkly. Frs (**Aa**) red at first, then usually **shiny** black or purple-red, rarely v dark red. Br Isles, vc; often highly invasive if unmanaged, therefore damaging habitats as much as well-known invasive plants like **Japanese Knotweed** (p 170 D); in scrub, wds, wa, hbs, etc. Fl 5–9.

B   AWI Stone Bramble, *R. saxatilis*, creeping per herb with long runners and erect ann fl-stems of 10–25 cm, both downy, scarcely prickly, bearing trifoliate lvs; end lft 2.5–8.0 cm, stalked, downy beneath. Fls only 8–10 mm across; petals narrow, white, 3–5 mm, no longer than sepals. Fr (**Ba**) scarlet, with fewer (2–6) and larger segments than in 379 and 381. Br Isles: Scot, c; Wales, N Eng, f-lc; Ire o-lf; S Eng vr; in wds, rocky gd, especially on basic soils. Fl 6–8.

C   Raspberry, *R. idaeus*, has erect rounded stems with straight slender prickles; lvs **pinnate**, with 3–7 lfts, green above, **white-woolly** below; fls (**Ca**) in panicles, with **erect, narrow**, white petals; ripe frs (**Cb**) **red, downy**. Br Isles, c in rocky or wet wds, hths. Fl 6–8.

D WO (NI) **Cloudberry**, *R. chamaemorus*, low per herb, 5–20 cm, with creeping rhizome; lvs few, simple, rounded, shallowly palmate, 5–7-lobed (resembling mulberry lvs), wrinkled, downy; fls solitary on ends of erect shoots, with 4–5 white petals 8–12 mm long and longer than sepals (a shy flowerer). Dioecious; fr (**Da**) orange when ripe, of few large segments. Br Isles: N Wales, r; N Eng, Scot, la N Ire vr; **S Eng abs**; on mt moors and upland bogs. Fl 6–8.

E AWI **Dewberry**, *R. caesius*, scrambling per with stems few-prickled, rounded, weak, with a waxy bloom; only 3 lfts per lf; fls always white. **Key difference** from A in **frs** (**E**) with **bluish waxy** bloom, with larger, fewer (to 20) segments per fr. Br Isles: Eng, Wales, SE Scot, f-c; N Scot r-abs; Ire lf; in damp wds, hths, scrub, hds mainly on basic soils. Fl 6–9.

Cinquefoils and **Barren Strawberry** (*Potentilla*) are herbs (rarely shrubs) with alt, lobed or compound lvs with lfy stipules; fls yellow, white or purple, with almost flat receptacle-cup bearing an epicalyx of 5 (less often 4) lobes and the same number of sepals and unjoined petals; many stamens; many separate tiny carpels, and a receptacle which does **not** swell up into a juicy fr. (Buttercups, in contrast, have **no** epicalyx, **no** cup below fl and **no** lfy stipules.) **Strawberries** (*Fragaria*) differ in lvs **always** trifoliate, fls **always** white, and in receptacle which **swells up** into a **juicy red** fr.

KEY TO CINQUEFOILS & STRAWBERRIES (*POTENTILLA* & *FRAGARIA*)

1 Ripe fr becoming red, fleshy, juicy, with tiny achenes embedded in it – lvs with 3 lfts only – fls always white **2**
Ripe fr remaining dry, scarcely fleshy, head of tiny achenes only – fls yellow (rarely white) -- lvs trifoliate, palmate or pinnate...**5**

2 Lfts ± hairless above – terminal lfts rounded at base – fls 20–35 mm across – frs 3 cm or more wide, achenes sunk in flesh..*Fragaria ananassa* (p 260 **K.1**)
Lfts hairy above – terminal lfts narrowed to base – fls 12–20 mm – frs less than 1.5 cm wide, achenes projecting **3**

3 Runners few or none – fl-stalk with spreading hairs – frs 15–25 mm – no achenes at base of fr ....*F. moschata* (p 260 **K**)
Runners many and long – fl-stalk with adpressed hairs – fls 12–18 mm – achenes all over fr
F. vesca (p 260 **J**)

4 Fls deep red – fr purple-red, but dry – lvs palmate above, pinnate below – tall marsh plant . to 45 cm ....*Potentilla palustris* (p 262 **A**)
Fls white or yellow – fr green or yellow **5**

5 Fls white **6**
Fls yellow **7**

6 Lvs trifoliate – low plant, to 15 cm only – widespread ....*P. sterilis* (p 260 **I**)
Lvs pinnate below – tall erect branched plant to 40 cm – vr ....*P. rupestris* (p 260 **G**)

7 Lvs pinnate **8**
Lvs digitate (5–7 lfts) or trifoliate (3 lfts) **9**

8 Branched shrub to 100 cm – lfts narrow, untoothed, grey below ....*P. fruticosa* (p 259 **C**)
Creeping herb with runners – lfts oval, toothed, silvery below ....*P. anserina* (p 260 **H**)

9 Lvs with undersides white-woolly ....*P. argentea* (p 259 **B**)
Lvs green on both sides...**10**

10 Fl-stems stiffly erect – fls terminal – lfts often 7 ....*P. recta* (p 259 **D**)
Fl-stems creeping or ascending – lfts 3–5 ....**11**

11 Fls in groups or clusters, five-petalled...**12**
Fls borne singly, four-five-petalled (if in groups, four-petalled)...**14**

12 Petals shorter than calyx – fls many, in terminal clusters – lvs trifoliate ....
P. norvegica (p 259 **D.1**)
Petals exceeding calyx – fls few, on branches from basal lf-rosette **13**

13 Stems creeping and rooting – stipules of rosette-lvs long, narrow – fls 10–15 mm wide ......
P. tabernaemontani (p 260 **F**)

Stems ascending, not rooting – stipules of rosette-lvs oval, blunt – fls 15–25 mm wide ....
P. crantzii (p 260 **F.1**)

14 Stems creeping to erect, never rooting – lf-stalks less than 5 mm or abs – fls four-petalled,
7–11 mm wide..*P. erecta* (p 259 **A**),
Stems creeping and rooting – lf-stalks 2.5 cm – fls five-petalled, 17–25 mm wide ....
P. reptans (p 259 **E**)

(Plants intermediate between the last two are either *P. anglica* (p 259 E.1) or,
if with sterile frs, hybrids of the two.)

▼ **A** **Tormentil**, *Potentilla erecta*, creeping and ascending per herb to 10 cm tall;
tufted branched stems, without the long rooting stolons of E below. Lvs shiny
deep green, almost hairless above, silky below, **without** stalks; all but basal lvs
have 3 lfts each but 2 stipules at lf-base make them appear digitate. Lfts deeply-
toothed, obovate. Fls on long stalks, 2–4 cm long, in lf-axils, loosely grouped, 7–11
mm across, with usually only 4 yellow petals. Br Isles, vc on gslds, hths, moors
(avoids chk). Fl 6–9.

▼ **B** NT **Hoary Cinquefoil**, *P. argentea*, similar to A, but more erect; stems **woolly**,
undersides of lfts (**Bb**) **silvery-white**, with dense down below; lfts 5, plus two small
untoothed stipules. Fls (**Ba**) yellow, 10–15 mm across, five-petalled. Br Isles: GB from
River Tay S, lf to E and SE, r to W; Wales vr; **Ire abs**; on dry sandy gslds. Fl 6–9.

▼ **C** **Ire NT ****Shrubby Cinquefoil**, *P. fruticosa*, shrub to 1 m tall; downy stems and lvs;
lvs **pinnate** with 3–7 (usually 5) lfts, 1–2 cm long, **untoothed**, elliptical, pointed,
greyish-downy especially below; stipules chaffy. Fls in loose infls, yellow, five-
petalled, 20 mm across. Br Isles: N Eng (Teesdale, Cumbria as native), W Ire vla, r
introd elsewhere; on damp rocky slopes and shingle by rivers and lakes. Fl 6–7.

▼ **D** **Sulphur Cinquefoil**, *P. recta*, erect hairy per herb, 30–70 cm tall; long-stalked
palmate lvs with 5–7 lfts, lfts to 7 cm long, deeply-toothed, hairy; fls pale yellow,
20–25 mm across; 5 petals, longer than sepals. Introd (C Eur); Br Isles: mostly S, E
Eng o; Scot, r; Ire abs; on wa, gsld. Fl 6–7. **D.1 Ternate-leaved Cinquefoil**, *P.
norvegica*, is shorter (20–50 cm) with lvs all trifoliate; deeper yellow fls, 10–15 mm
across; sepals equalling petals. Introd (E Eur); GB r; NE Ire vr; on wa. Fl 6–9.

▼ **E** **Creeping Cinquefoil**, *P. reptans*, resembles A, but has wholly creeping stems
rooting at nodes; **long-stalked** palmate lvs (no trifoliate lvs), with 5–7 lfts to 3 cm
long, and untoothed stipules; fls five-petalled, 17–25 mm across, solitary on **long**
stalks. Br Isles, vc (except N Scot, r) in lowland hbs, gslds, wa. Fl 6–9.
E.1 Trailing Tormentil, *P. anglica*, appears intermediate between A and E, but is
not a hybrid; stems trail and root as in E, but has ascending stems with **mostly
trifoliate lvs** (some palmate lvs also), hairy below; lower lvs with **short stalks** (1–2
cm), upper lvs with **even shorter** (0.5 cm) stalks, and ± untoothed stipules. Fls
borne singly on long stalks in lf-axils, mostly four- (**a few five-**) petalled, 14–18
mm across, yellow. Br Isles (except N Scot), f in W, o in E; on paths and wd borders
on moderately acid soils, hths etc. Fl 6–9. **Do not confuse** E.1 with the hybrid
between E and E.1 *P. x mixta*, which is **much more c** than E.1. This hybrid usually
has plants with **both** four-petalled and five-petalled fls like E.1 but lf-stalks **>1 cm**,
± all the same size (unlike E.1) and is nearly sterile.

F ****Spring Cinquefoil**, *Potentilla tabernaemontani* (*P. neumanniana*), resembles an early-flowering hairy A; but fls are in definite terminal infls, with 5 petals; stems mat-forming, prostrate, rooting at nodes, and arise from a thick branched stock covered with old lf-base fibres. Lvs palmate, the lower stalked, 0.5–2.0 cm long, with 5 lfts, upper lvs stalkless with 3 lfts; stipules narrow and long. Fl-stems, obviously downy and hairy, in axils of basal rosette-lvs; fls 10–15 mm across. Br Isles: throughout to central Scot, o-vl and scattered; **Ire abs**; on calc gslds and limestone rocks in sunny places. Fl 4–6. **F.1** ****Alpine Cinquefoil**, *P. crantzii*, resembles F, but branches never root or form mats; stipules **oval**; fls **15–25 mm**, **orange-yellow**. Br Isles: N Wales, N Eng, Scot Highlands, vl; **Ire abs**; on calc mt rocks. Fl 6–7.

G Sch 8 EN ****Rock Cinquefoil**, *P. rupestris*, tall erect per herb, 20–50 cm; hairy **pinnate** basal lvs-lvs (**Ga**) 7–15 cm, trifoliate upper stem-lvs. Fls in loose forked infls, white, five-petalled, 15–25 mm across. Br Isles: Scot, mid Wales only, vr; on basic rocks. Fl 5–6.

H **Silverweed**, *P. anserina*, creeping per herb with lvs **pinnate, silvery-silky** (especially below), in rosettes and on long rooting runners; lfts toothed; fls long-stalked in lf-axils, yellow; 5 petals, 15–20 mm across. Br Isles, vc on wa, rds, mds, hollows of sand dunes. Fl 6–8.

I AWI **Barren Strawberry**, *P. sterilis*, per herb 5–15 cm tall, with short runners; lvs, dull **bluish**-green, 0.5–2.5 cm long, trifoliate, with **terminal tooth of lfts shorter** than those on either side (**Ia**). Fls white, 10–15 mm across; petals 5, scarcely longer than sepals which are clearly visible **between** each petal; ripe frs do not become fleshy and red. Br Isles (except fens & N Scot) vc; in scrub, wds, hbs. Fl 2–5. **I.1** VU **Sibbaldia**, *Sibbaldia procumbens*, low per herb, resembles I, but lfts wedge-shaped, three-toothed at tip only, purple-tinged below; petals yellow, v narrow, shorter than sepals or abs. Br Isles: N Eng to Scot, lf; on mt gslds, rocks. Fl 7–8.

J AWI **Wild Strawberry**, *Fragaria vesca*, per herb 5–30 cm tall, is v like I, but with long arching runners; lvs long-stalked, bright **glossy** green, hairy; lfts ± unstalked, lfts 1–6 cm long, oblong, toothed, acute, **terminal tooth longer than side-teeth** (**Ja**); fls white, 12–18 mm across; petals 5, **exceeding sepals** in length; fl-stalks with adpressed hairs. Fr (**Jb**) bright red, fleshy, 1–2 cm long, like tiny garden strawberry. Br Isles, vc in wds, scrub, basic gslds. Fl 4–7. **Do not confuse** with I, the **key difference** being in the lfts, with terminal tooth of the lfts **shorter** than those either side in I and longer in J.

K **Hautbois Strawberry**, *F. moschata*, is taller than J, with runners; **stalked** lfts; fl-stalks much exceeding lvs, with **spreading** hairs. Fls 15–25 mm; fr (**Ka**) purplish-red, oblong, without achenes at base. Introd (C Eur); Br Isles r; in wds, scrub, hds, rds. Fl 4–7. **K.1 Garden Strawberry**, *F. ananassa*, (*F.* x *ananassa*) v variable, generally has lfts ± hairless above, terminal lft rounded to base; fls 20–35 mm across; frs, 3 cm or more wide, have achenes embedded in flesh. Introd from cultivation; Br Isles, f as a gdn escape. Fl 5–7

A

B Ba Bb

C

D

E

F

Ga G

H

I Ia

J Ja Jb

K Ka

A Marsh Cinquefoil, *Potentilla palustris*, per herb to 45 cm, with woody creeping rhizome; lower lvs pinnate, with 5 or 7 lfts; lfts 3–6 cm, oblong, strongly-toothed; upper lvs palmate to trifoliate. Fls (**Aa**) in loose terminal cymes, with 5 narrow purple petals and 5 purplish sepals longer and wider than petals; frs purple, not fleshy. Br Isles, o-la, more c in N Eng, Scot, Wales, Ire; in marshes and wet pools, avoids v limy water. Fl 6–7.

B *Mountain Avens, *Dryas octopetala*, low creeping shrub to 8 cm tall; oblong lvs lobed like tiny oak-lvs, dark green above, white-woolly below; white fls 2.5–4.0 cm across, with usually 8 petals, many golden stamens. Sepals oblong with sticky hairs, no epicalyx. Styles form a feathery head (**Ba**) in fr. Br Isles: N Wales, N Eng r; Scot Highlands lf; N and W Ire la; Arctic-alpine plant of limestone or basic rocks on mts, down to sea level in W Ire, N Scot. Fl 5–7.

C Lady's-mantle, *Alchemilla vulgaris* agg, complex of microspecies; spreading or erect per herbs with palmately-lobed lvs (**Ca**), pleated with strong veins. Lf-shape and type and presence or not of hairs is crucial for correct identification of the various spp. Terminal infls (cymes) of many small yellow-green fls with 4 sepals, no petals, 4–5 stamens, and 4 epicalyx lobes (**Cc**). Fls (**Cb**) usually 3–5 mm across; fr a tiny single achene. Br Isles, widespread but l in mds (most of the microspecies are upland plants). Fl 6–9. There is a useful scan of *Alchemilla* lvs, comparing many spp, in the *New Flora*.

D **Ire Alpine Lady's-mantle, *A. alpina*, has distinctive **compound palmate** lvs (**D**) with **separate** elliptical lfts cut **to the base**, **silvery-silky** below. Br Isles: N Eng, Scot, la on acid mts rocks; Ire vr. Fl 6–8. **Do not confuse** with the gdn plant *A. conjuncta* (often called Alpine Lady's-mantle) which has v similar lvs but lfts **joined** at the base.

E Garden Lady's-mantle, *A. mollis*, is the most widespread *Alchemilla* sp and it is worth learning to distinguish it from the native spp. **Key difference** from **all** native *Alchemilla* spp is epicalyx-segments (**E**) **as long as sepals** (rather than shorter than sepals as in C, D) ; consistently covered with outwards-sticking to **reflexed** hairs and v robust plant (other *Alchemilla* spp are sometimes similar in lf hairiness). Introd (SW Eur) apparently increasing gdn escape throughout GB, N Ire (Eire abs), c on rds, riverbanks, wa. Fl 5–9.

F Parsley-piert, *Aphanes arvensis*, is like a miniature of C, ann, 2–20 cm; lvs fan-shaped, short-stalked, with 3 deeply-toothed main lobes; lvs only 2–10 mm long. Fls (**Fa**) minute (less than 2 mm) in dense clusters in lf-axils, surrounded by cups formed by lf-stipules (**Fb**); 4 sepals, epicalyx, but no petals; stamens usually one. Fr (**Fc**). Br Isles, vc on ar, bare gd in gsld, on paths on dry or chalky soils. Fl 4–10.

G Slender Parsley-piert, *A. australis* (*A. microcarpa*), is even smaller, with **oblong** (**not triangular**) stipule-lobes (**Ga**) **when in fr** and **inclined** (**not erect**) sepals (sepal tips closer together than at their base). Fr (**Gb**). Br Isles, c on acid soils. Fl 4–10.

H Agrimony, *Agrimonia eupatoria*, erect hairy per herb, to 60 cm; stems often reddish; lvs pinnate, with smaller lfts between the pairs of main ones, largest lfts to 6 cm long, strongly-toothed and acute, not scented or scarcely so. Fls many in long spikes, 5–8 mm across; 5 yellow petals, usually not notched, 5 sepals, no epicalyx. Receptacle (**Ha**) below fl enlarges in fr to form a **grooved top-shaped** cup covered with hooked spines, the outermost **horizontal**, sometimes to 90° but **not** reflexed. A vr fragrant form occurs in N Eng. Br Isles f-c (except N Scot); on hbs, rds, gslds, scrub, especially on chky soils. Fl 6–9.

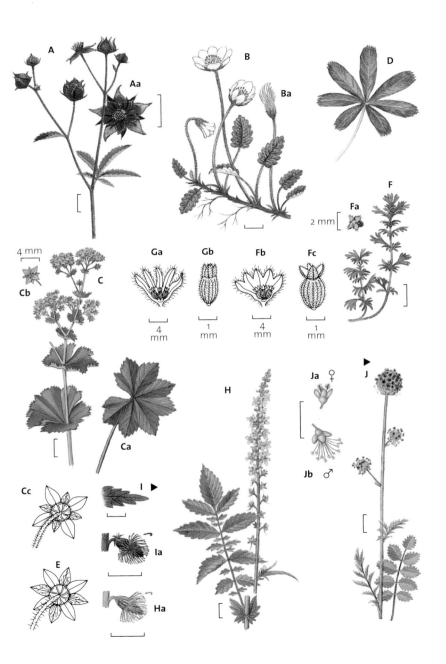

A

Aa

B

Ba

D

F

Fa

2 mm

4 mm

Cb

C

Ga   Gb   Fb   Fc

4
mm

1
mm

4
mm

1
mm

Ca

Cc

I ▶

Ja ♀

J

E

Ia

Jb ♂

H

Ha

▲ **I** **Fragrant Agrimony**, *Agrimonia procera* (*A. odorata*), similar, but more robust (to 1 m), never reddish-tinged; lfts (**I**) more pointed, bearing strongly lemon-scented glands; petals usually notched. **Key difference** from H in fruiting cups (**Ia**) **ungrooved, bell-shaped**, with the outermost hooked spines **reflexed**. Br Isles, more I than H, more c to S; always on acid soils, especially on wd borders, hbs. Fl 6–8.

▲ **J** **Salad Burnet**, *Sanguisorba minor* ssp *minor* (*Poterium sanguisorba*), ± hairless erect, branching per herb, 15–40 cm tall; basal rosettes of pinnate lvs, with rounded but deeply toothed lfts, smell of cucumbers when crushed and can be used in a salad. Stem-lvs similar but smaller. Infl 7–12 cm long of globular heads of tiny greenish fls; lower fls (**Jb**) male or bisexual with many long stamens, upper female (**Ja**) with 2 red-purple feathery stigmas; all with 4 green or purplish sepals, no petals, no epicalyx; fruiting-cup 4 mm long, four-angled, ridged, oval, netted. Br Isles: Eng, Wales f-la: S Scot r; S Ire l; on calc gslds. Fl 5–8. **J.1 Fodder Burnet**, *S. minor* ssp *muricata* (*Poterium polygamum*), more robust introd sp, with more deeply-toothed lfts, infl 10–15 mm long and larger (6 mm long) fruit-cup, four-**winged**, with **pitted rough** surface. Previously cultivated for fodder, sometimes included in 'wildflower' seed mixes. Br Isles, f on wa, rds. Fl 6–8.

A   FPO (Eire) **Ire Great Burnet**, *S. officinalis*, similar to J but much larger (to 1 m tall); lfts to 4 cm long with cordate bases; fl-heads oblong, dull crimson, 1–2 cm long, each fl (**Aa**) **bisexual** with 4 short stamens **and** a pale undivided stigma. Br Isles: Eng, Wales f-lc (except S r); S Scot vr; Ire vr; in wet mds, fens. Fl 6–9.

B   **Meadowsweet**, *Filipendula ulmaria*, tall (to 120 cm) per herb; lower lvs (**Ba**) stalked, pinnate, 30–60 cm long, with 2–5 pairs of main lfts (to 8 cm long) oval, pointed, sharp-toothed, dark green and hairless above, white-woolly or pale green and only downy below; pairs of tiny (1–4 mm) lfts alternate with larger ones; freshly broken lf-stalk smells v strongly of germaline (as does C below); stipules lfy, rounded. Fls, in dense irregular umbel-like infls to 15 cm across, are fragrant, creamy, 4–10 mm across; usually 5 sepals, 5 petals, many stamens; fr (**Bb**) of 6–10 green carpels twisted together, hairless. Br Isles, vc in swamps, mds, fens, wet wds, rds ditches and by rivers on less acid soils. Fl 6–9.

C   **Ire Dropwort**, *F. vulgaris*, per herb of similar habit to B, but usually **only 10–50 cm tall**; lvs (**Ca**) with 8–20 pairs of larger lfts, 5–15 mm long and 5 mm wide, oblong, deeply **pinnately**-cut into **narrow**, toothed lobes, green both sides, shiny, hairless. Fl-head more flattened, of **fewer, larger** (10–20 mm across) fls, cream, tinged **reddish** outside; petals **usually 6**; carpels (**Cb**) **erect, downy**. Br Isles: Eng, Wales f-lc; E Scot vr; Ire vl to W; on calc gslds. Fl 5–8.

D   **Wood Avens**, *Geum urbanum*, downy per herb to 60 cm tall; pinnate basal lvs (**Da**) with large, usually three-lobed blunt-toothed terminal lft, and 2–3 pairs of smaller unequal side-lfts 5–10 mm long; large, lfy stipules at base; upper lvs trifoliate or undivided. Fls 1.0–1.5 cm across, long-stalked, erect, with 5 **spreading yellow** 5–9 mm unstalked, **rounded** petals. Fr-head (**Db**) of many hairy achenes (**Dc**), with 5–10 mm-long zigzag-shaped hairless styles; in fr, lower part of style persists as a hook for animal dispersal. Br Isles, vc in wds, hbs, scrub on less acid soils. Fl 5–8.

E   AWI **Water Avens**, *G. rivale*, per herb with basal lvs (**Ea**), normally with a large, **unlobed, rounded** terminal-lft (**Ea**) up to 10 cm across, with small **sharp** teeth; **drooping** fls with **orange-pink**, **erect** petals, 10–15 mm long, with obvious stalks and

A

Aa

4 mm

Dc

D

C

Cb

Ca

B

Ba

Da

E

Ea

Db

shallow-notched tips, **purple** (not green) sepals and epicalyx. Fr-head stalked, styles hairy above. Br Isles: Scot, N Eng, c; elsewhere f-lc, but SW Eng r, **SE Eng abs**; Ire o-lf; in wet mds, marshes, moist wds, rock ledges on mts. Fl 5–9. Hybridizes with D (*G.* x *intermedium*) c where parents grow together, producing erect orange-pink fls or drooping yellow fls, with intermediate lf shapes; often fertile so a range of intermediate plants occur.

PEA FAMILY   *Fabaceae*

A v distinctive family, also known as the *Leguminosae*, easily recognized by its pea-like five-petalled fls, with wide, often erect **standard** at top, the two **wing** petals at the sides, and the two lower petals forming a ± boat-shaped **keel**. Stamens 10, often joined into a tube below; 1 style, 1 carpel; fr an elongated pod, often splitting into two **valves** when ripe to release several seeds, but sometimes not opening (then usually one-seeded). Sepals 5, forming a calyx-tube below. Ovary superior. Stipules usually conspicuous and lfy. Lvs various, usually trifoliate or pinnate in herb spp, sometimes reduced to spines, or one lft, in wdy shrub spp.

1 Fl-stalk
2 Standard petal
3 Wing petals
4 Keel petals
5 Calyx

PEA FLOWER

KEY TO PEA FAMILY

1 Trees – or shrubs, sometimes small and creeping..**2**
  Herbs (some may be wdy at extreme base) **8**

2 Trees with pendulous racemes **3**
  Shrubs with erect infls **4**

3 Fls white – lvs pinnate...*Robinia pseudoacacia* (p 270 **E**)
  Fls yellow – lvs trifoliate ....*Laburnum anagyroides* (p 270 **D**)

4 Lvs trifoliate, simple, or replaced by spines – fls in heads or lfy racemes, always yellow **5**
  Lvs pinnately or palmately compound – fls in erect lfless racemes.................................**7**

5 Stems round, 1–3 mm thick, 10–50 cm tall – lvs simple..*Genista* (p 268)
  Stems round, 3–6 mm thick, 100–200 cm tall – lvs all simple..
                                                    Spartium junceum (p 268 **G.2**)
  Stems grooved or angled lengthwise in upper green part – lvs trifoliate, simple, or absent on mature stems **6**

6 Mature shoots with branched spines and no lvs ....*Ulex* (p 268)
  Mature shoots without spines and with simple or trifoliate lvs *Cytisus scoparius* (p 268 **G**)

7 Lvs palmate – pod hairy, narrow, hard ....*Lupinus arboreus* (p 270 **H**)
  Lvs pinnate – pod hairless, inflated, bladder-like ....*Colutea arborescens* (p 270 **D.1**)

8 Some lvs trifoliate **9**
  None of the lvs trifoliate **13**

9 Fls solitary ....*Ononis* (p 270)
　 Fls 2 to many, in heads, racemes or spikes **10**

10 Pods spirally-coiled or sickle-shaped ....*Medicago* (p 288)
　 Pods straight...**11**

11 Lvs apparently trifoliate, but with 2 extra lfts at base of lf-stalk as well as a pair of narrow
　 stipules – fls in heads, yellow – pods to 3 cm long ....*Lotus* (p 290)
　 Lvs truly trifoliate – stipules different shape from lvs – pods 7 mm or less **12**

12 Infl a loose elongated raceme with many fls ....*Melilotus* (p 288)
　 Infl a dense head, or with few fls arranged like spread-out fingers ....*Trifolium* (key p 272)

13 Lvs palmate, lfts 6 or more – fls in spikes ....*Lupinus* (p 270)
　 Lvs pinnate, simple, or none **14**

14 Lvs simple, like grass blades ....*Lathyrus nissola* (286 **F**)
　 Lvs composed of a tendril only, with pair of lfy stipules at base ....*L. aphaca* (p 286 **G**)
　 Lvs pinnate...**15**

15 Lfts one to many pairs, but with no terminal lft, ending with or without a tendril..................
　　　　　　　　　　　　　　　　　　　　　Vicia and *Lathyrus* (see key p 280)
　 Lfts one to many pairs, with terminal lft present **16**

16 Fls in heads or umbels **17**
　 Fls in racemes **22**

17 Heads usually in pairs, each pair with lfy bract below – calyx inflated, white-woolly – fls
　 yellow (sometimes other colours) ....*Anthyllis vulneraria* (p 292 **B**)
　 Heads solitary – calyx hairless or only sparsely hairy, not inflated, not woolly **18**

18 Heads 10–20-fld – fls pink to purplish – calyx-teeth equal ....*Securigera varia* (p 292 **D**)
　 Heads 2–10-fld – fls yellow or white with red veins..**19**

19 Fls white with red veins, 3–4 mm long – pods pointed, hairy ....
　　　　　　　　　　　　　　　　　　　　　Ornithopus perpusillus (p 272 **H**)
　 Fls yellow or orange, over 6 mm long **20**

20 Lvs appearing trifoliate, but with lfts at base of lf-stalk resembling stipules – actual
　 stipules v narrow – pod straight, without joints ....*Lotus* (p 290)
　 Lvs with more than 5 lfts – pods curved, and jointed into segments – plant ± hairless **21**

21 Plant hairless – fls 10 mm long – fl-head stalks longer than lvs – pod segments shaped
　 like tiny horseshoes...*Hippocrepis comosa* (p 292 **A**)
　 Plant almost hairless – fls 6–8 mm long – fl-head stalks not longer than lvs – pod-
　 segments barrel-shaped – SW Eng, vr ....*Ornithopus pinnatus* (p 272 **H.1**)

22 Stipules papery, brown – fls bright pink or red ....*Onobrychis viciifolia* (p 292 **C**)
　 Stipules green – fls blue, purple, or yellow **23**

23 Plant hairless, robust **24**
　 Plant hairy...**25**

24 Plant creeping and ascending – fls greenish-yellow – infl-stalk much shorter than lf
　　　　　　　　　　　　　　　　　　　　　Astragalus glycyphyllos (p 270 **F**)
　 Plant erect – fls white or mauve – infl-stalk at least as long as lf ....
　　　　　　　　　　　　　　　　　　　　　Galega officinalis (p 292 **E**)

25 Keel of fl blunt – fls blue or purple ....*Astragalus danicus* or *A. alpinus* (p 272 **G, G.1**)
　 Keel of fl pointed – fls purple or yellow – Scot, r ....*Oxytropis* (p 272)

Whins (*Genista*), **Gorses** (*Ulex*), **Broom** (*Cytisus*) are ± evergreen shrubs with yellow fls that explode when insects (bees) alight on the lip, breaking the join between keel petals and thus releasing the coiled spring-like stamens that shoot out a cloud of pollen onto the insect. **Gorses** (*Ulex*) have no normal flat lvs on mature plants, and they are always spiny with branched spines. The calyx is yellow, divided into 2 lips to the base; upper lip has 2 tiny teeth at the top. **Whins** and **Greenweeds** (*Genista*) have simple, oval or oblong lvs; spines, if present, simple; the green or brown calyx is five-toothed, with 2 teeth on upper lip, 3 on lower; but the 2 lips not divided more than 1/2 way down.

A   NT **Petty Whin**, *Genista anglica*, low straggling shrub with sparse spreading spines, lfy when young, with also oval pointed **hairless** waxy-green lvs, 2–8 mm long. Fls in short terminal racemes, yellow, 8 mm long; pods **swollen, pointed, hairless**. Br Isles: GB f-lc; **Ire abs**; on dry hths. Fl 5–6.

B   NT ****Hairy Greenweed**, *G. pilosa*, spineless straggling shrub with oval lvs, 3–5 mm long, v **downy** below (**Bb**); fls (**Ba**) in short racemes, yellow, 10 mm long; calyx, corolla, stalk all downy; pods **flat, downy, pointed** at tips. Br Isles: Cornwall and Wales vl and r; on hths, limestone gsld, sea cliffs. Fl 4–6.

C   **Dyer's Greenweed**, *G. tinctoria*, shrub of 20–50 cm, ± erect; oblong lanceolate pointed lvs, with hairs on margins only. Infls of long racemes, mostly terminal; fls hairless, about 15 mm long; pods **flat, blunt, hairless**. Br Isles: Eng, Wales, f; S Scot o; **Ire abs**; in gsld, scrub on heavy soils. Fl 6–8.

D   **Gorse**, *Ulex europaeus*, **spiny** evergreen shrub to 2 m; has trifoliate lvs only when a v young plant. Fls yellow 2 cm across, coconut-scented, have bracteoles (**Da**) 3–5 mm long at bases of calyx; calyx (**Da**) 10–17 mm long; spines, 1.5–2.5 cm long, deeply furrowed. Br Isles, c (except high mts and the Fens); on rough gslds, hths, mostly on acid soils. Fl 1–12, peak 4.

E   **Dwarf Gorse**, *U. minor*, usually much smaller than D, but up to 1 m; its weaker, often ± prostrate stems have spines about 1 cm long, with only faint furrows; fls, 10–12 mm long, deeper yellow, have bracteoles (**Ea**) 0.5 mm only, **calyx** (**Ea**) **length < 9mm**. Br Isles: SE, S Eng f-la; Scot, Ire r introd; **SW Eng, S Wales abs** (distribution in Eng overlaps v little with that of F); on acid hths. Fl 7–9.

F   **Western Gorse**, *U. gallii*, in habit is intermediate between D and E, but not a hybrid; **key difference** from E is **calyx** (**Fa**) **length > 9 mm** (measure at least 10 fls if some calyces exactly 9 mm); spines more rigid than in E, 1.0–2.5 cm long, and faintly furrowed (as in E). Fls 10–12 mm, deep yellow, bracteoles (**Fa**) about 0.6 mm. Br Isles: W of a line Dorchester – Nottingham – Edinburgh, and near coast of E Anglia, vc; SE Eng vr; on hths, hill gslds. Fl 7–9.

G   **Broom**, *Cytisus scoparius* ssp *scoparius* (*Sarothamnus scoparius*), **spineless** erect shrub with long green straight hairless **five-angled** stems, borne ± in clusters. Stalked lvs, with 1–3 lfts each, occur on younger stems. Fls golden yellow, 20 mm long, on stalks to 10 mm; calyx (**Ga**), fl-stalks hairless; pods black, hairy. Br Isles, c on scrub on dry acid soils. Fl 5–6.   **G.1** NT***Prostrate Broom**, *C. scoparius* ssp *maritimus* is **procumbent** with young branches more densely silky hairy than G. Ire, Wales, SW Eng, CI, r on sea-cliffs. Fl 5–7. **G.2** **Spanish Broom**,

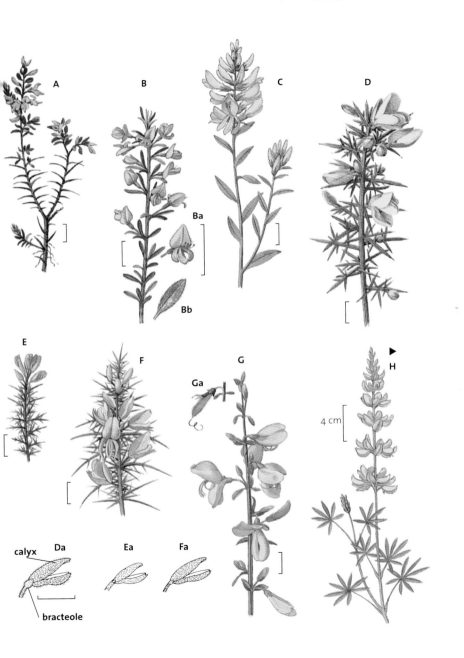

A

B

Ba

Bb

C

D

E

F

G

Ga

H

4 cm

calyx   Da   Ea   Fa

bracteole

Spartium junceum, resembles G, but has **cylindrical**, smooth stems and **simple** lanceolate lvs; fls yellow. Introd (Mediterranean); S Eng, o; elsewhere r; Ire abs; on wa, rds, cliffs on light soils. Fl 4–6.

▲ **H Tree Lupin**, *Lupinus arboreus*, erect shrub to 3 m, with palmate compound lvs with 7–11 radiating lfts to 3 cm long, silky below, hairless above. Fls in long racemes, yellow or white, 1.5–2.0 cm long. Introd (California); Br Isles, o-la, especially on rds, dunes. Fl 6–9. **H.1 Garden Lupin**, *L. polyphyllus* agg, complex group of gdn escapes; per herb with 9–16 radiating lfts, up to 10 cm; long spikes with fls of various colours. Originally from N America. Fl 5–7.

Restharrows (*Ononis*) are low shrubs or herbs with trifoliate lvs or with 1 lft; fls solitary in lf-axils, pink or yellow; plants ± sticky-hairy; all 10 stamens joined into a tube round the ovary. **Milk-vetches** (*Astragalus*) have pinnate lvs with terminal lfts, no tendrils, blunt keeled fls in heads or short spikes.

A Common Restharrow, *Ononis repens*, creeping to ascending **rhizomatous** per herb; stems 30–60 cm long to 30 cm tall, usually (but not always) spineless, rooting at the base; stems **hairy all round** (**Aa**). Lvs with downy stalks 3–5 mm long, large stipules, and blunt-toothed lfts 10–20 mm long, all sticky-hairy with a harsh, sweat-like odour. Fls 10–15 mm long, pink, short-stalked, **wings equalling keel**; pods erect, shorter than calyx. Br Isles: Eng, Wales, S Scot, S Ire f-c especially on calc gsld, scrub, dunes; NW Scot, N Ire r on mts. Fl 6–9.

B Spiny Restharrow, *O. spinosa*, erect per shrub **without** rhizomes, 30–40 cm tall; usually with straw-coloured spines on stem, which is hairy in **two opp vertical lines** (**Ba**); lfts narrower and more pointed than in A, fls deeper purple-pink, **wings shorter than keel**. Pods erect, longer than calyx. Br Isles: Eng f-lc; Scot vr; SW Eng, CI, IoS, Ire abs; on mds, gsld on heavy soils. Fl 6–9.

C Sch 8 ****Small Restharrow**, *O. reclinata*, erect downy **ann**, only 4–8 cm tall, sticky, without spines; fls pink, 7 mm or less, pods (**Ca**) **hanging downwards** when ripe, by bending of fr-stalks (only 6–7 mm long, enclosed in calyx). Lfts 3–5 mm, narrow, blunt. Br Isles: S Devon, CI, S Wales, S Scot vr; on rocky turf on limestone by sea, dunes. Fl 6–7.

D Laburnum, *Laburnum anagyroides*, tree to 7 m with trifoliate lvs, oval lfts; fls yellow, 20 mm long, in pendulous racemes 10–20 cm long; pods downy, 3–5 cm, **seeds v poisonous**. Introd (mts C Eur) and increasing; Br Isles, c (except Ire, o); on wds, gslds, railway banks. Fl 5–6. **D.1 Bladder-senna**, *Colutea arborescens*, shrub to 4 m; pinnate lvs with oval lfts, silky-hairy below; erect racemes of yellow fls with red veins; yellow-green **balloon-like** oval pods about 5 cm long. Introd (S Eur); Br Isles: o-f, especially in SE Eng; N Ire, Scot vr; in wa, on railway banks. Fl 5–7.

E False-acacia, *Robinia pseudoacacia*, tree to 25 m, with suckers; v furrowed bark; stipules on twigs, thorny; lvs pinnate, about 10 cm long; racemes pendulous, fls white, fragrant; pods flattened. Introd (E of N America); GB, lc; Ire, Scot vr; especially on railway banks, also wds, rds, wa on light soils. Fl 6.

F Wild Liquorice (Milk-vetch), *Astragalus glycyphyllos*, ± hairless robust sprawling per herb; stems 60–100 cm long, zig-zag; lvs 10–20 cm, pinnate, lfts elliptical,

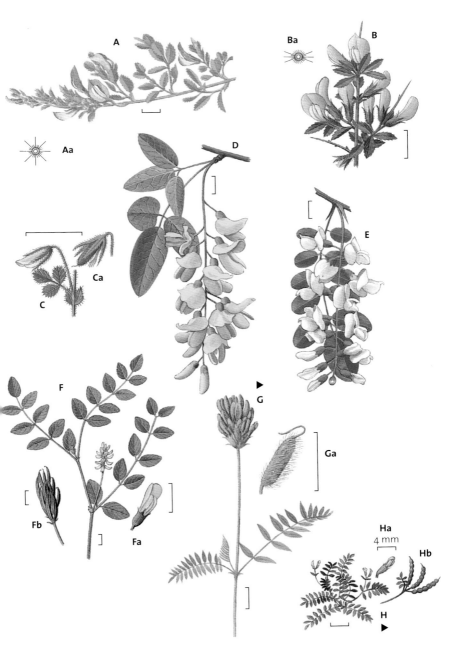

15–40 mm; stipules arrow-shaped, 2 cm; infl 2–5 cm, on stalks shorter than lvs; racemes of greenish-cream fls (**Fa**) 10–15 mm long, with a greyish tinge; pods (**Fb**) 25–35 mm, banana-shaped, rounded, pointed. Br Isles: SE and central Eng 0-la in calc gsld, scrub; N Eng, Wales, SE Scot vr; **Ire abs**. Fl 7–8.

▲ **G**   EN FPO (Eire) ****Ire Purple Milk-vetch**, *Astragalus danicus*, more like **Horseshoe Vetch** (p 292 A) in habit; low hairy per herb, 5–20 cm; hairy pinnate lvs 3–7 cm, lfts 5–12 mm; stipules joined below; racemes on stalks **much larger** than lvs, fls in heads, **bluish-purple**, erect, each about 15 mm long. Swollen pods (**Ga**) are dark brown, with crisp white hairs. Br Isles: Eng (N of Hants to Sutherland), E Scot, 0–vlf on calc gslds to S, dunes to N (**Wales, SW and SE Eng abs**); Ire (Aran Is only), W Scot vr; on dunes and calc gslds. Fl 6–7.   **G.1**   VU ****Alpine Milk-vetch**, *A. alpinus*, has free stipules, spreading, bluer fls, short infl-stalks. Scot vr on calc mt rocks. Fl 7.   **G.2** ****Purple Oxytropis**, *Oxytropis halleri*, has short heads of purple fls and hairy pinnate lvs as in G; but **silky-hairy** rather than downy; lfless infl-stalk stout, erect, from **base** of plant, **much longer** than lvs; fls 2 cm long; pod downy, 2.5 cm long, with a **pointed** keel (keel-tip blunt in other British *Astragalus* spp). N Scot only; vr but la on calc mt ledges, dunes. Fl 6–7.   **G.3**   VU ****Yellow Oxytropis (Milk-vetch)**, *O. campestris*, resembles G.2 in pointed keel to fl and silky-hairy lvs; but fls **pale yellow** or ± purple-tinged, on infl-stalk **shorter** than lvs. Scot only; vr on calc mt rocks. Fl 6–7.

▲ **H**   ****Ire Bird's-foot**, *Ornithopus perpusillus*, prostrate downy ann, stems 2–40 cm long. Fls in small heads, 3–6 together, with a pinnate bract below fl-head; fls (**Ha**) 3 mm or larger, short-stalked, creamy with red veins; pods (**Hb**) spreading like a bird's foot, curved, 10–20 mm, downy, constricted into many one-seeded joints. Br Isles: Eng, Wales f-la; S and E Scot lf; Ire, r; on dry sandy and stony gd, gslds, dunes. Fl 5–8.   **H.1** ****Orange Bird's-foot**, *O. pinnatus*, similar to H, but fls wholly orange; no bracts; pods have only slight constriction. **IoS, CI only**, vr on sandy turf, especially near sea. Fl 4–8.

Clovers (*Trifolium*) form a large genus of ann or per herbs with small fls in heads; trefoil lvs with stipules; fls with five-toothed calyx; wing petals longer than keel, and straight short pods enclosed in the withered calyx.

KEY TO CLOVERS (*TRIFOLIUM*)

1 Fls yellow **2**
   Fls white, pink or purple **6**

2 Fl-heads 2–3 cm long, sulphur-yellow ....*Trifolium ochroleucon* (p 274 C)
   Fl-heads 1 cm long or less, bright yellow **3**

3 Standard petal flat, bent forward...**4**
   Standard folded like a roof-ridge...**5**

4 Stipules oval – fls 4 mm long, pale yellow ....*T. campestre* (p 276 **A**)
   Stipules v narrow – fls 5 mm long, golden ....*T. aureum* (p 276 **A.1**)

5 Fl-stalks shorter than calyx-tube – standard not notched – 10–25 fls per head ....
                                                                                                           T. dubium (p 276 **B**)
   Fl-stalks equal to calyx-tube – standard notched – 2–10 fls per head *T. micranthum* (p 276 **C**)

6 Creeping pers – rooting runners – fl-heads on long stalks, all in lf-axils **7**
Plants not creeping – no rooting runners..**9**

7 Calyx bladder-like, pink and hairy in fr – fls pink ....*T. fragiferum* (p 274 **F**)
Calyx not bladder-like or hairy –fls white or pinkish **8**

8 Lf-stalks hairless – lfts >10 mm long ....*T. repens* (p 274 **D**)
Lf-stalks slightly hairy – lfts < 10 mm, usually 6–8 mm long ....*T. occidentale* (p 274 **D.1**)

9 Fls in lax heads, 6 or less together, spreading like fingers – plants ± prostrate, ann **10**
Fls many, crowded in dense heads **11**

10 Plant hairless – fls pink – pods longer than narrow calyx ....*T. ornithopodioides* (p 278 **C**)
Plant downy – fls creamy – pods enclosed in swollen calyx ....*T. subterraneum* (p 278 **B**)

11 All fl-heads terminal..**12**
Some fl-heads in lf-axils **15**

12 Fl-head cylindrical – calyx with long silky adpressed brownish hairs ....*T. incarnatum* (p 278 **F**)
Fl-head globose – calyx with long silky spreading white hairs **13**
Fl-head globose – calyx hairless, or with few long hairs **14**

13 Stipules ± triangular, abruptly narrowed to the tip – Hants & Sussex only, vr ....
T. stellatum (p 274 **F.1**)
Stipules gradually tapered to long, narrow tip ....*T. squamosum* (p 276 **D**)

14 Fl-heads obviously stalked – stipules v narrow, tapering to a green point....................................
T. medium (p 274 **B**)
Fl-heads scarcely stalked – stipules triangular, brown bristle-pointed ..*T. pratense* (p 274 **A**)

15 Fl-heads stalked – plants ± erect, 10–20 cm tall..**16**
Fl-heads stalkless – small ± prostrate plants of dry sandy gd **18**

16 Fl-heads cylindrical, v woolly – lvs downy ....*T. arvense* (p 278 **D**)
Fl-heads globular, not woolly – plant hairless **17**

17 Fls 8–10 mm, white to pinkish – calyx scarcely ribbed – stipules untoothed ....
T. hybridum (p 274 **E**)
Fls 4–5 mm, bright pink – calyx strongly ribbed and angled – stipules toothed – Cornwall,
r ann ....*T. strictum* (p 278 **G**)

18 Plants downy, at least on lf-undersides – calyx-tube strongly ribbed, with spine-like teeth **19**
Plants hairless – fl-heads dense, rounded, 0.5–1.0 cm across **21**

19 Lfts hairless above, downy below – fl-heads paired – fls white – Cornwall, vr ....
T. bocconei (p 278 **E**)
Lfts hairy both sides...**20**

20 Calyx-tube inflated, veins reddish, teeth erect – fls pink...............................*T. striatum* (p 276 **G**)
Calyx-tube not inflated, veins dark, teeth bent outwards in fr – fls white ....*T. scabrum* (p 276 **F**)

21 Lvs held erect on long, 2–3 cm stalks, above prostrate stems – corollas white but hidden
in calyx ....*T. suffocatum* (p 276 **A**)
Lvs spreading, on 1–3 cm stalks from slightly ascending stems – corollas pink, projecting
from calyx ....*T. glomeratum* (p 276 **E**)

A Red Clover, *Trifolium pratense*, hairy per, 10–40 cm tall, with grey-green elliptical to ovate lfts, 10–30 mm, often with a whitish crescent-shaped spot across them; **stipules triangular, bristle-pointed** (**Ab**), purple-veined. Fl-heads oval or globose, to 30 mm long, often paired, ± stalkless, with a pair of lvs close below. Fls (**Aa**) pink-purple (rarely white), 12–18 mm long, stalkless; **calyx-tube hairy**. Cultivated forms (var *sativum*) are larger than wild ones, with untoothed lfts, hollow instead of solid stems, and pale pink fls. Br Isles, vc in mds, rds, hbs, etc. Fl 5–9.

B Zigzag Clover, *T. medium*, is v like A, but has ± **zigzag** (not straight) stems; lfts 15–40 mm, narrower than in A, deep green (not grey-green), with only a faint whitened spot; fl-heads clearly stalked; ± **hairless calyx-tube**; fls deep purple-red; lf-stalks shorter (to 8 cm only). **Key difference** from A in **stipules tapering** (**Ba**), green to tip (not brown bristle-pointed). Br Isles f-lc; on scrub, hbs, wd borders, mds. Fl 6–9.

C NT *****Sulphur Clover**, *T. ochroleucon*, has habit of A, but fls **sulphur-yellow** to whitish-yellow, fading brown, and lfts **without** white spots; stipules oblong (not bristle-pointed); 2 lvs close below globose, sessile fr-heads (2–4 cm long). Eng, E Anglia to Lincs, vl-lf on rds, hbs, wd borders, gsld, mostly on calc clay or limestone. Fl 6–7.

D White (Dutch) Clover, *T. repens*, **creeping rooting** per herb; lfts oval, **usually > 10 mm** and with a white spot; **hairless**, long lf-stalks (to 14 cm). Fl-heads globular on long (5–20 cm) stalks from lf-axils on creeping stems; fls white, 8–10 mm, sometimes ± pink- or purple-tinged, scented. Calyx-teeth narrow, about 1/2 length of tube. Br Isles, vc in gslds on all soils except v acid ones. Fl 6–9.
D.1 *****Western Clover**, *T. occidentale*, also with creeping stems, is smaller, has tiny ± circular **lfts usually < 10 mm**, without spots, red stipules and calyx, lf-stalks slightly **hairy**; no scent. SW Eng, Wales, Ire, r – vlf; CI, IoS vc; on dry gd near sea. Fl 3–5.

E Alsike Clover, *T. hybridum*, similar to D but **not** rooting at nodes: plant nearly **hairless**, has **white** fls (8–10 mm) flushed with **pink** below, in globular heads, on stalks to 15 cm, all in lf-axils (none terminal); oblong, untoothed, oval **long-pointed stipules**; calyx-teeth broad-based, triangular, about twice length of tube. Br Isles, f (except S Ire, NW Scot r) but declining; escape from cultivation on ar, wa. Fl 6–9.

F Strawberry Clover, *T. fragiferum*, **creeping** ± hairless per with habit of D, but fls always pink; calyx-tubes **hairy**, swell in fr to form **pinkish, netted, downy bladders**, 5–6 mm long x 3–4 mm wide, hence resemblance of the 15–20 mm globular fruiting-heads to pink strawberries. **Key difference** from D when not flg or frg is side veins on lfts are thickened and curved back at tips in F. Br Isles: S, E Eng f-la; Wales, Ire, r; Scot vr; in damp mds, gsld near sea; rarer inland, on heavy soils. Fl 7–9. **F.1 Starry Clover**, *T. stellatum*, erect hairy ann with fl-head globose; calyx with long silky spreading white hairs; each fl 12–18 mm long, white becoming pink; sharply toothed stipules; fr heads are distinctive, the long calyx-lobes reflexed, star-like, to give the plant its common name. Fr-heads similar to **Sea Clover** (p 276 D) but stipules ± triangular, **abruptly** narrowed to the tip. Probably native, vr (but sometimes vla) in Hants and Sussex on shingle, wa, gdns near the sea. Fl 5–6.

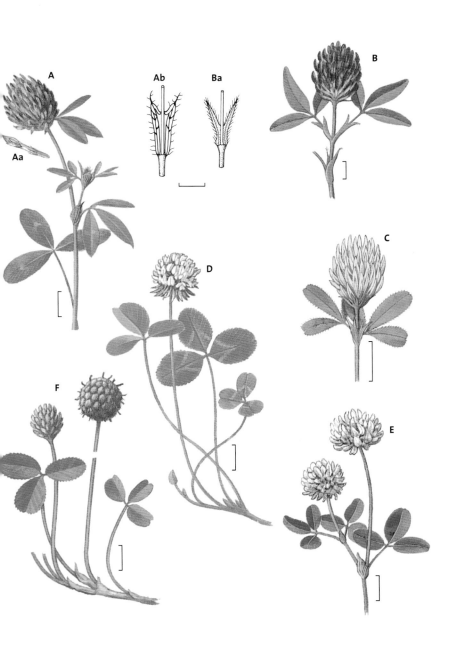

A

Aa

Ab

Ba

B

C

D

F

E

A Hop Trefoil, *Trifolium campestre*, sparsely hairy ann, 10–30 cm tall, ± erect; terminal lft has stalk longer than lateral lfts; globular fl-heads, 10–15 mm across, of >25 fls, each 4–7 mm long; fl-stalks 1/2 length of calyx-tube; standard petals of fls **broad and flat,** bending **forwards** and **downwards** over the pods, turning pale brown in fr, so that fl- and fr-heads are like tiny **hop-cones** in form. Br Isles, c on dry gslds, rds, hbs, especially on calc soils. Fl 6–9. **Do not confuse** large versions of B with A; also **Black Medick** (p 289 E), can be confused with B and A on this page, but it is v downy, its lfts have **minute points** in centres of tips, and frs are black and curved, not straight.   **A.1 Large Trefoil**, *T. aureum*, taller version of A, with narrower lfts, **all unstalked**; larger (5–6 mm long), richer, golden-yellow fls without stalks; stipules **linear** (not oval); oblong two-seeded (not oval one-seeded) pods. Introd (C Eur); GB vr; on ar, wa, gslds. Fl 7–8.

B Lesser Trefoil, *T. dubium*, close to A, but smaller, almost **hairless**; terminal lft has stalk longer than lateral lfts; fl-heads (5–7 mm across) of mostly <25 yellow fls, each 3–4 mm long; fl-stalks shorter than calyx-tube; standard petals of fls **fold** down on **either side** of mid-line over the pods like a **ridged roof**, turning brown in fr (so that fl- and fr-heads **not** hop-like). Br Isles, vc on dry gslds, rds, hbs and bare gd. Fl 5–9.

C Slender Trefoil, *T. micranthum*, resembles B, but has lfts of lvs ± **stalkless**; tiny 2.0–2.5 mm long fls, only 2–10 per fl-head, on **stalks** to 2 mm long. Fls (**Ca**) fold down mid-line as in B, but standard is **deeply notched** at top. Br Isles: Eng, Wales f (E and SE Eng c); Ire (E and S), Scot r; on gslds on light soils. Fl 5–7.

D *Sea Clover, *T. squamosum*, ± erect, ± downy ann, 10–30 cm, with lfts 10–25 mm x 3–4 mm, broader above; stipules **gradually** tapered to long, narrow tip; terminal, oval, short-stalked fl-heads 10 mm long, subtended by a pair of lvs; many tiny pink fls; in fr (**Da**) strongly-ribbed (5–7 mm-long) tough bell-shaped calyx-tubes, with rigid star-like spreading triangular green teeth. Br Isles: S Eng, CI r and l, decreasing (lf Thames Estuary); on dry banks near sea, especially on clay. Fl 6–7.

E   FPO (Eire) **Ire ***Clustered Clover**, *T. glomeratum*, hairless, ± prostrate or ascending ann, with stems 5–25 cm long. Lfts 5–7 mm long, widened above, toothed; stipules oval, long-pointed; lf-stalks 10–15 mm, **spreading**. Fl-heads, terminal and in lf-axils, 5–10 mm, unstalked, globular, each contain many dense-set fls (**Ea**); tiny pink corollas (length to three times calyx-tube); calyx-tubes, hairless, green, bell-shaped, with triangular, green sharp-pointed, spreading teeth. Br Isles: S Eng, Cornwall to Norfolk r; lf to E; CI, f; SE Ire vr; in dry open short turf on sand or gravel soils, especially near sea. Fl 6–8.

F Rough Clover, *T. scabrum*, downy all over, is close to G, but fls have calyx-tube **bell-shaped**, not inflated (**Fa**), with dark ribs, and rigid **outward-curving** teeth; petals white. Br Isles: Eng, Wales o-lc; E Ire, E Scot vr; on dry open sandy or gravelly gsld, especially by the sea, r inland. Fl 5–7.

G Knotted Clover, *T. striatum*, ann of similar habit to E, but usually taller, and v **downy** throughout; fl-heads egg-shaped, unstalked, 10–15 mm; calyx-tube **downy, inflated** (**Ga**), with strong reddish ribs and sharp **erect** teeth; petals pink. Br Isles: Eng, Wales f-lc; E Scot, E Ire r; on dry open sandy and gravelly gsld, by sea and inland. Fl 5–7.

A *****Suffocated Clover**, *Trifolium suffocatum*, prostrate hairless ann, rather like E (p 276); dense globular heads of fls **unstalked** on prostrate stems, but set in axils of **long-stalked** (20–30 mm) trifoliate lvs which are held **erect** like tiny parasols. Corollas of fls tiny (2 mm long), and white, but soon fading, and almost hidden in calyx-tubes (cylindrical, ribbed, with pointed spreading teeth). Br Isles: Eng, coastal, Cornwall to Yorks, IoM r (but lf); on dry sandy and gravelly gslds near sea, vr inland. Fl 4–8.

B   FPO (Eire) ******Ire **Subterranean Clover**, *T. subterraneum*, also a prostrate ann, **hairy** like F and G (p 276); long (2–5 cm) stalked dark green lvs, but fls are creamy, ± unstalked, **only 4–6 together**, with long (8–12 mm) narrow corollas (**Ba**) spreading **fan-wise** from their short common stalk; long curved bristle-like calyx-teeth. Some fls **sterile**, without corollas, but with rigid, palmate-lobed, enlarged calyx-teeth. Frs globular, 2.5 mm long, are forced downwards on elongating stalks when ripe. Br Isles: Eng (S and SE only) o-lf; IoS, CI c; Wales r; Ire vr; in dry open sandy and gravelly places. Fl 5–6.

C   **Bird's-foot Clover (Fenugreek)**, *T. ornithopodioides*, is close to B in habit; a prostrate ann, but has light green and **hairless** lvs; **fls 2–4**, held **fan-wise** on short-stalked heads, white or pale pink, 6–8 mm long; pods straight, much longer than calyx-tube; calyx-teeth narrow, straight. Br Isles: Eng, Wales o, lf to SE and E, CI, IoS c, elsewhere r; Ire (SE) vr; on dry open sandy and gravelly gd, especially near sea. Fl 5–9. **Key differences** from B (if frs are abs) are no **sterile** fls but **bracts** present (none in B).

D   **Hare's-foot Clover**, *T. arvense*, tall usually erect ann, 5–20 cm, with spreading lvs; lfts 10–15 mm x 3–4 mm, stipules oval, bristle-pointed; fl-heads **terminal** and **axillary**, oval to cylindrical, to 25 mm long, v **downy** with **soft white hairs**, set on stalks to 20 mm; fls white or pink, ± hidden in the hairs; calyx bell-shaped, with long hairy teeth. Br Isles: Eng f-lc, especially E and SE, rarer to N and in Ire; on sandy gsld, dunes etc, commonest near sea, but also inland. Fl 6–9.

E   VU ****Twin-headed Clover**, *T. bocconei*, ± erect, downy ann, 5–10 cm tall, most like a small G (p 276), with lfts 5–13 mm, **hairless** above, **downy** beneath; long bristle-like stipules; fl-heads oval, to 10 mm, unstalked and usually in pairs, with strongly **ribbed, dark** (not inflated) calyx with spine-like erect teeth; corolla (**Ea**) persistent, obvious, longer than calyx, white, then pink. Near Lizard, Cornwall, vr; on short gsld on cliffs. Fl 5–6.

F   VU ****Long-headed Clover**, *T. incarnatum* ssp *molinerii*, ± erect silky-hairy ann, 5–20 cm, with oval lfts; **cylindrical** fl-heads, all terminal, 10–20 mm long, of pale pink or cream fls, ± hidden among long, silky, buff-coloured hairs on calyx-tubes (cylindrical with long bristle-like teeth spreading starwise in fr). Lizard, Cornwall, CI vr; on short gsld on cliffs. Fl 5–6.   **F.1 Crimson Clover**, *T. incarnatum* ssp *incarnatum*, is like F, but 20–50 cm tall, with fl-heads 15–40 mm; fls **deep crimson**, rarely paler colours. Introd from cultivation; Eng, Wales r on field margins, ar, wa. Fl 6–7.

G   VU ****Upright Clover**, *T. strictum*, erect **hairless** ann, 3–15 cm, with narrow elliptical (5–15 mm) lfts; **oval, toothed**, whitish stipules; terminal and axillary globular heads of pale pink fls; calyx-tubes (**Ga**) strongly ribbed with spine-like teeth. Lizard, Cornwall, CI, vr on short gsld on cliffs; Wales vr on rocks inland. Fl 5–6.

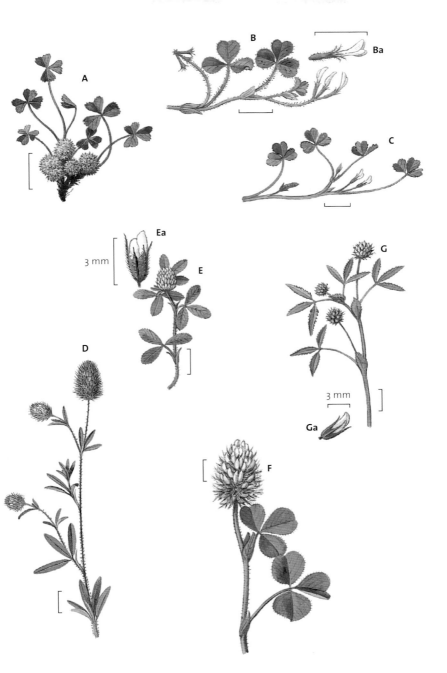

3 mm

3 mm

The two genera, **Vetches** (*Vicia*), pp 282–4, and **Peas and Vetchlings** (*Lathyrus*), pp 284–7, are closely related, and best considered together. The technical differences are not easy to see. In practice, *Vicia* spp **usually** have 2 to many pairs of lfts, ridged or angled (rarely winged) stems and round styles **hairy all over** (like a bottle-brush) **or hairless**. Despite what many floras say, *Vicia* stems are often only round **internally**, in cross-section. *Lathyrus* spp have either none, 1–5 pairs of lfts, angled or winged stems, and flattened styles **hairy on upper side only** (like a toothbrush). Both genera are non-spiny herbs with pinnate lfts in 1 to many pairs, or none; the terminal lft found in other Pea-fls is either abs or replaced by a tendril or a little point. Pods are straight, not spiny, oblong, two-valved and pea-like.

ID Tips Vetches, Vetchlings & Peas
• **What to look at: (1)** whether tendrils abs or present **(2)** lfts with parallel or pinnate veins **(3)** stipule size and shape **(4)** fl colour **(5)** whether calyx teeth all equal in length or not **(6)** fr-length, hairiness and seed characters.
• The **Tare-vetches** and **Spring Vetch** have tiny fls compared to all other *Vicia* spp, often around 5 mm long or less.
• **Meadow Vetchling** may not always have an unwinged stem but its yellow fls and strongly compressed frs help distinguish it.

KEY TO VETCHES, VETCHLINGS & PEAS (*VICIA* & *LATHYRUS*)

1 Lvs grass-like or abs **2**
Lvs not grass-like, lfts present **3**

2 Lvs of grass-like blades – stipules tiny, <1 cm long – tendrils none – fls red .... *Lathyrus nissolia* (p 286 **F**)
Lf-blades abs – stipules huge, >1 cm long, triangular, lf-like – tendrils long – fls yellow .... *L. aphaca* (p 286 **G**)

3 Tendrils abs or lf ending in short point **4**
Tendrils present **8**

4 Lfts with parallel veins .... *L. linifolius* (p 284 **E**)
Lfts with pinnate veins **5**

5 Per plant – most raceme-stalks (ie. stalk below all fls, not individual fl-stalks) > 3 cm – fls lilac white with purple veins .... *Vicia orobus* (p 284 **A**)
Ann plant – raceme-stalks abs or <2 cm **6**

6 Procumbent to weakly climbing hairy plant – lfts <2 cm long – frs <3 cm long **7**
Erect hairless plant – at least some lfts >2 cm long – frs >5 cm long .... *V. faba* (p 284 **C**)

7 Seeds with minute bumps (use hand lens) .... *V. lathyroides* (p 282 **E**)
Seeds smooth .... *V. sativa* (p 284 **D**)

8 Lfts with pinnate main veins **9**
Lfts with parallel main veins **24**

9 Lfts 1 pair only per lf **10**
Lfts 2 or more pairs per lf **12**

10 Stipules triangular, c. 10 mm long with broad, regular-teeth throughout stipule-margins
– calyx-teeth almost equal – pod hairy ....*V. bithynica* (p 284 **J**)
Stipules half-arrow shaped, not toothed – calyx teeth unequal – pod hairless or not
developing **11**

11 Fls <20 mm long, standard c. 1 cm long when flattened ....*L. tuberosus* (p 286 **A**)
Fls >25 mm long, standard c. 4 cm long when flattened ....*L. grandiflorus* (p 286 **C.1**)

12 Fls in stalked racemes with raceme stalk ± equal to or longer than individual fl **13**
Fls in stalkless racemes or raceme stalk shorter than individual fl **20**

13 Fls showy, usually >10 mm long, or more, if <10 mm then > 8 per raceme...............................**14**
Fls small, 8 mm or less – < 4 per raceme **18**

14 Prostrate blue-green hairless seashore plant – lvs broad elliptical – fls usually 20 mm or
more long ....*L. japonicus* (p 287 **F**)
Bright green plant – lvs various – fls usually <20 mm long **15**

15 Lfts 2–3 pairs only – fls usually 1 or 2 only .... *V. bithynica* (p 284 **J**)
At least some lfts in more than 3 pairs – racemes elongated with 4 to many fls **16**

16 Claw c. twice as long as limb of standard petal – calyx swollen at base with bulge on
upperside, the calyx-base projecting backwards beyond where the fl-stalk joins the calyx
V. villosa (p 282 **B**)

Claw equal to or shorter than limb of standard petal – calyx not swollen at base, the
calyx-base not or hardly projecting beyond where the fl-stalk joins the calyx **17**

17 Fls purplish-blue – at least some lvs slightly hairy – pods brown, 10-25 mm ..*V. cracca* (p 282 **A**)
Fls white or lilac-flushed, purple-veined – lvs ± hairless – pods black, 25-30 mm ....
V. sylvatica (p 284 **B**)

18 Calyx-teeth equal – frs hairy with 2 seeds ....*V. hirsuta* (p 282 **F**)
Calyx-teeth unequal – frs ± hairless 3–8 seeds **19**

19 Seeds with wide scar >2x as long as wide (x20) ....*V. tetrasperma* (p 282 **G**)
Seeds with scar little longer than wide ....*V. parviflora* (p 282 **H**)

20 Fls 9–30 mm long – seeds smooth **21**
Fls 5–9 mm long – seeds with minute bumps (x20) ....*V. lathyroides* (p 282 **E**)

21 Lfts 2–3 pairs only – stipules c.10 mm long...*V. bithynica* (p 284 **J**)
At least some lfts in more than 3 pairs or if less than 3 pairs then stipules <8 mm long **22**

22 Fls yellow, 1–2 in lf-axils – pod densely hairy ....*V. lutea* (p 284 **I**)
Fls pink, purple or blueish – pod sparsely hairy **23**

23 Plant per – calyx-teeth unequal, lower longer than upper (p 283 Ca) – lfts generally widest
below middle – fls dull purple, usually more than 2 together.....................*V. sepium* (p 282 **C**)
Plant ann – calyx-teeth almost equal (p 283 Da) – lfts generally widest at or above middle
– fls bright pink or purple, usually only 1-3 together ....*V. sativa* (p 282 **D**)

24 Stem square, or sometimes with small wings – fls yellow ....*L. pratensis* (p 284 **D**)
Stem with large wings – fls pink, purple or cream **25**

25 Lfts > 1 pair – pod flattened – hairless erect fen plant.....................................*L. palustris* (p 286 **E**)
Lfts 1 pair only **26**

26 Fls 1-3 only per raceme – pods hairy ....*L. hirsutus* (p 286 **D**)
Fls > 3 per raceme – pods hairless **27**

27 Lfts oval – stipules nearly as wide as stem – lowest calyx-tooth c. 8 mm *L. latifolius* (p 286 **C**)
Lfts lanceolate – stipules less than ¹/2 width of stem – lowest calyx-tooth c. 4 mm ....
L. sylvestris (p 286 **B**)

A Tufted Vetch, *Vicia cracca*, ± downy climbing **per** herb, 60–200 cm; lfts 10–25 mm long, in 8–13 pairs; branched tendrils; infl in **long** racemes 2–10 cm long on 2–10 cm stalks, 10–40 fld, dense; fls 8–12 mm, purplish-blue; claw ± **equal** to limb of standard petal; calyx-teeth (**Aa**) unequal, pod 10–25 mm, hairless, brown, two-six-seeded. Br Isles, c on hds, wd borders, scrub. Fl 6–8.

B Fodder Vetch, *V. villosa*, v close to A but **key differences** are **ann** plant with v **long fls**, the claw ± twice as long as the limb of standard petal; calyx (**B**) with a swollen base. Introd (S Eur); Br Isles: S Eng, especially London area, o-lf; N Eng, Scot r; Ire abs; possibly increasing on wa, ar, dry gslds. Fl 6–7.

C AWI **Bush Vetch**, *V. sepium*, climbing or spreading, ± hairless **per**, 20–50 cm; lfts 3–9 pairs, 10–30 mm long, oval, generally widest at base, base cordate, tip blunt and often notched; tendrils branched, stipules half-arrow-shaped, sometimes with a dark spot. Racemes short (1–2 cm), rounded, on short stalks, **2–6-fld**; fls 12–15 mm long, dull pink-purple; **lower calyx teeth longer than upper** (**Ca**); pod 20–25 mm, black. Br Isles, vc in wds, hbs, scrub. Fl 4–8.

D Common Vetch, *V. sativa*, **v variable** erect, trailing or scrambling downy **ann**, 15–40 cm; lfts 10–20 mm, 4–8 pairs, generally widest at middle or above, tendrils simple or branched (sometimes abs in young plants); stipules half-arrow-shaped, at least lower toothed, usually with a dark spot. **Fls 1–3**, 10–30 mm long, in lf-axils, bright pink or purple; **calyx teeth almost equal** (**Da**); pod (**Db**) hairy or hairless, 4–12-seeded, seeds **smooth**. Br Isles, vc in hds, hbs, gsld, scrub. Fl 5–9. There are 2 wild forms: ssp *nigra* is slender with narrow lfts of **different sizes**, upper lfts **narrower** than lower, fls 10–16 mm long, deep pinkish-purple; ssp *segetalis*, fls usually two-coloured and lfts **± the same**. Cultivated form (spp *sativa*) now r; v robust, with broader lfts 20–30 mm long and ± the same; fls paler, pod to 80 mm. **Key difference** between ssp *sativa* and ssp *segetalis* is subtle – pods of ssp *sativa* are often hairy and yellowish to brown; ssp *segetalis* pods usually hairless, brown to black.

E ******Ire **Spring Vetch**, *V. lathyroides*, slender prostrate downy **ann**; stems 5–20 cm long; lfts 2–3 pairs, 4–10 mm long, narrow, with mucronate-points in notched tips or sometimes with simple tendrils; lower stipules almost entire; fls solitary, 5–9 mm, lilac-pink; pod hairless, black, 15–25 mm. **Do not confuse** with small, young plants of D without tendrils. The **key difference** is D has smooth seeds; E has seeds with **minute bumps**. Br Isles: GB o-lf in dry sandy gd, especially by sea, mostly to E; Ire vr to E. Fl 4–5.

F Hairy Tare, *V. hirsuta*, slender trailing ± hairless ann, 20–30 cm; lfts 4–10 pairs, linear-oblong; racemes 1–9-fld, on slender stalks 1–3 cm long; fls (**Fa**) mauve-whitish, 4–5 mm; **calyx-teeth longer than calyx-tube, ± equal**; pod (**Fb**) 10 mm, oblong, **downy**, usually two-seeded. Br Isles generally c but rarer NW Scot, Ire; on hbs, gslds on less acid soils. Fl 5–8.

G Smooth Tare, *V. tetrasperma*, hairless ann with lfts 10–20 mm long, in 3–6 pairs, racemes 1–2-fld; fls (**Ga**) deep lilac, 4–8 mm. **Key difference** to distinguish G & H from F if no pods are present is **calyx-teeth unequal**, **upper 2 shorter than calyx-tube** in G and H; pod (**Gb**) 12–15 mm, oblong, **hairless**, four-seeded. Br Isles: Eng, Wales, f; Scot, Ire vr; in gsld, hbs on heavier soils. Fl 5–8.

H VU ***Slender Tare**, *V. parviflora* (*V. tenuissima*), is v close to G, but fls larger (8–9 mm); lfts longer (to 25 mm), fewer; infl-stalks longer than lvs; pod (**H**) 5–8-seeded. **Key difference** between G & H is the size of seed scars (see key) which is easy to see at x20 under a microscope (less so under a hand lens). S Eng r and l, declining; on gslds on heavier soils. Fl 6–8.

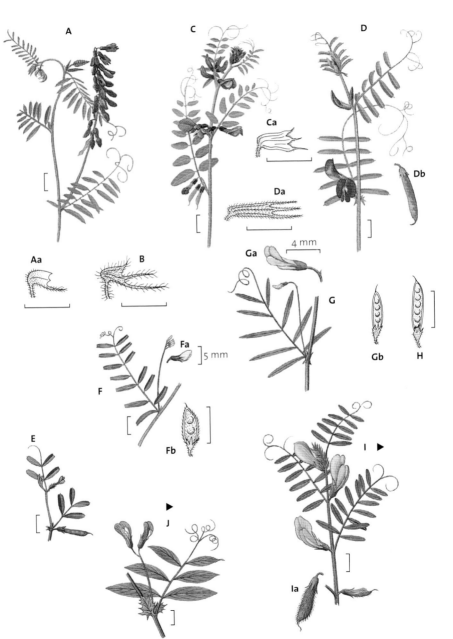

▲ **I** NT *****Yellow-vetch**, *Vicia lutea*, prostrate, hairy (sometimes almost hairless) ann; 3–10 pairs of lfts; tendrils on lf-tips; fls 1–3, pinky-white fading to pale greyish-yellow, 20–35 mm long; pod (**Ia**) v hairy. S Eng, vlc; Scot, Ire vr; mainly on shingle beaches near sea; also scattered, vr as introd inland. Fl 6–9. **Do not confuse** with introd, **yellow-fld** Vetches eg **Hungarian Vetch**, *V. pannonica*, erect and with v hairy lvs; **key difference** from I is **hairs** on back of standard petal.

▲ **J** VU *****Bithynian Vetch**, *V. bithynica*, climbing or trailing ± hairless **ann** herb, 30–60 cm tall; stem angled, ± winged; lvs with 2–3 pairs (sometimes 1 pair) of elliptical lfts, 20–50 mm long, lower blunt, upper narrower and pointed; **v large, oval-triangular**, spreading-toothed **stipules**, 1 cm or more long; branched tendrils. Fls 20 mm long, 1–3 on a common stalk (long **or** v short); standard-petal lilac-purple, wings and keel cream. Pod 30–40 mm, beaked, hairy. Br Isles: S Eng, Cornwall to Essex, r and l; S Wales, Scot, Galloway vr; on scrub, gsld, especially on heavy soils and near sea. Fl 5–6. **Do not confuse** with **Hairy Vetchling** (p 286 D) which has **v narrow, untoothed stipules**; smaller fls with **crimson-purple** standards.

A FPO (Eire) ******Ire NT *****Wood Bitter-vetch**, *V. orobus*, spreading to erect ± downy per, 30–60 cm; stem round; lvs have **no** tendrils, but a minute point at tip and 6–11 pairs of 10–15 mm, elliptical-mucronate lfts; stipules toothed, half-arrow-shaped; racemes 6–20-fld with **long stalks**, most >3 cm; calyx with short teeth; fls 12–15 mm, lilac-white, flushed with pink above, and with purple veins to standard and wings. Pod 20–30 mm, oblong, pointed, hairless. Br Isles: W of GB, Cornwall to Sutherland, r but vlf (Wales, f); W Ire r; in mds, scrub, rocks in hilly districts. Fl 6–9.

B AWI **Wood Vetch**, *V. sylvatica*, climbing hairless per, 60–200 cm; stems round. Lvs with 6–10 pairs of 5–20-mm elliptical mucronate lfts and much-branched **tendrils**; stipules lanceolate, with many narrow teeth. Racemes 1–7 cm long, on stalks to 10 cm, 6–20-fld, loose, one-sided; fls 15–20 mm, white with pencilled **purple veins**. Pod 25–30 mm, pointed, black when ripe, hairless. Br Isles, scattered, o-lf (Ire r; E and SE Eng vr-abs); in old open wds and wd borders, cliffs by sea. Fl 6–8.

C **Broad Bean**, *V. faba*, erect **hairless ann** to 1m; lvs elliptical, 5-10 cm; **no tendrils**; fls 1.6-3 cm, fl colour varies, white to lilac, darker wing-petals. Fr pod large, 5–20 x 1–2 cm. S Eng, f escape from cultivation, r-o elsewhere; on rds, wa, ar. Fl 5–6.

D **Meadow Vetchling**, *Lathyrus pratensis*, scrambling per herb, 30–120 cm long, usually with angled stems, each lf with 1 pair lanceolate grey-green lfts 1–3 cm long, with parallel veins, tendrils, and lf-like, arrow-shaped, 1.0–2.5 cm long stipules. Racemes on stalks longer than lvs, 5–12 fls, **yellow**, 15–18 mm; fr pod 25–35 mm, black when ripe. Br Isles, vc in hbs, gslds, scrub. Fl 5–8.

E AWI **Bitter-vetch**, *L. linifolius* (*L. montanus*), erect hairless per, 15–40 cm, with tuberous rhizome; lvs with 2–4 pairs of 1–4 cm long, narrow lanceolate to elliptical, pointed or blunt lfts; **no tendrils**; stipules toothed. Raceme stalked, with 2–6 fls; fls 12 mm, crimson-red, turning blue or green; pod 3–4 cm, rounded, hairless, brown. Br Isles f-lc, but abs E Anglia and C Ire; on hilly wds, hbs, scrub, hths on more acid soils. Fl 4–7.

F **Grass Vetchling**, *L. nissolia*, resembles a grass superficially when not in fl but lf-base does not enclose stem. Erect, hairless ann, 20–30 cm, with lvs composed of

A

B

D

C

G

▶

E

F

Fa

phyllodes (broad, grasslike midribs only) with no lfts; stipules tiny at base, tendrils abs. Fls bright crimson-red, 15 cm, erect, 1–2 together on long stalks; pod (**Fa**) 30–50 x 2–3 mm, straight, hairless, pale brown when ripe. S Eng: Devon to Lincs o-vla especially near sea in SE; r scattered elsewhere; Ire abs; gsld, scrub on ± basic, heavy soils. Fl 5–7.

▲ **G** ᵛᵁ *Yellow Vetchling, *Lathyrus aphaca*, hairless, **waxy grey-green**, scrambling hairless ann, to 40 cm; no lfts in mature plants, but only simple tendrils, that have **large**, 1–3 cm long, broad-**triangular stipules** paired at their bases and functioning as lvs. Fls on long stalks, yellow, erect, solitary, 10–12 mm long, with long calyx-teeth; pod 20–30 mm, ± curved. S Eng: Cornwall to E Anglia, CI r; in dry gslds. Fl 6–8.

A Tuberous Pea, *L. tuberosus*, scrambling ± hairless per herb, with tuberous roots and angled stem. Lvs with 1 pair of lfts, 1.5–3.0 cm long, elliptical; tendrils simple or branched; stipules narrow, 8–20 mm long, half-arrow-shaped. Long-stalked racemes with 2–7 fls, crimson, 12–20 mm long. Pod 25 mm, hairless, rounded. Introd (E Eur); GB r-o; Ire abs; in ar, wa. Fl 7.

B ᴬᵂᴵ **Narrow-leaved Everlasting-pea**, *L. sylvestris*, tall climbing hairless per herb, to 3 m. Stem **broadly-winged**; only **one pair** lfts, 7–15 cm long, **narrow-lanceolate**; tendrils branched; stipules to 2 cm long, less than 1/2 width of stem, lanceolate. Racemes on stalks 10–20 cm long, with 3–8 fls; fls 12–20 mm long, buff-yellow flushed with rose-pink above on each fl; calyx-teeth shorter than tube, lowest calyx tooth c. **4 mm** long. Br Isles: Eng, Wales o-lc; SW Scot r; Ire abs; in open wds, scrub, hds. Fl 6–8.

C Broad-leaved Everlasting-pea, *L. latifolius*, is close to B in habit, but has lfts **oval** and **blunt**, stipules more than 1/2 width of winged stem; tendrils branched; fls 15–30 mm long, 5–15 on a stalk, of vivid **magenta**-pink (or white), calyx-teeth longer than tube, lowest calyx tooth c. **8 mm** long. Introd (S Eur); S of GB, o-lc on railways banks, wa, scrub. Fl 6–8. **C.1 Two-flowered Everlasting-pea**, *L. grandiflorus*, easily confused with C but **stem** much more **slender, square** not winged; stipules 3–6 mm long; fls **1–2** (rarely more) on a stalk. Pods not usually formed in Br Isles, plants mostly spreading vegetatively. Introd (Mediterranean); Br Isles, r but sometimes vla, rampant gdn escape on railways, wa, hds. Fl 5–7.

D Hairy Vetchling, *L. hirsutus*, ± hairless, climbing ann, 30–100 cm, with winged stem; linear-oblong lfts in 1 pair per lf; tendrils branched; stipules 1.0–1.5 cm, v narrow, half-arrow-shaped, untoothed. Fls 7–12 mm long, 1–3 together on a long stalk. Standard petal red-purple, wings pale blue, keel cream. Pod (**Da**) 30–50 x 6–8 mm, densely hairy, brown when ripe. Introd (S Eur); Br Isles: Eng, N Ire vr, probably native on gsld on clay in Essex. Fl 6–8.

E ᵂᴼ ⁽ᴺᴵ⁾ **Ire NT *Marsh Pea**, *L. palustris,* erect, climbing per herb, 60–120 cm, with winged stem; lvs with 2–5 pairs narrow-lanceolate lfts 3.5–7.0 cm long; branched tendrils; narrow half-arrow-shaped stipules. Racemes long-stalked, with 2–6 fls, clear light purple, 12–20 mm long. Pod 3–5 cm, flattened, hairless. Br Isles: Eng, Wales r-lf (mostly E Anglia to SE Yorks); also Somerset, S Wales, vr; C Ire r; in calc fens and marshes. Fl 5–7.

F FPO (Eire) ****Ire ***Sea Pea**, *L. japonicus*, prostrate, **grey-green** ± hairless per herb, with angled stems to 100 cm long; lvs with 2–5 pairs lfts 2–4 cm long, broad-elliptical (sometimes narrow-lvd in E Scot, Suffolk), blunt, rather **fleshy**; tendrils sometimes abs; stipules broad-triangular, 2 cm. Racemes short-stalked, with 5–15 fls; fls 14–20 mm long, purple, fading blue with pale keel; pod 3–5 cm, hairless, swollen and garden pea-like in form and flavour of seeds. Eng: S and E coasts Cornwall to Suffolk o-vla; rest of Br Isles vr and sporadic on shingle beaches by sea. Fl 6–8.

Melilots (*Melilotus*) are close to Clovers (*Trifolium*), with trifoliate lvs; but infl is a loose, many-fld, erect raceme. Pods straight, short, oval, never spiny, rarely open. **Medicks** (*Medicago*) differ from both Clovers and Melilots in their **sickle-shaped** or **spirally-coiled** frs (bearing spines in some spp).

KEY TO MELILOTS (*MELILOTUS*)

1 Fls white, 4–5 mm ....*Melilotus albus* (p 289 **C**)
Fls yellow **2**

2 Fls about 2 mm – pod 2–3 mm, olive ....*M. indicus* (p 289 **D**)
Fls 4–6 mm – pod 3–6 mm **3**

3 Pods hairy, pointed, netted, black when ripe – petals all equal ....*M. altissimus* (p 288 **B**)
Pods hairless, blunt, ridged, brown when ripe – keel petal shorter......*M. officinalis* (p 288 **A**)

KEY TO MEDICKS (*MEDICAGO*)

1 Tall, ± erect plants (30–40 cm) – racemes 20–40 mm – fls 7–9 mm **2**
Low sprawling plants (less than 20 cm tall) – racemes 3–8 mm – fls 2–5 mm **4**

2 Fls purple – fl-stalks shorter than calyx-tube – pod smooth, spiral of 2–3 turns......................
Medicago sativa ssp *sativa* (p 290 **A**)
Fls yellow or various colours – fl-stalks longer than calyx-tube – pod smooth, sickle-
shaped or spiralled <2 turns **3**

3 Pod sickle-shaped – fls yellow ....*M. sativa* ssp *falcata* (p 289 **F**)
Pod curved or spiralled – fls often yellow and purple on same plant or various colours ......
M. sativa ssp *varia* (p 289 **F.1**)

4 Fl-heads 10–50-fld, compact – pod without spines, kidney-shaped, black when ripe ....
M. lupulina (p 289 **E**)
Fl-heads 1–5-fld, loose – pod with spines, brown when ripe...**5**

5 Pod hairy, 3–5 mm across ....*M. minima* (p 290 **D**)
Pod hairless, 4–6 mm across **6**

6 Lfts with dark spot – pod globular, dense spiral of 3–5 turns ....*M. arabica* (p 290 **C**)
Lfts without dark spot – pod flat – open spiral of $1^{1}/_{2}$–3 turns ....*M. polymorpha* (p 290 **B**)

A   Ribbed Melilot, *Melilotus officinalis* (*M. arvensis*), ± erect hairless bi, 60–120 cm; lvs trifoliate, lfts 15–20 mm; racemes **elliptical**, 20–50 mm, loose; fls 5–6 mm, **yellow**, wing and standard petals **equal**, but **longer** than keel; pod (**Aa**) 3–5 mm, **hairless**, oval, with **transverse ridges**, blunt, brown when ripe, style soon falling. Br Isles: widely introd (SE Eur), especially in S and E Eng, f; Ire o; in wa, gslds, scrub, ar, dunes. Fl 6–9. **Key differences** between *Melilotus* spp are in pod shape and surface characteristics (see Aa, B, C, D p 289).

B   Tall Melilot, *M. altissimus*, is v close to A, but has upper lfts ± parallel-sided; wing, standard and keel petals **all equal**; pod (**B**) **5–6 mm, hairy, oval**, pointed, with **netted** surface, **black** when ripe, style **persisting** in fr (frs are best distinction from A). Br Isles: GB N to S Scot; C and SE Eng c; Ire r; on hbs, gslds, open wds, especially on heavy soils. Fl 6–8.

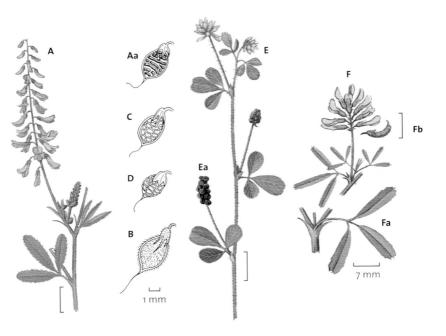

1 mm

7 mm

C White Melilot, *M. albus*, is close to A in **hairless** frs, brown when ripe, but pod (**C**) netted as in B, and fls **white**, 4–5 mm. Introd (SE Eur); Br Isles: S and E of GB, f; Ire vr; on wa, dunes. Fl 6–8.

D Small Melilot, *M. indicus*, has **yellow** fls, **only** 2 mm long, wings and standard equal, shorter than standard petal; pod (**D**) 2–3 mm, hairless, **netted, olive-green** when ripe. Introd (SE Eur to India); GB, o-lf to SE, r to W; CI f; Ire vr; on wa. Fl 6–8.

E Black Medick, *Medicago lupulina*, ascending, **downy** per, 5–25 cm; lfts 5–20 mm, toothed with **mucronate** tip; fl-heads stalked, compact, 3–8 mm, 20–40-fld; pods (**Ea**) 2 mm across, kidney-shaped, coiled, netted, **black** when ripe. Br Isles, vc on gslds, rds, especially on base-rich soils. Fl 4–8. **Do not confuse** with **Lesser Trefoil** (p 276 B) or **Hop Trefoil** (p 276 A) which both have lvs with few, scattered hairs and no mucronate tip.

F *Sickle Medick, *M. sativa* ssp *falcata* (*M. falcata*), hairless ascending or erect per, 30–40 cm; lfts (**Fa**) linear-lanceolate; fls in short racemes, 10–25 mm, clear **yellow**; fl-stalks longer than calyx-tube. Pod (**Fb**) curved or sickle-shaped, smooth, ± hairless, 2–5-seeded, 1–2 cm long. Eng: native in E Anglian Breckland, !f; introd elsewhere, o and declining on coastal gsld and hths; on calc gsld, rds. Fl 6–7.
F.1 Sand Lucerne, *M. sativa* ssp *varia*, is a fertile hybrid of F and A (p290), pod shape close to A (usually ± spiralled **up to 1.5** turns) but with a **range of fl colours**, often with both yellow and lilac/purple fls on the same plant. More c than F, S Eng o, Scot and W Ire r; on sandy soils in wa, urban areas, gsld. Fl 6–9.

A **Lucerne**, *Medicago sativa* ssp *sativa*, is close to F (p289), but more erect, with longer racemes (to 40 mm) of **purple** or **lilac** fls, with fl-stalks shorter than calyx-tube; pod (**Aa**) smooth, ± hairless, but a **spiral** with **2–3 turns**. Introd (S Eur); GB, f; Ire, r; in wa. Fl 6–9.

B ***Toothed Medick**, *M. polymorpha*, similar to C, but has a few hairs on stem, no blotches on lvs; jagged stipules; fls 3–4 mm; hairless pods (**Ba**) 4–6 mm across, in a flat spiral of 1¹/₂–3 turns, **strongly** netted, with a double row of hooked spines. Fl-head stalks as long as lf-stalks. GB, native S and SE coasts, o-lf; Ire vr; introd inland; on dry sandy gd near sea. Fl 5–8.

C **Spotted Medick**, *M. arabica*, prostrate ann, ± hairless; toothed trifoliate lfts **dark-blotched**; short-toothed stipules. Fl-heads 5–7 mm, few-fld, stalks shorter than lf-stalks; fls (**Ca**) 4–7 mm. Hairless pods (**Cb**) 4–7 mm across, **globular**, densely spiral, with 3–5 turns, faintly netted, with hooked spines in a **double** row on edge of spiral. Br Isles: Eng, Wales, lc to SE; Scot r casual; SE Ire r; on dry open sandy gd, especially near sea. Fl 4–9.

D vu ***Bur Medick**, *M. minima*, v downy ann, smaller than B, with **untoothed** stipules; pods (**Da**) like those of B in shape, with a double row of spines, but v **downy**, only 3–5 mm across. SE Eng: Sussex – Norfolk vlf; CI f; on beaches, dunes, Breckland hths. Fl 5–7.

Bird's-foot-trefoils (*Lotus*) are low herbs with lvs that appear trifoliate, but have a lower pair of lfts arising from base of stalk, like stipules; the true stipules are brown, minute, hard to see. Fls few together in one-sided heads, on long stalks in lf-axils. Pods elongated, spreading stiffly from common stalk like the toes of a bird from its leg.

E **Common Bird's-foot-trefoil**, *Lotus corniculatus*, low ± creeping, ± hairless per herb; stems **solid**, 10–40 cm long; lfts oval, infl stalks to 8 cm long; 2–8 fls per head; fls 15 mm long. **Calyx-teeth** (**Ea**) **erect in bud,** the two upper with an **obtuse** angle between them; buds flattened, deep red; open fls deep yellow often tinged red or orange. Br Isles, vc in gslds, rds etc, except on v acid soils. Fl 6–9. **Do not confuse** with the introd, sown variety (*L. corniculatus* var *sativus*), often in 'wildflower' seed mixes, which is **erect** and has **hollow** upper stems; open fls all yellow, not tinged red or orange.

F **Narrow-leaved Bird's-foot-trefoil**, *L. glaber* (*L. tenuis*), is close to E, but more slender, often more erect; **linear to lanceolate** lfts, usually 3 mm wide or less; **pale** yellow fls, 2–4 per head, about 10 mm long; 2 rear upper calyx teeth **converge**. Br Isles: Eng o-lf, especially in SE and E; Scot vr; Wales, Ire **abs or casual**; in dry gslds, especially on heavy clay soils. Fl 6–8.

G **Greater Bird's-foot-trefoil**, *L. pedunculatus*, close to E, but taller (to 60 cm), ± erect, usually **v hairy**; stem usually (but not always) **hollow**, stout; lfts (15–20 mm long) are **broadly oval, blunt**; infl-stalks to 15 cm long; 5–12 fls per head; **key difference** between E and G is the **calyx-teeth** (**Gb**) **reflexed in bud**, the two upper ones separated by an **acute** angle. Fls (**Ga**) clear yellow, not normally with any orange or red tinge. Br Isles f-la (except NW Scot, C Ire r); in damp mds, marshes, fens. Fl 6–8.

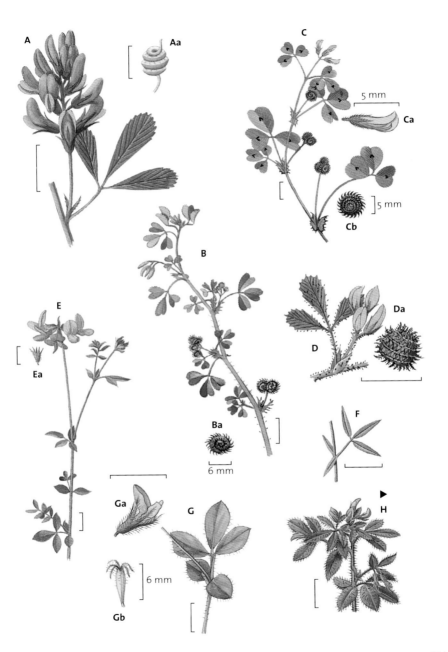

▲ **H** FPO (Eire) **Ire ***Hairy Bird's-foot-trefoil**, *Lotus subbiflorus* (*L. hispidus*), low ± spreading ann or sometimes per, with stems 3–30 cm long. Lfts to 20 mm, narrow-oval, **v hairy**; fl-heads, each of 2–4 small yellow fls only about 8 mm long, on stalks about 2 cm long, longer than lvs. Pods 6–12 mm long. Br Isles: Eng, Cornwall to Hants; IoS, CI, S Wales, S Ire vr but vlf in dry gsld close to sea. Fl 7–8. **H.1** NT ***Slender Bird's-foot-trefoil**, *L. angustissimus*, is close to H, but always **ann**, less hairy; fl-heads, one-two-fld, on stalks **shorter** than lvs; pods 20–30 mm. Br Isles: S Eng, Cornwall to Kent, CI r, vr to E; on dry gsld, usually near sea. Fl 7–8.

A Horseshoe Vetch, *Hippocrepis comosa*, spreading ± hairless per herb, 5–20 cm tall, with woody rootstock, bearing **pinnate** lvs with **4–5 pairs of side lfts** and a terminal lft, all narrow-oblong, 5–8 cm, long, with 2 narrow, pointed stipules at lf-base; no tendrils. Fl-heads have stalks 5–10 cm, bearing whorls of 5–8 fls (clear yellow, 8–10 mm long). Pods (**Aa**) arranged as in *Lotus*, but sinuous, about 30 mm long, breaking into horseshoe-shaped segments. Br Isles: Eng N to Teesdale f-la; Wales r; Scot, Ire abs; on old, short calc gsld and cliffs. Fl 5–7. **Do not confuse** A with *Lotus* spp, which have v similar fls but lvs with 5 lfts only (actually 3 lfts and 2 stipules), not pinnate.

B Kidney Vetch, *Anthyllis vulneraria*, prostrate to erect silky-hairy per, up to 30 cm tall; pinnate lvs 3–6 cm long, silky-white below, green above; terminal lft much larger (**Ba**) (elliptic on lower lvs); on upper lvs all lfts linear-oblong and equal (lowest lf sometimes one lft only). Fl-heads rounded, ± paired, 2–4 cm wide, of many fls (**Bb**) with **inflated white-woolly** calyx-tubes; petals usually yellow (white, pink, cream, purple or crimson in some coastal sites). Pod 3 mm long, rounded, flattened. Br Isles, c on calc gslds, sea cliffs, dunes. Fl 6–9. Divided into several poorly defined sspp but these include 2 introd sspp, of which ssp *polyphylla* is sometimes sown with 'wildflower' seed mixes (see the *New Flora* for key to ssp).

C NT **Sainfoin**, *Onobrychis viciifolia*, downy, erect (sometimes prostrate) per herb, 20–40 cm; lvs (**Ca**) pinnate, lfts 6–12 pairs, with oblong, pointed, **papery stipules**: fl-spikes many-fld, **conical**, on stalks longer than lvs; fls 10–12 mm, **salmon pink** with **red veins**; calyx-teeth longer than hairy tube; pods (**Cb**) not opening, 6–8 mm, oval, netted and warted, one-seeded. Introd (C Eur) formerly cultivated for fodder; S Eng, lf; N Eng, o-r; Scot, Ire abs; on rds, gslds on calc soils (smaller, more prostrate form probably native locally in old calc gsld in SE Eng). Fl 6–8.

D Crown Vetch, *Securigera varia* (*Coronilla varia*), straggling hairless per herb, stems 30–60 cm; pinnate lvs 5–10 cm long, with terminal lfts; fl-heads globular, 10–20-fld, on stalks longer than lvs; fls 12 mm, parti-coloured with pink and lilac, purple tips to pointed keels; pods four-angled, 4–5 cm long, breaking into one-seeded segments. Introd (S and C Eur); GB scattered throughout to SE Scot o-lf; Ire vr; in wa, scrub, especially on calc soils. Fl 6–8.

E Goat's-rue, *Galega officinalis*, erect hairless per herb, 60–100 cm; pinnate lvs with lfts broader (to 1 cm wide) than in C; erect oblong racemes (their stalks longer than lvs) of **white or mauve** fls with bristle-like calyx-teeth; pods cylindrical, rounded, 2–3 cm. Introd (E Eur); Br Isles: c in SE Eng (vc in London area), o-lf scattered elsewhere in C & N Eng; Scot, Wales, vr; Ire abs; concentrated in urban areas on wa, rds, disused quarries. Fl 6–7.

A

Aa

B

Bb

Ba

D

Cb

5
mm

C

Ca

E

SEA-BUCKTHORN FAMILY   *Elaeagnaceae*

Trees and shrubs with simple untoothed lvs covered with silvery scales; sepals 2 or 4; fr a coloured drupe.

A *Sea-buckthorn, *Hippophae rhamnoides*, thorny densely-branched shrub 1–3 m tall, spreading by suckers. Lvs 1–8 cm, alt, untoothed, linear-lanceolate, stalkless, covered with **silvery scales**. Fls v small, dioecious, borne in short spikes in lf-axils; 2 sepals, no petals; fr a bright **orange** berry. Eng, native and vla on E coast, SE Scotland to E Sussex, planted elsewhere on coasts and inland; on dunes. Fl 3–4; fr 9.

WATER-MILFOIL FAMILY   *Haloragaceae*

ID Tips   Water-milfoils

- Lf characters (length and number of lf-segments) refer to **mid-stem** lvs only.

- Always count lf-segments on **both sides** of lf.

- Plants with mid-stem lvs 8-26 mm long with all mid-stem lvs with 13-18 segments may be **either Alternate** or **Spiked Water-milfoil** (see *Plant Crib*).

B **Spiked Water-milfoil**, *Myriophyllum spicatum*, submerged aquatic herb often with reddish tinge; lvs rigid, 18–31 mm long with 13-41 segments, sometimes **slightly longer** than internodes. Fls tiny, **whorled**, in spikes in **tiny** bract-axils, emerge from water; lowest female, middle bisexual, uppermost male. Tiny petals are reddish; stamens yellow, 4 carpels per fl. Br Isles, c in still or slow, fresh or brackish water, tolerating nutrient-enriched water. Fl 6–8.

C **Alternate Water-milfoil**, *M. alterniflorum*, is more slender than B; lvs 3–26 mm long with 6–18 segments; upper fls **alt** (not whorled) in short spikes in axils of tiny bracts; petals yellow, red-streaked. Br Isles, vlf; in S Eng mainly in peaty water but in range of habitats in Scot including basic or calc water. Fl 5–8.

D   vu **Whorled Water-milfoil,** *M. verticillatum*, usually larger than B or C, lvs 15–45 mm long with 15–31 segments, 4 or 5 in a whorl; lvs often **much longer** than internodes; male, female and usually also bisexual fls in whorls in axils of **pinnate lf-like** bracts. Br Isles, lf in basic fresh water. Fl 7–8.   **D.1 Parrot's-feather,** *M. aquaticum*, differs from the native spp B–D by generally having **emergent** and submerged shoots and **key difference** in lvs, which are covered with **dense stalkless glands** (unlike D, having sparse glands; or B or C without glands) which give the emergent lvs a **blue-grey** colour. Whitish fls (**female only** in UK and Eire) in whorls of 4–6 from the lf-axils. Introd (central S America) from discarded aquarium or gdn pond plants, increasing and invasive; vla in S Eng, CI; in fresh water in reservoirs, canals, ditches or ponds. Fl 5–8.

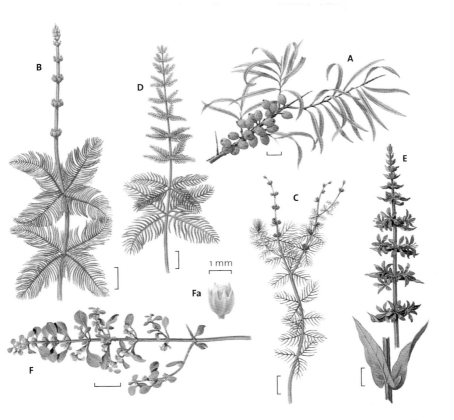

1 mm

Fa

PURPLE-LOOSESTRIFE FAMILY

Lythraceae

Herbs with opp or whorled lvs; fls radially symmetrical, usually with parts in sixes; ovary superior, within calyx-tube; fr a capsule.

E Purple-loosestrife, *Lythrum salicaria*, downy erect per herb, 60–120 cm; lvs sessile, oval-lanceolate, pointed, untoothed, in opp pairs or in whorls of 3 below, alt above. Infl dense, spike-like, 10–30 cm long; fls 10–15 mm across, in whorls in bract-axils, with downy calyx-tubes, 12 stamens, 6 red-purple petals. Br Isles, c (but o-r N Scot); by rivers, lakes, swamps, fens. Fl 6–8. **E.1** EN Sch 8***Grass-poly**, *L. hyssopifolium*, ann, 10–20 cm, with narrow linear lvs and pale pink fls 4 mm across, solitary in lf-axils. Br Isles: scattered throughout Eng, vr (but may be vla, eg. E Gloucs), SE Ire, Scot vr; CI r; on bare gd flooded in winter. Fl 6–7.

F AWI **Water-purslane**, *L. portula*, (*Peplis portula*), hairless **creeping** ann; stems 5–20 cm long, **square** and rooting at nodes; opp **oval** lvs widest above, 1–2 cm long; fls 1 mm across, **stalkless** and **solitary** in lf-axils, calyx-tube (**Fa**) with 6 pointed teeth, 6 petals hardly visible; fr a globular capsule 2 mm across. Br Isles f-lc especially in S; on damp bare gd on hths, wdland paths, by ponds, on acid soils flooded in winter. Fl 6–10.

SPURGE-LAUREL FAMILY   *Thymelaeaceae*

Small shrubs with alt undivided lvs without stipules; fls in short racemes or umbels, bisexual, with long calyx-tube bearing 4 spreading sepals; petals abs; stamens 8; ovary one-celled; fr a drupe.

A AWI **Spurge-laurel**, *Daphne laureola*, erect hairless **evergreen** shrub to 1 m tall; little-branched stems bear leathery lanceolate lvs gathered towards tops of shoots, reminiscent of Rhododendron. Fls (**Aa**) **green**, 8–12 mm long, in clusters in lf-axils; calyx four-lobed, fragrant; fr (**Ab**) 12 mm long, oval, black, fleshy. Br Isles: Eng f-lc; Wales vl; Scot, Ire introd, r; in wds (especially beech) especially on calc soils. Fl 2–4.

B VU AWI *****Mezereon**, *D. mezereum*, deciduous shrub of similar habit; lvs (**Bb**) pale green, thinner than in A; fls rich **rosy-pink**, strongly-scented; fr (**Ba**) **red**. Eng, S Wales r; **abs rest of Br Isles** except as r introd gdn escape; in wds on calc soils. Fl 2–4.

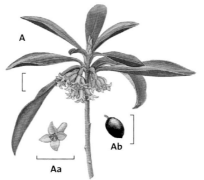
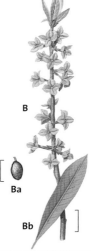

WILLOWHERB FAMILY   *Onagraceae*

Herbs with lvs in opp pairs, undivided; fl-parts in twos or fours (4 sepals, 4 petals, 8 stamens; 4 carpels in **Willowherbs**, *Epilobium*; parts in twos in **Enchanter's-nightshades**, *Circaea*). Ovary **inferior**.

KEY TO WILLOWHERB FAMILY

1 Petals present **2**
Petals abs – stamens 4 – creeping aquatic plant ....*Ludwigia palustris* (p 302 **D**)

2 Petals 2, white, but each deeply cleft – stamens 2 – fr a bristly nut ....*Circaea* (p 300)
Petals 4 – stamens 8 ....**3**

3 Shrub – fr a berry ....*Fuchsia magellanica* (p 302 **E**)
Herb – fr a capsule **4**

4 Fls yellow, large – seeds not plumed ....*Oenothera* (p 300)
Fls pink (or white) – seeds with plumes of hairs **5**

5 Lvs spirally arranged – petals not equal ....*Chamerion angustifolium* (p 300 **J**)
Lvs opp, at least below – petals all equal ....*Epilobium* (key below)

ID Tips   Willowherbs

- **Willowherbs** (*Epilobium*) are sometimes difficult to separate because the spp are so variable and hybrids occur. If a plant does not key out below, it may be a hybrid.

- **What to look at**: **(1)** shape of the stigma, whether club-shaped (p299 Ca) or 4-lobed (p299 Ba) **(2)** types of hairs **(3)** shape of the lvs **(4)** ripe capsules and seeds.

KEY TO WILLOWHERBS (*EPILOBIUM*)

1 Stem wholly creeping, rooting at nodes – lvs circular ....*Epilobium brunnescens* (p 300 **I**)
Stem at least partially erect **2**

2 Stem with long spreading hairs, many without glands – stigma four-lobed **3**
Stem hairless, or with adpressed hairs and/or spreading glandular hairs only (use hand lens) – stigma various (four-lobed or club-shaped) **4**

3 Petals over 1 cm long, purple-pink – lvs 6–12 cm, clasping stem...............*E. hirsutum* (p 298 **E**)
Petals less than 1 cm, pale pink – lvs 3–7 cm, not clasping stem ....*E. parviflorum* (p 298 **F**)

4 Stigma with 4 lobes in a cross (see Ba p 299) **5**
Stigma undivided, club-shaped (see Ca p 299) **6**

5 Lf-stalks short – lf-base rounded – fls pink in bud, pink when developed......................................
E. montanum (p 298 **A**)
Lf-stalks long – lf-base tapered into stalk – fls white in bud, pink when developed ....
E. lanceolatum (p 298 **B**)

6 Stem cylindrical, without ridges running down it – lvs usually less than 1 cm wide...............
E. palustre (p 298 **G**)
Stem with 4 (sometimes only 2) ridges running down it – lvs wider than 1 cm across **7**

7 Mid-stem lvs with stalks 0.5 cm or more...*E. roseum* (p 298 **D**)
Mid-stem lvs with stalks less than 0.5 cm, or none **8**

8 Stems wholly erect – fl-buds erect **9**
Stems creeping, then ascending – upper fls drooping – alpine only **11**

9 Stem and ovary with short spreading gland-tipped hairs as well as crisped adpressed ones ....*E. ciliatum* (p 298 **C**)
Stem without glandular hairs **10**

10 Lvs strap-shaped – no glandular hairs anywhere – capsule over 6.5 cm – no runners ....
E. tetragonum (p 298 **H**)
Lvs oval-lanceolate – some glandular hairs on calyx-tube – capsule 4–6.5 cm – runners in summer..*E. obscurum* (p 300 **H.1**)

11 Stem 1–2 mm across – lvs lanceolate, 1–2.5 cm – obvious runners – fls 4–5 mm across – in mt flushes in N of GB, vl ....**Alpine Willowherb** *E. anagallidifolium*
Stem 1.5–3 mm across – lvs oval, 1.5–4.0 cm long – runners ± underground – fls 8–9 mm across – in mt flushes in N GB, vl; NW Ire vvr ....
FPO (Eire) **Ire **Chickweed Willowherb** *E. alsinifolium*

A Broad-leaved Willowherb, *Epilobium montanum*, erect per herb, 20–60 cm, **round** stem (**Ab**), nearly hairless below, or with curved hairs above, lvs (**Ac**) opp, oval to lanceolate (sometimes in whorls of 3), toothed, nearly hairless, with **rounded** bases and **short** stalks. Fls 6–9 mm across, in terminal racemes; petals rosy-pink, notched; stigma (**Aa**) of 4 white lobes; capsule 4–8 cm, downy with curved hairs. The most widespread *Epilobium* spp in Br Isles, vc in wds, hbs, walls, rocks etc. Fl 6–8.

B Spear-leaved Willowherb, *E. lanceolatum*, has the stem (**Bb**) characteristics and **four-lobed** stigma (**Ba**) of A, but the nearly hairless lvs are generally alt above, **elliptical**-lanceolate, sparsely-toothed except at base, with **long lf stalks** (3–10 mm long) into which lvs **gradually narrow**. Fls white in bud, becoming pink. S Eng, S Wales lf; in dry habitats, open wds, walls, hbs, wa, rocks. Fl 7–9.

C American Willowherb, *E. ciliatum* (*E. adenocaulon*), per herb, with short-stalked hairless oval-lanceolate lvs (**C**) and habit of A, but with 4 **raised ridges** on the stem (**Cb**), which has many **crisped** and **short, spreading, glandular hairs** above and on the ovaries (**Cc**). Stigma (**Ca**) **club-shaped**, undivided. Eng, Wales, introd since 1891 from N America, vc on disturbed gd, walls, gdns, rds, wa. The commonest *Epilobium* spp in much of S Eng, especially in urban areas, and increasing but less c than A in N Eng, Scot and Ire. Fl 6–8.

D Pale Willowherb, *E. roseum*, has nearly hairless, elliptical pointed lvs (**D**) with **impressed veins**, **taper gradually** into a **long stalk** up to 20 mm long, much as in B. Fls white in bud, becoming pink like C but **club-shaped** stigmas (**Da**), capsule with sticky hairs. Br Isles: Eng f; Scot r; Ire vr shady, damp, disturbed places in wds, wa, hbs, gdns. Fl 7–8.

E Great Willowherb, *E. hirsutum*, tall per herb, 80–150 cm, with round stems; densely downy on stem and lvs, with **spreading** hairs, both gland-tipped and without glands; lvs opp, lanceolate, clasping, pointed, stalkless, 6–12 cm long; fls deep purple-pink, 15–23 mm across; stigma (**Ea**) with 4 stout creamy lobes that arch back. Br Isles (except NW Scot and some islands), vc; in fens, marshes, river banks and damp wa, sometimes drier areas, preferring neutral or basic soils. Fl 7–8.

F Hoary Willowherb, *E. parviflorum*, a smaller version of E, having round hairy stem, hairy lvs, and **four-lobed** stigma (**Fa**), as in E, but less tall (30–90 cm), less stout; hairs throughout much shorter; lf-bases do not clasp or run down stem as in E; fls (**F**) only 5–9 mm across, with pale pink petals. Stigma-lobes **upright**, not arched back. Br Isles (except Scot Highlands), vc; on streamsides, marshes, damp places but also drier wa. Fl 7–8.

G Marsh Willowherb, *E. palustre*, erect per herb, 15–60 cm, much more **slender** than all the previous spp. Stem (**Gb**) cylindrical without ridges (though sometimes 2 rows of crisp hairs); lvs usually stalkless, strap-shaped, 0.4–1.0 cm wide, narrowed to apex and base. Small fls (4–6 mm across) have **club-shaped** stigmas (**Ga**). Br Isles, f-lc in **acid** marshes, bogs, ditches although apparently much declined in S Eng due to habitat loss. Fl 7–8.

H Square-stalked Willowherb, *E. tetragonum*, resembles C in its 4 v conspicuous raised stem-ridges (**Hb**) and club-shaped stigmas (**Ha**), but lvs (**H**) strap-shaped, blunt, narrow (2–7 cm long x 0.3–1.0 cm wide), **almost parallel-sided**, almost stalkless. Fr capsules 6.5–10 cm long, rarely shorter; gland-tipped hairs abs, plants

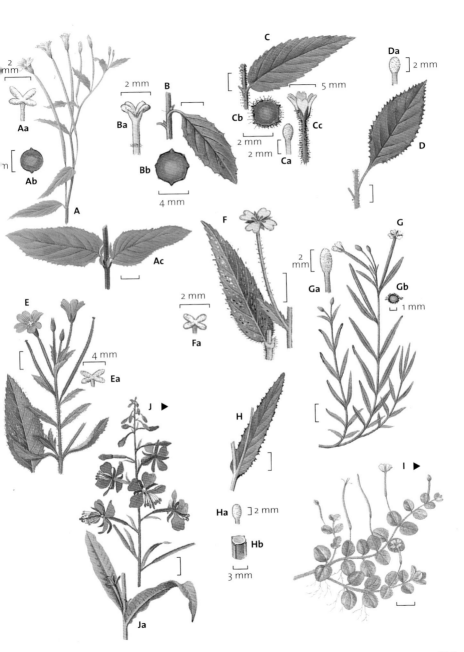

adpressed downy above. Br Isles: Eng, Wales f-lc; Scot abs to vr (expert confirmed records needed), Ire vr; in damp wd glades, hbs, streamsides, wa. Fl 7–8.

H.1 Short-fruited Willowherb, *Epilobium obscurum*, closely resembles H in its stems with 4 raised ridges, its stalkless, almost hairless lvs and club-shaped stigma; but has elongated **summer runners**, broader lanceolate lvs (3–7 x 0.8–1.7 cm), some **glandular hairs** on the calyx, and **shorter fr capsule** (4–6.5 cm long). Br Isles, c but less so in E Eng; moist wds, marshes, often in same habitats as H. Fl 7–8.

▲ **I   New Zealand Willowherb**, *E. brunnescens*, prostrate per herb, forming mats with stems **rooting at nodes**; lvs small, opp and ± **circular**. Fls 5–7 mm across on slender stalks arising from **lf-axils**; petals white or pale pink; stigmas **club-shaped.** Introd (New Zealand); Br Isles, c in N and W (except C Ire, o); o elsewhere; on moist open gd, gravel, stream-sides, tracks, walls. Fl 5–10.

▲ **J   Rosebay Willowherb**, *Chamerion angustifolium* (*Chamaenerion angustifolium*), tall erect per herb to 120 cm, nearly hairless; lanceolate alt lvs (**Ja**) **spirally** arranged up stems. Fls rose-purple, 2–3 cm across, borne in spikes, spread out horizontally; 2 upper petals **broader** than the 2 lower. Stigma (four-lobed) and stamens bend down eventually. Br Isles: vc-vla (except W Ire o-lf); wa, wds, gdns, railways etc. Fl 7–9.

A   Enchanter's-nightshade, *Circaea lutetiana*, per herb with creeping root-stock and **erect stems** 20–70 cm tall, v sparsely downy; lvs opp, oval, **rounded** at bases, tapering to tips, 4–10 cm long, with v small shallow teeth. Infl spike-like, elongated, held well above lvs; fls (**Aa**) each with 2 petals 2–4 mm long that are divided halfway, 2 long stamens, 2 stigma-lobes. Fr covered with hooked white bristles, not bursting open when ripe. Br Isles, vc (N Scot and Scot islands r); in shady wds, hbs, on base rich soils. Fl 6–8.

B   *Alpine Enchanter's-nightshade, *C. alpina*, smaller almost hairless per herb, 10–20 cm, with a **tuberous** rootstock; lvs 2–6 cm long, **cordate** at base, with **strong** but spaced-out teeth, are v thin, shiny, almost translucent. Fls in **tight cluster**, infl elongating after petals have fallen. Frs with soft, less hooked bristles. Br Isles: SW Wales, N Eng, SW Scot, r and l; in hilly rocky mt wds. Fl 7–8.   **B.1** AWI **Upland Enchanter's-nightshade**, *C.* x *intermedia*, the hybrid of A and B, is ± intermediate between them, with A's elongated fl-spikes and B's cordate strongly-toothed lvs. More c than B in Wales, N Eng, Scot, N Ire; in shady places on streamsides and wet rocks. Fl 7–8.

Evening-primroses (*Oenothera* agg) are a v difficult group as plants of C, C.1 and C.2 often form hybrid swarms, with a mix of different characters. It is likely that 'pure' plants (except for C.3), as described here, are uncommon. All are introd (N America, C.3 Chile).

C   Large-flowered Evening-primrose, *Oenothera glazioviana* (*Oe. erythrosepala*), tall bi herb with downy lfy-stems; top of stem **reddened**; **red bulbous-based** hairs on stem; lvs elliptic and often twisted; sepals red-striped; **petals 35–50** mm long and **style longer** than filaments. Br Isles, c in wa, rds, quarries on sandy soils. Fl 6–9.   **C.1 Common Evening-primrose**, *Oe. biennis*, **key difference** from C and C.2 being **no** red bulbous-based hairs on green part of stem; other useful characters

include: stems **green** at top; sepals green; petals yellow, 15–30 mm, wider than long; style ± equal to filaments.   **C.2 Small-flowered Evening-primrose**, *Oe. cambrica* (*Oe. parviflora*), **key difference** from C and C.1 is **glandular hairs** only top of stem and **style shorter** than filaments. Petals 20–30 mm long, c. as wide as long.   **C.3 Fragrant Evening-primrose**, *Oe. stricta*, is easier to identify for certain, has **v narrow** wavy-edged lvs, no red bulbous-based hairs; sepals red-striped; petals 25–40 mm, yellow with red spot at base first, turning orange-red; styles ± equal to filaments. The rarest *Oe*. sp, S Eng, CI only, r on sandy gd, dunes usually near the sea or casual inland (all *Oe*. spp are in fact fragrant in the evening). Fl 6–8.

▲ **D** ****Hampshire-purslane**, *Ludwigia palustris*, per **reddish**-suffused aquatic herb, creeping on mud or floating in water, with hairless stems rooting at nodes below; lvs 1.5–3.0 cm long x 0.5–2.0 cm wide, in opp pairs, oval to elliptical, short-stalked, blunt, hairless. Fls (**Da**) 3 mm across, solitary in lf-axils; ovary inferior, 4 sepals, no petals, 4 stamens, and a four-lobed stigma. S Eng vr, only native in and around ponds, swamps in New Forest; introd plants in SE Eng are likely to be a horticultural hybrid. Fl 6–8.

▲ **E** **Fuchsia**, *Fuchsia magellanica*, shrub to 3 m; lvs opp, elliptical 25-55 mm long, toothed; fls distinctive, hanging down on long fl-stalks comprising 4 bright red-pink sepals and 4 violet petals, 8 stamens; fr a black berry. Introd (S America), plants ± corresponding to this sp are planted as hedging and natd; c in Ire and SW Eng, o elsewhere in hds, scrub, stream-sides and on walls. Fl 6–10.

DOGWOOD FAMILY   *Cornaceae*

Shrubs with opp oval untoothed lvs; fls in umbels, with 4 petals; ovary inferior; fr a berry.

A **Dogwood**, *Cornus sanguinea* (*Thelycrania sanguinea*), deciduous shrub to 4 m tall; twigs purplish-red; lvs opp, 4–8 cm long, oval, pointed, rounded at base, without stipules, **untoothed**, slightly downy on both sides; the 3–5 main veins on each side curve round towards apex of lf; lf-stalk 8–15 mm. Lvs turn red in autumn. Infl a flat-topped stalked umbel of many creamy-white fls, 4 petals 4–7 mm long with 4 tiny sepals; 4 stamens alt with petals; two-celled inferior ovary

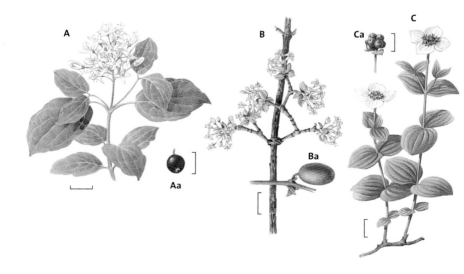

forms a black globular drupe in fr (**Aa**). Eng, N to Morecambe Bay and Tyne c-la; Wales o-lf; Scot, Ire r; in wds, scrub on calc soils. Fl 5–7. **A.1 Red-osier Dogwood,** *C. sericea*, similar shrub to A; lvs with tapering (not abrupt) point; petals smaller than A, 2–4 mm; fr white berry. Introd (N America), often planted in parks, rds etc for its red stems in winter; can be invasive in wet wds. Br Isles, f natd by suckers to wds, wa, hds, rds.

B Cornelian-cherry, *C. mas*, shrub or small tree to 8 m tall; twigs yellowish; oval lvs 4–10 cm long; umbels ± unstalked, with 4 yellow-green bracts at base; fls **yellow**, 3–4 mm across, produced before lvs expand; fr (**Ba**) scarlet, elliptical, 1.0–1.5 cm long. Introd (C Eur); Br Isles o-r gdn escape; on scrub, wds, on calc soils. Fl 2–3.

C NT **Dwarf Cornel**, *C. suecica* (*Chamaepericlymenum suecicum*), low creeping per herb, with lf shape and fl structure as in A. Erect fl-stems 6–20 cm tall; lvs opp, 1–3 cm; infls dense terminal heads, resembling fls, with 8–25 tiny (2–4 mm) blackish-purple fls, each surrounded by 4 large (5–8 mm) oval, white, petal-like bracts; frs (**Ca**) red, globular, 5 mm across. Br Isles: central Highlands Scot (**not in islands**) f-lc; S Scot, Yorks, vr; **S Eng, Ire abs**; in mts on hths, gsld. Fl 7–8.

SANDALWOOD FAMILY   *Santalaceae*

***Bastard-toadflax**, *Thesium humifusum*, prostrate semi-parasitic hairless per herb, 10–20 cm long, with yellow-green fleshy-looking stems and lvs. Lvs linear, alt, one-veined; infl terminal, loose, spike-like; fls, each with 3 bracts, like tiny white stars, 3–4 mm; a five-lobed calyx of spreading triangular sepals, white within, green-yellow outside, and a short tube; ovary inferior, oval, ribbed, in fr (**a**) twice length of calyx-lobes. Br Isles: S and E Eng, CI, N to Lincs o-lf; on dry calc gsld, dunes. Fl 6–8.

a

3 mm

MISTLETOE FAMILY

Mistletoe, *Viscum album*, woody parasitic shrub; repeatedly forked branching: narrow-elliptical lvs, blunt, leathery, widest near tip. Fls dioecious, in small tight clusters, with 4 tiny petals; fr a sticky white berry. Br Isles: Eng, E Wales, N to Yorkshire, lc in SE, and W Midlands; N Ire r; Scot vvr; parasitic on various trees, especially apple, lime. Fl 2–4; fr 11–12.

SPINDLE FAMILY

Shrubs with simple, usually opp lvs; stamens **alt** with petals.

AWI **Spindle**, *Euonymus europaeus*, deciduous shrub or small tree, 2–6 m tall, hairless; twigs green, four-angled; lvs opp, **oval-lanceolate**, 3–13 cm long, only **finely** toothed, orange in autumn. Fls (**a**), in stalked forking cymes in lf-axils, 3–10 together each fl 8–10 mm across, with 4 greenish-white petals, 4 short alt stamens; fr (**b**) four-lobed, 10–15 mm wide, bright **coral-pink**, opening by slits to expose the seeds, each with a bright orange sheath. Br Isles: GB N to Forth, vc to S, r to N; CI, vr (introd); Ire f; in wds, scrub, hbs on calc soils. Fl 5–6; fr 9–10.

HOLLY FAMILY   *Aquifoliaceae*

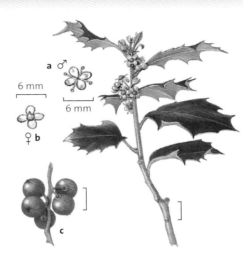

AWI **Holly**, *Ilex aquifolium*, small tree or shrub, usually 3–15 m, rarely to 20 m, with smooth thin bark and green twigs. Lvs alt, dark glossy green, leathery, oval, with wavy margins bearing large spine-pointed teeth; lvs at top of tree often spineless. Fls with 4 white petals joined below, 6 mm across, dioecious; male fls (**a**) have 4 stamens, female (**b**) 4 carpels; fr (**c**) a scarlet berry, 7–12 mm across. Br Isles vc generally, **abs Caithness, much of Scot Highlands**; in wds, scrub, hbs, on drier soils. Fl 5–8; fr winter.

BOX FAMILY   *Buxaceae*

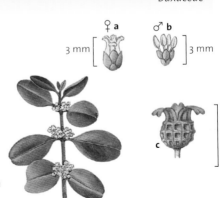

Box, *Buxus sempervirens*, **evergreen shrub or small tree, 2–5 m, rarely to 10 m, with green, **angled**, downy stems; lvs 1.0–2.5 cm, oval, shiny, leathery, opp, blunt, short-stalked. Fls in clusters in lf-axils, each with a terminal female fl (**a**) of 4 sepals and three-celled ovary, and several male fls (**b**) of 4 sepals, 4 stamens. Fr (**c**) a capsule. Br Isles, widely planted (native SE Eng, vr but vla); on steep hillsides on calc soils in scrub or wdland. Fl 4–5. **Do not confuse** with **Wilson's Honeysuckle** (p 426 A) which has v similar opp, leathery, dark green lvs but **round, glandular-hairy** stems.

SPURGE FAMILY   *Euphorbiaceae*

Members of the Spurge family in Br Isles are all herbs, with alt simple lvs and acrid **latex**, which is milky in *Euphorbia* and watery in *Mercurialis*. **Spurges** (*Euphorbia*) have fls in open umbel-like terminal heads, with primary branches in an umbel, further branches forked. Bracts conspicuous, often wider than lvs. Infl of perianth-like cups of 4–5 small teeth, alt with 4–5 conspicuous rounded or crescent-shaped **glands**; each cup surrounding a group of several tiny one-stamened male fls and 1 central female fl comprising a stalked three-celled ovary (capsule) with 3 stigmas.

ID Tips   Spurges

- **What to look at**: **(1)** structure of the fls **(2)** the shape of the **glands** surrounding the fls **(3)** shape of the lvs and whether lvs stalked or not **(4)** whether frs have warts or not and the shape of these warts.

1 Flower	
2 Upper bracts	
3 Stamens	
4 Gland	
5 Stigmas	
6 Fruit capsule	

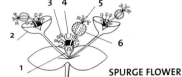

SPURGE FLOWER

- **Umbel** bracts are the bracts below the **whole** infl.

- **Upper** bracts are the bracts below **each group** of fls.

KEY TO SPURGES (*EUPHORBIA*)

1 Lvs opp **2**
Lvs alt...**3**

2 Plant prostrate, purple-suffused – stipules present – lvs not different from umbel bracts, unequal-sided – extinct but may reoccur – on seashore.................*Euphorbia peplis* (p 310 **D**)
Tall erect wdland & wa plant – no stipules – lvs strap-shaped in 4 rows *E. lathyris* (p 310 **A**)

3 Glands surrounding fl-clusters rounded both sides **4**
Glands with crescent-shaped horns on outer side..**8**

4 Plants many stemmed – oblong untoothed lvs, tapered to base **5**
Plants single stemmed – lvs toothed above **6**

5 Upper bracts rounded at base, yellow – glands yellowish – lvs hairy below ....
E. hyberna (p 310 **B**)
Upper bracts truncate at base, green – glands soon purple – lvs ± hairless....................................
E. dulcis (p 310 **B.1**)

6 Bracts and lvs blunt, tapered to base – capsule smooth ....*E. helioscopia* (p 308 **D**)
Bracts and lvs pointed – lvs cordate below – capsule with warts **7**

7 Capsule with hemispherical warts...*E. platyphyllos* (p 308 **F**)
   Capsule with cylindrical warts ....*E. serrulata* (p 310 **C**)

8 Tall hairy plant – lf rosettes raised above gd ....*E. amygdaloides* (p 307 **A**)
   Plant hairless **9**

9 Ann – on ar, wa **10**
   Bi or per – not on ar but sometimes wa **11**

10 Lvs oval, stalked...*E. peplus* (p 308 **E**)
   Lvs linear, stalkless ....*E. exigua* (p 308 **G**)

11 Plants grey-green, fleshy – lvs oval to obovate, less than 2 cm – coastal.......................................**12**
   Plants not fleshy – lvs linear to lanceolate, mostly more than 2 cm – inland **13**

12 Lvs v fleshy, no pointed tip, midrib obscure – seeds smooth.......................*E. paralias* (p 308 **B**)
   Lvs leathery, pointed tip, midrib prominent below – seeds pitted......*E. portlandica* (p 308 **C**)

13 Lvs of fl-stems linear, 2 mm or less wide ....*E. cyparissias* (p 308 **I**)
   Lvs of fl-stems linear-lanceolate, 4 mm or more wide..**14**

14 Lvs tapering to pointed tip, broadest at or below middle – lvs 4–5 mm wide ....
   E. x *pseudovirgata* (p 308 **H**)
   Lvs tapering to narrow base, broadest above middle with blunt lf-tip – lvs usually
   5–10 mm wide ....*E. esula* agg (p 308 **H.1**)

A AWI **Wood Spurge**, *Euphorbia amygdaloides*, downy per herb, 30–80 cm tall; tufted over-wintering stems, 10–20 cm tall, bear terminal rosettes (**Ab**) of strap-shaped, downy, dark green short-stalked lvs, 3–8 cm long; fl-shoots arise from rosettes. Umbel 5–10-rayed, with oval umbel bracts, upper bracts (**Ac**) kidney-shaped, joined in pairs, yellow. Glands around fls crescent-shaped (**Aa**); capsule rough. Br Isles: Eng, Wales, c in S, vr N of S Lincs; Ire, Scot vr (introd); in woods on basic-neutral soils. Fl 4–5.

5 mm

B Sea Spurge, *Euphorbia paralias*, hairless per, 20–40 cm tall, with several erect stems. Lvs 0.5–2.0 cm, alt, close-set on stems, v thick, **fleshy**, waxy-green, untoothed, oval, **blunt**, stalkless, midrib **obscure**. Umbel 3–6-rayed, upper bracts oval; glands crescent-shaped; seeds **smooth**. Br Isles: GB coasts N to Galloway and to Norfolk lf; Ire ± all round coasts; on dune sands, beaches. Fl 7–10.

C Portland Spurge, *E. portlandica*, is close to B, with glands crescent-shaped, but more slender; lvs rather thinner, obovate, with **pointed** tips (**Ca**), tapered to base; **key differences** in lvs with **midribs prominent** beneath; seeds **pitted**. Br Isles: GB on S and W coasts, Sussex to Galloway; Ire ± all round coasts vlc; on dune sands, sea cliffs. Fl 5–9.

D Sun Spurge, *E. helioscopia*, hairless ann 10–30 cm tall; single erect stems; lvs 1.5–3.0 cm, obovate, v blunt, tapered to base, toothed above. Umbel five-rayed, with 5 large obovate **toothed**, yellow-green umbel bracts (**Db**); 4 fl-glands, **oval**, green; fr (**Da**) smooth. Br Isles, vc on dry, disturbed gd on ar, wa. Fl 5–10.

E Petty Spurge, *E. peplus*, hairless ann, smaller version of D, but all lvs stalked, untoothed, green, oval, blunt; bracts shorter, triangular, narrower than in D, green; 4 glands of fl **crescent-shaped** (**Ea**) with long horns. Umbel usually three-rayed, umbel bracts in a whorl of 3, oval, pointed, untoothed. Br Isles, vc on ar, wa, pavements. Fl 4–11.

F Broad-leaved Spurge, *E. platyphyllos*, like D in habit and habitat, differs from it in having warted frs like C p 310, but warts hemispherical (not cylindrical) and frs rounded. Fr (**F**), seed (**Fa**) Umbel bracts (like lvs) elliptic-oblong, but upper bracts all shortly cordate, unlike the umbel bracts. S Eng r-o; on ar, wa, on heavy soils. Fl 6–9.

G NT **Dwarf Spurge**, *E. exigua*, slender grey-green ann, 5–15 cm, lvs **narrow**, **linear**, **unstalked**, untoothed. Bracts triangular-lanceolate with cordate bases; glands crescent-shaped. Br Isles: S, E Eng f-lc but declining especially in SE Eng; Wales, S Scot, Ire r; on ar, especially on calc soils. Fl 6–10.

H Twiggy Spurge, *E.* x *pseudovirgata* (*E. uralensis*), hairless per, 30–50 cm, with creeping rhizomes and erect branched stems, patch-forming; lvs numerous, alt, linear, pointed, stalkless, widest **at or below** middle, tapering (not cordate) to base, **pointed lf-tips**, 3–8 cm long x **4–5 mm wide**. Umbel bracts lanceolate (**Ha**), 12–35 mm long; upper bracts (**Hb**) triangular, yellow-green; glands (**Hc**) crescent-shaped. Introd (E Eur); GB, o-lf especially in SE Eng; on wa, gslds, tracks, wds. Fl 5–7. **H.1 Leafy Spurge**, *E. esula* agg, **key differences** from H in lvs widest **above** middle, tapering to base with **blunt** lf-tips and stalkless; usually **5–10 mm wide**. Introd (S Eur); GB r; in same habitats as H. Fl 5–7.

I Cypress Spurge, *E. cyparissias*, patch-forming hairless per herb, 10–30 cm, with tufted stems; fl-stems often with side branches above, taller than infl. Lvs v many, alt, **narrow-linear**, 1.5–3.0 cm x up to **2 mm wide**, stalkless, untoothed. Umbel 9–15-rayed, with oblong umbel bracts (**Ia**); upper bracts 3–6 mm, triangular, yellow, turning red. Glands crescent-shaped. Eng (possibly native in S, r gdn escape elsewhere); on calc gsld and scrub. Fl 5–7.

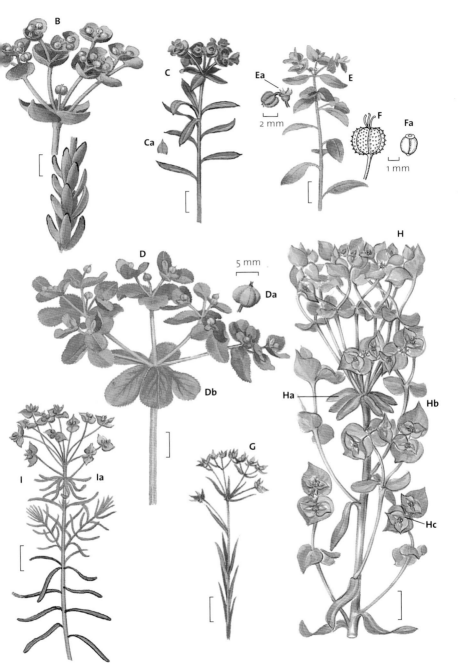

B

C

Ca

Ea

E

2 mm

F

Fa

1 mm

D

5 mm

Da

Db

H

Ha

Hb

Hc

I

Ia

G

A **Caper Spurge**, *Euphorbia lathyris*, tall erect (30–120 cm) hairless bi, with lvs (**Aa**) 4–20 cm, **opp**, in **4 vertical rows, strap-shaped**, blunt, **waxy** grey-green. Umbel wide, 2–6-rayed, umbel bracts triangular, lanceolate; upper bracts cordate; glands crescent-shaped. Fr (**Ab**) **v large**, 8–20 mm long, smooth. Eng: probably native in S, r; in wds on basic soils; more c as gdn escape in wa, ar, rds. Fl 6–7.

B VU ****Irish Spurge**, *E. hyberna*, per herb, with many erect simple stems 30–50 cm, lvs 5–10 cm, alt, oblong to elliptical, blunt, tapered to base, ± stalkless untoothed, hairless above, downy on midrib below. Umbel 4–6-rayed, umbel bracts (**Ba**) 3–6 cm, elliptic, **yellow**; upper bracts 8–30 mm, oval, base ± **cordate, yellow**. Glands 5, yellow, kidney-shaped; fr 5–6 mm, hairless, with cylindrical warts. Eng: Cornwall to Somerset r; Ire lc in SW; in open wds, scrub, gsld, not on calc soils. Fl 4–7.
B.1 **Sweet Spurge**, *E. dulcis*, is close to B, but more slender; lvs 3–5 cm, ± hairless; umbel bracts to 2–3 cm; upper bracts truncate at base, green; glands rounded, turning purple; frs with long warts. Introd (C Eur); GB r; on rds, wd, scrub.

C ****Upright Spurge**, *E. serrulata* (*E. stricta*), erect hairless ann, to 50 cm; stems red; lvs (**Ca**) oval-oblong, fine-toothed; umbel bracts with **clasping** bases, upper bracts (**Cb**) **progressively** shorter and more cordate. Glands (**Cc**) **rounded**; fr (**Cd**) three-angled, with **long, cylindrical** warts; seed (**Ce**). Eng, round Forest of Dean as a native vr, r introd elsewhere in Eng; in open rocky wds on limestone. Fl 6–9.

D EX ****Purple Spurge**, *E. peplis*, prostrate hairless ann; stems crimson, branches forked; lvs 3–10 mm, opp, waxy-grey, stalked, oblong, blunt, with a large round lobe on one side of base of each lf; forked stipules, bracts (**Da**) like lvs; glands undivided; fr smooth. SW Eng, CI **extinct** but could re-occur so included here; sandy beaches. Fl 7–9.

E AWI **Dog's Mercury**, *Mercurialis perennis*, **hairy** per herb, 15–40 cm tall, with creeping rhizomes; unbranched stems bear pairs of short-stalked opp oval-elliptical lvs, 3–8 cm long, with small teeth. Sexes on separate plants (dioecious); male fls (**Eb**) in erect, catkin-like spikes in lf-axils, each fl with 3 green sepals 2 mm long, and 8–15 stamens; female fls in groups of 1–3 on 3 cm-long stalks, each with 3 sepals, and rounded ovary that forms a two-celled **hairy** fr (**Ea**) 6–8 mm wide. (Lvs on female plants wider than on males.) Br Isles: GB c-la (except N Scot vr); Ire native one site in Clare, o scatted introd elsewhere; in wds on basic or calc soils, rocks on mts. Fl 2–4.

F **Annual Mercury**, *M. annua*, ann, almost **hairless** plant; stems branched; lvs only 2–5 cm long, fresh **shiny** green, with stalks 2–15 mm long; plant dioecious; male fls in catkins (**Fb**), female fls (**Fa**) almost stalkless with **bristly** frs 3–4 mm wide. Br Isles: S Eng c, rarer to N and hardly into Scot; E Ire o; in ar, gdns, wa, on fertile, often nutrient-enriched soils. Fl 7–10. **Key differences** from E are lvs and stem ± hairless and female fls almost stalkless (in clusters on stalks > 1 cm in E).

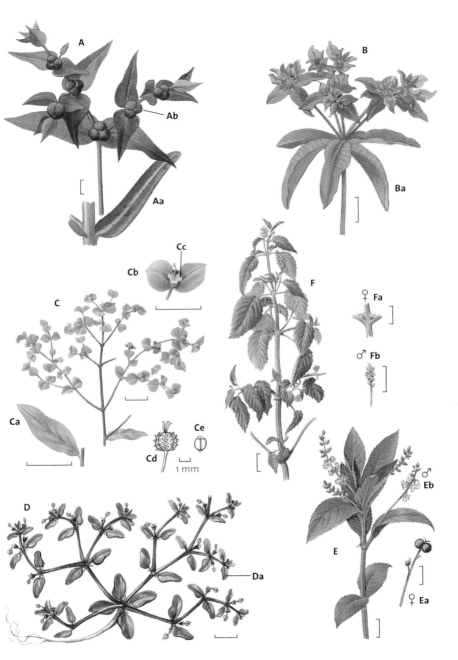

BUCKTHORN FAMILY   *Rhamnaceae*

Trees and shrubs with simple lvs; fls small, in small clusters in lf-axils; 4 or 5 petals, sepals, stamens; calyx tubular, ovary superior; stamens **opp** petals; fr a berry.

A Buckthorn, *Rhamnus cathartica*, shrub or small tree to 6 m tall, usually with some thorns; lvs **opp**, deciduous, oval, 3–6 cm long, with side veins curving up towards tip (as in Dogwood p 302, but A's lvs **toothed**, and twigs grey or brown, **not purple**); buds with scales. Fls in small clusters on short spurs below this year's lvs, green, four-petalled, 7 mm across; monoecious. Fr a berry (**Aa**) 6–10 mm across, green then **black**. Br Isles: Eng, Wales f-lvc in S and SE, far N r, **SW abs**; Scot vr; Ire o; in scrub, wds on calc soil and on fen peat. Fl 5–6.

B WO (NI) AWI **Alder Buckthorn**, *Frangula alnus*, shrub or small tree, with deciduous oval lvs, **alt, untoothed**, broadest near tips, 2–7 cm long, with veins **not curved towards tip**; buds **without scales**, v hairy; fls (**Ba**), 3 mm across, in small clusters in axils of upper lvs, have **5** greenish-white petals. Fr a berry 6–10 mm wide, green then **red**, finally black. Br Isles: Eng, Wales f-lc; Scot abs as native; Ire o-r; in fen carr, in damp wds, hths on acid soils (r on calc soils). Fl 5–6.

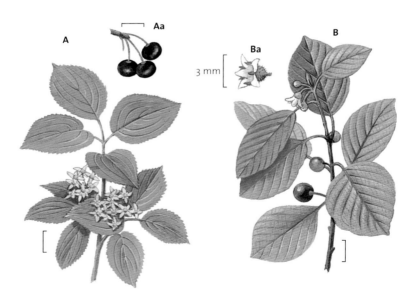

FLAX FAMILY   *Linaceae*

Flaxes are hairless erect herbs with narrow, unstalked lvs; fls in loose, repeatedly forked (dichasial) cymes, with 5 sepals, 5 stamens; five-celled ovaries forming dry globose capsules.

A Fairy Flax, *Linum catharticum*, erect ann herb, 5–25 cm; oblong, **blunt**, **opp**, lvs; open, forked infl of white fls (**Aa**), only 4–6 mm across. Br Isles, vc in calc gsld, dunes, flushes, rock ledges, fens. Fl 6–9.

B Pale Flax, *L. bienne*, ann or per, usually many stemmed 20–30 cm, with alt three-veined narrow pointed lvs, **pointed** sepals, and pale lilac-blue fls, petals 8–12 mm. Stigmas c. as high as anthers. Br Isles: Eng, Ire, CI; If dry gsld, especially near sea. Fl 5–9.

C End *****Perennial Flax**, *L. perenne*, resembles B, but is always per, many stemmed, usually taller (30–60 cm), with **one-veined** lvs; **blunt**, shorter, inner sepals, and sky-blue fls (**C**), 25 mm across. Stigmas either higher or lower than anthers. Eng r on calc gsld in E only. Fl 6–7. **C.1 Flax**, *L. usitatissimum*, resembles C in stature and fl size, but is much more c, ann, stems usually 1, with three-veined lvs, sepals all pointed, and pale blue fls with petals 12–20 mm. **Key difference** from C is sepals c. as long as ripe capsule and stigma c. as high as anthers. Br Isles, o-lf (introd escape from cultivation). Fl 6–9.

D NT AWI **Allseed**, *Radiola linoides*, tiny ann, 2–8 cm tall, with bushy forked stems; opp elliptical lvs 2–3 mm long; numerous tiny fls 1 mm across, each with 4 tiny white petals equalling the 4 sepals, and 1 mm-wide globose capsules. Br Isles: r (but If in S and W Eng, IoS, W Ire, CI); on damp bare acid sandy gd on hths, dune-slacks and on paths in wds. Fl 7–8.

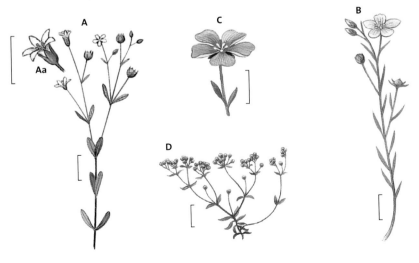

MILKWORT FAMILY *Polygalaceae*

Milkworts (*Polygala*) are low hairless per herbs with narrow-lanceolate lvs without stipules; small irregular fls have 3 tiny outer sepals and 2 large, coloured, lateral, petal-like ones (**inner sepals**); 3 true petals, v small, joined together into a little whitish fringed tube. 8 small stamens are joined into a tube with petals. Fls are blue, purple, mauve, pink or white, borne in spikes. Fr a notched two-celled compressed capsule with winged edges.

ID Tips   Milkworts

- All **Milkwort** (*Polygala* spp) fls can be in various shades of blue, pink, or white.

- **Heath milkwort** (*P. serpyllifolia*) has **opp** lower lvs; **Common Milkwort** (*P. vulgaris*) has **alt** lower lvs.

- The lower lvs often wither or get eaten; to see if these are opp or alt, look for the position of lf-scars on the stem.

A   Common Milkwort, *Polygala vulgaris*, has **alt** lvs, lower lvs **smaller**, lvs becoming larger and longer up the stem; fls 6–8 mm long; pointed inner sepals with little veins joining the bigger veins together laterally (**anastomosing veins**) near the edges of the inner sepal. Br Isles, c in gsld on basic soils. Fl 5–9.

B   Chalk Milkwort, *P. calcarea*, resembles and often difficult to tell from A, but the basal lvs ± form a **rosette** (although may be eaten and withers later in summer) **above** the lower, unbranched stem, which is **slender and lf-less**; lower lvs **larger** than upper, alt lvs. Fls 6–7 mm long in a cluster, arranged like outer spokes of a wheel (unlike A, with fls spaced out); inner sepals blunt, **without** anastomosing veins. Blue forms have bright **gentian-blue** fls, usually brighter than blue forms of A, but paler whitish-blue fls also occur. Br Isles: S, SE Eng N to Lincs only, vla; on short calc gsld. Fl 5–6 only.

C   **Dwarf Milkwort, *P. amarella*, the name in the *New Flora* for 2 closely related, but in the author's opinion, distinct, spp. *P. austriaca* (illustrated) has **basal lf-rosettes** like B but **at or v near** the base (no slender, lf-less stem below rosette) and all lvs broadest above middle, but tiny fls, 2–5 mm across, white, lilac, pink or pale (never bright) blue; inner sepals only ¹/₂ width of and shorter than capsule. Kent and North Downs area, vr; on chk gsld. Fl 5–7. **C.1** *P. amara* is taller; stem-lvs pointed; sepals 2/3 width of and longer than capsule. Yorks, on damp mt limestone gsld, vr. Fl 6–7.

D   Heath Milkwort, *P. serpyllifolia*, is close to A, but has at least some **opp lower** lvs, more diffuse growth, and fls more deeply (but not so brightly) coloured. Br Isles, vc on acid gslds, hths. Fl 5–9.

A

B

C

D

HORSE-CHESTNUT FAMILY

Hippocastanaceae

Horse-chestnut, *Aesculus hippocastanum*, deciduous tree to 25 m, with broad, dense crown, and stout twigs curving up at ends. Bark dark grey-brown, finally flaking. Buds stout, to 3.5 x 1.5 cm, oval, red-brown, **sticky**. Twigs hairless. Lvs **palmate**, divided into 5–7 lfts, 8–20 cm long, obovate, tapered to base, pointed at tip, toothed, hairless and dark green above, downy below when young, long-stalked. Infls large erect panicles, terminal on twigs, 20–30 cm long; fls 2 cm across, with 4 predominantly **white** petals with pink spots below; stamens arching down. Fr (**a**) to 6 cm across, globular, green, prickly (then brown). Seeds glossy brown, 1–2, globular, 3–4 cm across. Introd (Greece); Br Isles, vc planted and self-sown. Fl 5–6.

a

MAPLE FAMILY   *Aceraceae*

Trees and shrubs with opp, usually palmate lvs, without stipules; infl a raceme; 5 sepals, 5 petals, 8 stamens; fr dry, of two carpels, each with a propeller-like wing. Several other *Acer* spp are f planted and may be r natd.

A   AWI **Field Maple**, *Acer campestre*, deciduous tree up to 20 m or sometimes (in hds) a shrub. Bark fissured, flaking off. Twigs **downy**; lvs opp, palmate, bluntly 3–5-lobed, lobes ± untoothed, 4–7 cm long and almost as wide; lvs downy below, hairless above when mature; lf-stalk produces **latex** when cut; infl (**Aa**) an **erect rounded panicle**, flg with lvs, fls 6 mm across, cup-shaped, yellow-green; fr (**Ab**) usually **downy**, with two **horizontally**-spreading propeller-shaped wings. Br Isles: Eng, vc in S, o-r to N and W; Scot, Ire, introd only and o-f; in wds on basic soils, scrub, hds, la on chk. Fl 5–6.

B   **Sycamore**, *A. pseudoplatanus*, deciduous tree to 30 m, with a wide crown. Bark smooth when young, later in flakes; stalked opp palmate lvs, 7–16 cm long, five-lobed, hairless, blunt-toothed. Lf-stalk **not** producing **latex** when cut; twigs **hairless**. Infl (**Ba**) a narrow **drooping** panicle, 5–20 cm long; fls as in A, with lvs but more yellow, **not cup-shaped**. Fr (**Bb**) **hairless**, with two wings at an **acute angle**. Br Isles: introd in 15th century, but now c in wds and hds on richer soil, as well as planted. Fl 4–6.

C   **Norway Maple**, *A. platanoides*, deciduous tree to 30m; bark grey with fissures, not flaking; lvs palmate with sharp teeth to lobes; lf-stalk produces **latex** when cut; hairless twigs; **erect** infls of yellow-green fls, appearing **before** the lvs; fr hairless, with two ± horizontal wings. Introd (mts of Eur), often planted in Br Isles and f natd on rough gsld, wa, hds, wds. Fl 3–5, about 3 weeks before B.

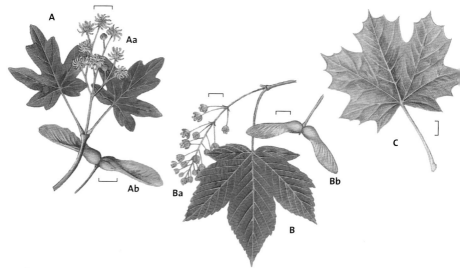

WOOD-SORREL FAMILY   *Oxalidaceae*

Per or rarely ann herbs, often with bulbs and/or rhizomes, with distinctive usually **trifoliate** lvs and 5 petalled-fls. Several other introd *Oxalis* spp occur, some quite widespread, but are omitted here as relatively difficult to identify (see the *New Flora* for line illustrations of many of these).

A AWI **Wood-sorrel**, *Oxalis acetosella*, creeping per herb, 5–10 cm, with long-stalked trefoil lvs like those of White Clover (p 274 D), but more delicate, with lfts drooping, purplish below, yellow-green above. Fls, 10–25 mm across, **solitary** on delicate 5–10 cm stalks, with 5 equal broad white **lilac-veined** petals, 10 stamens and 5 styles; fr a hairless five-angled capsule, 3–4 mm. Br Isles, c in shady dry wds and among rocks (avoids wet or v chky soils). Fl 4–5.

B Procumbent Yellow-sorrel, *O. corniculata*, has **several** fls per infl; fl (**Ba**) with narrower yellow petals, 8–10 mm, 10 stamens **all** with anthers and creeping **rooting stems** with alt trefoil lvs, tiny stipules and fr-stalks reflexed (**Bb**). Introd (native range unknown); S Eng c, elsewhere o; in gdns, wa etc. Fl 6–9. **B.1 Upright Yellow-sorrel**, *O. stricta* (*O. europaea*), is erect, **not** rooting, with lvs in whorls, and no stipules; yellow petals 4–10 mm; fr-stalks erect. Introd (N America); S Eng f, elsewhere o; in wa and gdns. Fl 6–9. **B.2 Least Yellow-sorrel**, *O. exilis*, is v close to B but smaller, v slender stem, only **1 fl** per infl; 10 stamens, 5 usually without anthers. Introd (Australia); GB o scattered; N Ire, o; on paths, short gsld, pavements. Fl 4–6. **B.3 Bermuda-buttercup**, *O. pes-caprae*, has lvs **all** rising from the base, yellow fls in **umbels**, and large petals 20–25 mm. Introd (S Africa); Br Isles: SW, S Eng, vr; IoS, CI vc; in ar. Fl 3–6.

C Pink-sorrel, *O. articulata*, has lfts ± downy with orange spots below (**Cb**); pink fls in umbels (**Ca**); petals 10–15 mm. Introd (S America); Br Isles: SW and SE Eng, f; o scattered elsewhere; in wa and gdns. Fl 5–10.

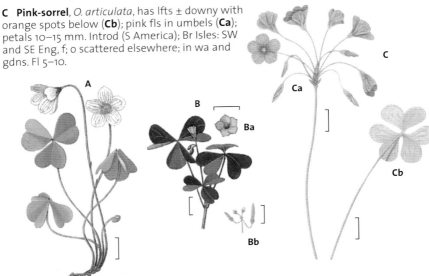

GERANIUM FAMILY   *Geraniaceae*

This family, in the Br Isles, is composed of herbs with alt lobed or compound lvs with stipules; fls solitary or in cymes; bisexual; with 5 separate sepals and 5 petals; 10 stamens, 5 of which lack anthers in **Stork's-bill** (*Erodium*). Ovary of 5 one-seeded cells, and a **long beak** topped with 5 stigmas; fr a five-lobed capsule. **Crane's-bills** (*Geranium*) have palmately-lobed lvs and 10 fertile stamens (except *G. pusillum*), and carpels whose beaks roll up to release seeds; **Stork's-bills** (*Erodium*) have pinnate lvs, only 5 of the stamens with anthers, and carpel beaks twisting spirally when ripe, the seeds remaining inside the carpel cells. If a plant does not key out using the key below, it may be one of many gdn *Geranium* spp which are often natd.

KEY TO CRANE'S-BILLS (*GERANIUM*) AND STORK'S-BILLS (*ERODIUM*)

1 Lvs with pinnate veins – only 5 of 10 stamens have anthers..............................*Erodium* (p 322)
   Lvs with palmate veins ....(*Geranium*) **2**

2 Petals usually 10 mm or more long or if <10 mm, then claw <¹/₃ length of petal **3**
   Petals (without claw) less than 10 mm **9**

3 Lvs kidney-shaped to round, lobes widened to tips – petals 10 mm or less, deeply-notched
                                                        Geranium pyrenaicum (p 320 **H**)
   Lvs with pointed lobes narrowed to tips – petals more than 10 mm **4**

4 Fls borne singly, 25–30 mm, purple-crimson (rarely white or pink) – petals rounded – lf-
   lobes not widened in middle ....*G. sanguineum* (p 320 **J**)
   Fls in pairs – not purple-crimson – lf-lobes widened in middle **5**

5 Petals blackish-purple, bent back, pointed...*G. phaeum* (p 322 **B**)
   Petals not blackish-purple, forming a cup, round or notched **6**

6 Petals notched, veins darker than petal ....*G. versicolor* (p 322 **C**)
   Petals rounded – veins not darker **7**

7 Fls pink – no sticky hairs ....*G. endressii* (p 320 **A**)
   Fls blue to violet – sticky hairs above **8**

8 Fls pale violet-blue, saucer-shaped – petals 15–18 mm – lf-lobes with pointed teeth 3–4
   times as long as wide ....*G. pratense* (p 320 **I**)
   Fls reddish-violet, cup-shaped – petals 10–15 mm – lf-lobes with blunt teeth 1 ½ times to
   twice as long as wide...*G. sylvaticum* (p 322 **D**)

9 Petals notched **10**
   Petals unnotched **12**

10 Lvs divided at least ⁵/₆ of length into linear lobes that are pinnately lobed again – frs and
   lvs hairy ....*G. dissectum* (p 320 **G**)
   Lvs divided to ³/₄ or less of length into lobes widened above **11**

11 All 10 stamens with anthers – fr hairless, wrinkled ....*G. molle* (p 320 **F**)
   5 stamens with anthers – fr hairy, unwrinkled ....*G. pusillum* (p 320 **E**)

12 Lvs dull green, downy – sepals spreading **13**
   Lvs shiny bright green – sepals erect...**14**

13 Lvs divided to 5/6 of length into linear lobes – fl-stalks 2 cm or more – frs hairless ....
G. columbinum (p 319 **B**)

Lvs divided to 1/2 of length or less into broad lobes – fl-stalks 0.5–1.5 cm frs hairy...................
G. rotundifolium (p 320 **C**)

14 Lvs rounded, 5–7-lobed to 1/2 of length or less – sepals ± hairless, keeled, forming a five-angled calyx ....*G. lucidum* (p 320 **D**)

Lvs triangular, lobed to base – sepals hairy, unkeeled, forming a rounded calyx **15**

15 Anthers yellow – pollen yellow – petals 5–9 mm long, not spreading ....
G. purpureum (p 319 **A.1**)

Anthers orange, purple, pink or red – pollen orange – petals 8–14 mm long, spreading at tips..*G. robertianum* (p 319 **A**)

A AWI **Herb-Robert**, *Geranium robertianum*, ann, 10–40 cm high, often over-wintering, with a strong smell; lvs palmate to base, sparsely hairy, shining bright green, the lfts pinnately-cut; stem and lvs often reddish-flushed. Petals 8–14 mm with claw, broadest part of petal 4–6 mm long, pink (or white with yellow anthers), unnotched; **anthers orange** or **purple**; frs netted. Br Isles, vc in wds, hbs, rocks, shingle and on dry, disturbed wa. Fl 4–9. **A.1** * **Little-Robin**, *G. purpureum*, has smaller lvs, petals **only 5–9 mm, yellow anthers**, and more wrinkled frs. SE and SW Eng, CI, S Ire r on stony soils, shingle and rocks near sea; inland on railway ballast. Fl 5–9.

B **Long-stalked Crane's-bill**, *G. columbinum*, slender, spreading to erect ann, 10–40 cm, with adpressed hairs; lvs (**Ba**) long-stalked below, **five-angled** in outline, divided to 5/6 of length into 5–7 pinnate lobes with narrow segments. Fls (**Bb**) on long, 2–6-cm, stalks; a few together on common stalks to 12 cm. Sepals bristle-tipped; **petals unnotched**, pink, 7–9 mm; frs (**Bc**) ± **hairless**, unwrinkled. Br Isles: Eng, Wales lf-lc; Scot r; Ire lf; on gsld, scrub on calc soil. Fl 6–7.

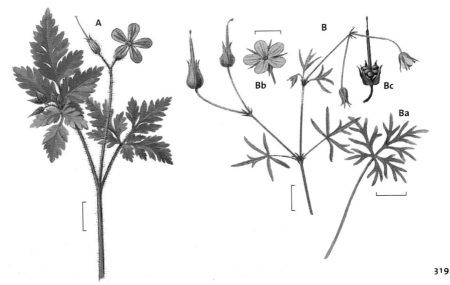

C Round-leaved Crane's-bill, *Geranium rotundifolium*, branched v hairy ann, 10–30 cm, with long-stalked, kidney-shaped downy lvs (**Ca**), cut for ¹/₂ of length or less into 5–7 two- to three-toothed blunt lobes, wider above; fls numerous in open cymes, fl-stalks 0.5–1.5 cm; sepals pointed (no bristle); petals **unnotched**, pink, 5–7 mm; frs (**Cb**) **downy**, wrinkled; seeds **netted**. Br Isles: S Eng, S Wales o-lf; N Eng, o-r; Scot vr; Ire o-r; on dry hbs, walls, gsld on limy soils. Fl 6–7.

D Shining Crane's-bill, *G. lucidum*, branched, ascending, ± hairless ann, 10–40 cm, with brittle fleshy stems, and **shining** green lvs, often red-tinged. Lvs long-stalked, 5–7-lobed for ¹/₂ of length or less, with **± hairless** lobes widened above, each with 2–3 blunt teeth; sepals ± hairless, **keeled**, netted, forming a five-angled calyx; petals 8–9 mm including claw, pink, **unnotched**. Br Isles: GB c (N Scot o); Ire o-lf; on calc rocks, walls, hbs. Fl 5–8.

E Small-flowered Crane's-bill, *G. pusillum*, branched spreading ann, with closely downy (not hairy) stems and lvs. Lvs round, with 7–9 lobes, widened above; sepals **hairy, not keeled**; petals 2–4 mm, notched, dingy **mauve**; **5 of 10 stamens lack anthers**; frs (**Ea**) **downy, smooth**; seeds smooth. Br Isles: Eng, Wales, c to E, o to W; Scot vlc in E; Ire r; on dry gsld, wa ar, especially on sand. Fl 6–9.

F Dove's-foot Crane's-bill, *G. molle*, resembles E, but has long 1–2 mm hairs on stems; petals 4–6 mm, **pink**, notched; all 10 stamens with rosy-pink anthers; **key difference** from E is in frs (**Fa**) **hairless, wrinkled**; seeds **smooth**. Br Isles, vc on dry gsld, ar, wa, dunes. Fl 4–9. **Do not confuse** with C, which has similar lvs but **netted** seeds.

G Cut-leaved Crane's-bill, *G. dissectum*, ann to 60 cm; lvs divided **almost to base** into linear lobes that are lobed again; petals 4.5–6 mm, pink, notched; frs and lvs hairy; seeds netted. Br Isles, vc (except N Scot, o-r) on ar, wa, hbs. Fl 5–8.

H Hedgerow Crane's-bill, *G. pyrenaicum*, rather like an erect, larger (25–60 cm) version of F, but has a per rootstock; stems and lvs hairy; lvs (**Ha**) more rounded than in F, basal lvs long-stalked, and 5–8 cm across; **petals 7–10 mm**, dull purple, notched; frs downy, smooth. Probably introd (mts S Eur); Br Isles: Eng c, especially in SE; rest of GB rarer W and N; Ire o, only c in E; on hbs, wa, scrub. Fl 6–8.

I Meadow Crane's-bill, *G. pratense*, erect hairy per herb, 30–80 cm; sticky-hairy above (including fl-stalks and calyx); basal lvs deeply 5–7-times palmately lobed, lobes pinnately-cut into narrow sharp teeth 3–4 times as long as wide; fls in **pairs**, saucer-shaped; petals 15–18 mm, **violet-blue** to **sky-blue**, unnotched, veins paler; fr hairy, unwrinkled, stalk bent down. Br Isles: GB (except N Scot and SE Eng, o-r) f but l (c mid Eng to S Scot); Ire r; on mds, rds, gsld, especially on calc soils. Fl 6–9.

J AWI **Bloody Crane's-bill**, *G. sanguineum*, bushy per herb, 10–40 cm; deeply 5–7-lobed lvs, with white adpressed hairs and long stalks below, the lobes trilobed and with almost parallel-sided segments; fls **solitary** on long stalks; petals usually **crimson**, 12–18 mm, not notched. Br Isles, lf on rocks, gsld, open wds, fixed dunes or calc or basic soils (**SE Eng, abs**). Fl 7–8.

▼ **A French Crane's-bill**, *G. endressii*, erect hairy per herb, 30–80 cm; palmately-lobed lvs cut more than ¹/₂ of their length; fls in pairs on long peduncles; petals about 16 mm, unnotched, deep pink, veins **not darker**; frs downy, unwrinkled. Introd (W French Pyrenees); GB, o; Ire, r; on rds, hbs. Fl 6–7.

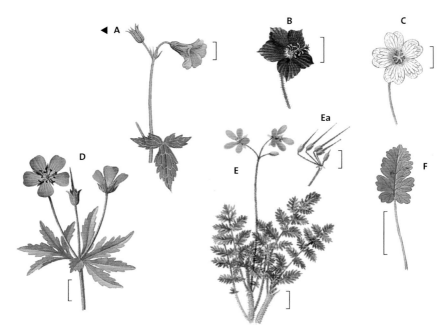

B   **Dusky Crane's-bill**, *Geranium phaeum*, has habit of D (below), a per herb, 30–60 cm; lvs far less narrowly divided than in D; fls (**B**) with blackish-purple pointed petals that are bent backwards. Introd (C Eur); Br Isles: GB o-vl, especially Eng, S Scot; N Ire o, elsewhere vr; shady, moist places on hbs, scrub, rds. Fl 5–6.

C   **Pencilled Crane's-bill**, *G. versicolor*, hairy per of habit like D, but with pale pink or white notched petals, 15–18 mm, with **darker**, violet veins, and cup-shaped fls (**C**). Introd (S, SE Eur); Br Isles: o gdn escape, especially in S Eng (c Cornwall); in hbs etc. Fl 5–9. **C.1 Knotted Crane's-bill**, *G. nodosum*, has fls as C, but, unlike it, has its stems swollen at the nodes and stem-hairs abs or adpressed; the 3–5 main lf-lobes merely toothed (not secondarily lobed). Introd (S Eur); r gdn escape in GB; vr Ire; on rough gsld, rds, usually near housing. Fl 5–9.

D   AWI WO (NI) **Wood Crane's-bill**, *G. sylvaticum*, is rather close to I (p 320), but differs in having broader lf-lobes with blunt teeth only 1½–2 times as long as wide; fls **reddish** or **pinkish-violet**, more cup-shaped; petals smaller, 10–15 mm; frs held erect. Br Isles: N Eng, Scot c; Wales o-r; SW and SE Eng abs; N Ire vr; on mds, hbs, damp wds, rocks. Fl 6–7.

E   **Common Stork's-bill**, *Erodium cicutarium*, spreading ann to 30 cm, usually v hairy, often sticky, with lvs pinnate, the lfts pinnately-cut **almost to** the mid rib; stipules whitish, conspicuous, pointed. Fls in loose umbels of 3–7, **fls mostly > 10 mm** across; petals 6–8 mm, often **unequal**, rose-purple to white, often with black spots; only 5 stamens have anthers. Fr (**Ea**) 5.0–6.5 mm, with a large **rimmed pit**

at apex; carpels hairy with beaks 15–40 mm, that twist spirally when ripe and remain attached to the separating carpels (which retain their seeds inside). Br Isles: GB, c to S, f to N; Ire f on coastal areas; along coasts on dunes, cliffs; inland on sandy wa, ar. Fl 6–9. **E.1** *****Sticky Stork's-bill**, *E. lebelii* (*E. glutinosum*), v close to E, a densely sticky-hairy plant, has equal pale pink petals, fls in umbels of 2–4, **fls** mostly **< 10 mm** across, unspotted; **key difference** from E is in fr, less than 5 mm long with a tiny **rimless pit** at apex (x 20). Br Isles: S and W coasts of GB; E and S coasts of Ire; r-vlf on dunes. **E.2 Musk Stork's-bill**, *E. moschatum*, smells of musk and is v sticky. Lfts of lvs only **shallowly** toothed, divided ≤ ¹/₂ way to midrib, stipules broad and **blunt** (not pointed). Br Isles: S Eng, Wales, Ire; r-lf; IoS, CI c; near coast on wa, rds and dunes; inland as casual on dry wa. Fl 5–7.

F   Sea Stork's-bill, *E. maritimum*, has **simple lobed** (not pinnately-cut) lvs (**F**), and the petals either none or falling v soon. Br Isles: S and W coasts of GB, E and S coasts of Ire, o-lc; IoS, CI c; on dunes, gsld, walls and pavements near sea. Fl 5–9.

BALSAM FAMILY *Balsaminaceae*

Balsams are hairless **ann** herbs, with fleshy stems, and oval stalked toothed lvs. Fls in spikes, irregular, five-petalled, with wide lip, smaller hood and spur; frs explosive oblong capsules.

A   Indian Balsam, *Impatiens glandulifera*, tall (1–2 m); reddish stem; lvs in **whorls** of 3, red-toothed; fls (**A**) 2.5–4.0 cm long, with short curved spurs, purple-pink (or white). Introd (Himalayas); Br Isles, c, invasive and increasing; on riversides, wa. Fl 7–10.

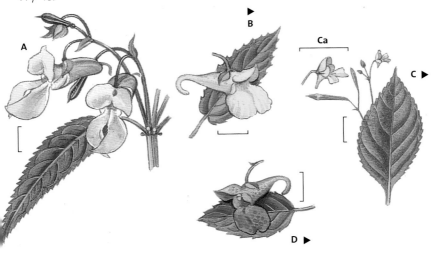

▲ **B** ***Touch-me-not Balsam**, *Impatiens noli-tangere*, 20–60 cm tall; stem nodes swollen, lvs **alt**, with 10–15 teeth (2–3 mm deep) on each side; fls bright yellow, 3.5 cm long, including gradually curved and tapered spur. Br Isles: N Wales, NW Eng, vlf as a native, often also introd elsewhere; in wet wds, riversides. Fl 7–9.

▲ **C** **Small Balsam**, *I. parviflora*, has alt lvs with more than 20 teeth each side, and small (5–15 mm) pale yellow fls (**Ca**). Introd (E Asia); GB, lf and increasing; Ire, vvr; in wds and on shaded riversides. Fl 7–10.

▲ **D** **Orange Balsam**, *I. capensis*, like B, but has less than 10 (shallower, 1–2 mm deep) teeth on each side of lf; **orange, red-spotted** fls with spur suddenly narrowed and bent into a hook. Introd (N America); Br Isles: S Eng, Wales, o-la and increasing; by rivers, ponds, canals. Fl 6–9.

IVY FAMILY

Araliaceae

A **Ivy**, *Hedera helix* ssp *helix*, evergreen woody climber, ascends to 30 m, and also forms carpets on wdland floors; stem clothed with sucker-like adhesive roots, young twigs downy. Lvs 4–10 cm long, hairless, glossy dark green with paler veins above, paler below; **juvenile** lvs (**Aa**) of non-fl stems palmately 3–5-lobed, usually < 8 cm across with triangular untoothed lobes and lf-underside with **whitish hairs**; **mature** lvs of fl-shoots oval or elliptic, untoothed. Fls (only borne on shoots in good light at top of tree or wall) in umbels, with 5 small sepals, 5 yellow-green 3–4 mm petals, 5 yellow stamens. Fr (**Ab**) a black globular berry. Br Isles, vc on gd in dry wds, and climbing trees, hbs, rocks, walls etc. Fl 9–11.

A.1 **Atlantic Ivy**, *H. helix* ssp *hibernica*, v close to A but **key differences** are lvs of creeping or climbing (**not flg**) stems usually lobed <$^1/_2$ way, > 8 cm across; lf-underside with **yellowish hairs**; more c than A in Ire, CI, W and SW GB, elsewhere o in same habitats as A. Fl 9–11.

CARROT (UMBELLIFER) FAMILY   *Apiaceae*

Also known as the *Umbelliferae* or Umbellifers, herbs with alt lvs, no stipules (except in **Marsh Pennywort**) but generally with **sheathing, inflated** bases to the lf-stalks; fls small, in umbels which may themselves also be in umbels (hence infls often umbrella-like). Fls with 5 **separate** petals (sepals 5, or **abs** in many *genera*), 5 stamens, 2 stigmas; ovary **inferior**, with 2 carpels or lobes that are variously ridged, winged or flattened in fr. Frs dry, dividing into 2 parts.

ID Tips   Umbellifers

- **What to look at: (1)** lf form **(2)** presence or absence of whorl of bracts (in key called 'bracts') at base of main umbel **(3)** presence or absence of whorl of bracts (in key called 'bracteoles') at base of upper umbels **(4)** shape of frs.

- Only 3 Umbellifers smell of aniseed when crushed: **Sweet Cicely**, **Fennel** and **Dill**.

- Only one Umbellifer is dioecious with entirely male or female fls, **Honewort** (*Trinia glauca*).

- Frs with hooked bristles are uncommon, found only in **Bur Chervil** (*Anthriscus caucalis*), **Hedge Parsleys** (*Torilis* spp); **Carrots** (*Daucus* spp) and **Sea-hollies** (*Eryngium* spp).

KEY TO UMBELLIFERS

Lvs undivided, or palmately-lobed, or spiny ....**A**
Lvs (at least lower ones) pinnately-lobed, pinnate or trifoliate ....**B**

A. Umbellifers with undivided (or divided ≤ ¹/₂ way to base), or palmately-lobed, or spiny lvs

1 Lvs spiny – fls in dense rounded to oval heads (resembling Thistles, but petals separate, not joined in a tube as in Thistles and Teasels) ....*Eryngium* (p 344)
Lvs not spiny **2**

2 Lvs undivided, linear or oval – fls yellow ....*Bupleurum* (p 340)
Lvs undivided or divided ≤ ¹/₂ way to base, circular, stalk attached to centre – fl in whorled spikes ....*Hydrocotyle* (p 344)
Lvs palmately-lobed – fls white or pink **3**

3 Bracts tiny – fls in loose compound umbels – frs bristly ....*Sanicula europaea* (p 336 **E**)
Bracts in a conspicuous toothed ruff below the dense, head-like simple umbel – frs smooth ....*Astrantia major* (p 336 **F**)

B. Umbellifers with at least lower lvs pinnately-lobed, pinnate or trifoliate

1 Plants growing in water – submerged lvs finely divided **2**
Plants without submerged fine-cut lvs **4**

2 Plants large – lvs 15–30 cm long **3**
Plants small, creeping – lvs 1–8 cm long...*Apium inundatum* (p 344 **D**)

3 Stout erect fl-stems present – submerged lvs with hairlike segments – aerial lvs with flat, deep-cut lfts – in still water ....*Oenanthe aquatica* (p 342 **B**)
Slender ascending fl-stems sometimes present – submerged lvs with flattened, linear-segmented lfts – aerial lvs twice-pinnate with broad oval lfts – in rivers and canals ....
O. fluviatilis (p 342 **C**)

4 Lower lvs only once pinnate, or pinnately-lobed ....**C**
Lower lvs 2–4 times pinnate ....**D**

C Umbellifers with lower lvs once pinnate only

1 Plants ± hairy, at least on lower lvs **2**
Plants hairless **5**

2 Fls yellow – lfts oval, coarse-toothed ....*Pastinaca sativa* (p 338 **A**)
Fls white or pink **3**

3 Bract and bracteoles abs – plant sparingly downy ....*Pimpinella saxifraga* (p 334 **B**)
Bracteoles present – plant roughly hairy **4**

4 Stout plant with no bracts – calyx minute ....
Heracleum sphondylium or *H. mantegazzianum* (p 336 **D.1**)
Slender plant, with both bracts and bracteoles – calyx-teeth half as long as petals – r ....
Tordylium maximum (p 332 **D**)

5 Lf-stalks and stems tubular, swollen – umbels dense, globular – on marshy gd ....
Oenanthe fistulosa (p 342 **D**)
Lf-stalks not swollen – umbels loose, flat..**6**

6 Lf-segments thread-like, many, in dense whorls – vl ....*Carum verticillatum* (p 332 **D**)
Lf-segments lanceolate-oval, not whorled **7**

7 Bracts abs **8**
Bracts present **9**

8 Umbels almost stalkless, lf-opposed or terminal umbels with stalks shorter than rays – in wet places ....*Apium* (p 334 and p 344)
Umbels terminal, on stalks longer than rays – in wds, hbs ....*Pimpinella major* (p 334 **C**)

9 Some bracts up to half length of shorter umbel-rays..**10**
All bracts much shorter than umbel-rays – stems much branched – lft-teeth shallow..........
Sison amomum (p 334 **A**)

10 Bracts bristle-like – stem solid, wiry – lft-teeth deep – in dry places..
Petroselinum segetum (p 332 **E**)
Bracts wider, sometimes divided – stem hollow, brittle – in wet places **11**

11 Lf segments 4–6 pairs – lfts finely-toothed – plants 1–2 m tall ......*Sium latifolium* (p 342 **A**)
Lf segments 7–10 pairs – lfts coarsely-toothed – plants 30–100 cm tall ....
Berula erecta (p 344 **B**)

D Umbellifers with at least lower lvs 2 or more times pinnate

Fls yellow or green-yellow ....**E**
Fls white or pinkish..**F**

E Umbellifers with yellow fls and pinnate lvs

1 Lf-segments fleshy, cylindrical – on seashores.........................*Crithmum maritimum* (p 338 **D**)
Lf-segments not fleshy, but flat or thread-like **2**

2 Lf-segments linear or thread-like, over 10 times as long as wide **3**
Lf-segments lobed or toothed, 2–6 times as long as wide **4**

3 Lfts long-narrow, linear, but flat and not thread-like – near sea – vr ....
Peucedanum officinale (p 338 **F**)
Lfts pinnately-divided into cylindrical thread-like segments ....*Foeniculum vulgare* (p 338 **E**)

4 Lvs mostly once pinnate only – lfts oval, coarse-toothed, v hairy ..*Pastinaca sativa* (p 338 **A**)
Lvs all 2 or more times pinnate **5**

5 Lower lf-segments large (3 cm or more), oval-oblong, toothed, blunt ....
Smyrnium olusatrum (p 338 **B**)
Lower lf-segments small, pinnatifid, with narrow pointed lobes.................................**6**

6 Lf-stalks with broad whitish margins – lfts crisped, edges not fine-toothed ....
Petroselinum crispum (p 334 **D**)
Lf-stalks with only narrow margins – lfts flat, edges with fine teeth *Silaum silaus* (p 338 **C**)

F Umbellifers with white (or pinkish) fls, and lvs 2 or more times pinnate

1 Stems purple-spotted **2**
Stems green, without purple spots **3**

2 Stems and lvs hairy..*Chaerophyllum temulum* (p 330 **F**)
Stems and lvs hairless...*Conium maculatum* (p 330 **G**)

3 Plants hairy, at least on lf edges or veins or on infl-stalks (use hand lens) **4**
Plants quite hairless throughout...**I**

4 Fl-stems hollow (collapses when squeezed)...**G**
Fl-stems solid (or almost so) ....**H**

G Umbellifers ± hairy, with hollow unspotted stems, lvs 2 or more times pinnate, and white or pinkish fls

1 Plant hairless (except for infl-stalks) – lfts narrow-lanceolate-oblong, v finely-toothed (use hand lens) – in fens ....*Peucedanum palustre* (p 346 **D**)
Plant downy, at least on lower stem or lvs...**2**

2 Lower lf-segments broad (at least 1 cm wide x 3 cm long), only shallowly-toothed – v large robust plants **3**
Lower lf-segments narrow (0.5 cm wide x 2.0 cm long), deeply cut **4**

3 Lvs with hairless, oval, evenly-toothed lfts – stems round, hairless or only downy below
Angelica (p 336 **A**, **B**)

Lvs with roughly hairy, irregularly-lobed lfts – stems ridged, roughly hairy ....
Heracleum sphondylium (p 336 **D**)

4 Plants smelling of aniseed – lvs fern-like – frs stout, 20–25 mm long ....
Myrrhis odorata (p 330 **B**)

Plants not smelling of aniseed...**5**

5 Frs 30–70 mm, long, slender-beaked – umbel only 1 or 2 rays – r on ar ....
Scandix pecten-veneris (p 332 **C**)

Frs short, 3–10 mm long – umbels 3–10-rayed – widespread **6**

6 Stems hairless – frs prickly, 3 mm long..*Anthriscus caucalis* (p 329 **B**)
Stems downy – frs smooth, 6 mm long..*A. sylvestris* (p 329 **A**)

H Hairy Umbellifers with white (or pinkish) fls, solid unspotted stems, and lvs 2 or more times pinnate

1 Stems with many straight and downward-adpressed hairs – frs oval, spiny ....*Torilis* (p 330)
Stems hairy, but hairs not straight or downward-adpressed...**2**

2 Hairs on stem tiny, crisped...**3**
Hairs on stem spreading **4**

3 Slender plant – bracts, bracteoles abs – calyx abs – umbel-rays ± hairless – no fibrous
stock at base – fr smooth...*Pimpinella saxifraga* (p 334 **B**)
Stout plant – bracts and bracteoles present – calyx conspicuous – umbel-rays downy –
many fibres crown rootstock at stem-base – fr hairy ....*Seseli libanotis* (p 346 **A**)

4 Bracts of umbel 7–13, large, pinnately-lobed, lfy – centre fl of umbel red – fr spiny.................
....*Daucus carota* (p 346 **C**)
Bracts of umbel 0–5, tiny, undivided – but bracteoles conspicuous – fr not spiny but long-
pointed..*Scandix pecten-veneris* (p 332 **C**)

I Hairless Umbellifers with lvs 2 or more times pinnate, unspotted stems, and white (or pinkish) fls

1 Bracteoles abs (or 1 only) **2**
Bracteoles 2 or more..**3**

2 Lower lvs persistent, twice trifoliate – lfts oval, toothed, not pinnatifid, 3–6 cm long –
many creeping fleshy rhizomes – c garden weed ....*Aegopodium podagraria* (p 334 **F**)
Lower lvs soon withering, 3 times pinnate – lfts deeply pinnatifid, small (to 1 cm long) –
root a round tuber ....*Conopodium majus* (p 330 **I**)

3 Lf-segments long, narrow (to 30 x 2 cm), strap-shaped, sharp-toothed, curved ....
....*Falcaria vulgaris* (p 332 **D.1**)
Lf-segments under 10 cm long, not strap-shaped **4**

4 Bracteoles conspicuous, long, pointing downwards.....................*Aethusa cynapium* (p 330 **H**)
Bracteoles small, spreading...**5**

5 Plant low (3–20 cm), erect – waxy-grey solid stem – rosette of waxy-grey fine-cut lvs –

persistent fibrous lf-bases – dioecious – frs oval – vr on limestone turf ....
....*Trinia glauca* (p 346 **B**)

Taller plant not as above **6**

6 Tall plants of fen, marsh, streamside or damp meadow..**7**

Plants of dry gslds, wds, dunes or rocks **12**

7 Lfts 3–10 cm long, 1 cm wide, narrow-lanceolate, toothed all along – no bracts – many
bracteoles – r in swamps..*Cicuta virosa* (p 336 **C**)

Lfts 1–2 cm long, pinnatifid..**8**

8 Lfts broad oval, 1 cm wide, with 2–3 blunt broad teeth, wedge-shaped at base – bracts,
bracteoles many – in ditches, river banks, marshy wds ....*Oenanthe crocata* (p 341 **A**)

Lfts narrow, less than 1 cm wide, divisions narrow-lanceolate **9**

9 Secondary umbels dense – fl-stalks shorter than fr – fr oblong, with conspicuous calyx-
teeth ....*Oenanthe* (key p 341)

Secondary umbels open – fl-stalks longer than frs – fr globular – no calyx **10**

10 Stem solid..*Selinum carvifolia* (p 346 **E**)

Stem hollow **11**

11 Bracts 4 or more – r in fens ....*Peucedanum palustre* (p 346 **D**)

Bracts none – vl on river banks, mds in N ....*Peucedanum ostruthium* (p 336 **F.1**)

12 Basal lvs 3–4 times pinnate, segments hairlike, whorled, v aromatic – rootstock bearing
fibrous bases of old lvs ....*Meum athamanticum* (p 332 **G.2**)

Basal lvs with flat lfts (not whorled, hairlike or aromatic) **13**

13 Stout plants – lower lvs 1–2 times trifoliate – lfts oval, ± unlobed, 3–5 cm long x 2 cm wide
– on dunes or rocks by sea in N ....*Ligusticum scoticum* (p 332 **F**)

Slender plants – stems narrowed to base – lower lvs 2–3 times pinnate – lfts linear-
lanceolate, 1–2 mm wide – root with rounded tuber...**14**

14 Bracts, bracteoles abs (or, rarely, 1) – styles of fr erect – stem hollow when in fr –
widespread in acid gsld, wds..*Conopodium majus* (p 330 **I**)

Bracts and bracteoles many – styles of fr recurved – stem always solid – r on chk in Eng ....
....*Bunium bulbocastanum* (p 330 **A**)

▼ **A Cow Parsley**, *Anthriscus sylvestris*, downy tall erect per herb, 60–100 cm; stems
hollow, furrowed, **unspotted**, downy below; lvs 2–3 times pinnate, fern-like,
scarcely downy, fresh green, lfts pointed; umbels to 6 cm across, 4–10-rayed, with
no bracts, but with bracteoles; fls pure white, 3–4 mm across, fl-stalks thin. Fr
(**Aa**) 6 mm long, smooth, oblong, short-beaked, turning brown-black. Br Isles, vc
in hbs, rds, wd borders (by nearly every Eng rdside in May except on v poor soils).
Fl 4–6.

▼ **B Bur Chervil**, *A. caucalis*, small version of A (to 50 cm), but ann; stems hollow,
hairless; lvs with tiny (5 mm) pale green lfts; fls only 2 mm across; frs (**B**) 3 mm
long, oval, on stout stalks and covered with **hooked spines**. Br Isles, lc in E Eng, CI;
r and scattered elsewhere; on sandy wa, hbs, mostly near sea. Fl 5–6.

C   Upright Hedge-parsley, *Torilis japonica*, July–Aug successor to A on rds; ann, slenderer than A; stems **solid, unspotted, rough**, with **adpressed** straight hairs; lvs narrower than in A, rough-hairy, dull green; umbels to 4 cm across, 5–12-rayed, bracts and bracteoles present; **frs** (**C**) oval, 2–2.5 mm (excluding spines), covered with **curved** but not hooked **spines**, styles recurved. Br Isles (except N Scot), vc on rds, hbs, wd borders. Fl 7–8.

D   EN BAP *Spreading Hedge-parsley**, *T. arvensis*, short (10–20 cm) spreading ann; 3–5 umbel-rays. **Key differences** from C are: bracts **none or 1**; frs (**D**) **3–4 mm**, with ± straight spines minutely **hooked** at tips. SE Eng. r-o; on ar. Fl 7–9.

E   Knotted Hedge-parsley, *T. nodosa*, prostrate or erect v slender ann, to 30 cm tall; stems solid; lvs under 10 cm long; umbels (**E**) small (0.5–1.0 cm), ± **stalkless, lf-opposed**; no bracts, but bracteoles; fls 1 mm across, pinkish; fr (**Ea**) 2–3 mm, covered with straight spines **and** warts. Br Isles: Eng, Wales, f-lc to E, o to W; SE Scot vr; Ire r; on dry grassy banks on sea coasts, ar inland in E Eng. Fl 5–7.

F   Rough Chervil, *Chaerophyllum temulum* (*C. temulentum*), resembles A and is its June-July successor on hbs and rds; stems (**Fc**) hairy, ridged, solid, **purple-spotted**; lvs **dull** dark **grey-green, v hairy**, with lfts **blunter** than in A; fl (**Fb**) as in A; frs (**Fa**) 5–7 mm, much as in A. Br Isles: GB (not N, W Scot) c; Ire vr (introd); in hbs etc. Fl 6–7.

G   Hemlock, *Conium maculatum*, resembles F in its **purple-spotted stems** (**Gb**), but is **hairless**; tall (to 2 m) erect unpleasant-smelling bi, with fine-cut fern-like lvs to 30 cm long; umbels 2–5 cm across, rays 10–20; bracts, bracteoles few; frs (**Ga**) 3 mm long, **globular**, with **wavy ridges**. Br Isles (except NW Scot), c by streams, on wa, rds. Fl 6–7. **Poisonous.**

H   Fool's Parsley, *Aethusa cynapium*, is another herb like A, but ann, to 50 cm tall, and hairless; bractless umbels have **conspicuous narrow bracteoles** (**H**) **1 cm long, hanging down** from the partial umbels; frs (**Ha**) 3–4 mm, **oval**, with broad ridges and no spines. Br Isles (except N, W Scot) c; on ar, wa. Fl 7–8.

I   AWI **Pignut**, *Conopodium majus*, slender erect per; smooth stem, 30–50 cm tall (stem becomes hollow after flowering). Basal lvs (**Ib**) soon wither; lvs twice pinnate, lfts deeply-cut, segments linear, 1–2 mm wide; umbel 3–7 cm across, 6–12 rays, bracts (usually) none, bracteoles few. Fr (**Ia**) 4 mm, narrow, oval, beaked, styles short, **erect**. The plant's name refers to its edible tubers (**Ic**). Br Isles, c; in dry wds, old gsld on non-calc soils. Fl 5–6.

▼ **A**   ****Great Pignut**, *Bunium bulbocastanum*, is v like I in habit and has similar tubers, but stems always solid; umbels (**Ab**) have many bracts and bracteoles; styles of fr (**Aa**) **recurved**. Eng: SE Midlands only, r and vl; on chk gslds, rds. Fl 6–7.

▼ **B**   **Sweet Cicely**, *Myrrhis odorata*, has aspect of A (p 329), but is bushier; when bruised smells strongly of **aniseed**. Stem hollow, ± downy; lvs fern-like, ± downy, to 30 cm long, often white-blotched at base; **frs** (**B**) **linear-oblong, v long and stout** (20-25 mm long x 8–10 mm wide), sharply-ridged with **pointed beak**. Br Isles: most c early-flowering Umbellifer of upland N Eng and S Scot; **vr** (as gdn escape only) **SE Eng**; N Scot, N Ire f, S Ire vr; rds, mds. Fl 5–6.

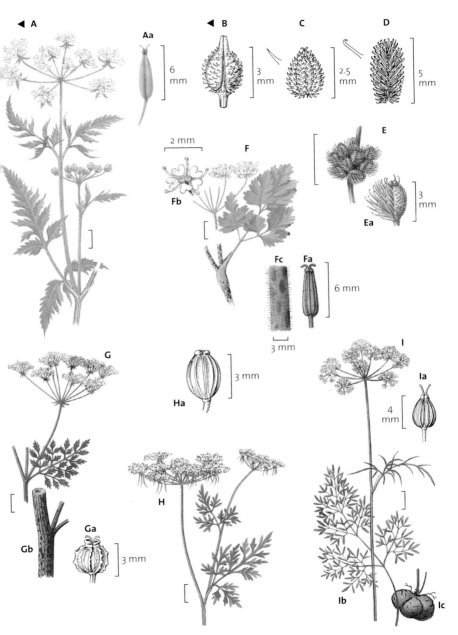

A

Aa

6 mm

B

3 mm

C

2.5 mm

D

5 mm

2 mm

F

Fb

E

Ea

3 mm

Fc Fa

6 mm

3 mm

Ha

3 mm

G

Gb

Ga

3 mm

H

I

Ia

4 mm

Ib

Ic

C CR BAP **Shepherd's-needle**, *Scandix pecten-veneris*, erect ann, 15–50 cm; stems ± downy, becoming hollow when old; lvs oblong, 2–3 times pinnate, with segments widened at tips. Umbels simple, or of only 2 short stout rays, spiny-edged bracteoles; fls small (1 mm across), white. **Key difference** from all other British Umbellifers is frs (**C; Ca** opening), 30–70 mm long, with v distinctive **long slender beaks**. Eng: SE, E Eng N to Yorks (and Shetland), formerly c weed of cornfields, now r; in ar, especially on calc soils. Fl 5–7

D **Hartwort**, *Tordylium maximum*, erect branched ann to 1 m tall; adpressed-hairy stem and roughly-hairy lvs give it resemblance to C (p 330), but lvs only **simply pinnate**, with lfts toothed, **basal lvs oval**, lanceolate higher up stem; umbels with bristly bracts, long bracteoles; fls (**Db**) white to pinkish; fr (**Da**) 5–8 mm, broadly oblong, bristly, with **thickened whitish hairless ridges**. Possibly introd; **Essex only**, vr on gsld on clay. Fl 6–7. **D.1** **Longleaf**, *Falcaria vulgaris*, also has linear-lanceolate lvs, **key difference** from D in **basal lvs**, once- or twice-trifoliate, long-stalked, with **linear, strap-shaped**, pointed, **sharply and finely-toothed, curved** lfts up to **30 cm** long, **grey-green**. Stem-lvs shorter, with 3–5 lfts, arranged like spread fingers. Umbels compound, as in **Cow Parsley** (p 329 A), with many **long** rays and small white fls; bracts, bracteoles present, linear. Introd (E Eur); Br Isles: S Eng, CI r; on gsld on calc soils. Fl 7.

E **Corn Parsley**, *Petroselinum segetum*, slender hairless bi, 30–100 cm, parsley-scented; stem round, solid; lvs (**Ec**) once-pinnate, linear-oblong; many pairs lfts, 0.5–1.0 cm long, oval, toothed or lobed; upper lvs with narrow lfts. Umbels irregular, few-rayed, 1–5 cm across, bracts and bracteoles bristle-like. Fls (**Eb**) white, 1 mm across, 3–5 only per partial umbel. Fr (**Ea**) egg-shaped, ridged, no calyx. S and E Eng, Wales lf near coast, o inland; **abs rest Br Isles**; on dry grassy banks, near sea and on hbs, rds inland. Fl 8–9. **Do not confuse** E with **Stone Parsley** (p 334), which has similar lvs but with a petrol smell when crushed.

F **Scots Lovage**, *Ligusticum scoticum*, **stout-stemmed** celery-scented per, 20–90 cm; stems often **purplish**, lvs (**F**) twice trifoliate, **dark shiny green**; lfts **oval, toothed in upper half**, 3–5 cm long; lf-stalks **inflated**, sheathing stem. Fls greenish-white, 2 mm across, in dense umbels (**Fb**); bracts and bracteoles linear. Fr (**Fa**) 4 mm long, oblong, with calyx-teeth and sharp ridges. Br Isles: coasts of Scot, N Ire, o-lf; among rocks or shingle by sea. Fl 7.

G **Whorled Caraway**, *Carum verticillatum*, erect slender hairless per, 30–60 cm, with hollow, little-branched stem; erect once-pinnate lvs (**G**); lfts divided deeply into linear **bristle-like** green lobes, appearing as if **in whorls** up lf-stalks. Umbels 2–5 cm across, rays 8–12; bracts and bracteoles many, narrow; fls 1 mm across, white. Fr (**Ga**) 2 mm long, oval, with sharp ridges. Br Isles: S, SW Eng, Wales, CI r-lf; W Scot lc; Ire N, SW vl; in damp acid mds, gslds, Fl 7–8. **G.1** EN *Caraway, C. carvi*, a hairless bi herb, infl similar to G but lvs alt not whorled, 2–3 pinnate with ultimate lobes linear; umbels with <10 rays; frs 3–4 mm with distinctive caraway smell when crushed. Introd (Eur) formerly cultivated, possibly native, now appearing as a casual only; Br Isles, r except Shetland where it is f; on dunes, wa, gsld. Fl 6–7. **G.2** NT *Spignel*, *Meum athamanticum*, like G has **apparently-whorled** hair-like lobes to lfts of the three-four times pinnate lvs; but a stouter plant, with larger triangular lvs, with few or no bracts; frs **elliptical**, 6–10 mm long. Br Isles: N Wales, N Eng, Scot; ol-lf; N Ire vvr (introd); on gslds, rds in mts and hills. Fl 6–8.

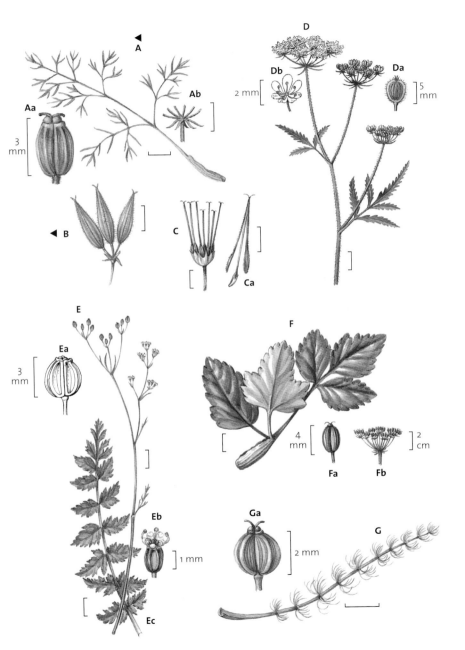

333

A **Stone Parsley**, *Sison amomum*, erect, **much-branched**, bushy, slender hairless bi, 50–100 cm, with unpleasant **petrol** smell; stem solid; basal lvs (**Ab**) once-pinnate, long-stalked, lfts 2–7 cm, **oval**, lobed and toothed; terminal lft **trifoliate**, lfts v narrow. Umbels both terminal and in lf-axils, 1–4 cm across, **few-rayed, few-fld**; bracts and bracteoles 2–4, bristle-like. Fls white, 1 mm across, petals notched; fr (**Aa**) globular, 3 mm long, ridges thin. S and E Eng, f-lc; N Wales r; **abs rest of Br Isles**; on gslds, rds, hbs on heavy clay soils. Fl 7–9.

B **Burnet-saxifrage**, *Pimpinella saxifraga*, erect slender **downy** per, 30–100 cm, with rough round stem. Basal lvs (**Bb**) usually once-pinnate, with **oval**, toothed lfts but may be **v variable** and vegetative lvs may appear pinnatifid (see illustrations in *BSBI Umbellifers Handbook* and *Plant Crib*); stem-lvs twice-pinnate, with **v narrow** lfts, no bracts or bracteoles. Fls 2 mm across, white, styles v short. Fr (**Ba**) 3 mm long, wide-oval. Br Isles, vc (but abs NW Scot); on dry mostly calc gslds. Fl 7–9.
B.1 ****Bladderseed**, *Physospermum cornubiense*, has lfts rather like those of B, but a **hairless woodland** herb to 75 cm; lvs **twice-trifoliate**, lfts deep-cut, dark green, with **wedge-shaped** bases, on long stalks; umbels of white fls have lanceolate bracts and bracteoles; frs **smooth**, 4–5 mm long, rounded, resembling tiny green **bladders**. S Eng: Bucks, Cornwall only, r in dry open wds. Fl 7–8.

C AWI **Greater Burnet-saxifrage**, *Pimpinella major*, erect hairless per, 50–120 cm tall, **not** v like B; stem (**Cb**) smooth, hollow, strongly ridged; **lvs all once-pinnate**, with **glossy**, dark green, **oval-elliptical** coarse-toothed lfts 2–8 cm long, lf-stalks sheathing. Umbels terminal, 3–6 cm across, flat, many-rayed; no bracts or bracteoles (**Cc**); fls 3 mm across, white (pinkish in N); styles long. Fr (**Ca**) 4 mm long, narrow-oval. Br Isles: Eng, lc in parts of Midlands, SE, abs in many parts; S Ire vlf; on rds, wd-borders, hbs, on heavier usually base-rich soils. Fl 6–7.

D **Garden Parsley**, *Petroselinum crispum*, stout erect hairless bi, 30–70 cm, has familiar parsley scent. Solid, ridged stem; lvs (**D**) thrice-pinnate, triangular in outline; lfts 1–2 cm, bases wedge-shaped, toothed at tips, often much **crisped**. Umbels flat-topped, 2–5 cm across, many-rayed; bracts and bracteoles with **white edges** and **sheathing bases**. Fls 2 mm across, yellowish. Fr (**Da**) 2.5 mm long, oval, finely-ridged. Br Isles o, especially in S (gdn escape); on wa, rocks, old walls. Fl 6–8.

E **Wild Celery**, *Apium graveolens*, stout erect hairless bi, 30–60 cm, has familiar scent of celery. Stems hollow, grooved; basal lvs (**Eb**) once-pinnate, with long sheathing stalks, lfts diamond-shaped to triangular, 0.5–3.0 cm long, dark shiny-green, lobed and toothed; upper lvs trifoliate, lfts narrower. Umbels both short-stalked (shorter than rays), terminal, and ± **stalkless in lf-axils**; rays unequal, bracts, bracteoles **abs**. Fls greenish-white, 0.5 mm across; fr (**Ea**) oval, ridged, 1.5 mm long. Br Isles: Eng, Wales, S Ire, lc near coasts, r inland; Scot vr (introd); in brackish mds, marshes, ditches, riversides by tidal water or near sea. Fl 6–8.

F **Ground-elder**, *Aegopodium podagraria*, erect hairless, carpet-forming per, 40–100 cm tall, with stout hollow grooved stem (**Fb**); lvs 10–20 cm, triangular in outline, long-stalked, once or twice trifoliate; lfts oval to elliptical, 2–7 cm long, toothed, ± stalkless, fresh light green; umbels terminal, 2–6 cm across, many-rayed, no bracts or bracteoles; fls white, 1 mm across. Fr (**Fa**) 4 mm long, egg-shaped. Br Isles, vc (except in mts) on rds, wa, hbs, wd borders, and a serious

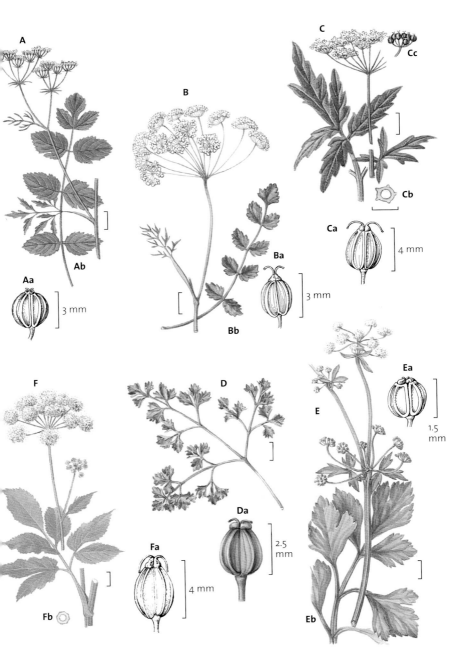

A

Aa
3 mm

Ab

B

Ba

Bb
3 mm

C

Cc

Cb

Ca
4 mm

D

Da
2.5 mm

E

Ea
1.5 mm

Eb

F

Fa
4 mm

Fb

garden weed because of far-creeping rhizomes. Fl 5–7. **F.1** NT *Masterwort*, *Peucedanum ostruthium*, tall per; twice-trifoliate lvs with broad, oval, toothed lfts as in F, but stems stouter with inflated sheathing lf-stalks, lfts downy below; umbels 5–10 cm across, bracteoles present; fls often pinkish; fr almost spherical. Introd (mts S Eur); Br Isles: N Eng, Scot, N Ire, r-vl; in damp places in mds, riversides, rds. Fl 7–8.

A **Wild Angelica**, *Angelica sylvestris*, ± hairless per to 200 cm tall, with v stout hollow **purplish** stems (**Ad**), downy only near base; lvs (**Af**) 30–60 cm, triangular in outline, 2–3 times pinnate; lfts 2–8 cm, oblong-oval, sharp-toothed; lf-stalks (**Ae**) broad, hollow, channelled on upperside, with broad, inflated, sheathing bases; upper lvs reduced to sheaths around umbels. Umbels terminal and in lf-axils, 3–15 cm across, many-rayed, no bracts, bracteoles few, narrow. Fls (**Ab**) 2 mm across, white or pink; no calyx; petals curved inwards; frs (**Aa**) 5 mm long, oval, v flattened (section **Ac**), with broad wings on edges. Br Isles, c in damp mds, fens, wds. Fl 6–9.

B **Garden Angelica**, *A. archangelica*, larger than A; stems and fls green, fls with calyx, fr (**B**) with corky wings. Introd (Denmark); GB, r and scattered but lc in London area; Ire vr; in wa, riversides etc. Fl 6–9.

C *Cowbane*, *Cicuta virosa*, erect hairless per, 50–130 cm tall, with ridged **hollow** stem (**Cb**); lvs (**C**) triangular in outline, to 30 cm long, with long hollow stalks, 2–3 times pinnate, with **linear-lanceolate**, strongly and **sharply-toothed** lfts 2–10 cm long. Umbels 7–13 cm across, convex, many-rayed, bracts none, bracteoles many and strap-shaped; fls white, 3 mm across, with oval calyx-teeth; fr (**Ca**) 2 mm long, globular, wider than long, with blunt ridges. Br Isles, v scattered and l (lc in Norfolk and central Ire); on fens, lake borders, ditches. Fl 7–8.

D **Hogweed**, *Heracleum sphondylium*, robust, roughly hairy bi to 200 cm; stems hollow (**Dc**), ridged, with downward-pointing hairs. Lvs 15–60 cm, once-pinnate, rough, grey-green, with clasping bases (**Dd**), and with oval- to oblong-lobed, pointed, coarse-toothed lfts to 15 cm long, lower ones stalked. Umbels 5–15 cm, stalked, many-rayed; bracts usually none, bracteoles bristle-like, down-turned. Fls (**Db**) white or pinkish, 5–10 mm across; petals notched, **unequal**. Fr (**Da**) 7–8 mm long, oval, whitish green, v flattened, smooth with club-shaped dark marks on sides. Br Isles, vc on rds, hbs, gslds, wds. Fl 6–9. **D.1 Giant Hogweed**, *H. mantegazzianum*, perhaps the largest herb wild in W Eur; like D, but to 5 m tall; stem red-spotted, to 10 cm across; lvs to 1 m long, with teeth sharper than in D. Umbels to 50 cm across; petals to 12 mm long. Introd (Caucasus); Br Isles, o-la on wa, riversides. Fl 6–7. **Poisonous** – sap causes severe blisters due to skin becoming hyper-sensitive to sunlight.

E AWI **Sanicle**, *Sanicula europaea*, hairless per, 20–60 cm tall, with long-stalked basal lvs (**Eb**) palmately-lobed (rather like Anemone (p 109) lvs, but lvs v **shiny** below), to 6 cm wide below, upper lvs smaller. Umbels few-rayed, bracts 3–5 mm, pinnatifid; bracteoles undivided. Fls pink or white. Fr (**Ea**) 3 mm long, oval, covered with hooked bristles (animal-dispersed). Br Isles, vc in wds on richer soils, especially beech wds on lime. Fl 5–8.

F **Astrantia**, *Astrantia major*, has simple umbels and is superficially more like a Scabious (p 432) than an Umbellifer. Erect hairless per, 30–75 cm tall, with long-stalked palmately-lobed basal lvs, 6–17cm across, with coarsely toothed lobes,

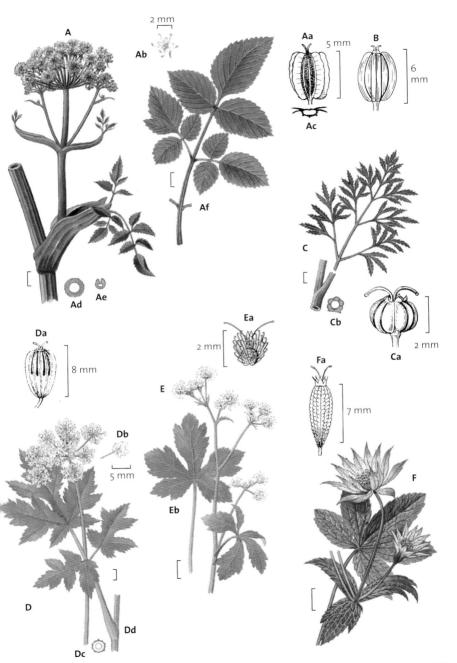

A

Ab 2 mm

Aa 5 mm B 6 mm

Ac

Af

C

Cb

Ca 2 mm

Da 8 mm

Db
5 mm

Ea 2 mm

E

Eb

Fa 7 mm

F

D

Ad Ae

Dd

Dc

veins stronger than in E. Umbels **simple**, dense, rounded, 1.5–2.0 cm across, with a ruff of long lanceolate bracts, green-purplish above. Fls white or pink, 2 mm across, fine-stalked; fr (**Fa**) 6–8 mm long, oval, with swollen wrinkled ridges. Introd (C Eur); GB, r in partially shaded places in wds, mds, rds. Fl 5–7.

A   Wild Parsnip, *Pastinaca sativa*, erect downy branched per, with hollow, furrowed, angled stems; rough once-pinnate lvs (**Ac**) with oval-pinnatifid and toothed segments. Umbels 5–10 cm across, 5–15-rayed; fls (**Ab**) yellow, 1.5 mm across; fr (**Aa**) oval, flattened, with narrow winged edges (the ancestor of the vegetable). Br Isles: S, central, E Eng c; Wales o-lf; Scot, Ire, vr introd; on rds, gsld, scrub, wa, especially on dry calc soils. Fl 6–8. **Poisonous** – sap causes severe blisters due to skin becoming hyper-sensitive to sunlight.

B   Alexanders, *Smyrnium olusatrum*, stout celery-scented bi with furrowed stem, solid until old, with branches often opp above; lvs dark green, shiny; basal lvs thrice trifoliate, to 30 cm long; lfts stalked, **oval** to diamond-shaped, lobed, toothed; upper lvs less divided; stalks with sheathing bases. Umbels terminal and in lf-axils, ± globular, many-rayed; few bracts and bracteoles. Fls 1–5 mm across, **yellow-green**; young fr (**Bb**), mature fr (**Ba**) 8–10 mm long, **oval, angled, black**. Br Isles: Eng, Wales, Ire, f-lc near coasts, r-o inland; Scot r; on cliffs, wa near sea, hbs. Fl 4–6.

C   Pepper-saxifrage, *Silaum silaus*, slender branched erect per to 100 cm, with solid, ridged stems. Lower lvs triangular, 2–3-times pinnate, with linear, often pinnatifid, finely-toothed, pointed lfts 1–5 cm long; smelling peppery when crushed. Umbels 2–6 cm wide, on long stalks; 5–10 rays; usually no bracts, bracteoles linear; fls (**Cb**) 1.5 mm, **sulphur-yellow**; fr (**Ca**) oblong-oval, nearly smooth, shiny, brownish-purple. Br Isles: Eng f-lc; Scot SE only, Wales r; **Ire abs**; on mds, gslds, rds on heavy clay soils. Fl 6–8.

D   Rock Samphire, *Crithmum maritimum*, **fleshy**, spreading per to 30 cm tall; stems solid, ridged; lvs triangular, short-stalked, once-twice-trifoliate, with fleshy rounded segments; lf-stalks sheathing. Umbels 3–6 cm wide, many-rayed, with many narrow bracts and bracteoles. Fls 2 mm across, yellow-green. Fr (**Da**) 6 mm long, corky, purplish, oval, round, thick-ridged. Br Isles: coasts of GB round clockwise from Suffolk to Hebrides; SW of GB c, to N r; Ire lc; on sea cliffs, shingle. Fl 6–8.

E   Fennel, *Foeniculum vulgare*, stout erect hairless per with solid stem; lvs repeatedly pinnate into long **thread-like** waxy-green lfts, not all in one plane. Crushed lvs smell of **aniseed**. Umbels 4–8 cm wide, many-rayed, no bracts or bracteoles; fls 1–2 mm wide, **bright yellow**. Fr (**Ea**) **oval**, rounded, stout-ridged. Br Isles: Eng, Wales, c on coasts, f inland; S, E Ire r; on cliffs, wa near sea; gdn escape inland on wa, rds. Fl 7–10. **E.1   Dill**, *Anethum graveolens*, similar gdn escape; **key difference** from E is frs **flattened**, with **winged** ridges. Scattered, r in GB, IoM, N Ire on rds, wa. Fl 7–8.

F   ****Hog's Fennel**, *Peucedanum officinale*, robust, hairless, erect per to 150 cm tall; stems stout, ridged, solid; lvs 4–6-times trifoliate, with **flat**, narrow-linear, untoothed lfts 4–10 cm long (**not rounded** as in E). Umbels 8–20 cm wide, many-rayed, bracts few or none, bracteoles bristle-like; Fls (**Fb**) 2 mm across, **sulphur-yellow**. Frs (**Fa**) 5–10 mm long, **elliptical**, ridged. Eng: Kent, Essex only, r but la; on clayey banks near tidal water and sea. Fl 7–9.

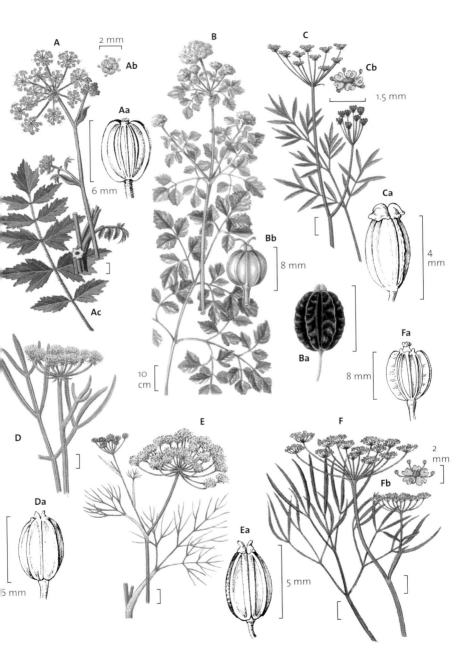

A

2 mm

Ab

Aa

6 mm

Ac

B

Bb

8 mm

Ba

10 cm

C

Cb

1.5 mm

Ca

4 mm

Fa

8 mm

D

Da

6 mm

E

Ea

5 mm

F

2 mm

Fb

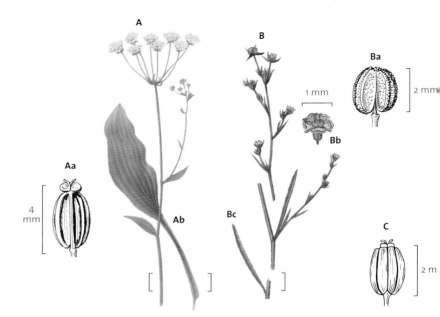

A <small>Sch 8</small> **Sickle-leaved Hare's-ear**, *Bupleurum falcatum*, erect hairless per, 50–100 cm, stem hollow; lvs **undivided**, untoothed (as in all genus *Bupleurum*), narrow-elliptical, curved; lower (**Ab**) stalked, upper clasping, 3–8 cm long. Umbels stalked, 2–6 cm wide, 5–11 rays, bracts and bracteoles lanceolate; fls golden-yellow, 1 mm wide; **fr** (**Aa**) oblong, ridged, **red-tipped**. Introd (S Eur); Eng: Essex, Yorks only vr; on calc gslds. Fl 7–9. **A.1** CR** **Thorow-wax**, *B. rotundifolium*, has oval lvs, the upper forming **rings** round stem, umbels simple or few-rayed; no bracts; bracteoles like upper lvs; fls yellow, 1.5 mm; fr smooth. S, E Eng vr, especially on calc soils. Fl 6–8.

B VU *****Slender Hare's-ear**, *B. tenuissimum*, more slender than A, has wiry stems and linear-lanceolate lvs (**Bc**) 1–5 cm. Umbels only to 5 mm across, in lf-axils, ± stalkless; bracts, bracteoles bristle-like. Fls (**Bb**) 1 mm, yellow; fr (**Ba**), 2 mm long, globular, **black, rough**. S, E Eng from Severn to Yorks, o-r; on grassy banks, saltmarsh edges usually near sea. Fl 7–9.

C <small>Sch 8</small> VU ** **Small Hare's-ear**, *B. baldense*, is tiny (2–10 cm), with sharp-pointed spoon-shaped lvs 1.0–1.5 cm; umbels 5–10 mm across, with few short rays, few fls, few bristle-like bracts. Oval, spine-tipped bracteoles hide yellow 2-mm fls; fr (**C**) smooth, 2–3 mm. Eng: S Devon, Sussex, CI vr; on dunes, cliff-tops. Fl 6–7.

Water-dropworts (*Oenanthe*) are hairless Umbellifers usually found in water or moist places; stems often ± swollen, ridged; lvs 2–4 times pinnate, of triangular outline, with lfts either linear, or wedge-shaped at bases, with sharp teeth above. Lf-stalks sheathing stem. Umbels many-rayed, mostly long-stalked, bracts and bracteoles many, v narrow. Fls white, 2 mm across, petals unequal. Frs strongly angled, oval, ridged, calyx-teeth prominent. **All are poisonous; A is deadly**.

KEY TO WATER-DROPWORTS (*OENANTHE*)

1 Umbels in lf-axils – some umbels with infl-stalks shorter than fl-stalks rays **2**
   Umbels not in lf-axils, terminal – all umbels with infl-stalks longer than rays **3**

2 Lvs often all submerged, twice-pinnate, lfts with wedge-shaped bases, cut into fine
   narrow segments – aerial lvs (if present) once-twice-pinnate, with oval, blunt, toothed lfts
   – fr 5–6.5 mm, oblong, 3 times length of styles ....*Oenanthe fluviatilis* (p 342 **C**)
   Submerged lower lvs 3-times pinnate, with hairlike segments – upper, aerial lvs with oval,
   fine-toothed lfts 5 mm long – fr 3.5–4.5 mm, oval, twice length of styles ....
   ....*O. aquatica* (p 342 **B**)

3 Upper lvs 3–4 times pinnate, lfts 2 x 1 cm, with wedge-shaped bases and coarse teeth at
   ends ....*O. crocata* (p 341 **A**)
   Upper lvs with narrow linear lfts **4**

4 Stems slender but hollow, inflated between lf-junctions – lf-stalks hollow, inflated, longer
   than pinnate blades – partial umbels form dense balls of frs after flowering............................
   ....*O. fistulosa* (p 342 **D**)
   Stems solid and ridged or hollow, not inflated between lf junctions – lf-stalks solid, not
   inflated, shorter than pinnate blades – partial umbels either open, or dense and flat in fr**5**

5 Basal lvs (best seen Apr-June, then withering) with lfts oval, bases wedge-shaped, with
   several pointed teeth – umbels dense in fr, flat-topped, rays stout (0.5–1.0 mm) – usually
   on acid soils in dry or damp mds, gslds...*O. pimpinelloides* (p 342 **E**)
   Basal lvs with lfts narrow, spoon-shaped or elliptical, blunt – umbels open in fr, convex,
   rays slender (0.25 mm) – in fens, marshes near sea ....*O. lachenalii* (p 342 **G**)
   Basal lvs with lfts narrow, linear-lanceolate, pointed, like those of upper lvs – umbels
   rather open in fr, flat-topped, rays v stout (1–2 mm) – on inland calc mds...................................
   ....*O. silaifolia* (p 342 **F**)

E   F   G

WATER-DROPWORT BASAL LEAFLETS

▼ **A Hemlock Water-dropwort**, *Oenanthe crocata*, robust erect hairless per to 150
cm; stems hollow, cylindrical, grooved, lvs triangular, 3–4 times pinnate; lfts oval,
bases tapered to stalk, with several deep teeth above; lf-stalks sheathing stem.
Umbels stalked, terminal, 5–10 cm across, many-rayed, bracts and bracteoles
many, linear, deciduous. Fls (**Ab**) white, 2 mm wide, petals unequal. Fr (**Aa**) 4–6
mm long, cylindrical, styles 2 mm, erect. Br Isles: S, W of GB vc; E, central Eng, N
Scot r – abs; Ire f-lc; in marshes, wet wds, ditches. Fl 6–7. **Poisonous**.

B Fine-leaved Water-dropwort, *Oenanthe aquatica*, rather bushy per, 30–100 cm tall, with runners; stem (**Bb**) **v swollen** in lower part, 3–4 cm thick, hollow, v finely-ridged, with **transverse** joints. Lvs 3 times pinnate, lfts hairlike on submerged lvs, lanceolate, finely-toothed, on aerial lvs (**Bc**). Umbels **short-stalked**, terminal and in lf-axils; 4–10 rays; bracts 0–1, bracteoles bristle-like. Fls white, 2 mm wide, petals ± equal. Fr (**Ba**) 3–4 mm long, oval, twice length of styles. Br Isles: Eng, f-lc in E, SE Midlands, r elsewhere in GB; Ire o; in ponds, ditches, sometimes in flowing water. Fl 6–9.

C River Water-dropwort, *O. fluviatilis*, similar to B, but often lacks infls; submerged lvs (**C**) twice-pinnate, with narrow linear lfts; lvs on aerial fl-shoots once- or twice-pinnate, lfts broad oval, blunt-toothed, 1–2 cm long. Fls as in B; fr (**Ca**) 5–6 mm, oblong, 3 times length of styles. S, E Eng vlc; Ire r; usually in fast-flowing water but also canals. Fl 6–9.

D VU **Tubular Water-dropwort**, *O. fistulosa*, erect slender hairless per, 30–60 cm, with slender but **hollow** stems, **inflated** between lf-junctions. Lvs (**Db**) once- (or twice-) pinnate; lfts of lower lvs oval-lobed, of upper lvs linear, untoothed. Lf-stalks longer than pinnate blades. Umbels few-rayed, rays stout, bracts none; partial umbels 1 cm, forming **dense balls** of frs after flowering; bracteoles many, fls white, 3 mm wide; fr (**Da**) 3–4 mm long, angled; styles spreading. Br Isles: Eng f-lc, r-o elsewhere; on marshes, pond edges. Fl 7–9.

E Corky-fruited Water-dropwort, *O. pimpinelloides*, erect hairless per, 30–80 cm; stems usually solid, cylindrical, ridged; basal lvs (**Ec**) twice-pinnate, with oval, deeply-toothed lfts as in A, but half size. Upper stems-lvs once-pinnate, lfts long, linear. Umbels (**E**) terminal, 2–6 cm wide, compact, **dense, flat-topped** with stout rays; **bracts**, bracteoles bristle-like; fls (**Eb**) 3–4 mm wide. Frs (**Ea**) 3 mm long, ribbed, **cylindrical**, with **swollen** corky bases; styles erect. Br Isles: Eng, lc S of Thames–Severn, vvr-**abs elsewhere**; on dry or damp mds, gslds on clay, often near sea. Fl 6–8.

F NT ***Narrow-leaved Water-dropwort**, *O. silaifolia*, v close to E, but with basal and stem-lvs twice-pinnate, all (**F**) with narrow-linear, pointed lfts; umbels flat-topped but ± open (not dense-topped) in fr; bracts usually abs; frs (**Fa**) cylindrical, corky-based, rounded at tops, narrowed suddenly at junction with stalks. SE, central Eng only, r in mds by inland rivers. Fl 6.

G Parsley Water-dropwort, *O. lachenalii*, is close to both E and F: differs from both in that the twice-pinnate basal lvs (**Gc**) have narrow, but **bluntly elliptical** or **obovate** lfts without teeth; differs from E in umbels (**Gb**) **open, convex**, especially in fr; from F in having **bracts, oval** (not corky-based) frs (**Ga**), and **slender** umbel-rays. Upper lfts usually narrower than shown in illustration. Br Isles: all round coast c (but **NE Scot abs**), vlc inland; in brackish mds, coastal marshes, inland fens. Fl 6–9.

▼ **A** EN BAP ***Greater Water-parsnip**, *Sium latifolium*, robust, erect, hairless per; stem thick, hollow, ridged, to 2 m tall; lvs to 30 cm long, **long-stalked**, once-pinnate, with 4–6 pairs **oval** to oblong, toothed lfts, 2–15 cm long, bright green; lf-stalks (**Ac**) hollow, sheathing stem. Umbels 6–10 cm wide, **terminal, flat-topped, long-stalked**, with many rays, 2–5 cm; bracts, bracteoles **large, lfy**. Fls (**Ab**) white, **4 mm**

A

Ab

2 mm

B

Bc

Da

4 mm

D

Db

Aa

5 mm

Ca

nm

Ba

Bb

4 mm

C

G

Gb

E

Ea

Eb

3 mm

mm

F

Ga

2 mm

Ec

Fa

3 mm

Gc

wide; sepals present. Fr (**Aa**) oval, **3 mm** long, ridges low. **Poisonous**. Br Isles: N to S Scot r, but lc in E Anglia; Ire vr; in fens, marsh ditches. Fl 7–8.

B   Lesser Water-parsnip, *Berula erecta*, smaller version of A, 30–100 cm tall, but less erect and has **runners**; lvs, once-pinnate, have lfts (**B**) more **deeply toothed**, oval, 2–5 cm long, in 7–10 pairs, dull **bluish**-green; stem-lvs much smaller than in A. Always has a **ring-mark** (sometimes also a pair of tiny lfts) on the stem just above the lf-axil (although A may also have this). Umbels only 3–6 cm wide, **short-stalked**, in **lf-axils** and terminal (**Bb**), with fewer, **shorter** (1–3 cm) rays. Bracts (**Bc**), bracteoles as in A. Fr (**Ba**) almost **spherical**, 2 mm long. Br Isles, f (lc to S and E); Scot, Ire o-lf; in fens, ditches, ponds. Fl 7–9. **Poisonous**.

C   Fool's-water-cress, *Apium nodiflorum*, rather close to B, but **never** has a **ring-mark** on the stem above the lf-axil; ± creeping, with ascending fl-stems to 80 cm. Lvs (**Cb**) once-pinnate, with 4–6 **opp** pairs of **shallowly, bluntly-toothed**, oval, **bright** green, **shining** lfts (**Cc**). Umbels ± stalkless in lf-axils; no bracts; but bracteoles present; fr (**Ca**) **broadly oval**, 1.5 mm long. **Do not confuse** with Water-cress (p 210 E), a Cabbage family plant which does not have a sheathing, inflated base to lf-stalks like most Umbellifers and has **alt** or only near-opp lvs. Br Isles: Eng, Wales, Ire, c in lowlands; Scot r (except W islands); in fens, ditches, ponds, springs. Fl 7–8.   **C.1**   EU Sch 8 VU BAP ****Creeping Marshwort**, *A. repens*, close to small plants of C but has procumbent stems **rooting** at ± all nodes; frs **‹1 mm** with **narrow ridges.** S Eng vr, Oxon & Essex, formerly scattered in E Eng & Scot; on disturbed gd in damp mds generally with seasonal flooding, ponds, ditches. Fl 7.

D   Lesser Marshwort, *A. inundatum*, slender, ± prostrate, hairless per, 10–50 cm long. Lvs pinnate and of 2 types: **submerged lvs** with deeply **pinnatifid, hair-like** lfts; **aerial lvs** with linear to narrow wedge-shaped **three-lobed** lfts. Umbels 1–2 cm wide, in lf-axils, on stalks 1–3 cm long; bracts none; bracteoles 3–6, lanceolate, blunt. Fls white, 1 mm wide; fr (**Da**) 2 mm, elliptical, low-ridged. Br Isles, o-lc but declining; on mud or in water of shallow ponds. Fl 6–8.

E   Marsh Pennywort, *Hydrocotyle vulgaris*, with no superficial resemblance to Umbellifers, is often put in separate family, *Hydrocotylaceae*. Hairless prostrate creeping and rooting per, with **circular**, shallow-toothed lvs (rather like those of Wall Pennywort, p 240 I), 1–5 cm wide, held erect like parasols, on stalks 1–15 cm, attached to **lf-centres**. Infls (hard to spot and often abs) 2–6 cm tall, with simple **whorls** of 2–5 stalkless pinkish-green fls 1 mm wide, with triangular bracts. Fr (**Ea**) ± circular, flattened, ridged, 2 mm wide. Fls have Umbellifer structure (5 petals, 5 stamens, inferior ovary of 2 carpels). Br Isles, c in bogs, fens etc. Fl 6–8. **E.1** Floating **Pennywort**, *H. ranunculoides*, similar to E but often free-floating with circular lvs divided ± $^1/_2$ way to base, that are **not** parasol-like, with the stem joining the **lf-edge**. Introd (N America), increasing and invasive aquarist throw-out; S Eng o-vla, Scot, Ire abs; in still or slow-moving canals, rivers, ditches. Fl 7–8 (rarely).

F   Sea-holly, *Eryngium maritimum*, branched per, 30–60 cm; lvs **waxy-grey** with thick margins, edged with long **spines**; basal lvs long-stalked, palmate-veined, rounded; stem-lvs stalkless. Umbels **simple** (more like teasel heads, p 430, than those of Umbellifers), oval, 2.5–4.0 cm long, with close-packed fls (8 mm wide) of bright clear blue; bracts of umbels whorled, **oblong**, spiny; bracts of each fl spine-

4 mm

◀ A

Ab

Bb

30 cm

B

Bc

Aa

Ac

Ba

2 mm

Ca

2 mm

C

Cb

D

Da

2 mm

Cc

2 mm

Ea

E

Ga

4 mm

F

Fa

5 mm

G ▶

like, three-lobed. Fr (**Fa**) to 5 mm long, oblong, covered with hooked bristles. Coasts of Br Isles (except N, E Scot), f-la on sandy or shingly beaches, dunes. Fl 7–9.

▲ **G**   Sch 8 CR ****Field Eryngo**, *Eryngium campestre* has **pinnate**, stalked basal lvs to 20 cm long, with strap-shaped, spiny, grey-green lfts; stem-lvs shorter, unstalked, with clasping bases. Umbels **1–2 cm long**, oval, with **narrow** spiny bracts, to 3 cm. Fls **mauve-white**, 2–3 mm wide, with individual bracts much longer than fls, spine-like. Fr (**Ga**) smaller than in F, 3 mm long. Probably introd; S Eng, CI vr; dry gslds, banks, dunes, rds. Fl 7–8.

A   NT ****Moon Carrot**, *Seseli libanotis*, per, is stout-stemmed, downy, and robust (to 1 m tall), like C below; stem **solid, downy, strong-ridged**; remains of old lvs usually persist at stem-base as tuft of fibres; lvs (**A**) downy, twice-pinnate, 10–20 cm long; lfts **stalkless**, oval, pinnatifid, with oblong, pointed lobes, **not all in one plane**; some lvs in **opp pairs**. Umbels 5–10 cm wide, terminal, dense, convex; rays many, **downy**; bracts and bracteoles many, narrow. Fls (**Ab**) 1–2 mm, white, with **long calyx-teeth**; fr (**Aa**) **oval, downy**, rounded, with strong ridges. Eng: Sussex, Beds to Cambs only, vr; on calc gsld, rocks, scrub. Fl 7–8.

B   ****Honewort**, *Trinia glauca*, low, bushy, erect, **waxy** grey-green per, 3–20 cm; stem grooved, branched from base. Lvs (**Bb**) 3 times pinnate, lfts narrow-linear, waxy-grey. Dioecious; umbels of male plants 1 cm wide, flat-topped, with 4–7 rays, equal, 5 mm long; umbels of female plants (**B**) to 3 cm wide, with longer (to 4 cm) unequal rays. Bracts none, bracteoles 2–3. Fls tiny, white; fr (**Ba**) **oval**, 2 mm long, **hairless**, with broad raised ridges. Eng: S Devon, Bristol, N Somerset, r and vl; among limestone rocks in sun. Fl 5–6.

C   **Wild Carrot**, *Daucus carota* ssp *carota*, erect, branched, **roughly hairy** bi, to 100 cm, with solid ridged stem; lvs (**Cc**) 3 times pinnate, with oval-lanceolate, pinnatifid lfts 5–8 mm long. Umbels 3–7 cm wide, rays many and bristly, bracts **many, large, lfy, pinnatifid**, forming a 'ruff' to umbel below. Fls 2–3 mm, white, but central fl usually **dark red**. Fr (**Ca**) 2.5–4.0 mm long, oval, downy, with **4 stout, spiny ridges** on each carpel; umbels concave in fr. Br Isles, c except in mt and moorland areas; on gslds, cliffs, rds, hbs, especially on calc soils and near sea. Fl 6–8. **C.1** ***Sea Carrot**, *D. carota* ssp *gummifer*, has fleshier, blunter lfts, **convex** or flat umbels in fr, spines on fr-ridges **webbed** together. Coasts S Eng, o-lf. Fl 6–7. The cultivated carrot, with a fleshy tap-root, is ssp *sativus*.

D   VU ***Milk-parsley**, *Peucedanum palustre*, ± hairless bi, to 150 cm; stems (**Dd**) ridged, hollow; lvs triangular in outline, 2–4 times pinnate, with lfts pinnatifid into oblong-lanceolate lobes **without** spiny tips (**Db**), 5 mm x 2 mm. Umbels 5–10 cm wide, many-rayed, rays **downy**; **several** bracts and bracteoles, lanceolate. Fls (**Dc**) 2 mm, greenish-white; fr (**Da**) 4–5 mm long, shortly oval, hairless, much flattened, with broad wings on edges and narrow ridges on faces. Eng: E, S, r (but la in E Anglia); **abs rest of Br Isles**; in calc fens. Fl 7–9.

E   Sch 8 VU ****Cambridge Milk-parsley**, *Selinum carvifolia*, v close to D, is tall, ± hairless per, with stems more ridged than in D. Lvs (**E**) 2–4 times pinnate; lfts have linear-lanceolate, v shortly **spine-tipped** lobes. Umbels of white fls (**Eb**) many-rayed, **without bracts** but with many linear bracteoles; fr (**Ea**) 3–4 mm long, long-oval; 2 side wings, 3 back wings on each carpel, **all thick**, prominent. **Cambs only** vr; in fens. Fl 7–10.

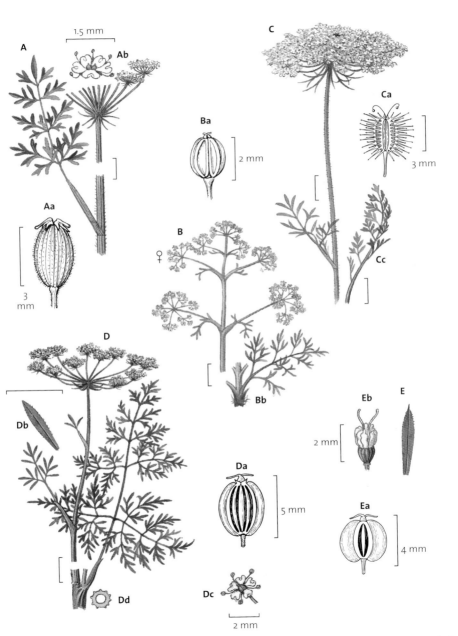

A

1.5 mm

Ab

Aa

3 mm

Ba

2 mm

B

♀

Bb

C

Ca

3 mm

Cc

D

Db

Da

5 mm

Dc

Dd

2 mm

Eb

E

2 mm

Ea

4 mm

GENTIAN FAMILY   *Gentianaceae*

Herbs ± hairless, with opp, untoothed, ± stalkless lvs without stipules; infls forked cymes with a fl in each fork (dichasia). Fls regular, with 4, 5 or more lobes, the corolla-lobes twisted round one another in bud. Stamens, on corolla-tube, alt with and same number as lobes. Ovary **superior** (similar-looking *Campanulaceae* have ovary inferior); 2 carpels, forming a capsule in fr.

KEY TO GENTIAN FAMILY

1 Fls yellow **2**
Fls pink, blue, purple or white...**3**

2 Fls 6–8-lobed, 10–15 mm across – oval lvs joined round stem in a ring..........................
Blackstonia perfoliata (p 350 **J**)
Fls 4–5-lobed, 3–5 mm across – linear lvs not joined round stem ......*Cicendia filiformis* (p 350 **I**)

3 Corolla pink – style on top of ovary distinct and threadlike, withering in fr **4**
Corolla blue, purple or white – style abs, or ovary gradually tapering into short stout style, with 2 stigmas remaining on top of fr ....*Gentiana* or *Gentianella* (key below)

4 Four-lobed calyx and corolla – anthers not twisting after flg – Guernsey only, r ....
Exaculum pusillum (p 350 **I.1**)
Five-lobed (rarely 4) calyx and corolla – anthers twisting after flg – widespread ....
Centaurium (p 350)

KEY TO GENTIANS (*GENTIANA* & *GENTIANELLA*)

1 Corolla-lobes fringed with white hairs along edges – calyx-tube and corolla-tube four-lobed, corolla to 40 mm across, lobes spreading, oval ....*Gentianella ciliata* (p 349 **B.2**)
Corolla-lobes not fringed with hairs **2**

2 Corolla with tiny lobes between large lobes, blue, throat without fringe of white hairs ......
(*Gentiana*) **3**
Corolla without tiny intermediate lobes, purple-mauve, throat with fringe of white hairs ..
(*Gentianella*) **5**

3 Plant with many basal rosettes of oval lvs, forming tufts or cushions..
Gentiana verna (p 349 **B**)
Plant without basal rosettes of lvs, lvs on stem only..**4**

4 Plant tall (10–40 cm) – lvs narrow-linear, blunt, 1.5–4.0 cm long – corolla-tube bell-shaped, 2.5–4.0 cm long, blue with green stripes down outside ....*G. pneumonanthe* (p 349 **A**)
Plant short (3–15 cm) – lvs oval, pointed above, 2–5 mm long – corolla-tube cylindrical, 1.0–1.5 cm long, lobes spreading, no green stripes – Scot mts – r ....*G. nivalis* (p 349 **B.1**)

5 Calyx four-lobed, 2 outer lobes much broader than 2 inner, overlapping them ....
Gentianella campestris (p 350 **E**)
Calyx four- or five-lobed, lobes ± equal, all narrow..**6**

6 Corolla 25–35 mm long, twice length of calyx or more – lower lvs oval ..*G. germanica* (p 350 **D**)
Corolla 9–22 mm long, less than twice length of calyx – lower lvs lanceolate **7**

7 Stems with 5–10(-14) pairs of lvs above the basal rosette – distance between two uppermost lf-pairs (ie. the terminal internode) typically forms 20% or less of the height of the plant .... *G. amarella* (p 349 **C**)
Stems with 0–3(-4) pairs of lvs above the basal rosette – distance between two uppermost lf-pairs (when present!) 40% or more of the height of the plant.......................... **8**

8 Sepal lobes nearly equal and ± adpressed to corolla – stems with 1–3(-4) pairs of lvs above the basal rosette – fl 4-6 ....*G. anglica* (p 349 **C.1**)
Sepal lobes often markedly unequal and often spreading away from corolla – stems with 0–2(-4) pairs of lvs above the basal rosette – fl 6-11 ....*G. uliginosa* (p 349 **C.2**)

▼ **A** **Marsh Gentian*, *Gentiana pneumonanthe*, erect hairless per, 10–40 cm tall, with blunt linear lvs 1.5–4 cm x 3–5 mm, **not** in rosettes at base. Fls 1–7 in a dense terminal group; corolla-tube 2.5–4.0 cm long, bright blue, with 5 green stripes on the outside, narrowing below, its oval lobes ascending (**not** spreading at 90°). Br Isles: Eng, S (Dorset–Sussex), and E (Norfolk–Yorks) vlf, Lancs – N Wales r; **Scot, Ire abs**; declining in wet hths, dune slacks. Fl 8–9

▼ **B**   Sch 8 ***Spring Gentian*, *G. verna*, hairless per with dense **basal rosettes** of oval-elliptical lvs 5–15 mm long. Stems 2–6 cm tall, with few pairs of smaller lvs. Fls solitary, terminal, on erect stems; corolla-tube 1.5–2.5 cm, deep rich blue, cylindrical, its oval lobes **spreading** out at 90° in a star shape, 1.5–2.0 cm wide. Eng, Teesdale vlf; Ire, Clare, Galway, la; on calc gslds, limestone crevices. Fl 4–6.
B.1   Sch 8 NT ***Alpine Gentian*, *G. nivalis*, slender ann; 3–15 cm-tall, ± branched stems; fls only 8 mm across spreading lobes. Br Isles: vr on ledges of mts in Scot. Fl 7–9.   **B.2**   Sch 8 CR ***Fringed Gentian*, *Gentianella ciliata* (*Gentiana ciliata*), is taller than B, without rosettes of lvs at base; blue corolla has 4 spreading lobes to 4 cm across, that bear fringes of white bristles on them (not in throat of fl). Br Isles: vr, **Bucks** only; on dry calc gslds. Fl 8–10.

▼ **C**   **Autumn Gentian**, *Gentianella amarella* ssp *amarella*, erect branched bi, 5–30 cm tall; stem-lvs pointed, oval-lanceolate, 1–2 cm long; stems with **5–10**(-14) **pairs** of lvs above the basal rosette, **distance between** two uppermost lf-pairs typically forms **20% or less** of the height of the plant. Corolla-lobes purple, 4- or 5-lobed, bell-shaped, 14–18(-20) mm long, with white fringe of hairs in throat and lobes triangular, ± spreading; tube **less than twice** length of calyx. Br Isles: Eng, Wales f; Scot, Ire lf; on calc gslds, dunes. Fl 7–11. End NT **ssp septentrionalis* has a creamy-white corolla, tinged purplish-red on outside; Br Isles: Yorks N to Scot only, r on dunes, calc gsld. Fl 7–8. End ssp *hibernica* has a corolla usually over 19 mm long or more; in **Ire only**.   **C.1**   EU Sch 8 End BAP **Early Gentian*, *G. anglica*, like C but only 4–15 cm tall; stems with 1–3(-4) pairs of lvs above the basal rosette and **distance between** two uppermost lf-pairs (when present!) **40% or more** of the height of the plant; sepal lobes nearly **equal** and ± **adpressed** to corolla. S Eng, S Wales r in calc gsld, dunes. **Fl 4–6**.   **C.2**   Sch 8 VU BAP ***Dune Gentian*, *G. uliginosa*, is v close to C.1, but stems with **0–2**(-4) **pairs of lvs** above the basal rosette; sepal lobes often markedly **unequal** and often **spreading** away from corolla. S Wales, N Devon, W Scot vr; in dune hollows. Fl 6–11.

D *__Chiltern Gentian__, *Gentianella germanica*, is close to C, but lvs broader-based (10 mm across or more) than in C; corolla-tube 25–35 mm long, twice (or more) length of calyx (**D**) and transversely wrinkled on outside; fls about 20 mm across lobes. Eng, N Hants to Hertford, vlf; on calc gslds. Fl 8–9.

E VU **Field Gentian**, *G. campestris*, differs from C and D mainly in its calyx (**E**), in which two **wide-oval** acute sepals overlap the two **narrow**-lanceolate inner ones. Corolla as in C, but blue-lilac. Br Isles: N Eng, Scot f but declining; S Eng, Wales r; Ire f; on neutral to acid gslds, dunes. Fl 7–10.

F **Common Centaury**, *Centaurium erythraea*, hairless erect ann, 10–40 cm, with basal rosette of obovate lvs over 5 mm wide, and 1 or more stems; **lvs oval-elliptical**, never parallel-sided. Fls ± unstalked, in ± dense forking cymes on top of stem, not umbel-like; **calyx <3/4 as long as** corolla-tube, corolla-lobes flat, oval, spreading, pink; fl 10–12 mm across; **fl-stalks 0-1 mm**. Br Isles, vc (except NE Scot r) c; in dry gslds, dunes, wd clearings. Fl 6–10. **F.1** EN **__Perennial Centaury__**, *C. scilloides* (*C. portense*), creeping **per**, with round, stalked lvs on creeping stems; fl-stems erect with oblong lvs; fls 16–18 mm across, **key difference** from all other British *Centaurium* spp is corolla-lobes larger, **>7 mm** long. Pembroke, Cornwall, vr (vr introd elsewhere); sea cliffs. Fl 7–8.

G WO (NI) *__Seaside Centaury__, *C. littorale*, close to F in erect stems, but basal lvs only to 5 mm wide, linear to spoon-shaped; stem-lvs **parallel-sided**, strap-shaped, blunt; fls in **dense** terminal **umbel-like** clusters; **calyx >3/4 as long as** corolla-tube; corolla 12–14 mm across, deeper pink and more concave than in F. Br Isles: N, W coasts of GB lf; N Ire vr; on dunes. Fl 7–8.

H FPO (Eire) **Lesser Centaury**, *C. pulchellum*, smaller, slenderer ann than F and G, 3–15 cm, with v open forked branching and **2-4 internodes**; fls in the forks and on shoot tips, no basal lf-rosette; lvs all elliptical; **fl-stalks 1-4 mm**. Corolla (**Ha**) dark pink, 4–8 mm across, 5- (or 4-) lobed (sometimes reduced to a tiny unbranched one-fld plant). Eng, Wales, o, lc in S; Ire vr; on bare, often damp, grassy open gd on calc and acid soils. Fl 6–9. **H.1** Sch 8 VU **__Slender Centaury__**, *C. tenuiflorum*, looks like a taller H (10–35 cm tall), but with **5-9 internodes**, strictly erect branches and dense infls and fl usually white. S coast Eng vr; on moist bare gd near sea. Fl 7–9.

I VU *__Yellow Centaury__, *Cicendia filiformis*, slender ann, 2–12 cm tall, with simple or few **erect** rather fleshy pinkish-green branches; wide-spaced pairs of linear lvs 2–6 mm long; fls (**Ia**) yellow or often tinged pink, long-stalked, terminal, only open in sun; four-lobed corolla; calyx-tube with 4 short triangular lobes. SW Eng, Wales, r but vlf; SW Ire vl; on damp open sandy gd on hths and in wds. Fl 8–9. **I.1** **Guernsey Centaury**, *Exaculum pusillum*, like I, but often smaller (although height variable in both this and I) with **spreading** branches; corollas **pink**, four-lobed, 2–3 mm across; calyx deeply four-lobed. **Guernsey only**, r-lf; on moist sandy gd nr coast. Fl 7–9.

J **Yellow-wort**, *Blackstonia perfoliata*, erect waxy **glaucous** ann, 15–40 cm, with basal rosette of obovate lvs; stem-lvs oval-triangular, **joined in pairs to form rings** round stem. Fls in loose forking cymes; corolla 10–15 mm across, 6–8-lobed, **yellow**. Br Isles: Eng c; Wales vlc; Scot vr-abs; Ire lc; on calc gslds, dunes. Fl 6–10.

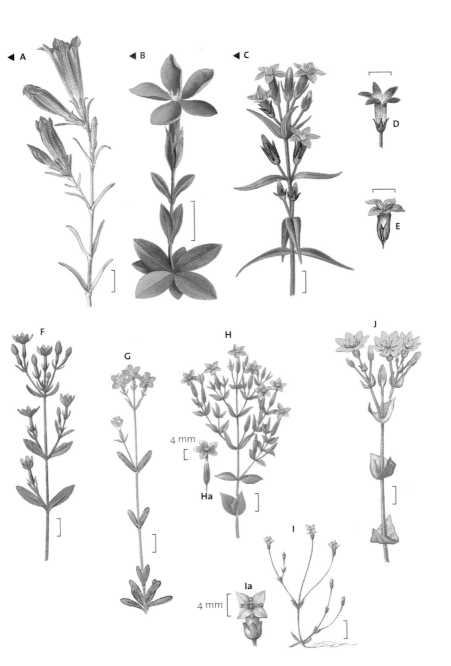

A

B

C

D

E

F

H

Ha

4 mm

G

J

I

Ia

4 mm

PERIWINKLE FAMILY

Apocynaceae

Creeping woody shrubs with opp evergreen oval untoothed lvs, and corollas with 5 spreading lobes, twisted in bud.

A Lesser Periwinkle, *Vinca minor*, ± prostrate evergreen shrub with **hairless**, elliptical, shiny dark green short-stalked lvs (**Aa**) in pairs. Calyx-teeth (**Ab**) lanceolate, **hairless**. Fls (**Ac**) usually solitary in lf-axils, with corolla of 5 almost flat blue-violet lobes, 25–30 mm across. Br Isles: GB c and probably native in S, rarer and planted to N; Ire o-f; in wds, hbs. Fl 3–5.

B Greater Periwinkle, *V. major*, v similar to A, but has broader, longer-stalked (1 cm), oval lvs (**Ba**); fls (**Bc**) 40–50 mm across; **key difference** from A in lf-margins with tiny **hairs** and calyx-teeth (**Bb**) narrow, **hairy**. Br Isles, c in S, f-lc in N, Ire; introd (Mediterranean) in wds, hbs. Fl 4–5.

NIGHTSHADE FAMILY

Solanaceae

Herbs and shrubs with alt lvs without stipules, often **poisonous**; fls regular, bell-shaped or wheel-shaped, with 5 sepals; corollas five-lobed, stamens projecting in a column in **Nightshades** and **Potato** (*Solanum*); 5 stamens; fr two- or sometimes four-celled, a berry or capsule.

C Duke of Argyll's Teaplant, *Lycium barbarum*, shrub with arching grey, often spiny, stems to 2.5 m; lanceolate grey-green alt lvs to 6 cm long; fls 1 to 3 in short shoots in lf-axils, brownish to rosy-purple; corollas funnel-shaped, with 5 lobes in a star 1 cm across; fr (**Ca**) a red oval berry. Introd (China); Br Isles: S half of GB f; Ire, Scot r; in hbs, wa etc. Fl 6–9.

D Thorn-apple, *Datura stramonium*, stout ann herb 50–100 cm tall, with alt oval coarsely-toothed lvs to 20 cm long on long stalks; white (rarely purple) trumpet-shaped fls, with long corolla-tubes, and 5 long-pointed corolla-lobes; corollas 6–8 cm across. Fr (**Da**) an oval **spiny** capsule 4–5 cm long. Introd (SE Eur); Eng, Wales o-f; Scot, Ire vr; previously cultivated for alkaloids to treat asthma; most c in hot summers in ar, wa. Fl 7–10.

E   vu **Henbane**, *Hyoscyamus niger*, tall ann or bi with sticky-hairy strong-smelling lvs and stems; lvs oval-oblong, toothed or not, upper clasping, 10–20 cm long, alt up stem. Fls in two rows on a forked cyme, calyx-tube hairy, swollen, with long rigid teeth; corolla funnel-shaped with 5 lobes, not all quite equal, dull yellow with purple veins and eye; fr (**Ea**) a large (1–2 cm long) capsule, opening at top. Eng, o-lf in S and E, r in rest of Br Isles; on wa, open grassy downs, sandy gd near sea. Fl 6–8. **Whole plant v poisonous**.

F Bittersweet (Woody Nightshade), *Solanum dulcamara*, downy woody per, scrambling over other plants to 2 m. Lvs alt, stalked, oval, often with two spreading lobes or lfts at base, tip pointed, to 8 cm long. Fls in loose cymes, 1 cm across; corolla with 5 pointed, arched-back, purple lobes, and a conical column of

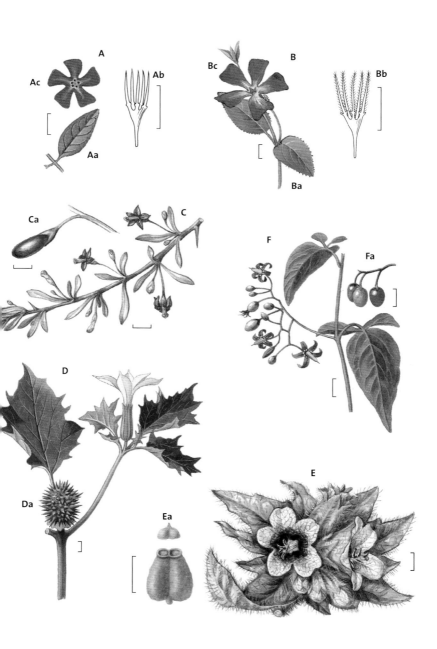

353

yellow stamens to 8 mm long. Fr (**Fa**) a **red oval** berry, to **1 cm** long. Br Isles: GB, N to Scot, NE Ire vc, N Scot, rest of Ire f; in wds, hbs, scrub, wa, beaches. Fl 6–9.

A   Black Nightshade, *Solanum nigrum*, ann, lower and more bushy than F (p 352) (to 50 cm), with erect or spreading stems; lvs oval, pointed, ± lobed, but **not** with lfts at base as in F; fls similar to F, 6–10 mm across, **white**. Fr (**Aa**) a **black round** berry, to 8 mm long. Br Isles: Eng, Wales c; Scot, Ire vr; on ar, wa. Fl 7–9.   **A.1 Potato**, *S. tuberosum*, with white fls, and tubers, and **A.2   Tomato**, *Lycopersicon esculentum*, with yellow fls, are related plants with similar fl-structure.

B   Deadly Nightshade, *Atropa belladonna*, tall (to 150 cm) stout, shrubby-looking, per herb; pointed oval lvs, to 20 cm long, narrowed into the stalks, alt or in unequal pairs. Fls 1 or 2 together, in lf-axils and in forks of branches; corollas bell-shaped with ± parallel sides, stalked, 25–30 mm long, drooping, dull brownish-purple or green, with 5 blunt lobes. Fr (**Ba**) a black glossy berry, 15–20 mm across, with the five-lobed calyx persisting below it. S and E Eng, native and o-lf, elsewhere in Br Isles r; in wds, scrub on calc soils, introd in wa near buildings. Fl 8–10. **Whole plant v poisonous**.

BINDWEED FAMILY   *Convolvulaceae*

Mostly creeping or climbing plants with alt lvs and no stipules; fls with 5 sepals, corolla trumpet-shaped, five-angled, 5 stamens inside tube of corolla; fr a two-celled capsule.

C   Field Bindweed, *Convolvulus arvensis*, creeping and climbing per herb, with ± hairless stems arising from stout fleshy undergd stems; lvs 2–5 cm long, oblong-arrow-shaped, stalked, alt; fls trumpet-shaped, white or pink, 30 mm across, five-angled at mouth; calyx five-lobed, **no epicalyx**. Br Isles, vc (except N Scot r); in wa, hbs, ar, rds, gsld near sea (a serious weed in gdns). Fl 6–9.

D   Hedge Bindweed, *Calystegia sepium*, of similar habit to C, but much larger, climbing to 3 m or more; lvs to 15 cm long and more rounded; fls white (rarely pink); an **epicalyx** of 2 bracteoles (**Da**), longer than calyx-lobes but **not (or scarcely)** overlapping each other, surrounds calyx. Bracteoles **10–18 mm** wide when flattened out. Br Isles, vc (except N Scot lf); on wa, hbs, scrub, wd borders, fens. Fl 7–9.

E   Large Bindweed, *C. silvatica*, similar to D, but **key difference** in the 2 epicalyx-bracteoles (**E**) **strongly inflated, overlapping** each other and ± concealing the sepals; bracteoles **18–45 mm** wide when flattened out; fl-stalks hairless. Introd (Mediterranean); Br Isles, vc (except N Scot, o-r) in wa, hds, etc. Fl 7–9. Hybridizes with D (*C. x lucana*) which is o-vlf but overlooked, has bracteoles intermediate between those of its parents: slightly overlapping with the sepal visible and may be highly fertile or sterile.   **E.1   Hairy Bindweed**, *C. pulchra*, close to E with large **overlapping** bracts but has pink or pink and white **striped fls** and **fl-stalks hairy** or at least slightly so. Introd, probably originally a gdn escape; Br Isles, f throughout near housing on wa, hds. Fl 7–9. **Do not confuse** with *C. sepium* spp

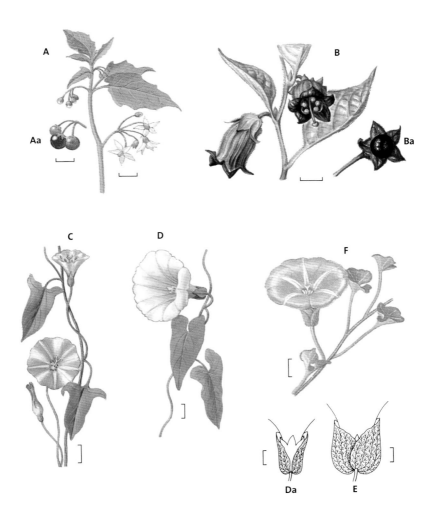

roseata which also has hairy fl-stalks and pink fls but **key difference** is the bracteoles **not** overlapping, like D.

F Sea Bindweed, *C. soldanella*, creeping hairless per with ± fleshy, **kidney-shaped** lvs; corolla-trumpets 25–40 mm across, **pink** with **white stripes**; epicalyx-bracts **shorter** than sepals, rounded. Br Isles: coasts of Eng, Wales, SW Scot, Ire f-lc; on sand-dunes. Fl 6–8.

DODDER FAMILY

Cuscutaceae

Ann to per parasitic herbs with v thin, rootless stems that twine over host vegetation and small, globose infls.

A vu **Dodder**, *Cuscuta epithymum*, parasitic climber without green chlorophyll; threadlike **red stems**; lvs reduced to small scales; fls (**Aa**) in dense clusters 6–10 mm across, bell-shaped, with projectıng stamens, pink five-lobed corollas 3–4 mm across, and 5 scales inside tubes. S Eng f-lc; rest of Br Isles r; **Scot abs,** declining throughout; parasitic on gorse, heather, clovers etc. Fl 7–9. **A.1** *****Greater Dodder**, *C. europaea*, larger, with fl-heads **10–15 mm** across; paler corollas **4–5 mm** across; stamens and styles **enclosed** in corolla-tubes, scales **smaller**, corolla-lobes **blunt**, not pointed. S Eng r; parasitic on nettles etc. Fl 8–9.

BOGBEAN FAMILY

Menyanthaceae

These are like members of the Gentian family, but are **water plants** with alt lvs.

B **Bogbean**, *Menyanthes trifoliata*, per with creeping stems, bearing alt, erect long-stalked, **trifoliate** lvs, with oval untoothed grey-green lfts 3–7 cm long; erect lfless fl-stems, 10–30 cm tall, bearing fl-racemes; corollas five-lobed, star-shaped, **pink outside, white within**, 15 mm across, conspicuously **fringed** with stout white hairs; fr (**Ba**) an oval capsule. Br Isles c (except central Eng o); in wet bogs, fens, ponds. Fl 5–7.

C *****Fringed Water-lily**, *Nymphoides peltata*, a floating aquatic, with lvs (**Ca**) like small Water-lily lvs (see p 99), **round to kidney-shaped,** purple below, 3–8 cm across, lower alt, upper in opp pairs; fls few together in lf-axils, long-stalked; corolla 3 cm across, with five **fringed yellow** lobes. Br Isles, o-lf but native only in E, C Eng; in ponds, rivers. Fl 6–9.

PHLOX FAMILY

Polemoniaceae

Herbs with 5 sepals, 5 equal-lobed corollas, 5 stamens and **three**-celled ovary. Fr a capsule.

D ******Jacob's-ladder**, *Polemonium caeruleum*, per herb 30–90 cm, with erect unbranched stem, bearing alt hairless **pinnate** lvs, lower long-stalked, upper stalkless, 10–40 cm long, with 6–12 pairs of oval-lanceolate lfts (**Da**). Infl a loose, terminal head; fls 2–3 cm across, drooping; corolla blue or white, with a short tube, and 5 equal spreading lobes; stamens golden, protruding, style long, with 3 stigmas. N Eng r, but vlf from Derby to Northumberland; on grassy slopes, rocks, screes on limestone, also gdn escape in wa, hbs. Fl 6–7. **Do not confuse** with the gdn form with **broader** lfts, which is a r introd casual (see illustration in *Plant Crib*).

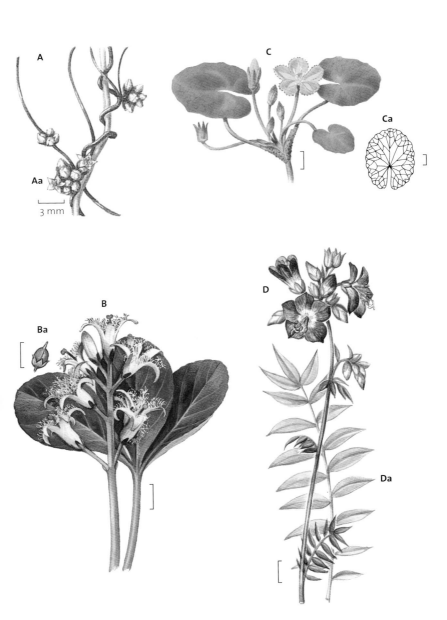

A

Aa

3 mm

C

Ca

Ba

B

D

Da

BORAGE FAMILY   *Boraginaceae*

A family of mostly hairy- or bristly-stemmed and -lvd herbs, rarely (**Oysterplant**, p 362 C) hairless. Lvs alt, undivided, without stipules; fls in curved or forked cymes. Calyx five-toothed, corolla five-lobed, wheel-, funnel- or bell-shaped, usually regular (with unequal lobes in **Viper's-bugloss**, p 362 B); stamens 5, attached to corolla-tube; the single style arises from **between** the 4 lobes of the deeply divided ovary (as in Deadnettle family). Fls often pink in bud, opening blue; sometimes yellow, purple or white. Fr of 4 separating one-seeded **nutlets**.

ID Tips   Borage Family

- The fr of **4 nutlets** is the key distinguishing character of the **Borage** family.

Do not confuse with members of :
- Figwort family, which are not bristly, have two-lipped fls, and fr an oval or rounded capsule with style on fr tip; or
- Dead-nettle family, which have opp lvs and the usually irregular two-lipped (not regular) corollas.

- **What to look at: (1)** frs, whether bristly or not **(2)** fl colour, shape and characters **inside** corolla, such as presence of scales or hairy folds **(3)** lf shape and width.

- **Viper's-bugloss** (*Echium*) is the only member of the Borage family with **2 stigmas**.

KEY TO BORAGE FAMILY

1 Plant hairless, waxy-blue or grey-green ....*Mertensia maritima* (p 362 **C**)
Plant hairy or bristly **2**

2 Nutlets of fr with hooked bristles – calyx-lobes spreading – fls red-purple...................................
Cynoglossum (p 361)
Nutlets of fr without bristles – calyx-lobes erect, hiding fr..**3**

3 At least some stamens protruding from corolla **4**
Stamens all enclosed inside corolla-tube **6**

4 Corolla funnel-shaped, its lobes unequal, with definite lip – stamens spreading.....................
Echium vulgare (p 362 **B**)
Corolla wheel-shaped, its lobes all equal, ± turned back – stamens protruding together in a close column, as in Bittersweet (p 352 F) **5**

5 Fls blue – stamens hairless – anthers 8–10 times as long as wide ....
Borago officinalis (p 364 **H**)
Fls purple – stamens hairy – anthers 2–3 times as long as wide ....
Trachystemon orientalis (p 364 **H.1**)

6 Corolla bell-shaped, hanging down, stigma projecting, with 5 v small lobes inside 1–2 mm long..*Symphytum* (p 363)
Corolla not bell-like, ± erect, with 5 larger spreading lobes that join to form a funnel or saucer-shape, stigma not projecting...**7**

7 Corolla orange-yellow ....*Amsinckia micrantha* (p 364 **I**)
  Corolla blue, red, purple or cream **8**

8 Corolla funnel-shaped, with lengthwise hairy folds in its throat **9**
  Corolla wheel-shaped, throat mouth closed by ring of 5 tiny rounded scales **10**

9 Calyx bell-like, divided at most only halfway down into 5 lobes – nutlets brown – lvs
  usually white-spotted..*Pulmonaria* (p 364)
  Calyx divided nearly to base into 5 v narrow long teeth – nutlets hard, whitish, shiny like
  porcelain ....*Lithospermum* (p 362)

10 Scales in corolla-tube hairless flat – fls blue, with yellow eye – plant downy or softly
  hairy **11**
  Scales in corolla-tube hairy oblong – fls blue or purple, with white eye – plant bristly **12**

11 All lvs stalked, oval-cordate, slightly downy only ....*Omphalodes verna* (p 360 **A.1**)
  Upper lvs stalkless, oblong, hairy ....*Myosotis* (key below)

12 Lvs oval – corolla-tube straight, shorter than its lobes....*Pentaglottis sempervirens* (p 362 **A**)
  Lvs narrow-oblong – corolla-tube curved, as long as its lobes ....*Anchusa* (p 361)

KEY TO FORGET-ME-NOTS (*MYOSOTIS*)

1 Hairs on calyx-tube adpressed (or almost abs)..**2**
  At least some hairs on calyx-tube stiff, hooked or curled **7**

2 Calyx-teeth forming equilateral triangle ....*Myosotis scorpioides* (p 360 **A**)
  Calyx-teeth forming isosceles triangle..**3**

3 Lower stem with ± projecting hairs **4**
  Lower stem with only adpressed hairs **5**

4 Fr-stalks 2$^{1}/_{2}$ –5x length of calyx ....*M. secunda* (p 360 **B**)
  Fr-stalks <2x as long as calyx ....*M. alpestris* (p 360 **D.1**)

5 Creeping runners with short spoon-shaped lvs from most lower lf-axils...................................
  M. stolonifera (p 360 **C.1**)
  No creeping runners **6**

6 Calyx with many adpressed hairs, teeth pointed – corolla flat – in wet places – c.................
  M. laxa (p 360 **C**)
  Calyx almost hairless, teeth blunt – corolla cup-like – Jersey only, vr......*M. sicula* (p 360 **C.2**)

7 Fr-stalks equal to or longer than calyx – tall ann or per (8–40 cm) plants – hairs
  spreading on stem throughout **8**
  Fr-stalks shorter than calyx – small (2–15 cm) ann plants of open dry gd – hairs on stem
  spreading below, adpressed above...**10**

8 Corolla not more than 5 mm across, cup-shaped ....*M. arvensis* (p 360 **E**)
  Corolla up to 8 mm across, flat **9**

9 Fr-stalks equal to calyx – nutlets black – lowest lvs long-stalked – fls fragrant in evening –
  in mts – vr ....*M. alpestris* (p 360 **D.1**)
  Fr-stalks 1$^{1}/_{2}$ –2 times length of calyx – nutlets brown – lowest lvs scarcely stalked – fls
  unscented – widespread ....*M. sylvatica* (p 360 **D**)

10 Open corolla yellow or cream, then turn pink to blue – corolla-tube twice length of calyx
  – calyx-teeth erect in fr..*M. discolor* (p 360 **F**)
  Open corolla always blue (rarely white) – corolla-tube shorter than calyx – calyx-teeth
  spreading in fr ....*M. ramosissima* (p 360 **G**)

A Water Forget-me-not, *Myosotis scorpioides*, creeping per, with runners, 15–30 cm tall; stem with hairs spreading below, adpressed above and on calyx; lvs oblong, 3–5 times as long as wide; fls (**Aa**) bright blue, up to 8 mm across (sometimes smaller); calyx-teeth (**Ab**) short, form **equilateral triangles**, 1/3 length of calyx-tube; fr-stalks once or twice length of calyx. Br Isles, c in marshes, ponds, streamsides. Fl 5–9. **A.1 Blue-eyed-Mary**, *Omphalodes verna*, has fls 10 mm across, v like those of A, but lvs **broad**, **rounded**, **cordate**, pointed at tips, 3–4 cm long x 3 cm wide, on **long** (to 10 cm) **stalks**; infls few-fld, loose; corolla has **white** eye; fr-stalks **bent** down, fr with **grooved**, hairy margins. Introd (SE Eur); Br Isles, r in wds near houses. Fl 3–5.

B Creeping Forget-me-not, *M. secunda*, like A in having runners; upper stem with adpressed hairs and **lower** stem with long **projecting** hairs; corollas up to 6 mm across with slightly notched lobes; calyx-teeth (**Ba**) long, forming **isosceles triangle** and **half or more** length of calyx-tube; fr-stalks **3–5 times** length of calyx; **lfy** bracts on lower part of infl. Br Isles, c in N, W, lf in SE, vr in E; in acid marshes and bogs. Fl 5–8.

C Tufted Forget-me-not, *M. laxa* (*M. caespitosa*), close to B with calyx-teeth (**Cb**) long, pointed, forming **isosceles triangle**; but **key differences** are **no** runners and stem with **adpressed** hairs **throughout**. Larger lvs 3 x as long as wide or more. Corollas (**Ca**) with notched lobes, usually smaller than B up to 5 mm across (but corolla-size not reliable). Br Isles, c in marshes, ponds, streamsides. Fl 5–8. **C.1** **Pale Forget-me-not*, *M. stolonifera* (*M. brevifolia*), also has hairs adpressed **throughout** stem but has **runners** from all lower lf-axils with short, **spoon-shaped** lvs, rarely >3 x as long as wide. Br Isles: N Eng, S Scot vlf, **elsewhere abs**; in wet places in mts. Fl 6–8. **C.2 Jersey Forget-me-not**, *M. sicula*, differs from C in almost **hairless** calyx with **blunt** teeth, and cup-shaped corolla 2–3 mm across. Jersey only, vr in dune hollows. Fl 4–6.

D AWI **Wood Forget-me-not**, *M. sylvatica*, downy **per**, 15–40 cm tall; no rooting runners; hairs on stem spreading, on calyx (**Da**) **stiffly curled** or **hooked**; corollas (**D**) **up to 8 mm** across, pale blue, flat; fr-stalks 1½–2 times length of calyx; infl in fr not longer than lfy part of stem; nutlets **brown**. Br Isles: GB c; N Ire o; **Eire** mostly **abs**; N Scot o; in wds on richer soils, also as **ann** gdn escape. Fl 4–8. **D.1** NT ***Alpine Forget-me-not*, *M. alpestris*, close to D, but usually much shorter; lower lvs **long-stalked**; fr-stalks **as long as** calyx; corolla up to 10 mm across; nutlets **black**; fls fragrant, especially in evening. Teesdale, Scot Highlands vr but la; on calc mt rocks. Fl 7–9.

E Field Forget-me-not, *M. arvensis*, close to D, except corolla (**E**) is only **up to 5 mm** (usually c. 3 mm) across, cup-shaped, with tube **shorter** (not as long or longer) than calyx. Br Isles, vc in dry wds, ar, hbs, rds, dunes. Fl 4–9.

F Changing Forget-me-not, *M. discolor*, **ann** 8–20 cm tall; stem with hairs **spreading** below, adpressed above; long lfless infls exceed lfy part of stem. Tiny fls, corolla 2 mm across, have curly-haired calyx **longer** than fl-stalk; fls (**Fa**) open **yellow or cream**, then turn pink to blue; corolla-tube **twice length** of calyx; calyx-teeth **erect** in fr. Br Isles, c in dry open sandy gslds. Fl 5–6.

G Early Forget-me-not, *M. ramosissima*, small slender **ann**; resembles F in hair patterns and fl-size etc, but is only 2–15 cm tall (rarely more); corollas (**Ga**) **blue** (rarely white), up to 3 mm across, tube **shorter** than calyx. In fr calyx-teeth **spreading**, and infl **much** longer than lfy part of stem. GB, c (except N, W Scot r); Ire o-r; on dry open gsld, especially on sand. Fl 4–6.

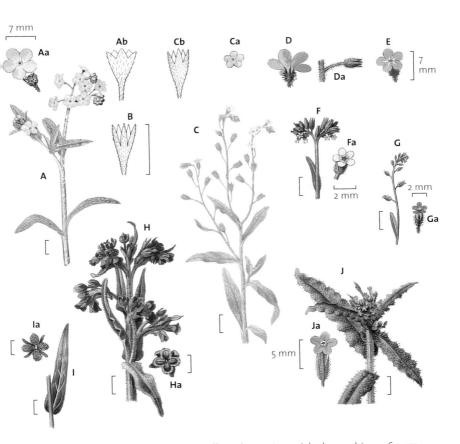

7 mm

Aa Ab Cb Ca D Da E 7 mm

B C F Fa G Ga

A H J Ja

Ia I Ha 5 mm

2 mm

H NT **Hound's-tongue**, *Cynoglossum officinale*, erect greyish-downy bi, 30–60 cm tall, smelling of **mice**; basal lvs 10–25 cm, stalked, elliptical, pointed; upper lvs stalkless, lanceolate. Infl a long forked cyme; fl-stalks mostly **<5 mm** long; calyx-lobes to 7 mm long, oblong, blunt, divided nearly to base. Corollas to 1 cm across, widely funnel-shaped, equally five-lobed, dull purplish-red; fr (**Ha**) of 4 flattened oval nutlets, each 5–6 mm wide, covered with **hooked bristles**, and with thick, **raised border**. Br Isles: Eng, Wales, f to E and SE, o to W, Scot, Ire, r and eastern, on gslds, wd edges on light sandy or calc soils, dunes. Fl 6–8.

I Sch 8 CR ****Green Hound's-tongue**, *C. germanicum*, close to H but **key differences** in lvs green (**I**), almost **hairless above**, fl-stalks mostly **>5 mm** long, and nutlets (**Ia**) **without thickened border**, but with longer bristles at edge. SE, S, mid Eng, vr; in wds, hbs, on calc soil. Fl 5–7.

J **Bugloss**, *Anchusa arvensis* (*Lycopsis arvensis*), erect, v bristly ann, 15–40 cm tall; bristles with bulbous bases; lvs to 15 cm long, lanceolate-oblong, wavy-edged, ± blunt, toothed, lower stalked, upper clasping. Infl often forked, elongating, with lfy bracts; calyx-teeth v narrow; corollas (**Ja**) 5–6 mm wide, blue, corolla tube **curved**, throat closed by 5 **hairy** scales; nutlets 4 mm wide, netted. Br Isles: GB c, but r in mts; Ire f in E; on sandy ar, wa, gsld, dunes. Fl 6–9. **J.1 Alkanet**, *Anchusa officinalis*, has lvs lanceolate, basal lvs **<5 cm wide**, bracts **not** wavy-edged; blue-purple fls, to 10 mm wide, with corolla-tubes **straight**, in long, curved cymes. Introd (Eur); Br Isles r; gdn escape in wa, sandy gsld, ar. Fl 6–7.

A Green Alkanet, *Pentaglottis sempervirens*, erect bristly per, 30–60 cm tall; lvs (**Aa**) **oval**, basal lvs **>5 cm wide**, **untoothed**, flat, often (but not always) with **blisters**, lower stalked, to 20 cm long. Infls dense, v bristly, long-stalked, each with a lfy bract below it; calyx-teeth v narrow; corollas bright blue, 10 mm wide, wheel-shaped with **short** tube, throat closed by 5 **downy** white scales. Nutlets netted on surfaces. Introd (SW Eur); Br Isles: GB vc; Ire o; in hbs, wa. Fl 4–7. **Do not confuse** non-flowering plants with **Oxtongues** (*Picris* spp) (p 478) which also have lvs with blisters.

B Viper's-bugloss, *Echium vulgare*, bristly bi; erect stem to 80 cm, dotted with **red**-based bristles; rosette of stalked, strap-shaped basal lvs to 15 cm long, with strong mid vein but no side veins **apparent**; stem-lvs shorter, stalkless, **rounded** at bases. Fls in **curved** cyme clusters in bract-axils up stem, forming a large panicle; buds pink, resembling clusters of tiny grapes; calyx-teeth narrow, short; corollas (**Ba**) bright blue, **funnel-shaped** with **unequal** lobes; 5 stamens, 4 of them long and protruding from fl; nutlets rough. Br Isles: GB c to S, E; N Scot r; Ire r; on dry gslds on sand and chk, dunes, cliffs. Fl 6–9. **B.1** *Purple Viper's-bugloss*, *E. plantagineum* (*E. lycopsis*), is similar to B, but **softly** hairy, not bristly; basal lvs **oval**, with side veins **obvious**; stem-lvs with **cordate** bases; corollas red, then **purplish**-blue; only 2 of stamens long-protruding. Cornwall, IoS, CI r-lf; vr scattered elsewhere in GB, Ire abs; on ocean cliffs, dunes. Fl 6–8.

C FPO (Eire) WO (NI) NT *Oysterplant*, *Mertensia maritima*, prostrate per herb; stem (30–60 cm) and lvs **blue-grey**, **hairless**, fleshy; lvs oval, lower stalked, uppersides dotted. Infls terminal, branched; fls long- (5–10 mm) stalked; calyx-lobes **oval**; corollas 6 mm wide, bell-shaped, pink, then pale blue, no scales in throat, but with 5 folds. Br Isles: coasts of NW Eng, W and N Scot, N Ire, r (to lf in N); on shingly, sandy shores. Fl 6–8.

D Common Gromwell, *Lithospermum officinale*, erect downy per to 60 cm, with alt lanceolate, **pointed**, **stalkless** lvs to 7 cm long up the stem, their side veins **conspicuous**; infls cymes, in lf-axils and terminal, forming a **V-shaped elongated spray** in fr (**Dc**); calyx (**Da**) with 5 narrow teeth; corollas creamy-yellow, wheel-shaped, with a short tube, 3–4 mm wide, throat with 5 hairy folds in it; nutlets (**Db**) oval, 3 mm long, **hard**, **white**, shiny like porcelain. Eng, Wales f-lc in S, E; Scot, Ire r; in open wds, scrub, hbs on calc soils. Fl 6–7.

E EN **Field Gromwell**, *L. arvense*, erect downy ann to 50 cm, differing from D in little-branched stem, lower lvs **blunt, stalked**, with side veins **not** obvious, and upper lvs strap-shaped; nutlets (**Ea, Eb**) hard, but conical, **grey-brown, warty**. Br Isles, r (but S, E Eng o); on ar. Fl 5–7.

F **Purple Gromwell**, *L. purpureocaeruleum*, creeping downy per; erect fl-stems to 60 cm tall; long, arching and rooting, lfy runners up to 90 cm long; lvs stalkless, narrow-lanceolate, dark green and rough above, pointed, to 7 cm long, side veins not obvious. Infls terminal forked lfy cymes; calyx-teeth (**Fb**) to 10 mm, narrow; corollas (**Fa**) red-purple, then bright **deep blue**, 12–15 mm across, funnel-shaped, with white scales in throat, twice length of calyx. Nutlets hard, white, shiny, as in D. SW Eng, Wales r and l as a native; vr introd elsewhere; in open wds, scrub on calc soils. Fl 4–6.

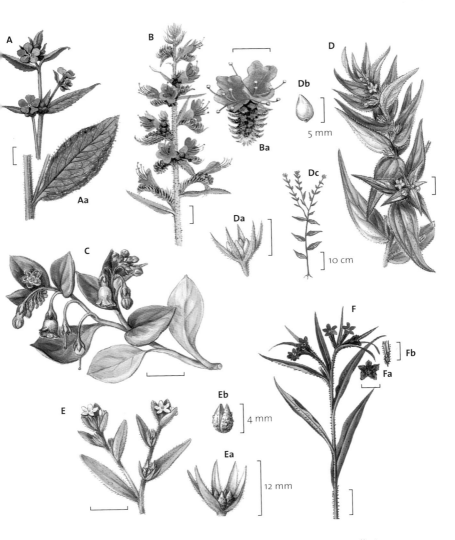

▼ **A** **Common Comfrey**, *Symphytum officinale*, erect bristly per to 1.5 m tall; stem
strongly **winged all the way from one lf-axil to the next** (**Aa**); basal lvs stalked, oval-
lanceolate, 15–25 cm long, soft-hairy; upper lvs narrower, stalkless, their bases running
down stem and clothed with long, down-pointing, tapering hairs. Infl of forked,
coiled cymes; calyx-teeth long (4 mm), **pointed** (**Ab**), narrow, twice length of tube;
corollas tubular to narrow bell-shaped, 8–20 mm long, with short (2 mm) triangular
bent-back teeth, cream, yellow or red-purple; 5 narrow scales inside corolla-tube;
nutlets black, shiny. Br Isles: GB, N to mid Scot c; N Scot, Ire o; by riversides, in marshes,
ditches, damp rds. Fl 5–6.

B Tuberous Comfrey, *Symphytum tuberosum*, differs from A in tuberous rhizomes, little-branched **unwinged** stems, **mid** lvs of stem longest and stalked; calyx-lobes 3 times length of tube, **pointed** not blunt (**B**); fls **yellow**. Br Isles: Scot, N Eng native, in wds, hbs; introd on hbs etc in rest of GB, and Ire, r-o. Fl 5–7.

C Creeping Comfrey, *S. grandiflorum*, is a per herb, stems decumbent with erect flg-stems to 30 cm, creeping by rhizomes, stems slightly winged or not; lvs generally with cordate base and long, winged stalks; **calyx-lobes blunt** (**C**) (unlike the similar-looking B); fls **cream or yellow**, reddish in bud; nutlets dark brown. Introd (Caucasus mts), a gdn escape which often spreads vigorously; S, C Eng f; o elsewhere, apparently abs Ire; flg earlier than A or B, **often the first comfrey to fl** along with D. Fl 3–5.

D White Comfrey, *S. orientale*, similar to C but **key differences** that distinguish D from other comfreys are **calyx-teeth < half length of tube** (**D**) and fls always **white**; stem unwinged, lvs softly downy; nutlets warty, brown. Br Isles: S Eng f ; Scot, Wales, Ire, r; introd in hbs, wa. Fl 3–5.

E Russian Comfrey, *S. x uplandicum*, hybrid of A and C, now **most c Comfrey**; stems v rough, only narrowly ± winged, bristly, upper lvs forming **short wings** (**Ea**) only down stem; fls often purplish-blue or deep red, but **v variable**; calyx (**Eb**). Br Isles, vc on rds, hbs, wa, wds. Fl 6–8.

F AWI ***Narrow-leaved Lungwort**, *Pulmonaria longifolia*, erect downy per, 20–30 cm tall; basal lvs (**Fa**) lanceolate, 10–20 cm, pointed, tapered to bases, white-spotted; stem-lvs broader, clasping. Infls short, in terminal cymes; calyx bell-shaped, teeth half length of tube, narrow; corollas 5–6 mm wide, pink then deep blue, funnel-shaped, with tufts of hairs alt with stamens in throat. Nutlets oval, **flattened**, shiny. Eng: E Dorset, SW Hants, IoW, vlc as a native; vr introd elsewhere in GB; in wds on clay or loam soils. Fl 4–5.

G Lungwort, *P. officinalis*, has lvs (**G**) normally white-spotted as in F, **oval-cordate, pointed, long-stalked**; calyx-teeth only 1/3–1/4 times length of more cylindrical calyx-tube; corollas 10 mm wide with frequent glandular hairs in fls; nutlets oval, **rounded**. Introd (C Eur); GB, c; Ire r; in wds, hbs, on chk, clay. Fl 4–5. **G.1** EN AWI ****Suffolk Lungwort**, *P. obscura*, has unspotted lvs (or rarely with faint light green spots) and fewer glandular hairs in the fls. Probably native (unlike G), vr, Suffolk only (but perhaps over-looked elsewhere) in coppiced, old wds. Fl 3–5.

H Borage, *Borago officinalis*, bristly erect ann to 60 cm tall; lower lvs oval, stalked, upper narrower, wavy-edged, stalkless. Fls in loose branched infls, bright blue, 20–25 mm wide; petals spreading, narrow; stamens hairless, in a purple column. Introd (S Eur); Br Isles: S Eng, CI, IoS, IoM f; N of GB, o; Ire r; in wa. Fl 6–8. **H.1 Abraham-Isaac-Jacob**, *Trachystemon orientalis*, similar to H, but per, with basal lvs large, **cordate, blunt, long-stalked**, blades to 30 cm long (10–20 cm in 8o2); fls as in H, but purple, with **hairy** stamens; anthers only 2–3 times as long as wide (8–10 times in H). Introd (SE Eur); Br Isles: Eng, o (most f in SE Eng); Scot, Ire r; in wds. Fl 4–5.

I Common Fiddleneck, *Amsinckia micrantha*, ann white-bristly herb to 20 cm; unstalked, linear to lanceolate lvs; fls **yellow** on a coiled cyme, corolla 3–5 mm long **without** hairy scales inside; warty nutlets. Introd (N America); E Eng, E Scot f-vla; C, S Eng o, Wales and Ire vr; on ar, wa and disturbed gd on light, sandy soils. Fl 4–8.

◄ A

Ab

B

Eb

E

Aa

C

Ea

D

H

F

G

Fa

I

VERBENA FAMILY   *Verbenaceae*

Very like Deadnettle Family but style **terminal** on fr.

Vervain, *Verbena officinalis*, erect hairy per herb, 30–60 cm; stems tough, with pairs of opp pinnatifid bristly lvs, 2–7 cm long, elliptical; infls terminal, in uppermost lf-axils, spike-like and lfless, slender, loose, 8–12 cm long. Fls (**a**) bluish-pink, 4 mm across; corolla-tube twice length of calyx, five-lobed, ± two-lipped; stamens 4; fr with style on top, separating into 4 nutlets. Br Isles: Eng, Wales, c to S, r to N; Ire f in S; **Scot abs**; in gsld, rds, wa, scrub, on dry often calc soils. Fl 6–9.

DEAD-NETTLE FAMILY   *Lamiaceae*

Also known as the *Labiatae*, a distinctive family of herbs with **square stems** and **opp lvs** without stipules; infls in axils of opp pairs of bracts (these often like ordinary lvs), cymose, but so dense as to appear whorled; calyx ± five-toothed, often two-lipped; corolla with a tube, 4- or 5-lobed, usually clearly two-lipped; stamens 4 (or 2 only), fixed to corolla-tube; ovary superior, its 4 lobes ripening in fr to **4 separate one-seeded nutlets**, style arising from **hollow** between lobes. Often scented or aromatic when crushed, containing various volatile oils.

ID Tips   Dead-nettle Family
• **Dead-nettles** look similar to members of the Figwort family. **Key differences** are that Figworts have ovary oval or globular, two-celled, style arising from its tip (not from a hollow between 4 lobes) and fr a many seeded capsule (not composed of nutlets).
• **Do not confuse** Dead-nettles with **Vervain** (above) which is v like a Dead-nettle in its opp lvs, square stem, and fr of 4 nutlets, but fls are in **slender spikes**, **not whorled**, and two-lipped corolla has 5 ± **equal, spreading** lobes.
• **What to look at: (1)** lf-shape and width **(2)** calyx-tube shape **(3)** corolla shape and colour **(4)** stamen number.
• **White Horehound** (*Marrubium*) is the only member of Dead-nettle family with **10** (not 5) **calyx-teeth**.

KEY TO DEAD-NETTLE FAMILY

1 Corolla-lobes 4, ± equal – corolla 5 mm wide or less **2**
   Corolla with well-developed lower lip, upper lip abs, or composed of 2 tiny teeth only **3**
   Corolla with 2 unequal, lobed or toothed lips..**4**

2 Stamens 2 – lvs pinnatifid, odourless ....*Lycopus* (p 371)
   Stamens 4 – lvs shallowly-toothed at most, strongly mint-scented ....*Mentha* (key p 369)

3 Corolla with a single five-lobed lower lip – corolla-tube hairless inside, with no teeth at
   top – lvs acrid-smelling, matt ....*Teucrium* (p 382)
   Corolla with three-lobed lower lip – corolla-tube with ring of hairs inside and 2 small
   teeth at top – lvs odourless, shiny ....*Ajuga* (p 382)

4 Lower lvs deeply palmately-lobed – upper lvs trilobed – fls white, in dense whorls spaced
   out in lf-axils up stem – tall (60–100 cm) downy plant ....*Leonurus cardiaca* (p 371 **A.1**)
   Lower lvs kidney-shaped, blunt-toothed, on stalks longer than lf-blades – fls violet (rarely
   pink) – creeping and ascending plant................................*Glechoma hederacea* (p 378 **J**)
   Lower lvs oval, oblong or cordate, ± pointed, with pinnate veins **5**

5 Calyx tubular, ten-veined, equally ten-toothed – plant erect, white-woolly – stamens
   hidden in corolla-tube – fls white ....*Marrubium vulgare*(p 380 **D**)
   Calyx with 5 teeth or less..**6**

6 Stamens obviously projecting from fls, spreading out, longer than corolla upper lip...........**7**
   Stamens shorter than corolla upper lip, lying beneath it, side by side **8**

7 Tall (30 cm or more) erect plant – fls in a broad (5–8 cm wide), rounded, branched
   terminal head – lvs oval, over 1 cm wide ....*Origanum vulgare* (p 376 **G**)
   Short (usually less than 10 cm tall) creeping and rooting plant – fls in small (1–2 cm wide)
   heads or short spikes – lvs small, 5 mm wide or less ....*Thymus* (p 371)

8 Calyx of 2 untoothed lips, upper with scale on back – erect, narrow-lvd ..*Scutellaria* (p 378)
   Calyx two-lipped, with lips toothed **9**
   Calyx ± equally five-toothed, not clearly two-lipped **14**

9 Stamens 2 – upper corolla-lip strongly-arched, hooded, often sickle-shaped ....*Salvia* (p 374)
   Stamens 4 – corolla never sickle-shaped..**10**

10 Calyx with tube rounded above, its upper lip with 3 long narrow teeth, lower lip with 2
   long teeth..**11**
   Calyx with tube v flattened above, its upper lip with 3 short (1 mm) broad-triangular but
   bristle-pointed teeth, lower lip with 2 long narrow teeth **13**
   Calyx bell-shaped, upper lip with 2–3 tiny blunt irregular teeth, lower with 2 small
   rounded lobes – erect plant with v large fls in lfy racemes – lower corolla-lip 1–2 cm long,
   white with purple patch..*Melittis melissophyllum* (p 372 **E**)

11 Calyx-tube straight – fls in tall spikes composed of pairs of loosely-branched, stalked infls
   in axils of opp pairs of lvs – fls pale, pinkish- or violet-mottled...
   Clinopodium ascendens, *C. menthifolium* or *C. calamintha* (p 376)
   Calyx-tube curved – fls in unstalked clusters in axils of opp pairs of lvs – on calc soils **12**

12 Fls violet, in loose, 3–8-fld whorls – plant creeping and ascending to 10 cm tall; lvs not
   more than 1.5 cm long x 7 mm wide – calyx with a pouch below ....
   C. acinos (p 376 **F**)
   Fls rosy-purple, in dense, many-fld whorls – plant erect, 15 cm tall or more – lvs at least 2
   cm long x 1.5 mm wide; calyx without a pouch below...................................*C. vulgare* (p 376 **A**)

13 Calyx-tube strongly flattened above and below, outer edges parallel (so outline from above rectangular) – fls in dense oblong heads – lvs unscented – corolla-tube straight – c ....*Prunella* (p 380)

Calyx-tube flattened only on upper side, outer edges diverging (so outline bell-shaped) – fls in whorls in lf-axils spaced along stems – lvs lemon-scented – corolla-tube curved upwards – not c ....*Melissa officinalis* (p 376 **E**)

14 Upper corolla-lip flat – 2 outer stamens shorter than inner 2 – lvs white-woolly below, green downy above – stem white-woolly – fls white with purple spots...*Nepeta cataria* (p 380 **E**)

Upper corolla-lip arched or hooded – 2 outer stamens longer than inner 2 ....**15**

15 Calyx funnel-shaped, teeth v short, broad – plant rough – fls dull purple ....*Ballota nigra* (p 380 **C**)

Calyx tubular or bell-shaped, with long teeth (2 mm or more long) **16**

16 Lower corolla-lip forked at tip, side-lobes tiny, ± reduced to teeth or flaps ..*Lamium* (p 372)

Lower corolla-lip at most notched at tip, side-lobes well-developed, spreading **17**

17 Corolla deep yellow – plant with lfy runners.........................*Lamiastrum galeobdolon* (p 374 **A**)

Corolla purple, pinkish, or creamy yellow – no runners...**18**

18 Calyx-teeth forming long (5–7 mm) spines – corolla with conical bosses at base of lower lip – upper lip forming hood flattened on each side ....*Galeopsis* (p 374)

Calyx-teeth long-pointed (3–4 mm), but not spine-like – corolla without bosses – upper lip arched forward, but not flattened on each side to form hood ....*Stachys* (key p 378)

ID Tips Mints

- **Mints** (*Mentha*) are difficult to separate because the spp are so variable and all forms can reproduce vegetatively; also frequent occurrence of hybrids and many naturalized cultivars makes matters even more complex.

- Most *Mentha* spp occur in two fl forms:
  Type A: **fertile** stamens **projecting** out of the corolla mouth – plants producing seed (nutlets) and pollen;
  Type B: **sterile** stamens **hidden** inside corolla tube – female plants, producing only seed.
- The exception is **Corsican Mint** (*M. requienii*) with stamens never projecting from corolla. Hybrids are nearly always sterile with B type stamens, or rarely with stamens protruding but anthers aborted.
- Rarely, both types of fl may occur on the **same** plant.

- Separating the *M. spicata* group (couplet 10 onwards in the key below) is particularly difficult. These plants include many garden escapes. Characters given here are a broad guide only. Leaf shape is often unreliable in separating *M. spicata* from its hybrids, and both may be hairy or hairless.

- Check whether your plant **produces seed** (nutlets) or not, as none is a good test for most **hybrids**. However, note that isolated female plants very rarely do not produce seed, if no male pollen bearing plants are near at hand.

Mints (*Mentha*) are aromatic herbs with creeping stems either below or above gd; small, lilac-purple fls have corolla divided into 4 nearly equal lobes, and 4 stamens.

KEY TO MINTS (*MENTHA*)

1 Small creeping plants – lvs less than 1 cm wide, ± untoothed – calyx-throat hairy, teeth unequal **2**
Larger, ± erect plants – lvs over 1.5 cm wide – calyx-throat hairless, teeth equal....................**3**

2 Tiny plants – stems creeping, thread-like – lvs rounded, 3–5 mm long – whorls 2–6-fld with stamens never projecting ....*Mentha requienii* (p 370 **G.1**)
Larger plants – stems ascending, stouter (2 mm thick or more) – lvs oval, 5–15 mm long – whorls many-fld with stamens sometimes projecting or not ....*M. pulegium* (p 370 **G**)

3 Whorls of fls all in lf-axils, not forming heads, spaced out along stem – stem-tip leafy......**4**
Whorls of fls close-set, forming terminal spikes or heads – stem-tip not leafy......................**7**

4 Calyx bell-shaped...**5**
Calyx tube-shaped **6**

5 Calyx-tube hairy, teeth triangular, short – stamens projecting from fl – abundant seed produced – scent usually peppery ....*M. arvensis* (p 370 **B**)
Calyx-tube hairless except for teeth, teeth narrow, twice as long as broad – stamens not projecting – usually no seed produced – scent minty, aromatic ....
Bushy Mint *M.* x *gracilis* (illustration p 370 **H**)

6 Calyx-tube hairy, calyx mostly < 3.5 mm – stamens usually not projecting – scent weakly peppery ....*M.* x *verticillata* (p 370 **C**)
Calyx-tube hairless except for teeth, calyx mostly >3.5 mm – stamens usually projecting – scent minty-aromatic – lvs ± reddish-purple **Tall Mint** *M.* x *smithiana* (illustration p 370 **I**)

7 Fls in rounded heads ....*M. aquatica* (p 370 **A**)
Fls in spikes **8**

8 Lvs stalked, lanceolate – usually with peppermint scent ....*M.* x *piperita* (p 370 **F**)
Lvs ± stalkless **9**

9 Lvs 2–4 cm, hairy, rounded, often slightly oblong, bluntly-toothed or teeth curved towards underside of leaf – abundant seed usually produced......**Round-leaved Mint** *M. suaveolens*
Lvs lanceolate to oval, usually > 3cm long, hairy or glabrous, sharply-toothed – seed produced or not **10**

10 Lvs with forward-pointing teeth unless lvs oval to ± rounded **11**
Lvs usually hairy, with teeth near lf-base pointing outwards and never oval to ± rounded **12**

11 Lvs lanceolate, usually glabrous and smooth, rarely both wrinkled and hairy – plant producing seed .... *M. spicata* (p 370 **E**)
Lvs ± oval, often both wrinkled and hairy, rarely glabrous or smooth – plant not producing seed ....*M.* x *villosa* (p 370 **D**)

12 Plant always hairy and usually producing seed – lvs ± parallel-sided – rarely found in the Br Isles ....**False Apple-mint**, *M.* x *rotundifolia*
Plant not producing seed – lvs elliptical to lanceolate-oblong, widest near middle ....
....**Sharp-toothed Mint**, *M.* x *villosonervata*

A   Water Mint, *Mentha aquatica*, downy, erect per herb, 15–60 cm tall; lvs 2–6 x 1.5–4.0 cm, opp, oval, ± hairy, with blunt tips and teeth; infl a terminal rounded **head**, 2 cm long, often with extra whorls below. Calyx-tube hairy (**Aa**) corolla (**Ab**) mauve; stamens projecting from fls; lvs **fresh mint**-scented. Br Isles, vc in marshes, fens, wet wds, by fresh water. Fl 7–10.

B   Corn Mint, *M. arvensis*, downy prostrate to erect per herb, 10–30 cm tall, 'peppery' mint scent when bruised; lvs stalked, 2–6 x 1–2 cm, rounded to elliptical, blunt-tipped, teeth ± hairy. Fls in **separated** dense **whorls** in **lf-axils**; calyx (**Ba**) **bell-shaped**, **v hairy**, with short, **blunt, triangular** teeth; corolla mauve, stamens projecting. Br Isles, c but possibly declining in ar, paths in wds, mds. Fl 5–10.

C   Whorled Mint, *M. x verticillata* is A x B, close to B but more robust, to 90 cm tall; calyx (**C**) **tube**-shaped, with **narrowly triangular** teeth twice as long as broad, calyx-tube hairy; stamens **not** projecting; scent like B, peppery. Br Isles, c in damp places. Fl 7–10.

D   Apple Mint, *M. x villosa* (*M. x niliaca*), stout, erect, v hairy per herb, 40–100 cm; lvs (**D**) 3–10 x 2–4 cm, stalkless, **oval** or oblong, pointed, sharply-toothed, **often wrinkled and hairy**, green, grey-woolly below, bases cordate. Infls terminal spikes, 3–8 cm long; bracts narrow, shorter than fls; calyx (**Da**) and fl-stalks hairy; corolla mauve, **hairy**, stamens **not** projecting. Eng, Wales, f on rds, ditches, wa. Fl 8–9.

E   Spear Mint, *M. spicata*, the usual plant for mint-sauce; v variable; erect per herb 30–90 cm, may be ± hairless or sparsely hairy; lvs (**E**) lanceolate, pointed, **stalkless**, unlike D rarely **both wrinkled and hairy**; infls in terminal spikes, 3–8 cm long; bracts v narrow, longer than fls; calyx (**Ea**) and fl-stalks usually hairless; corolla mauve, usually hairless, stamens **projecting**. Smells of spearmint. Br Isles, introd (often as discarded gdn rubbish), f in ditches, rivers, rds, wa. Fl 8–9. Previous records for **Horse Mint**, *M. longifolia*, are now thought mostly to be errors for hairy plants of E.

F   Peppermint, *M. x piperita* is A x E and has **stalked** lvs, usually with a peppermint smell; calyx (**F**). Br Isles, f in damp gd, wa, ditches, rds. Fl 7–10.

G   Sch 8 WO (NI) FPO (Eire) BAP EN *Pennyroyal, *M. pulegium*, **prostrate, creeping** downy per herb to 30 cm long; lvs 1–2 cm long x 0.5–1.0 cm wide, oval, short-stalked, finely downy, scented, v blunt-tipped and -toothed; fls in distant whorls in lf-axils; calyx and fl-stalks downy; corolla mauve, hairy outside only. Eng, Wales, vr as a native and mostly in S (also r introd); Ire vr; on moist bare sandy gd by ponds and on commons. Fl 8–10. **G.1   Corsican Mint**, *M. requienii*, is much **smaller** than G; tiny **creeping**, rooting, mat-forming per herb; stems thread-like; lvs tiny, **round**, ± hairless, **strong mint**-scented, 3–5 mm long; fls few in each lf-axil, corollas tiny, pale mauve, enclosed in calyx. Introd (Corsica); scattered throughout Br Isles (Scot vr-abs), on cultivated gd, on paths in wds, wet moors. Fl 6–8. **Do not confuse** with **Mind-your-own-business** (p 122) which has v similar habit and lvs but is not mint-scented.

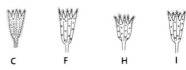

C   F   H   I

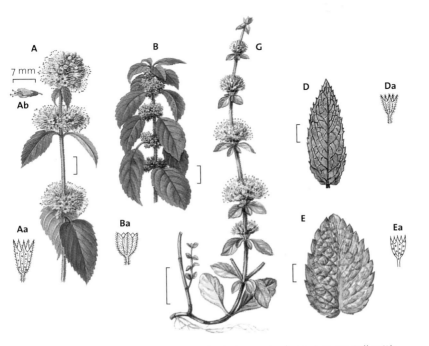

▼ **A Gypsywort**, *Lycopus europaeus*, erect ± hairy per herb, 30–100 cm tall, with ascending branches. Lvs 5–10 cm, elliptical-lanceolate, short-stalked, **deeply pinnately-lobed**, with pointed tips and lobes. Fls (**Aa**) in well-spaced dense whorls in lf-axils; calyx bell-shaped, hairy, equally five-spine-toothed; corolla (**Ab**) white, 3 mm or less long and wide, with 4 ± equal lobes, lower lip purple-dotted; stamens only 2, slightly protruding. Br Isles (except NE Eng, E and N Scot r), vc; on stream banks, in ditches, fens. Fl 6–9. **A.1 Motherwort**, *Leonurus cardiaca*, also erect downy per, 60–80 cm, fls in whorls in lf-axils; lvs deeply-lobed, lower lvs **palmately**-lobed, with cordate bases, upper lvs deeply trilobed and narrowed into stalks; fls have white, **downy, two-lipped** corollas, 12 mm long. Introd (Eur); Br Isles, r; Ire abs; on wa, hbs. Fls 7–9.

▼ **B Large Thyme**, *Thymus pulegioides*, shortly creeping low shrub-like per with ascending fl-shoots to 20 cm, **strongly** thyme-scented. Lvs oval-elliptical, blunt, short-stalked, hairy only at bases, slightly folded upwards, 6–10 mm x 3–6 mm; Fl-stems (**Ba**, cut across) four-angled, **long-hairy on angles**, the 2 broader faces **hairless**, the 2 narrower shortly **downy**; fls in **interrupted spikes** of few-fld whorls; calyx two-lipped, purplish, hairy; corolla (**Bb**) to 5 mm long, two-lipped, rose-purple, stamens **far protruding**. Br Isles: S, E Eng f-lc; Scot, Ire abs; on dry calc or hthy gslds, banks. Fl 7–8.

▼ **C Wild Thyme**, *T. polytrichus* (*T. drucei*), close to B, but lower, more mat-forming, far-creeping; only **faintly** thyme-scented; lvs 4–8 mm long, **flat**; fls in dense heads; fl-stems (**C**) bluntly four-angled, **v hairy** on 2 opp faces, ± hairless on other 2. Br Isles vc (except E Anglia o-lf); on dry gslds, hths, dunes, rocks. Fl 5–8.

D **Breckland Thyme, _Thymus serpyllum_, close to C, but smaller; fl-stems (**D**) almost round, **equally hairy all round**; lvs erect, shortly oval, blunt; fls in short heads. Eng: Breckland only, vlf; on sandy hths, gslds. Fl 7–8.

E AWI VU *Bastard Balm, _Melittis melissophyllum_, erect hairy-stemmed per, 20–50 cm; lvs 5–8 cm long, oval, stalked, bluntly-toothed. Fls in few-fld whorls in lf-axils; calyx **bell-shaped, two-lipped**, upper lip with 2–3 small teeth, lower with 2 **rounded lobes**; corolla **v large**, 2.5–4.0 cm long, cream-coloured or pink; lower lip **to 2 cm long**, with a large rosy-purple blotch. S, W Eng, Wales (from Sussex W), r but vlf; in open wds, hbs, not on chk. Fl 5–7.

Dead-nettles (_Lamium_) have a tubular to bell-shaped calyx, with 5 ± equal, long, **triangular-pointed** teeth; corolla two-lipped, lower lip's 2 side-lobes **v small**, bearing small teeth, mid-lobe divided into **2 large teeth**; fls in dense, separated whorls in lf-axils. **Hemp-nettles** (_Galeopsis_) are like Dead-nettles, but all ann, have 5 **spine-like** ± equal calyx-teeth; corolla has lower lip with 2 narrow side-lobes not **much** shorter than mid-lobe, and 2 cone-shaped **projections** at its base. **Claries** and **Sages** (_Salvia_) have a strongly arched upper corolla-lip and only 2 stamens, the single anther of each on a **hinged** stalk, attached by its centre to the main stamen-stalk.

F **Red Dead-nettle**, _Lamium purpureum_, ± erect downy ann; stem 10–30 cm, branched from base, ± purplish; lvs (**Fa**) and bracts 1–5 cm long, **cordate-oval**, coarse-toothed, teeth**< 2 mm** long. Infl dense; calyx downy, teeth equalling tube, spreading in fr. Corolla 10–15 mm, pink-purple, tube longer than calyx, with ring of hairs inside. Br Isles, vc on ar, wa, hbs. Fl 3–10.

G **Cut-leaved Dead-nettle**, _L. hybridum_, close to F, but more slender; lvs (**G**) and bracts triangular-oval, **truncate** at base, **deeply-toothed** with many teeth **> 2 mm** long, corolla shorter, few or no hairs inside. Br Isles, f (but c E Eng); on ar, wa on dry, fertile soils. Fl 3–10.

H **Henbit Dead-nettle**, _L. amplexicaule_, also close to F; downy ann 5–25 cm, branched from base; lower lvs 1.0–2.5 cm, **rounded, blunt**, long-stalked (3–5 cm); base truncate, teeth large but blunt; bracts **stalkless**, ± **clasping** stem in pairs. Calyx-tube (**Ha**) with **dense white spreading** hairs, teeth shorter than tube, **erect** in fr; corollas mostly **15 mm**, pink-purple, with **long** tube **hairless** inside, but also some **v short corollas** enclosed in calyces. Br Isles: S, E of GB f-lc (but Wales, W Scot, o-lf); Ire r; dry places on ar, wa, walls. Fl 4–8.

I **Northern Dead-nettle**, _L. confertum_ (_L. moluccellifolium_), is close to H, but stouter; has bracts **not** clasping, lower bracts **stalked**; calyx-tube (**I**) has **adpressed** hairs and long **spreading** teeth in fr. Scot, IoM f; Ire o; N Eng, Wales r; on ar, wa. Fl 5–9.

J **White Dead-nettle**, _L. album_, erect hairy tufted per, 20–60 cm; lvs and bracts 3–7 cm long, oval, cordate-based, pointed, coarse-toothed and nettle-shaped but **stingless**; fls in dense, well-spaced whorls; corollas (**Ja**) 2 cm long, **white**, tiny side-lobes of lower lip with 2–3 small teeth; upper lip long, **hairy**; tube curved backwards in an arch and narrowed suddenly near base. Br Isles: GB vc (except N Scot r); Ire lc in E only; on hbs, rds, wa. Fl 5–12. **J.1 Spotted Dead-nettle**, _L. maculatum_, has habit and size of J, but lvs often bear large white blotches, and fls (same size as in J) **pink-purple**, with only one tooth on side-lobes of lower lip. Introd (Eur); GB f; Ire o-r; in open wds, rds, wa. Fl 4–10.

A

Bb

B

Ba

C

5 mm

Aa

Ab

D

3 mm

Ha

I

E

F

H

J

Fa

G

Ja

A AWI **Yellow Archangel**, *Lamiastrum galeobdolon* (*Galeobdolon luteum*), ± hairy erect per, 20–60 cm tall, with **long creeping lfy runners**, especially after fl-time. Lvs 4–7 cm long, oval, pointed, rounded at base, stalked, coarsely-toothed. Corolla 2 cm long, **bright yellow** with red-brown streaks; lower lip ± equally three-lobed, mid-lobe **untoothed**. Br Isles: Eng, Wales, N to Yorks c, lf to N; Scot, Ire, r; in wds, hbs, on richer or calc soils. Fl 5–6. A gdn form with **variegated lvs**, *L. galeobdolon* ssp *argentatum*, is increasingly natd and may become invasive.

B FPO (Eire) BAP CR *****Red Hemp-nettle**, *Galeopsis angustifolia*, erect ± branched downy ann; stem 10–60 cm tall (often stunted on shingle), joints **not** swollen; lvs 1.5–8.0 cm x 0.5–0.8 cm, green, narrow-lanceolate, pointed, narrowed to stalk, downy, with **few** small teeth. Calyx hairy, tubular; corolla 1.5–2.5 cm long, rosy-purple, white spots on lip, tube much **longer** than calyx-teeth. Br Isles: declining throughout GB: Eng o-r in S, E, r to W, N; Wales vr; Scot, vr-abs; Ire vr but apparently expanding its range; on disturbed gd on calc soils, ar, shingle beaches. Fl 7–10.

C **Common Hemp-nettle**, *G. tetrahit*, erect branched ann; stem 10–80 cm, branched, bristly, with sticky hairs, joints swollen; lvs 2.5–10.0 cm long, oval, pointed, coarse-toothed, hairy, narrowed to stalk; calyx bristly, teeth long, spine-pointed; corolla (**Ca**) 13–20 mm long, pink or white, with purple markings on lower lip, reaching about ½ way to lip-edge; tube ± **equalling** calyx-teeth. Br Isles, vc on ar, hbs, wds, fens. Fl 7–9.

D **Bifid Hemp-nettle**, *G. bifida*, similar to C but with darker markings on lower lip reaching close to lip-edge; lower lip end lobe curved, appearing ± notched, with rolled-under sides. Br Isles, c in similar habitats to C although preferring ar and disturbed gd. Fl 7–9.

E VU **Large-flowered Hemp-nettle**, *G. speciosa*, resembles C, also ann but stouter, stem bristly throughout; corollas (**E**) 30 mm long, pale yellow with **large violet spot** on lower lip, tube **twice length** of calyx. Br Isles: GB, c to N, r to S; Ire lc to N, vr to S; on ar, wa. Fl 7–8.

F **Wild Clary**, *Salvia verbenaca* (*S. horminoides*), downy, little-branched per, 30–80 cm tall, slightly sage-scented; rosette of basal lvs, each 4–12 cm long, oval-oblong, cordate, blunt, wrinkled, stalked, with **deep, jagged teeth**; stem-lvs few, stalkless above; upper stem-lvs and calyces purplish-blue. Infl of whorled spikes; calyx 7 mm, sticky-downy, two-lipped, with **long white hairs**, upper lip's teeth v short, converging, mid tooth **shorter**; corollas of 2 sizes, larger 7 mm long, violet-blue with 2 white spots on lower lip, upper lip ± arched; smaller corollas ± hidden in calyx, not opening. Br Isles S, E Eng lf; scattered elsewhere in GB, Ire r; on dry gslds, dunes, hbs, rds, usually on base-rich soils. Fl 5–8. **F.1** **Whorled Clary**, *S. verticillata*, is close to F, but upper calyx-lip has conspicuous straight ± **equal** teeth, and corolla has **hairs** inside; upper lvs stalked. Introd (S Eur); Br Isles, r on wa. Fl 6–8.

G Sch 8 NT *****Meadow Clary**, *S. pratensis*, downy per, 30–100 cm, sage-scented. Basal lvs 7–15 cm long, in a rosette, oval to oblong, blunt, long-stalked, cordate-based, wrinkled, **doubly**-toothed, but **not** deeply-cut as in F; upper lvs stalkless. Calyx 6.5 mm, **without** long white hairs; corollas (**Ga**) of larger fls 15–25 mm long, brilliant blue, upper lip **strongly** arched like a **sickle** and forming a hood; style **long-projecting** (in F style short). Eng: native in S, r, introd N to Lincs vr; on calc gslds. Fl 6–7.

375

A Wild Basil, *Clinopodium vulgare*, erect, hairy, weakly-scented per herb, 10–40 cm, little-branched; lvs 2–5 cm, oval, blunt, rounded at base, scarcely-toothed, stalked, v like those of Marjoram (G) but veins more impressed; fls in **dense** whorls in lf-axils (**Aa**); calyx tubular, curved, hairy, ± two-lipped, 13-veined, teeth of lower lip longer than of upper; corolla 15–20 mm, bright rosy-purple, two-lipped, hairless within. Br Isles: GB, c to S and E, o to N, **abs N W Scot**; Ire, vr introd; on calc scrub, hbs, gslds. Fl 7–9.

B Common Calamint, *C. ascendens* (*Calamintha ascendens*), erect, tufted hairy per herb, 30–60 cm; lf-stalks green, 1 cm; lvs 2–4 x 1–3 cm, oval, blunt, bases truncate, with adpressed hairs, teeth 5–8 on each side, shallow and blunt; infls in long spikes of distant whorls in lf-axils; whorls (**Ba**) ± **loose**, obviously branched; calyx (**Bb**) 6–8 mm long, tubular, **straight**, five-toothed, hairy on the 13 veins and within throat; calyx-teeth with **long hairs**, 2 teeth of lower lip **curved** up, much **longer** than **spreading** teeth of upper lip; corolla (**Bc**) 10–15 mm long, pinkish-white, with small purple spots on lower lip. Br Isles: Eng lf to S, r to N; Wales vlf; Ire o to S; **Scot abs**; on hbs, wd borders, scrub on dry or calc soils. Fl 7–9.

C Sch 8 CR ****Wood Calamint**, *C. menthifolium* (*Calamintha sylvatica*) close to B, but lvs 3.5 to 6.0 cm long; corolla (**C**) **15–22 mm long, rosy-pink**, lower lip (**Ca**) 6 mm long x 8 mm wide, side-lobes shallow, purple blotches larger than in B. Eng: **IoW only**, vr; on hbs, scrub on calc soils. Fl 8–9.

D VU ***Lesser Calamint**, *C. calamintha* (*Calamintha nepeta*), has tufted habit and height of B, but v aromatic-scented, stems and lvs (**D**) grey-downy; lvs only 1–2 cm long, scarcely-toothed, lf-stalks 5 mm or less; **key difference** from B is infls looser, **branched 1–3x** (unlike B with upper infl unbranched); calyx (**Da**) 4–6 mm, **all** teeth ± **straight**, ± of **equal** length, with **v short or no** hairs; corolla (**Db**) 10–15 mm, pale mauve, scarcely spotted. Hairs **protrude** from calyx-throat in fr, not confined inside as in B and C. Br Isles: E and S Eng, CI only, o-lf; on hbs, rds, dry gslds on sand or gravel. Fl 7–9.

E FPO (Eire) VU **Basil Thyme**, *C. acinos* (*Acinos arvensis*), creeping and ascending ann, to 15 cm tall; stems hairy; lvs 0.5–1.5 cm, oval-elliptical, ± hairless, shallow-toothed, bases wedge-shaped; infls much shorter than in B, C and D, whorls only 4–6 fld; calyx-tube (**Ea**) **curved**, with **pouched** base, hairy inside and out, 13-veined; corolla (**Eb**) 7–10 mm, violet with white blotches on lower lip. Br Isles: S, E Eng f, r to N; Wales, Scot, vr-abs; Ire r; on ar, gslds, rocks on calc soils. Fl 5–9.

F Balm, *Melissa officinalis*, erect branched tufted per herb with habit of B, C and D, but lvs **lemon-scented, deeply-toothed**, with strongly **impressed** veins; calyx two-lipped, bell-shaped, v hairy upper calyx-lip almost **flat** with 3 v **short, broad, triangular shortly-bristle-pointed** teeth; corolla (**Fa**) two-lipped, **white** (buds (**Fb**) bright yellow), 12 mm long, **curved** upwards. Introd (Mediterranean); Br Isles: S Eng, Wales, S Ire f; elsewhere o-r; in hbs, wa. Fl 8–9.

G Wild Marjoram, *Origanum vulgare*, erect, tufted, sparsely downy per herb, 30–60 cm; lvs (**Gb**) 1.5–4.5 cm long, oval, stalked, ± untoothed, sweetly-scented. Fls in **dense, rounded**, terminal **panicles**; lower bracts like lvs, upper purple; calyx-tube hairy inside, with 5 short equal teeth; corolla (**Ga**) rose-purple, 6–8 mm, two-lipped, tube longer than calyx, stamens **protruding** from corolla. Br Isles: S of GB c to l vla, f to N, vr-abs N, W Scot; Ire f; on gslds, scrub, hbs on calc soils. Fl 7–9.

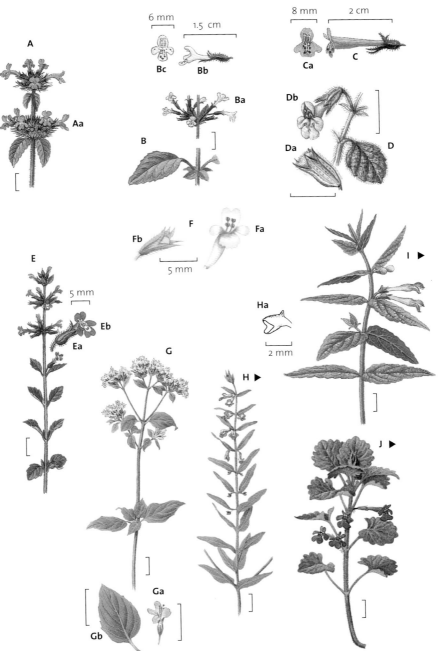

▲ **H** AWI **Lesser Skullcap**, *Scutellaria minor*, erect ± bushy, hairless per herb, 10–15 cm tall; lvs **1–3 cm** long, oval-lanceolate, cordate-based, pointed, untoothed; fls in pairs in lf-axils; calyx (**Ha**) two-lipped, lips **untoothed**, but with small **scale** on back of upper lip; corolla-tube **straight**, 6–10 mm long, 2–4 times length of calyx, two-lipped, **pink**, lower lip purple-spotted. Br Isles: S, W Eng, Wales, f-lc; E, N Eng r; Scot to W, only lc; Ire, S and W only, lc; only near coast; on damp hths, paths in hthy wds. Fl 7–10.

▲ **I** **Skullcap**, *S. galericulata*, resembles H in habit, but is taller; lvs 2–5 cm long, **shallow-toothed**; calyx as in H; corolla-tube 10–20 mm, slightly **curved, blue-violet**. Br Isles: GB, c (except NE Scot r); Ire f; on riverbanks, canals, streamsides, fens, mds, wet wds. Fl 6–9. (The hybrid of H and I (*S. x hybrida*) resembling H in lf size, but with fls violet, occurs occasionally, especially with H).

▲ **J** **Ground-ivy**, *Glechoma hederacea*, ± softly hairy creeping and rooting per herb, with ascending lfy fl-stems 10–20 cm tall; lvs 1–3 cm wide, **kidney-shaped, blunt-tipped**, toothed, long-stalked; fls in 2–4 fld whorls in lf-axils; calyx tubular, ± two-lipped; corollas 15–20 mm, pale violet (or, rarely, pink) with purple spots on hanging, three-lobed lower lip; upper lip flat; tube straight, tapered to base. Br Isles, vc (but NW Scot r); in wds on all except poorest soils, scrub, hbs, gslds. Fl 3–5.

ID Tips   Woundworts

Woundworts (*Stachys*) are rather like **Dead-nettles** and **Hemp-nettles**, but the **key differences** are:

- Unlike Dead-nettles, Woundworts have a calyx **without** spine-like teeth; corolla lower lip with mid-lobe **oblong, much longer** than side-lobes and **convex** in shape.

- Unlike Hemp-nettles, the lower lip of Woundwort corollas do not have conical knobs at the base.

KEY TO WOUNDWORTS (*STACHYS*)

1 Lvs mostly in basal rosette – stem-lvs few, in distant pairs – fls mostly in a dense, short, oblong, red-purple spike ....*Stachys officinalis* (p 380 **A**)
Lvs mostly in pairs up stem – fls in ± elongated whorled spikes **2**

2 Ann weed of ar, to 20 cm tall – corolla 7 mm long or less, little longer than calyx – fls pink, purple-streaked ....*S. arvensis* (p 380 **B**)
Creeping or tufted per or bi herbs, 30–80 cm tall – corolla 12 mm long or more, nearly twice as long as calyx...**3**

3 Stem and lvs felted with dense white silky hairs **4**
Stem and lvs green, hairy, but not felted...**5**

4 Lower lvs with cordate bases .... *S. germanica* (p 379 **C.1**)
Lower lvs with wedge-shaped bases ....*S. byzantina* (p 380 **C.2**)

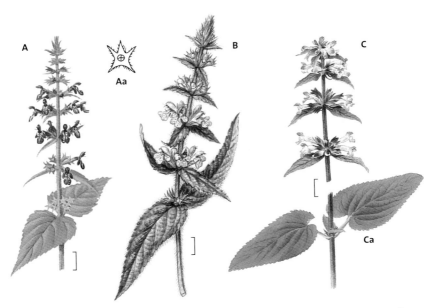

A

Aa

B

C

Ca

5 Bracts of fls nearly as long as calyx – whorls of fls dense – corollas pinkish-red with yellow eye – calyx-teeth ¹/₂ length of tube – plant tufted, no creeping stems – vr ....
S. alpina (p 379 **C**)

Bracts of fls minute, bristle-like, at most ¹/₄ length of calyx – whorls of fls looser – corollas pink or purple, no yellow eye – calyx-teeth more than half length of tube – creeping stems – c **6**

6 Lvs lanceolate-oblong, odourless, v short (5 mm)-stalked or stalkless above, bases rounded – fls pink-purple ....*S. palustris* (p 379 **B**)

Lvs oval, stalks all long (1.5–7.0 cm), cordate-based, with harsh smell; fls beetroot colour with white pattern ....*S. sylvatica* (p 379 **A**)

A AWI **Hedge Woundwort**, *Stachys sylvatica*, harsh-smelling bristly per, with creeping rhizome and erect stems, 30–80 cm; lvs **oval-cordate**, 4–9 cm long, with **stalks 1.5–7.0 cm**, ¹/₄ or more of length of lf and stalk together; infl a loose terminal spike; fl-bracts minute; calyx (**Aa**) with rigid triangular teeth; fls beetroot-red with white pattern on lip. Br Isles, vc in wds, hbs. Fl 7–9.

B Marsh Woundwort, *S. palustris*, odourless bristly per, with creeping rhizome and erect stems, 30–80 cm, lvs **lanceolate or oblong**, 5–12 cm long, with lvs on flg stem **short-stalked** (5 mm) below, **stalkless** above; fls pink-purple with white pattern on lip. Br Isles, c on streamsides, marshes. Fl 7–9. Hybridizes with A (*S. x ambigua*), often looking like a large version of B but lvs on flg stems with **short** upper lf-stalks (up to 1/5 length of lf and stalk together) and plant sterile.

C Sch 8 **Limestone Woundwort**, *S. alpina*, softly hairy per with tufted stems 40–80 cm, none creeping; lvs (**Ca**) **oblong-cordate**, 4–16 cm long, stalks 3–10 cm; infl a long interrupted spike of dense close whorls; fl-bracts linear, ± length of calyx; fls pinkish-red with **yellow eye**. GB: Glos, Denbigh vr; in open wds on calc soil. Fl 6–8.

C.1 Sch 8 VU ****Downy Woundwort**, *S. germanica*, per or bi herb, 30–80 cm; differs from C in plant densely felted with **white silky hairs**; at least lower lvs **cordate** at base; green lf-surface can be seen at least on upper surface of lvs; fls in dense terminal spike; corolla 20 mm, pale rose-pink; bracts linear, ± length of calyx. Eng: Oxon only, vr; on gslds, rds, hbs on calc soils, especially after disturbance. Fl 7–8. **Do not confuse C.1** with **C.2 Lamb's-ear**, *Stachys byzantina*, a popular gdn plant that is also white-woolly but **key difference** is in lvs **wedge-shaped** (not cordate) at base and **densely** white-woolly so that the green lf-surface can not be seen. Introd (Turkey); o gdn escape on wa, rds, quarries. Fl 6–8.

A FPO (Eire) WO (NI) AWI **Betony**, *S. officinalis* (*Betonica officinalis*), sparsely hairy per herb, 10–60 cm; basal **rosette** of long-stalked, oblong, cordate lvs, 3–7 cm long; stem-lvs (**Aa**) in few distant pairs, upper sessile; all lvs coarsely-toothed and blunt; infl a short oblong whorled spike, lowest bracts like the lvs; calyx 7–9 mm, with bristle-pointed teeth; corolla (**Ab**) 15 mm, red-purple, tube longer than calyx. Eng, Wales c (except E Anglia o); Scot, Ire r; in open wds, hths, gslds on both calc and acid soils. Fl 6–9.

B NT **Field Woundwort**, *S. arvensis*, ann with spreading and ascending hairy stems to 20 cm tall; lvs 1.5–3.0 cm, oval, blunt, toothed; infl whorls few-fld; corollas small (6–7 mm long) pale pink, purple-streaked. Br Isles: S of GB f-lc Scot r; Ire lf; on ar on non-calc soils. Fl 4–11.

C **Black Horehound**, *Ballota nigra*, roughly hairy per herb, 40–80 cm tall, with **harsh** resinous smell; lvs (**Ca**) oval-cordate, stalked, rough, coarse-toothed, 2–5 cm long; infls **spikes** of many-fld whorls in lf-axils; calyx (**Cb**) 1 cm long, **funnel-shaped**, with 10 veins and 5 short, oval, pointed, ± equal teeth; corolla two-lipped, **dull purple**, hairy all over, 12–18 mm long, with short tube and concave upper lip. Br Isles: GB, c N to SE Scot, r N Scot; Ire r; on hbs, rds. Fl 6–10.

D *White Horehound**, *Marrubium vulgare*, erect per herb 30–60 cm, stem **white-felted**; lvs 1.5–4.0 cm, rounded-oval, **wrinkled, blunt-toothed**, white-woolly beneath, grey-downy above, lower stalked; infl in dense whorls in lf-axils up stem; calyx (**Da**) with **10** small, ± equal, **hooked** teeth; corolla 1.5 cm long, **white**, upper lip flat, two-lobed, hairless. S Eng, S Wales r; rest of Br Isles vr-abs, decreasing; on rds, hbs, gslds, especially near coasts. Fl 6–10.

E VU **Cat-mint**, *Nepeta cataria*, minty-scented per herb with grey-downy erect stems 40–80 cm; lvs 3–7 cm, oval-cordate, stalked, **coarse-toothed**, grey-woolly below, downy but greener above; infl an oblong head of dense whorls; calyx (**Ea**) downy, tubular, with 5 **straight** teeth; corolla 12 mm, two-lipped, tube **curved**, hairless, white with purple dots, upper lip flat, rounded. Eng, IoM o in S and E; rest of Eng r; Scot, Ire vr-abs; on hbs, rds on calc soil. Fl 7–9.

F **Selfheal**, *Prunella vulgaris*, sparsely downy per herb with creeping runners and erect fl-stems to 20 cm tall; lvs **stalked**, oval, **widest at base**, **dull**, ± untoothed, pointed, bases rounded or wedge-shaped; infl a dense oblong **head**, with hairy, ± purplish bracts; purplish calyx (**Fa**) with 3 teeth of flattened upper lip **v short**, bristle-pointed, the **outer 2 diverging**; lower lip with 2 long narrow teeth; edges of calyx-tube parallel (so outline **rectangular**); corolla 10–14 mm long, violet, rarely white or pink, upper lip v concave. Br Isles, vc in gslds, wds, wa. Fl 6–10. **F.1 Cut-leaved Selfheal**, *P. laciniata*, close to F, but fls normally cream, upper lvs pinnate or pinnatifid, v downy; upper calyx-lip teeth **parallel**. Possibly native; Eng: S, SE only, S Ire r; in gslds on calc soils. Fl 6–10. Hybridises with F (*P.* x *intermedia*).

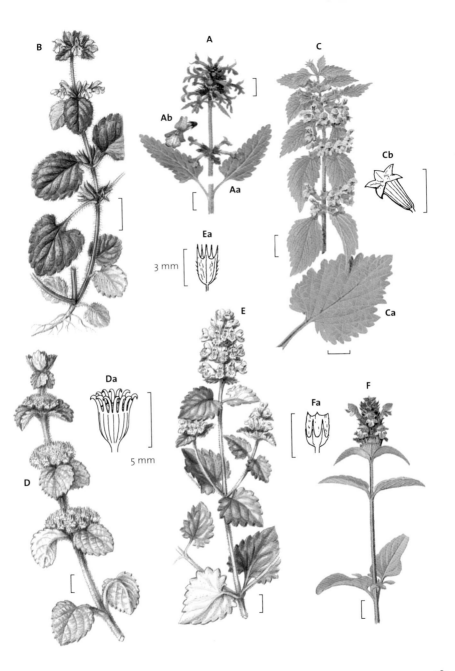

B

A

Ab

Aa

C

Cb

Ea

3 mm

E

D

Da

5 mm

Fa

F

A Bugle, *Ajuga reptans*, per herb with long lfy rooting runners; erect fl-stems 10–30 cm tall, stem ± **hairy** on **2 opp sides** only; basal lvs in a rosette, 4–7 cm long, **shiny**, **obovate**, ± untoothed, narrowed into long stalks; stem-lvs in few pairs, shorter, **unstalked above**; infl a spike of whorls of fls in lf-axils; upper bracts **shorter** than fls, bluish-green; calyx bell-shaped, 5–6 mm, with short teeth; corolla blue (rarely pink or white), lower lip with white streaks. Br Isles, vc in damp wds, hbs, mds. Fl 4–6.

A.1 WO (NI) VU **Pyramidal Bugle*, *A. pyramidalis*, has no runners above gd; stem hairy all round and tufted, not solitary; oval basal lvs form a dense rosette, infl dense, and its **lfy** bracts hairy and all **much longer** than fls, untoothed and **bright purple**. Br Isles: NW Eng vr; N, W Scot o-lf; W Ire vl; calc rocks. Fl 5–7.

B Sch 8 EN ****Ground-pine**, *A. chamaepitys*, hairy ann, often branched from base, 5–20 cm tall, with a pine-resin smell when bruised; stem-lvs each divided into 3 narrow, linear, blunt lobes like conifer lvs; fls 2 per whorl, shorter than lvs, corolla **yellow**, lower lip 5–7 mm, red-dotted. SE Eng, r and decreasing; vr introd elsewhere in Eng; on ar and open stony gd on calc soils. Fl 5–9.

C Wood Sage, *Teucrium scorodonia*, downy erect tufted per herb, 15–60 cm tall; lvs 3–7 cm, oval, cordate-based, wrinkled, blunt-toothed, pointed, stalked; infls loosely-branched lfless spikes; fls (**Ca**) in opp pairs; bracts v small; calyx with uppermost tooth rounded, broader than other teeth; corolla pale greenish-yellow, lip 5–6 mm long, stamens red. Br Isles, vc (but E mid Eng, central Ire r); in dry wds, gslds, hths, dunes, not on v calc soils. Fl 7–9.

D Sch 8 **Cut-leaved Germander**, *T. botrys*, erect, soft-hairy ann or bi, 10–30 cm tall; lvs (**Db**) 1.0–2.5 cm, **deeply pinnatifid** into ± lobed blunt segments 1–2 mm wide; fls (**Da**) in 2–6-fld whorls up stem in lf-axils; calyx tubular, ± equally five-toothed, net-veined, pouched below; corolla **bright pink**, lip 8 mm long. Possibly introd from cultivation; SE Eng, SE from Cotswolds, vr; on ar, open stony gslds on calc soils. Fl 7–9.

E Sch 8 EN ****Water Germander**, *T. scordium*, softly hairy per with runners, and erect fl-stems to 40 cm; lvs 1–5 cm, oblong, grey-green, **stalkless**, bases rounded, coarse-toothed; side-veins strong. Fls 2–6 together in distant whorls up stem, in axils of lvs much **longer** than fls. Corolla **pink**, lip to 10 mm long. Eng: N Devon, E Anglia, CI vr; Ire, by R Shannon and in Clare vlf; in wet dune-hollows, swamps. Fl 7–10.

F Wall Germander, *T. chamaedrys*, spreading and ascending, ± hairy, tufted per herb, woody at base; fl-stems 10–20 cm tall. Lvs oval, blunt, narrowed to short stalk, shiny dark green, with deep but rounded teeth; infls 5–10 cm, spike-like, upper bracts shorter than fls; calyx with 5 ± equal teeth; corolla deep pinkish-purple, lip 8 mm long. Introd (W Eur); GB: Eng, Sussex coast, probably native in chk gsld, vr; scattered elsewhere in Br Isles; on old walls, rocks. dry banks. Fl 6–9.

MARE'S-TAIL FAMILY   *Hippuridaceae*

G Mare's-tail, *Hippuris vulgaris*, **aquatic** unbranched per herb, with stout spongy round stems bearing strap-like blunt lvs in **whorls**; stems both erect emergent, and trailing submerged. Fls tiny, petalless, green; male and female fls separate at base of lvs on emergent stems, male with one reddish anther. Br Isles, lf in base-rich waters of lakes and slow streams. Fl 6–7.

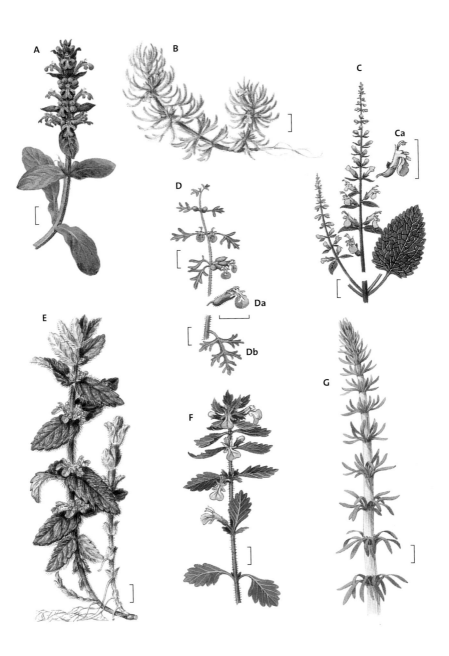

WATER-STARWORT FAMILY   *Callitrichaceae*

Water-starworts (*Callitriche*) are herbs of fresh water or damp mud, also occasionally in pools and rivers at the upper edge of the saline limit. They have slender stems and **opp** pairs of **untoothed**, linear or oval lvs **without stipules**; in many spp upper lvs are close-set to form a floating **terminal rosette**. Fls tiny, male and female separate or occasionally together in leaf axils, lack perianths; male of 1 stamen, female of an oval, 4-lobed ovary with 2 styles. Fr 1-2 mm long, the 4 lobes **winged** in some spp. The key (below) includes all spp of Br Isles, and uses both lf and fr characters; however plants **lacking fr cannot be named** with confidence.

ID Tips   Water-starworts

- **What to look at: (1)** Water-starworts (*Callitriche* spp) cannot be identified without ripe frs **(2)** pollen colour can help with id and yellow anthers can often be seen among the lvs in floating rosettes **(3)** upper and lower lf shape is also useful but often v variable (lvs vary in form according to whether they are floating, submerged, on mud, or in still or flowing water).

- Frs can usually be found if you look carefully.

- Shoots with yellow anthers in the rosette are often the best to check for frs. The ripest frs will be found towards the base of shoots.

1 Male flower
2 Female flower
3 Bracts
4 Filament
5 Anther
6 Fruit
7 Stigmas

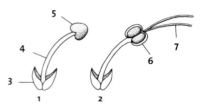

KEY TO WATER-STARWORTS (*CALLITRICHE*)
by R. V. Lansdown

1 Frs without a wing that is clearly different from the seed when held up against the light **2**
Frs with a wing that is clearly different from the seed when held up against the light **3**

2 Anthers yellow – submerged lvs ± parallel-sided, floating lvs expanded, opaque, opposed lvs joined at the base – often with floating rosette, floating lvs usually ridged along veins
Callitriche obtusangula (p 386 **D**)
Anthers translucent, difficult to locate – lvs never expanded, translucent, opposed lvs not joined at the base – upper lvs never forming a floating rosette................*C. truncata* (p 385 **B**)

3 Ripe frs blackish..**4**
Ripe frs not black but generally brown or grey...**5**

4 Ripe frs winged only in the upper parts, the wing tapering towards the base of the frs –
stigmas minute, erect – anthers pale yellow ....*C. palustris* (p 386 **F**)
Ripe frs winged from base to apex, of ± even width throughout – stigmas folded down
against the side of the frs, often emerging below the top of the frs – anthers translucent,
difficult to locate...*C. brutia* (p 386 **G**)

5 Anthers colourless, difficult to locate – lvs translucent, opposed lvs not joined at the base
C. hermaphroditica (p 385 **A**)
Anthers yellow – lvs opaque, joined at base **6**

6 Most lvs broadly rounded in outline – wing of fr broad.............................*C. stagnalis* (p 386 **C**)
Upper and rosette lvs elongate spoon-shaped, lower lvs often ± parallel-sided – wing of fr
narrow ....*C. platycarpa* (p 386 **E**)

A Autumnal Water-starwort, *Callitriche hermaphroditica* ssp *hermaphroditica*,
has all lvs **narrow**, translucent, ± parallel-sided or tapering slightly from base to
apex, leaf rosettes **never** formed; anthers **colourless**, difficult to locate; frs (**Aa**
p 386) **to 1.7 mm** long, almost stalkless, **brown to dark brown** when ripe, lobes
divergent in the form of a cross from above, with conspicuous **wings** as wide as
the seed. Br Isles: Eng from midlands north, o; lakes and occasionally rivers. **A.1**
ssp *macrocarpa*, has frs (**Ab** p 386) **1.5–3 mm** long. Br Isles, N from Anglesey, o;
lakes, ponds and occasionally rivers. Fl 5–9.

B FPO (Eire) *****Short-leaved Water-starwort**, *C. truncata*, resembles A in all lvs
narrow, translucent, more or less parallel-sided; rosette lvs **never** formed; anthers
colourless, difficult to locate; frs (**Ba** p 386) **1–2 mm** long, almost stalkless or
stalked to 1 mm, brownish when ripe, lobes divergent in the form of a cross from
above, the lobes **unwinged**. Br Isles: Eng and Wales, o; SW Ire, vr; in lakes, canals
and occasionally rivers. Fl 5–9.

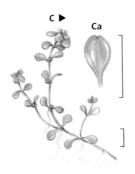

C ▶   **Ca**

▲ **C** **Common Water-starwort**, *Callitriche stagnalis*, unlike A or B most lvs (**Ca**) broadly **spoon-shaped** to **oval**, occasionally lower lvs ± parallel-sided, upper lvs often in floating rosette; anthers conspicuous **bright yellow**; stigmas erect; frs (**Cb**) 1–2 mm long, almost stalkless, **greyish brown** when ripe, flattened, the lobes broadly **winged**. Br Isles, c; in ponds, tracks, ditches and lake margins, often on wet mud. Fl 4–10.

D **Blunt-fruited Water-starwort**, *C. obtusangula*, lower lvs ± parallel-sided, **upper lvs angular spoon-shaped**, with longitudinal ridges; anthers conspicuous **bright yellow**; stigmas erect; frs (**Da**) 1–1.7 mm, almost stalkless, brownish when ripe, flattened, the lobes **unwinged**. Br Isles, c; in rivers, streams and ditches, occasionally lakes and wet mud. Fl 5–8.

E **Various-leaved Water-starwort**, *C. platycarpa*, lower lvs ± parallel-sided, upper spoon-shaped; anthers conspicuous **bright yellow**; frs (**Ea**) 1.4–1.8 mm, almost stalkless, flattened, brownish when ripe, the lobes clearly **winged**. Br Isles, c; in lakes, rivers, streams and ditches, occasionally on wet mud. Fl 5–9.

F EN ****Narrow-fruited Water-starwort**, *C. palustris*, lower lvs ± parallel-sided, upper spoon-shaped; anthers inconspicuous v small, **pale yellow**; stigmas erect; frs (**Fa**) **0.8–1.1 mm**, almost stalkless, flattened, **black when ripe**, the lobes winged **only** in the upper half. Br Isles, vr; one site W Ireland, one site SW Scotland; in ponds, tracks, ditches and lake margins, often on wet mud. Fl 5–9.

G **Intermediate Water-starwort**, *C. brutia* (*C. intermedia*) ssp *brutia*, lower lvs ± parallel-sided, upper spoon-shaped; anthers **translucent**, difficult to locate; **stigmas folded down** against the side of the frs, often emerging below the top of the fr; frs (**Ga**) 1–1.5 mm, stalkless or on stalk up to 12 mm, **blackish** when ripe, flattened, the lobes narrowly **winged**. Br Isles, f; in streams, pools and ditches, often wet mud. **G.1** ssp *lacustris* (*C. hamulata*), **key difference** from G in lf tips often **spanner-shaped** and frs **always** almost stalkless. Br Isles, f in lakes, rivers, streams and ditches, occasionally on wet mud. Fl 5–9.

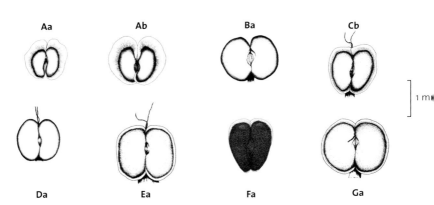

Aa Ab Ba Cb

1 mm

Da Ea Fa Ga

PLANTAIN FAMILY

Plantaginaceae

Herbs with lvs usually all in **basal rosettes**; infls in heads or spikes on long **lfless stalks**; fls tiny, with the scarious calyx and corolla both four-lobed, 4 long conspicuous stamens, and 2–4-celled ovary forming a capsule in fr. **Plantains** (*Plantago*) are land plants with no runners and bisexual fls; **Shoreweed** (*Littorella*) is ± aquatic, on sandy lake-shores, with stolons, and male and female fls on separate stalks.

A Ribwort Plantain, *Plantago lanceolata*, rosette-forming per herb; lfless, silky-hairy fl-stems 10–40 cm; lvs all basal, lanceolate or oval-lanceolate, spreading or erect, scarcely toothed, with 3–5 strong, ± parallel veins, narrowed to short stalk; infl-stalk deeply **furrowed**; oblong infl of many small fls, each with a tiny pointed bract; fls 4 mm, calyx green, corolla brownish, its 4 bent-back lobes with brown midribs; stamens long, white. Br Isles, vc on gslds, wa, rds. Fl 4–10.

B Greater Plantain, *P. major*, differs from A in ± hairless, **broad-oval** to **elliptical**, many-veined lvs (**B**), all narrowed into **stalks as long as lvs**; infl-stalk **unfurrowed**, 10–15 cm, about as long as lvs; corolla-lobes dirty-white, with no brown midribs; stamens short, yellow. Br Isles, vc on ar, wa, rds, especially trodden places. Fl 6–10.

C Hoary Plantain, *P. media*, differs from B in **greyish-downy**, 5–9 veined elliptical or diamond-shaped lvs, gradually narrowed into **short** stalks, mostly forming **flat rosette** on gd; unfurrowed infl-stalks bear oblong spikes, 2–6 cm long, of **scented** fls; corollas white, stamens **pinkish** with **purple** stalks. Pollinated by insects (unlike other Plantains, which are wind-pollinated). Br Isles, r or vl (but Eng f-lc); on calc gslds. Fl 5–8.

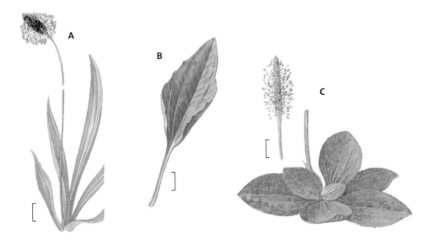

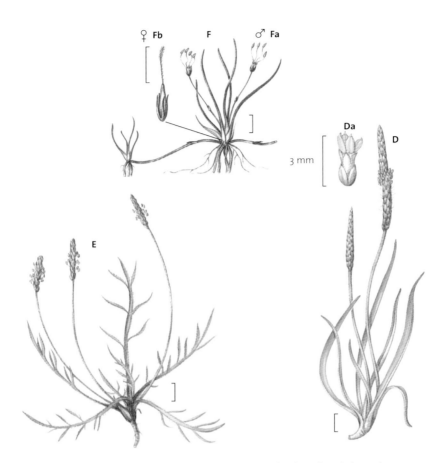

♀ Fb   F   ♂ Fa

Da   D

3 mm

E

D **Sea Plantain**, *Plantago maritima*, ± hairless per herb with wdy base; lvs 5–20 cm long, ± erect, **fleshy**, 3–5-veined, ± untoothed, **narrow-linear**; infl-spike 2–6 cm long, on a long erect unfurrowed stalk. Fls (**Da**) 3 mm long, corollas brown with darker midribs; stamens pale yellow. Br Isles: coasts o; mts of N Wales, N Eng, Scot, lc; inland limestone areas of W Ire; on salt marshes, wet mt cliffs. Fl 6–8.

E **Buck's-horn Plantain**, *P. coronopus*, usually downy bi herb; flat rosette of deeply **pinnatifid** lvs, 2–6 cm long, segments linear, one-veined, often again deeply-lobed (but lvs sometimes narrow-linear, unlobed, merely toothed; sometimes erect). Bracts of fl-spike with usually long, spreading points; corolla brownish, stamens yellow. Br Isles, c near coasts (also inland in S, E Eng, f-lc); on rocks, sea-cliffs, dry sandy or gravelly gslds near sea, on commons inland. Fl 5–7.
E.1 Branched Plantain, *P. arenaria* (*P. indica*), **key difference** from other Plantains

in erect **branched** stems bearing along them **opp**, narrow-linear lvs to 10 cm long; and in long-stalked infls in lf-axils; fl-heads egg-shaped, 1 cm long. Introd (S Eur); S Eng vr; on wa, dunes. Fl 7–8.

F Shoreweed, *Littorella uniflora*, hairless turf-forming per herb of shallow water or lake shores; creeping and rooting runners form **rosettes** or erect tufts of **linear** lvs; lvs, 2–10 cm long, flat one side, rounded on other (thus half-cylindrical in shape), **spongy** inside, not pointed, have sheathing bases. Male fls (**Fa**) solitary on stalks 5–8 cm long, have 4 small scale-like sepals, 4 tiny whitish petals, and 4 long-stalked (1–2 cm) stamens; female fls (**Fb**) almost stalkless, 1–3 at base of male fl-stalk, with style 1 cm long. Scot, N Eng, Wales, Ire lc; S, mid Eng r-vlf, submerged (to 4 m deep) in shallow non-calc lakes, and exposed on their sandy or gravelly shores as water falls in summer. Fl 6–8.

BUTTERFLY-BUSH FAMILY *Buddlejaceae*

Deciduous to semi-evergreen **shrubs** with alt or opp lvs without stipules; small fls in dense, **showy panicles**, fragrant; sepals 4 fused into a tube; **4** corolla-lobes and **4 stamens** with a 2-celled superior ovary; fr a 2-valved capsule.

Butterfly-bush, *Buddleja davidii*, shrub to 5 m; opp lvs lanceolate to ovate, usually toothed, white-downy below; infl in dense spike-like panicle, fls mauve-purple, 4 petals fused into a tube with 4 stamens. Introd (China), very invasive and, in effect, damages butterfly habitats by shading out caterpillar food plants. Br Isles, vc but restricted to lowland; on wa, railways, rds, urban areas, preferring dry disturbed gd. Fl 4–8.

OLIVE FAMILY

Oleaceae

Trees and shrubs with **opp** lvs; fls with **4 sepals**, usually **four-lobed** corolla, 2 stamens; superior ovary of 2 joined carpels.

A Ash, *Fraxinus excelsior*, tall tree (to 30 m), deciduous, with pale grey bark, ridged with age; twigs **smooth, grey**; buds **black**, 5–10 mm long; lvs (**Ac**) opp, 20–30 cm long, pinnate, with 3–6 pairs of oval, pointed, shallow-toothed lfts. Fls (**Aa**), in panicles, appear before lvs; no perianth; 2 purple stamens, and a narrow ovary that forms a flat, oblong fr (**Ab**), 3 cm long, winged at tip like a propeller-blade. Br Isles, c-la in wds, hds, on moister, base-rich soils. Fl 4–5.

B Lilac, *Syringa vulgaris*, deciduous, suckering shrub to 7m; opp lvs are oval, untoothed, cordate at base, ± hairless; infl in large terminal pyramidal panicles; fl with 4 petals white, lilac or purple forming a tube below, 2 stamens; fr a capsule. Introd (SE Eur); gdn escape natd throughout Br Isles on rds, railway-banks, wa, rough gsld. Fl 4–5.

C Wild Privet, *Ligustrum vulgare*, semi-evergreen shrub with opp elliptical-lanceolate minutely **hairy** 1st year twigs, 3–6 cm long; fls in terminal panicles; corolla white, 4–5 mm across, four-lobed, scented, the 2 stamens projecting from corolla-tube. Fr (**Ca**) a black shiny berry, 6–8 mm across. Br Isles, c (except N Scot r); in hbs, scrub, gsld, especially on calc soils. Fl 6–7. **C.1 Garden Privet**, *L. ovalifolium*, the sp usually planted, has oval **hairless** 1st year twigs. Br Isles, c near housing on wa, hds, railway banks. Fl 7.

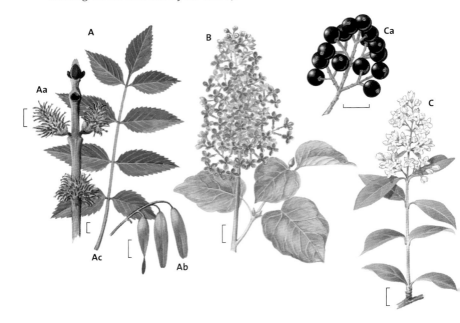

FIGWORT FAMILY   *Scrophulariaceae*

Herbs (in Br Isles); lvs with no stipules, and often alt above, opp below (or all opp). Fls bisexual; calyx five- or four-lobed; corolla irregular, usually two-lipped, five-lobed (sometimes four-lobed); stamens 5 or 4 (or 2 in *Veronica*). Ovary two-celled, with style **on top** of the two lobes; fr a capsule with 2 cells, each several-seeded.

ID Tips   Figwort Family

- **What to look at**: **(1)** number of stamens **(2)** presence or not of a two-lobed swelling projecting from lower lip (**palate**) **(3)** calyx **(4)** corolla shape and colour.

- **Do not confuse** with the Deadnettle family which has frs of 4 separating one-seeded nutlets, not a many-seeded capsule as in Figwort family.

- **Mullein** (*Verbascum*) is the only member of the Figwort family with **5 stamens** (v rarely 4).

- **Mulleins** often hybridize; hybrids have seed capsules which readily collapse to dust when rubbed between finger and thumb.

KEY TO FIGWORT FAMILY

1 Stamens 5 – corolla five-lobed, almost regular, lower petal only slightly longer than others – lvs alt – fls in long racemes ....*Verbascum* (p 392)
Stamens 2 or 4 ....**2**

2 Stamens 2 – corolla ± with 4 lobes – lvs opp...*Veronica* (key p 398)
Stamens 4 ....**3**

3 Corolla two-lipped, pouched or spurred below mouth of tube, closed by a two-lobed swelling projecting from lower lip ('palate') – Snapdragon-like fls.................................**4**
Corolla not pouched or spurred, no 'palate' on lower lip **8**

4 Corolla pouched below, but without spur...............................*Antirrhinum* & *Misopates* (p 396)
Corolla with a pointed hollow spur projecting from base of lower lip **5**

5 Fls in terminal racemes or spikes – lower lvs whorled, upper alt ....*Linaria* (p 394)
Fls in lf-axils along stem – lvs nearly all alt **6**

6 Lvs linear, strap-shaped – plant erect ....*Chaenorhinum minus* (p 396 **A**)
Lvs long-stalked, broad, round or arrow-shaped – plant ± creeping **7**

7 Lvs pinnately-veined, oval or arrow-shaped ....*Kickxia* (p 396)
Lvs palmately-veined, ivy-shaped ....*Cymbalaria muralis* (p 396 **I**)

8 Plant small, creeping, rooting at intervals – corollas tiny (under 5 mm), solitary, with v short narrow corolla-tubes **9**
Plant ± erect, lfy, not creeping and rooting – corollas with long or cup-shaped tubes – fls grouped in infls **10**

9 Lvs oblong-linear, untoothed – tiny plant on mud ....*Limosella aquatica* (p 398 **H**)
Lvs kidney-shaped, long-stalked, toothed – in moist wds ....*Sibthorpia europaea* (p 398 **I**)

10 Lvs opp (except bracts sometimes alt in infl) **11**
Lvs alt **17**

11 Calyx clearly five-toothed **12**
Calyx four-toothed..**13**

12 Stems square in section – corolla-tube like a small rounded cup, with 2 small (2–3 mm-long), usually red-brown upper lobes, 3 greenish lower lobes ....*Scrophularia* (p 394)
Stems round in section – corolla with tube 10–45 mm long, with 5 large spreading yellow lobes..*Mimulus* (p 396)

13 Calyx-tube inflated, flattened on each side, like a tiny oval purse, narrowed to four-toothed tip – corolla yellow, with 2 violet teeth on upper lip ....*Rhinanthus* (p 407)
Calyx-tube not inflated; but cylindrical or bell-shaped..**14**

14 Upper corolla-lip flattened on each side to form roof-like ridge – mouth of corolla-tube almost closed – lower lvs ± untoothed, or toothed only near their bases ....
Melampyrum (p 404)
Upper corolla-lip not flattened at sides – mouth of corolla-tube open – all lvs and bracts toothed along all edges..**15**

15 Upper corolla-lip ± erect, two-toothed – lower lip with purple lines ....*Euphrasia* (p 405)
Upper corolla-lip unlobed, arched forwards..**16**

16 Fls purple, over 1 cm long, with oval purple bracts – in mts ....*Bartsia alpina* (p 406 **C.1**)
Fls yellow, over 1 cm long, with lanceolate green bracts ....*Parentucellia viscosa* (p 406 **C**)
Fls pink, 4–8 mm long, with lanceolate green bracts ....*Odontites vernus* (p 405 **B**)

17 Corolla clearly two-lipped, lower lip with 3 spreading lobes, only 2–3 times length of tubular inflated five-toothed calyx – fls in short racemes – lvs pinnate ..*Pedicularis* (p 406)
Corolla tubular to bell-shaped, with 5 lobes, not clearly two-lipped **18**

18 Corolla bell-shaped much longer than deeply five-lobed calyx – fls in long one-sided racemes – lvs unlobed, oval-lanceolate ....*Digitalis purpurea* (p 394 **D**)
Corolla-tube narrow forming 5 petal-like lobes, scarcely longer than calyx – fls in terminal clustered raceme – lvs obovate or lobed ....*Erinus* (p 394 **E**)

A **Great Mullein**, *Verbascum thapsus*, tall (to 2 m) stout white-woolly bi, with round, usually unbranched stem; broad oval-elliptical alt lvs with winged stalks running down stem (decurrent). Infl a dense terminal spike, sometimes with small side spikes from upper lf-axils. Fls bright yellow, with 5 ± equal spreading lobes, lowest lobe rather larger; 5 stamens, stalks of **upper 3 only** clothed densely with white or yellow hairs; 3 **transverse-fixed** anthers (fig 1) and 2 **oblique** (or ± decurrent)-**fixed** anthers (figs 2, 3). **Key differences** from all other British *Verbascum* spp are: anthers **<2 mm long**; stigma knob-like; 2 lower filaments ± hairlesss; and upper-stem lvs decurrent. Fr an oval capsule. Br Isles, c except N and W Scot r, Ire o-f; in open wds, dry hbs, wa. Fl 6–8. By far the most widespread British *Verbascum* sp.

B **Dark Mullein**, *V. nigrum*, differs from A in angled stem, lvs dark green above, only thinly downy, lower ones with cordate bases and long stalks, upper stalkless; also in dense **purple hairs** on stalks of **all** stamens (**Ba**); all 5 anthers transverse-fixed (fig 1). Corolla yellow, 12–20 mm wide, purple-spotted at base. Br Isles: Eng, Wales o, but lc in S and E; Scot, Ire casual and r; on dry hbs, wa, rds, gsld on calc or sandy soils. Fl 6–9.

C ****White Mullein**, *V. lychnitis*, has angled stem; lvs short-stalked, dark green, nearly hairless above, white powdery-downy on undersides. Infl branched like a

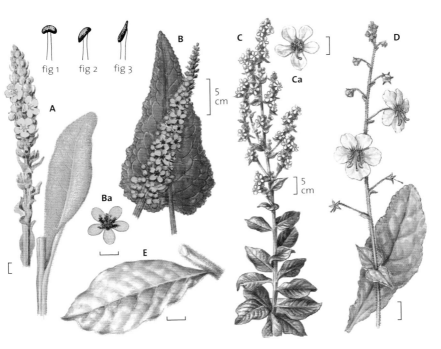

candelabrum, side branches almost erect and parallel with main stem; stalks of all stamens with dense **white hairs**; all 5 anthers transverse-fixed (fig 1). Fls (**Ca**) usually **white** in Eng, but yellow in Somerset. Br Isles: GB r (except SE lc); **Ire abs**; in open wds, scrub, dry banks, on calc soil. Fl 7–8.

D Moth Mullein, *V. blattaria*, has stem and lvs nearly hairless, but upper part of stem sticky-hairy; lvs dark shiny green, toothed, wrinkled, lower narrowed to base, upper cordate at base, stalkless. Fl-stalks mostly **longer** than calyx, **single fls** in axils of each bract; stamens all clothed with purple hairs as in B, but fls usually **white** (sometimes yellow), 20–30 mm wide. **Key difference** from other *Verbascum* spp (except D.1): 2 **decurrent-fixed** anthers (fig 3) and 3, v small **transverse-fixed** anthers (fig 1). Fr a **globular** capsule. Introd (Eur); Eng, IoM, CI o-r; in wa. Fl 6–9. **D.1 Twiggy Mullein**, *V. virgatum*, is v close to and with anthers fixed as D, but more glandular throughout; **key differences** are fls in clusters of 1–5 in axils of bracts (not single), on fl-stalks mostly **shorter** than calyx. Br Isles: possibly native Devon, Cornwall, If (introd elsewhere o-r); on gsld, wa, hbs. Fl 6–8.

E *Hoary Mullein, *V. pulverulentum*, resembles A, but lvs (**E**) thickly clothed on both sides with a **mealy** white wool that rubs off easily; upper lvs with **cordate** bases. Infl branched to form a pyramidal panicle, branches **not** erect as in C, but more spreading; all stamens equally white-hairy. All 5 anthers transverse-fixed (fig 1). Br Isles: Eng, E Anglia as native only, vlf; r introd elsewhere; on rds, scrub, gsld on calc soil. Fl 7–8.

A AWI **Common Figwort**, *Scrophularia nodosa*, erect per, 40–80 cm, hairless except for sticky hairs on infl; stem square, but ± without wings; lvs opp, oval, **pointed**, coarsely-toothed but unlobed, short-stalked. Sepals 5, oval, blunt, green, with scarcely visible white edges; corollas (**Aa**) 1 cm long, green, cup-shaped, with 2 red-brown upper and 3 green lower lobes; stamens 4, plus an upper stamen without anther (staminode). Br Isles, c (except N Scot r); in wds, hbs, wa. Fl 6–9.

B **Water Figwort**, *S. auriculata* (*S. aquatica*), is close to A, but stems (**Ba**) **four-winged**; lvs (**B**) **bluntly** tipped, **bluntly** toothed; sepals have **broad** (0.5–1.0 mm) **white borders**; staminode rounded as in F. Br Isles: Eng, Wales c; Scot vr; Ire f; in moist wds, marshes, by fresh water. Fl 6–9. **B.1 Green Figwort**, *S. umbrosa*, has stems even more broadly winged than in B, but is quite hairless; lvs pointed as in A, sharply-toothed; bracts in infl longer, lf-like; staminode forked into 2 lobes. Br Isles o, only vlf (SW Eng abs); in moist wds, by fresh water. Fl 7–9.

C **Balm-leaved Figwort**, *S. scorodonia*, has square stem of A, but whole plant **grey-downy**; lvs (**C**) wrinkled, cordate, shortly-pointed, deeply doubly-toothed; bracts lf-like; fls purple; staminode rounded. SW Eng, lf; IoS, CI c; S Eng, Wales r; in scrub on sea cliffs, hbs. Fl 6–8. **C.1 Yellow Figwort**, *S. vernalis*, with yellow fls and pointed green sepals, occurs introd (S & C Eur) o in GB (f in SE Scot), Ire abs; usually in shade on hbs, wdland clearings, wa. Fl 4–6.

D **Foxglove**, *Digitalis purpurea*, tall downy bi to 150 cm; unbranched stem; lvs green, oval-lanceolate, softly downy, bluntly-toothed, alt, 15–30 cm, with winged stalks. Fls in long erect raceme, corolla 40–50 mm, tubular to narrow bell-shaped, pink-purple, with dark purple spots on a white ground inside tube; calyx much shorter than corolla, cup-like with pointed lobes. Fr an oval capsule. Br Isles vc (except in the Fens and on chk); in wds, hbs, mt rocks, on acid soils. Fl 6–8. **Do not confuse** Foxglove lvs with **Ploughman's-spikenard** (p 444 D) lvs, which look v similar; **key difference** is habitat: Ploughman's-spikenard is restricted to calc soils whereas Foxglove usually prefers acid soils.

E **Fairy Foxglove**, *Erinus alpinus*, a tufted per herb to 20 cm; hairy obovate or spoon-shaped lvs in basal rosette with few alt stem-lvs; fls purple to deep pink (sometimes white) with 5 notched petals fused into a tube about as long as the calyx; calyx 5 linear lobes; 4 stamens; fr an ovoid capsule. Introd (mts of C Eur); Br Isles: N Eng, Scot, SW Eng o-lf, scattered elsewhere; natd on walls, stony gd, especially on limestone or lime mortar. Fl 5–8.

Toadflaxes (*Linaria*) and **Fluellens** (*Kickxia*) have spurred corollas, with their throats closed by a two-lobed swelling on the lower lip, called the **palate**.

F **Common Toadflax**, *Linaria vulgaris*, erect grey-green-lvd, ± hairless per, 30–80 cm; lvs 3–8 cm, numerous, ± whorled below, alt above, linear-lanceolate. Fls (**Fa**) in long dense terminal racemes; sepals **oval**, pointed; corollas 15–25 mm long, **yellow** with **orange** palate and long ± **straight** spur. Br Isles: GB c (but NW Scot r); Ire o; on gsld, ar, wa, hbs, railways. Fl 7–10. **F.1 Prostrate Toadflax**, *L. supina*, has yellow fls like F, but corollas only 10–15 mm long; stems creeping then ascending, only **5–20 cm** tall; lvs **waxy-grey, blunt**; sepals **linear**, blunt. Introd (SW Eur); SW Eng, IoM vr; on dunes, wa, on calc soils. Fl 6–9.

G Purple Toadflax, *L. purpurea*, is like F, but fls in much longer racemes; corollas **violet, unstriped**, 8 mm long, with **long** (1/2 rest of corolla) curved spur. Br Isles, vc in S Eng, elsewhere o-lc; gdn escape on old walls, wa. Fl 6–8.

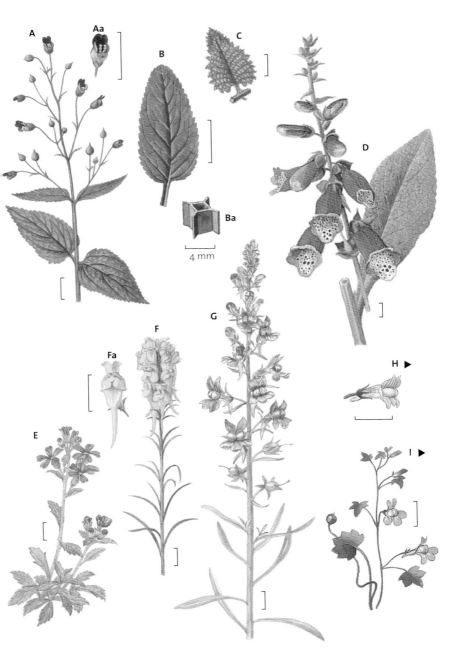

▲ **H Pale Toadflax**, *Linaria repens*, erect per with habit of G, but corollas (**H**) 7–14 mm long, **white** or **lilac, striped** with **purple**; spurs **straight**, only ¼ length of rest of corolla. Br Isles: Eng, Wales o-lf; Scot, Ire r; on dry calc gslds, hbs. Fl 6–9.

▲ **I Ivy-leaved Toadflax**, *Cymbalaria muralis*, **creeping** hairless per; lvs **long-stalked, palmate-veined, ivy-shaped, thick, alt**, 2.5 cm long; fls **solitary** in lf-axils; corollas 8–10 mm, lilac or white, with yellow spot on white palate; spur short, curved. Br Isles, vc (but NW Scot r); on walls, rocks. Fl 5–9.

A Small Toadflax, *Chaenorhinum minus*, erect sticky-downy ann, 8–25 cm; lvs alt, linear-oblong, blunt, short-stalked; fls (**Aa**) axillary, **long-stalked**; corollas 6–8 mm long, pale purple, spurs short, blunt, curved. GB, c but declining (except N, W Scot vr); Ire f; usually on calc soils on wa, ar, railways. Fl 5–10.

B Round-leaved Fluellen, *Kickxia spuria*, ± prostrate sticky-hairy ann, 20–50 cm long; lvs (**B**) oval to ± round, 1–5 cm, long-stalked, hairy; fls in lf-axils, with **long, woolly stalks**; corollas 8–11 mm, **yellow**, upper lips **dark purple, spur curved**. S, E Eng f-lc; Wales r; **rest of Br Isles abs**; on ar, wa especially calc. Fl 7–10.

C Sharp-leaved Fluellen, *K. elatine*, close to B, differs in **triangular**, less hairy, **hastate, pointed** lvs (**C**), more slender habit, **pale** purple upper lips to corollas, **straight** spurs, **hairless** fl-stalks. Br Isles: S, E Eng, IoS, CI f-lc; Wales o; S Ire, IoM vr; **Scot abs**; on ar, wa. Fl 7–10. **Do not confuse** young plants of C with B, as young lower lvs of C often will not have developed the characteristic hastate shape.

D FPO (Eire) VU **Weasel's-snout (Lesser Snapdragon)**, *Misopates orontium* (*Antirrhinum orontium*), erect ann 20–50 cm, sticky-downy above; lvs 3–5 cm, linear, untoothed; fls in loose lfy racemes; **calyx-lobes** (**Da**) **linear, as long as corollas**; corollas (**Dc**) pink-purple, 10–15 mm, without spurs, but pouched below. Fr (**Db**) a capsule, glandular-hairy, opening by 3 pores, apparently looking like a weasel's snout. Br Isles: Eng, Wales, IoS, CI f-lc; Scot, Ire vr; on ar on sand. Fl 7–10.

E Snapdragon, *Antirrhinum majus*, the familiar gdn plant, found on old walls as a gdn escape; lvs lanceolate; corollas **3–4 cm** long, unspurred, various colours; calyx-lobes (**Ea**) **oval, v short**; fr (**Eb**). GB o-lf, Ire o; on walls, wa. Fl 7–9.

F Monkeyflower, *Mimulus guttatus*, erect per, creeping with long stolons, hairless below, 20–50 cm; lvs 2–7 cm, **opp**, oval, toothed, upper clasping. Fls in lfy racemes; calyx and fl-stalks **downy**, calyx tubular, five-toothed; corolla 2.5–4.5 cm, yellow with **small** red spots in throat (not petals), strongly two-lipped with three-lobed lower lip much longer than upper, throat nearly closed by palate of 2 boss-like swellings. Introd (N America); Br Isles, f-lc on streamsides, mds. Fl 7–9.
F.1 Blood-drop-emlets, *M. luteus*, v like F, but calyx and fl-stalks hairless; the scarcely two-lipped corollas have **open** throat and **large** (2–3 mm) red to purplish-brown blotches. Mainly in Scot; introd (S America) o by streams. Fl 6–9.
F.2 Hybrid Monkeyflower, *M.* x *robertsii*, a hybrid of F and F.1, often close to F and difficult to distinguish therefore as a **broad guide only**: has short stolons; 2 boss-like swellings v small, throat ± open; petals (not only throat) marked with orange, red or purplish spots; may be fertile and then v close to F. Br Isles, f (N Eng, N Ire lc); often confused with, and in the same habitats as, F. Fl 6–9. Other *Mimulus* hybrids occur and are difficult to distinguish from F.2.

▲ **G Musk**, *Mimulus moschatus*, smaller than F, ± **prostrate, sticky-hairy** all over; yellow unspotted corolla only 1–2 cm long, 1.0–1.5 cm wide. Introd (western N America); Br Isles, o in wet places. Fl 7–8.

▲ **H** FPO (Eire) WO (NI) ***Mudwort**, *Limosella aquatica*, low hairless ann, creeping by runners, which produce rosettes of lvs 5–15 mm long, upper with elliptical blades and long stalks, lower linear. Fls (**Ha**) v small, long-stalked, in lf-axils; petals shorter than sepals, star-like, 2.5–3 mm across, white or lilac with bell-shaped tube and pointed lobes, scentless. GB r, Ire vr; on mud on pond edges. Fl 6–10.
H.1 Sch 8 **** Welsh Mudwort**, *L. australis*, close to H but lvs not having distinct blade and stalk; fls larger, 3.5–4 mm across, scented. Possibly introd; Wales only, vr; saltmarsh pools and mudflats.

▲ **I** AWI ***Cornish Moneywort**, *Sibthorpia europaea*, small creeping and rooting hairy per; lvs alt, long-stalked, kidney-shaped, blunt-toothed, 0.5–2.0 cm wide; short-stalked fls 1–2 mm wide, with almost equally five-lobed spreading corollas; 2 upper lobes yellow, 3 lower pink; stamens 4. Br Isles: Sussex, SW Eng, IoS, CI, S Wales vlf; Ire vl; by streams, wet gd in wds, avoiding calc soils. Fl 7–10.

ID Tips   Speedwells

- All **Speedwells** (*Veronica* spp) have only **2 stamens**.

- **What to look at**: (**1**) lf-shape, hairiness and whether stalked or not (**2**) whether plant creeping and rooting at nodes or erect (**3**) fl colour (**4**) length of fl-stalks (**5**) whether stem hairs in 2 lines or all the way round (**6**) fr shape and hairiness.

KEY TO SPEEDWELLS (*VERONICA*)

1 Fls solitary in lf-axils...**A**
　Fls in racemes, heads or clusters **2**

2 Main stems ending with infl, so infl not in lf-axils ....**B**
　Infls in lf-axils ....**C**

A   Speedwells with fls solitary in lf-axils

1 Lvs with palmate veins, deeply 5–7-lobed, ± ivy-shaped – fls pale lilac – sepals with cordate bases..*Veronica hederifolia* (p 402 **G**)
　Lvs with pinnate veins, not all arising together from base of lf – fls blue, or blue with white lip..**2**

2 Lvs kidney-shaped, v rounded – stems rooting at nodes...........................*V. filiformis* (p 402 **F**)
　Lvs oval or oblong – stems not rooting at nodes (or doing so only near base).........................**3**

3 Fr-lobes widely spreading to 90° – fls large (8–12 mm wide) – lower petal white...................
　　　　　　　　　　　　　　　　　　　　　　　　　　　　　　　V. persica (p 402 **B**)
　Fr-lobes spreading to 45° only – fls small (4–8 mm wide) **4**

4 Sepals oval, pointed – fr with long glandular and short non-glandular hairs – corolla all blue..*V. polita* (p 402 **C**)
　Sepals oblong, blunt – fr with glandular hairs only – lower petal white ....
　　　　　　　　　　　　　　　　　　　　　　　　　　　　　　　V. agrestis (p 402 **D**)

B   Speedwells with fls in racemes or heads terminating main stems, not in lf-axils

1   Fls in long, dense lfless, spike-like terminal raceme – lvs elliptical, blunt-toothed, downy – corolla-lobes narrow, deep Oxford blue – r ....*V. spicata* (p 400 **E**)
Fls in shorter or looser, lfy terminal racemes or heads – lower bracts lf-like, upper narrower **2**

2   Stem creeping at base, then ascending – lvs scarcely toothed, ± hairless **3**
Stem erect – lvs strongly toothed or cut, obviously downy – ann **5**

3   Stem woody at base – fls few in a loose head – corolla 10 mm wide, bright blue with a red eye – on mt rocks – vr ....*V. fruticans* (p 400 **C.2**)
Stem not woody at base – corolla 7 mm wide or less, white, pale or dull blue **4**

4   Fls white or pale blue, with dark blue lines, in a long raceme – fl-stalks longer than calyx – frs wider than long, ± equalling calyx – style as long as fr – c............*V. serpyllifolia* (p 400 **C**)
Fls dull blue, in oval head – fl-stalks shorter than calyx – fr longer than wide, far longer than calyx – style v short – on mt rocks – r ....*V. alpina* (p 400 **C.1**)

5   Lvs deeply 3–7-lobed into segments longer than width of central part of lf – E Eng – r **6**
Lvs oval, merely coarsely-toothed **7**

6   Lvs pinnately 3–7-lobed – fr broader than long – fl-stalks v short (1 mm)*V. verna* (p 402 **E.1**)
Lvs palmately 3–7-lobed into segments like radiating fingers – fr as long as wide – fl-stalks 5–8 mm, longer than calyx..*V. triphyllos* (p 402 **E.2**)

7   Fl-stalks v short (1 mm) – upper bracts (4–7 mm) longer than fls plus fl-stalks – calyx longer than corolla or fr – widespread ....*V. arvensis* (p 402 **E**)
Fl-stalks long (5 mm) – upper bracts (3–4 mm) and calyx both shorter than fl-stalks – calyx shorter than corolla or fr – E Eng – r ....*V. praecox* (p 402 **E.3**)

C   Speedwells with infls (racemes) in lf-axils

1   Plants of wet ground or in water – lvs hairless (rarely hairy in *V. scutellata*)...........................**2**
Plants of dry ground – lvs hairy **5**

2   Racemes always in opp pairs (in axils of 2 opp lvs) – lvs oval to broad-lanceolate **3**
Racemes always in axil of only 1 of a pair of opp lvs – lvs linear-lanceolate – fls white, pink or pale blue ....*V. scutellata* (p 400 **F**)

3   Lvs stalked, blunt, oval – fls deep blue, 7–8 mm wide ....*V. beccabunga* (p 402 **A**)
Lvs stalkless, pointed, lanceolate – fls pale blue or pink, 5–6 mm wide..**4**

4   Petal veins almost to edge of petals – petals 3–3.5 mm wide – fr-stalks erect or spreading to 45° – fr scarcely notched..*V. anagallis-aquatica* (p 400 **G**)
Petal veins much shorter, not to the edge of petals – fr-stalks spreading to 90° – fr clearly notched ....*V. catenata* (p 400 **H**)

5   Lvs oval, under 1 cm wide, with v shallow teeth – fl-stalks only to 2 mm, shorter than bracts ....*V. officinalis* (p 400 **D**)
Lvs oblong to oval-triangular, mostly over 1 cm wide, coarsely-toothed – fl-stalks 4 mm or more, longer than bracts **6**

6   Stem hairy in 2 opp rows only (sometimes hairy all around stem in shade) – bright blue with white eye – fr shorter than calyx ....*V. chamaedrys* (p 400 **A**)
Stem equally hairy or downy all round – fls mauve-lilac to dull pink – fr longer than calyx
V. montana (p 400 **B**)

A   Germander Speedwell, *Veronica chamaedrys*, creeping and ascending hairy per, fl-stems to 20 cm tall; stem (**Aa**) with two **opp rows of long white hairs**, hairless between (sometimes hairy all around stem in shade); lvs 1.0–2.5 cm long, oval-triangular, **± stalkless** (sometimes short-stalked), blunt, cordate-based, coarsely- and bluntly-toothed, hairy. Racemes in axils of one (or both) of a pair of upper lvs, long-stalked, loose, many-fld; corollas (**Ab**) *c*. 1 cm wide, bright blue with white eye. Fr **shorter than** lanceolate calyx-lobes, wider above. Br Isles, vc in wds, hbs, gsld. Fl 3–7.

B   AWI **Wood Speedwell**, *V. montana*, like A, creeping, hairy per; lvs large (2–3 cm long), coarse-toothed; long racemes in lf-axils; but stem (**Ba**) hairy **all round**, lvs have **lf-stalks** 5–15 mm long, and are pale green; corollas only 7 mm wide, lilac; fr almost round, **longer than** oval calyx-lobes. Br Isles, c (N Scot r); in wds on less acid soils. Fl 4–7.

C   Thyme-leaved Speedwell, *V. serpyllifolia*, creeping per, but ± hairless; lvs ± **untoothed**, hairless, 1.0–1.5 cm long; fls in ± erect **terminal** racemes; bracts oblong, longer than fl-stalks. Corollas (**Ca**) white or pale blue with dark blue lines, 5–6 mm wide; fl-stalks longer than calyx; fr broader than long, ± equalling calyx. Alpine forms ± prostrate, with larger (6–7 mm wide) blue fls (ssp *humifusa*). Br Isles, c on gslds, hths, open wds, wa. Fl 3–10. **C.1** *Alpine Speedwell**, *V. alpina*, and **C.2** NT *Rock Speedwell**, *V. fruticans*, are r plants of Scot mts; see key (p 399) for details.

D   Heath Speedwell, *V. officinalis*, creeping, racemes in lf-axils as in A and B, per herb, with hairy lvs only 1–2 cm, oval, v **shallowly**-toothed; lowest lvs stalked; fls in **dense** racemes; fl-stalks only 2 mm (4 mm in A and B), shorter than bract and calyx; corollas (**Da**) only 6 mm wide, pale lilac. Br Isles, vc in gslds, hths, open wds. Fl 5–8.

E   Sch 8 *Spiked Speedwell**, *V. spicata*, erect downy per, 8–60 cm; lowest lvs oval, stalked, upper lanceolate, unstalked, shallowly-toothed. Infl a lfless, dense, many-fld, terminal spike-like raceme. Fls (**Ea**) v short-stalked, corollas deep Oxford-blue, narrow-lobed; fr rounded. Eng: E Anglia vr (small form to 30 cm only, lvs sparingly-toothed); NW Eng, Bristol, Wales r on rocks (to 60 cm, lvs toothed all round); on dry calc gslds, dunes, calc rocks. Fl 7–9.

F   Marsh Speedwell, *V. scutellata*, hairless (or rarely downy) per, with **slender** creeping stems, ascending to 30 cm tall. Lvs 2–4 cm, opp, **linear-lanceolate**, pointed, stalkless, clasping, ± untoothed, yellow-green or purplish. Infls open; racemes few-fld in axils of only **one** of a pair of lvs; fl-stalks twice length of the linear bracts; sepals oval; corollas (**F**) 6–7 mm wide, white, or pale blue with purple lines. Frs flat, wider than long, longer than calyx. Br Isles, f-lc in bogs, wet mds, ponds. Fl 6–8.

G   Blue Water-speedwell, *V. anagallis-aquatica*, more robust than F, with **fleshy**, creeping and ascending stems to 30 cm tall, lvs **oval-lanceolate, pointed**, clasping, scarcely-toothed. Infls (**G**) many-fld racemes in axils of **both** of a pair of lvs. Fl-stalks equalling or **longer** than **linear** bracts; sepals **oval-lanceolate**; corollas (**Ga**) 5–6 mm wide, **pale blue**; petal veins almost to edge of petal; frs round, ± inflated, scarcely notched, **fr-stalks erect or spreading** to **45°**. Br Isles, f-lc in ponds, streams, rivers, wet wds. Fl 6–8.

H   Pink Water-speedwell, *V. catenata*, differs from G in **usually** (but not always) **pink** corollas (**H**), with petal-veins that do not reach the petal edge; fl-stalks **shorter** than **lanceolate** bracts, sepals **oblong**, fr more notched, **fr-stalks** spreading

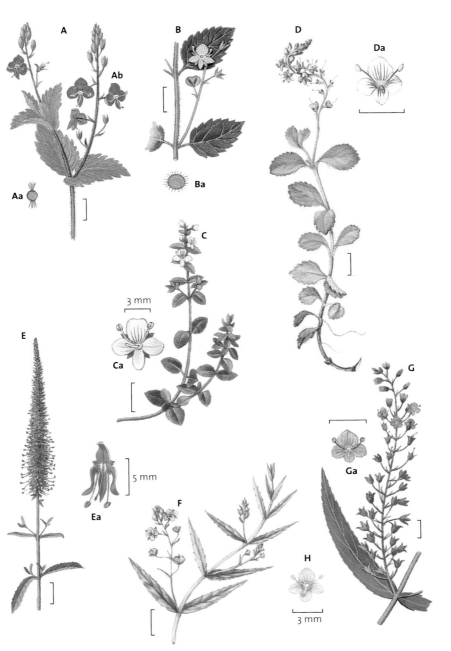

Aa

Ab

A

B

Ba

D

Da

C

Ca

3 mm

E

Ea

5 mm

F

G

Ga

H

3 mm

to **90°**. Br Isles: Eng, Wales, Ire f-lc; Scot vr; in fresh water, mds. Fl 6–8. Hybridizes with G (*V.* x *lackschewitzii*) with infl increasing in length throughout summer (not stopping growth when frs ripen) and plants sterile or slightly fertile; overlooked and probably f in S and E Eng.

A   Brooklime, *Veronica beccabunga*, hairless per, stems creeping and rooting, then ascending to 30 cm, fleshy; lvs 3–6 cm long, thick, **oval, blunt**, shallow-toothed, **short-stalked**. Infls loose, many-fld racemes in axils of **both** of a pair of lvs; sepals oval, pointed. Corollas (**Aa**) 7–8 mm wide, **blue**; capsule **rounded**. Br Isles, c in ponds, rivers, streams, mds, marshes. Fl 5–9.

B   Common Field Speedwell, *V. persica*, branched spreading hairy ann, fresh green, 10–30 cm long; lvs 1–3 cm, oval-triangular, coarsely-toothed, short-stalked, hairy underneath. Fls **solitary** in **lf-axils**, on stalks longer than lvs; calyx-lobes 5–6 mm, oval, hairy, pointed, spreading in fr; corolla (**Bb**) 8–12 mm wide, bright blue, lower lip white; fr (**Ba**) with 2 **spreading, sharp-edged** lobes, twice as wide as long, with long hairs. Introd (Asia); Br Isles, vc and much more c than C or D in ar, wa. Fl 1–12. **Key differences** between B, C and D are characters found in the frs: B has lobes spreading to 90° whereas C and D have lobes only spreading to ± 45°.

C   Grey Field Speedwell, *V. polita*, has habit of B, but smaller; lvs grey-green 5–15 mm long; fl-stalks equalling or shorter than lvs; sepals **oval, pointed**. Corollas (**C**) 4–8 mm wide, **wholly** bright blue; fr (**Ca**) broader than long, but lobes **erect, rounded** (**not** spreading or sharp-edged), with short **curled** hairs and some glandular hairs. Br Isles, f-lc in S, r to N; on ar. Fl 3–11.

D   Green Field Speedwell, *V. agrestis*, close to C, but sepals **oblong**, blunt; corollas (**D**) 4–8 mm wide, but pale blue with **white** lower lips as in 716, fl-stalks **short** as in C; fr (**Da**) with **erect, rounded** lobes as in C, but with **glandular** hairs **only** (**no** short, curled hairs). Br Isles, f-lc on ar. Fl 3–11.

E   Wall Speedwell, *V. arvensis*, **erect** downy ann, 5–20 cm, ± branched at base. Lvs oval-triangular, coarsely-toothed, to 1.5 cm long, upper stalkless; racemes long, terminal, bracts longer than fls, narrow above, grading into ordinary lvs below. Fl-stalks much shorter than calyx; corollas (**Ea**) blue, 4–5 mm wide; fr hairy, as long as wide. Br Isles, vc in ar, dry open gslds, walls. Fl 3–10.   **E.1** EN ** **Spring Speedwell**, *V. verna*, **E.2** Sch 8 EN **Fingered Speedwell**, *V. triphyllos*, and **E.3   Breckland Speedwell**, *V. praecox*, are erect small ann plants related to E, vr on sandy gslds and rds in E Anglian Breckland, see key (p 399).

F   Slender Speedwell, *V. filiformis*, **creeping**, mat-forming per; lvs stalked, **kidney-shaped, rounded**, pinnate-veined; fls (**F**) blue, 8–10 mm wide, with **white** lower lips, stalks 2–3 times length of lvs. Introd (Caucasus); Br Isles, c; in gslds, gdns, wa, churchyards. Fl 4–6.

G   Ivy-leaved Speedwell, *V. hederifolia*, hairy **spreading** ann; lvs stalked, **palmately**-lobed, ivy-shaped, to 1.5 cm long, with large blunt teeth. Fls in lf-axils, stalks shorter than lvs; sepals oval, cordate-based; corollas shorter than calyx, 4–5 mm wide, pale lilac; fr globular, hairless. Br Isles, c (but NW Scot r); in wds, ar, hbs, wa. Fl 4–5.

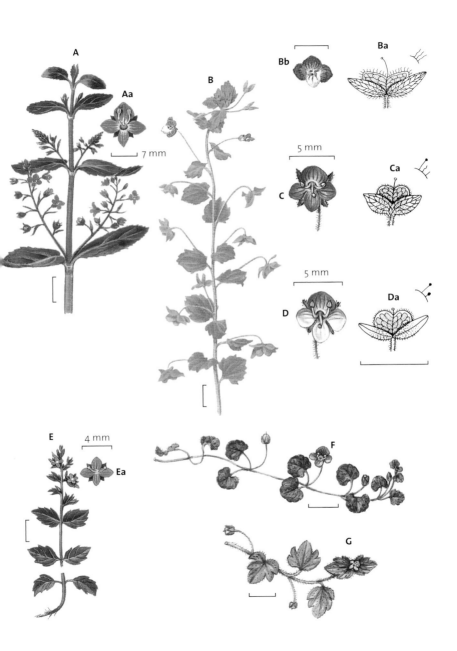

A

Aa

7 mm

B

Bb

Ba

C

Ca

5 mm

D

Da

5 mm

E

Ea

4 mm

F

G

Cow-wheats (*Melampyrum*) are erect ± hairless ann herbs, partially parasitic on other plants; lvs opp, lower ± untoothed; fls in lfy terminal spikes; calyx tubular, four-toothed; corolla two-lipped, upper lip shaped like a roof-ridge, mouth ± closed by arched-up lower lip; fr flattened, with large seeds.

A AWI **Common Cow-wheat**, *Melampyrum pratense*, variable, ± hairless erect ann, 8–40 cm, branches spreading; lvs 1.5–8.0 cm, ± stalkless, oval to linear-lanceolate, lower untoothed; infl a loose spike, fls in pairs in axils of lf-like, ± toothed bracts, both fls turned to one side. Corolla (**Aa**) usually pale **yellow**, 11–17 mm long, **much longer** than calyx; calyx-teeth erect, narrow, long; corolla-mouth ± **closed**, its lower lip **straight**. Br Isles, f-lc (but o-r E Eng); in wds, scrub, hths, upland moors. Fl 5–9.

B WO (NI) BAP AWI EN *****Small Cow-wheat**, *M. sylvaticum*, like A, but corolla (**B**) always **deep** yellow, only 5–8 mm long; tube ± equal to calyx, lower lip **bent down**, corolla-mouth **open**; calyx-teeth **spreading**. GB: N Eng, Teesdale vr, Scot Highlands r; in hthy pine or birch wds. Fl 6–8.

C AWI VU ******Crested Cow-wheat**, *M. cristatum*, ± downy ann, 20–40 cm; differs from A in its **dense**, **four-angled** fl-spikes, with **arched-back**, **cordate-based** bracts. Bracts, **rosy-purple**, pinnatifid, have 2 mm-long, sharp teeth at bases, but are long-pointed, untoothed, green towards tips. Corollas yellow, purple-flushed, 12–16 mm long. E Eng only, r-vlf; on wd edges on calc soils. Fl 6–9.

D Sch 8 **Field Cow-wheat**, *M. arvense*, has fls in dense spikes with coloured bracts as in C, but infls rounded, less dense; bracts **erect**, bright **rosy-red**, pinnatifid **all the way up sides**, with 8 mm-long teeth. Corolla **pink**, 20–24 mm long, with **yellow** throat. Possibly introd; S Eng, IoW vr; on field-margins, rds, wa, cliffs. Fl 6–9.

Eyebrights (*Euphrasia* agg), are short ± branched ann semi-parasitic herbs; lvs mostly opp, ± oval, deeply-toothed, hairy or hairless, often bronze-tinted, upper often alt; stem with curled white hairs. Fls in lfy spikes; calyx bell-shaped, four-toothed; corolla small, 4–11 mm long, white (more rarely yellow or purple-flushed), two-lipped; upper lip two-toothed, curved forward, lower flatter, three-lobed, with purple lines and a yellow blotch. Some 21 spp recorded in Br Isles. They are difficult to distinguish except by specialists; only the most c sp of the Br Isles is given here, as an example, plus one v distinct but vl sp.

A Common Eyebright, *Euphrasia nemorosa*, erect, 15–20 cm tall, stem and lvs purplish; lvs oval, hairless above, sharp-toothed, 6–12 mm; bracts oval, pointed, sharp-toothed. Corolla (**Aa**) 5.0–7.5 mm long, white, lower lip longer than upper; fr hairy. Br Isles: Eng, Wales c; Scot, Ire o; on gslds, downs, hths, open wds. Fl 7–9.
A.1 Irish Eyebright, *E. salisburgensis*, is distinctive: slender, with v **narrow** oblong lvs and bracts, both with few, long teeth; corolla narrow, white; capsule hairless. W Ire only, lf; on dunes, limestone rocks. Fl 7–8.

B Red Bartsia, *Odontites vernus*, slender, erect, branched downy ann, to 50 cm tall; lvs oblong-lanceolate, few-toothed, unstalked, opp, 1–3 cm long. Infl a long lfy branched raceme with bracts like the lvs; calyx four-toothed, bell-shaped; corollas (**Ba**) 9 mm, pink, two-lipped, upper lip curved forward, ± undivided; lower three-lobed, mid-lobe 4 mm long, throat open, tube equals calyx. Br Isles, c on ar, wa, trampled gslds, rds. Fl 6–7.

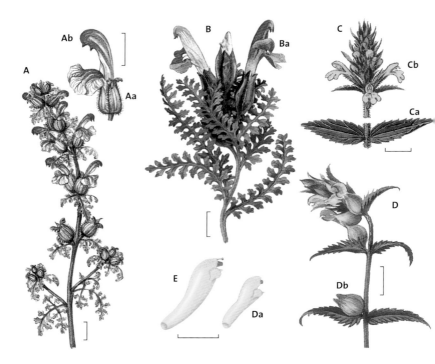

A **Marsh Lousewort**, *Pedicularis palustris*, ± hairless ann, to 60 cm tall, with single erect branched stem; lvs 2–6 cm, **alt**, oblong in outline, deeply pinnatifid, lobes toothed; bracts similar. Fls in loose lfy spike; calyx (**Aa**) tubular, ± inflated below, reddish, many-ribbed, with 2 short but lf-like lobes, **downy**; corollas (**Ab**) pink-purple, two-lipped; upper lip arched, flattened on each side, **four-toothed**, lower three-lobed, spreading downwards. Br Isles: N, W of GB, c; SE, central Eng r and declining; Ire c but declining; in fens, valley bogs, mds. Fl 5–9.

B **Lousewort**, *P. sylvatica*, per, shorter than A; **many** stems **spread** from base and then rise to 10–20 cm tall; lvs 2–3 cm; spikes fewer-fld; calyx **stouter**, five-angled, **hairless** (but **downy** in Ire and NW Scot Is), with 4 small lf-like lobes; corollas (**Ba**) **pink**, upper lips **two-toothed**. Br Isles, f-c (except in v cultivated areas of mid Eng); on damp hths, gslds, avoids chky soils. Fl 4–7.

C **Yellow Bartsia**, *Parentucellia viscosa*, erect, v sticky, hairy ann, 10–50 cm, ± unbranched; lvs (**Ca**) 1.5–4.0 cm long, lanceolate, pointed, opp, stalkless, with few blunt teeth. Fls in long terminal raceme with bracts like the lvs; calyx tubular, ribbed, with 4 spreading teeth; corollas (**Cb**) bright **yellow**, 2 cm long, the lower, three-lobed lip **much longer** than upper hooded ± untoothed one. Br Isles: S, W coastal areas of GB N to W Scot, r (but lc in SW); Ire o; on damp dune-slacks, moist hthy gslds. Fl 6–10. **C.1** ** **Alpine Bartsia**, *Bartsia alpina*, has the opp unstalked lvs, erect stems, and fl shape of C; but shorter (10–20 cm), quite unbranched; lvs **oval**, blunt-toothed, 1–2 cm long; infls short, with conspicuous **purple bracts**; corollas dull **purple**, 2 cm long, **upper** lip much **longer** than lower. N Eng, Scot Highlands r in mds, on calc rock ledges in mts. Fl 6–8.

D Yellow-rattle, *Rhinanthus minor*, erect ann, parasitic, ± hairless; stem to 50 cm, black-spotted; opp pairs of stalkless, narrow-lanceolate coarse-toothed lvs. Fls in short lfy spikes; calyx v flattened on each side, inflated and bladder-like in fr (**Db**), hairless except for margins, four-toothed; corolla (**Da**) yellow, two-lipped; upper lip flattened on each side, with 2 **short** (1 mm) violet teeth, lower three-lobed. Fr flattened; large seeds 'rattle' inside calyx when ripe. Br Isles, c in gslds, dunes. Fl 5–8.

E Sch 8 **Greater Yellow-rattle**, *R. angustifolius* (*R. serotinus*), more robust, more branched than D, with **key differences** the violet teeth on upper lip of corolla (**E**) **1–2 mm** long or more, longer than wide, and tube **curved** upward (straight in D). Possibly introd; GB vr and scattered; in gslds, dunes. Fl 6–9.

BROOMRAPE FAMILY   *Orobanchaceae*

A family of total parasites on roots of various plants; lvs reduced to scales, no chlorophyll; fls in spikes, with two-lipped corollas, as in Figwort family; stamens 4, ovary one-celled (two-celled in Figwort Family), with a single style and two-lobed stigma; fr a capsule with v many tiny seeds; stems and fls usually similarly coloured. **Toothworts** (*Lathraea*) have creeping scaly rhizomes, an equally four-lobed calyx, and a ± tubular corolla with non-spreading, parallel lips. **Broomrapes** (*Orobanche*) have no rhizomes, two-lipped calyx, two-lipped corolla with spreading lips, lower three-lobed, upper ± two-lobed. Most fl 6–7; G, below, fl 5–6.

ID Tips Broomrapes

- **What to look at: (1)** distribution of hairs and sticky glands on stamen-stalks or filaments **(2)** colour and degree of separation or joining of the 2 stigma-lobes **(3)** whether corolla has a straight or curved back.

- Be wary of making **assumptions** about the **plant host** – eg often Common Broomrape *appears* with ivy but in fact is growing on the roots of a sp some distance away, even on gdn plants.

- **Do not confuse Ivy Broomrape** (*O. hederae*) with slightly purplish stigmas with (the much more c) **Common Broomrape** (*O. minor*). Ivy Broomrape corolla lobes are more pointed (not rounded) and have a distinct pinch behind the mouth. Other usual (but **not** diagnostic) characters include: fls more spread out along flg-spike with some fls near to the gd; and top of fl-spike has cluster of buds of unopened fls.

KEY TO BROOMRAPE FAMILY

1   Parasitic on hazel, elm, other trees – rhizome present – calyx with 4 equal lobes – corolla with lips parallel, not spreading ....*Lathraea* p 408 **J**
    Parasitic on herbs, or on shrubs of pea family – no rhizome – calyx two-lipped (each lip often has 2 teeth so 4 projections surround the corolla base) – corolla with spreading upper and lower lips ....(*Orobanche*) **2**

2 Each fl of spike with 1 large bract below it and 2 smaller bracts on each side (which look like part of the calyx but are separate from the calyx-lobes) – fls dull purplish-blue (rarely yellowish-purple or lilac) – stigma-lobes white – 15–45 cm tall – on yarrow – S and E Eng, CI, r ....VU ***Yarrow Broomrape** *Orobanche purpurea* **I**
Each fl of spike has single bract only – stigma-lobes various colours but not white **3**

3 Stigma-lobes yellow..**4**
Stigma-lobes purple, red, or brown (not yellow) **6**

4 Robust plants, usually over 40 cm tall – stems >8 mm wide – corollas long, 18–25 mm, yellowish **5**
Smaller plants, usually 10–40 cm – stems purple <8 mm wide – corollas short, usually <18 mm long (rarely to 22 mm), cream, veined with purple – filaments (**Hb**) ± hairless below, with glands above – stigma-lobes (**Ha**) partly joined – on ivy – S Eng, Wales, Ire, o-lc ....WO (NI) AWI **Ivy Broomrape** *O. hederae* **H**

5 Upper corolla-lip untoothed – filaments (**Eb**), hairless below, attached <2 mm from base of corolla-tube, few glands above – stigmas (**Ea**) with lobes yellow, separate – on broom and gorse – Eng, Ire, Wales, r and decreasing everywhere ....
NT ***Greater Broomrape** *O. rapum-genistae* **E**
Upper corolla-lip (**Dc**) finely-toothed; filaments hairy below, attached 3–6 mm from base of corolla-tube, with glands throughout (**Db**) – stigmas (**Da**) yellow, with lobes touching – on Greater Knapweed – S, E Eng, f-lc on calc soils..............**Knapweed Broomrape** *O. elatior* **D**

6 Fls with conspicuous sweet smell **7**
Fls without smell or foul-smelling **8**

7 Stem and corolla purplish-red – corolla 15–20 mm long – filaments (**Cb**) ± hairy below, with glands above – stigma-lobes touching, reddish (**Ca**) – plant 8–25 cm tall – on thyme – Cornwall, Yorks, W Scot, Ire r, on rocky slopes especially by sea ....
***Thyme Broomrape** *O. alba* **C**
Stem yellowish or pale pink – corolla large, 20–32 mm long, bell-shaped, back curved – lobes of lower lip ± equal, crisped, toothed, pink or creamy yellow – deliciously clove-scented – filaments (**Gb**) hairy below, with glands above – stigma-lobes (**Ga**) well separated, purple – plant 15–40 cm tall – on bedstraw – Kent coast, r and l on dunes, chk gsld ....Sch 8 NT ***Bedstraw Broomrape** *O. caryophyllacea* **G**

8 Corolla with dark purplish glands – stem yellow to purple – corollas yellow with purple margins, upper lip notched, lower lip squarish – filaments (**Fb**) attached near base of corolla-tube, with a few hairs below, and a few glands above – stigma-lobes (**Fa**) touching, dark purple – on thistles – Yorks vr, in gslds ....Sch 8 NT ** **Thistle Broomrape** *O. reticulata* **F**
Corolla without dark glands – filaments hairless or hairy below but with no glands above **9**

9 Plant usually purplish – corolla yellowish, purple-veined, its back arched in smooth curve, upper lip straight, lower lip rounded – filaments (**Ab**) and stigma-lobes (**Aa**) very variable – on various hosts – S half Br Isles f, in N r-abs..**Common Broomrape** *O. minor* **A**
Plant very pale, ± ivory – corolla upper lip upkinked, lower lip squarish – stem and corollas ivory to creamy yellow – filaments (**Bb**) v hairy below, ± hairless above – stigma-lobes (**Ba**) just touching, purple – on Hawkweed Oxtongue – S Eng, vr ± restricted to calc sea cliffs ....
Sch 8 EN ****Oxtongue Broomrape** *O. picridis* (*O. artemisiae-campestris*) **B**

J AWI **Toothwort**, *Lathraea squamaria*, has stout stem, white or pale pink, 8–30 cm; fls in one-sided spike with scaly bracts; calyx sticky-hairy, tubular below, yellowish-white; corolla pink, nearly tubular, little longer than calyx. Parasitic mostly on hazel and elm. GB N to mid Scot o-lf; E Ire o; in wds on richer or calc soils. Fl 4–5.

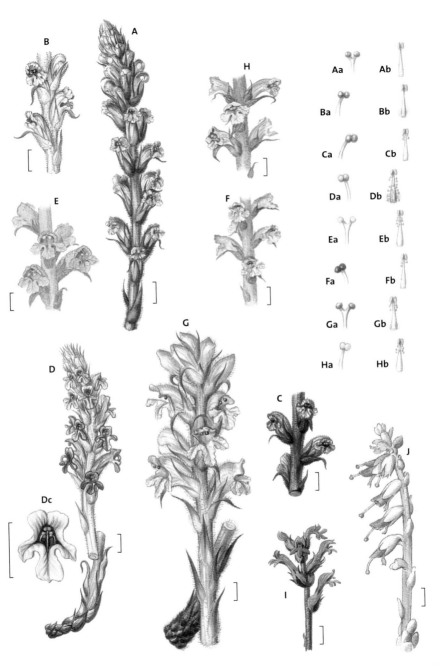

BUTTERWORT FAMILY   *Lentibulariaceae*

A small family of per insect-catching herbs of fresh water or wet places. Calyx five-lobed, corolla two-lipped, spurred, five-lobed; stamens 2, attached to corolla; carpels 2, forming one-celled fr-capsule with many tiny seeds. They over-winter either as rootless buds, or as lf-rosettes (**Ca**). **Butterworts** (*Pinguicula*) have undivided lvs in basal rosettes; fls solitary on long erect lfless stalks; sticky glands all over the lf-surfaces trap insects, whose soft parts are then digested. **Bladderworts** (*Utricularia*) are rootless aquatic plants with long, lfy floating stems. Lvs deeply divided into v narrow green segments, of which some or all bear tiny bladders that trap small animals by a vacuum mechanism and digest their soft parts for nitrogen compounds. Fls in short racemes on erect lfless stems, rising above water; corollas yellow, two-lipped, spurred, with a **palate**.

A   Common Butterwort, *Pinguicula vulgaris*, with rosette of spreading oval-oblong, ± pointed yellow-green lvs, 2–8 cm long, resembling a yellowish starfish. Fl-stalks 5–15 cm tall; calyx two-lipped; lobes blunt, oval. Corolla (**A**) **violet**, 10–15 mm long x 12 mm wide, with white patch in throat; lower lip lobes **well-separated**, **flat**, longer than broad; spur 4–7 mm long, back-pointing, **tapered to a point**. Br Isles: uplands of Wales, N Eng, Scot, c; mid, S Eng vr; Ire f-c; in (especially calc) bogs, fens, on wet rocks. Fl 5–7.

B   Large-flowered Butterwort, *P. grandiflora*, close to A, but fl-stalks 8–20 cm tall; corollas (**B**) violet, 15–20 mm long x 25–30 mm wide; lower lip lobes ± **overlapping**, **wavy**, broader than long; white throat patch **longer**, more **purple-streaked**; spur 10 mm or more long. Introd (mts of SW Eur); SW Ire lc; Cornwall; in bogs, wet rocks. Fl 5–7.

C   Pale Butterwort, *P. lusitanica*, has lvs of rosette (**Ca**) 1–2 cm long, **olive**, with **purple veins**, oblong, blunt. Fl-stalks only 3–10 cm tall; corollas (**C**) **lilac-pink**, 5–8 mm wide, with **yellow** throat, lobes of upper lip **rounded**, 2–3 mm long; spur **cylindrical**, **blunt**, 2–4 mm long, bent **downwards**. Br Isles: Eng, Hants to Cornwall, lf-lc; S Wales vr; W, N Scot f-lc; Ire lf; in Sphagnum bogs, wet hths. Fl 6–10.

D   Common Bladderwort, *Utricularia vulgaris*, has floating stems, from 10 – 100 cm long, bearing **pinnately**-divided lvs (**Da**) 2.0–2.5 cm long, oval; green; threadlike lf-segments bear groups of bristles and oval bladders to 3 mm long. Infl 10–20 cm tall, 2–8-fld; corollas 12–18 mm long, bright yellow; upper lip only **as long as** palate of lower lip; **margins** of lower lip **turned down**; fl-stalk 8-15 mm; spur conical. Br Isles, o-lf, in Eng r away from Norfolk Broads and Somerset; in still, often peaty and calc fresh water. Fl 7–8.

E   Bladderwort, *U. australis* (*U. neglecta*), **v close** to D, but **key difference** in corolla-lips; lower lip (**E**) **flat**, spreading out ± **horizontally**, upper lip usually twice length of lower lip's palate; fl-stalk **elongating after flg** to 10–30 mm. Br Isles, more c than D: Eng, S Scot, Ire o; E Eng vr–abs (but lf Essex); in more acidic water than D. Fl 7–8 (flg vr in Scot).

F   Lesser Bladderwort, *U. minor*, smaller (3–30 cm long) slender plant than D and E; lvs, only 3–6 mm long, palmately-divided into thread-like **untoothed** segments

without bristles except at lf-tips; lvs bearing bladders 2 mm long. Shoots buried in the peaty substratum have fewer lvs, many bladders. Infl 4–15 cm tall, 2–6-fld. **Key difference** from other Bladderwort spp being **corollas 6–8 mm** long, pale yellow, with short **spurs 1–2 mm** long. Br Isles o but lf; in peaty bog pools. Fl 6–9.

G   Intermediate Bladderwort, *U. intermedia* agg, a group of 3 v similar spp, is near to F in size, (stem 5–40 cm long) but **key difference** from other Bladderwort spp is plants with **two distinct types of shoots**: some horizontal, 10–25 cm long, bearing palmately-divided lvs (**G**), 3–6 mm long, with segments **flat, toothed**, 1–2 mm wide, and **no** bladders; others, descending into the peat substrata, have no lvs, but **bear bladders** 3 mm long (**Ga**). Fls r; infl 9–16 cm tall, 2–4-fld; corollas 8–12 mm long, **bright yellow** with reddish lines; spurs **conical**. Br Isles: Dorset, New Forest vr; Wales vr; N Eng, Scot, Ire o-lf; in shallow peaty pools in bogs, fens. Fl 7–9.

BELLFLOWER FAMILY	*Campanulaceae*

A family of herbs with **alt, undivided** lvs without stipules; fls either small, in dense heads or spikes, or large, showy, ± bell-shaped in loose infls; calyx five-lobed, lobes green, not spiny; corolla **regular** (irregular in *Lobelia*), with 5 **equal lobes or irregular** of 5 lobes joined below (*Lobelia*); 5 stamens alt with corolla-lobes; ovary **inferior**, 2–5-celled, with 2–5 stigmas; fr a capsule.

ID Tips   Bellflower Family

• **Do not confuse Bellflowers** with members of the **Gentian** family (p 348), which have a superior ovary and opp lvs.

• **Rampions** and **Sheep's-bit** can be confused with Scabiouses (in Teasel family, p 430), but in latter lvs **opp**, calyx has 5 long obvious **spine-like** teeth, and an outer cup-like **epicalyx** encloses each fl.

• **What to look at: (1)** whether plant hairless or hairy **(2)** corolla irregular or regular **(3)** size of fls **(4)** shape of corolla-tube, whether almost flat or bell- or funnel-shaped **(5)** plant creeping or erect.

KEY TO BELLFLOWER FAMILY

1   Fls small, narrow (under 5 mm wide) with linear petals, in dense heads or spikes **2**
    Fls at least 7 mm wide, mostly much more, solitary, or in open spikes or panicles **3**

2   Plants hairless – fl-buds curved – petals at first joined together at their tips – stigma-lobes v narrow ....*Phyteuma* (p 416)
    Plants hairy – fl-buds straight – petals never joined at tips – stigma-lobes short, stout........
    Jasione montana (p 417 **D**)

3   Corolla irregular – filaments fused to form tube around style ....*Lobelia* (p 417)
    Corolla regular – filaments free (although clustered around style) **4**

4   Ovary and fr cylindrical – corolla-tube v short, lobes ± flat, forming a wheel shape – ann plant of chalky ar ....*Legousia hybrida* (p 416 **H**)
    Ovary and fr oval or globular – corolla with bell or funnel-shaped tube – creeping or tufted plants of gslds, wds, hbs, etc **5**

5   Plant small, creeping – stems thread-like – all lvs tiny (1 cm wide), rounded, ivy-shaped, long stalked – fls 1–2 together, on long stalks, bell-shaped, 6–10 mm long x 7 mm wide
    Wahlenbergia hederacea (p 416 **A**)
    Plant ± erect – fl-stems bearing spikes or panicles of bell-or funnel-shaped fls – upper lvs stalkless, always smaller than lower ....*Campanula* (key p 413)

KEY TO BELLFLOWERS (*CAMPANULA*)

1 Epicalyx of broadly cordate, bent-back lobes present between the 5 calyx-teeth – lvs oval-lanceolate, toothed – corolla large (4–5 cm long), inflated-bell-shaped – stigmas 5 – robust, bristly plant – S Eng, introd in wa, on railway banks ....
Canterbury-bells *Campanula medium*

No epicalyx present – calyx of 5 lobes only – stigmas 3 ....**2**

2 Lvs on middle of stem ± rounded or oval (length 2–4 times breadth) **3**
Lvs on middle of stem linear, lanceolate, or narrow-oblong **8**

3 Fls in dense terminal heads, often additional fls down stem – all fls erect, stalkless – corolla hairless, purplish blue – basal lvs stalked, downy, cordate-oval ....*C. glomerata* (p 414 **D**)
Fls in loose spikes, racemes, or panicles – all fls with their own stalks – corolla-lobes ± hairy ..**4**

4 Calyx-teeth spreading or bent back in fl – fls (in a ± lfless narrow raceme) drooping to one side, corolla only 20–30 mm long, funnel-shaped, lobes equalling tube – plants erect 30–60 cm – Br Isles, o introd – on gsld, hbs, wa – fl 7–9 ....
Creeping Bellflower *C. rapunculoides*

Calyx-teeth erect to slightly spreading in fl **5**

5 Plant procumbent, often ± creeping stems <30 cm **6**
Plant erect, stems usually >50 cm **7**

6 Corolla funnel-shaped (width of fl much less than length) – corolla-lobes cut to c. ¼ (sometimes more) way to base – mid-stem lvs mostly ± circular in outline without strongly cordate bases...*C. portenschlagiana* (p 414 **F**)
Corolla bell- or star-shaped (width of fl ± equal length) – corolla-lobes cut over ½ to base – mid-stem lvs mostly oval with cordate bases ....*C. poscharskyana* (p 414 **G**)

7 Plant softly downy – stem bluntly-angled – basal lvs tapering into stalk, stem-lvs stalkless – fls in an unbranched lfy raceme – corolla 35–55 mm long ....*C. latifolia* (p 414 **B**)
Plant bristly-hairy – stem sharply-angled – basal lvs cordate-based, not tapering into stalk, stem-lvs stalked – fls in loose clusters in a branched lfy panicle – corolla 25–35 mm long ..
C. trachelium (p 414 **A**)

8 Calyx with spreading, lanceolate lobes (3 mm wide at base) – infl a tall, loose raceme, fls 1–2 together – stem-lvs hairless, linear-lanceolate (to 10 cm long) with small distant teeth – corolla widely cup-shaped with v short lobes – GB introd, f – fl 6–9 ....
Peach-leaved Bellflower *C. persicifolia*
Calyx with erect linear lobes (not over 1 mm wide at base) – infl with spreading branches ..**9**

9 Plant ± hairless, smooth – basal lvs long-stalked, rounded-cordate, about 1 cm wide – lower stem-lvs stalked – fls drooping on hair-like stalks, bell-shaped, lobes v short ....
C. rotundifolia (p 414 **C**)

Plant rough to touch – basal lvs oblong-ovate, narrowed to short stalks – all stem-lvs stalkless – fls erect, corolla funnel-or widely bell-shaped, lobes widely spreading **10**

10 Basal lvs gradually narrowed to base, running down stalk – infl with several long, spreading branches – fl-stalks with bracts on middle – corolla to 40 mm wide, with spreading lobes as long as tube, purple ....*C. patula* (p 414 **E**)
Basal lvs suddenly narrowed to stalk – infl narrow, fls clustered below – fl-stalks with bracts at base – corolla 10–15 mm wide, lobes half as long as tube, pale mauve – GB vr and scattered (Ire, Wales abs) – on hbs, wa – fl 5–8 ....EN ****Rampion Bellflower** *C. rapunculus*

A AWI **Nettle-leaved Bellflower**, *Campanula trachelium*, erect robust bristly-hairy per herb, 50–100 cm tall, with sharply angled stem; basal lvs long-stalked, oval-triangular, **cordate-based**, blades to 10 cm long; stem-lvs short-stalked, oval, pointed, all lvs **coarsely** sharp-toothed; infl a lfy **panicle** with fls in groups of 1–4 on branches; fls with stalks to 1 cm, spreading or erect; calyx-teeth erect, 1 cm long, pointed-triangular; corolla **25–35 mm long**, bell-shaped, pale purplish-blue, with short lobes, hairy outside. Fr globular. Br Isles: Eng, Wales, N to R Humber, c to S, o to N; SW Eng, Scot r introd; Ire SE only as native, r; in wds, hbs, on calc soils. Fl 7–9.

B AWI **Giant Bellflower**, *C. latifolia*, ± downy per herb, 50–100 cm tall; differs from A in **bluntly**-angled stem, basal lvs (**B**) **tapering** into stalks, 10–20 cm long, long-pointed, **less deeply-toothed** than in A; stem-lvs stalkless; infl a lfy **raceme**, rarely branched; fl-stalks 2 cm; corolla **40–50 mm long**, paler than in A, often white; fr oval. Br Isles: GB from S mid Eng to Ross-shire, r to S, lc in N Eng, o-lf in Scot; N Ire o introd; in wds, hbs on calc or richer soils. Fl 7–8.

C **Harebell**, *C. rotundifolia*, slender, erect hairless per herb, 15–40 cm tall, stems shortly creeping then ascending; basal lvs (**Ca**) 1 cm wide, **rounded, cordate-based**, long-stalked; stem-lvs narrow-linear, lower stalked, upper stalkless; fls few in a loose panicle, **drooping** on long thread-like stalks; calyx-teeth spreading, bristle-like; corolla 15 mm long, bell-shaped, with short oval lobes, pale blue. Fr globular. Br Isles: GB c (but r SW Eng); Ire, c in N, o in S; on dry gslds, hths, hbs, dunes. Fl 7–9.

D **Clustered Bellflower**, *C. glomerata*, erect, closely downy per herb, 3–30 cm; basal lvs (**Da**) long-stalked, oval, blunt, **cordate-based, downy**, with small blunt teeth; upper stem-lvs stalkless, clasping, oval-lanceolate; fls in a **dense terminal head**, often with extra fls lower down stem, all erect, **stalkless**; calyx-teeth narrow-triangular, pointed; corolla 15–20 mm long, narrow-bell-shaped, rich purplish blue, hairless, lobes ± equalling tube; style not protruding from corolla. Br Isles: S, E Eng, f-lc, o-lf to N; Wales, E Scot r; **Ire abs**; on calc gslds. Fl 6–10.

E AWI EN *****Spreading Bellflower**, *C. patula*, slender erect per herb, 20–60 cm tall, stems and lvs rough; basal lvs 4 cm, obovate, blunt, weakly-toothed, running down into short stalk; stem-lvs stalkless, small, narrow, pointed; infl **much branched**, v open, spreading; fls erect, on slender 2–5 cm stalks, each with a little bract in the middle; calyx-teeth 1 cm, bristle-like; corolla (**E**) funnel- or widely bell-shaped, lobes spreading, oval, pointed, as long as the tube, bright **rosy-purple**. GB: S, W midland Eng, E Wales, r and scattered; vr introd elsewhere; on riverbanks, hbs, rds. Fl 7–9.

F **Adria Bellflower**, *C. portenschlagiana***,** procumbent to trailing finely downy per herb to 30 cm; mid-stem lvs mostly ± circular, coarsely toothed, without strongly cordate bases; long stalks on lower lvs, upper ± stalkless; fl (**F**) bluish-purple with long **funnel-shaped** tube (width of fl much less than length) and 5 lobes cut c. ¹/₄ (sometimes more) way to base; fr a capsule. Introd (Balkans); much grown gdn escape. Br Isles: Scot, Ire, o; S Eng, Wales, IoM c in urban areas on walls, pavement cracks, rocky gd, wa. Fl 6–10.

G **Trailing Bellflower**, *C. poscharskyana*, bluish-purple fls (**Gb**) superficially similar to F but plant grey-hairy; mid-stem lvs (**Ga**) mostly oval with cordate bases, finely toothed; corolla **bell-shaped** with longer petals which look like stars from above,

415

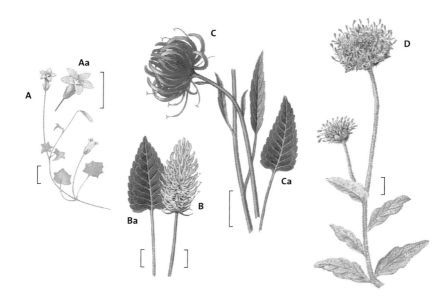

width of fl ± equals length, corolla-lobes cut **over ¹/2 way** to base and the corolla-tube much less distinct than F. Br Isles: Scot, Ire, o; S Eng, Wales, IoM c. Same habitats and flg time as F.

▲ H **Venus's-looking-glass**, *Legousia hybrida*, erect or sprawling bristly ann, with stems 5–20 cm; lvs oblong, wavy-edged, stalkless, 1–3 cm long; fls erect, in loose terminal clusters; calyx-teeth 5, oblong-lanceolate, half length of **cylindrical** ovary; corolla **wheel-shaped**, mauve to dull purple, five-lobed, 5–10 mm across, only opening in sunshine, lobes only half length of calyx-teeth; fr a three-angled linear-oblong capsule 15–25 mm long, opening by pores below calyx. S, E, mid Eng, f-lc; S Scot vr; abs rest of Br Isles; on ar on calc soils. Fl 5–8.

A AWI NT **Ivy-leaved Bellflower**, *Wahlenbergia hederacea*, delicate, **creeping**, patch-forming, hairless per herb; with stems to 20 cm, weak, **thread-like**; lvs all long-stalked, alt, rounded, palmately-lobed, rather **ivy-shaped**, pale green, 5–10 mm long and wide, bases ± cordate; fls on fine stalks 1–4 cm long, in lf-axils, erect or inclined; calyx-teeth narrow, 2–3 mm long, longer than ovary; corolla (**Aa**) bell-shaped with 5 short lobes, **pale sky-blue**, 6–10 mm long x 5–8 mm wide. S Eng, Kent to Cornwall o-lf; Wales f; NW Eng, W Scot r; S, W Ire r; on moist hthy tracks in wds, boggy moorland rivulets. Fl 7–8.

B Sch 8 EN ****Spiked Rampion**, *Phyteuma spicatum*, erect hairless per herb, 30–80 cm tall; lower lvs (**Ba**) long-stalked, cordate-based, oval, blunt-toothed and -tipped, 3–7 cm long; upper lvs narrower, stalkless. Infl a dense **spike** 3–8 cm long; fls with calyx-teeth narrow, spreading in fr, corollas creamy-yellow, 1 cm long, curved in bud, their narrow lobes at first **joined at tips**; stigmas slender; frs in a dense cylindrical **spike**. Eng: E Sussex only, r; in open wds, hbs. Fl 5–7.

C ***Round-headed Rampion**, *P. tenerum* (*P. orbiculare* in the *New Flora*), slender erect hairless per herb, 8–40 cm tall; lvs as in B, but basal lvs (**Ca**) less cordate, more oblong, all stem-lvs stalkless, linear-lanceolate, few; infl a dense **rounded head** 1.0–2.5 cm wide, of deep **Oxford-blue** fls. S Eng, Sussex to Wilts only, f-l va; on calc gslds. Fl 7–8. *P. orbiculare*, is close to C, but is the C Eur plant, in the author's opinion not found in the Br Isles, having **broad oval**-lanceolate (not narrow-lanceolate) bracts below head, **longer** than the outer fls.

D **Sheep's-bit**, *Jasione montana*, erect or sprawling **downy** bi herb, 5–30 cm tall, with rosette of short-stalked, **strap-shaped**, **wavy-edged**, weakly-toothed basal lvs to 5 cm long, stem lvs shorter, stalkless; fls in dense rounded **heads** 10–30 mm across; corollas 5 mm long, **pale sky-blue**, their narrow lobes **not** joined at tip, and 2 **stout** stigmas on tip of style. Br Isles: GB, f-lc to W and S, o-r to E, vr-abs N Scot (except Shetland, f); Ire lc; on cliffs, dry gslds, hths, dunes, on acid sandy or stony soils. Fl 5–8.

E vu ****Heath Lobelia**, *Lobelia urens*, ± hairless erect per herb, 20–60 cm tall; stems angled, lfy; basal lvs (**Ea**) **obovate**, linear-oblong above, toothed, shiny dark green, scarcely stalked; fls in a loose branched raceme; bracts linear, ± equalling calyx; calyx-teeth (**Eb**) long, v narrow, spreading; **fl-stalks <1 cm** long; corolla 10–15 mm long, bluish-purple, with 2 narrow lobes to upper lip, 3 narrow teeth on lower lip; **corolla-lobes <2 mm** wide. S, SW Eng, r but vlf; on grassy hths, open hthy wds. Fl 8–9. **E.1 Garden Lobelia**, *L. erinus*, a slender, trailing, usually **ann** version of E, fls various colours, the most familiar being the deep blue cultivar 'Sapphire'. **Key difference** from E **fl-stalks 1-2 cm** long and lower 3 **corolla-lobes >2mm** wide. Introd (S Africa); Br Isles: gdn escape o natd on pavements, wa, urban areas. Fl 5–7.

F **Water Lobelia**, *L. dortmanna*, hairless erect per, 20–60 cm, with rooting runners; rosettes of submerged, **linear**, blunt, untoothed, arched back, fleshy lvs (**Fa**) 2–4 cm long; lvs with **2 tubes** in cross section and exuding milky **latex**. Infl-stem, emerging from water, lfless except for small scales, bears a few-fld raceme of drooping fls with stalks to 1 cm long, and blunt bracts much shorter than fl-stalks; calyx-teeth oval, blunt, short, erect; corolla 15–20 mm, **pale lilac-mauve**. Br Isles: Wales, NW Eng, Ire, lf; W, N Scot, c; on stony bottoms of shallow lakes of non-calc fresh water, especially in mts. Fl 7–8.

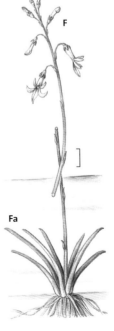

F

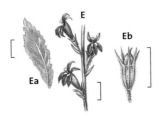

E

Eb

Ea

BEDSTRAW FAMILY   *Rubiaceae*

Herbs (rarely woody below) with lvs apparently in **whorls** of 4–12 along the **four-angled** stems (technically, lvs are in opp pairs, with lf-like stipules making up the rest of each whorl). Fls small, in clusters, terminal or in lf-axils; calyx small or abs; corolla 4–5-lobed; stamens 4–5, alt with corolla-lobes; inferior ovary of 2 rounded cells; fr of 2 nutlets, or a berry. Many members of the family lacking whorled lvs (eg, Coffee plant, *Coffea*) exist in the Tropics, but not in Br Isles.

KEY TO BEDSTRAW FAMILY

1 Scrambling woody evergreens to 150 cm long – lvs elliptical, rigid, leathery, dark green, one-veined, prickly on their edges, 2–6 cm long, 4–6 in a whorl – corollas yellow-green, five-lobed – fr a black berry ....*Rubia peregrina* (p 422 **C**)
Non-evergreen herbs – corollas four- (or three-) lobed – fr 2 dry nutlets **2**

2 Fls with distinct 4–6 toothed calyx – corollas mauve-pink, funnel-shaped – fls in small dense heads with a ruff of bracts below each head ....*Sherardia arvensis* (p 422 **E**)
Fls with calyx abs, or reduced to a small rim on ovary top **3**

3 Corolla funnel-shaped, with tube as long as, or longer than, the 4 lobes **4**
Corolla wheel-shaped, with v short tube and 4 ± flat lobes ....
(most *Galium* spp & *Cruciata*, see key below)

4 Plant prostrate to ascending – infls loose, few-fld – fls with corollas white inside, pinkish outside – lvs linear, 4–6 per whorl, ascending at an angle – frs warted ....
Asperula cynanchica (p 423 **F**)
Plant erect, 15–30 cm tall – fls white, in terminal umbel-like heads – lvs in spreading whorls up stems – lvs elliptical-lanceolate, 6–8 per whorl – corolla four-lobed – frs covered with hooked bristle ....*Galium odoratum* (p 422 **D**)

KEY TO BEDSTRAWS (*CRUCIATA* & *GALIUM* EXCEPT *G. ODORATUM*)

1 Fls yellow – fr smooth **2**
Fls white, cream or greenish – fr wrinkled, warted or bristly **3**

2 Lvs linear, mucronate, one-veined, hairless, at least 10 times as long as wide, 8–12 in a whorl...*Galium verum* (p 422 **B**)
Lvs oval-elliptical, blunt, three-veined, hairy, not more than 3 times as long as wide, 4 in a whorl ....*Cruciata laevipes* (p 422 **A**)

3 Lvs three-veined, elliptical-lanceolate, blunt, widest at or below middle, rarely more than 5 times as long as wide, 4 to a whorl – fls white – fr with hooked bristles ....
G. boreale (p 420 **G**)
Lvs one-veined, mostly widest above middle, 5–10 times as long as wide, 4–8 per whorl ..**4**

4 Lvs blunt or slightly pointed, never mucronate at tips – in wet places **5**
Lvs mucronate at tips **6**

5 Lvs linear-lanceolate to broad-lanceolate, 1–3 cm long – tiny backward-pointing prickles on lf-edges – panicle of fls loosely pyramidal with wide-spreading branches – fls 3.0–4.5 mm across ....*G. palustre* (p 420 **E**)

Lvs narrow-linear, 0.5–1.0 cm long – tiny forward-pointing prickles on lf-edges – panicle of fls V-shaped, with ascending branches – fls 2.5 mm across ....*G. constrictum* (p 420 **F**)

6 Stems v rough with backward-pointing prickles on angles **7**
Stems ± smooth on the angles..**11**

7 Lvs with forward-pointing prickles on edges, only 3–10 mm long – fls 0.5 mm across, reddish outside, whitish inside – fr 1 mm long, warted, hairless – in dry places – r ....
G. parisiense (p 420 **B.1**)
Lvs with backward-pointing prickles on edges – fls 1–3 mm across, white or greenish (not reddish) **8**

8 Frs hairless, but ± wrinkled and warted – fls white or cream **9**
Frs covered with hooked bristles – fls greenish-white **10**

9 Fls white, 2.5–3.0 mm across – frs 1 mm long, wrinkled, not warted – fr-stalks straight – erect plant of wet places – f ....*G. uliginosum* (p 420 **D**)
Fls cream, 1.0–1.5 mm across – frs 3–4 mm long, v warted – fr-stalks strongly arched down – sprawling plant of dry ar – vr...*G. tricornutum* (p 422 **H.2**)

10 Fls 2 mm across – fr 4–6 mm long, purplish when ripe, its hooked bristles with swollen bases – often scrambling in hds – vc ....*G. aparine* (p 420 **H**)
Fls 1 mm across – fr 1.5–3.0 mm long, blackish when ripe, its hooked bristles without swollen bases – on ar – vvr...*G. spurium* (p 422 **H.1**)

11 Lvs of fl-shoots lanceolate-obovate, not more than 5 times as long as wide – prickles on lf-edges forward-pointing or abs...**12**
Lvs of fl-shoots linear to narrow-lanceolate, more than 5 times as long as wide – prickles on lf-edges backward-pointing **13**

12 Plant robust – lvs 8–25 mm long – fls in large terminal panicles – frs wrinkled (not warted) ....*G. mollugo* (p 419 **A**)
Plant slender, more creeping – lvs 5–10 mm long – fls in small clusters on stem-tips and in lf-axils – corolla-lobes not long-pointed – frs with pointed warts ....
G. saxatile (p 419 **B**)

13 Plant ± creeping – short ascending fl-stems with close-set lf-whorls – frs with pointed warts...*G. sterneri* (p 420 **C.1**)
Plant ± erect – lf-whorls distant – frs with low dome-shaped warts ..*G. pumilum* (p 420 **C**)

▼ **A Hedge Bedstraw**, *Galium mollugo*, robust per herb; stems (**Ad**) sprawling to erect and scrambling, to 100 cm long, downy or not, four-angled, smooth, solid; lvs 8–25 mm long, 6–8 per whorl, lanceolate-obovate, one-veined, mucronate, prickles (**Aa**) on edges **forward**-pointing. Infls are large loose terminal panicles; corollas (**Ab**) white, 3–4 mm across, with sharp points to lobes; fr (**Ac**) hairless, wrinkled, 1–2 mm long. Br Isles: Eng vc; Wales, Scot, Ire, r-o; in hbs, scrub, open wds on richer soils. Fl 6–9.

▼ **B Heath Bedstraw**, *G. saxatile*, mat-forming per herb, with ascending fl-shoots 10–20 cm tall; stems (**Bc**) hairless, four-angled, smooth, v branched; lvs 5–10 mm long, 6–8 per whorl, lanceolate and broadest above middle on fl-shoots, obovate on prostrate stems, all mucronate and one-veined, with prickles (**Ba**) on edges **forward**-pointing. Infls open lfy cylindrical panicles; corollas white, 3 mm across, lobed (not sharp-pointed); fr (**Bb**) 1.5–2.0 mm long, hairless, with pointed warts.

Br Isles, c (except mid Eng o); on hths, gslds, wds on acid soils. Fl 6–8. **B.1** VU **Wall Bedstraw*, *Galium parisiense*, sprawling slender herb with stems 10–20 cm long, rough on the 4 angles, with small **down-pointing** prickles; **narrow** lvs only 3–10 mm long, with **forward-pointing** prickles as in B; fls in small open clusters in lf-axils, forming a narrow panicle; corollas only 0.5 mm across, reddish outside, whitish within; fr 1 mm long, warted, hairless. Br Isles: E, S Eng only, r; on old walls, sandy gsld. Fl 6–7.

C EN **Slender Bedstraw*, *G. pumilum*, slender ± erect herb, 20–30 cm tall, with ± smooth four-angled stem; lvs (**C**) linear-lanceolate, 14–18 mm long, with mucronate tips and a few **backward**-pointing prickles on edges, in distant whorls of 5–7. Fls cream, 3 mm across, in a long open terminal panicle; fr 1.5 mm long, hairless, with **dome-shaped** warts. Br Isles, S Eng only, vr; on calc gslds, rocks. Fl 6–7. **C.1 Limestone Bedstraw**, *G. sterneri*, is close to C, but of more compact, mat-forming habit, with shorter (10–20 cm) ascending fl-shoots; crowded whorls of mucronate lvs with **more**, **backward-pointing**, prickles; fr 1.3 mm, hairless, with **pointed** warts. Br Isles, hill areas of N Eng, Wales, Scot, Ire, lf; on calc gslds, rocks. Fl 6–7.

D **Fen Bedstraw**, *G. uliginosum*, slender herb with ascending weak stems 10–50 cm long, **v rough** on the 4 angles, with **backward-pointing** prickles (**Db**); lvs one-veined, linear-lanceolate, wider above, **mucronate**, also with **backward-pointing** prickles (**Da**) on their edges, 6–8 per whorl; panicle narrow; corollas white, 2.5–3.0 mm across; fr 1 mm long, hairless, wrinkled. Br Isles: GB f-lc (except **NW Scot abs**); Ire o: in calc fens. Fl 6–8.

E **Marsh-bedstraw**, *G. palustre*, variable slender or robust herb with creeping stems and also erect fl-stems to 100 cm tall, four-angled, hairless, smooth or rough; lvs (**Ea**) to 3.5 cm long, one-veined, obovate-lanceolate, ± blunt, (but **never** with mucronate tip like D), 4, 5 or 6 in a whorl; prickles on lf-edges **backward**-pointing. Fls 3.0–4.5 mm across, white; fr wrinkled, hairless, 1–2 mm long, in **pyramidal**, **wide-spreading** panicles. Br Isles, vc in marshes, fens, ditches, ponds. Fl 6–8.

F ****Slender Marsh-bedstraw**, *G. constrictum* (*G. debile*), is smaller, more slender and prostrate than E; stems (**Fc**) smooth or slightly rough, lvs only 0.5–1.0 cm long, 4–6 per whorl, linear, one-veined, **never mucronate**, with prickles on edges **forward-pointing** (**Fa**); fls 2.5 mm across, in **V-shaped** panicles with **ascending** branches; fr (**Fb**) 1 mm long, hairless, warted. Br Isles: New Forest vlf, S Devon, CI, vr; Ire, Scot abs; in damp hollows in gsld and around ponds on acid soils. Fl 6–7.

G **Northern Bedstraw**, *G. boreale*, erect herb 20–40 cm tall, with rigid smooth four-angled lf-bearing stems; lvs (**G**) **elliptical-lanceolate**, **three-veined**, **blunt**, widest **at** or **below** middle, rarely over 5 times as long as wide, in whorls of 4; fls white, 4 mm across, in lfy pyramidal terminal panicles; fr 2.5 mm long, with many **hooked bristles**. Br Isles: N Eng lf; Scot f-lc; Wales vr; N, W Ire o; in mt mds, wds, mt ledges, rocks, dunes. Fl 7–8.

H **Cleavers (Goosegrass)**, *G. aparine*, sprawling or ascending ann herb, often climbing over other plants to 100 cm long; four-angled stems (**Hb**) v rough, with large **backward-pointing** prickles; lvs 12–50 mm long, 6–8 to a whorl, linear-lanceolate or elliptical, mucronate, one-veined, ± bristly, edges with **backward**-

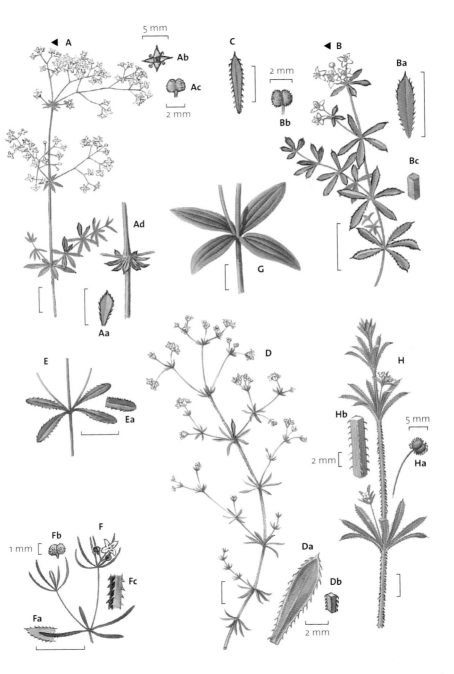

pointing prickles. Fls in 2–5-fld axillary cymes, whitish-green, 2 mm across; fr (**Ha**) 4–6 mm long, purplish when ripe, covered with **hooked bristles** with **swollen** bases. Br Isles, vc on hbs, stony slopes, ar, wa, beaches, moist wds, etc. Fl 6–8. **H.1 False Cleavers**, *Galium spurium*, is close to H, but ± prostrate; fls 1 mm across; fr only 1.5–3.0 mm long, **blackish** when ripe, bases of its hooked bristles **not swollen**. Introd from cultivation; S half of Eng vvr; on ar. Fl 7. **H.2** BAP CR ****Corn Cleavers**, *G. tricornutum*, in appearance like H and H.1, has similar prickly stems, but frs hairless, 3–4 mm long, v warted; fr-stalks **strongly arched downwards**; fls cream, 1.0–1.5 mm across. S, E Eng, vr, decreasing; on ar on calc soils. Fl 6–9.

A   Crosswort, *Cruciata laevipes* (*G. cruciata*), per herb with spreading, then erect hairy stems 15–60 cm tall; lvs to 2.5 cm long, **4 in a whorl**, **oval**-elliptical, **three-veined**, v hairy, yellow-green. Fls in dense axillary clusters **shorter** than lvs, 2–3 mm across, clear **yellow**, **honey-scented**; fr 1.5 mm long, hairless, smooth, black when ripe. Br Isles: GB, c (but r near W coast); **N**, **W Scot, Ire abs**; in open wds, scrub, gslds, rds, hbs. Fl 5–6.

B   Lady's Bedstraw, *G. verum*, per herb with creeping stems at base and fl-stems ± erect, ± four-angled, ± hairless, 15–60 cm tall; lvs **linear**, mucronate, 6–25 mm x 0.5–2.0 mm, one-veined, dark green above with margins rolled back, 8–12 to a whorl. Panicles lfy, terminal; corollas 2–3 mm across, **golden-yellow**, with pointed lobes; fr 1.5 mm long, hairless, smooth, black when ripe. Br Isles, vc in gslds, hbs, dunes. Fl 7–8. Hybridizes with Hedge Bedstraw (p 419) (*G.* x *pomeranicum*) to produce plants with intermediate lf-size and shape (see lf scan in the *New Flora*) and pale yellow fls.

C   Wild Madder, *Rubia peregrina*, evergreen per, woody below, with scrambling hairless four-angled stems, prickly on angles, to 150 cm long; lvs 4–6 in a whorl, rigid, leathery, dark green, elliptical-lanceolate, one-veined, shiny, 2–6 cm long, with curved prickles on edges and on midrib below. Fls (**Ca**) in spreading cymes in lf-axils; corollas 5 mm across, yellow-green, the 5 lobes long-pointed; fr (**Cb**) a black berry, 4–6 mm across. Br Isles: S and SW Eng, IoS, CI, Wales, f-la near coast (but r E to Kent and inland); S Ire r but lf; on scrub, bushy cliffs, wds, rocks, hds, especially near sea. Fl 6–8.

D   AWI **Woodruff**, *Galium odoratum*, erect per, 15–30 cm tall, vanilla-scented when bruised, stems unbranched, four-angled, only hairy below the distant, spreading whorls of 6–8 elliptical-lanceolate, pointed, hairless lvs, that have forward-pointing prickles on edges. Fls in umbel-like heads; corollas (**Da**) 4–6 mm long, white, **funnel-shaped**, four-lobed halfway; fr (**Db**) 2–3 mm long, rough with hooked, black-tipped bristles. Br Isles: GB c (but E Anglia o, N Scot o-r); Ire f; in wds on calc or richer soils. Fl 5–6.

E   Field Madder, *Sherardia arvensis*, hairless ± prostrate ann with stems 10–30 cm long; lvs 4 to a whorl below, 5–6 to a whorl above, elliptical, pointed, 5–20 mm long, with backward-pointing prickles on edges. Fls 4–8 together in dense terminal heads, each head with a ruff of bracts below it; sepals 4–6, green, enlarging in fr; corollas mauve-pink, funnel-shaped, 2–3 mm across, 4–5 mm long, with long tube; fr dry, 4 mm long, with sepals persisting on top, bristly. Br Isles: GB, f (c in S and E, r to N); Ire f; on ar, wa. Fl 5–10.

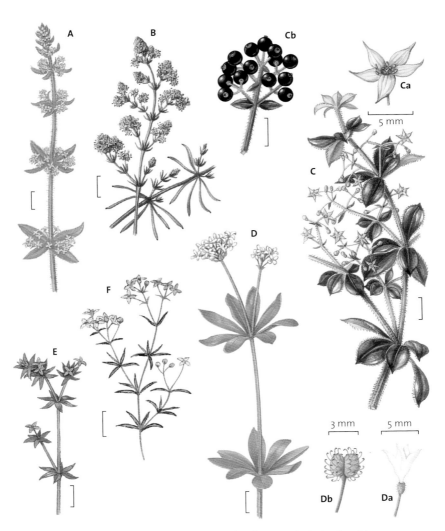

F Squinancywort, *Asperula cynanchica*, per hairless prostrate herb, with branched four-angled ascending shoots 5–20 cm long; lowest lvs elliptical, others linear, 4 to a whorl, 6–20 mm long, often **unequal** and at **acute angle** to stems; infls few-fld, long-stalked clusters; corollas funnel-shaped, 6 mm long x 3–4 mm wide, white within, pink outside, four-lobed; fr 3 mm long, warted. Br Isles: S, SE Eng, S Wales f-lc, rest of Eng N to Yorks and Westmorland r; **Scot abs**; Ire vlf in W; on calc gslds, dunes. Fl 6–7.

HONEYSUCKLE FAMILY   *Caprifoliaceae*

Shrubs, or more rarely herbs, with **opp** lvs; fls in heads or umbels; ovaries **inferior**; petals joined into a five-lobed corolla-tube; 5 stamens; 2–5 carpels; fr **fleshy** (in Br Isles), except in **Twinflower** (*Linnaea*). **Elders** (*Sambucus*) have pinnate lvs; in other genera lvs undivided. In **Guelder-rose** and **Wayfaring-tree** (*Viburnum*) fls are in umbel-like cymes; in **Honeysuckles** (*Lonicera* and *Leycesteria*) and **Snowberries** (*Symphoricarpos*) they are in whorls, spikes or pairs or clusters in lf-axils, trumpet-shaped and two-lipped.

A   Elder, *Sambucus nigra*, shrub or small tree, to 10 m tall, producing erect straight suckers from base; bark deeply-furrowed and corky. Lvs opp, pinnate; lfts 3–9 cm long, oval or elliptical, toothed, ± hairless; infl umbel-like, flat-topped, much-branched, 10–20 cm across; corolla 5 mm across, creamy, five-lobed; stamens 5, creamy; stigmas 3–5; fr (**Aa**) a black edible berry-like drupe 6–8 mm wide; fruiting heads **drooping**. Br Isles, vc on nutrient-enriched soils in wds, scrub, wa (disliked by rabbits so common near their colonies). Fl 6–7; fr 8–9, edible.

B   Dwarf Elder, *S. ebulus*, robust hairless per **herb**, 60–120 cm tall; stems stout, erect, grooved; lvs as in A, but strong-smelling and with lfts 5–15 cm long, narrower and more sharply-toothed, stipules at base of lvs oval, **conspicuous** (tiny or abs in A); infl similar, but 7–10 cm across; fls (**Ba**) often pink-tinged outside; stamens **purple**; fr black, **poisonous**; fr heads **erect** (**Bb**). Introd (Eur); Br Isles, o-f (but Scot vr); on rds, wa. Fl 7–8; fr 8–9.

C   Red-berried Elder, *S. racemosa*, hairless shrub to 4 m; lvs and fls as in A, but infls **oval panicles** (**not flat-topped**), and frs (**Ca**) bright **red**, 5 mm wide. Introd (C Eur); Br Isles: Eng, Ire vr; Scot lf; in wds. Fl 4–5; fr 6–7.

D   AWI **Guelder-rose**, *Viburnum opulus*, deciduous shrub 2–4m tall, with grey hairless angled twigs, and **scaly** buds; opp **palmately-lobed** lvs, 5–8 cm long, lobes sharply-toothed, dark green, hairless above, ± downy below; infls flat umbel-like heads, inner fls fertile, 6 mm across, outer fls sterile (no stamens or stigmas), 15–20 mm across, all fls white; fr (**Da**) 8 mm long, globular, shiny red. Br Isles, c (but N Scot r-o); in wds, scrub, hds, especially on damper soils. Fl 6–7; fr 9–10 **Poisonous**.

E   AWI **Wayfaring-tree**, *V. lantana*, deciduous shrub 2–6 m tall, with pale brown downy rounded twigs, and buds **without** enclosing scales; opp, undivided, oval, wrinkled pointed finely-toothed lvs, 5–10 cm long, sparsely downy above, densely grey-downy below. Infls flat umbel-like heads, 6–10 cm across, but fls **all alike**, fertile, 5–6 mm wide, creamy. Fr (**Ea**) 8 mm long, oval, flattened, first red, then black. S Eng: c to W to S Devon and N and E to Cambridge, S Lincs and Worcs; introd and r to N and W and in rest of Br Isles; in scrub, hds, wds on calc soils. Fl 5–6; fr 7–10. **Poisonous**.

F   AWI **Honeysuckle**, *Lonicera periclymenum*, twining shrub, climbing trees etc, to 6 m, and then flowering, also carpeting gd in dry wds (but then without fls); lvs opp, 3–7 cm long, oval to elliptical, grey-green, ± downy or not, **untoothed**, pointed, the lower short-stalked; fls in whorled terminal **heads**, 5–6 cm across, **stalkless**, with **tiny** bracts; corollas 4–5 cm long, trumpet-shaped and two-lipped, upper lip four-toothed, curved up, lower untoothed, curved down, creamy-yellow within, yellow to purplish-pink outside, v fragrant; style and the 5 stamens all long-protruding. Fr (**Fa**) a red globular berry. Br Isles, vc in wds, hds, scrub, rocks, mostly on acid soils. Fl

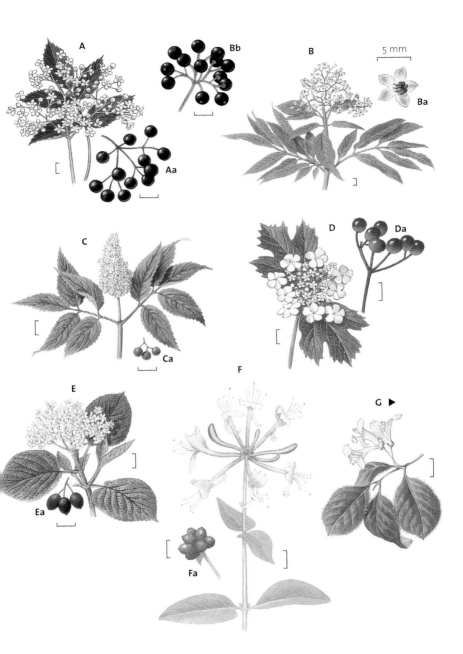

A

Bb

B

5 mm

Ba

Aa

C

Ca

D

Da

E

Ea

F

Fa

G ▶

6–9; fr 8–9. **Poisonous**. **F.1 Japanese Honeysuckle**, *Lonicera japonica*, similar twining shrub to F but semi-evergreen; **key difference** from F is fls in **pairs** on **stalks** in **lf-axils** (rarely terminal like F); corolla yellow, 3-5 cm long, hairy on outside. Introd (E Asia); Br Isles: o-vla gdn escape in S Eng, IoM, CI in wds, scrub, hds, wa. Fl 5–7. **F.2 Perfoliate Honeysuckle**, *L. caprifolium*, twining shrub close to F, but lvs and the large **bracts** to fl-heads are **joined in pairs** (ie perfoliate) round stems. Introd (SE Eur); GB r; in wds, hds. Fl 5–6.

▲ **G Fly Honeysuckle**, *L. xylosteum*, erect bushy **non-climbing** shrub, 1–3 m tall; twigs **downy**; lvs opp, grey-green, downy, oval, stalked, pointed, 3–6 cm long; fls in **pairs** on a shared stalk in upper lf-axils; corollas as in F, but only 1–2 cm long, yellow, **downy**; berries red. Eng: Sussex only, vr possibly native in wds on chk; r but widespread introd elsewhere; in wds, scrub. Fl 5–6.

A Wilson's Honeysuckle, *L. nitida*, evergreen shrub to 2.5m with arching branches and **round glandular-hairy** stems, (may be confused with Box (p 305) which has similar lvs but 4-angled hairless or downy stems); opp oval lvs; bracteoles fused into a cup-like organ below fls; fls (**Aa**) usually in pairs in lf-axils, 5–7mm long, white with 5 lobes; glands covering calyx, corolla and lvs; fr a violet berry. Introd (China) and increasing gdn escape. Br Isles: SW Eng, Ire c; elsewhere f-vlc on scrub, wds, wa, hds, rds. Fl 3–4 (although often not flg).

B Himalayan Honeysuckle, *Leycesteria formosa*, shrub to 2m; opp lvs 5–17 cm long, oval with pointed tip, entire or toothed; infl in long hanging spikes surrounded by large purple-green bracts; fls (**Ba**) reddish, white or pink-violet; fr (**Bb**) a purplish-black berry. Introd (Himalayas); S Eng f gdn escape; elsewhere o-r; on wa, rds, wds. Fl 7–9.

C Snowberry, *Symphoricarpos albus*, suckering shrub to 2m; lvs 2–5 cm generally oval to almost circular but some usually variously lobed, dark green above, paler below, very finely downy or hairless; infl mostly in terminal spike, few fl in lf-axils; fls rosy-pink, whitish on inside, bell-shaped with 4 or 5 corolla lobes; fr (**Ca**) a white berry. Introd (N America); Br Isles, vc gdn escape, sometimes planted as game cover in wds, scrub, rds, wa. Fl 6–9.

D BAP ***Twinflower**, *Linnaea borealis*, is a tiny creeping shrub with oval lvs 5–10 mm long, with bell-shaped pink fls 8 mm long, **hanging in pairs** on erect stalks 3–7 cm tall; fr a dry nutlet. Scot only as a native (N Eng vr introd); r in mossy pine forests. Fl 6–8.

MOSCHATEL FAMILY   *Adoxaceae*

E WO (NI) AWI **Moschatel (Town-hall Clock)**, *Adoxa moschatellina*, erect hairless per herb, 5–12 cm tall, with creeping scaly rhizomes. Not closely related to other flowering plants, so in a family on its own. Basal lvs long-stalked, twice-trifoliate, lfts 5–10 mm long, with 2–3 blunt lobes, each tipped with a **tiny spine**; lvs on fl-stem 2 only, opp, short-stalked, smaller, trifoliate; all lvs **fleshy**, **pale green**. Infl (**E**) a long-stalked head of 5 fls; 4 face outwards at right angles (like faces of a clock-tower) and have three-lobed calyx, five-lobed corolla, 10 stamens; fifth fl faces upwards, with two-lobed calyx, four-lobed corolla, 8 stamens; all fls pale yellow-green with golden stamens and inferior ovaries. Fr green, fleshy, drooping. Br Isles: GB N to mid Scot f-lvc; N Scot r; Ire, E coast only, vr; in wds on richer soils, mt ledges. Fl 4–5.

VALERIAN FAMILY   *Valerianaceae*

A family of herbs with **opp** lvs, no stipules; infls **head-** or **umbel-like cymes** of **small** fls, sometimes dioecious; calyx **small**, like a rim on top of ovary, toothed or not; corolla funnel-shaped, sometimes with a spur, and with 5 blunt lobes; stamens 3 or 1; ovary **inferior, three-celled** (but only 1 cell producing a seed); fr a **nutlet**. Valerians (*Valeriana*) are erect pers with pinnate lvs, corolla-tubes swollen below, 3 stamens, and a feathery pappus on top of fr, as in Dandelion (p 470 A), but much smaller; **Spur-valerians** (*Centranthus*) are tall herbs with unlobed lvs, spurred corollas, 1 stamen per fl, and a pappus; **Cornsalads** (*Valerianella*) are small anns with forked branches, tiny mauve fls in heads, 3 stamens and no pappus.

A ᴬᵂᴵ **Common Valerian**, *Valeriana officinalis*, erect per herb, 30–120 cm tall, stem hairy below; **all lvs** opp, pinnate, to 20 cm long, lower long-stalked, upper stalkless; lfts lanceolate, ± bluntly-toothed, ± hairless, side veins conspicuous; infl a terminal umbel-like head, 5–12 cm wide; corollas (**Aa**) funnel-shaped, pinkish-white, five-lobed, tube swollen at base; fr 4 mm long, oblong, with a white feathery pappus. Br Isles, c in fens, riversides, wet wds, also as a smaller form in dry calc gslds and scrub. Fl 6–8.

B **Marsh Valerian**, *V. dioica*, erect dioecious herb, 15–30 cm tall, with creeping runners at base; basal lvs (**Ba**) long-stalked, blades oval-elliptical, blunt, **undivided**, 2–3 cm long; stem-lvs ± stalkless, **pinnatifid**. Infls in terminal heads; on male plants 4 cm across with fls 5 mm wide; on female plants only 1–2 cm across, with florets 2 mm wide; all fls pinkish. Frs as in A, but smaller. Br Isles: GB N to central Scot f-lc; **N Scot, Ire abs**; in fens, wet mds. Fl 5–6.

C **Red Valerian**, *Centranthus ruber*, erect per herb, 30–80 cm tall; lvs **grey-green**, opp, oval-lanceolate, **undivided**, untoothed, 5–10 cm long, lower stalked; infls terminal panicles; corollas (**Ca**) 5 mm wide, deep pink, scarlet, or white, with slender tube 8–10 mm long bearing a pointed **spur** 3–4 mm long; stamen 1 only, protruding; fr with a pappus. Introd (Mediterranean); Br Isles, vc in S, r to N; on old walls, cliffs, rocks, wa. Fl 6–8.

D **Common Cornsalad**, *Valerianella locusta*, slender, ± hairless ann, 7–30 cm tall, with repeatedly forked branches, much more compact in dry places (**Db**); lvs blunt, ± untoothed, lower spoon-shaped, upper oblong. Fls in dense terminal heads, 1–2 cm across; corollas (**Dc**) pale mauve, 1–2 mm across, five-lobed; stamens 3; fr (**Da**) a nutlet 2.5 mm long x 2.0 mm wide, without a pappus, with a corky bulge on the cell with the seed in it. Br Isles, f-lc (except N Scot, NW Ire r); on dunes, sandy gslds, ar, banks, rocks. Fl 4–6.

This is the commonest Cornsalad; others (**E–H**) are of v similar appearance, but **key differences** are in their **frs**; use key p 430 to separate them from D.

1 mm [

Da   E   F   G   H

A

Aa

3 mm

B

♀

Ba

C

Ca

2 mm

Dc

Db

D

KEY TO CORNSALADS (*VALERIANELLA*)

1 Calyx above fr minute, scarcely visible **2**
Calyx above fr larger, obvious **3**

2 Fr (**Da**) flattened on 2 sides (chestnut-shaped), 2.5 mm long x 2 mm wide, almost round in side view – cell with seed inside has swollen corky wall ....*Valerianella locusta* (p 428 **D**)
Fr (**E**) almost square in cross-section, oblong in side view, 2.0 mm long x 0.75 mm wide – cell with seed inside has thin (not swollen, not corky) wall – on old walls, rocks – Eng o-lf to S; Scot abs; Ire vr – fl 4–6 ....**Keeled-fruited Cornsalad** *V. carinata* (illustration p 428 **E**)

3 Fr (**H**) hairy (rarely hairless), oblong, 1.2 mm long x 0.9 mm wide – calyx above fr with 5–6 teeth – S Eng, introd and r – on old walls, wa – fl 6–7 ....
                                            Hairy-fruited Cornsalad *V. eriocarpa* (illustration p 428 **H**)
Fr usually hairless, 1 calyx-tooth **4**

4 Fr (**G**) inflated, broadly egg-shaped, not flattened, 2.0 mm long x 1.5 mm wide – calyx teeth all short, entire – S Eng, S Ire, r, decreasing – on ar on calc soil – fl 6–7..............................
                    BAP EN* **Broad-fruited Cornsalad** *V. rimosa* (illustration p 428 **G**)
Fr (**F**) not inflated, pear-shaped, flat on one side, 2 mm long x 1 mm wide – calyx with one much longer tooth, calyx-teeth with small sub-teeth – Eng f-vlc in S and E, rest of Br Isles r – on ar on calc soil – fl 6–7 ....
                    EN **Narrow-fruited Cornsalad** *V. dentata* (illustration p 428 **F**)

TEASEL FAMILY   *Dipsacaceae*

A family of herbs with erect stems, **opp** lvs, and fls in dense **heads** seated on a common receptacle-disc, with calyx-like whorl of bracts below each head. Fls small, often asymmetrical; each fl surrounded by a tubular **epicalyx**. Calyx tubular, cup-like, or divided into 4–8 long bristle-like teeth; corolla tubular, 4–5-lobed, lobes ± equal or else fl ± two-lipped; stamens **long-protruding**, separate, 2 or 4; style long; stigma simple or two-lobed; **ovary inferior**; fr an achene.

A   **Wild Teasel**, *Dipsacus fullonum*, stout bi herb, in first year producing a lf-rosette of short-stalked, oblong-lanceolate lvs, with swollen-based prickles; stem to 2 m, **hairless**, **v prickly**, branched, arises in second year with opp, long narrow lanceolate lvs (**joined into a cup** at base of each pair that collects rainwater) with few teeth and **prickles only on underside of midribs**. Fl-heads 3–8 cm long, **egg-shaped**, erect; bracts below head linear, rigid, spiny, 5–9 cm long; bracts among florets linear, spiny, **longer** than pink-purple corollas 5–7 mm long. Br Isles: Eng, Wales, c; Scot, Ire r; in open wds, stream banks, rds, gslds, especially on clay soils. Fl 7–8.

B   AWI **Small Teasel**, *D. pilosus*, erect bristly bi herb to 120 cm, much more slender than A, with angled **weakly**-prickly branched stems; basal lvs (**Ba**) oval, **hairy**, **long-stalked**, prickly **only** on midrib below; stem-lvs oval-elliptical, ± toothed, hairy (not prickly, **not** joined round stem). Fl-heads **globular**, 2.0–2.5 cm wide, on weak prickly stalks; bracts below head hairy, narrow-triangular, spine-tipped, **shorter** than fl-head; bracts among florets equalling them, obovate, hairy, spine-tipped; corollas 6–9 mm, white or pinkish. Br Isles: Eng, Wales, o-lf; **Scot, Ire, abs**; in damp open wds, by rivers and streams, hbs. Fl 7–8.

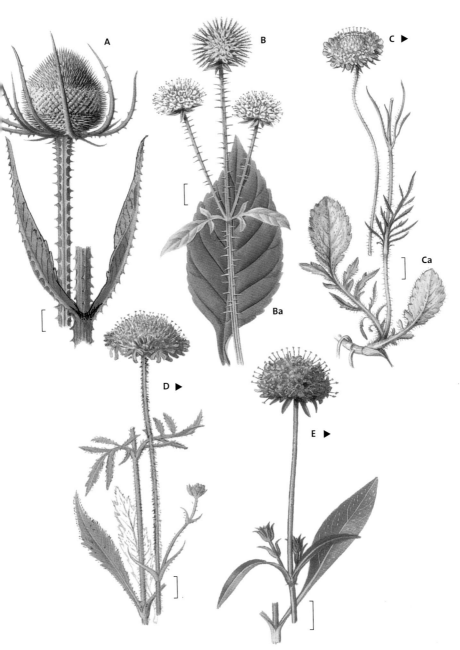

A

B

C ▶

Ba

Ca

D ▶

E ▶

▲ **C Small Scabious**, *Scabiosa columbaria*, slender erect per, 15–70 cm; rosette-lvs (**Ca**) long-stalked, obovate, toothed, or ± pinnatifid with large end lft; stem-lvs pinnate, ± downy, upper ones stalkless, with **linear** lobes; fl-heads 2–3 cm, on long downy stalks; bracts of head linear, shorter than florets; outer florets much larger than central ones; 5 long, blackish, bristle-like calyx-teeth; corollas blue-violet with **5 unequal lobes**. Br Isles: Eng, Wales, f-lc; Scot vr; **Ire abs**; on calc gslds. Fl 7–9.

▲ **D Field Scabious**, *Knautia arvensis*, a **stouter**, more **roughly-hairy** version of C; basal lvs larger, roughly-hairy, usually **unlobed**, but often blunt-toothed; stem-lvs deeply pinnatifid, with **coarse** hairy segments and **elliptical** end-lft; fl-heads 3–4 cm wide, stalks **stout** (2–3 mm); bracts of head oval; calyx-teeth 8; corollas blue-violet, with **4 unequal lobes**; epicalyx four-angled. Br Isles, c (but N, W Scot r); on dry gslds, hbs, rds. Fl 7–9.

▲ **E Devil's-bit Scabious**, *Succisa pratensis*, **key difference** from C and D in its **undivided**, opp, obovate-lanceolate, ± hairy, ± untoothed lvs; stem lvs narrower than basal lvs, otherwise v similar. Fl-heads 1.5–2.5 cm wide, outer florets (unlike those in C and D) **no larger than inner**; calyx with 5 bristle-teeth; bracts among florets lfy, **longer** than calyx-teeth; corollas mauve to dark purplish-blue, with 4 ± **equal lobes**. Br Isles, vc in gslds, mds, fens, damp wds, on acid and calc soils. Fl 6–10.

ID Tips Teasel Family

• Members of the **Teasel** family are similar to Daisy family (*Asteraceae*), but have long-protruding free stamens (in *Asteraceae* joined in a tube and not, or little-protruding), in the persistent epicalyx round each fl (and, later, fr), and in the long calyx-teeth.

• **Do not confuse Scabiouses** with **Rampions** (*Phyteuma*) and **Sheep's-bit** (*Jasione*) in the Bellflower family (p 412). Scabiouses have stamens protruding from the fls, calyx-teeth are long (not short, as in Bellflowers) and there is an epicalyx cup.

• **What to look at: (1)** whether stems spiny or not **(2)** lf-shape, whether entire, toothed or pinnatifid **(3)** number of corolla-lobes and whether equal or unequal.

KEY TO TEASEL FAMILY

1 Stems spiny – long spiny bracts below fl-heads, shorter spine-tipped bracts among narrow florets ....*Dipsacus* (p 430)
Stems hairy but not spiny – bracts of fl-heads and florets not spine-tipped – corollas broad-lobed **2**

2 Stem-lvs pinnatifid – outer florets of head larger than central ones – corollas all unequally 4- or 5-lobed **3**
Stem-lvs untoothed or only weakly so – all florets equal in size – corollas all equally four-lobed ....*Succisa pratensis* (p 432 **E**)

3 4 corolla-lobes – no bracts among florets – calyx with 8 long bristle-teeth – epicalyx four-angled ....*Knautia arvensis* (p 432 **D**)
5 corolla-lobes – bracts among florets – calyx with 5 long bristle-teeth – epicalyx a cylindrical funnel..*Scabiosa columbaria* (p 432 **C**)

DAISY FAMILY   *Asteraceae*

The world's largest family of flowering plants, with c. 23,000 spp, and in evolution one of the most advanced. Also known as the *Compositae* or Composites. In Br Isles, herbs (many shrubs elsewhere); no stipules to lvs; small fls (**florets**) in dense heads, the florets seated unstalked on a flat or rounded disc (**receptacle**) on the end of a stem. The disc is surrounded by a calyx-like series of bracts of varied form, in one or more rows, called the **involucre**. Individual florets may, or may not, have undivided scale-like bracts between them, attached to the receptacle-disc; in different spp florets may have both stamens and ovary together, stamens only, ovary only, or may be sterile. Ovary, when present, is always **inferior**. Petals are always joined into a corolla-tube with (normally) 5 teeth or lobes; corolla is of 2 main types, either **tubular** and ± regular, tipped with 5 small ± equal teeth, or **strap-shaped**, composed of a long petal-like strip extending out to one side of the floret with 5 (or 3) small teeth at its tip. The 5 stamens are joined into a tiny tube attached to the inside of the corolla-tube; the single style is forked into 2 stigmas above; ovary produces a single seed. Calyx may consist of: (1) a parachute-like **pappus** of one or more rows of simple or pinnate-feathered hairs, for wind dispersal of the small achene-like frs; (2) small scales sometimes joined in a cup; or (3) calyx may be abs.

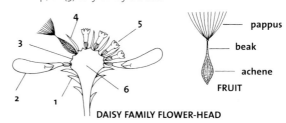

1 Involucre-bract
2 Ray-floret
3 Achene pit
4 Receptacular scale
5 Disc-floret
6 Receptacle
7 Fruit (achene)

pappus
beak
achene
FRUIT

DAISY FAMILY FLOWER-HEAD

ID Tips   Daisy Family

- **What to look at: (1)** florets tubular, or strap-shaped, or of both shapes **(2)** arrangement, length and shape of the bracts surrounding fl-heads (**involucre**) **(3)** presence or absence of pappus on fr (**achene**), and whether its hairs, if present, are simple or feathered **(4)** shape of fl-heads **(5)** presence or absence of scales among florets (6) fr shape and whether beaked or not.

- **Do not confuse** with:
    Members of the **Teasel** family which have stamens all **separate, long-stalked**, distinctly **protruding** from florets, bristle-like calyx-teeth, and an epicalyx to each floret (abs in *Asteraceae*); or
    Sheep's-bit (*Jasione montana*) and **Rampions** (*Phyteuma* spp) in the Bellflower family (p 416), but in these latter stamens are separately stalked and short, and calyx has 5 narrow green teeth; or
    Valerians (*Valeriana*) also have a pappus, but their stamens are 3 in number (5 in *Asteraceae*).

KEY TO DAISY FAMILY

Note: go directly to section F for all *Asteraceae* not agreeing with headings of Keys A–E, especially for *Asteraceae* with tiny heads, that are not obviously daisy-, dandelion- or thistle-like.

1 Florets of fl-heads all strap-shaped (as in a dandelion), all either yellow or blue – stems with milky latex when cut...**A**
At least central florets of fl-heads tubular, regular – stems without milky latex **2**

2 Heads with florets all tubular – no strap-shaped florets visible......................................**3**
Heads with at least some strap-shaped outer florets (perhaps v short, erect), as well as tubular inner florets (as in a daisy) **5**

3 Florets in thistle-like heads, either with florets projecting, ± like a shaving brush, well beyond the elongated cup of spiny or scale-like bracts (or, if florets not projecting, bracts hooked or spine-tipped) ....**E**
Florets not in thistle-like heads, but either in button-like discs or in narrow, few-fld heads, variously grouped or clustered **4**

4 Florets in well-separated, button-like, rounded or flattish heads – heads broader than deep, with many (20–60) densely-packed yellow florets ....**D**
Florets in heads longer than broad, with few (5–20) florets – heads themselves often clustered in heads, panicles or racemes...**F**

5 Ray and disc-florets yellow ....**B**
Ray-florets white, blue, or purple – disc-florets yellow (sometimes white or greenish-grey)..**C**

A. Dandelion-like Asteraceae, with florets of fl-heads all strap-shaped, either yellow, blue or purple; stems with milky latex

1 Fls blue, purple or mauve **2**
Fls yellow or orange **4**

2 Fls bright blue – heads 2.5–4.0 cm across – pappus of scales only (not of hairs) ....
Cichorium intybus (p 470 **D**)
Fls pale blue, purple or mauve – pappus of hairs...**3**

3 Achenes narrowed into a beak above...*Tragopogon porrifolius* (p 470 **B.1**)
Achenes not narrowed into a beak – lvs with large terminal lobe, small side-lobes ....
Cicerbita (p 472)

4 Pappus abs on frs...*Lapsana communis* (p 473 **D**)
Pappus of hairs on frs **5**

5 Pappus with at least its inner hairs feathery **6**
Pappus wholly of unbranched hairs...**11**

6 Stem-lvs tiny or abs – lvs nearly all in basal rosette...**7**
Stem-lvs large, conspicuous (at least below) **8**

7 Lanceolate scarious scales present between florets on receptacle-disc *Hypachoeris* (p 474)
No scales between florets on receptacle-disc..*Leontodon* (p 474)

8 Stem and lvs bristly – stem-lvs oblong – outer bracts of fl-heads ± spreading in a ruff, different from adpressed inner ones – pappus soft ....(*Picris*) **9**
Stem and lvs ± woolly when young, ± hairless when mature – stem-lvs all narrow – bracts of fl-heads all similar, adpressed to head – pappus with many rays, rigid..............................**10**

9 Outer ruff of fl-head bracts widely cordate – bristles on lvs with white swollen bases – achenes with long (6 mm or more) beaks...*Picris echioides* (p 478 **H**)
Outer ruff of fl-head bracts narrow, lanceolate, spreading – bristles on lvs without swollen bases – achenes with short (1–2 mm) beaks ....*P. hieracioides* (p 478 **G**)

10 Bracts of fl-heads long, narrow, in a single, equal row – stem-lvs grass-like, sheathing, 10 cm or more – basal lvs similar, but longer ....*Tragopogon pratensis* (p 470 **B**)
Bracts of fl-heads oval-oblong, in many overlapping rows – stem-lvs flatter – basal lvs elliptical-lanceolate, stalked...*Scorzonera humilis* (p 470 **C**)

11 Lvs all in a basal rosette – infl-stem hollow, lfless, with 1 fl-head ..*Taraxacum* agg (p 470 **A**)
Stems lfy, at least on creeping runners – infl-stem solid or hollow...**12**

12 Achenes ± cylindrical, not flattened...**13**
Achenes strongly flattened, lens-shaped in cross-section...**15**

13 Plants with lfy runners at base – stem-lvs abs or few – rosette-lvs unlobed, untoothed – achenes unbeaked, toothed at tips, below brownish brittle pappus ....*Pilosella* (p 480)
Plants with no runners at base – stem-lvs usually numerous – rosette-lvs and stem-lvs ± toothed or lobed – achenes not toothed at tips **14**

14 Involucre-bracts many, overlapping, unequal – frs never beaked – pappus brownish, brittle..*Hieracium* (p 478)
Involucre-bracts in two rows, inner row of longer equal ones outer row of shorter adpressed or ± spreading bracts – frs sometimes beaked – pappus white, soft.........................*Crepis* (p 476)

15 Frs with no beak, not narrowed upwards – stems stout, hollow – fl-heads 1.5 cm wide or more, with many florets per head, in loose umbels..*Sonchus* (p 480)
Frs beaked – stems slender, ± solid – fl-heads under 1 cm wide, with only 5–12 florets per head, in panicles or racemes..**16**

16 Side-branches of infl spreading at 90° to main stem – florets only 5 per fl-head – bracts around fl-heads in an inner row of equal long ones, plus outer row of v short bracts............
Mycelis muralis (p 472 **F**)
Side-branches of infl ascending at acute angle to main stem – florets more than 5 per fl-head – bracts around fl-heads in many unequal overlapping ranks..................*Lactuca* (p 472)

B Asteraceae with yellow tubular disc-florets in centre of head, and yellow strap-shaped ray-florets round edge of head; latter may be v short, or erect (rather than spreading and daisy-like)

1 Stem-lvs (mostly) opp or at least lower lvs opp **2**
Stem-lvs (mostly) alt, or all lvs from base of stem only **3**

2 Lower lvs with tapering bases – fl-heads not more than 3 cm across (including rays) – fr crowned by 2–4 persistent barbed spines ....*Bidens* (p 448)
Fl-heads more than 3 cm across, up to 30 cm or more – fr flattened without barbs ....
Helianthus annuus (p 442 **C**)

3 Fl-heads solitary, terminal **4**
Fl-heads not solitary **5**

4 Stems scaly lfless (in early spring) – lvs (appearing later) all basal, polygonal in outline – pappus of white hairs ....*Tussilago farfara* (p 450 **E**)
Stems with spiralled lvs – lvs ovate with cordate bases – pappus o or of scales ....
Helianthus annuus (p 442 **C**)

5 Heads arranged in elongated racemes or pyramidal panicles – lvs ± stalkless, simple – pappus of white hairs ....*Solidago* (p 442)
Heads not in elongated racemes or pyramidal panicles **6**

6 Fr without hairy pappus **7**
Fr with pappus of hairs **8**

7 Lvs grey-green, hairless, toothed or pinnately-lobed, with broad segments – no scales between disc-florets...*Chrysanthemum segetum* (p 448 **E**)
Lvs bright green, hairy below, deeply twice-pinnatifid into narrow segments – scales present between disc-florets..*Anthemis tinctoria* (p 455 **E.1**)

8 Involucre-bracts in 1 (or 2) rows, all bracts in each row of equal length, erect – also a few small bracts at involucre-base...**9**
Involucre-bracts in several overlapping rows, outer bracts progressively shorter.................**10**

9 Fl-heads not over 3.5 cm across, in loose, umbel-like infls – involucre-bracts in 1 equal row – also a few small bracts at involucre-base ....*Senecio* & *Tephroseris* (p 440–1)
Fl-heads 4 cm or more across, few per fl-stem, long-stalked, not in a loose, umbel-like infl – involucre-bracts in 2 equal rows ....*Doronicum pardalianches* (p 442 **D**)

10 Fr with a row of small scales on top, surrounding base of hairy pappus ......*Pulicaria* (p 444)
Fr with no scales surrounding pappus-base ....*Inula* (p 444)

C Asteraceae with daisy-like heads, disc-florets yellow or whitish, but ray-florets white, purple, or bluish (not yellow)

1 Lvs opp – heads under 1 cm across – ray-florets white, v short, only 4 or 5*Galinsoga* (p 450)
Lvs alt, or all in a basal rosette **2**

2 Frs with a hairy pappus...**3**
Frs without a hairy pappus **6**

3 Ray-florets v short, narrow, erect or ascending at an angle **4**
Ray-florets broad, spreading, their length at least equal to width of fl-head disc **5**

4 Fl-heads 12–18 mm across, in a loose umbel – ray-florets purple, ascending ....
Erigeron acer (p 446 **B**)
Fl-heads 3–5 mm across, many, in a long panicle – ray-florets whitish, erect, v short..............
Conyza (p 446)
Fl-heads 10–15 mm across, vanilla-scented, appearing late winter, spreading on long stalks in a spike-like raceme – ray-florets pale lilac, erect, short – lvs long-stalked, kidney-shaped, 10–20 cm wide ....*Petasites fragrans* (p 452 **G**)

5 Lower lvs mostly linear, >3x as long as wide, often lobed or toothed but without 1 pair of large lobes – infl-bracts either all green or more green in terminal half ....*Aster* (p 446)
Lower lvs obovate, mostly <2x as long as wide, with 1 pair of large lobes – infl-bracts more green in basal half ....*Erigeron karvinskianus* (p 446 **C**)

6 Fl-heads solitary on lfless unbranched stalks – lvs all in a basal rosette, undivided, obovate to spoon-shaped ....*Bellis perennis* (p 450 **D**)
Fl-heads on lfy, ± branched stalks – lvs not all in a basal rosette **7**

7 Disc-florets whitish or greenish-white (never bright yellow) – ray-florets white or pink – fl-heads small (4–18 mm across), in umbel-like heads ....*Achillea* (p 452)
Disc-florets yellow – ray-florets always white – fl-heads larger (15–50 mm across) **8**

8 Lvs coarsely-toothed, or pinnatifid with broad segments at least 3–4 mm wide ....
Leucanthemum & *Tanacetum* (p 448)
Lvs twice pinnatifid into narrow (1 mm wide) linear or hairlike segments.....................................
Mayweeds etc. (key p 453)

D Asteraceae with yellow disc-florets in button-like rounded or flattish heads; heads not more than twice as long as broad, with many (20–60) densely packed florets (resembling daisy-heads with the ray-florets missing); bracts to heads not spiny

1 Stem-lvs in opp pairs – fr crowned with 2–4 barbed spines – no hairy pappus ......*Bidens* (p 448)
Stem-lvs alt..**2**

2 Lvs simple (sometimes v shallowly lobed), toothed or untoothed.................................**3**
Lvs deeply pinnately lobed or divided..**6**

3 Lvs hairless ....*Aster linosyris* (p 446 **F**), and rayless forms of *A. tripolium* (p 446 **A**)
Lvs hairy **4**

4 Lvs white-woolly – no pappus on fr – on beaches – vr..............*Otanthus maritimus* (p 444 **A**)
Lvs hairy, but not white-woolly – hairy pappus on fr – widespread **5**

5 Fl-heads to 1 cm across, in irregular umbel-like infls – ray-florets abs or v short, disc-florets dark yellow – lvs oval-oblong, downy (like foxglove lvs) ....*Inula conyzae* (p 444 **D**)
Fl-heads 3–5 mm across, many in a long panicle – disc-florets pale yellow – minute white erect ray-florets often present, but not always easy to see – lvs linear-lanceolate, hairy ......
Conyza (p 446)

6 Hairy pappus present – bracts of involucre in one row ....*Senecio vulgaris* (p 440 **F**)
No hairy pappus **7**

7 Disc-florets with corolla-tube white, but with 4 yellow lobes – fl-heads solitary, dome-shaped – plant aromatic – r ....*Cotula coronopifolia* (p 455 **F.1**)
Disc-florets with five-lobed corollas wholly yellowish **8**

8 Lvs 2–3 times pinnately-divided into hairlike segments – fl-heads conical, greenish-yellow, pineapple-scented, terminating lfy branches ....*Matricaria discoidea* (p 455 **F**)
Lvs once-pinnately-divided into flat, lanceolate, toothed segments – fl-heads flat-topped, numerous, in umbel-like flat-topped corymbs................................*Tanacetum vulgare* (p 448 **D**)

E Asteraceae with thistle-like fl-heads (either with florets projecting in a shaving brush-like tuft from the involucre cup of bracts, or, if florets shorter, fl-head has hook-tipped or spiny bracts)

1 Florets yellow..**2**
Florets purple, red, blue or white...**3**

2 Lvs and outer bracts of fl-head spiny, thistle-like – inner bracts of fl-head scarious, yellow, spreading in fr..*Carlina vulgaris* (p 470 **F**)
Lvs with hairs or fine bristles on margins, but without strong spines ....
Centaurea solstitialis (p 467 **D**)

3 Florets blue, radiating outwards in all directions, in spherical fl-head 4–6 cm wide – each floret with whorl of outer bristle-like bracts, and inner pinnate bracts – lvs spine-tipped – tall thistle-like plant...*Echinops sphaerocephalus* (p 468 **B**)
Florets not (or only shortly) projecting beyond bracts – bracts of fl-head involucre narrow, rigid, spreading, hook-tipped – head in fr forming a round burr that sticks to clothes or hair ....*Arctium* (p 468)
Florets projecting thistle-like in a tuft well beyond fl-head bracts – bracts of fl-head broader, erect, not hook-tipped **4**

4 Bracts of fl-head involucre each formed of 2 segments, the lower pale or greenish, the upper brown, black or silvery, scarious or spiny – pappus (if present) not longer than achenes ....*Centaurea* (p 466)

Bracts of fl-head not composed of 2 separate segments – pappus always longer than achenes **5**

5 Lvs and fl-heads quite devoid of spines or prickles...**6**
Lvs spiny, or at least fringed with fine prickles ....(Thistles) **7**

6 Pappus feathered – lvs unlobed, but ± toothed, white-cottony below – fl-heads few, in a dense terminal cluster – on mts – r-lf ....*Saussurea alpina* (p 468 **D.1**)
Pappus simple – lower lvs pinnatifid, upper lvs ± unlobed but fine-toothed, hairless below – fl-heads in a v open terminal cluster ....*Serratula tinctoria* (p 468 **D**)

7 Pappus of feathered hairs ....*Cirsium* (p 462)
Pappus of simple hairs **8**

8 Lvs hairless, shiny dark green, with strong milk-white veins above – stems without spiny wings ....*Silybum marianum* (p 468 **C**)
Lvs ± hairy, without strong white veins above – stems with spiny wings **9**

9 Both sides of lvs, and the winged stem wholly whitish-grey with cottony hairs – achenes four-angled ....*Onopordum acanthium* (p 470 **E**)
Upper sides of lvs green, not cottony – lower sides of lvs and stem ± cottony – achenes rounded ....*Carduus* (p 465)

F Asteraceae with tubular florets only; florets rarely bright yellow, 5–40 together in narrow elongated fl-heads, not daisy-, dandelion- or thistle-like

1 Stem-lvs opp, palmately-lobed, with 3–5 elliptical segments, or else simple, oval-lanceolate – fl-heads small, tassel-like, of 5–6 pinkish-purple florets, grouped into large terminal panicles or corymbs – robust plant to 120 cm tall ....*Eupatorium cannabinum* (p 453 **C**)
Lvs large, all from root, long-stalked, cordate or kidney-shaped – infl-stalk with alt scale-like lvs only **2**
Stem-lvs well-developed, alt – basal lvs (if present) ± similar to stem-lvs **3**

2 Fl-heads gathered into racemes on stout lfless but scale-bearing stalks, produced in winter or spring – basal lvs large (10 cm or more wide), cordate, often appearing later than infls – florets pink or white – widespread..*Petasites* (p 450)
Fl-heads solitary, terminal, on slender lfless but scale-bearing stalks – basal lvs kidney-shaped, small (to 4 cm wide), appearing with infls – florets purple – on mts of Scot – vr ....
Homogyne alpina (p 450 **E.1**)

3 Lvs twice pinnately-divided – fl-heads v small (2–4 mm wide), bell-shaped or oval, grouped into long, usually much-branched panicles ....*Artemisia* & *Seriphidium* (p 456)
Lvs undivided, narrow, strap-shaped or obovate, ± woolly – fl-heads ± woolly **4**

4 Long lfy runners arising from basal rosette of obovate lvs, green above and white below – fl-heads woolly, in an umbel-like head – plants dioecious – male fl-heads with spreading white scarious tips to bracts – female fl-heads with erect pointed pink-tipped bracts ....
Antennaria dioica (p 450 **A**)
Plants without long lfy runners at base **5**

5 Fl-heads grouped into dense rounded or egg-shaped clusters in forks of repeatedly branching stems, along stem branches and at branch tips – outer bracts of fl-heads woolly, inner bracts scarious – lvs strap-shaped, untoothed, woolly – woolly ann plants with no lf-rosette at fl-time..*Filago* (p 459)
Fl-heads either in spike-like racemes or in terminal looser umbel-like clusters surrounded by lvs (not in tight clusters as in *Filago*) – all bracts of fl-heads scarious ....
Gnaphalium (p 459)

KEY TO RAGWORTS, GROUNDSELS & FLEAWORTS
(*SENECIO* & *TEPHROSERIS*)

1 At least some lvs deeply pinnatifid **2**
All lvs unlobed, oval-lanceolate **10**

2 Lvs thickly white-felted below – ray-florets spreading......................*Senecio cineraria* (p 440 **E**)
Lvs ± hairless (or merely cottony or downy) below **3**

3 Ray-florets v short (5 mm or less) or abs..**4**
Ray-florets long, spreading, conspicuous **6**

4 Heads, when young, stalkless, in dense clusters, later stalked – ray-florets usually abs..........
S. vulgaris (p 440 **F**)
Heads always stalked, in loose umbel-like clusters – ray-florets present, short, erect or
recurved **5**

5 Stems and lvs v sticky-hairy – about 20 bracts per fl-head, the outermost 2 or 3 at least $\frac{1}{3}$
as long as others – heads 10–12 mm long x 8 mm wide – achenes hairless ....
S. viscosus (p 440 **H**)
Stems and lvs ± cottony, not sticky – about 13 bracts per fl-head, the outermost v short,
scale-like – heads 7–9 mm long x 5 mm wide – achenes downy ....*S. sylvaticus* (p 440 **G**)

6 Fl-heads in dense umbel-like corymbs on erect stems **7**
Fl-heads in loose, spreading long-branched infls **8**

7 Fls golden-yellow – lvs bright green, ± hairless below – end-lobes of lower stem-lvs and
basal lvs broad, blunt – outer bracts of heads v short ($\frac{1}{4}$ length of inner ones) – achenes
hairless ....*S. jacobaea* (p 440 **A**)
Fls pale clear yellow – lvs grey-green, cottony, especially below – end-lobes of all lvs v
narrow, pointed – outer bracts of heads $\frac{1}{2}$ length of inner, longer ones – achenes hairy ....
S. erucifolius (p 440 **C**)

8 Stem-lvs with large oval terminal lobe – basal lvs elliptical, often without side-lobes – fl-
heads 2.5–3.0 cm wide – in wet places...*S. aquaticus* (p 440 **B**)
Stem-lvs with all lobes narrow – fl-heads 2 cm wide or less – on wa, walls, rds **9**

9 Ray-florets >8 mm long .... *S. squalidus* (p 440 **D**)
Ray-florets < 8 mm long ....*S. cambrensis* (p 440 **D.1**)

10 Involucre with some short outer bracts forming a cup round the long equal inner ones –
lvs sharp-toothed, hairless above **11**
Involucre with no short outer bracts, bracts all long and equal – lvs scarcely toothed,
cottony both sides ....*T. integrifolius* (p 442 **B**)

11 Lvs v cottony below, shiny, hairless above – fl-heads 3–4 cm wide, each with 10–20 ray-
florets ....*S. paludosus* (p 442 **A**)
Lvs hairless both sides – fl-heads 2–3 cm wide, each with 6–8 ray-florets................................
S. fluviatilis (p 442 **A.1**)

Ragworts and **Groundsels** (*Senecio*) are a genus of ± cottony-hairy herbs (in Eur),
with alt lvs and fl-heads in loose or dense umbel-like infls; bracts of fl-heads in **one**
long **equal** row, with mostly a few **short** outer bracts at base of head. Disc- and ray-
florets all yellow (latter sometimes abs); achenes cylindrical, ribbed, not beaked;
pappus white, of simple hairs only; no scales between florets.

A Common Ragwort, *Senecio jacobaea*, erect **bi**, 30–100 cm; stems furrowed, ± hairless, lfy; basal rosette-lvs pinnatifid, with large oval terminal lobes, non-flg rosettes looking very cabbage or kale-like from above. Stem-lvs 1–2 times pinnatifid, with short, often blunt, toothed terminal lobes, all green, ± cottony below. Fl-heads in a branched terminal umbel-like corymb; heads 15–25 mm across, bright yellow; outer bracts less than ¼ length of the long equal inner ones; ray-florets spreading, 5–8 mm long, their outer achenes **hairless**, with simple pappus; disc-floret achenes (**Aa**) **hairy**. Br Isles, vc on neglected gslds, rds, wa, dunes. Fl 6–10. **Key difference** between A and B is in achenes; however lf-shape and habit is useful.

B Marsh Ragwort, *S. aquaticus*, differs from A in elliptical-oval undivided basal lvs (**Ba**), stem-lvs with **large**, **oval** end-lobes (**Bb**), and in the **spreading**, **loosely-branched** infls, larger golden-yellow fl-heads 25–30 mm across; achenes all **hairless**. Br Isles, c in mds, marshes. Fl 7–8.

C Hoary Ragwort, *S. erucifolius*, has erect habit and umbel-like infls of A, but stem and lvs **grey-green** with cottony hairs, especially on lf-undersides; all lvs (**Cb**) have both side- and end-lobes **narrow**, **pointed**, with curved-back margins; fl-heads (**Ca**) 15–20 mm across, their outer bracts half length of the long inner equal ones; florets all **pale** clear yellow, achenes all **hairy**. Br Isles: Eng, Wales, c in lowlands; Scot vr introd; E Ire only, r; on gslds on clay or calc soils. Fl 7–9.

D Oxford Ragwort, *S. squalidus*, usually **ann** (or per), more spreading and bushy than A–C; may be woody at base; lvs almost hairless, 1–2 times pinnatifid (although variable and may be much less deeply cut), with **all** lobes narrow, pointed, non-flg rosettes not cabbage-like; fl-heads 15–20 mm across, bright yellow, **ray-florets** usually **>8 mm** long; all achenes usually **hairy**. Introd (S Italy); Br Isles: Eng, Wales vc; Scot, Ire o; on wa, railways, walls, rds. Fl 5–12. **D.1** End NT ****Welsh Groundsel**, *S. cambrensis*, similar lvs to F and fls to D but with **ray-florets 4–7 mm** long. Br Isles: Wales, Scot only as native, vr on disturbed, open gd on wa, tracks, rds and walls. Fl 5–10.

E Silver Ragwort, *S. cineraria*, low shrub, 30–60 cm; lvs twice-pinnatifid, **white-felted** below (**Ea**), green on uppersides; infl in corymbs, 8–12 cm across, fls bright yellow; bracts woolly; achenes hairless. Introd (Mediterranean); Br Isles: Eng, r but lc; Ire vr; on sea cliffs, beaches, wa. Fl 6–8.

F Groundsel, *S. vulgaris*, ann weed, with weak erect stems to 40 cm; lvs ± cottony, pinnatifid, with short, oval-oblong, blunt lobes, upper lvs clasping, lower stalked. Fl-heads 4 mm across, ± stalkless, in dense clusters at first, later stalked; short outer bracts often black-tipped, inner narrow, also black tipped, 8–10 mm; ray-florets usually abs; achenes (**Fa**) hairy. Br Isles, vc in disturbed places on ar, wa, pavements. Fl 1–12.

G Heath Groundsel, *S. sylvaticus*, taller (30–70 cm) plant than F; lvs pinnatifid, longer-lobed than in F, cottony at first, but later hairless; infl much looser than in F, with long-stalked heads; fl-heads (7–9 x 5 mm) more conical, with **v short** outer bracts, and ± **sticky-hairy**, long inner bracts **not** black-tipped. Ray-florets present but v short, recurved, achenes green, hairy. Br Isles, f-lc on open wds, hths, gslds on acid sandy soils. Fl 7–9.

H Sticky Groundsel, *S. viscosus*, v close to G in habit, lf-form and fl-heads, but **v sticky-hairy** all over stem, lvs and involucres. Fl-heads (**H**) **longer**, **wider** (10-12 x 8 mm) than in G (7–9 x 5 mm), with more long, equal bracts; short outer bracts nearly

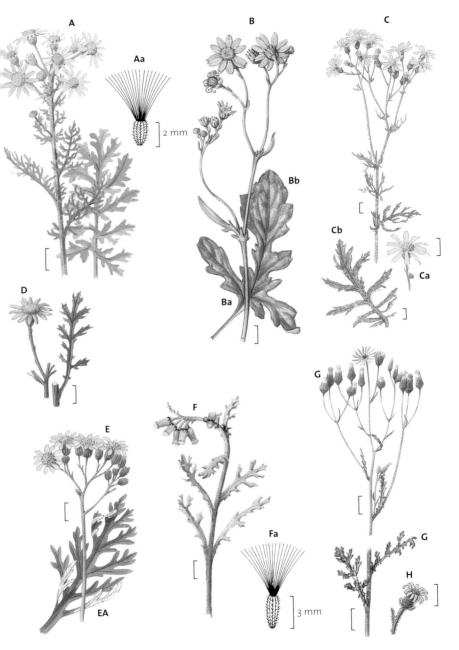

A

Aa

] 2 mm

B

Bb

Ba

C

Cb

Ca

D

E

EA

F

Fa

] 3 mm

G

G

H

441

half length of inner. Ray-florets as in G, but longer; achenes **hairless**. Introd (C Eur); Br Isles: GB f-lc; Ire o; on beaches, wa, railways, rds, walls. Fl 7–9.

A Sch 8 CR ****Fen Ragwort**, *Senecio paludosus*, tall per (to 200 cm); stems and **undersides of lvs cottony**; lvs (**A**) shiny, narrow-lanceolate, sharp-toothed, erect, stalkless, **hairless above**; fl-heads 3–4 cm wide, bright yellow, each with 10–20 ray-florets in a dense erect compound umbel-like corymb. Eng, E Anglia vr in the fens. Fl 7–8. **A.1 Broad-leaved Ragwort**, *S. fluviatilis*, close to A, but lvs **hairless both sides**; fl-heads 2–3 cm wide, with 6–8 ray-florets only. Introd (C Eur); Br Isles, r; in fens, by streamsides. Fl 7–9.

B Field Fleawort, *Tephroseris integrifolia* (*Senecio integrifolius*), erect per, 7–30 cm tall; stem and lvs all ± cottony, especially above; rosette-lvs ± **untoothed**, **round-oval**, **short-stalked**, **flat on gd**, 3–5 cm long; stem-lvs few, lanceolate, stalkless, ± clasping. Fl-heads 1–10, orange-yellow, 1.5–2.5 cm across; involucre 5–8 mm long, bracts with **tuft of hairs** at tip; receptacle surface rough; pappus of disc-florets **equalling** corollas; achenes hairy. Plants of Pennine and Anglesey colonies are taller (to 60 cm) stouter, fleshier. Br Isles: S, E Eng, o-lf (EN *ssp *integrifolia*); Anglesey vr (Sch 8 End VU **ssp *maritima*); on short turf in calc gslds. Fl 5–6.

C Sunflower, *Helianthus annuus*, ann with erect usually unbranched stems, 1.5–4m tall; bristly hairy, most lvs alt, ovate, stalked, with lower lvs opp and lf-bases **cordate**. Solitary fls up to 30cm or more across, yellow with yellow or red to purple centre; involucre bracts ovate, **abruptly contracted** (not gradually tapering) to a **mucronate** point. Achenes (**Ca**) 1–1.5 cm, flattened, often striped. Introd (N America); Br Isles, o preferring fertile neutral to calc soils in open areas on wa, rds. Fl 7–9. Several other introd *Helianthus* spp occur as r casuals.

D Leopard's-bane, *Doronicum pardalianches*, erect per, 30–90 cm tall, with hairy stems and alt lvs; rosette-lvs long-stalked, **broadly cordate**, untoothed; stem-lvs stalked below, upper clasping, all **cordate**, pale green, thin, hairy, ± untoothed. Fl-heads terminal on stem and branches, 4–6 cm across, bright yellow; bracts on saucer-shaped involucre in 2–3 equal rows. Achenes black, pappus simple. Introd (W Eur); Br Isles: Eng, Wales f, E Scot f-lc; Ire r; in shady places in wds, hbs. Fl 5–7.

E AWI **Goldenrod**, *Solidago virgaurea*, per herb with erect little-branched lfy stems 5–70 cm tall; basal lvs obovate, 2–10 cm long, short-stalked, ± hairless, weak-toothed; stem-lvs narrower, pointed, ± untoothed. Fl-heads in a **raceme** or **panicle**, each 6–10 mm across, with yellow ray- and disc-florets; bracts narrow, unequal, many, overlapping, rough-edged. Achenes (**Ea**) brown, downy; pappus of white hairs. Br Isles, c in dry wds, gslds, cliffs, dunes. Fl 7–9.

F Canadian Goldenrod, *S. canadensis*, much taller plant (60–200 cm) than E**,** with **downy stems** (at least in top ¹/₂) and strap-shaped **hairy lvs**; infl with long, dense side-branches forming a **pyramidal** panicle 10–15 cm wide; fl-heads only 5 mm across, ray-florets v short. Introd (N America); Br Isles, f-lc; in wa, on railways. Fl 8–10. **F.1 Early Goldenrod**, *S. gigantea*, close to F but has **lvs hairless** or hairy only on lower-side veins and ± **hairless stems**. Introd (N America); GB f-lc; Ire vr-abs; natd on a wide range of soils in wa, rds, railways, rough gsld. Fl 7–9. Several other garden *Solidago* spp are r natd.

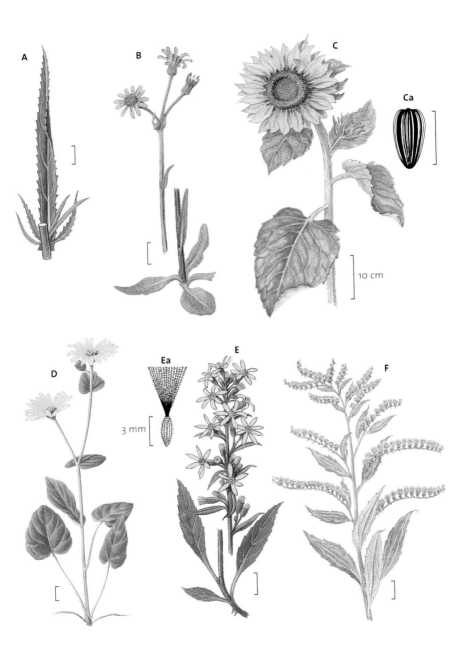

A

B

C

Ca

10 cm

D

Ea

3 mm

E

F

443

A FPO (Eire) EX **Cottonweed**, *Otanthus maritimus*, a striking plant; erect tufted per, with stems and lvs all densely **white-woolly**, lvs oblong-lanceolate, shallow-toothed, 2–3 cm long; fl-heads in a dense umbel-like corymb, each 6–9 mm across; tube-florets bright yellow; **ray-florets abs**; involucre of many overlapping woolly bracts. Achenes hairless, pappus abs. **GB abs**; SE Ire, vr but vla; on sandy, shingly seashores. Fl 8–9.

Fleabanes (*Pulicaria*) have alt downy lanceolate lvs; both ray- and disc-florets yellow; bracts to fl-heads in **overlapping** rows; achenes hairy, with a **simple** hairy pappus and also a row of **scales** on top. *Erigeron* are close to *Aster*, but fl-head bracts are **equal**, and ray-florets are **narrow** and numerous. **Ploughman's-spikenard** etc (*Inula*) is close to *Pulicaria*, but there is no row of scales on top of achene around pappus.

B Common Fleabane, *Pulicaria dysenterica*, erect hairy branched **per** herb with runners, 20–60 cm; lvs alt, downy, oblong, ± untoothed, 3–8 cm long, upper lvs **cordate**-based, clasping. Fl-heads 1.5–3.0 cm across, in a loose corymb; ray-florets **twice as long** as disc-florets, all florets golden-yellow; bracts of head narrow, pointed, sticky-hairy; achenes (**Ba**) hairy, pappus surrounded by a scaly cup. Br Isles: Eng, Wales, c in lowlands; Scot, S only, r; Ire f; in mds, rds, ditch-sides, on clay or wet soils. Fl 8–9.

C Sch 8 CR ****Small Fleabane**, *P. vulgaris*, erect ± sticky-hairy branched **ann**, 8–40 cm; stem-lvs as in B, but rounded- **not** cordate-based, 2.5–4.0 cm long; fl-heads (**Ca**) many, *c.* 1 cm across, outer bracts **spreading**; ray-florets **short**, ± equal to length of disc-florets, erect, all florets **pale** yellow. Achenes as in B, but scales on top are **separate**, **not** forming a cup. S Eng, decreasing, vr (but New Forest f); on open, well-grazed hollows or tracks on commons, pond edges ± flooded in winter, ± dry in summer. Fl 8–10.

D Ploughman's-spikenard, *Inula conyzae*, bi or per herb, erect, downy, 20–80 cm; lower lvs foxglove-like, oval-oblong, stalked, pointed, toothed, downy; upper narrower, ± stalkless. Fl-heads 1 cm across, many, in an umbel-like corymb; outer bracts green, downy, spreading; inner scarious, ± purplish; ray-florets **v short**, **or abs**; all florets dark yellow. Achenes (**Da**) dark-brown, hairy, with pappus of simple rosy hairs, but **no scales** around it. Br Isles: Eng, Wales only, f-lc; Scot vvr; Ire vr introd; on gsld, scrub, rocks on calc soils. Fl 7–9.

E ***Golden-samphire**, *Inula crithmoides*, fleshy tufted erect per herb 15–80 cm; lvs many up stems, alt, **linear**, **fleshy**, hairless, stalkless, three-toothed at tip or untoothed, 2.5–6.0 cm long, ± clustered above in lf-axils. Fl-heads few, 2.5 cm across, in a loose umbel-like corymb; bracts to heads hairless, erect; ray-florets many, golden, **twice as long as longest bracts**; disc-florets orange; achenes downy, pappus white. Br Isles: S Eng, Wales, S Ire only, o-lf on coast; on upper drier salt-marshes, shingle banks, sea cliffs. Fl 7–9. **E.1 Elecampane**, *I. helenium*, differs from E and from most other yellow-rayed daisy-like *Asteraceae* in its stout stems 60–150 cm tall; in its v large lvs, the basal lvs elliptical, long-stalked, 25–40 cm long, the stem-lvs stalkless, clasping, oval-cordate, all ± hairless above, woolly below; and in its **v large**

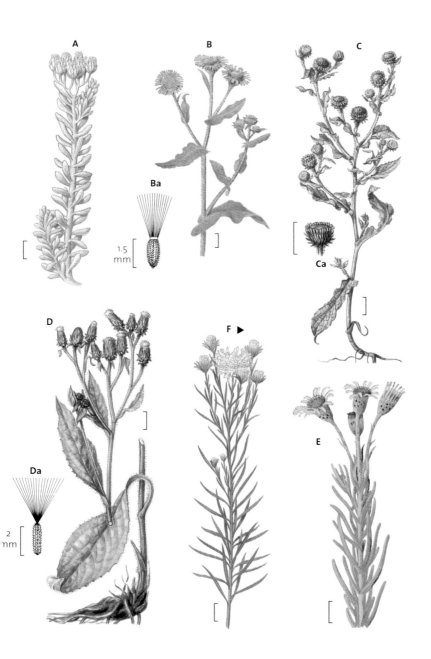

A

B

Ba

1.5 mm

C

Ca

D

Da

2 mm

F ▶

E

fl-heads, 6–8 cm across, with many narrow rays; stout cup-like involucres with hairy overlapping oval spreading bracts; achenes hairless, pappus reddish. Introd (W Asia); Br Isles, o on wa, rds. Fl 7–8. **E.2** FPO (Eire) **Irish Fleabane**, *I. salicina*, is close to E in habit and fl-head size; erect per 25–50 cm tall; stem-lvs, many, alt, **elliptical**-lanceolate, pointed, 3–7 cm long, are firm but **not fleshy**, ± untoothed, hairless above but fine-**bristly** on edges and veins below. Fl-heads 2.5–3.0 cm across, 1–5 in a loose head as in E, ray- and disc-florets yellow; outer bracts of fl-heads **hairless**, lanceolate; achenes hairless. **GB abs**; Ire, by Lough Derg, vr-vlf and declining; in fens, by stony lake shores. Fl 7–8.

F ****Goldilocks Aster**, *Aster linosyris*, slender erect hairless per herb, 10–40 cm tall; stems bear many alt, **narrow-linear**, pointed lvs 2–5 cm long, untoothed but rough-edged; fl-heads in a dense umbel-like corymb, each head 12–18 mm across, its bracts v **narrow**, many, **overlapping**, outer spreading, inner adpressed; **no ray-florets**; disc-florets bright yellow, longer than bracts, in a brush-like tuft. Achenes downy; pappus reddish, simple. GB: SW Eng, Wales, Cumbria only, vr limestone rocks, cliffs. Fl 8–9.

A **Sea Aster**, *Aster tripolium*, is closer to a fleshy-lvd Michaelmas Daisy than to F. Erect branched hairless per, 15–100 cm; lvs 7–12 cm long, **fleshy**, ± untoothed; basal lvs stalked, obovate; upper lvs linear-oblong. Fl-heads in loose umbel-like corymbs, 8–20 mm across, involucre bracts **blunt**, adpressed, overlapping, scarious-tipped; ray-florets long and spreading, **mauve** or white, or **abs** (var *discoideus*); disc-florets yellow. Achenes (**Aa**) brown, hairy, with brownish simple pappus. Br Isles, lc all around coasts, estuaries on salt-marshes, sea cliffs, rarely inland. Fl 7–10. **A.1 Michaelmas-daisies**, *A. novi-belgii* agg, is a group of v similar spp that resemble A in blue, reddish (or white) ray-florets and yellow disc-florets, but **lvs** thin, **not fleshy**, involucre-bracts long and pointed. Many spp introd (N America) occur as gdn escapes in wa, etc in Br Isles. Fl 7–10.

B WO (NI) **Blue Fleabane**, *Erigeron acer*, resembles a shorter, narrow-lvd version of D (p 444) with purple ray-florets, but bracts to fl-head are **equal** and not overlapping. Erect hairy ann or bi, 8–40 cm tall, with many alt linear-lanceolate, untoothed clasping stem-lvs and basal rosette of stalked, obovate-lanceolate lvs. Fl-heads 12–18 mm across, in a loose panicle; ray-florets pale purple, narrow, erect, many, little longer than yellow disc-florets. Achenes (**Ba**) yellow, hairy, pappus long, reddish-white. Br Isles: Eng, Wales, c to E, f to W; Ire r; Scot r introd; on calc gslds, hbs, wa. Fl 7–8. **B.1** Sch 8 VU ****Alpine Fleabane**, *E. borealis*, close to B, but **short** (7–20 cm), unbranched; lvs and stem much hairier; fl-heads only 1–3, 20 mm wide, with purple ray-florets **much longer** than disc-florets. Scot, Perthshire to Aberdeenshire, vr; on calc mt rock-ledges. Fl 7–8.

C **Mexican Fleabane**, *E. karvinskianus*, sprawling per herb to 50 cm, often procumbent; lvs obovate, upper stem lvs usually entire, unlobed; lower lvs with 1 pair of large lobes and regularly toothed; infl 1.5–3 cm across; fl white or pale mauve, often deeper pink beneath, usually several per stem. Introd (Mexico); Br Isles, f (SW Eng and CI c); on walls, rds, wa, stony gd, urban areas. Fl 4–10. **Do not confuse** with **Daisy** (p 450 D) which has **no** stem-lvs and **one fl-head** per stem.

D **Canadian Fleabane**, *Conyza canadensis*, ann, 8–80 cm tall, resembles a narrow-lvd, slender, **lime-green**, less hairy version of B, with ± bristly hairs; many more, much smaller (3–5 mm across) fl-heads (**Db**) in long panicles; ray-florets erect, narrow, **v short, whitish**; disc-florets yellow with **4 lobes**; **bracts** on involucre **hairless** to v slightly hairy; achenes (**Da**) downy, pale yellow with simple **yellowish** pappus. Introd (N America); SE Eng c; rest of Eng, Wales o; Scot, Ire r; on wa, rds, dunes. Fl 6–10. **D.1 Guernsey Fleabane**, *C. sumatrensis*, is v close to D but more densely covered with **grey-green soft** hairs; brittle stems that (unlike D) snap easily; infl more pyramidal in shape than D; ray florets often **abs**; v tiny disc-florets with **5 lobes** (x20) and **bracts** on involucre **densely hairy**. Introd (S America); Br Isles: S Eng, CI f-vla; E Ire vvr; one of the commonest plants around London; chiefly urban areas on wa, railways and pavements but also wds. Fl 6–10.

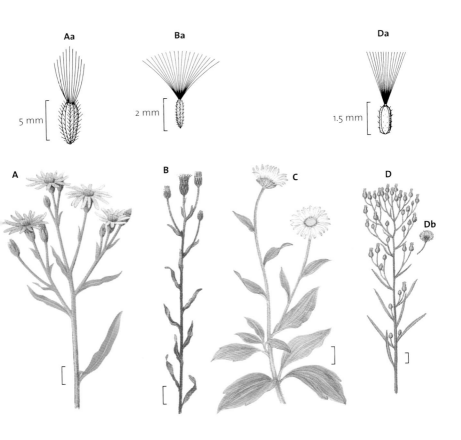

Aa

Ba

Da

5 mm

2 mm

1.5 mm

A

B

C

D

Db

A Oxeye Daisy, *Leucanthemum vulgare* (*Chrysanthemum leucanthemum*), erect, slightly hairy little-branched per herb, 20–70 cm tall; basal rosette of long-stalked, ± spoon-shaped, toothed lvs (**Aa**), **abruptly contracted** to broadly rounded base, usually some lobed. Stem lvs stalkless, clasping, oblong, alt, deeply-toothed, all dark green. Fl-heads long-stalked, daisy-like, 2.5–6 cm across, ray-florets **white**, disc-florets **yellow**; bracts oblong, overlapping, purple-edged. Achenes (**Ab**) hairless, ribbed, without a pappus. Br Isles, c in gslds, mds, rds, on fertile soils. Fl 5–9.

B Shasta Daisy, *L. x superbum* (name used for gdn hybrids and variants of *L. lacustre*), close to A, fls usually larger 6–10 cm across; lf-bases of basal lvs (**B**) **gradually** contracted to v narrow, rounded base and toothed, never lobed. Achenes (**Ba**). Introd fertile hybrid of gdn origin. Br Isles, c in disturbed rds, wa, gsld. Fl later than A, 7–9.

C Feverfew, *Tanacetum parthenium* (*Chrysanthemum parthenium*), erect, downy, strongly **aromatic** per herb, 25–60 cm, much branched above; lvs oval in outline, **pinnate** with toothed segments, the lower long-stalked; fl-heads in a loose, ± flat-topped umbel-like corymb; heads 12–20 mm across, with **short**, **broad**, **white** rays, **yellow** discs; bracts to heads overlapping, oblong, **downy**, with **pale** scarious edges. Achenes without pappus, ribbed, hairless. Br Isles, c (except S Ire f; NW Scot o-r); on hbs, wa, rds, walls. Fl 7–9.

D Tansy, *T. vulgare* (*Chrysanthemum vulgare*), erect **strong-smelling** hairless per herb, 30–100 cm; lvs, 15–25 cm long, up the stems, are alt, pinnate, their lfts deeply pinnatifid. Fl-heads in a dense umbel-like corymb 7–15 cm across, each head 7–12 mm across, yellow, button-like, **without** ray-florets; overlapping **hairless** bracts with pale scarious edges; achenes (**Da**) ribbed, hairless, without a pappus, but with a toothed **cup** on top. Br Isles: GB c; Ire f; on rds, hbs, wa, riverbanks. Fl 7–10.

E vu Corn Marigold, *Chrysanthemum segetum*, ± erect hairless ann, 20–50 cm; lvs grey-green, ± fleshy, deeply-toothed, oblong; fl-heads large (3.5–6.5 cm), solitary, with both **disc** and rays **golden-yellow**; bracts of heads overlapping, with broad pale brown scarious margins; achenes (**Ea**) **cylindrical**, hairless, without pappus. Br Isles, f but declining on ar on acid or sandy soils. Fl 6–10. **E.1 Pot (Garden) Marigold**, *Calendula officinalis*, also has large (4–5 cm across) heads, but with both rays and disc **orange**; lvs downy, oblong-obovate ± untoothed; achenes **boat-shaped**, without pappus. Br Isles, f gdn escape. Fl 5–10.

F Nodding Bur-marigold, *Bidens cernua*, erect ann, 8–60 cm; lvs **opp**, **undivided**, lanceolate, pointed, coarsely-toothed, stalkless, ± **hairy**; fl-heads few, long-stalked, **drooping**, 15–25 mm across, button-like, dull yellow, usually **without** ray-florets (rarely with few, golden-yellow, broad rays, 12 mm long). Achenes (**Fa**) straight-sided, flattened, usually with 3–4 barbed spines on top. Br Isles: Eng, Wales, Ire f-lc to S, o to N; Scot vr; by muddy pond edges, on grassy commons, where water stands in winter but not in summer. Fl 7–10.

G Trifid Bur-marigold, *B. tripartita*, is similar to F, but hairless or downy only; lvs (**G**) **trifoliate**, with toothed, lanceolate lobes and short winged stalks; fl-heads ± **erect**; ray-florets v rarely present. Achenes (**Ga**) **obovate**, v flattened, with 2–4 barbed spines (rarely extra shorter ones) on top. Br Isles: Eng, Wales, f-lc to S, o to N; Scot vr, Ire o; habitats as for 911. Fl 7–10.

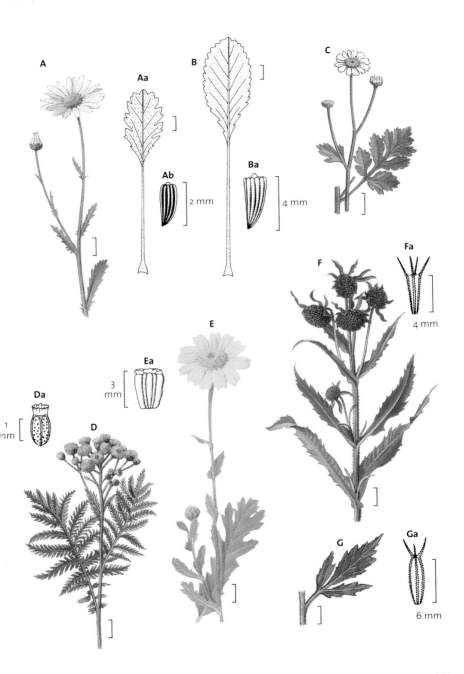

A

Aa

Ab

2 mm

B

Ba

4 mm

C

Da

1
mm

D

Ea

3
mm

E

F

Fa

4 mm

G

Ga

6 mm

A   Mountain Everlasting, *Antennaria dioica*, per herb with creeping woody stem that produces lfy rooting runners, and erect unbranched, lfy white-woolly fl shoots 5–20 cm tall; rosette- and runner-lvs obovate to **spoon**-shaped, blunt, mucronate, stem-lvs erect, narrow, pointed; all lvs alt, **green** above, **white**-woolly below. Fl-heads white-woolly, in a close terminal umbel; dioecious; female heads (**Ac**) 12 mm across, with their woolly bracts narrow, pointed, **erect**, with **pink** scarious tips; male heads 6 mm across, with bracts obovate, **blunt**, **spreading** like ray-florets, usually **white**-tipped; florets all tubular, pink; achenes of female plants (**Aa**) with slender white pappus hairs, achenes of male plants (**Ab**) with a few pappus hairs, club-shaped above like **butterfly antennae**. Br Isles: S, E Eng vr; Wales, N Eng o-lf; Scot f, vc in N; Ire o-lc; on hths, dry gslds, mt ledges. Fl 6–7.

B   Gallant-soldier, *Galinsoga parviflora*, erect ± hairless ann, 10–70 cm tall; lvs **opp**, oval, pointed, stalked, few-toothed; forked cymose infls with a fl-head in each fork; fl-heads 3–5 mm across, bracts few, oval; ray-florets usually only 5, 1 mm long x 1 mm wide, white; disc-florets yellow; **scales** on receptacle (**Ba**) **3-lobed**; achenes (**Bb**) black, bristly, three-angled, oval, with a pappus of 8–20 silvery lanceolate **scales** without bristle-tips. Eng f, but lc in SE (London area vc); Scot, N Ire vr; on wa, ar. Fl 5–10.

C   Shaggy Soldier, *G. quadriradiata* (*G. ciliata*), v like B, but clothed with spreading hairs; usually 4 narrower ray-florets; **key difference** is **scales** on receptacle (**Ca**) **unlobed** or with 2 small side lobes and pappus-scales of achenes (**Cb**) shorter, tapering into **bristle**-tips. Eng f, but lc in SE (London area vc); Scot, Ire vr; in wa, rds, pavements. Fl 5–10.

D   Daisy, *Bellis perennis*, per herb with short-stalked, obovate, blunt-tipped, blunt-toothed downy lvs 2–4 cm long, **all in a basal rosette**. Fl-stems 3–12 cm tall, **lfless**, hairy, each with **solitary** fl-head 16–25 mm across; bracts many, green, blunt, hairy, oblong; ray-florets many, white, spreading; disc-florets yellow; achenes (**Da**) obovate, pale, flattened, downy, pappus **abs**; receptacle beneath achenes **conical**. Br Isles, vc in short gslds, mds. Fl 3–10.

E   Colt's-foot, *Tussilago farfara*, per herb with stout scaly runners, and solitary fl-heads on erect scaly lfless stems 5–15 cm tall; fl-heads 15–35 mm across, rays and disc florets yellow, bracts of head many, ± purplish, narrow, in **one** equal row forming a long narrow cup, with a few broader outer ones; fl-stem in fr lengthens to about 30 cm; achenes (**Eb**) hairless, pale, pappus of long simple white hairs. Lvs (**Ea**) appearing **after** fls, long-stalked, rounded-triangular, cordate-based, 10–20 cm across, downy above, white-felted below, with small sharp teeth. Br Isles, c on clay on wa, ar, cliffs, landslips, dunes, scree etc. Fl 3–4. **E.1** Sch 8 EN ****Purple Colt's-foot**, *Homogyne alpina*, creeping per; like E in lfless scaly fl-stems with solitary fl-heads; but **kidney-shaped** glossy dark green lvs, purple below, 4 cm wide, appear with fls. Stems bearing heads are 10–30 cm tall in fl; fl-heads of purple **tubular** florets only. Scot: Glen Clova, vr on moist open places in mts. Fl 6–9.

F   Butterbur, *Petasites hybridus*, per herb with stout rhizomes, producing fl-heads in **stout**-stalked racemes, 10–40 cm tall, in spring; fl-stems have green lanceolate scales only; plants ± dioecious; fl-heads reddish-pink, male 7–12 mm long, all florets **tubular**; female fl-heads shorter (3–6 mm long), florets narrower. Basal lvs (**Fa**), appearing after fls, long-stalked, rounded, distantly-toothed, deeply cordate, 10 cm wide at first, but becoming huge (60 cm across or more) in summer, green above, grey-downy below. Achenes (**Fb**) pappus white. Br Isles f-lc (except **N Scot abs**); on

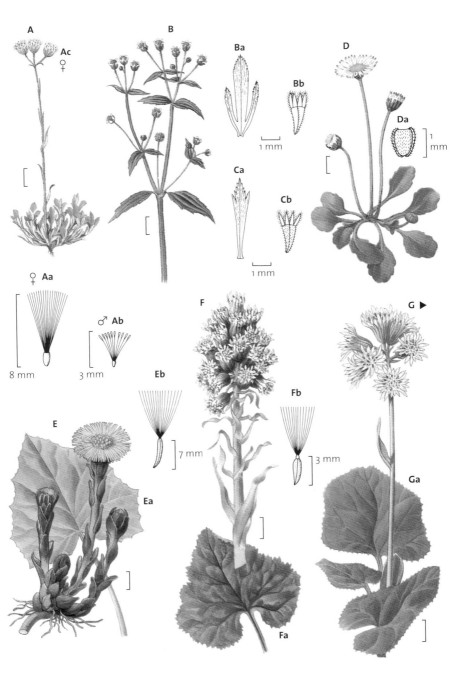

A

Ac ♀

Aa ♀   Ab ♂

8 mm   3 mm

B

Ba   Bb

1 mm

Ca   Cb

1 mm

D

Da

1 mm

E

Eb

Ea

7 mm

F

Fb

3 mm

Fa

G ▶

Ga

wet mds, rds, streamsides, wet copses; male plants widespread, but female plants lc only in N Eng. Fl 3–5. **F.1 Giant Butterbur**, *P. japonicus* has **creamy** fls in **branched** flattish panicles, with broad, dense pale green bract-scales on their stalks, appearing like cauliflowers from a distance; lvs v large (to 1 m across), rounded-cordate. Introd (Japan); GB, o; Ire vr; in damp places on rds, riversides. Fl 3–4. **F.2 White Butterbur**, *P. albus* also has **white** fl-heads, but smaller (to 30 cm wide) rounded-cordate lvs, **white**-woolly beneath. Introd (mts Eur); GB, N Ire, o-r (E Scot f); in moist wds, wa, gsld.

▲ **G Winter Heliotrope**, *P. fragrans*, **slenderer** plant than F, with 10–25 cm-tall racemes of few, **lilac**, **vanilla-scented** fl-heads with short erect **rays** to outer florets; lvs (**Ga**), **appearing with** fls (in winter), are **rounded**, kidney- to heart-shaped, 10–20 cm wide, evenly-toothed, **green both sides**. Introd (SW Eur); Br Isles, vc to S; o to N (except Scot r-o) Ire f; on wa, banks, rds, streamsides. Fl 12–3.

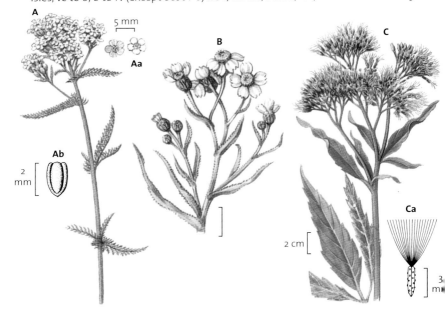

A Yarrow, *Achillea millefolium*, erect tufted downy per, 8–40 cm tall, strongly-scented, with runners; lvs lanceolate in outline, 5–15 cm long, 2–3 times pinnate into short narrow **linear** segments, lower lvs stalked, upper stalkless, shorter. Fl-heads (**Aa**) many, 4–6 mm across, in dense terminal umbel-like corymbs; involucre oval, of many overlapping oblong, blunt, keeled bracts with scarious edges; ray-florets usually 5, three-toothed at tip, as long as **wide**, white or pink; disc-florets **dirty white** or **creamy**. Achenes (**Ab**) flattened, hairless, **without** pappus. Br Isles, vc in mds, gslds, rds, hbs. Fl 6–8.

B Sneezewort, *A. ptarmica*, erect tufted per herb, 20–60 cm, little-branched, hairy on upper parts; lvs **unlobed**, linear-lanceolate, 1.5–6.0 cm long, ± hairless, **finely** and sharply toothed; fl-heads larger, fewer than in 920, 12–18 mm across; ray-florets (4 mm) longer than in A, oval, white; disc-florets greenish white. Br Isles, f-lc but r in chky or v cultivated areas; on wet mds, hthy gslds, scrub, on acid soils. Fl 7–8.

C Hemp-agrimony, *Eupatorium cannabinum*, tall downy per to 120 cm; basal lvs obovate, stem-lvs ± stalkless, trifoliate, each of 3 (or 5) elliptical-lanceolate toothed lfts 5–10 cm long; all lvs **opp**. Fl-heads v small (and superficially more Valerian- than Daisy-like) with only 5–6 reddish-pink tubular florets and 8–10 oblong purple-tipped bracts, but grouped into large, rounded, much-branched panicles to 15 cm across; achenes (**Ca**) black, angled with white simple pappus-hairs. Br Isles: S Eng, Wales c, f elsewhere, r-o N Scot; in fens, marshes, wet wds. Fl 7–9.

Mayweeds and **Chamomiles** (*Tripleurospermum, Matricaria, Anthemis, Chamaemelum*) have daisy-like fl-heads with disc-florets yellow and ray-florets (mostly) white (or abs); alt lvs are 2 or more times **pinnate** into almost **hairlike** segments. See key to **Mayweeds** etc below.

ID Tips Mayweeds etc

- **What to look at: (1)** smell of lvs when crushed **(2)** solid or hollow receptacle **(3)** presence or not of scarious, lanceolate **scales** at base of disc-florets on the receptacle (best seen by breaking fl-head in half and looking amongst the florets) **(4)** achene shape and surface characteristics.

- **Do not confuse** non-flg plants with **Yarrow** (p 452), which also has finely segmented lvs, or **Tansy** (p 448).

- In *Anthemis* and *Chamaemelum* each disc-floret has a **scale** at its base.

- *Tripleurospermum* and *Matricaria* have **no** scales among disc-florets.

KEY TO MAYWEEDS AND CHAMOMILES etc with fine lf-segments divided > ³/₄ way to midrib

1 Ray-florets abs ....*Matricaria discoidea* (p 455 **F**)
Ray-florets present **2**

2 Scales in fl-head – plant hairy **3**
No scales in fl-head – plant ± hairless **5**

3 Per plant – base of corolla-tube of individual disc-floret enlarged, forming a pouch, completely enclosing achene (**Gc** p 454) ....*Chamaemelum nobile* (p 455 **G**)
Ann plant – base of corolla-tube of individual disc-floret not pouched **4**

4 Receptacle scales linear – plant with strong foetid smell – achenes with warts ....
Anthemis cotula (p 455 **D**)
Receptacle scales broad – plant with no or scarce smell – achenes without warts.................
A. arvensis (p 455 **E**)

5 Receptacle hollow ....*M. recutita* (p 455 **C**)
   Receptacle solid **6**

6 Lvs fleshy – lf-segments blunt – achene with elongated oil glands (**Aa** below)..........................
   Tripleurospermum maritimum (p 455 **B**)
   Lvs not fleshy – lf-segments bristle-pointed – achene with circular oil glands (**B** below) ....
   T. inodorum (p 455 **A**)

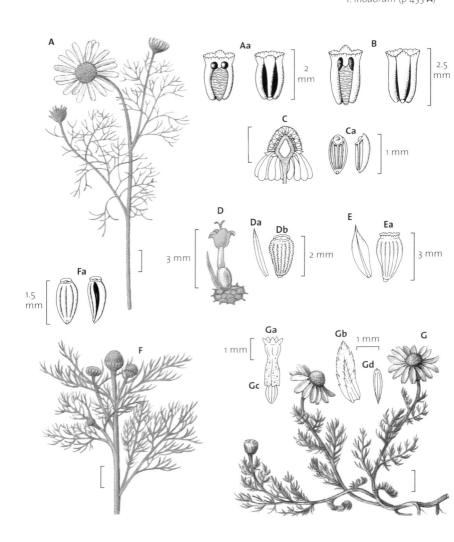

A Scentless Mayweed, *Tripleurospermum inodorum*, ann or per herb 10–50 cm tall, erect or sprawling, ± **scentless**; lvs alt, hairless, 2–3 times pinnate into narrow segments, thread-like, fine-pointed . Fl-heads in loose infls, 2.0–4.5 cm across, with many spreading white ray-florets and yellow disc-florets; no scales among disc-florets; receptacle **solid**, **dome-shaped**; involucres cup-like, of blunt, oblong, overlapping bracts with brown scarious edges (dark edges often absent in winter); achenes (**Aa**) top-shaped, flattened, with 2 **circular black spots** (oil glands) near top on one side and no pappus, but a narrow rim on top. Br Isles, c in ar, wa. Fl 7–9.

B Sea Mayweed, *T. maritimum*, v close to A but lvs blunt, cylindrical, ± fleshy; **key difference** is in the frs: **achenes** (**B**) with 2 **elongated** oil glands on one side. Br Isles, c on sandy or shingly shores, rocks, by sea. Fl 7–9.

C Scented Mayweed, *Matricaria recutita*, v near to A but pleasantly **aromatic**; receptacle of fl-head **hollow**, **conical** (**C**); achenes (**Ca**) obliquely truncate above, have **no** black spots (oil-glands) near top; ray-florets soon bend back. Br Isles: Eng, Wales, f-lc; Scot, Ire, o-r; on ar, wa, on sandy soils. Fl 6–7.

D VU **Stinking Chamomile**, *Anthemis cotula*, looks v like A, is also ± hairless, but has unpleasant smell, and narrow **scales** (**Da**) occur on the conical receptacle among **inner** disc-florets (**D**); achenes (**Db**) ribbed, **warty**, no pappus. Br Isles, f-lc to S, r to N, Ire r; on ar, wa, especially on clay. Fl 7–9.

E EN **Corn Chamomile**, *A. arvensis*, resembles D in most points, but **downy** or woolly, not hairless; **aromatic**, **not** nasty-scented; scales (**E**) on receptacle **lanceolate**-mucronate, **not** linear; ray-florets have styles (abs in D's rays); achenes (**Ea**) strong-ribbed, but **not** warty. Br Isles: Eng o-f but declining; rest of GB r; Scot, Ire vr; on ar on calc soils. Fl 6–7. **E.1 Yellow Chamomile**, *A. tinctoria*, is distinctive in its fl-heads 2.5–4.0 cm across, both ray- and disc-florets **yellow**; structure and lvs as in E, but looks more like Corn Marigold (p 448 E). Introd (C Eur); Br Isles, r in wa, r; on ar, rds, wa. Fl 7–8.

F Pineappleweed, *Matricaria discoidea* (*M. matricarioides*), erect ann with habit and fine lf-segments of A, and without scales between disc-florets, but **lacks** ray-florets, has **conical**, **hollow** receptacles to fl-heads, greenish-yellow disc-florets, and smells strongly of **pineapples** when crushed. Achenes (**Fa**). Br Isles, vc on wa, trackways, rds, ar. Fl 5–11. **F.1 Buttonweed**, *Cotula coronopifolia*, has similar fl-heads to F, **aromatic** hairless creeping ann, lvs alt, linear-pinnatifid, but with lobes **flat**, **not hairlike**, bases **sheathing** stem; fl-heads yellow, **button-like**, single on long stalks; disc-florets with **white** tubes and **4 yellow** corolla-lobes; pappus abs. Introd (S Africa); Eng, r-lf (vla along Thames in E London); on mud in marshy places and wa, dunes. Fl 7–10.

G VU **Chamomile**, *Chamaemelum nobile*, **creeping** downy **per** herb, strongly chamomile-scented; stems 10–30 cm long; lvs 1.5–5.0 cm, 2–3 times pinnate, with short linear ± hairy segments; fl-heads 18–25 mm across, solitary on short erect stems; ray-florets white, spreading; disc-florets (**Ga**) yellow, with oblong blunt **scale** (**Gb**) at base of each, and bases of corolla-tubes **swelling** to cover top of achenes in a hood (**Gc**) (use hand lens); receptacle conical; bracts of fl-head blunt, overlapping, scarious-edged. Achenes (**Gd**). S Eng, S Wales, lf; N Eng, Scot, vr; Ire r (lc in SW); on sandy gsld on commons. Fl 6–8.

Mugworts and **Wormwoods** (*Artemisia* & *Seriphidium*), are tall herbs with lvs alt, **pinnatifid**; long, much-branched **panicles** of tiny fl-heads, each with rather few **tubular** florets only. Involucre globular or bell-shaped, of many hairy, scarious-edged, overlapping bracts; florets yellowish or reddish; pappus abs. **Cudweeds** (*Gnaphalium, Filago*) have similar tiny ± woolly fl-heads, but lvs are **simple**, strap-shaped.

KEY TO MUGWORTS AND WORMWOODS
(*ARTEMISIA* & *SERIPHIDIUM*)

1 Plant only 3–6 cm tall – fl-heads only 1–3, nodding, 12 mm wide – lvs silky-white, ± pinnately-lobed into narrow 3–5-toothed, wedge-shaped segments – Scot mts – vr ....
Artemisia norvegica (p 456 **C.1**)
Plant over 20 cm tall – many small fl-heads in long panicles **2**

2 Terminal lf-lobes narrow-linear or hair-like, 1 mm wide or less **3**
Terminal lf-lobes flat, linear-lanceolate, 2 mm wide or more...**4**

3 Plants unscented, ± hairless or slightly silky – on hths – r.....................*A. campestris* (p 458 **D**)
Plants strongly scented – lvs white-woolly below – in salt marshes – lc ....
Seriphidium maritimum (p 458 **E**)

4 Lf-lobes ± hairless above, white-woolly below – fl-heads narrow-oval **5**
Lf-lobes whitish-hairy both sides – fl-heads bell-shaped ....*A. absinthium* (p 456 **C**)

5 Plant tufted – stems ± hairless – only the larger lf-veins translucent – infl-branches all erect – fls in summer – vc ....*A. vulgaris* (p 456 **A**)
Plant with stems borne singly, on long runners – stems downy – all veins of lvs translucent – infl-branches v lfy, arching outwards – fls late autumn, if at all – on wa – r (except London area)..*A. verlotiorum* (p 456 **B**)

A Mugwort, *Artemisia vulgaris*, erect tufted ± downy, aromatic per, 60–120 cm tall; grooved, reddish stems with white pith inside occupying **c. 4/5** of total stem diameter (**Ac**). Lvs 5–8 cm long, dark green, hairless above, white-cottony below, twice-pinnatifid; lower lvs stalked, upper unstalked, clasping and less divided, with lanceolate-oblong pointed segments, 3–6 mm wide; only **main** veins translucent. Infls of erect racemes in a lfy panicle; fl-heads (**Aa**) 3–4 mm long x 2–3 mm wide, oval, erect; their bracts lanceolate, woolly, scarious-margined; florets all tubular, red-brown; achenes (**Ab**) hairless. Br Isles, vc (except Scots mts r); on wa, rds, hbs, scrub. Fl 7–9.

B Chinese Mugwort, *A. verlotiorum*, close to A, but stems **downy** (**not** tufted), on creeping rhizomes, and pith occupying c. **1/3** of total stem diameter (**Bb**); **all** lf-veins translucent; infl-branches v lfy, **arching** outwards, do not produce fls till late autumn (if then); fl-head (**Ba**). Introd (China); SE Eng lf, r elsewhere; in wa. Fl 10–12.

C Wormwood, *A. absinthium*, erect, v aromatic per, 30–90 cm tall; **key differences** from A in its **blunt** lf-segments, lvs **silky-hairy both** sides; fl-heads (**Ca**) cup-shaped, 3–5 mm wide, drooping, with blunt silky-hairy bracts and yellow florets. Br Isles: Eng, Wales, f-lc; Scot, Ire, r; on wa, rds, hbs. Fl 7–8. **C.1** BAP VU ****Norwegian Mugwort**, *A. norvegica*, small aromatic alpine per herb, 3–6 cm tall, v **unlike** other Mugworts. Basal lvs 2 cm long, stalked, palmately-lobed, with 3–5 toothed wedge-shaped lobes; stem-lvs pinnate, stalkless; all lvs silky white. Fl-heads only 1–3, 12 mm across, nodding on erect stems; bracts with green midribs and broad dark brown scarious edges. Scot, mts of W Ross, vr in alpine moss-hth. Fl 7–9.

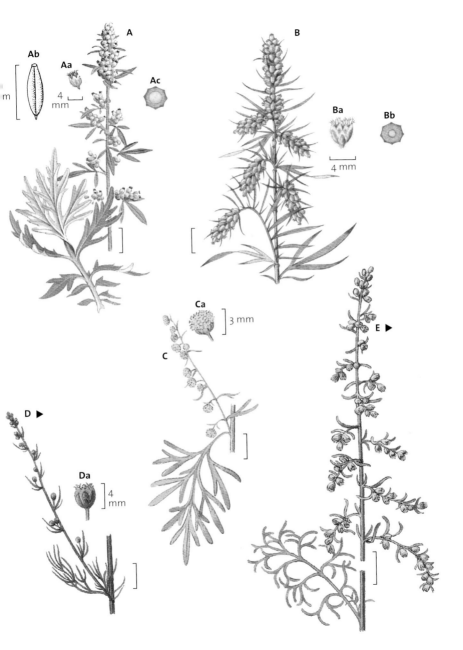

A

Aa

Ab

m

Ac

4
mm

B

Ba

Bb

4 mm

Ca ⌉ 3 mm

C

D ▶

Da ⌉ 4
mm

E ▶

457

▲ **D** Sch 8 VU ****Field Wormwood**, *Artemisia campestris*, scentless, ± hairless ascending per, with shoots 20–60 cm long, woody below. Lower lvs 2–3 times pinnate, stalked, upper lvs stalkless, undivided; all with segments linear, pointed, to 1 mm wide, becoming hairless. Panicles loose, erect; fl-heads (**Da**) oval, 3 mm wide; bracts oval, hairless, green with scarious edges; florets yellow or red. E Anglian Breckland, S Wales only, vr; on sandy grassy hths. Fl 8–9.

▲ **E Sea Wormwood**, *Seriphidium maritimum* (*A. maritima*), strongly aromatic per, with spreading to erect fl-stems 20–50 cm long, and also separate lf-rosettes. Lvs 2–5 cm, twice-pinnate, segments linear, blunt, 1 mm wide; lower stalked, upper not; all **white**-woolly both sides. Fl-heads many, 1–2 mm wide, oval, in panicles, side-branches and heads usually drooping, sometimes erect; bracts of heads oblong, downy, scarious-edged; florets yellow or reddish. Br Isles: GB, coasts N to E Scot, f-lc in SE, o to N; Ire r; drier salt marshes, sea walls. Fl 8–9.

Cudweeds (*Filago* and *Gnaphalium*) are ± woolly all over, with narrow strap-shaped untoothed lvs. Fl-heads small, few-fld, of tubular florets only, and frs with a hairy pappus, bear no superficial resemblance to Daisies, Dandelions or Thistles. *Filago* fl-heads are gathered into dense stalkless globular, oval and conical clusters in lf-axils, stem-forks, and at branch-tips, always with lvs at base of each cluster; involucres have **outer** bracts **woolly** grey-green, **inner** bracts **scarious**, ± straw-coloured; fl-head receptacles conical, **scaly-edged** under the florets. *Gnaphalium* fl-heads, separately stalked, are in spikes, racemes, panicles or terminal umbel-like clusters; **bracts all scarious-edged**, receptacles flat, **without** scales at edges.

ID Tips   Cudweeds

- **What to look at**: (**1**) number of fl-clusters per fl-head (**2**) number of fls per cluster (**3**) number of lvs at base of fl-cluster (**4**) whether lvs overtop the fl-heads (**5**) lf-shape .

- Look at fresh, young plants as all plants go brown when old and sometimes characters used in this key are lost.

- When counting the **number** of **fl-clusters** per fl-head, only look at the **oldest** fl-head at the base of the main stem branches because other fl-heads are often atypical and smaller.

- Get to know **Marsh Cudweed** (*Gnaphalium uliginosum*), the most common Cudweed sp, preferring damp places often with winter flooding, where *Filago* spp generally will not grow.

KEY TO CUDWEEDS (*FILAGO & GNAPHALIUM*)

1 Fl-heads in dense, globular or oval clusters, in lf-axils, stem-forks and branch-tips – outer bracts of heads woolly, inner scarious – receptacle scales present (use hand lens)(*Filago*) **2**
Fl-heads in spikes, racemes, or purely terminal dense or umbel-like clusters, fl-heads brownish to pale yellow, in spikes, or in dense umbel-like clusters – all bracts of heads scarious-tipped – receptacle scales abs...(*Gnaphalium*) **6**

2 Clusters of fl-heads globular, 8–40 heads per cluster – bracts bristle-pointed, erect in fr ..**3**
Clusters of fl-heads narrow, ± oval, 2–8 heads per cluster – bracts not bristle-pointed, spreading star-wise in fr...**5**

3 Lvs wavy-edged, without mucronate points at tips – clusters of fl-heads longer than the lvs at their bases – 20–40 fl-heads per cluster ....*Filago vulgaris* (p 459 **A**)
Lvs ± flat-edged, with mucronate points at tips – clusters of fl-heads shorter than the lvs at their bases – 8–20 heads per cluster...**4**

4 Each fl-head cluster with 3–5 lvs at its base – outer bracts of each fl-head with yellow, slightly curved outwards (easier to see x20), bristle-points – plants branched from base, spreading, white-woolly ....*F. pyramidata* (p 460 **B**)
Each fl-head cluster with 1–2 lvs at its base – outer bracts of each fl-head with bright red-tips (look on young fl-heads), erect bristle-points – plants erect, branched only above, yellowish-woolly ....*F. lutescens* (p 460 **C**)

5 Lvs narrow-linear, upper far longer than fl-head clusters just above them ..*F. gallica* (p 460 **D.1**)
Lvs lanceolate, upper shorter than fl-head clusters just above them......*F. minima* (p 460 **D**)

6 Ann plants with fl-heads in dense terminal clusters – bracts of fl-heads without dark edges...**7**
Tufted or creeping pers (rarely ann) with fl-heads in spikes or racemes – bracts of fl-heads with scarious dark brown edges..**8**

7 Lvs narrow-oblong, ± pointed – terminal clusters of fl-heads surrounded by long, over-topping lvs resembling petals – bracts of fl-heads brown – fls brownish-yellow with pale stigmas – short branched plant – c in damp places ... .*Gnaphalium uliginosum* (p 460 **E**)
Lower lvs broadly oblong and blunt, upper clasping, pointed – terminal clusters of fl-heads without petal-like lvs around them – bracts of fl-heads wholly pale yellow – fls yellow with red stigmas – erect little-branched plant – vr in dry sandy places .. ..
....*G. luteoalbum* (p 460 **E.1**)

8 Short (under 12 cm) v tufted plant with 1–7 fl-heads in short compact terminal spikes – lvs woolly both sides – on Scot mts ....*G. supinum* (p 460 **F.2**)
Taller (12–60 cm) erect plants with fl-heads >10 in elongated spikes **9**

9 Lvs linear-lanceolate, one-veined, hairless on uppersides, woolly below – lvs progressively shorter above..*G. sylvaticum* (p 460 **F**)
Lvs lanceolate, three-veined, woolly both sides – lvs all ± equal in length ....
G. norvegicum (p 460 **F.1**)

▼ **A** NT **Common Cudweed**, *Filago vulgaris* (*F. germanica*), erect ann, 50–30 cm, stems and lvs v woolly, stems ± branched at base, then forking into 2–3 branches below terminal fl-head clusters. Lvs erect, 1–2 cm long, strap-shaped, wavy, blunt-tipped; fl-heads in dense globular clusters of 20–40 together in branch-forks and terminating branches; lvs below clusters shorter than clusters; bracts of fl-heads linear, five-rowed, **straight**, keeled, the outer woolly, the inner **yellow**-bristle-pointed, **erect**. Br Isles: GB lf (but decreasing) to S, r to N; Ire r; on sandy gslds, ar, wa rds, on dry ± acid soils. Fl 7–8.

B Sch 8 BAP EN *Broad-leaved Cudweed, *Filago pyramidata* (*F. spathulata*), white-woolly **spreading** ann, close to A, but branched from base, stems 6–30 cm long, lvs mucronate, obovate; fl-heads 8–15 per cluster, clusters terminal and **along** stems, the 3–5 lvs at each cluster base are longer than cluster; outer fl-head bracts have **yellow**, slightly **outward-curving** bristle-points. S, E Eng r and declining (**rest of Br Isles abs**); on ar on calc soils. Fl 7–8.

C Sch 8 BAP EN *Red-tipped Cudweed, *F. lutescens* (*F. apiculata*), also close to A, but with erect **yellow**-woolly stems, **mucronate** non-wavy lvs; fl-heads 10–20 per cluster (**C**) with 1–2 overtopping lvs; fl-head bracts with **bright-red-tipped**, **erect** bristle-points. SE Eng, r (**rest of Br Isles abs**); sandy ar. Fl 7–8.

D FPO (Eire) **Small Cudweed**, *F. minima*, slender erect grey-woolly ann 5–15 cm tall, with ± erect branches above middle; lvs 5–10 mm long, linear-**lanceolate**; five-angled fl-heads 3 mm long, in oval clusters of 3–6, at tips and in forks of stems, **longer** than lvs at their bases; bracts of fl-heads (**Da**) woolly-based with hairless scarious **blunt** yellowish tips, spreading, resembling a star in fr. GB, o-lc (**but NW Scot abs**); Ire r; on sandy gslds, wa, railways, hths. Fl 6–9. **D.1** EW **Narrow-leaved Cudweed**, *F. gallica*, close to D, but all lvs **narrow-linear**, 8–20 mm long; fl-heads, 4 mm long, are in clusters of 2–6 in forks and at stem tips, and are **much exceeded** by lvs at their bases; bracts of fl-heads with sharp tips. Br Isles: CI, vr; on gravelly gslds, wa. Fl 7–9.

E **Marsh Cudweed**, *Gnaphalium uliginosum*, v woolly ann 4–20 cm tall, much-branched from base; lvs woolly both sides, 1–5 cm long, narrow-oblong, narrowed to base, ± pointed; fl-heads 3–4 mm long, in dense **terminal** clusters of 3–10 surrounded by **long lvs** resembling **petals**; fl-head bracts woolly, pale below, with **dark** scarious hairless **tips**; florets **brownish**-yellow; stigmas **pale**. Achenes (**Ea**). Br Isles, vc on damp ar, gslds, commons, paths, on acid soils. Fl 7–9. **E.1** Sch 8 **Jersey Cudweed**, *G. luteoalbum*, little-branched above; lower lvs broad-oblong, blunt; upper lvs clasping, pointed, wavy, all lvs woolly both sides; fl-heads 4–5 mm long, in dense terminal **lfless** clusters of 4–12; bracts elliptical, scarious, shiny, pale yellow; florets **bright yellow**; stigmas **red**. Eng: Norfolk, Kent, CI vr as native (vr introd elsewhere) but may be vla; on sandy hths, dunes. Fl 6–8.

F FPO (Eire) AWI EN **Heath Cudweed**, *G. sylvaticum*, per herb with short lfy runners and erect unbranched lfy fl-shoots 8–60 cm tall; rosette and lower stem-lvs stalked, lanceolate, pointed, 2–8 cm long, upper lvs progressively shorter and narrower, stalkless, all lvs one-veined, **hairless** above, **white-woolly** below. Infl a long spike, sometimes with short side-branches forming over half height of plant; fl-heads (**Fa**) 6 mm long, separate or in open clusters; bracts of fl-heads with green centre and broad scarious **dark brown** edges; florets pale brown; achenes (**Fb**) hairy, pappus-hairs reddish. Br Isles: GB o-r (N Scot f); Ire r and declining; in open wds, hths, gslds on acid soils. Fl 7–9. **F.1** *Highland Cudweed*, *G. norvegicum*, has habit of F, but only 8–30 cm tall; lvs three-veined, all ± **equal** in length, v woolly **both sides**; spike short, ± compact, only ¼ height of plant. Scot, central and N Highlands r; on acidic mt rocks. Fl 7–8. **F.2** NT **Dwarf Cudweed**, *G. supinum*, dwarf tufted per with unbranched ± erect woolly fl-stems 2–12 cm tall; lvs lanceolate below, linear and pointed above, 0.5–1.5 cm long, woolly **both sides**; fl-heads 1–7 in a short, ± compact terminal spike; bracts elliptical with woolly greenish central stripe and **broad**, **brown**, **scarious** margins. Scot Highlands f-lc; on mt rocks, mt hths (alpine). Fl 7–8.

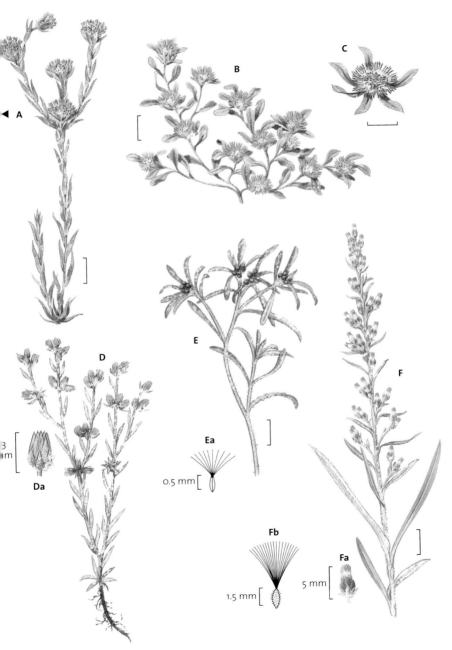

A

B

C

D

Da

3 mm

E

Ea

0.5 mm

F

Fa

5 mm

Fb

1.5 mm

> **ID Tips   Thistles**
>
> • **Plume-thistles** (*Cirsium*) differ from other Thistle genera (*Carduus, Silybum, Onopordum*) in having a pappus of **feathered** hairs to their achenes (in the other genera, the pappus is of **simple**, unbranched hairs).
>
> • *Carduus* has stems always ± **spinous-winged**; lvs spine-edged, green above, ± cottony beneath; and pappus of fr is of **simple** hairs.
>
> • **What to look at**: **(1)** whether plant has creeping runners or not **(2)** If characters including: lvs simple or pinnate; white woolly below; purple-flushed or not **(3)** whether stem wings, if present, continuous or interrupted **(4)** size, colour and number of fl-heads.

KEY TO PLUME-THISTLES (*CIRSIUM*)

1 Plant consisting of a spreading rosette of wavy-and spiny-edged, pinnatifid lvs, with, at its centre, 1 ± unstalked fl-head (rarely 2–3, and rarely stalked to 10 cm) ....
Cirsium acaule (p 463 **E**)
Plants with tall, lfy stems **2**

2 Lf-edges with only soft short prickles (not sharply spiny) – bracts of fl-heads never strongly spine- tipped – fl-heads only 1–3 on each rounded cottony stem **3**
Lf-edges with strong sharp spines – bracts of fl-heads sometimes strongly spine-tipped – fl-heads many, on each much-branched, round or winged stem **5**

3 Lvs deeply twice pinnatifid, green both sides, only slightly cottony beneath – no creeping runners ....*C. tuberosum* (p 464 **G.1**)
Lvs elliptical-lanceolate, ± toothed, but at most shallowly-lobed, green above, white-cottony beneath – creeping runners present **4**

4 Lvs thickly white-felted below, hairless above, over 4 cm wide – upper lvs with broad cordate clasping bases – fl-heads large (3.5–5.0 cm long x 3–5 cm wide) ....
C. heterophyllum (p 463 **F**)
Lvs only white-cottony (not felted) below, hairy above, under 3 cm wide – upper lvs narrowed to bases, scarcely clasping – fl-heads smaller (2.5–3.0 cm long x 2.0–2.5 cm wide)..*C. dissectum* (p 464 **G**)

5 Stems with spiny wings or flanges **6**
Stems without spiny wings...**7**

6 Lvs shiny, without prickles on upper surface, ± purple-flushed – stem-wings continuous fl-heads in crowded clusters, each head 1.5–2.0 cm long x 1.0–1.5 cm wide – outer bracts purplish, with erect, only shortly-pointed tips ....*C. palustre* (p 463 **C**)
Lvs dull, with hairs and prickles on upper surface, not purple-flushed, often white-downy below – stem-wings interrupted – fl-heads few, in loose clusters, each head 3–5 cm long x 2–4 cm wide – outer bracts green, with long, arched-back spine-tips *C. vulgare* (p 463 **A**)

7 Stout plant – lvs deeply pinnatifid, prickly-hairy above, the lobes tipped with stout long spines – involucres globular, to 4 cm long x 4–7 cm wide – bracts thickly webbed with white cotton – florets red-purple ....*C. eriophorum* (p 463 **B**)
Slender plant – lvs shallow-pinnatifid, ± hairless and shiny above, lobes tipped with slender short spines – fl-heads with oval involucres, 1.5–2.5 cm long x 1 cm wide – bracts purplish, ± hairless – florets mauve or white.......................................*C. arvense* (p 463 **D**)

A **Spear Thistle**, *Cirsium vulgare*, erect bi 30–150 cm, branched above; stems cottony, with interrupted spiny **wings**; basal lvs 15–30 cm long, short-stalked, deeply pinnatifid, wavy-edged and toothed, lobes and teeth with long **stout spines**; stem-lvs stalkless, smaller, with long terminal lobes; all lvs prickly-hairy above, not shiny. Fl-heads few, in loose clusters or solitary; heads 2–5 cm long x 2.5–4 cm wide, ± **cottony**, outer bracts green, with long, **arched-back yellow** spine-tips; florets pink-purple. Pappus of fr (**Aa**) feathered. Br Isles, vc on wa, rds, gslds, open wds. Fl 7–10.

B **Woolly Thistle**, *C. eriophorum*, stout erect bi 60–150 cm, branched above; stems v cottony, round, unwinged, not prickly; basal lvs to 60 cm long, short-stalked, deeply pinnatifid into narrow lanceolate two-lobed segments tipped with strong stout spines, one lobe arched up, one down, making lf three-dimensional; stem-lvs similar, but stalkless; all lvs prickly-hairy and green above, white cottony below. Fl-heads solitary, **v large**; involucres 3–5 cm long x 4–7 cm wide, bracts spine-tipped (with a purple wing just below tip), **thickly cobwebbed with white wool**; florets red-purple. Mid, S, E Eng, N to Durham, f-lc: SE Eng, S Wales, r; **rest of Br Isles abs**; on calc gsld, scrub, rds. Fl 7–9.

C **Marsh Thistle**, *C. palustre*, erect bi, with continuously **spiny-winged** hairy stem 30–150 cm, branched above; lvs ± deeply pinnatifid, with wavy spine-tipped and -edged lobes; lvs shiny, but hairy above, dark green and often **purple-flushed**, like stem; basal lvs lanceolate, stalked, less deeply lobed. Fl-heads in crowded clusters, each head 1.5–2.0 cm long x 1.0–1.5 cm wide; outer bracts purplish, with erect, shortly-pointed tips; florets dark red-purple or sometimes white. Br Isles, vc in marshes, mds, damp gslds, open wds, hbs. Fl 7–9.

D **Creeping Thistle**, *C. arvense*, creeping per herb with erect, branched, spineless, furrowed, **unwinged** fl-shoots 30–90 cm tall, ± hairless. Lvs oblong-lanceolate, ± pinnatifid, with strong slender spines on their wavy and toothed edges; upperside ± **hairless**, grey-green, ± cottony beneath; lower lvs stalked, upper clasping stem. Fl-heads in open clusters, 1.5–2.5 cm long x 1 cm wide; involucre-bracts purplish, ± hairless, oval, with spreading spine-tips; florets **mauve** or white. Br Isles, vc in gslds, rds, wa, ar. Fl 7–9.

E **Dwarf Thistle**, *C. acaule*, per herb with spreading rosette of deeply pinnatifid, wavy-edged, stoutly spine-edged lvs 10–15 cm long, ± hairless above, ± hairy beneath. Fl-heads usually 1, sometimes 2–3, **stalkless** from centre of rosette (rarely stalked to 10 cm or more), 3–4 cm long, oval; florets bright red-purple. S, E Eng, N to Yorks, S Wales, f-lc; **rest of Br Isles abs**; on short calc gslds. Fl 6–9.

F WO (NI) AWI **Melancholy Thistle**, *C. heterophyllum*, erect **spineless** per herb with runners, 45–120 cm tall; stems grooved, ± unbranched, cottony, unwinged. Basal lvs 20–40 cm long, 4–8 cm wide, elliptical-lanceolate, stalked, with fine soft-prickly (but hardly **spine-like**) teeth; upper lvs unstalked with **cordate clasping** bases, prickly-edged, mostly **unlobed**; all lvs **green**, **hairless** above, thickly **white-felted** below. Fl-heads **solitary** (or rarely 2–3), 3.5–5.0 cm long x 3–5 cm wide; involucre oval, with oval short-pointed bracts, purple-tipped, adpressed. Florets red-purple. Br Isles: N Wales, N Eng to N Scot, f-lc; **S Eng**, **S Wales abs**; Ire vr; in upland mds, gslds, rds, open wds. Fl 7–8.

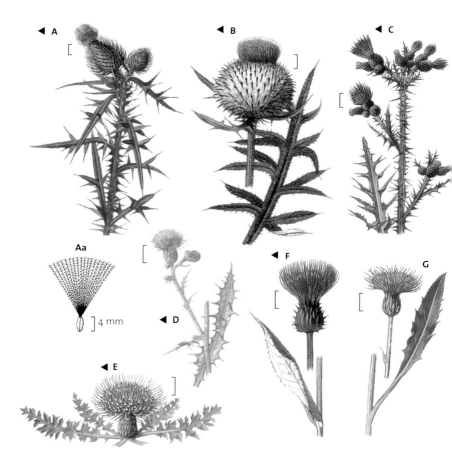

G **Meadow Thistle**, *Cirsium dissectum*, erect per herb with runners, 15–50 cm tall, resembling a slender version of F in its round grooved unwinged cottony stem and elliptical-lanceolate, scarcely-lobed, spineless lvs; but lvs narrower (under 3 cm wide) and white-**cottony**, (**not felted**) beneath, **green and hairy** above, narrowed to bases and scarcely clasping; lf-edges wavy-toothed with soft prickles. Fl-heads solitary, sometimes 2–3 on long branches, 2.5–3.0 cm long x 2.0–2.5 cm wide; bracts of heads lanceolate, adpressed, outer shortly spine-tipped, cottony, purplish; florets dark red-purple. S, SW Eng, E Eng S Wales, N to Yorks, f-lc, **abs to N**; Ire c; in fens and less acid peat bogs. Fl 6–8. **G.1** NT ****Tuberous Thistle**, *C. tuberosum*, close to G in fl-heads and cottony wingless stem, but **no runners**, and lvs mostly deeply twice **pinnatifid**, green **both sides**, only **slightly** cottony below, edges bristly but scarcely spiny. Eng: Wilts, Dorset r; Cambs, Gloucs, Glamorgan vr; in calc gslds. Fl 6–7.

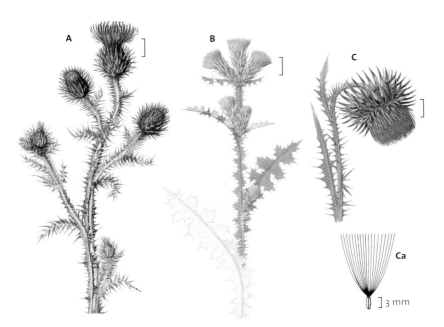

A **Welted Thistle**, *Carduus crispus* (*C. acanthoides*), erect v branched bi 30–120 cm tall, with narrow, **wavy**, **spiny-edged wings** ± continuous along stems. Lvs pinnatifid, with lobes, each itself three-lobed, oval, spiny, dull green, cottony below; lower lvs elliptical in outline, stalked, upper narrower, running down stem, more deeply cut. Fl-heads 3 cm long x 2 cm wide, many, in **clusters** of 3–5; involucre oval, ± cottony, bracts v narrow, ± spreading, **weakly** spine-tipped, ± purplish; florets **two-lipped**, red-purple. Achenes with **simple** pappus hairs. GB, c (except N, W Scot r); Ire o mostly in E only; in open wds, scrub, hbs, wa, mostly on ± calc soils. Fl 6–8.

B **Slender Thistle**, *C. tenuiflorus*, bi, 15–100 cm tall; close to A, but more erect and narrowly branched; spinous stem-wings **broader** (to 5 mm); stems and lvs more greyish-cottony; fl-heads **narrower**, oblong-cylindrical, 1.5 cm long x 0.8 cm wide, in **dense** terminal clusters; bracts oval-lanceolate, hairless, tipped with **outward-curved** spines; florets **paler**, pink, in a **narrower** brush, **equally** five-lobed. Br Isles: GB, f-lc near coasts in S, r inland and in N Scot; Ire o-lf; on dry banks, wa, beaches, rds, especially near sea. Fl 6–8.

C **Musk (Nodding) Thistle**, *C. nutans*, bi, 20–100 cm tall; stem erect, cottony, branched above, with spinous wings interrupted and abs for some way below fl-heads. Lower lvs wavy-edged, upper deeply pinnatifid with lobed spine-tipped segments; all lvs shiny but ± hairy above, woolly on veins below. Fl-heads **solitary**, rounded, **large** (4–6 cm across), well-separated on long stalks, **drooping**, with involucre of long, bent-back, lanceolate, spine-tipped, **purple-red** bracts; florets in a red-purple brush much narrower than involucre, v fragrant, tubes **two-lipped**. Achenes (**Ca**) with **simple** pappus hairs. Br Isles: GB, N to R Forth f-lc, N, W Scot, Ire vr introd; on calc gslds, wa, rds. Fl 5–8.

Knapweeds and **Star-thistles** (*Centaurea*) have **non-spiny** but ± bristly lvs and thistle-like fl-heads, with distinctive **involucre-bracts**. These each consist of two segments: the lower is green or pale; the upper is **either** flat, pale to blackish brown, **papery**-textured, its margins shortly-toothed or fringed with long, branched, bristle-like teeth, **or** consists of one or more stout, woody, pale yellow **spines**. In some, fl-heads are surrounded with a crown-like circlet of enlarged, sterile, trumpet-shaped outer tube-florets, each with several corolla-lobes 5–10 mm long. Achenes have either a short hairy pappus or none.

KEY TO KNAPWEEDS & STAR-THISTLES (*CENTAUREA*)

1 Tips of involucre-bracts each bearing one or more yellowish spines **2**
  Tips of involucre-bracts spineless, ± toothed, silvery to dark brown **3**

2 Florets purple-pink – lateral spines pinnate ....*C. calcitrapa* (p 466 **C**)
  Florets yellow – lateral spines palmate ....*C. solstitialis* (p 467 **D**)

3 Outer florets enlarged, bright blue, inner purple **4**
  All florets red-purple or pink...**5**

4 Basal lvs and non-flg shoots abs when plant flg – fl-head 1.5–3 cm across....................................
                                                             C. cyanus (p 467 **E**)
  Basal lvs and non-flg shoots present when plant flg – fl-head 6–8 cm across ....
                                                          C. montana (p 467 **E.1**)

5 Brown tip of involucre-bracts horseshoe-shaped ....*C. scabiosa* (p 466 **B**)
  Brown tip of involucre-bracts triangular-shaped ....*C. nigra* (p 466 **A**)

A   **Common Knapweed**, *Centaurea nigra*, v variable erect rough-hairy per 15–60 cm tall, stems grooved, branched above; lower lvs stalked, upper stalkless, usually unlobed, sometimes pinnatifid, oblong to linear-lanceolate. Fl-heads 2–4 cm across, terminal, solitary; involucre globose; tips of bracts brown, ± **triangular**, fringed with long, branched, bristle-like teeth; lower parts of bracts oblong, pale. Florets red-purple, outer enlarged into a crown-like whorl in few local populations only. Achenes (**Aa**) hairy, with pappus of v short bristly hairs. Br Isles, vc on gslds, rds, wa. Fl 7–9.

B   **Greater Knapweed**, *C. scabiosa*, erect downy per 30–80 cm tall, stems grooved, branched above; rosette-lvs stalked, 10–25 cm, upper stalkless, stouter, all ± **deeply pinnatifid**, lobes ± toothed. Fl-heads 3–6 cm across, terminal, solitary, often with enlarged outer florets, florets all bright purple-red; involucre globose, bracts with oval **green** lower parts and dark brown **horseshoe**-shaped tips, **enclosing** tip of green part and fringed with bristle-like teeth. Achenes (**Ba**) with pappus of long bristly hairs. Br Isles: Eng, Wales, f-lvc; Scot r but lf; mid-Ire lf; on gslds, rds, hbs, scrub on calc soils. Fl 6–8.

C   CR ****Red Star-thistle**, *C. calcitrapa*, bushy bi, with hairless, ascending or erect, grooved stems 15–60 cm tall, branching below fl-heads; lower lvs deeply pinnatifid, lobes narrow, upper ± unlobed, stalkless, all lvs hairy, bristle-toothed. Fl-heads 1–3 cm across, purplish-pink; involucre oval, hairless, 1 cm wide, bract-tips each with a **stout yellow spine** to 2.5 cm long, and shorter **pinnate**

side-spines at base. Eng, probably native near S coast, Thames estuary, vr; introd elsewhere vr; on dry calc gslds, wa. Fl 7–9.

D Yellow Star-thistle, *C. solstitialis*, differs from C in **yellow** fls and **palmately** arranged, spiny bract-tips with middle spine only 1–2 cm. Introd (S Eur); S Eng, vr; on ar and wa. Fl 7–9.

E BAP **Cornflower**, *C. cyanus*, cottony ann with erect branches 20–50 cm tall; basal lvs and non-flg shoots **abs** when flg; lower lvs stalked, pinnatifid, 10–20 cm, upper stalkless, linear-lanceolate. Fl-heads **1.5–3.0 cm** across, solitary, long-stalked; bright **blue**, inner red-purple. Involucre-bracts with green, **dark-margined centre**, and **silvery**, **jagged-toothed** edges. Achenes (**Ea**) with pappus of long bristly hairs. Br Isles, formerly f-lvc, now vr as a native, probably only in Kent and IoW, its rarity status blurred by f records originating from sown 'wildflower' seed; on ar, wa. Fl 6–8. **E.1 Perennial Cornflower**, *C. montana*, resembles E in enlarged blue outer florets; but **key differences** are per, stouter; basal lvs and non-flg shoots **present** when flg; **oblong-lanceolate** lvs **run down stem**; fl-heads **6–8 cm** across. Introd (mts C Eur); GB f, much more widespread than E; in wa, rds. Fl 6–8.

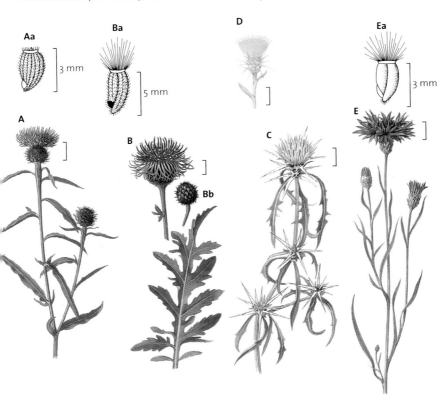

Burdocks (*Arctium*) are robust, downy, branched herbs; lvs large oval-cordate, non-spiny; fl-heads oval-globose, with red-purple florets; **involucre-bracts** many, overlapping, narrow, with long spreading **hooked** tips, forming burrs in fr that attach to clothes etc; pappus of rough hairs.

A   Lesser Burdock, *Arctium minus*, 60–130 cm tall; stems furrowed, ± woolly; **basal** lf-stalks **hollow**, lvs longer than wide, to 40 cm long; fl-heads generally 1.5–3.0 cm wide, **short-stalked** (< **2 cm**), in **racemes**; corolla often glandular-hairy; involucre-bracts (**Aa**) may be shorter than, or ± equal to, or exceed corolla. Achenes (**Ab**). Br Isles, vc in open wds, scrub, hbs, rds, wa. Fl 7–9. **A.1 Wood Burdock**, *A. nemorosum*, is v difficult to distinguish from A as many intermediate forms occur, which are probably best regarded as ssp of A; (see the *New Flora* for details of how these may be distinguished). **A.2 Greater Burdock**, *A. lappa*, has **basal** lf-stalks **solid**; lvs as wide as long; fl-heads generally 3–4 cm wide, **long-stalked** (> **2.5 cm**), in a flat-topped **corymb**; involucre-bracts **longer** than corolla by at least 1mm. S half Eng, S Wales, f-lc; in wds, rds, hbs on clay soils. Fl 7–9.

B   Glandular Globe-thistle, *Echinops sphaerocephalus*, tall (50–200 cm) thistle-like per herb; lvs oval-oblong, pinnatifid, green and sticky-hairy above, white-woolly below, ends of lvs and their lobes strongly spine-tipped. Fl-heads **globular**, 4–6 cm wide; actually composed of **many**, **one-fld** fl-heads attached to and radiating from a central point, each with an outer involucre of bristle-like bracts and an inner one of fringed scales; florets tubular, pale blue; achenes (**Ba**) with a cup of partly fused hairs on top. Introd (C, S Eur); Eng r casual, on wa, scrub. Fl 6–9. Other similar *Echinops* spp are also r introd.

C   Milk Thistle, *Silybum marianum*, erect ann or bi, 40–100 cm; stem grooved, ± branched, unwinged, ± cottony; lvs oblong-lanceolate, wavy-lobed or pinnatifid, spiny-edged, hairless, shiny green, and netted with conspicuous **milk-white veins** above; lower lvs narrowed to base, upper clasping. Fl-heads 4–5 cm long x 4 cm wide, solitary, long-stalked, florets red purple. Bracts hairless, with oval bases and long, **triangular**, **spine-edged** appendages, each tipped with a **stout yellow spine**. Achenes (**Ca**) blackish, obovate, with **simple**, long white pappus surrounded by a yellow basal ring. Most f in E and SE Eng, possibly native near sea, r-lf; r and introd rest of Br Isles; on dry banks near coasts, wa. Fl 6–8.

D   AWI **Saw-wort**, *Serratula tinctoria*, spineless, hairless per herb, with wiry, branched, erect, grooved stems, 20–80 cm tall. Lvs 12–20 cm, hairless, edges finely **bristle-toothed**, vary from ± undivided, lanceolate, to deeply **pinnatifid** with narrow lobes; lower stalked, upper not. Fl-heads 1.5–2.0 cm long in loose lfy infls, florets red-purple, involucre **narrow-oblong** with bracts adpressed, oval, pointed, purplish, **not spiny**. Achenes (**Da**) with **simple** yellowish pappus. Br Isles: Eng, Wales, S Scot, f-lc (r in E Eng); **Scot, Ire abs**; on hths, calc gslds, fens, open hthy wds. Fl 7–9. **D.1**   WO (NI) **Alpine Saw-wort**, *Saussurea alpina*, spineless, erect, thistle-like per, 7–40 cm tall, near to D; but stem **cottony**, lvs **undivided** but ± toothed, oval to lanceolate, ± hairless on uppersides, **white-cottony** beneath; lower lvs stalked. Fl-heads 1.5–2.0 cm long, narrow as in 956, but in a **dense terminal** cluster, unstalked; florets purple; bracts oval, blunt, **hairy**. Achenes brown, 4 mm, pappus **feathered**. Br Isles: mts of Scot f; NW Eng, Wales, Ire, vr on ± calc alpine cliffs. Fl 7–9.

A

Ab

Aa

7
mm

B

Ba

8 mm

C

Ca

7
mm

D

Da

5 mm

♂

E ▶

Ea

5
mm

F ▶

Fa

2
mm

▲ **E Cotton Thistle**, *Onopordum acanthium*, erect robust bi, 45–150 cm; stem **white-cottony**, **continuously** and **broadly spiny-winged**, branched above; lvs, oblong, stalkless, wavy-lobed, are edged with **strong spines**, v white-cottony both sides, running down into stem-wings. Fl-heads solitary, involucre globose, 3–5 cm wide, bracts green, lanceolate, cottony, ending in strong ± spreading yellow spines; florets purplish-pink. Achenes (**Ea**) wrinkled, grey-brown, pappus **simple**, long, pale reddish. Br Isles: GB, f in E and S, r to N; Ire vvr; on dry banks, rds, wa. Fl 7–9.

▲ **F Carline Thistle**, *Carlina vulgaris*, erect **spiny** bi, 10–60 cm, ± cottony and ± purple-flushed; lvs oblong-lanceolate, wavy-lobed, cottony below, fringed with weak spines, clasping stem. Fl-heads solitary or clustered, conical in bud, opening to 3–4 cm wide in fl and fr; outer bracts spine-edged, lanceolate, green, cottony; inner **straw-yellow**, linear, **spreading** like ray-florets in fr; florets brownish-**yellow**. Achenes (**Fa**) with rusty hairs and long, **feathered** pappus. Br Isles (except N Scot vr), f-lc; on calc gslds. Fl 7–10.

A Dandelion, *Taraxacum* agg, comprises some hundreds of closely similar 'microspecies', too complex to separate here. In general, a per herb with tap-root and **basal rosette** of lanceolate-obovate, sparsely hairy, ± sharply-lobed and -toothed lvs; fl-heads 2–6 cm across, solitary on **lfless**, **hollow**, **unbranched** stems, several per plant; **milky latex abundant**, florets bright yellow; no scales among florets; involucre-bracts in 2 erect inner rows, and an outer row of shorter ones that may be addressed to head, or spreading, or arched back. Achenes (**Aa**) ribbed, beaked, with white pappus of simple hairs. Br Isles, vc in mds, gslds, rds, wa, dunes, mt rocks. Fl 3–10.

B Goat's-beard, *Tragopogon pratensis* agg, erect bi with milky latex, 30–100 cm tall, little-branched, ± woolly when young, later hairless. Lower lvs 10–30 cm long, linear-lanceolate, keeled lengthwise, grass-like, grey-green, long-pointed, untoothed, hairless, with white midrib, wider and sheathing at base; upper lvs shorter, erect. Fl-heads to 5 cm wide, solitary, long-stalked, only opening in **morning sunshine**, otherwise closed (**Ba**). Bracts of heads **equal**, 8 or more, lanceolate, pointed, ± hairless, **in one row**. Achenes (**Bb**) rough, long-beaked, pappus with main rays almost woody and radiating, interwoven across like a spider's web with fine white side-hairs. GB 0 (except N, W Scot, and mts, vr); Ire 0; on rds, gslds, wa. Fl 5–7. **Do not confuse** with B.1 as closed fls of B may have a pale purplish tinge on the outside. **B.1 Salsify**, *T. porrifolius*, closely resembles B, but stems **v swollen** just below fl-heads, and florets mauve-**purple**, ± equalling the bracts; achenes 4 mm long only. Introd (Mediterranean); S, E Eng, mostly by lower Thames, o-lf; rest of Br Isles vr; on rough gsld, cliffs, sea walls, rds. Fl 5–7.

C Sch 8 VU ****Viper's-grass**, *Scorzonera humilis*, slender erect per rather like B in habit, but only 7–50 cm, usually unbranched, more persistently cottony in lf-axils; basal lvs **elliptical**-lanceolate, **flat** at tips, sheathing at base, untoothed as in B; stem-lvs narrow, **flat**, **short**, but ± clasping at bases. Fl-heads 2.5 cm wide, solitary, florets deep yellow, involucre-bracts woolly-based, 2.0–2.5 cm long, **oval**-lanceolate, **unequal**, **overlapping**, often **orange-flushed**. Achenes (**Ca**) smooth-ribbed, **beakless**, pappus as in B. Eng: Dorset, S Wales, vr; **rest of Br Isles abs**; in moist mds. Fl 5–7.

D Chicory, *Cichorium intybus*, per herb with erect, tough, grooved ± hairy, stiff-branched stems 30–100 cm; basal lvs stalked, lanceolate, unlobed, pointed, clasping with pointed basal lobes. Fl-heads 2.5–4.0 cm wide, on thick, v short stalks, in clusters in lf-axils; florets bright blue; 2 rows of bracts, inner bracts

Aa

Bb

B

Da
3 mm

D

Ba

4 mm

A

E ▶

Fa
4 mm

F ▶

Ca
8 mm

C

longer, erect, outer row shorter, ± spreading. Achenes (**Da**) angled, **without pappus**, but with **toothed scales** on top. Eng f-lc but declining: rest of Br Isles r-lf; on rds, gslds, banks, ar on ± calc soils. Fl 7–10.

▲ **E Common Blue-sowthistle**, *Cicerbita macrophylla*, erect per herb to 2m with strong rhizomes; **pale blue** fl-heads 30 mm wide; plant spreading and loosely, bushy-branched; lvs with **cordate end-lobes** and usually only 1 pair side-lobes, clasping stem, **all** sticky-hairy, at least on mid rib and veins of lf-underside. Introd (E Eur); GB f, Ire o; on wa, rds, river banks. Fl 7–9. **E.1** Sch 8 VU ****Alpine Blue-sowthistle**, *C. alpina*, erect bristly per, 50–200 cm; stem with dense red sticky hairs above, little-branched; **lower lvs hairless**, lyre-shaped, pinnatifid, with large, **triangular-pointed end-lobes**; upper smaller, less lobed, with cordate clasping bases. Fl-heads in a close **short** panicle 7 cm wide x 10 cm long, at top of tall stem, each head 20 mm wide, florets **pale-violet-blue**; involucre sticky-hairy, bracts in 2 unequal rows, purplish. Achenes linear, beakless, pappus white, of simple hairs. **E Scot** only, vr on moist mt rocks. Fl 7–9.

▲ **F Wall Lettuce**, *Mycelis muralis*, erect hairless per herb 25–100 cm; lower lvs **lyre-shaped**, pinnatifid, stalk winged, lobes triangular, end-lobes larger and three-lobed; upper lvs stalkless, smaller, less lobed, clasping with **arrow-shaped** bases; all lvs **thin**, ± **reddish-tinged**. Fl-heads small, 1 cm wide or less, florets yellow, usually only 5, in an **open** panicle with **branches at 90°** to main stem; involucre narrow, bracts in **2 distinct rows**, inner bracts **erect**, linear, **equal**, outer **v short**, spreading, all blunt. Achenes (**Fa**) short-beaked, spindle-shaped, black, pappus of simple white hairs, the inner longer than outer. Br Isles: Eng, Wales, f-lc; Scot r-lf; Ire o; in wds (especially beech) on calc soils, walls, calc rocks. Fl 6–9. **Do not confuse** with *Lactuca* spp, which have pappus hairs equal (not unequal like F) and involucre bracts overlapping, not in 2 distinct rows.

A Prickly Lettuce, *Lactuca serriola*, erect ± hairless bi 30–120 cm; stems often reddish, much milky latex. Lvs **erect**, oblong-lanceolate, often pinnatifid (especially lower lvs), all waxy grey-green, **thick**, with fine **spines** along edges and on undersides of **whitish** veins, clasping stem with spreading **arrow-shaped** basal lobes (**Ab**). Infl a loose panicle with branches at **acute angle** to main stem; fl-heads 11–13 mm wide, with 7–12 yellow florets; involucre narrow, 8–12 mm long, bracts **unequal**, erect, overlapping, **not** in 2 distinct rows, lanceolate, grey, **purple**-tipped. Achenes (**Aa**) **olive**-grey, **bristly** at tips, **beaked**, pappus white, of **equal** simple hairs. GB, vc to S and E, vr to N; Ire vr; on wa, railways, rds. Fl 7–9.

B Great Lettuce, *L. virosa*, bi, close to A, but **taller** (to 200 cm), stouter, stem and lvs more **purple-flushed**; lvs less divided, more **spreading**, basal clasping-lobes **rounded**, adpressed to stem; ripe achenes (**B**) **purple-black**, **without** bristles at tip; pappus as in A. Br Isles: Eng. f-lc to SE, E; rest of GB r-vr; **Ire abs**; on wa, rds, railways. Fl 7–9.

C Sch 8 EN ****Least Lettuce**, *L. saligna*, ann, much more slender than A and B, 30–70 cm tall: upper stems-lvs ± **vertical, linear**-lanceolate, wavy-edged, ± untoothed; bases (**Ca**) clasping, arrow-shaped; lower lvs pinnatifid with narrow lobes, or unlobed; midribs broad, **white**; all lvs grey-green, hairless. Fl-heads in axils of arrow-shaped bracts, in a narrow, spike-like raceme or panicle. Bracts of **narrow**-cylindrical involucre erect, **linear**, blunt, greenish with **white** edges; florets few, pale yellow, little longer than the involucre. Achenes **pale**, 3–4 mm, long-beaked; pappus white, simple. SE Eng, mostly near coast, r and decreasing; **rest of Br Isles abs**; on dry banks near sea, shingle and estuaries. Fl 7–8.

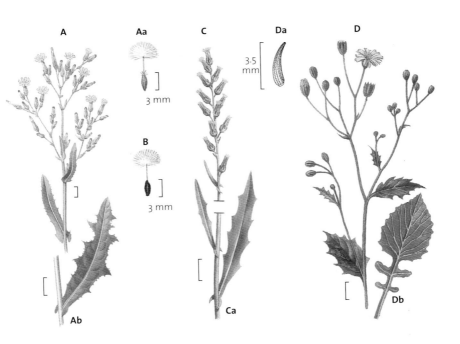

A · Aa · C · Da · D

3.5 mm

3 mm

B

3 mm

Ab · Ca · Db

D Nipplewort, *Lapsana communis*, erect **lfy-stemmed** ann, **much-branched** and hairless above, hairy below. Basal lvs (**Db**) lyre-shaped, pinnatifid with large, oval. toothed end-lobes; upper lvs oval to diamond-shaped, with few large teeth; all lvs **thin**, ± hairy. Fl-heads 1.0–1.5 cm wide, in a loose panicle; only 8–15 florets, pale yellow, short; involucre an oval **cup** of one row of narrow equal bracts and few tiny outer ones. Achenes (**Da**) ribbed, **without beak or pappus**. Br Isles, vc (except mt areas); on hbs, wds, rds, walls, wa. Fl 7–9.

ID Tips   Some yellow dandelion look-a-likes

- **Cat's-ears** (*Hypochaeris*) may be confused with **Hawkbits** (*Leontodon*) as both have basal lf-rosettes, lfless infl-stems, unbranched or little-branched, and feathered pappus to frs.

- **Cat's-ears** have narrow papery **scales** on receptacle between the strap-shaped florets (as in Aa p 475); **Hawkbits** have **no** scales between the florets (as in Ea p475).

- When looking for scales, look at fully open, **mature** fl-heads only as sometimes scales are not developed in young plants.

- **Nipplewort** (*Lapsana*) has frs with **no** pappus.

A Cat's-ear, *Hypochaeris radicata*, per herb with basal rosette of oblong-lanceolate, bristly, wavy-toothed lvs, 5–12 cm long. Fl-stems, 20–40 cm, ± hairless, simple or branched 1 or 2 times only, are ± lfless but have a few scale-like dark-tipped bracts along them, and **enlarge** towards fl-heads. Fl-heads 2–4 cm across, florets bright yellow, outer greyish beneath. Involucre, narrow bell-shaped, suddenly narrowed to stalk, is of many unequal erect overlapping purple-tipped bracts, hairless except for bristles on midribs. **Receptacle** with **scales** (**Aa**) among florets, visible when fl-head broken in ½ and florets pulled off. Achenes (**Ab**) rough, beaked, with both feathered and simple pappus-hairs. Br Isles, vc in mds, gslds, rds, dunes, not usually on v calc soils. Fl 6–9.

B WO (NI) VU **Smooth Cat's-ear**, *H. glabra*, 10–20 cm tall, smaller than A; lvs shorter, usually ± hairless, glossy ± red-tinged. Fl-heads 1–1.5 cm across; **florets scarcely longer than** purple-tipped **involucre-bracts**, their yellow straps only **twice as long as wide**, only spreading in **sunshine**. Inner achenes (**Ba**) usually with beak, outer achenes (**Bb**) without beak. Br Isles: Eng, Wales, lf to E, and CI, r to W; Scot, N Ire, vr; on sandy gslds, ar, hths, dunes. Fl 6–9.

C VU ****Spotted Cat's-ear**, *H. maculata*, per herb with bristly-hairy lvs 7–15 cm long, all in a basal rosette as in A, but usually blotched with **dark purple spots** and with red midribs, wavy-toothed. Infl-stems **bristly**, unbranched or once-branched, **hardly** enlarged upwards, with few or no scale-like bracts. Fl-heads 3–5 cm across, florets **lemon-yellow**, much longer than v hairy unequal blackish-green, lanceolate involucre-bracts. Outer achenes (**Ca**) wrinkled, inner short-beaked; pappus **wholly** of feathered hairs. E Eng, Cornwall, CI, Cumbria, N Wales, vr; **rest of Br Isles abs**; on gslds, grassy sea cliffs, on calc soils and on serpentine rock. Fl 6–8.

D Rough Hawkbit, *Leontodon hispidus*, **v hairy** per herb to 60 cm, with basal rosette of lanceolate coarse wavy-toothed lvs, narrowed to base; infl-stems **lfless, unbranched**, 10–40 cm, **v hairy throughout**, lf-undersides and stems (**Da**) with hairs **forked** (use hand lens); fl-heads 2.5–4.0 cm across, involucre narrowing **suddenly** into stalk; florets golden-yellow, far longer than **hairy unequal** involucre-bracts. Achenes (**Db**) unbeaked; **dirty-white** pappus of **both** feathered **and** simple hairs; **no scales** among florets. Br Isles: GB c (except N Scot vr); Ire f to S, **abs to N**; on calc gslds, mds, fens. Fl 6–9.

E Autumn Hawkbit, *L. autumnalis*, per herb to 60 cm, ± hairless, lvs **deeply pinnatifid**; fl-stems **branched** 2 or 3 times, involucre **long-tapered** into stalk. Achenes (**Eb**) ribbed, with **white** pappus of feathered hairs only. Receptacle (**Ea**) **without** scales, as in D and F. **Key difference** from other British *Leontodon* spp in having lf-undersides with **unforked** hairs. Br Isles, vc in mds, gslds, rds, not usually on v calc soils. Fl 6–10.

F Lesser Hawkbit, *L. saxatilis* (*L. taraxacoides*), resembles D in wavy-toothed rosette-lvs, lf-undersides with forked hairs, and solitary fl-heads on unbranched lfless infl-stalks; but infl-stalks shorter (8–40 cm), **bristly** below, **hairless** above; involucre-bracts, **hairless** except for few bristles on midribs, in one inner longer **equal** row with a **few** overlapping outer ones. Fl-heads 2.0–2.5 cm across. Inner achenes (**Fa**) 5 mm, v short-beaked, pappus of **both** feathered **and** simple hairs; outer achenes scaly-topped, **without** pappus. Br Isles c (except N Scot vr); on dunes, dry sandy or calc gslds. Fl 6–9.

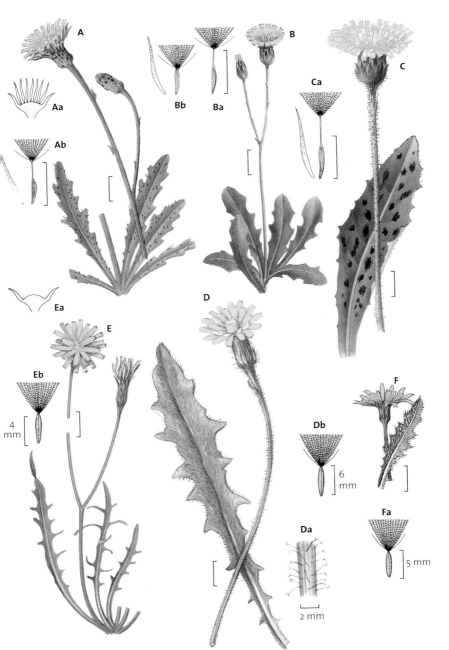

Hawk's-beards (*Crepis*) are erect, branched, usually v **lfy-stemmed** (except F.1), florets all yellow, strap-shaped, inner involucre-bracts in **one erect**, **equal** row, outer **shorter**, spreading (but ± adpressed in C and E); achenes cylindrical, with **simple**, **white** (except in E) soft pappus. **Oxtongues** (*Picris*) are also erect, branched, lfy-stemmed, with similar (inner erect, outer spreading) involucre-bracts; but plants **v bristly**, achenes **curved**, ribbed, wrinkled, and pappus of **feathered** hairs.

ID Tips   Hawk's-beards

- **Do not confuse** Hawk's-beards (*Crepis* spp) with Hawkweeds (*Hieracium* spp). *Crepis* have involucre bracts in 2 distinct inner and outer rows: an erect, equal inner row, outer shorter, spreading (but ± adpressed in C and E).

- **Smooth Hawk's-beard** (*C. capillaris*) is the only common *Crepis* that is **± hairless** (except for hairs on the involucre bracts).

- **Beaked Hawk's-beard** (*C. vesicaria*) is the only *Crepis* with **all** achenes having beaks.

KEY TO HAWK'S-BEARDS (*CREPIS*)

1 Stem lfless, except for small lfy bracts at bases of infl-branches – outer fl-head bracts adpressed ....*Crepis praemorsa* (p 478 **F.1**)
Stem lfy **2**

2 Lvs pinnatifid **3**
Lvs toothed but unlobed, elliptical-lanceolate **6**

3 Frs beaked – lvs roughly hairy **4**
Frs unbeaked – lvs hairy or not **5**

4 Fl-heads drooping in bud – plant smelling of prussic acid – outer achenes short-beaked, inner long-beaked – r ....*C. foetida* (p 477 **B**)
Fl-heads erect in bud – plant not smelling of prussic acid – all achenes long-beaked – c ....
C. vesicaria (p 476 **A**)

5 Plant tall (40–100 cm), bristly-hairy – outer fl-head bracts spreading ....*C. biennis* (p. 477 **D**)
Plant shorter (20–60 cm), ± hairless, with shiny lvs – outer fl-head bracts adpressed to heads ....*C. capillaris* (p 477 **C**)

6 Stem-lvs elliptical or oblong, hairless, strongly-toothed, clasping with arrow-shaped bases – pappus dirty white, brittle ....*C. paludosa* (p 478 **E**)
Stem-lvs oblong, ± hairy, scarcely-toothed, clasping with cordate bases – pappus pure white, soft ....*C. mollis* (p 478 **F**)

A Beaked Hawk's-beard, *Crepis vesicaria*, ± erect, **downy**, branched bi, 15–70 cm tall, stem furrowed, bristly, ± purple-flushed below; lower lvs lyre-shaped, ± deeply pinnatifid, terminal lobes ± lobed; upper less lobed, clasping, all downy. Fl-heads **erect** in bud, 15–25 mm across, long-stalked, gathered into loose corymbs; involucres 8–12 mm long, cylindrical, downy, outer bracts **spreading** in a ruff; florets golden-yellow, outer **orangey-red-striped** outside. Achenes **all** long-beaked (**Aa**). Br Isles: Eng, Wales, vc to S and E, r to N; Scot vr; Ire f; on gslds, banks, rds, wa. Fl 5–7

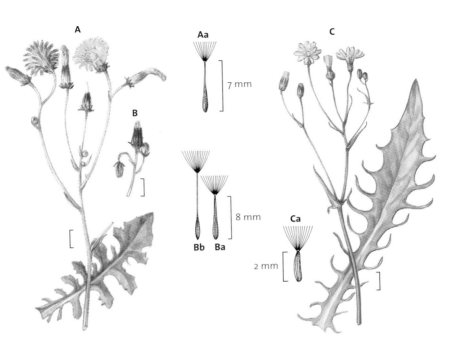

B Sch 8 BAP EW ****Stinking Hawk's-beard**, *C. foetida*, like A downy with beaked achenes; but unlike A **smells of prussic acid** (bitter almonds) when bruised, is shorter (10–40 cm), more branching from base, has stalked rosette-lvs **v hairy**, with large triangular end-lobes, and fl-heads (**B**) droop in bud. Inner involucre-bracts each **wrap round** an outer achene; outer achenes (**Ba**) ± **un-beaked**, the inner ones (**Bb**) **long-beaked** (to 10 mm). Re-introd by conservationists; Eng: Kent, Sussex and Somerset only, vr; on shingle beaches, wa. Fl 6–8.

C **Smooth Hawk's-beard**, *C. capillaris*, ± **hairless**, **glossy-lvd** ann or per, ± branched from base; basal lvs 5–15 cm, lyre-shaped, pinnatifid with narrow ± toothed lobes; upper lvs lanceolate, pointed, stalkless, clasping, bases arrow-shaped, all lvs ± shiny and hairless. Fl-heads 10–13 mm across, **erect** in bud, slender-stalked, in loose corymbs. Involucre **flask-shaped**, 5–8 mm long, **all** bracts adpressed, **hairless** on inner faces, downy and ± bristly on outer. Achenes (**Ca**) curved, **beakless**, pappus white. Br Isles, vc except in mts; on gslds, wa, hths, walls, rds, hbs. Fl 6–9.

▼ **D** **Rough Hawk's-beard**, *C. biennis*, more rough-hairy, more erect, taller (to 120 cm) than A; stem-lvs clasping, but **not** arrow-shaped at bases. Fl-heads in a loose corymb, 20–35 mm across, with inner involucre-bracts **dark-hairy** outside, **downy inside**, outer **spreading** in a ruff; florets clear **pale** yellow (not red-striped outside); achenes (**Da**) **unbeaked**, pappus white. SE Eng: Kent, Surrey, Herts, f-lc and native; rest of Br Isles f introd; on gslds, scrub, hbs, rds, mostly on calc soils. Fl 6–7.

E Marsh Hawk's-beard, *Crepis paludosa*, erect ± hairless per 30–90 cm tall, branched above only; lower lvs elliptical-lanceolate with **short** winged stalks, upper lvs clasping with long, **back-pointed basal lobes** forming **arrow-shape**; all lvs shiny, thin, sparsely wavy-toothed, hairless. Fl-heads few in a loose corymb, 15–25 mm across, erect; florets golden yellow, involucre-bracts woolly with **black sticky-tipped hairs**, outer shorter, **adpressed**. Achenes (**Ea**) beakless, pappus of **brittle**, **brownish** hairs. Br Isles: Wales and N Eng to N Scot f-lc; **S Eng abs**; Ire o-lf; in moist mds, damp wds on richer soils, in hilly or mt areas. Fl 7–9.

F EN ****Northern Hawk's-beard**, *C. mollis*, rather like E, but stem-lvs clasping with **rounded** bases (**Fb**), wavy-toothed or untoothed, ± hairless; basal lvs lanceolate, **blunt**, narrowed into **long** winged stalks; fl-heads 20–30 mm across, as in E but less **hairy**; achenes unbeaked, pappus **white, soft**. N Eng, Scot, r and declining; **rest of Br Isles abs**; in moist wds, by streams in mts. Fl 7–8. **F.1** EN ****Leafless Hawk's-beard**, *C. praemorsa*, is related to F but rosette-lvs scarcely toothed; downy stem, 15–40 cm, **lfless except for bracts**, and bears few, pale yellow heads in a short terminal **raceme**. Achenes **unbeaked**, pappus **white**, **soft**. Probably native, discovered 1988; Cumbria only, vr on limestone gsld. Fl 5–7.

G Hawkweed Oxtongue, *Picris hieracioides*, erect bi or per with stout **bristly** furrowed stem 15–70 cm, branched above; lower lvs 10–20 cm, lanceolate, blunt, **short-stalked**, upper narrower, clasping at base, all wavy with shallow teeth, bristly. Fl-heads 20–30 mm across, in a loose umbel-like corymb; involucre oval, inner bracts **narrow**, equal, erect, outer shorter, often **spreading**, bristly on midribs; florets bright yellow. Achenes (**Ga**) curved, v **short-beaked**, wrinkled, pappus creamy, of **feathered** hairs. Eng and Wales, c to SE, o to N and W; **Scot abs**; Ire vr; on gslds, scrub, cliffs, rds, wa, on calc soils. Fl 7–10.

H Bristly Oxtongue, *P. echioides*, of habit similar to G, but ann or bi more thickly covered with blister-like **bristles** with **swollen**, **whitish bases**; stem-lvs lanceolate, with clasping **cordate** bases; 3–5 outer involucre-bracts are sepal-like, **broadly cordate**, 2 x as wide as inner; all v bristly. Achene beaks as long as achenes (**Ha**). Br Isles: Eng, Wales, f-lc, mostly in S and E; Scot, Ire, vr; in hbs, mds, gslds, banks on clay soils or chk. Fl 6–10. **Do not confuse** lvs with those of **Green Alkanet** (p 362), which also often has lvs with blister-like, white bristles.

I Hawkweeds (*Hieracium* agg) form a large genus (over 260 'microspecies' described for Br Isles alone) of great difficulty. Because they reproduce non-sexually, the ovules forming seeds without pollination, many v similar, self-reproducing (but not cross-breeding) forms occur. These cannot be considered in detail here. Hawkweeds are all erect, **± lfy-stemmed** pers **without** runners; all have yellow strap-shaped florets **only**; fl-heads (**Ia**) with overlapping, **unequal** rows of bracts. Hawkweed achenes (**Ib**) are important to distinguish the group, cylindrical, 10-ribbed, beakless, with a **ring** on top, and pappus of **simple**, **brittle**, **brownish**, **unequal** hairs. The illustration (**I**) is an example of a hawkweed.

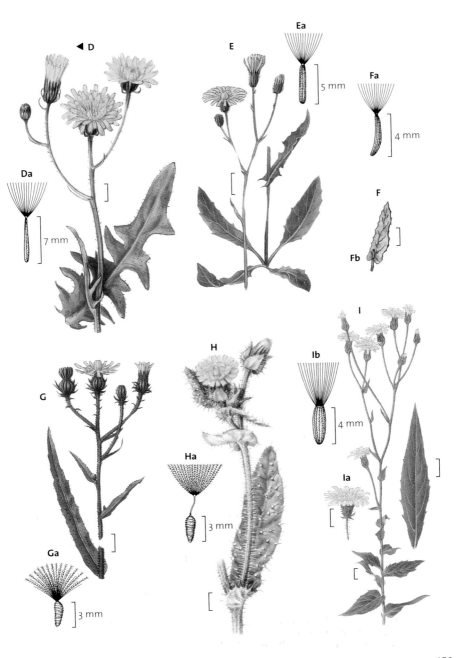

◄ D

Da

7 mm

E

Ea

5 mm

Fa

4 mm

F

Fb

G

Ga

3 mm

H

Ha

3 mm

I

Ib

4 mm

Ia

Mouse-ear-hawkweeds, *Pilosella* (formerly in the genus *Hieracium*), are short plants; **lvs all** (or **mostly**) in a **basal rosette**, usually with **lfy runners**; achenes only 1.5–2.0 mm long, and with 10 **teeth** at their tips; all pappus hairs of ± **equal** length. Several other *Pilosella* spp occur as r native plants but are v difficult to distinguish, so omitted here.

A   Mouse-ear-hawkweed, *Pilosella officinarum*, per herb with **long**, **lfy**, **runners**; basal rosettes of obovate ± blunt untoothed lvs, 3–8 cm long, stalked, **white-felted** below, **green** but with scattered **stiff long white hairs** above; infl-stems, 5–30 cm tall, lfless, unbranched, bear **solitary** heads 15–25 mm wide of **pale** yellow florets (**Aa**); involucres oval, with narrow bracts that have both sticky, black-based, and also white hairs; florets **red-striped** beneath, styles yellow, achenes purple-black. Br Isles, vc in gslds, hths, banks, rocks, walls. Fl 5–8.   **A.1** End ****Shetland Mouse-ear-hawkweed**, *P. flagellaris* ssp *bicapitata*, similar to A but has **2 fls** per fl-stalk; **Shetland only**, vr on rocky coastal gsld. Fl 5–8. (ssp *flagellaris* is a v similar gdn plant, usually with 2-6 fls per fl-stem; r found natd on rds, banks, wa).

B   Fox-and-cubs, *Pilosella aurantiaca*, differs from A in having a **few stem-lvs** as well as basal rosette, in height usually **over 20 cm**, and plant (but not lvs) **densely long-hairy**; several fl-heads of **orange florets** in a **close-umbellate cluster**. Introd (C Eur); Br Isles, f-lc; Ire o; in wa, banks, rds, railway banks. Fl 6–7.

Sow-thistles (*Sonchus*) are tall, robust; stems, **stout**, **hollow**, containing **abundant** milky latex in canals, bear lvs ± deeply pinnatifid with clasping bases; fl-heads in umbels; florets strap-shaped, yellow; achenes **unbeaked**, **flattened**, ribbed, with a pappus of white, **simple** hairs.

C   *Marsh Sow-thistle, *Sonchus palustris*, tall robust per 90–300 cm tall, with stout **hollow** angled stem (**Cb**), hairless below, sticky-hairy above. Basal lvs lanceolate-oblong, with deeply clasping **arrow**-shaped bases (**Cc**), and pinnatifid with **few narrow** side-lobes and large lanceolate end-lobes, edges and teeth finely spine-tipped; upper lvs **lanceolate**, untoothed, with long-pointed clasping basal lobes; all lvs wavy, grey-green, ± hairless. Fl-heads in large open umbel-like corymbs, each head **30–40 mm** across; all stalks and bracts densely covered with **blackish-green** sticky hairs; involucres **oval**, 12–15 mm long, florets **pale** clear yellow, 25 mm long (**Ca**); achenes **yellow**. Pappus white, of simple hairs. S Eng, E Anglia lf; **rest of Br Isles abs**; in reed-swamps by tidal rivers, ditches, fens. Fl 7–9.

D   Perennial Sow-thistle, *S. arvensis*, **per** herb 60–150 cm tall; stem much as in C; stem-lf bases **rounded**, clasping (**Da**); all lvs ± pinnatifid, with side-lobes **broader**, **shorter** than in C, **shiny green**, edged with fine spines. Fl-heads **40–50 mm** across, florets **deep** yellow; branches in infl (less branched, looser, less umbel-like than in C) and involucres all covered with **yellow** sticky gland-tipped hairs (**Da**). Involucres **bell-shaped**, 13–20 mm long. Achenes **brown**. Br Isles, vc on ar, wa, hbs, fens. Fl 7–10.

E   Smooth Sow-thistle, *S. oleraceus*, erect **ann** (or rarely bi) 20–150 cm tall; stout **hairless** stems; lvs hairless except when young, with **pointed spreading** basal lobes, grey-green, pinnatifid into **broad**-triangular, toothed, but **non**-spiny lobes. Fl-heads, 20–25 mm across, in a loose umbel; florets **pale** yellow, involucre short (10–15 mm), usually hairless, **not** sticky. Achenes yellow, **wrinkled**. Br Isles, vc on ar, wa, hbs, open wds. Fl 6–10.

F   Prickly Sow-thistle, *S. asper*, ann, close to E; lvs (**F**) less pinnatifid, lobes narrow, but **deeply doubly-toothed**, **crisped**, **spiny** on edges like thistle lvs; basal clasping lobes **rounded**, **adpressed** to stem (**Fb**). Fl-heads as in E, but florets **deep golden** yellow; achenes (**Fa**) yellow, **smooth**. Br Isles, c on ar, wa, open wds, hbs. Fl 6–10.

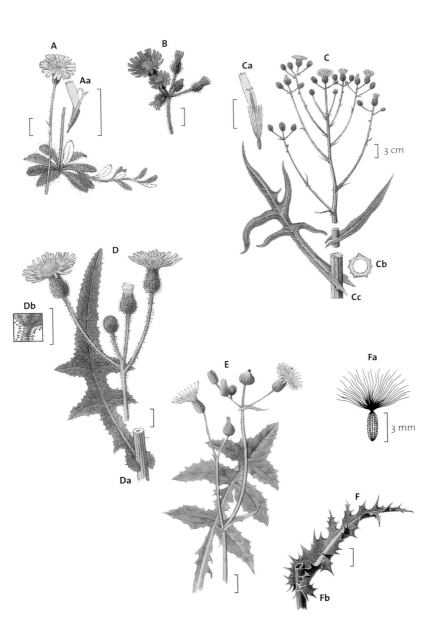

A

Aa

B

Ca

C

] 3 cm

Cb

Cc

D

Db

Da

E

Fa

] 3 mm

F

Fb

481

PLANT DESCRIPTIONS – MONOCOTYLEDONS

FLOWERING-RUSH FAMILY   *Butomaceae*

A family of **aquatic** herbs. Lvs untoothed, parallel-veined; fls have 3 sepals, 3 petals, as in most Monocotyledons. The carpels are oblong and form **many-seeded follicles** in fr, opening at tips to release seeds.

A   Flowering-rush, *Butomus umbellatus*, hairless erect per to 150 cm, with **linear**, pointed, **three-angled**, twisted lvs ± as long as fl-stems, all from base; infl-stems round, with terminal **umbel** of many, unequally- and long-stalked fls; fls (**Aa**) 2.5–3.0 cm across; 3 sepals smaller than 3 petals, but **all** rosy-pink; stamens 6–9; carpels 6–9, **free** except at base, dark red, oblong, to 7 mm long, each **many-seeded**, style on top. Br Isles: Eng f-lc; Wales, Scot, Ire or; in still and slow moving fresh water on muddy substrates. Fl 7–9.

WATER-PLANTAIN FAMILY   *Alismataceae*

A family of **aquatic** herbs. Lvs untoothed, parallel-veined; fls have 3 sepals, 3 petals, but the carpels, **numerous**, **free** from one another, in whorls or heads, ripen into **achenes** v like those of Buttercups; sepals **green**, unlike the petals; stamens 6 (or more); ovary **superior**.

KEY TO WATER-PLANTAIN FAMILY

1 Fls in several, usually simple, successive whorls up erect infl – separate male fls with many stamens – female fls with carpels in a dense round head ....
Sagittaria sagittifolia (p 484 **E**)
All fls bisexual, with 6 stamens and 6 to many carpels **2**

2 Fls usually solitary, long-stalked, arising from creeping or floating stems....................................
Luronium natans (p 484 **D.1**)
Fls in erect, whorled infls **3**

3 Lvs cordate-based, at least some floating on long stalks – ripe carpels 6 (-10), 12 mm long, long-beaked, in a spreading, star-like whorl ....*Damasonium alisma* (p 484 **C.2**)
Lvs narrowed to, or rounded at, base (if cordate, lvs erect and not floating) – ripe carpels many, 1–2 mm long, without long beaks...**4**

4 Infl either a simple umbel or of 2 successive simple whorls – carpels oval, pointed, in a crowded head..*Baldellia ranunculoides* (p 484 **D**)
Infl of several successive whorls, which are themselves whorled-branched – carpels flattened, blunt, style attached to one side, in a single whorl ....*Alisma* (p 484 **B**, **C**)

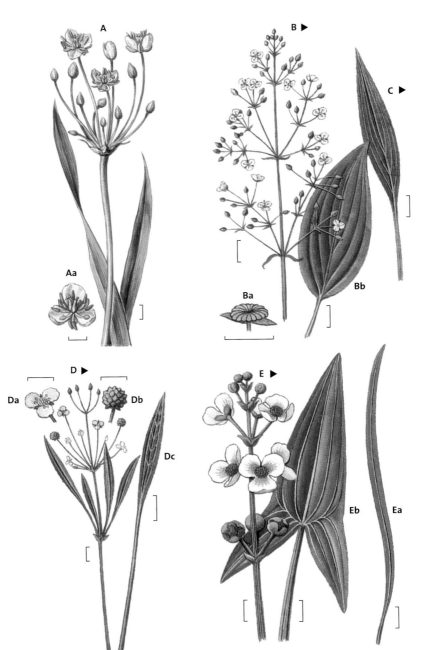

▲ **B Water-plantain**, *Alisma plantago-aquatica*, erect hairless per, 20–100 cm; lvs (**Bb**) long-stalked, 8–20 cm long, ± erect, oval, rounded at base (or occasionally tapering). Infl of several tiers of whorled branches which again branch in whorls; fls to 1 cm across, open in the afternoon, sepals 3, blunt, green, petals 3, rounded, pinkish, stamens 6; carpels many in a flat single whorl (**Ba**). Br Isles, c (except N Scot r); on mud in and by fresh waters. Fl 6–8. **Key difference** between B and C is not clear, although one study suggests that the lf-tips of B are acute, whereas those of C are acuminate.

▲ **C Narrow-leaved Water-plantain**, *A. lanceolatum*, differs from B only in its usually (but not always) **narrow**-lanceolate lvs (**C**), **tapered** to stalk (although may be rounded at base and v close to B), fls open in the morning; styles all **erect** in a ring. Br Isles: Eng, Wales, o-lf; Scot vr; Ire r; habitat as for B. Fl 6–8. **C.1** Sch 8 BAP CR [**] **Ribbon-leaved Water-plantain**, *A. gramineum*, close to C, lvs either submerged or emergent; **key difference** is styles **recurved in a ring**. One site in E Eng, Worcs; vr; in ponds etc. Fl 6–8. **C.2** Sch 8 BAP CR [**] **Starfruit**, *Damasonium alisma*, low ann, 5–20 cm tall, with blunt **cordate-based** ± **floating lvs**, 3–5 cm long, in a basal rosette; infl of a few superimposed whorls of white fls 6 mm across; carpels 6(-10), **12 mm long**, **long-beaked**, ± two-seeded, in a **spreading star-like** whorl. SE Eng, vvr, almost **extinct** apart from reintroductions in Bucks and Surrey; in or beside shallow sandy or gravelly-bedded ponds. Fl 6–8.

▲ **D** NT **Lesser Water-plantain**, *Baldellia ranunculoides*, erect per 5–20 cm tall, like a small version of B or C, but sometimes with runners; infl either a simple umbel or of 2 successive simple whorls; lvs (**Dc**) 2–4 cm long, mostly basal, **linear-lanceolate**, pointed, tapered into long stalks and smelling of **coriander** when crushed. Fls (**Da**) 15 mm wide, pale pink; carpels many, in a **crowded head** (**Db**), oval, with **pointed** tips. Br Isles: GB, o to S; N Scot r; Ire f; in fens, ponds, ditches, usually on calc peat. Fl 6–8. **D.1** EU Sch 8 BAP [*]**Floating Water-plantain**, *Luronium natans*, has **blunt elliptical** floating lvs 1.0–2.5 cm long, on long stalks, and **linear** submerged lvs to 10 cm long x 2 mm wide; fls usually **solitary**, long-stalked, arising from lf-axils, 12–15 mm wide, white with **yellow spots** at bases of their 3 petals; carpels in a crowded head as in D. **Key difference** from submerged, non-flowering forms of B (which may have ± linear leaves) and D is that D.1 has **runners** (stolons) and, unlike D, does not smell of coriander when crushed. Mid and E Eng, Wales, r-lf; Scot, Ire vr; in lakes, rivers, streams, canals and temporary pools. Fl 7–8.

▲ **E Arrowhead**, *Sagittaria sagittifolia*, erect hairless per, over-wintering by detached submerged buds; submerged lvs **linear**, translucent (**Ea**), floating lvs **oval**-lanceolate, aerial lvs (**Eb**) ± erect, long-stalked, **arrow-shaped**, 5–20 cm long; infl a simple erect whorled raceme, 30–80 cm tall, fls 2 cm or more across, 3–5 per whorl; lower fls **female** with many carpels in a **dense head**; upper fls **male**, larger and longer-stalked, with **many stamens**; all petals white with **purple spot** at base. Br Isles: Eng f-lc; Wales, Scot vr; Ire o-lf; on muddy substrates in still or slow-moving fresh waters. Fl 7–8.

FROGBIT FAMILY
Hydrocharitaceae

A family of **aquatic** herbs. Fls with 3 sepals, 3 petals, but **ovary inferior, one-celled**, producing several seeds; fls 1–3 together, in bud enclosed in a **spathe** of 1 or 2 bracts; fls unisexual.

A VU **Frogbit**, *Hydrocharis morsus-ranae*, floating herb with long runners, not rooted in substrate, over-wintering by detached buds in mud; lvs 3 cm across, floating, **rounded-kidney-shaped**, long-stalked, with large papery basal stipules; fls long-stalked, 2 cm across, arising from water; petals 3, **crinkly**, white, with yellow spot at base; sepals green, smaller; male fls 2–3 together in a spathe of 2 bracts, female solitary. Br Isles: Eng o-lc but declining; Wales, Scot vr; Ire o; over mud or peat substrates in ponds, ditches, fens. Fl 7–8.

B NT ****Water-soldier**, *Stratiotes aloides*, aquatic herb with runners not rooted in substrate, ± submerged, but rising partly out of water to flower; lvs in a large ascending **crown-like** rosette, 15–40 cm long, rigid, lanceolate, pointed, **spine-edged** like **aloe** lvs, but **translucent**, many-veined, **brownish-green**; in Br Isles female plants only; fls erect, solitary, on stalks 5–8 cm long, with 2 large bracts below fl; fls 3–4 cm across, three-petalled, white; Eng o, but lf in E; o-r introd rest of Br Isles; in ponds, ditches, fens. Fl 6–8.

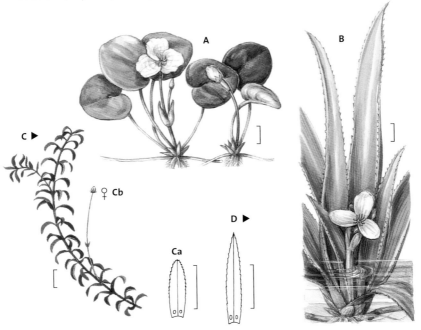

▲ **C Canadian Waterweed**, *Elodea canadensis*, submerged per aquatic herb with stalkless, oblong **± blunt**, dark green translucent lvs (**Ca**), **widest at middle** of lf, to 17 mm long x 3–4 mm wide, in whorls of 3–4 (rarely 5 or 2 on lower stem) along stems, lvs bearing minute untoothed green scales on upper sides near base and **minute teeth** to lower **lf-margins**. Dioecious but female plant only in Br Isles; female fls (**Cb**) 5 mm across, **floating** at surface on **v long** thread-like stalks, with 3 sepals and 3 whitish-reddish petals, 3 styles and inferior ovary; fl-stalk with **two-toothed** tubular **sheath** 1–2 cm long surrounding its base. Introd (N America); Br Isles, f-lc in still or slow-moving fresh water. Fl 5–10.

▲ **D Nuttall's Waterweed**, *E. nuttallii*, similar to C but with lvs (**D**) **tapering** to an acute **point** (not ± blunt), **widest at base** of lf, some lvs recurved or twisted, with minute teeth. Introd (N America); probably more c than C in S Eng; Ire, Scot o; same habitats as C. Fl 5–10. **Do not confuse** with vu ****Esthwaite Waterweed** (*Hydrilla verticillata*) which has lvs in whorls of 3 but lf-tip pointed, teeth **throughout** the lf-margin and 2 minute brown **finger-like** (not entire as in *Elodea* spp) scales at base; vvr in Scot and Ire only.

Several other aquatic plants similar to Waterweeds are now natd in Br Isles, mostly thrown out from aquaria etc; **see key to these on p 81 C**.

RANNOCH-RUSH FAMILY *Scheuchzeriaceae*

A ****Rannoch-rush**, *Scheuchzeria palustris*, erect per herb 10–20 cm tall; creeping rhizome clothed with papery, straw-coloured remains of old lf-bases, swollen at joints; lvs present up stem, alt, linear, ± flat, to 20 cm long, some overtopping fls; lf-tips blunt, with **conspicuous pore** at tip (**Aa**); infl **short**, **v loose**, of **few** (3–8) fls on long stalks each in axil of a long lf-like green bract with **inflated sheathing base**. Fls yellow-green, 4 mm across, with 6 narrow perianth segments, 6 stamens, and (usually) 3 carpels joined only at base and in ripe fr oval, 4–5 mm long, glossy, pointed. Scot, Rannoch Moor only, vr but vlf; in pools and wet hollows of ancient undisturbed *Sphagnum* bogs. Fl 6–8.

ARROWGRASS FAMILY *Juncaginaceae*

Marsh or aquatic herbs with **linear**, sheathing, mostly basal lvs; fls **small**, **greenish**, in **erect spikes** or **racemes**; fl-parts in threes, but carpels either **3 or 6**, joined into a **superior** ovary; fr a capsule. **Arrowgrasses** (*Triglochin*) have long narrow racemes of many fls, with 6 carpels.

B Marsh Arrowgrass, *Triglochin palustre*, slender erect per herb 15–40 cm tall **without runners** (stolons); pleasant aromatic smell when bruised. Lvs linear, 10–20 cm long, rounded on lower side, deeply **grooved** on upper; raceme many-

fld. Fls (**Ba**) with 6 green, ± purple-edged perianth segments 2 mm long; style short, white-woolly; raceme elongating in fr; frs (**Bb**) **club-shaped**, 10 mm long x 2 mm wide near tip, adpressed to stem, opening **from below** when ripe into 3 valves pointed at base to form **arrow-shape** (**Bc**). Br Isles, f-lc, especially to N, except in v cultivated regions; in damp gsld usually on calc soils, fens, mds, calc springs. Fl 6–8.

C Sea Arrowgrass, *T. maritimum*, like B, but **with runners** (stolons), stouter, with half-cylindrical **fleshy** lvs **not** furrowed above, not v aromatic; raceme denser, and more like that of Sea Plantain (p 388); fls (**Ca**) **fleshier**; frs (**Cb**) **oval** when ripe, 4–5 mm long x 2 mm wide, the opening valves **not** forming arrow-shape, but falling off completely. Br Isles, c on coasts, vr inland; in salt marshes, mostly near sea. Fl 7–9.

PONDWEED FAMILY   *Potamogetonaceae*

Fresh water floating or submerged aquatic herbs with opp or alt lvs in 2 opp ranks; infls stalked **spikes** in lf-axils (although may appear terminal). Fls small, each with 4 perianth segments, 4 stamens and separate carpels ripening to small nutlets, usually 4 or less per fl; membranous scales (**stipules**) (2, below) usually present in axils of lvs.

ID Tips   Pondweeds

- **What to look at**: **(1)** whether floating lvs or submerged or both types of lvs are present **(2)** mature lf-shape, toothing and texture **(3)** whether lf-base clasping stem, sessile or petiolate **(4)** whether longitudinal veins translucent or not **(5)** whether or not stipules (2, below) are fused to lf-base **(6)** length and texture of stipules **(7)** fr-size (measurements include the beak).

- When looking at submerged lvs, **only** look at **mid- or lower**-stem lvs.

- Some narrow-lvd Pondweeds can **only** be identified if you look at a **cross-section** of a **young** stipule (ie. stipule nearest tip of plant) under a microscope (x 20 is enough). See pp 67-68 of the *BSBI Pondweeds Handbook* for an explanation of how to do this. It is fiddly but not difficult to see as long as you cut the stipule v thinly.

- Some narrow-lvd Pondweeds (with lvs <2 mm wide) can be confused with other narrow-lvd water plants. Narrow-lvd Pondweeds have **alt** lvs; **stipules**; and no auricles or teeth along lf-margins.

1 Stem 2 Stipule 3 Lf-blade 4 Lf-sheath	PONDWEED STEM 

KEY TO PONDWEEDS (*POTAMOGETON & GROENLANDIA*)

1 Lvs in opp pairs ....*Groenlandia densa* (p 493 **J**)
  Lvs alt (except lvs immediately below infl opp) **2**

2 Lf-margin with conspicuous teeth, visible to naked eye – lvs oblong, strongly wavy-edged and crisped ....*Potamogeton crispus* (p 492 **F**)
  Lf-margin entire or with v small teeth not visible to naked eye **3**

3 Some or all lvs elliptical, with convex sides **4**
  All lvs linear or thread-like, with parallel sides **15**

4 Lvs all floating apparently without submerged lvs..**5**
  Some or all lvs submerged..**7**

5 Floating lvs with a buff or discoloured, flexible hinge-like joint where stalk joins blade – lvs when held to light with fine, translucent longitudinal veins ....*P. natans* (p 490 **A**)
  Floating lvs with no buff or discoloured, flexible hinge-like joint – lvs when held to light with longitudinal veins ± darker than lf **6**

6 Floating lvs leathery – If cross-veins dark, opaque, difficult to see – frs 1.9–2.6 mm long ....
P. polygonifolius (p 490 **B**)
Floating lvs thin, almost translucent – If cross-veins easy to see – frs 1.5–1.9 mm long ....
P. coloratus (p 492 **C**)

7 Submerged lvs stalkless and clasping stem **8**
Submerged lvs with stalk or stalkless but not clasping the stem...**9**

8 Submerged lvs clasping stem ≥ ½ way round – lf-tip not or slightly hooded – at least
some stipules <1 cm long, inconspicuous, usually present only on youngest lvs – frs 2.6–4
mm long ....*P. perfoliatus* (p 492 **E**)
Submerged lvs clasping stem < ½ way round – lf-tip markedly hooded – stipules
conspicuous, >1 cm long, usually persistent – frs 4.5–5.5 mm long (frs larger than any
other broad-lvd Pondweed spp) – lf in N and W Br Isles, o-r elsewhere in lakes, rivers,
canals and fenland drains...NT **Long-stalked Pondweed** *P. praelongus*

9 Submerged lvs linear with parallel sides – vr Outer Hebrides in lochs, r introd in canals in
N Eng ....VU ****American Pondweed** *P. epihydrus*
Submerged lvs elliptical to oblong, without parallel sides **10**

10 Most or all submerged lvs without stalks **11**
All submerged lvs with stalks at least 1 mm long **12**

11 Stems unbranched – lf-margin of submerged lvs entire, lf-tip rounded or pointed but not
mucronate – f in N Eng, Scot, N Ire in lakes and ditches; r in S Eng, S Ire in canals, ditches,
streams, flooded gravel/sand pits ....**Red Pondweed** *P. alpinus*
Stems usually much branched – lf-margin of submerged lvs with minute teeth, lf-tip
mucronate – lf in Scot, N Eng, Ire; r in S Eng, Wales, SW Ire in lakes, fenland ditches, canals,
flooded gravel/sand pits..**Various-leaved Pondweed** *P. gramineus*

12 Lf-margin with minute teeth, persisting on older lvs – stipules with 2 ribs on lower side
which are often strongly winged for at least half their length ....*P. lucens* (p 492 **E**)
Lf-margins entire or toothed only on v young lvs and falling off older lvs – stipules with 2
ribs on lower side but no wings **13**

13 Floating lvs translucent and similar in texture to the submerged lvs ..*P. coloratus* (p 492 **C**)
Floating lvs leathery and v different in texture to submerged, translucent lvs **14**

14 Submerged lvs 16–28 cm long – lf-margin with minute teeth on young lvs only – frs
2.7–4.1 mm long (although vr-never fr in Br Isles) – r in calc rivers in S Eng ....
VU ** **Loddon Pondweed** *P. nodosus*
Submerged lvs 6–16 cm long – lf-margin entire – frs 1.9-2.6 mm long ....
P. polygonifolius (p 490 **B**)

15 Lvs all linear, long, like lf-stalks, without a blade – stipules large, 40–170 mm long..................
P. natans (p 490 **A**)
Lvs with midrib and blade – stipules 4–55 mm long **16**

16 Lf-blade pulls away from the stem with the stipule apparently attached to the lf-sheath,
looking like a grass ligule **17**
Stipule and lf are separate and originate from the same point at the junction of the lf-
sheath and lf-blade **18**

17 Lf-sheath open along its entire length – mature frs 3.3–4.7 mm long ....*P. pectinatus* (p 492 **H**)
Lf-sheath closed at base when young – mature frs 2.2–3.2 mm long......*P. filiformis* (p 493 **I**)

18 Most lvs >2 mm wide..**19**
Most lvs <2 mm wide **24**

19 Lvs with 3–5 main veins and many short, interrupted, longitudinal strands between these (seen when lf held up to light) – stems flattened or strongly compressed, at least 2.5 x as long as wide in cross-section...**20**

Lvs with 3–5 main veins only – stems round or compressed or grooved....................................**21**

20 Fl-stalks 28–95 mm long – fls with mostly 2 carpels – lvs with 5 main veins – o in C and E Eng; vr in E Scot in canals, lakes ....BAP EN **Grass-wrack Pondweed* P. compressus

Fl-stalks 5–20(-30) mm long – fls with 1 carpel – lvs with 3 main veins – r in S and E Eng in grazing-marsh ditches or ponds................................CR ** **Sharp-leaved Pondweed** P. acutifolius

21 Lf-margin toothed – stem with shallow groove running down both or one of the broad sides...P. crispus (p 492 **F**)

Lf-margin entire or if with minute teeth then no groove running down stem **22**

22 Lvs with mucronate tip – stipules closed at base when young – o-lf (lc in Shetland & Orkney, Norfolk broads) especially in canals, also ditches, rivers, lakes..

NT **Flat-stalked Pondweed** P. friesii

Lvs without mucronate tip – stipules open **23**

23 Stems forming dense fan-like sprays – lvs often tinged pink or reddish-brown – stipules with (8-)10–17 veins between ribs – infl with 6–8 fl – o-vla throughout Br Isles (vr SW Eng, N and W Scot) in ponds, lakes, ditches, canals .... **Blunt-leaved Pondweed** P. obtusifolius

Stems not forming dense fan-like sprays – lvs without any pink or reddish-brown tinge – stipules with (4-)5–8(-9) veins between ribs – infl with 2–4 fl ....P. berchtoldii (p 492 **G**)

24 Stipules closed and fused into a tube around stem...**25**

Stipules open, with margins overlapping around stem..**26**

25 Lf-tip gradually tapering to v fine point, almost bristle-like – lvs rigid – vl in Scot only ....
....BAP ** **Shetland Pondweed** P. rutilus

Lf-tip pointed – lvs not rigid – widespread ....P. pusillus (p 492 **G.1**)

26 Lvs tapering to fine point – lvs stiff out of water due to broad mid-rib, occupying 30-70% of lf-width near base – lvs mostly <1 mm wide – fls with 1–2 carpels – o-lf in Eng (Scot, Wales r; Ire abs) in range of habitats, tolerating eutrophic water ....**Hairlike Pondweed** P. trichoides

Lf-tip rounded – lvs flaccid out of water due to mid-rib only occupying 10-20% of lf-width near base – lvs mostly >1 mm wide – fls rarely with 3, usually 4 or more carpels ....
P. berchtoldii (p 492 **G**)

A Broad-leaved Pondweed, *Potamogeton natans*, aquatic herb with floating lvs (**A**) and often with submerged grass-like structures (**phyllodes**) which are actually modified lf-stalks (but no submerged lvs with blades); stems cylindrical, little branched, 1–2 m long or more; floating lvs dark green, leathery, opaque with **translucent** longitudinal veins, oval to elliptical-lanceolate, 5–10 cm long, pointed at tips, rounded at base. **Stipules** large, **4–17 cm** long. **Key difference** from all other Pondweeds, **when present**, is **floating** lvs with **buff or discoloured flexible joint** just below top of the long lf-stalk. Fl-spikes dense, cylindrical, 3–8 cm long, stout-stalked, emerging from water. Frs (**Aa**) 4–5 mm long x 3 mm wide, olive, obovoid, ± flattened, both margins convex; beak short, straight. Br Isles, c in still and slow moving fresh water. Fl 5–9.

B Bog Pondweed, *P. polygonifolius*, v variable but often close to A in habit and shape of floating lvs (**Ba**), but also has submerged lvs (**Bb**) (at least when young). All lvs **stalked**, with lanceolate-oval blades; floating lvs leathery, 4–10.5 cm long,

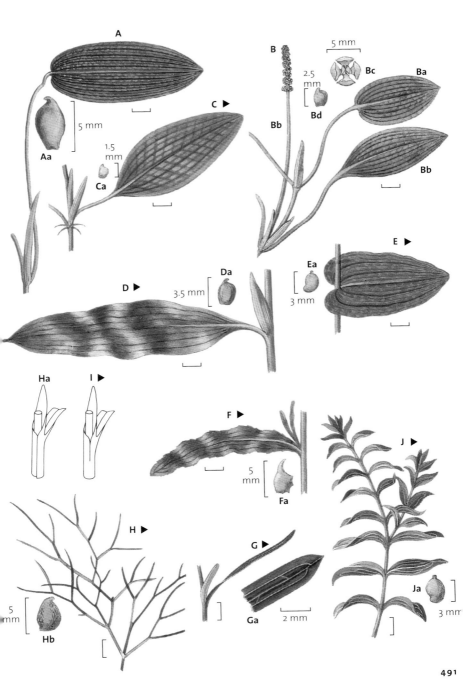

either tapered into stalk or cordate-based, but **without** any buff-coloured flexible **joint** at stalk-top and longitudinal veins **darker than lf, opaque** (not translucent like A); **stipules 1–5 cm,** but only **2-ridged** not winged (like D); submerged lvs translucent and more **linear-lanceolate** than floating lvs. Fl-spikes (**B**) dense, as in A but smaller; fls (**Bc**) with 4 perianth segments, 5 mm across. Frs (**Bd**) 1.9–2.6 mm long x 1.4–1.9 mm wide, reddish, beak tiny. Br Isles c (but mid, E Eng, and v cultivated regions, o-lf only); in shallow peaty acid water, bog pools. Fl 5–10.

▲ **C** **Fen Pondweed*, *P. coloratus*, also similar to A, but also has submerged lvs (**C**); all lf-stalks **shorter** than floating lvs, which are **thin**, **translucent**, olive to reddish-brown, **tapered** into stalks, with a fine **network** of **cross-** and longitudinal-veins; like B has **no joint** at top of lf-stalk. Floating lvs oval-elliptical, 5–10 cm long, submerged lvs narrow. **Key difference** from A and B in **smaller frs** (**Ca**) 1.5–1.9 mm long x 1–1.3 mm wide, green or greenish-brown, ovoid, flattened, inner edge ± straight, beak v short. Br Isles: GB, N to Edinburgh, and in Hebrides, r (but lf E Anglia to E Yorks); Ire o-lf; in calc peaty ditches, fens. Fl 6–7.

▲ **D** Shining Pondweed, *P. lucens*, **wholly submerged** aquatic herb; lvs (**D**) all **similar, translucent** yellow-green, **wavy, shiny,** oblong-lanceolate, 10–20 cm long x 2.5–6.0 cm wide, with dense network of longitudinal and cross-veins; lf edges **minutely-toothed**; lf-stalks **v short**, 1–12 mm long; **stipules winged**, 3–8 cm; infl-stalks thickened above. Frs (**Da**) 3.5 mm long x 2.4–3 mm wide, brown, ovoid, not flattened, inner edge ± straight, beak v short. Br Isles: S, mid, E Eng, f-lc; rest of GB o-r; Ire o; in still and slow-moving fresh, especially calc, water. Fl 6–9.

▲ **E** Perfoliate Pondweed, *P. perfoliatus*, submerged aquatic herb, **stalkless**, ± oval, **translucent** lvs (**E**), all **submerged**, 2–8 cm long, ± **completely clasping** stem-bases; stipules often present **only** on **youngest** lvs. Frs (**Ea**) 3.0–4.0 mm long x 2.5–3.0 mm wide, olive-green, ovoid, swollen, inner side concave near base, hardly keeled. Br Isles widespread and f, to lc in S and E Eng; in still or flowing fresh water. Fl 6–9.

▲ **F** Curled Pondweed, *P. crispus*, submerged aquatic herb with linear-oblong lvs (**F**), narrower than A–E; lvs translucent dark green, stalkless, with blunt tips and strongly **waved** or **crisped** margins and **grooved stem**. Infls loose, stalks slender. **Key differences** from all other Pondweeds (excepting hybrids) are **strongly toothed** (almost hack-saw-like) lvs and frs with **beak** $^1/_2$ as long or more as fr. Frs (**Fa**) 2–5 mm long (without the long curved beak), olive, oval, flattened, with outer edge keeled, with one **tooth**. Br Isles, c (except in mts); in still and flowing fresh water. Fl 5–10.

▲ **G** Small Pondweed, *P. berchtoldii* (illustrated) and **G.1 Lesser Pondweed** (*P. pusillus*) are v v close and can only be separated by looking at stipule cross-sections under a microscope. Both are **slender** submerged aquatic herbs; lvs (**G**, **Ga**) **narrow-linear**, flat, generally < 2 mm wide (G may be wider), stalkless. **Key difference** between these spp is stipules **open** (*P. berchtoldii*) or **closed** (*P. pusillus*). Br Isles, both spp f-lc in still and slow-moving fresh water. Fl 6–9.

▲ **H** Fennel Pondweed, *P. pectinatus*, submerged aquatic, long narrow-linear lvs, generally < 2 mm wide, each lf composed of **2 slender parallel tubes**. **Key difference** from all other narrow-lvd Pondweeds (except I) is that stipules (**Ha**) join lf-base, so that when you pull the lf, the sheath and stipule come away, resembling a grass sheath and ligule. Lf-sheath open. Infls open, v loose, on long, slender stalks; stems branching mainly in the **upper part** of plant. Frs (**Hb**) 3–5 mm long x 2–4 mm wide, olive-brown, ovoid, flattened, beak short, curved. Br Isles: GB, N to mid Scot, c; N Scot, Ire, o; in still and flowing fresh water, especially in lowlands and near coasts (only sp of *Potamogeton* c in brackish ditches near sea). Fl 5–9.

▲ **I** *****Slender-leaved Pondweed**, *P. filiformis*, another narrow-lvd sp with a leaf sheath structure like H. **Key difference** from H is lf-sheath **closed** at the base when young and plant branched mainly from its base. Br Isles: Ire, Scot o; Orkney & Shetland c; **abs in Eng**. Fl 5–9.

▲ **J** FPO (Eire) VU **Opposite-leaved Pondweed**, *Groenlandia densa*, submerged aquatic with **all** lvs in **opp** pairs (in *Potamogeton* lvs **alt**, except those with infls in axils); lvs **without** stipules (except those with infls in axils), oval-triangular to lanceolate, green-olive, **unstalked**, minutely-toothed on edges, often lengthwise folded, **densely-set**. Fls in tiny, few-fld **heads**, (not spikes); frs (**Ja**) 3 mm long x 2 mm wide, olive, shortly oval, flattened, both edges keeled and convex; beak v short, central on fr-tip. Br Isles: Eng f-lc but declining; Scot, Wales, vr; Ire o; in running or still fresh water. Fl 5–9.

TASSELWEED FAMILY *Ruppiaceae*

A Beaked Tasselweed, *Ruppia maritima*, lvs to 10 cm, only 0.5 mm wide, may be **alt** or **opp**, with small **teeth** (use hand lens) at tip. Fls in cluster appearing stalkless, each fl with 2 stamens and several carpels; frs in small umbel-like heads of 3–5, on a common **fr-stalk to 2.6 cm** long; fr (**Aa**) 2–2.8 mm long, ovoid, v asymmetrical, with short beak. Br Isles: o to N, W, and in Ire; lc in S, E, Eng; in shallow brackish, often stagnant ditches, pools, lakes near sea, sometimes inland in areas with natural salt deposits. Fl 7–9.
A.1 NT *****Spiral Tasselweed**, *R. cirrhosa* (*R. spiralis*), close to A, but **key differences** are: common **fr-stalk** is **4 cm or more** and usually coiled in a spiral in fr; frs ovoid, but nearly **symmetrical** and **2.7–3.4 mm** long. Br Isles: o S and E coasts; r in Ire; r in NW Scot, Shetland and Orkney only; in similar habitats to A but often in deeper, more salty water. Fl 7–9.

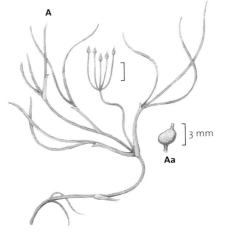

A

Aa

3 mm

NAIAD FAMILY
Najadaceae

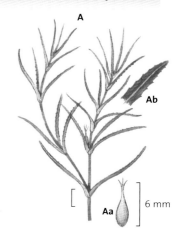

A

Ab

Aa

6 mm

A EU Sch 8 FPO (Eire) BAP ***Slender Naiad**, *Najas flexilis*, slender submerged aquatic herb with smooth stems to 30 cm long; lvs (**Ab**) linear, **minutely-toothed**, 20–25 mm long x 0.5–1.0 mm wide, in **opp** pairs or whorls of 3, translucent, pointed, with few-haired sheathing bases; fls 1–3 together in lf-axils: male with a tiny sheath, two-lipped perianth and 1 stamen; female fls naked with three stigmas forming a narrow oval fr (**Aa**) 4–6 mm long including beak and stigmas. Br Isles: NW Eng, W Scot, Hebrides, r; W Ire r; in lakes of acid to slightly base-rich water. Fl 7–9.
A.1 Sch 8 BAP VU ****Holly-leaved Naiad**, *N. marina*, has broader, **strongly spiny** lvs, and a few teeth on the stiff, brittle stems. Lf-sheaths hairless; plant dioecious; frs **twice** as long as in A. Eng: Norfolk Broads only, vr but lf in water over silty or peaty soils. Fl 7–8.

HORNED PONDWEED FAMILY
Zannichelliaceae

B Horned Pondweed, *Zannichellia palustris*, submerged aquatic with much branched slender shoots to 50 cm long; lvs opp, linear, 1.5–5.0 cm long x 0.5–2.0 mm wide, flat or sometimes hairlike, fine-pointed, translucent, with few parallel veins (like narrow-lvd *Potamogeton* spp) with stipules. Tiny fls in ± **unstalked clusters** in lf-axils, usually 1 male fl of 1 stamen and 2–6 female together in a tiny cup-shaped sheath). Frs (**Ba**) distinctive **banana** shape, 2–6 together, green, 2–3 mm long, usually short-stalked, often with toothed edge, beak to 2 mm long. Br Isles: widespread, Eng, lc to E and S, o-lf to W; Scot, Wales, Ire, o; in still or slow moving fresh brackish water, chk streams or nutrient-rich lakes or ponds. Fl 5–8.

EELGRASS FAMILY
Zosteraceae

ID Tips   Eelgrasses
• **Eelgrasses** (*Zostera* spp) are the only narrow-leaved aquatic plants with a **congested infl**, **enclosed** in the lf-sheath and **alt** lvs.
• Eelgrasses are restricted to maritime habitats, growing submerged in the sea and are often found as fragments washed up on beaches.

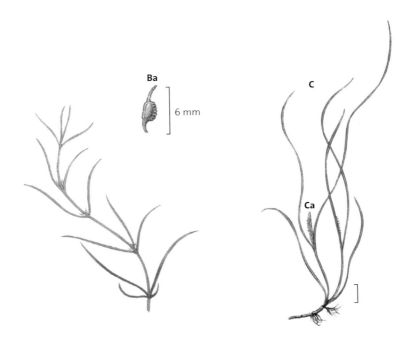

Ba

6 mm

C

Ca

C NT **Eelgrass**, *Zostera marina*, (includes plants previously called *Z. angustifolia*) lvs ribbon-like, **15–50 cm** (or more) long x 2–10 mm broad, with 1, 3 or more veins, with rounded mucronate or notched tips; fl-stems much **branched**, fls (**Ca**) in **spikes** on one side of a flattened axis, ± enclosed inside lf-sheaths; fls of two sexes separate but male and female fls alt within spikes, without bracts or perianth; stigma twice length of style; seeds 2.5–3.5 mm long, ribbed. Br Isles, lf around most of coast, but decreased; in sea from spring tide low water down to a vertical depth of 4 m, on sand or mud. Fl 6–9. **C.1** VU *****Dwarf Eelgrass**, *Z. noltei*, **key differences** from C are fl-stems **unbranched**; lvs one-veined only, **6–12 cm long**, under **1 mm wide**, with notched tips, only one vein and lf-sheaths **open**; seeds 2 mm long, smooth. Br Isles, r-lf round coasts. Fl 6–10.

ARUM FAMILY   *Araceae*

This family has tiny fls crowded into a dense, spiked infl (the **spadix**). In *Arum*, spadix is **enclosed** in a large sheathing bract (**spathe**), the whole resembling a large one-petalled fl with the naked, club-shaped spadix tip visible above the rolled-up cup-like lower part of the spathe. Fls have **no** perianths and are **unisexual**; lower part of spathe contains, at **base** of spadix, whorls of **female** fls (with tiny yellow ovaries and long bristle-like stigmas), then above them the **male** fls (with tiny short red-brown stamens), finally a ring of **sterile** bristle-like fls which close the cup's **top**. Small flies, attracted by the meaty scent, push down past sterile fls to enter cup, are at first trapped, and, if carrying pollen from another fl, pollinate the female fls. Later, the flies pick up pollen released by the male fls, and, when the sterile fls wither, escape and may pollinate another plant. In *Arum*, lvs are **triangular** to arrow-shaped, and (rare in Monocotyledons) **long-stalked, net-veined**. Frs are red berries in a dense spike. **Sweet-flag** (*Acorus calamus*) has **no spathe,** spadix is borne on **side** of stem, and fls each have 6 tiny **perianth segments,** 6 stamens **and** an ovary; lvs, **linear,** sword-like, smell of tangerines when crushed.

A   Lords-and-Ladies (Cuckoo-pint), *Arum maculatum*, erect hairless per, 30–50 cm tall, from a tuber. Lvs arrow-shaped to triangular, 7–20 cm long, net-veined, ± wrinkled, thin, v shiny bright green, often purple-spotted, on **long stalks**, appearing Feb–March. Spathe (**A**) with cup-shaped, rolled-up, basal part, and cowl-shaped, **erect, pointed** upper part 15–25 cm long, pale yellow-green, **purple-edged**, sometimes purple-spotted. Spadix (**Aa**) 7–12 cm long, upper part cylindrical club-shaped, 3–4 cm long, naked, **chocolate-purple**, rarely yellow; lower part (with fl-spike) hidden in cup. Frs (**Ab**) red berries 5 mm wide, in a dense spike 3–5 cm long. Br Isles: Eng, Wales, vc; S half Scot, Ire, f; Scot r-lf (abs Scot mts); in wds, hbs, mostly on calc or richer soils. Fl 4–5, fr 7–8.

B   NT *ⁱ**Italian Lords-and-Ladies**, *A. italicum* ssp *neglectum*, close to A, but some lvs appear in **autumn**, and are blunter, **thicker**, more **leathery-textured, never wrinkled or spotted**; midribs of the **triangular spring** lvs ± **paler** yellow-green (not dark green); upper part of spadix **always bright yellow; spathe** (**B**) up to **40 cm** long, wholly yellow-green, **3 times as long** as whole of **spadix** (only **twice** as long in A); tip of spathe **bends forward** and **hangs down** over spadix when mature; frs scarlet berries 10 mm wide, in a spike to 14 cm long. S Eng, S Wales, IoS, CI o-vlf as a native; vr introd elsewhere in Eng; in wds, scrub near sea, but in Hants, W Sussex inland on moist calc screes at foot of steep wds on chk. Fl 6, fr 8–9. **B.1** *A. italicum* spp *italicum* has **broad-triangular dark green** lvs with **cream-white** veins. Introd (S Eur); S Eng, f and more widespread than B.

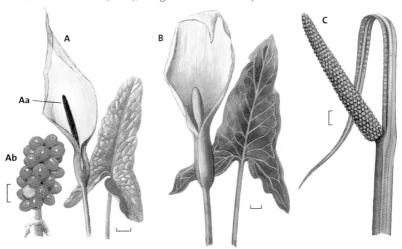

C Sweet-flag, *Acorus calamus*, stout hairless erect per herb to 1 m tall; lvs bright green, sword-shaped, narrowed to tips, 1–2 cm wide, with thick midrib, and wavy, often **crinkled** edges, smelling strongly of **tangerines** or **dried oranges** when crushed. Infl (**C**), borne **laterally** at angle of 45° on the flattened winged stem, is a stout cigar-shaped spike, tapering to blunt tip to 8 cm long; stem **continues** as a long lfy point above spike. Fls densely packed, **each** with 6 tiny **yellowish-green** perianth segments, 6 stamens, and a 2–3-celled ovary; **no spathe** or bract to infl. Introd (unknown origin); Br Isles: E, central, SE Eng, f-lc; Wales, Scot, Ire, r; in shallow fresh water of ponds, canals, rivers, ditches. Fl 5–7.

DUCKWEED FAMILY	*Lemnaceae*

Tiny floating plants that form green carpets on the surface of fresh water. No obvious division into stem and lvs, but with merely a rounded or elliptical **thallus**. In *Lemna*, **single** roots are present, hanging down from thallus; in *Wolffia*, roots **abs**; in *Spirodela*, **several** roots per thallus. Fls, minute, hard to see and rarely produced in most spp, consist of tiny male fls with 1–2 stamens, and female fls with minute ovary, borne in a pocket on thallus edge or upper surface.

ID Tips Duckweeds

- These floating pad-like plants are unique but **do not confuse** with **Water Fern** (*Azolla filiculoides* p 499 F) which is also a surface-floater.

- **What to look at**: well-grown spring or summer plants; those over-wintering are atypical, often with fewer veins or roots.

- Often veins are not visible. **Least Duckweed** (*L. minuta*) is blue-green and elliptical, like a grain of rice in shape; **Common Duckweed** (*L. minor*) is bright green and oval to rounded.

KEY TO DUCKWEED FAMILY

1 Plant thallus ovoid or globular, rootless ....*Wolffia arrhiza* (p 498 **D**)
   Plant thallus ± flattened above, with roots ....*(Lemna & Spirodela)* **2**

2 Plants floating just below surface of water, translucent, elliptical, stalked, several plants attached together ± in a branched cluster ....*Lemna trisulca* (p 498 **E**)
   Plants floating on surface of water, opaque, round or oval, unstalked, plants separate........**3**

3 Each thallus with several roots, flat, 1.5–8 mm wide ....*Spirodela polyrhiza* (p 498 **B**)
   Each thallus with one root only **4**

4 Thallus convex above, sometimes v swollen and spongy below ....*L. gibba* (p 498 **C**)
   Thallus ± flat above and below **5**

5 Thallus with 3–5 veins – green, ovate, ± opaque..*L. minor* (p 498 **A**)
   Thallus with 1 vein – blue-green, elongate, ± translucent ....*L. minuta* (p 498 **A.1**)

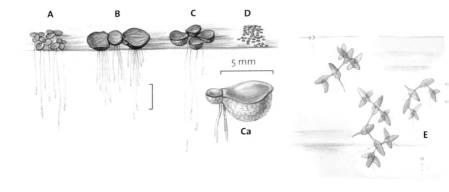

A **Common Duckweed**, *Lemna minor*, plants flat both sides, 1–8 mm long, each with a **single** v long root hanging down into water and **3–5 veins**. Br Isles, vc in fresh, still or slow-moving water. Fl 6–7 (vr flg). **A.1** **Least Duckweed**, *L. minuta*, v close to A but plants 0.8–4 mm long with **1 vein** only (but difficult to see unless stained with red ink under microscope). Thallus **blue-green** and usually slightly translucent, elliptic, symmetrical whereas A is usually opaque, obovate and asymmetrical. Introd (N&S America) first discovered 1977 and probably overlooked. Br Isles: S Eng c, elsewhere in Eng, Wales, Scot, Ire o-r, same habitat as A, often with other *Lemna* spp. Fl 5–7 (vr flg).

B **Greater Duckweed**, *Spirodela polyrhiza* (*L. polyrhiza*), plants flat both sides as in A, 1.5–10 mm long, often red-purple below, each with **several** long roots. Br Isles: Eng f-lc; Ire o-lf; Wales, r; Scot, abs-vr; in still fresh or slow-flowing, usually calc water. Fl 7 (not or vr flg).

C **Fat Duckweed**, *L. gibba*, plants 3–5 mm long, green and **convex** above, either flat (as A) or white, **spongy** and v swollen below (**Ca**), each with **one** long root; (**Ca** also shows a plant forming a daughter thallus by budding). **Key difference** from A is when thallus held up to light, air spaces (lacunae) can be seen overlapping cells (use hand-lens). Br Isles: Eng f-lc; Wales, Ire, r; Scot vr; in still or brackish nutrient-rich water. Fl 6–7 (vr flg).

D VU *****Rootless Duckweed**, *Wolffia arrhiza*, plants like tiny ovoid or ellipsoid grains (so that it can be **rolled** in your fingers, unlike all other *Lemnaceae*), 0.5–1.5 mm long, **without roots**; infl of 1 male and 1 female fl. S Eng only, r-lf in still fresh water. **Not flg** in Br Isles. The smallest flowering plant in the world!

E **Ivy-leaved Duckweed**, *L. trisulca*, differs from other Duckweeds being slightly **submerged** with elliptical-lanceolate, translucent thalli, which taper when mature into stalks at the base, and are usually joined up at **right angles** into branched colonies, the terminal triplets of thalli looking ± like tiny ivy lvs. Br Isles: Eng c; Ire o-lf; Wales, S half Scot only, r; in still fresh or brackish water. Fl 5–7 (vr flg).

WATER FERN FAMILY   *Azollaceae*

F **Water Fern**, *Azolla filiculoides*, is an aquatic floating **fern** but is included here as it resembles a Duckweed in habit and habitat. **Key differences** from Duckweeds are its tiny oval closely **overlapping** lvs along short **branched stems**. Fronds turn **red** in later summer. Introd (N & S America); Br Isles: S Eng, f-lc; Ire, Wales, N Eng, Scot o-r; in fresh water.

PIPEWORT FAMILY   *Eriocaulaceae*

G ****Pipewort**, *Eriocaulon aquaticum* (*E. septangulare*), slender aquatic herb; creeping rootstock produces at intervals erect to spreading tufts of linear, submerged translucent basal lvs 5–10 cm long, flattened at sides, tapered to fine points, and with internal cross-partitions. Fl-stems arising from lf-rosettes, lfless, **erect**, **6–8 (usually 7)-angled** (**Gb**), 20–60 cm tall, twisted, bearing button-like heads 0.5–2.0 cm across, of tiny close-packed fls at tips, each head with tiny grey bracts around it in a tightly adpressed whorl. Fls (**Ga**) 4 mm long, with 2 grey sepals with hairy tips, 2 whitish black-tipped petals, 2–4 stamens and stigmas; infls resemble whitish-headed knitting-needles. Br Isles: Scot, only in Skye, Coll, W Argyll, vla; Ire, Donegal to Cork near W Coast, vla; in shallow lakes and pools of acid water on peaty substrates, avoids limestone. Fl 7–9.

For a **key** to **water plants** with ± linear, quill-like **lvs in rosettes**, see p 82.

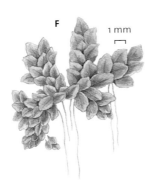

F

1 mm

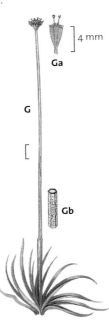

4 mm

Ga

G

Gb

BUR-REED FAMILY   *Sparganiaceae*

Per aquatic herbs with creeping rhizomes, simple or branched stems, and linear lvs **broadside-on** to stems; fls tiny, in unisexual **globular heads**, the female heads at base of infl or its branches, the male heads above; perianth of 3–6 papery spoon-shaped scales; female fls have one-celled stalkless ovary with elongated style at its tip, male 3 or more stamens. Fruiting-heads are globular and spiky; fr a nut.

KEY TO BUR-REEDS (*SPARGANIUM*)

1 Infl branched, each branch with male heads above, female heads below – perianth segments black-tipped ....*Sparganium erectum* (p 500 **A**) 2
   Infl unbranched, only one group of male heads above per plant – perianth segments wholly pale **3**

2 Fr with an obvious 'shoulder' (p 501 **Aa**), upper part dark, flattened, (3-)4–6(-7) mm wide at shoulder ....*S. erectum* ssp *erectum* (p 501 **Aa**)
   Fr with an obvious 'shoulder', upper part dark, but domed and wrinkled below style, 2.5–4.5 mm wide at shoulder...*S. erectum* ssp *microcarpum* (p 501 **Ab**)
   Fr ellipsoidal, tapered evenly into style, no 'shoulder', 2.0–3.5(-4.5) mm wide..............................
   ....*S. erectum* ssp *neglectum* (p 501 **Ac**)
   Fr ± spherical, tapered into style, 4–7 mm wide................*S. erectum* ssp *oocarpum* (p 501 **Ad**)

3 Stem-lvs keeled at base in cross-section – male heads 3 or more, clearly separate ....
   ....*S. emersum* (p 500 **B**)
   Stem-lvs flat or inflated but not keeled at base – male heads 1–3(-4), close together..........**4**

4 Lf-like bract below lowest female head 10–60 cm, twice length or more of whole infl – male heads usually 2–3 ....*S. angustifolium* (p 501 **D**)
   Lf-like bract below lowest female head <10 cm, barely longer than whole infl – male head usually 1 ....*S. natans* (p 500 **C**)

A   Branched Bur-reed, *Sparganium erectum*, erect **branched** hairless per 50–150 cm; stem-lvs **keeled** at base, broadside on to stems, all usually erect (may develop submerged or floating lvs in deep or fast-flowing water which are keeled to tip). Each infl branch has male heads above, female below. Perianth segments 6, black-tipped. **For ssp see key above**: most are c in S Eng; in still and moving fresh water. Fl 6–8.

B   Unbranched Bur-reed, *S. emersum*, erect plant 20–60 cm tall, v like A but smaller and more often with **both** erect stem-lvs keeled at base and ± flat to slightly keeled, **ribbon-like** basal **floating**-lvs. **Key difference** from A is infl an **unbranched** raceme, therefore only **one** group of male heads (not several groups) per plant. Male heads 3 or more; perianth segments wholly pale (**Bb**); fr (**Ba**) elliptical, narrowed into style. Br Isles, f (except NW Scot r); in still and moving fresh water, especially on mud substrates. Fl 6–7.

C   Least Bur-reed, *S. natans* (*S. minimum*), small floating plant rooted in substrate with stems 6–80 cm long; **all lvs unkeeled**, 2–6 mm wide, hardly inflated at bases; infl emerging from water, **unbranched**, usually with 1 male fl-head and 2–3

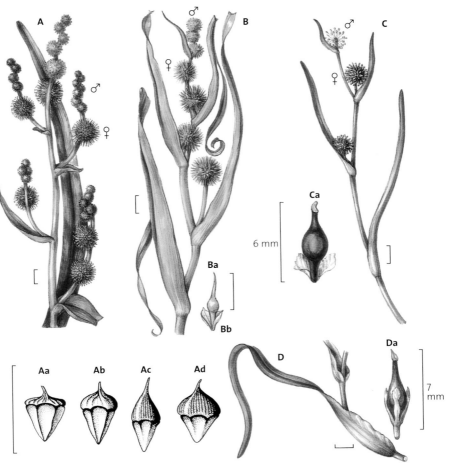

± **stalkless** female heads, all under 1.5 cm wide; bract of lowest female fl-head **not or barely longer** than whole infl, <10 cm. Fr (**Ca**) **obovate**, 3.5–4.5 mm long (excluding style on top). Br Isles: Scot, Ire, f-lc; N Eng o; S Eng, Wales, r; in peaty fen and bog pools, ditches. Fl 6–7.

D   Floating Bur-reed, *S. angustifolium*, close to and can not be separated vegetatively from C; unbranched infl has usually 2–3 male fl-heads and 2–4 female ones; bract of lowest female fl-head is **twice or more** length of whole infl; fr (**Da**) **elliptical** in outline, 7–8 mm long (including style). Br Isles: N, NW Scot, N Wales f-lc; NW Eng, Ire, o-vlf; **S, E Eng, abs** (except New Forest vr); in peaty lakes, pools. Fl 8–9.

BULRUSH FAMILY
Typhaceae

Bulrushes (*Typha*) are tall erect stout-stemmed herbs, growing from rhizomes in shallow water or mud; lvs mostly basal, linear, flat, **grey-green**, on **two opp sides** of stem, with sheathing bases; infl a **cylindrical spike** of densely packed tiny fls; its **lower** part **stout**, **brown**, of many female fls each with a one-celled stalked ovary surrounded by densely-packed brown hairs; its **upper** part **narrow**, **yellow**, of many male fls each with 2–5 stamens.

A Bulrush (Common Reedmace), *Typha latifolia*, 1.5–2.5 m tall; lvs **16–18 mm wide**; infl with **no gap** (rarely < 0.5 mm) between male and female parts of spike; female part of spike **3–4 cm wide**; **no** bracts to female fls. Br Isles: Eng c; Wales, S, E Scot, o-lf; N Scot vr; Ire f; in reed swamps on mud or silt in still or slow-moving fresh water. Fl 6–7.

B Lesser Bulrush, *T. angustifolia*, as tall as A, but more slender; lvs **only 4–6 mm** wide, curved on the backs; **a gap** (**Ba**) 2–9 cm long between male and female parts of spike; female part of spike **1.5–2.0 cm wide**; female fls have tiny **bracts**. Br Isles: Eng f-lc; Wales, S Scot, Ire, r; **N Scot abs**; in reed swamps, often on more peaty soils than A, and more c in ditches near sea. Fl 6–7. Hybrid of A and B (*T.* x *glauca*) is probably f Eng (more r Ire, Scot), overlooked, intermediate in lf width and has a **small gap** (< 15 mm) between the male and female parts of the spike and plant sterile.

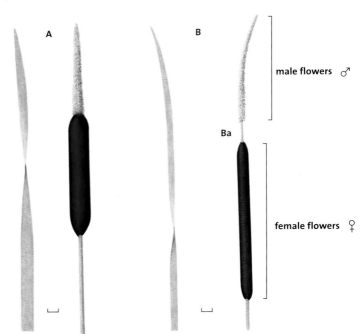

A   B

male flowers ♂

Ba

female flowers ♀

LILY FAMILY   *Liliaceae*

A large family of herbs with bulbs or tubers and a few shrubs, of diverse form, but all with 6 equal, similar perianth segments (**May Lily** has fl-parts in fours), in some joined into a tube, in others separate; usually 6 stamens (3 in **Butcher's-broom**; 4 in **May Lily**; 8 in **Herb Paris**), a three-celled **superior** ovary (except Daffodils, Snowdrops & Snowflakes sometimes placed in a separate family, *Amaryllidaceae*), forming in fr either a dry capsule or a berry. In **Orchid** family **ovary is inferior**; one petal is enlarged, v different from rest of perianth segments, and forms a **lip to fl**, usually on the **lower side**; and there is usually only one, stalkless, anther. **All** plants in the Lily and Orchid families (except Herb-Paris) have **parallel-veined**, **unlobed**, **untoothed** lvs, like most Monocotyledons.

KEY TO LILY FAMILY

1 Much-branched low shrub – true lvs reduced to tiny papery scales – oval, tough, evergreen, spine-tipped (flattened lf-like stems **cladodes**) in axils of the scale-lvs, bear fls on their upper sides – berry red..*Ruscus aculeatus* (p 506 **B**)
Much-branched herb – true lvs reduced to tiny papery scales – bright green, needle-shaped cladodes in axils of scale-lvs – fls in axils of scale-lvs on normal stems – berry red
Asparagus (p 506)

Herbs with aerial stems unbranched, or branched only in infls – true lvs present, either linear, grass-like or ± oval, parallel-veined, at base of, and/or up the stems **2**

2 Ovary superior **3**
Ovary inferior..**16**

3 Infl a terminal umbel..**4**
Infl a terminal raceme **5**
Fls solitary on stem-tips **13**
Infl a branched panicle, fls white inside, purplish-pink outside, 15–20 mm across – perianth segments 5–7-veined – stamen stalks white-woolly ....*Simethis* (p 508)
Fls in clusters of 1–3 in axils of elliptical or lanceolate lvs up lfy stem – perianth tubular or bell-shaped, six-lobed – fr a berry..*Polygonatum* (p 504–5)

4 Fls many per umbel (sometimes partly or wholly replaced by small bulbils) – papery spathe enclosing infl in bud, splitting into 1 or more bracts in fl – plants with garlic or onion smell when bruised ....*Allium* (see key p 513)
Fls few (2–5) per umbel, always yellow, never mixed with bulbils – 1–3 lf-like green bracts at base of, but not enclosing infl – no garlic smell when bruised..........*Gagea lutea* (p 512 **G**)

5 Lvs linear, all flattened and folded in one plane, edgeways-on to stem (as in a gdn Iris) – lvs mostly, but not all, basal – in bogs **6**
Lvs variously shaped, broadside-on to stem, not all flattened in one plane **7**

6 Fls creamy – stamens hairless – styles 3..*Tofieldia pusilla* (p 508 **B**)
Fls yellow – stamens orange-woolly – style 1 ....*Narthecium ossifragum* (p 511 **F**)

7 Fl-stem lfy **8**
Fl-stem lfless (but sometimes with scales) – lvs all from root **9**

8 Lvs cordate-based, alt, 2–3 only – infl a dense raceme only 2–5 cm long – fls small, white, 2–5 mm wide, with 4 free perianth segments only............*Maianthemum bifolium* (p 506 **A**)
Lvs elliptical-lanceolate, not cordate-based, whorled or alt, many – infl a loose raceme 20–30 cm long – fls large, various colours, 30 mm wide or more, with 6 free perianth segments, often arched back ....*Lilium* (p 508)

9 Lvs oval-lanceolate, in pairs, stalked – rootstock creeping – perianth white, cup-like, six-lobed – fr a red berry..*Convallaria majalis* (p 505 **E**)
Lvs linear or grass-like, stalkless – rootstock a bulb **10**

10 Perianth-tube inflated in middle, narrowed to mouth, its lobes v short........*Muscari* (p 508)
Perianth not inflated, not narrowed to mouth, its 6 segments free almost to base **11**

11 Fls white – the 6 free perianth segments each with a green central stripe on the outside
Ornithogalum (p 510)
Fls blue or purplish (rarely white) – perianth segments never with a green stripe on outside **12**

12 Infls with no bracts (or one only) below each fl – perianth segments 8 mm or less long, free to base, forming an open cup- or star-shape...*Scilla* (p 512)
Infls with 2 bracts below each fl – perianth segments 10 mm or more long, joined at base, forming a bell-shape...*Hyacinthoides* (p 510)

13 Fls green, with 4 sepals and 4 petals – lvs broad, oval in whorl of 4 near top of stem ....
Paris quadrifolia (p 506 **D**)
Fls various colours, never green – lvs narrow, lanceolate or linear, not in whorls of 4 ....**14**

14 Fl-stalk erect, lfless, pink, arising directly from ground in September, no lvs at base – thick lvs, as in gdn hyacinth, produced in spring..................................*Colchicum autumnale* (p 508 **E**)
Fl-stalk bearing lvs **15**

15 Fls drooping, bell-shaped, 3–5 cm long, chequered pink and purplish or sometimes pure white, with 6 free segments, with a shiny nectary at base of each inside – lvs linear ....
Fritillaria meleagris (p 506 **E**)
Fls erect, yellow, bell-shaped, 3–5 cm long, no nectaries – lvs lanceolate ....
Tulipa sylvestris (p 506 **F**)
Fls erect, white, cup-shaped, 1 cm long, nectary present – lvs almost thread-like......................
Lloydia serotina (p 508 **E.1**)
Fls erect, yellow, star-shaped, 1–2 cm long, nectary abs – lvs thread-like ....
Gagea bohemica (p 512 **G.1**)

16 Corolla of 6 equal yellow or cream perianth-segments, with inner trumpet-shaped tube
Narcissus (p 516)
Corolla of 3 white outer perianth-segments, with 3 smaller, inner perianth segments often tinged green ....*Galanthus nivalis* (p 516 **B**)
Corolla of 6 equal perianth segments, all white tinged green.........................*Leucojum* (p 516)

Solomon's-seals (*Polygonatum*) have lfy stems bearing alt or whorled, **parallel-veined** lvs with tubular or bell-shaped whitish fls, in clusters in lf-axils.

A AWI **Solomon's-seal**, *Polygonatum multiflorum*, hairless per herb 30–80 cm tall, with **round** (**Aa**) **arching** stems; lvs alt, 5–12 cm long, untoothed, oval-elliptical, pointed, stalkless, ± spreading on each side of stem; fls in stalked clusters of 2–4 in lf-axils, pendulous; perianth 9–15 mm, tubular, six-toothed, **narrowed** in middle, greenish-white; stamen-stalks **downy**; fr (**Ab**) a **blue-black** berry. GB, f ; N Ire o; in dry wds on calc or sandy soils. Fl 5–6.

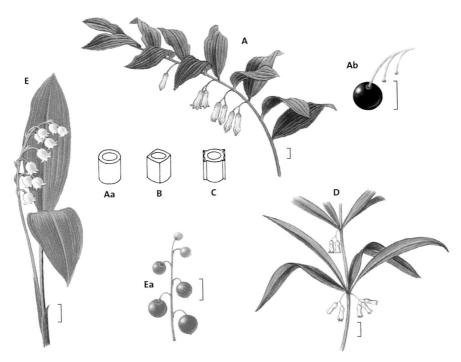

B Garden Solomon's-seal, *P.* x *hybridum*, is **t**he hybrid A x B and is the usual gdn plant, sometimes escaping, much more widespread than, and often confused with, A. This hybrid has the arched stem and drooping fls of A but with a 2– 4 **ridged stem** (**B**) and is partly sterile.

C AWI *****Angular Solomon's-seal**, *P. odoratum*, like A, but shorter (15–30 cm), with fls longer (18–22 mm), **not** narrowed in middle but **narrow bell-shaped, fragrant**; stamen-stalks **hairless**; stem (**C**) distinctly **angled** in cross-section. Eng, r but vlf; Wales vr; Scot, Ire, IoM vr-abs introd; in wds on limestone and on calc sands. Fl 6–7.

D Sch 8 VU ******Whorled Solomon's-seal**, *P. verticillatum*, **erect** per herb 30–80 cm, with ± angled stem and **linear-lanceolate** lvs in **whorls** of 3–6, 5–12 cm long; fls 1–4 per cluster, pendulous; perianth 6–8 mm long, **narrowed** in middle as in A; stamen-stalks **rough**; fr a **red berry** (unlike that of A and B). Scot, Perthshire vr as a native; Staffs vr introd; in mt wds. Fl 6–7.

E AWI **Lily-of-the-valley**, *Convallaria majalis*, hairless per with creeping rhizomes; lvs **long-stalked**, in pairs from rhizomes, **oval-elliptical**, untoothed, parallel-veined, 8–20 cm long x 3–5 cm wide; infl-stalks **lfless**, 10–20 cm tall, with fls in **one-sided racemes**; fls **white**, 5–8 mm long, **globose**, six-toothed, v fragrant; fr (**Ea**) a red berry. Br Isles: Eng f-la; Wales, Scot, vr; Ire vr, in dry wds on calc or sandy soils. Fl 5–6.

A AWI VU ****May Lily**, *Maianthemum bifolium*, erect per, 8–20 cm tall, hairless except on upper stem; lvs deeply **cordate**-based, oval, pointed, basal lvs long-stalked, stem-lvs 2, alt, v short-stalked, all 3–6 cm long; fls in a **dense** raceme 2–5 cm long, each 2–5 mm across, unscented, white, with **4 free** perianth segments; fr a red berry. Lincs, Yorks, Lancs, Norfolk vr but la as a native; vvr introd elsewhere; in dry wds on acid sandy soils. Fl 5–6.

B AWI **Butcher's-broom**, *Ruscus aculeatus*, erect hairless, much-branched dioecious evergreen shrub 25–100 cm tall, with creeping rhizomes; lvs reduced to small papery scales; lf-like **cladodes** (flattened stems) in their axils are **oval**, **tough**, **evergreen**, **spine-tipped**, 1–4 cm long; fls, 1–2 on upper sides of cladodes in axils of tiny bracts, are 5 mm across, six-petalled, greenish-white; fr (**Ba**) a red berry 1 cm wide. Br Isles: S Eng, S Wales, native f-lc; N to central Scot, Ire, introd r; in dry wds. Fl 1–4, fr 10–5.

C **Garden Asparagus**, *Asparagus officinalis*, per hairless herb, erect to 100 cm; stems stout, ann, much-branched; lvs tiny papery whitish scales, with clusters of green needle-shaped lf-like **cladodes** (really modified stems) in axils of some. **Average** length (measure **several**) of **longest cladodes 10–32 mm**; cladodes flexible and usually green (not blue-green). Fls (**Ca**) 1–2 together on stalks in axils of some scale-lvs on larger stems, 3–6 mm long, with bell-shaped greenish-yellow six-lobed perianth; male and female fls separate; male larger, stamens 6. Fr (**Cb**) a red berry. Eng, Wales, o-la; Ire, Scot vvr; in wa, dunes; often an escape from cultivation, but probably native on some dune areas. Fl 6–9. **C.1** FPO (Eire) BAP EN ****Wild Asparagus**, *A. prostratus*, v close to C but **key differences** are plant **procumbent** to slightly erect with creeping underground stem; cladodes usually blue-green and rigid; **average** length of **longest cladodes 2–16 mm**. SW Eng, S Wales, SE Ire, CI, r and vl; on grassy sea cliffs. Fl 6–9.

D AWI **Herb-Paris**, *Paris quadrifolia*, erect hairless per herb with naked stems 15–40 cm tall, topped by a parasol-like **whorl** of usually 4 (sometimes 3, 5 or more) obovate to diamond-shaped, **net-veined** pointed lvs 6–12 cm long, tapered to bases. Fl in centre of lf-whorl, **solitary**, long-stalked, with 4 (rarely more) green lanceolate sepals 2.5–3.5 cm long, 4 narrower green petals, 8 long narrow stamens, and a superior 4(or 5)-celled purple ovary. Fr (**Da**) a black berry. Br Isles: Eng f-lc (Devon, Cornwall abs); Wales, E, S only, r; Scot, S, E only, r; **Ire abs**; in damp wds on calc soils. Fl 5–7.

E VU ***Fritillary**, *Fritillaria meleagris*, erect hairless herb 20–40 cm, with few narrow-linear alt stem-lvs; fls **solitary** (rarely 2), **drooping**; perianth segments 6, 3–5 cm long, oblong, forming a parallel-sided cup **chequered pink** and **brownish-purple** (sometimes all white), with a glistening **nectary** at base of each segment inside; stigmas 3, long, narrow; capsule globose, three-angled. GB: S central Eng, E Anglia, r but la, decreasing; in mds, especially those flooded in winter. Fl 4–5.

F **Wild Tulip**, *Tulipa sylvestris*, hairless herb to 50 cm; lvs linear-**lanceolate**, to 18 mm wide; **erect**, **yellow** fls; buds hanging; **no nectaries** on base of pointed perianth segments inside; stigmas 3, v short. Fr a capsule. Introd (Eur); GB, scattered, vr; Ire vvr-abs; in wds, mds, hbs, grassy banks. Fl 4–5.

A FPO (Eire) **Kerry Lily**, *Simethis planifolia*, erect hairless per herb 15–45 cm tall; lvs linear, **grey-green**, grass-like, all from base, as long as infl, ± curved; infl a loose panicle of fls (**Aa**) each 20 mm across, with short bracts; 6 perianth segments, **white inside**, **pink outside**; 6 stamens, **white-woolly**-stalked; style undivided; capsule globose. Ire, Kerry vr but lf; hths, rocky gd. Fl 5–7.

B **Scottish Asphodel**, *Tofieldia pusilla*, erect hairless per herb 5–20 cm tall; lvs mostly basal, **all flattened in one plane, edgeways on** to stem (like a miniature gdn Iris). Infl a **short**, **dense** raceme; fls with short (1 mm) stalks, short **three-lobed** bracts, and 6 **blunt**, oblong greenish-cream perianth segments 2 mm long; styles 3; capsule short, oval. GB: Teesdale r but lf; Scot Highlands f-lc in springy mt bogs on basic soils. Fl 6–8.

C VU ****Grape-hyacinth**, *Muscari neglectum* (*M. atlanticum*), erect hairless per herb 10–25 cm tall, arising from a bulb; lvs linear, half-cylindrical, grooved on inner sides, 15–30 cm long x 1–3 mm wide, all basal; fls in a **dense cylindrical** terminal raceme 2–3 cm long, **plum-scented**; **dark blue** perianth-tubes 3–5 mm long, **oval**, inflated, with 6 white teeth at the narrow mouth; upper fls smaller, paler, sterile. Capsule globular. Eng: E Anglia, Oxford, Cotswolds r; also introd widely; on dry calc gslds. Fl 4–5 **C.1 Garden Grape-hyacinth**, *M. armeniacum*, the most widespread *Muscari* and usual garden plant, v close to C but has **bright blue**, more ± globular **scentless** flowers (see *Plant Crib*).

D **Tassel Hyacinth**, *M. comosum*, taller (20–50 cm); lvs **6–15 mm wide**; **long**, **loose infl** of long (1–2 cm) stalked fls, lower fls brown, fertile, **oblong**, spreading, upper fls sterile, **purple**, in an **erect tassel**. Introd (S Eur); Eng, Wales, IoS, CI r; on dunes, wa. Fl 4–7.

E FPO (Eire) AWI NT **Meadow Saffron**, *Colchicum autumnale*, hairless herb with underground corm producing erect fls arising directly from gd **in autumn**, **without lvs**; apparent **white** stalk of fl, 5–20 cm tall, is actually its perianth-tube, bearing at its tip 6 oblong **rosy-pink** perianth segments 3.0–4.5 cm long, resembling a Crocus fl, but with 6 stamens (Crocuses have only 3 stamens). Styles 3, v long, separate up tube from below gd level. Lvs, oblong-lanceolate, 12–30 cm long x 1.5–4 cm wide, thick, glossy-green (like lvs of a **gdn Hyacinth**), are produced **in spring** with frs from **previous** year's fls; capsule oval, 3–5 cm long, on short stalk with lvs below. Eng, E Wales, lf in a belt from N Dorset N to Shropshire and E to Oxfordshire; r-o introd elsewhere; in mds, damp wds, rds on clayey soils. Fl 8–10, fr 4–6. **E.1** Sch 8 VU ****Snowdon Lily**, *Lloydia serotina*, erect hairless herb 5–15 cm tall, with a small bulb in gd; lvs linear, almost **thread-like**, basal 15–25 cm long, stem-lvs few, much shorter; fls **solitary**, erect, bell-shaped, perianth segments 6, oblong, 1 cm long, **white** with **purple veins**, pointed, **crocus-like**. Fr a three-angled capsule. N Wales, Snowdon range only, vr on mt ledges on basic rocks. Fl 5.

F **Martagon Lily**, *Lilium martagon*, erect, robust per herb to 100 cm; stem rough; lvs oval-lanceolate, 5–15 cm long, in **whorls** of 5–10 up stem, uppermost alt; fls in a raceme, each 4–5 cm across, with 6 oblong, **strongly curved-back**, **red-purple** perianth segments with darker spots. Anthers red, long-stalked. S Eng, Kent to Wye Valley, r, scattered but la, possibly native; r introd elsewhere in GB; in wds on calc soils. Fl 6–7. **F.1 Pyrenean Lily**, *L. pyrenaicum*, lvs **alt**, fls **yellow**, black-dotted. Introd (S Eur mts); Br Isles: SW Eng, E Scot, Wales o; elsewhere vr; natd in wds, rds, hbs. Fl 5–7.

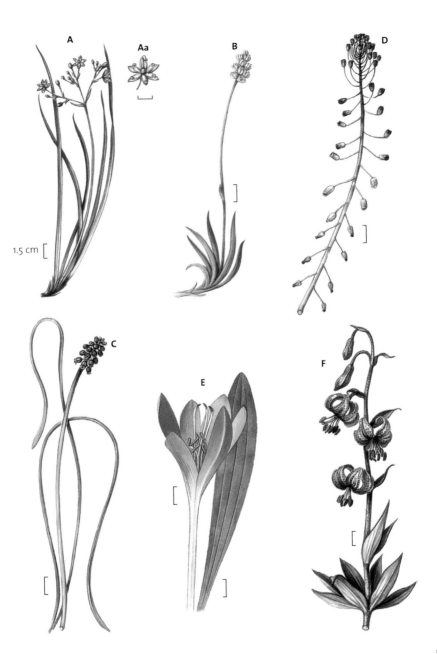

1.5 cm

A

Aa

B

C

D

E

F

Bluebells (*Hyacinthoides*) grow from undergd bulbs; all lvs basal; racemes of blue, bell-shaped fls each with 2 bracts and 6 perianth segments **joined at the base**. **Stars-of-Bethlehem** (*Ornithogalum*) have racemes of **white** fls; green stripe down back of each of 6 perianth segments; stamens with broad flattened hairless stalks. **Yellow Stars-of-Bethlehem** (*Gagea*) have erect **yellow** fls with a green stripe down back of each perianth segment; fls in **umbels**, with a spathe of lf-like bracts **below umbel**. **Squills** (*Scilla*), bulbous plants, are close to Bluebells, but with no bracts or 1 only in each fl; perianth is **wide open**, cup- or saucer-(not bell) shaped, its segments **separate** to base.

A Sch 8 AWI **Bluebell** (Hyacinth in Scot), *Hyacinthoides non-scripta* (*Endymion non-scriptus*), hairless herb 20–50 cm tall; lvs linear, all basal, glossy-green with hooded tips, 20–40 cm x 1–2 cm; infl a **one-sided** raceme of **drooping** fls, each with **2 blue bracts** at base. Fls sky-blue (or white), **cylindrical-bell-shaped**; 6 perianth segments, **parallel** below, **curled back** at tips, are joined at **base** and bear filaments with **cream** anthers (**Aa**); capsule (**Ab**) oval, 15 mm long. Br Isles, vc except high mts, fens; **Orkney, Shetland, abs**; in wds, hbs, gslds, sea cliffs in W. Fl 4–5.

B **Hybrid Bluebell**, *H. x massartiana*, similar to A but the commonly grown gdn plant, usually fertile and sometimes mistaken for C; with broad lvs 1–3.5 cm wide and infl not one-sided, **spiralled around** stem but perianth segments **slightly** curled back (**Bb**); anthers blue (or cream in white forms); 3 **outer** filaments fused to perianth-lobes **at** or **just below the middle** of the lobe (**Ba**). Br Isles, f-lc gdn escape on wa, rds, wds. Fl 4–6.

C **Spanish Bluebell**, *H. hispanica*, similar to B, but fls **erect, widely bell-shaped**, perianth segment tips **not** curled or arched back, flared outwards; anthers **blue** and 3 **outer** filaments fused **well below** the middle (**C**). Introd (SW Eur); Br Isles, probably r and much recorded in error for B, as most gdn escapes are of the hybrid; in wa, wds. Fl 4–6. All Bluebell spp sometimes have pink or white fld forms.

D *****Spiked Star-of-Bethlehem**, *Ornithogalum pyrenaicum*, erect hairless per with **stout** stem 50–100 cm tall; lvs, linear, grey-green, all basal, 30–60 cm x 3–12 mm, wither before fls open; raceme **dense**, many fld; bracts of fls whitish, short; perianth segments 6–10 mm long, separate, narrow, **spreading starwise**, greenish-**cream** with green stripes, filaments lanceolate, **narrowed** to anther; capsule **oval**, 8 mm long. Eng: E Somerset to Wilts and Berks, f-la; Bedford, r but lf; r introd elsewhere in Eng; in wds, hbs, on calc or clay soils. Fl 6–7.

▼ **E** **Star-of-Bethlehem**, *O. angustifolium* (*O. umbellatum*), has lvs **shorter** (15–30 cm), **narrower** (6 mm) than in D, with white stripes down midrib, < 35 lvs per bulb. Fls in a short, **umbel-like** raceme, lower fls on long ascending stalks, nearly **level** with upper ones; fls **erect, cup-shaped, larger** than in D; perianth segments 15–20 mm long, **white**, with **conspicuous** green stripe on back; capsule **obovate**, six-angled; bracts **2–3 cm long**. Br Isles: GB o-lc, probably native in E Anglia, otherwise introd; E Ire vr; in dry gslds, hbs, wa. Fl 4–6. **E.1** **Greater Star-of-Bethlehem**, *O. umbellatum*, is now recognised as a separate sp and is probably more c than E. **Key differences** are outer perianth segments **20–30 mm long** and lvs < **10 per bulb**. Introd (S Eur); Br Isles distribution unclear due to confusion with E but probably f-c in S Eng and scattered elsewhere; on grassy places, rds. Fl 4–6.

E.2 Drooping Star-of-Bethlehem, *O. nutans*, like D has fls in long racemes, but **all turn to one side**; separate fls as in E, but **larger, drooping**; perianth segments 2–3 cm long, **bell-shaped**, with green stripe on back v broad, white edges narrow. Filaments v broad, **two-toothed at tips, anther in the notch**; capsule **pendulous**. Introd (Balkans); Mid, E Eng, r-lf; in wa, gslds. Fl 4–5.

▼ **F Bog Asphodel**, *Narthecium ossifragum*, creeping hairless per with erect fl-stems 10–40 cm tall; lvs mostly basal, ± curved, 5–20 cm long, **all flattened in one plane** as in gdn Iris; stem-lvs shorter, sheathing. Infl an erect raceme; bracts of fls 5–10 mm, equalling fl-stalks; fls **star-like**, with 6 **bright yellow** lanceolate perianth segments 6–8 mm long; 6 **orange-red** anthers, filaments **orange-woolly** (**Fa**); capsules elliptical, 12 mm long, turning orange like rest of plant in fr. Br Isles, uplands of W, N, f-lvc, hth areas of S Eng f; **S mid, E Eng** (except Norfolk vlf), ± **abs**; in wet hths, *Sphagnum* bogs. Fl 7–8.

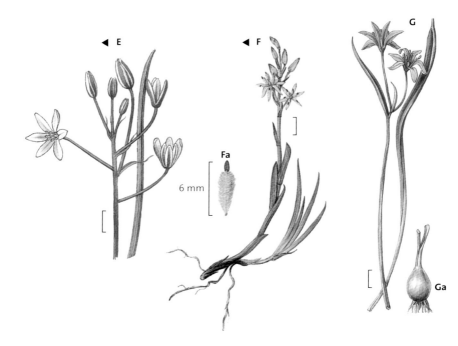

G AWI **Yellow Star-of-Bethlehem**, *Gagea lutea*, almost hairless, 8–25 cm tall, stem arising from a bulb (**Ga**); the **solitary** linear basal lf, 15–40 x 0.5–1.2 cm, is like those of A p 510, but hooded tip ± **curled** over like a crook, with 3–5 ridges on back; **2–3 bracts**. Fls **golden-yellow** inside, **star-like** when fresh, fading **pale** yellow and partially closing up to show green stripe down outside of each of 6 narrow oblong perianth segments that are 10–15 mm long. Br Isles: GB: from central S Eng and E Anglia to E Scot, o but vla; **SE, SW Eng, Wales, Ire, abs**; in wds on basic soils. Fl 3–5. **G.1** Sch 8 VU ****Early Star-of-Bethlehem**, *G. bohemica*, similar to G but **key difference** is **2 basal lvs** thread like and wavy; usually **4-6 bracts. Wales only**, vr on ledges and cracks on calc rocks. Fl 1–3.

▼ **A** ***Autumn Squill**, *Scilla autumnalis*, ± hairless per 4–20 cm tall; lvs all basal, narrow-linear, 4–15 cm long x 1–2 cm, appearing with the fls; infl a dense erect 4–20-fld raceme. Fls stalked, **without bracts**; 6 spreading oblong **dull purple** perianth segments 4–6 mm long; anthers **purple**. S Eng, Cornwall to Essex, r; on dunes, sea cliffs, dry gravelly gsld inland. Fl 7–9.

▼ **B** **Spring Squill**, *S. verna*, close to A, but fl **in spring** and linear curly lvs appear **before** fls: infl a 2–12-fld raceme, dense, round-topped, 5–15 cm tall. Fls each with 1 **bluish-purple bract** longer than fl-stalk; perianth segments **pointed, pale violet-blue**, 5–8 mm long; anthers **violet-blue**. GB: W coast from IoS, Cornwall, S Devon N to Hebrides, and in Shetland, f-la; NE Scot, Northumb, N, E, Ire, lf; in short gsld on rocky sea cliffs. Fl 4–5.

Garlics, Leeks, Onions (*Allium*), erect per herbs arising from bulbs; **onion smell** when bruised. Fls in terminal **umbels**, a **spathe** of one or more papery or greenish **bracts** below umbel; **bulbils** mixed with fls in some spp; perianth segments and stamens 6 each, ovary **superior**.

KEY TO GARLICS, LEEKS, ONIONS etc (*ALLIUM*)

1 Lvs elliptical, long-stalked – fls white..*Allium ursinum* (p 515 **C**)
Lvs narrow, grass-like or ± cylindrical, stalkless – various colours..**2**

2 Infl-stalk three-angled **3**
Infl-stalk round **4**

3 Infl without bulbils among the white fls – lvs 2–5.....................................*A. triquetrum* (p 515 **F**)
Infl with many bulbils among the white fls or bulbils only – one lf only ....
A. paradoxum (p 515 **F.1**)

4 Lvs flat, solid **5**
Lvs ± round or half-cylindrical **9**

5 Robust plants 60–200 cm tall – stems stout – lvs 12–35 mm wide – spathe of one bract only, longer than umbel – umbel 7–10 cm across..**6**
Slender plants 30–80 cm tall – stems thin – lvs 4–15 mm wide – spathe of 1–4 bracts – umbel 5–6 cm across **7**

6 Umbel globular, many-fld, with no or with few small bulbils........*A. ampeloprasum* (p 515 **E**)
Umbel irregular, few-fld, with many large bulbils ....
A. ampeloprasum var *babingtonii* (p 515 **E.1**)

7 Spathe of 2 long-pointed bracts much longer than umbel – fls pink (or bulbils only) – stamens long-protruding – o introd – in wds, on river banks ......**Keeled Garlic** *A. carinatum*
As above, but fls whitish – stamens not protruding ....*A. oleraceum* (p 515 **D.1**)
Spathe of 2 or more short bracts – stamens not protruding **8**

8 Lvs with keel on back, rough-edged – perianth segments abs (bulbils only) or 5–8 mm, red-purple ....*A. scorodoprasum* (p 515 **D.2**)
Lvs without keel, smooth-edged – perianth segments 10–12 mm, bright pink (fls with or without bulbils) – introd S Eng, o (IoS, CI c) in wa, ar ....**Rosy Garlic** *A. roseum*

9 Umbels without bulbils among fls **10**
Umbels with bulbils among fls or of bulbils only **12**

10 Lvs cylindrical, grey-green, – perianth segments 7–14 mm long, spreading, pale pink-purple – fl-stalks shorter than fls – stamens shorter than perianth segments ....
A. schoenoprasum (p 515 **G**)
Lvs half-cylindrical, grooved, green – perianth segments 3–6 mm long, erect – fl-stalks ± equalling fls – stamens at least as long as perianth segments **11**

11 Spathe of 1 bract ....*A. vineale* (p 515 **D**)
Spathe of 2 lf-like pointed bracts ....*A. sphaerocephalon* (p 516 **G.1**)

12 Stamens shorter than perianth segments – filaments of stamens undivided, slender ....
A. oleraceum (p 515 **D.1**)
Stamens at least as long as perianth segments – filaments of stamens broad, divided into 3 teeth, anther on the shorter middle tooth **13**

13 Spathe of 1 bract ....*A. vineale* (p 515 **D**)
Spathe of 2 lf-like pointed bracts ....*A. sphaerocephalon* (p 516 **G.1**)

C AWI **Ramsons**, *Allium ursinum*, per herb to 45 cm; 2–3 lvs, **elliptical-oval**, 10–25 cm x 4–7 cm, bright green, pointed, on **long stalk** twisted through 180°; stem 10–45 cm tall, weakly three-angled; spathe of 2 papery bracts shorter than fls; umbel 6–20 fld, ± flat-topped, **bulbils abs**; perianth segments **white**, 8–10 mm long, lanceolate; stamens with short, narrow stalks. Br Isles, c (except NE Scot, Ire, f); in moist wds, hbs, especially on calc or richer soils. Fl 4–6.

D **Wild Onion (Crow Garlic)**, *A. vineale*, the commonest narrow-lvd *Allium*, v variable per herb to 120 cm; lvs **half-cylindrical**, grooved above, **hollow**, smooth, 20–60 cm x 2 mm; stem 30–80 cm tall, round; spathe of 1 papery bract, no longer than fls, 1-valved; umbel has oval, green-purple **bulbils** (**Da**) usually mixed with fls (**Db**), sometimes bulbils only (var *compactum*), rarely fls only; fl-stalks 1–2 cm long, fls with pink or greenish-white oblong perianth segments 5 mm long; stamens at least **as long as** perianth segments, filaments of inner 3 three-toothed, with anther on tip of middle tooth which is only half as long as the others. Br Isles: GB, c to S, r to N; Ire o; on dunes, rds, ar. Fl 6–7. **D.1** VU **Field Garlic**, *A. oleraceum*, differs from D in its spathe of **2 v long-pointed** lf-like **bracts**, **much longer** than umbel of bulbils and fls (rarely bulbils only); fls long-stalked; stamens not protruding from perianth, their stalks untoothed. Br Isles: GB, mid, N Eng, f (elsewhere r); Ire vr; on dry gslds. Fl 7–8. **D.2 Sand Leek**, *A. scorodoprasum*, rather like D, but lvs **flat**, **solid**, linear, keeled **rough-edged**, spathe of **2** bracts **shorter** than umbel; few **reddish-purple** fls, with stamens as in D, but **enclosed** in perianth; purple bulbils. GB: N Eng to central Scot, o; scattered elsewhere, vr; dry gslds, rds. Fl 5–8.

E *Wild Leek**, *A. ampeloprasum*, tall (60–200 cm) **robust** per herb with **stout** cylindrical stem; lvs **flat**, but hollow, **keeled**, 15–60 cm x 12–35 mm, rough-edged, waxy grey-green. Spathe of **1** papery bract, falling as fls open; umbels 7–10 cm across, globose, many-fld, bulbils none or few; perianth (**Ea**) bell-shaped, pale purple, 6–8 mm long; stamens slightly protruding; style (**Eb**) protruding. GB: S Eng, Cornwall, Somerset, S Wales, IoS, CI vr; on dunes, hbs, rocky gd near sea. Fl 7–8. **E.1** *A. ampeloprasum* var *babingtonii*, resembles E, but umbels **irregular**, with **many** short-stalked **large bulbils** and **few** fls, some of these in clusters in stalked secondary umbels. Br Isles: Cornwall, IoS, NW Ire; on rocks and dunes near sea. Fl 7–8.

F **Three-cornered Garlic**, *A. triquetrum*, has **linear, flat to keeled** lvs 12–20 cm x 5–10 mm, and erect strongly **three-angled stem** 20–50 cm tall; spathe of 2 narrow papery bracts **shorter** than fls; umbel of **drooping** long-stalked fls, **no bulbils**; perianth segments, 12–18 mm long, **white** with **green** stripe outside, form a bell longer than stamens. Introd (Mediterranean) Br Isles: SW Eng, IoS, CI vc; r and scattered elsewhere in Eng, Wales; Scot vr; Ire o especially near coasts; on wa, hbs, rds, scrub. Fl 4–6. **F.1 Few-flowered Garlic**, *A. paradoxum*, differs from F in having only **one** lf, and **few fls**, but **many bulbils** (or bulbils only), per umbel. Introd (Caucasus); GB: o-vla throughout; central Scot lf; N Ire r; on rds, wds, riverbanks, wa. Fl 4–5.

G FPO (Eire) *Chives**, *A. schoenoprasum*, tufted per herb with **cylindrical hollow** lvs all from base, 10–25 cm x 1–3 mm; stem 15–40 cm, **cylindrical**, with **dense** umbel of pale purplish-pink fls and **no bulbils**; perianth segments 7–14 mm, spreading, **longer** than stamens, rather **thin and papery**; spathe of **2 short bracts**. GB:

Cornwall, S Wales, Wye Valley, N Eng, W Ire (Mayo) r as a native; rest of Br Isles o introd gdn escape; on rocky calc gslds or rock outcrops in rivers. Fl 6–7. **G.1** Sch 8 VU ****Round-headed Leek**, *A. sphaerocephalon*, taller (30–80 cm), with lvs **half**-cylindrical, hollow, but **grooved** on upper side; umbel dense, **± spherical**, 2.0–2.5 cm across, with spathe of 2 short bracts, 2-valved; fls many, bulbils **abs** as in G; perianth segments **5 mm long**, **red-purple**; stamens at least **as long as** perianth segments. Eng: Avon Gorge near Bristol, CI as a native, vr; vvr introd elsewhere in S Eng; on limestone rocks, coastal dunes. Fl 6–8.

Daffodils, **Snowdrops** & **Snowflakes** are bulbous herbs, sometimes placed in their own family (*Amaryllidaceae*), because ovary always **inferior**, unlike other *Liliaceae*; fls (or fl-clusters) in bud enclosed in a **spathe**; perianth segments 6, stamens 6, style single, ovary three-celled. In *Narcissus* perianth consists of 6 segments around a central **trumpet**.

A   AWI **Daffodil**, *Narcissus pseudonarcissus* ssp *pseudonarcissus*, hairless per herb with flat, linear, erect grey-green lvs 12–35 cm x 5–12 mm, and **solitary** fls, turned to one side, on ± flattened stalks 20–35 cm tall; fls 5–6 cm wide, with outer whorl of **6 pale yellow** perianth segments and inner **trumpet-shaped** tube 25–30 mm long, **golden yellow** with many small blunt lobes at its mouth; capsule obovate, 12–25 mm long. Eng, Wales, scattered but f-la especially in W and S; Scot, r; Ire, vr-abs; in wds, old gslds, on clay and loam soils. Fl 3–4. **A.1** Tenby Daffodil, *N. pseudonarcissus* ssp *obvallaris* (*N. obvallaris*), has ± erect fls, **both** trumpet **and** outer perianth segments **deep yellow**, and trumpet **expanding** to mouth, with **spreading** edges. Wales: Pembroke, vla; in mds. Fl 4. **A.2 Pheasant's-eye Daffodil**, *N. poeticus*, **sweet-scented**; outer whorl of perianth segments 5.5–7.0 cm wide, **white**, ± **flat**; trumpet **v short** (3 mm long), yellow, with **crisped red edge**. GB, introd (S Eur) o; in parks, wds Fl 4–5. **A.3   Primrose-peerless**, *N.* x *medioluteus* (*N.* x *biflorus*), has fls as A.2 in size and shape, but outer perianth segments **creamy-yellow**, trumpet short, **deep yellow**, and fls **2 together**. Eng, Wales, S Ire, introd (gdn origin); o in parks, mds. Fl 4–5.

Many **gdn Daffodil** varieties, often hybrids and usually with larger fls, occur as gdn escapes.

B   AWI **Snowdrop**, *Galanthus nivalis*, small bulbous herb 15–25 cm tall; lvs keeled, grey-green, linear, 4 mm wide, all basal; fls **solitary, drooping**; narrow oblong spathe notched into 2 lobes, green with broad scarious edges; the 3 outer perianth segments 14–17 mm long, obovate, blunt, pure white; the 3 inner half as long, deeply-notched, with **green spot near tip**; capsule oval. GB, f-vla but generally regarded as introd (perhaps native in Welsh Borders); in damp wds. Fl 1–3.

C   Summer Snowflake, *Leucojum aestivum*, **taller** (30–60 cm), **stouter** than B; with **3–7 fls** in a loose **umbel**; spathe **undivided**, 4–5 cm long, green; perianth segments all **alike and equal**, obovate, white with a green spot near tip, 14–18 mm long; capsule pear-shaped; fls 30–50 cm x 10–15 mm. The native plant is *ssp *aestivum* having 2 sharp stem-edges with **minute teeth**, whereas the introd gdn plant is ssp *pulchellum* which has **entire** stem-edges. Br Isles: o throughout, more c in S Eng, CI, IoS in swamps, wet mds, winter-flooded alder or willow wds by both fresh and tidal rivers. Fl 4–5.   **C.1 Spring Snowflake**, *L. vernum*, has perianth

segments all alike and equal as in C, but they are **larger** (20–25 mm long), fls are usually **solitary**, and the spathe **bilobed** at tip. Eng: W Dorset, Somerset, vr, possibly native; v scattered elsewhere as r introd; **Ire abs**; in wds, scrub, hbs. Fl 2–4.

IRIS FAMILY   *Iridaceae*

Iris family has fls with an **inferior** ovary; only **3 stamens**, and perianth segments joined into a tube below; style **three-lobed**.

D WO (NI) **Blue-eyed-grass**, *Sisyrinchium bermudiana*, hairless erect per herb with short rhizome; lvs linear, 1–3 mm wide, all basal; **flattened**, **winged** stem; fls 2–4 in a loose raceme with 2–3 **lf-like** bracts below; perianth segments 7 mm long, blue, obovate, with **mucronate** points; stamens joined below; capsule globular. W Ire, Kerry to Donegal, o-lf as a native; vr introd in GB; on marshy mds, lake shores. Fl 7–8.

A **Yellow Iris**, *Iris pseudacorus*, per herb with erect stems 40–150 cm, arising from a stout rhizome; lvs sword-shaped, ± equalling infl, 15–25 mm wide; infl ± branched, with fls 2–3 in each papery-tipped spathe; fls 8–10 cm across, bright **yellow**; falls oval to oblong, **unbearded**, veins purple; standards spoon-shaped, v short (2 cm); styles yellow; seeds brown. Br Isles, vc in fens, marshes, wet wds, by and in fresh water. Fl 5–7.

B AWI **Stinking Iris**, *I. foetidissima*, per **tufted** herb with erect stems 30–80 cm, and clumps of sword-shaped **evergreen** lvs ± equalling infl, 10–20 mm wide, smelling of **fresh meat** when crushed (hence local name of Roast-beef plant); fls in infls arranged as in A, but 6–8 cm across, falls obovate-lanceolate, **unbearded**, **dull grey-purple** at margins, **brownish-yellow** to centre, veins purple; standards spoon-shaped, brown-yellow, shorter (3–4 cm) than falls. Styles coloured as standards; capsule (**Ba**) club-shaped, 4–5 cm long, splitting into 3 segments when ripe to expose **orange-red** seeds. Br Isles: Eng and Wales, c in S, r to N, Ire, Scot o introd; in wds, hbs, scrub, on sea cliffs and inland, mainly on calc soils. Fl 5–7. **B.1 Bearded Iris**, *I. germanica*, has fls of various colours (natd form purple); falls **yellow-bearded** above, standards **erect**, **longer** than falls, **incurved**, oval: fls 10 cm or more wide; lvs 5 cm wide, grey-green; gdn escape, o-lf in various colour varieties in Br Isles, o (S Ire, vvr); on walls, wa. Fl 5–6.

C Sch 8 ****Wild Gladiolus**, *Gladiolus illyricus*, slender erect hairless per herb 40–90 cm, growing from a corm with **unbranched** stems; lvs at base and up stem, to 30 cm long x 1 cm wide, tapered to tips, all **flattened in one plane**, grey-green; **3–8 fls** in a long one-sided raceme, each fl with a spathe of 2 green, purple-tipped bracts; perianth segments 2.5–3.0 cm long, obovate, long-stalked, v unequal, the 3 lower narrower and pointing down to form a lip, 3 upper broader, forming a loose hood to fl, bright crimson-purple with paler streaks. Capsule obovate, three-angled; anthers only **half length** of their filaments; 3 stigmas suddenly swollen towards tips. Br Isles, New Forest only, lf; under bracken on hths and in open wds on acid loamy soils. Fl 6–7 (brief). **C.1** **Eastern Gladiolus**, *G. communis* ssp *byzantinus*, much more f than C, **key differences** are stems often **branched**; having more fls, **10–20**. Introd (Mediterranean), the most c cultivated European *Gladiolus* sp; SW Eng, CI, IoS f on ar, wa, rds. Fl 6–8.

D **Montbretia**, *Crocosmia* x *crocosmiiflora*, like C, but **densely** tufted and has runners; fls 2.0–5.0 cm across, deep orange, funnel-shaped. Introd (hybrid of gdn origin); Br Isles, vc to S and W; f-lc to E and N on rds, wds, hbs, cliffs. Fl 7–10. Several other *Crocosmia* are r–o natd.

E Sch 8 VU ****Sand Crocus**, *Romulea columnae*, small crocus-like herb growing from a corm; basal rosette of wiry, curly lvs 5–10 cm long x 1–2 mm wide, and 1 (to 3) fls on an erect stem 3–5 cm tall. Fls **star-shaped**, only open in sun, 7–10 mm wide; perianth segments **all alike**, **purplish-white inside** with darker veins and yellow spot at base, **greenish outside**, pointed; stamens 3. Eng: Dawlish, S Devon, S Cornwall vr but lf; in short turf on sea cliffs, dunes. Fl 4. **Crocuses** (*Crocus* spp) are sometimes natd and have erect showy cup- or funnel-shaped fls to 5 cm long; long pale stalks are in fact perianth-tubes arising from gd; lvs narrow, green with **white midrib**; **stamens 3**, **ovary inferior**. **E.1** **Autumn Crocus**, *Crocus*

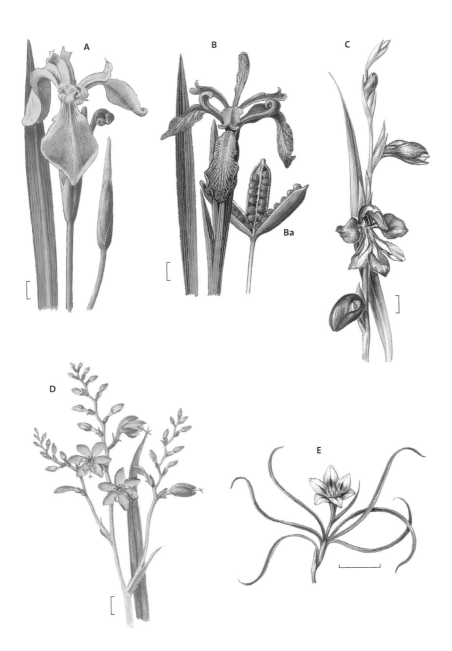

nudiflorus, can be confused with **Meadow Saffron** (p 508 E), as fls appear in 9–10, **lvs in spring only**, but Meadow Saffron has **6** stamens and is **pink**-fld (**not** purple as in E.1). **E.2 Spring Crocus**, *C. vernus* (*C. purpureus*) has **purple fls** in **spring**, appearing with the lvs. Introd (mts C Eur); GB, o-lf natd gdn escape in gsld, churchyards, hbs. Fl 3–4.

YAM FAMILY *Dioscoreaceae*

A mostly tropical family of creeping or twining and climbing monocotyledonous herbs with tuberous rhizomes and **cordate**, **net-veined** lvs; fls small, in racemes, **dioecious**, with **3 sepals and 3 petals all alike** and joined in a tube below; ovary **inferior**, 6 stamens in male fls. Fr (in Br Isles) a berry.

ᴬᵂᴵ **Black Bryony**, *Tamus communis*, climbing herb, twining clockwise, **without** tendrils; lvs **cordate**, pointed, 3–10 cm x 2.5–10.0 cm, hairless, long-stalked, v **glossy** dark green, with **palmate main veins** and **netted side-veins** (unlike most Monocotyledons except Herb-Paris (p 506 D), and Arum family (p 495). Fls in racemes in lf-axils, male (**b**) stalked, female ± stalkless, 4–5 mm across, bell-shaped, the 6 perianth lobes narrow, yellow-green; frs (**a**) red **poisonous** berries. Br Isles: Eng, Wales, vc (except in mt areas) N to S Cumbria and Durham; elsewhere vr-abs introd; in open wds, scrub, hbs, on richer or calc soils. Fl 5–7. **Do not confuse** with **White Bryony** (*Bryonia dioica* p 190), also a dioecious, climbing herb **but** with palmately-**lobed** lvs, **tendrils**, and **five**-petalled fls; frs also red berries.

ORCHID FAMILY   *Orchidaceae*

One of the largest and most distinctive families of flowering plants, with *c.* 18,500 spp worldwide; in tropics many are **epiphytes**, growing on trees, but in Br Isles all grow in the ground. Per herbs, with rhizomes or tuberous roots, and erect fl-stems; lvs undivided, untoothed, parallel-veined. Some orchids are **saprophytes**, ie, plants without, or with v little green **chlorophyll**, which, with lvs either **abs** or **reduced** to scales, feed on vegetable humus in the soil through the partnership of a fungus whose filaments penetrate their roots. Fls, in spikes or racemes, have perianth segments in 2 whorls: the **outer** of 3 sepals, ± alike, and usually coloured; the **inner** of 3 petals, but with one v different from the others, forming a **lip** to the fl, called the **labellum**. This is usually on lower side of fl, but sometimes on upper, or pointing upwards; it may be much lobed, sometimes furry, insect-like, sometimes pouch-like, often of different colours from rest of fl. Ovary **inferior**, in effect forming part of fl-stalk. Except in **Lady's-slipper**, (with 2 anthers) there is only 1 stalkless anther at back of fl, which forms two stalked, or stalkless, pollen-masses called **pollinia**. These normally have sticky bases which can adhere to heads of insects visiting fls. Below the anther and above the concave, stalkless stigma a projection, called the beak or **rostellum**, normally prevents self-pollination – but in some spp beak is abs and fls can pollinate themselves if not visited by insects. Insects are attracted to the fls by colour, scent, shape, or by **nectar** often secreted in the hollow **spur** developed in many spp at rear of labellum. There is evidence that male bees, wasps etc may mistake the bizarre insect-like labella of some orchids for females of their own species, and attempt copulation, so removing pollen from the fl and carrying it to another, effecting fertilization of the ovules. Fls of some spp are so specialized in shape, size and attractiveness, that only one sp (or few) of flying insect are able to extract nectar from their spurs and so cross-pollinate them. After fertilization the ovary ripens into a one-celled capsule containing a large number of minute, dust-like seeds, which can be wind-dispersed for long distances, and may account for the unexpected appearance of orchid colonies in totally new places.

Many orchids are v rare; **none** should ever be picked, only photographed or painted in the field. Some only produce fl-stems (or even lvs) at intervals, so may incorrectly be supposed to have disappeared from a locality. Many are found only in chky soils, but some in bogs, fens, on acid hths and moors, or in the deep peaty humus of pine or beech wds.

A note on Orchid classification and names
Recent genetic work on the classification of orchids has resulted in some species being moved into different genera to reflect their evolutionary relationships, and several new species being recognized. These new names and additional species are not included in this book as the classification follows the *New Flora*, simply because this is the most widely available and used text. The alternative genetic classification is used in *Orchids of the British Isles* (2005) by M. Foley & S. Clarke, which includes keys.

ID Tips   Orchids

- Orchids are the only European Monocotyledons with **inferior** ovaries and **irregular** fls with a definite lip, except *Gladiolus*, in which lip formed of 3 (not 1) perianth segments.

- **Do not confuse** the non-green saprophytic orchids, such as **Bird's Nest Orchid**, with either parasitic **Broomrapes** (*Orobanche* spp)(p 407), or with **Yellow Bird's-nest** (*Monotropa hypopitys*)(p 230):
  Broomrapes have a superior ovary, a **tubular** corolla, all in one piece, with lip formed of 3 corolla-lobes and only 2 lobes arching over it, and 4 long-stalked stamens.
  Yellow Bird's-nest has **regular** fls and corolla-tube divided into 4 or 5 **equal** lobes and **no lip**.

- Some members of Deadnettle and Figwort families have rather orchid-like fls with elaborate lips, but these can be distinguished by their **superior** ovaries, their usually **broad**, **net-veined** lvs, **long-stalked** stamens and **long** styles.

- The **Helleborines** (*Epipactis* & *Cephalanthera*) are the only orchids **without** basal lvs in rosettes.

- This key does not include the **Tongue-orchids** (*Serapias* spp) that are possibly native but vvr in CI, Cornwall and Suffolk only (see *Plant Crib*).

1   **Fl-stalk**
2   **Ovary**
3   **Spur**
4   **Labellum**
5   **Lateral sepals**
6   **Dorsal sepal**
7   **Lateral petals**
8   **Pollinia**
9   **Rostellum**
10 **Stigma**

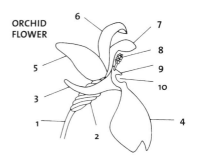

ORCHID FLOWER

KEY TO ORCHID FAMILY

1   Fls v large, solitary or only 2 on stem – lip pouch-shaped, 3 cm or more long, hollow, yellow – other perianth segments 6–9 cm long – vvr............*Cypripedium calceolus* (p 528 **A**)
    Fls in spikes or racemes, rarely 1 alone – lip not pouch-shaped, not hollow – other perianth segments not over 2 cm long **2**

2   Plants without green lvs (though stem and fls may be greenish) – lvs reduced to small scales up stem..**3**
    Plants with well-developed green lvs at base of, or up stem **5**

3   Lip three-lobed, mid-lobe cordate, pink, on upper side of fl, spurred – other perianth segments banana-flesh-coloured – vr...*Epipogium aphyllum* (p 530 **E**)
    Lip pointing downwards, on lower side of fl **4**

4 Plant wholly honey-brown – lip unspurred, two-lobed, to 12 mm long – other perianth segments forming a hood – fls to 10 mm wide ....*Neottia nidus-avis* (p 530 **D**)
Plant with yellow-green stem – lip unspurred, three-lobed, to 5 mm long, white, side-lobes v small – other perianth segments forming a loose hood, yellow-green – fls to 6 mm wide...*Corallorhiza trifida* (p 532 **A**)

5 Fls with spur or pouch extending back (or down) from rear of lip **6**
Fls without spur or pouch at back of lip **13**

6 Tall (30–90 cm) plant with mid-lobe of lip 3–5 cm long, coiled in bud like a spring, ribbon-like, twisted in fl – fls in long spikes, smelling strong and unpleasant ....
....*Himantoglossum hircinum* (p 540 **A**)
Lip of fl under 2 cm long, not coiled or twisted **7**

7 Lip long strap-shaped, unlobed, to 15 mm long, greenish-white – spur 15–30 mm long – fls sweetly-scented – stem with 2 oval-elliptical lvs at base but sometimes v small upper lvs on stem also...*Platanthera* (pp 538–40)
Lip toothed or lobed, not long strap-shaped..**8**

8 Spur v short (3 mm or less), blunt, pouch-like – lip short (6 mm or less), unspotted, except sometimes at its base – fl-spikes dense, perianth segments forming a close hood to fl **9**
Spur conical or cylindrical, usually over 5 mm long (if under 2 mm, lip white, three-lobed, bearing red spots on lobes)...**10**

9 Fls yellow-green to reddish-brown – lip rectangular in outline, hanging down, with 3 v short teeth at tip – widespread but o ....*Coeloglossum viride* (p 540 **E**)
Fls greenish (or pinkish-white) – lip pointing forward, with 2 side-lobes shorter than three-toothed mid-lobe, pink or whitish, sometimes purple-spotted at base of lip only – Ire, IoM only, r ....*Neotinea maculata* (p 540 **B**)
Fls creamy-white – lip with 3 ± equal lobes, curved downward ....*Pseudorchis albida* (p 540 **D**)

10 Spur 12 mm long or more, cylindrical, v slender – lip ± equally three-lobed, fls wholly pink or reddish, unspotted (as are lvs) **11**
Spur under 10 mm long, conical, stout – lip variously lobed but never equally three-lobed, fls either spotted or lined with darker markings, or lip with a whitish patch on a darker ground...**12**

11 Lip divided nearly to base into 3 ± equal lobes, and with 2 ± erect ridges running out onto its base – fls smell foxy – lvs grey-green, arranged spirally up stem ....
....*Anacamptis pyramidalis* (p 536 **H**)
Lip only divided less than halfway into 3 ± equal lobes (or outer lobes longer), and with no ridges – fls v fragrant – lvs shiny bright green, arranged in 2 opp ranks on lower part of stem – fl-spike long, cylindrical..*Gymnadenia* (p 538)

12 Fl-spike in bud emerging from gd wrapped round with spathe-like lvs – bracts of fls narrow (1–2 mm wide), short, membrane-like – lvs arranged spirally around base and up stem ....*Orchis* (key p 534)
Fl-spike in bud emerging from gd naked – bracts of fls broad (3–5 mm wide), long, like small lvs – lvs in 2 ± opp ranks at base of, and up stem ....*Dactylorhiza* (key p 541)

13 Plants with 2 ± equal lvs in an opp pair at base of, or above base of stem – no (or only 1–2 tiny) stem-lvs above this pair **14**
Plants with lvs all alt, never in equal opp pairs...**15**

14 Plants with the opp pair of lvs spreading at a wide angle from a point on the stem several cm above gd – lips of fls long, forked, on lower side of fl ....*Listera* (p 532)
Plants with the opp pair of lvs spreading at an angle up to 45° from base of stem – lips of fls unlobed, on upper side of scentless fl – in fens....................................*Liparis loeselii* (p 532 **D**)

Plants with the opp pair of lvs spreading ± flat on gd from base of stem – 1 or 2 smaller alt stem-lvs usually present – lips of fls v short, three-lobed, on lower side of honey-scented fl ....*Herminium monorchis* (p 538 **A**)

15 Plants with at least some lvs arising at (or beside) base of stem, often in a rosette **16**
Plants with no lvs arising from base of stem, stem ± scaly, but lfless at base, largest lvs well above stem-base **19**

16 Tiny plant (usually under 12 cm) – 3–5 blunt obovate lvs under 1 cm long, one above the other at base of stem – fls green, measure 7 mm or less top to bottom, flat-faced, lips on uppersides – in sphagnum bog pools – r......................................*Hammarbya paludosa* (p 532 **E**)
Taller plants (usually over 12 cm) – lvs, all over 3 cm long, at base of (or in a rosette adjacent to base of) stem – lips on lower sides of fls **17**

17 Lips of fls velvety-brown, often convex on upper side, insect-like, with a ± complex pattern of lines and patches of different colours – fls few (2–6, rarely more) in a v loose spike..........
Ophrys (pp 545–6)
Lips of fls like a tiny figure, with 2 arms and 2 legs, hanging down from lower side of fl, hairless, yellowish, often brownish-edged – the other perianth segments forming a close green (often red edged) helmet – fls many, in a dense spike ....
Aceras anthropophorum (p 536 **G**)
Lips of fls small, undivided, forward-pointing, (not insect or man-shaped) – fls whitish, downy, small, in 1 or more spirally-twisted rows, the fl-stalks together resembling a plait of green hair...**18**

18 Plant with obvious creeping runners – lvs stalked, oval, their cross-veins conspicuous between main parallel veins – lips of fls with untoothed edges, spout-shaped ....
Goodyera repens (p 530 **C**)
Plant without creeping runners – lvs stalkless, oval-lanceolate, cross-veins not obvious – lips of fls with frilled edges, curved downwards ....*Spiranthes* (p 530)

19 Fls large (2 cm or more long), white (rarely rosy-red), not opening widely, held erect or ascending – ovaries twisted – few (3–15) fls per spike – lip formed of an inner and an outer joint (but inner joint not forming a cup) ....*Cephalanthera* (p 528)
Fls smaller (under 1 cm long), greenish to brick-red, ± wide open, held horizontal or drooping – ovaries not twisted – many (10–100) fls per spike – lip of 2 joints, but inner joint forming a cup, outer triangular and ± bent downwards ....*Epipactis* (key p 525)

Helleborines (*Epipactis*) have **no basal rosette-lvs**, only stem-lvs, of which those on mid-stem are larger than those above and below. Fls are usually many (sometimes only few), in long dense spikes. Each fl has 3 pointed outer perianth segments and 2 inner ones, which are usually all ± alike, and wide-spreading in most spp (**not** spreading in F p 526), and a lip on the lower side of each fl. The lip is jointed into 2 parts, an inner, usually ± cup-shaped **hypochile** usually containing glistening nectar, and an outer, triangular or cordate **epichile**, which is often turned downwards in a flap. The ovary below fl is straight, not twisted, and pear-shaped in fr; the two **pollinia** are **unstalked**, and lie side-by-side on top of the **beak**, but in D, E and F (p526) there is **no** beak in the mature fl, so the pollen-masses can fall on to the hollow **stigma** below and effect self-pollination. In A, B, C and G, the projecting beak **persists** below the anther, preventing self-pollination, so that these spp are dependent on wasps (or, in G, on bees) for cross-pollination.

ID Tips   Helleborines

- **What to look at: (1)** whether lvs spiral or in 2 ranks **(2)** lf-shape, length in relation to internodes and colour **(3)** colour of perianth segments **(4)** colour and shape of lip-epichile.

- **Green-flowered, Narrow-lipped** and **Young's Helleborine** are all self-pollinated. Therefore if you insert a pencil tip into a newly-opened fl, the pollinia are not drawn out stuck to it in one piece; instead, they crumble and can not be easily pulled out whole.

KEY TO HELLEBORINES (*EPIPACTIS*)

1 Fen and marsh plant (also calc gsld) – fls brownish-grey outside, purplish-red within – lip-epichile large (to 10 mm long x 7 mm wide), white, frilly-edged, blunt-tipped, with yellow spot at base, flexibly hinged to lip-hypochile ....*Epipactis palustris* (p 528 **G**)
Wdland, scrub and rock-crevice plants – fls green-purplish outside and within – lip-epichile smaller (5 mm wide or less), greenish-reddish, edges untoothed, never frilly, tip pointed (but sometimes curved under), never yellow-spotted, fixed ± rigidly to lip-hypochile **2**

2 Fls wholly brick- to purple-red – lip-epichile wider than long, with wrinkled, red swellings near base – stem reddish above but with dense whitish down – ovary v downy – beak present ....*E. atrorubens* (p 526 **B**)
Fls greenish or yellowish to purplish-flushed only – lip-epichile with no (or only smooth) swellings near base – stem hairless or slightly downy, not reddish above – ovary ± hairless..**3**

3 Lvs spirally arranged up stem...**4**
Lvs in 2 opp ranks (but not in opp pairs) up stem **5**

4 Stems clustered – lvs oval-lanceolate to lanceolate, ascending at an acute angle, grey-green above, purple-flushed below and on stem – hypochile-cup of lip violet-mottled – perianth segments whitish-green, lip-epichile pinkish-white ....*E. purpurata* (p 526 **C**)
Stems single or up to 3 together – lower lvs broad-oval, ± horizontally spreading, upper narrower, all dull green both sides – hypochile-cup dark red or green, not mottled – perianth segments green or purplish, lip-epichile often purple ....*E. helleborine* (p 525 **A**)

5 Lvs often shorter than internodes, fresh green, with v weak veins – upper stem within infl ± hairless – fls drooping, not opening fully – lip-hypochile almost flat...................................
....*E. phyllanthes* (p 526 **F**)
Lvs usually longer than internodes, yellowish-green, with strong, ± pleated veins – upper stem within infl hairy – fls ± horizontal – lip-hypochile cup-like...........................(*E muelleri*) **6**

6 Fls widely open, lip-epichile longer than wide, ± flat, narrow-triangular point, greenish-yellow – in wds on calc soils ....*E. muelleri* var *muelleri* (p 526 **D**)
Fls not opening widely, lip-epichile wider than long, tip curved under, greenish-white – in dune hollows ....*E. muelleri* var *dunensis* (p 526 **E**)

▼ **A** AWI **Broad-leaved Helleborine**, *Epipactis helleborine*, robust erect per, 25–80 cm, with only 1–3 stems arising together; stem ± downy above; no lf-rosette at base, lowest lvs are no more than sheathing **scales**; largest lvs on mid-stem 5–17 cm long x 2.5–10.0 cm wide, **broad-oval**, ± **horizontally** spreading, upper smaller and narrower, all strongly veined, **dull green** both sides, spirally arranged up stem. Infl dense raceme 7–30 cm long with 30–100 **wide-open** fls, held **horizontal** or ±

inclined; bracts narrow, lowest longer than fls; outer perianth segments (**Aa**) 1 cm long, oval-lanceolate, green-purplish; lip-hypochile **cup-shaped**, dark red or green within, **not mottled**; epichile (**Aa**) cordate, **broader than long**, green to purple, tip turned **under**, swellings on lip **smooth**; beak **well-developed**; ovary drooping in fr, ± hairless. Br Isles: GB, N to mid Scot f-lc; NW Scot r; Ire o-lf; in wds, dune hollows. Fl 7–9.   **A.1**   Sch 8 End BAP ****Young's Helleborine,** *Epipactis youngiana*, a sp only described in 1982 (but now regarded as a var of A), differs from A in having upper lvs of **2-ranks** (not spirally arranged), and from D in having **pinkish** (not green) perianth segments and stigma with 2 basal bosses forming a **3-horned** shape. Br Isles: Wales, S Scot, Northumberland and Lanarks only, vr; in wds on heavy soils. Fl 7–8.

B   ***Dark-red Helleborine**, *E. atrorubens*, shorter than A (only 15–40 cm); stem **reddish** above, but with **dense white down**; lvs in two opp ranks, largest oval-elliptical, pointed, ± reddish-tinged, more folded than in A; raceme dense, 8–20-fld, fls (**Ba**) **wholly red-purple** or **brick-red**, less wide-open than in A; hypochile cup **red-spotted** within; epichile with tip turned under, swellings on lip **wrinkled**, **bright red**; beak well-developed; ovary drooping in fr, **v downy**, **purplish**. Br Isles: N Eng, N Wales, NW Scot, W Ire, r-lf; in wds, rocks, screes, on limestone; on chk downs, open wds on limestone. Fl 5–8.

C   AWI **Violet Helleborine**, *E. purpurata*, robust like A, but many **stems clustered** together; lvs (**C**) in spiral up stem as in A, but **narrower** (oval-lanceolate), **ascending** at an angle, **grey-green** above, ± **purple-flushed** below (and on stem), **strong-veined**; fls (**Ca**) wide-open, horizontal, in dense racemes; outer perianth segments **greenish-white**, lanceolate; hypochile-cup **violet-mottled** within; epichile cordate, **dull whitish**, with **purple streak** in centre; lip swellings smooth; ovary rough, ± hairy, horizontal in fr; beak well-developed; bracts longer, purplish. Eng: SE, S central, o-lf; in wds (mostly beech) on clay, often over chk. Fl 8.

D   AWI ***Narrow-lipped Helleborine**, *E. muelleri* var *muelleri* (*E. leptochila*), tall like A, but more slender; upper stem within infl hairy; lvs in 2 **opp ranks**, oval-lanceolate, 10 cm long, usually longer than internodes, **yellow-green**, **strong-veined**; fls (**D**) **wide open**, horizontal, **greenish-yellow**, lip-epichile longer than wide, **straight**, **narrow-triangular** point; hypochile with a cup; beak scarcely developed. Eng: Kent to Chilterns and Cotswolds, r but lf; in beechwds on calc soils. Fl 6–7.

E   **Dune Helleborine**, *E. muelleri* var *dunensis* (*E. dunensis*), close to D in colour, lvs, and lack of beak to fls; but **key differences** in fls (**E**) **less open**, lip-epichile **wider than long** with tip **curved under**, greenish-white, ± pink-tinged. GB: Lancs, Northumberland, Anglesey, r-lf; in dune hollows, dune pine wds. Fl 6–7.

F   WO (NI) AWI ***Green-flowered Helleborine**, *E. phyllanthes*, v variable slender plant 20–40 cm; upper stem within infl ± hairless; lvs in 2 **opp ranks**, **fresh green**, oval-elliptical, **weakly-veined**, to 6 cm long, often shorter than internodes. Fls (**Fa**) **drooping**, **rarely widely open**, **yellow-green**, lip-hypochile with cup scarcely hollowed, epichile **flat**, longer than broad, **greenish**, **without swellings**; **beak abs**; v variable; fls in some forms never open. Br Isles: Eng, Wales, N to Cumbria, r, but lf in S Eng and on Lancs coast; Ire vr; in beechwds on calc soils, dune hollows etc. Fl 7–9.

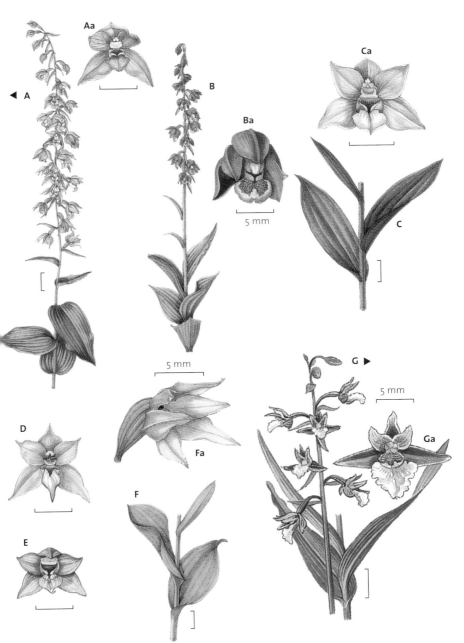

Aa

◀ A

B

Ba

5 mm

Ca

C

G ▶

5 mm

Ga

5 mm

D

Fa

E

F

▲ **G** WO (NI) **Marsh Helleborine**, *Epipactis palustris*, per herb 15–50 cm, with runners; stem downy above; lvs oblong-oval to lanceolate, upper ascending at an angle to 15 cm long, concave above and folded, veins strong and pleated; infl 7–15 cm, with 7–14 fls in a **loose** raceme; bracts lanceolate, not longer than fls; fls (**Ga**) wide open, to 15 mm wide, horizontal in fl, drooping in bud; perianth segments **brownish-grey**, **downy** outside, **purplish-red** within; lip-epichile to 10 mm long, cordate-oval, blunt, white, **frilly-edged**, with a **yellow spot at base**, hypochile joined to it by narrow **hinge**, its cup white with **red** veins; beak well-developed. Br Isles: Eng, Wales, Ire, r-lf (lc in Norfolk); Scot, E only, vr; in calc fens, calc dune-hollows and calc gsld that is probably wetter in winter. Fl 7–8.

A EU Sch 8 BAP CR ****Lady's-slipper**, *Cypripedium calceolus*, a striking plant; per herb, 15–40 cm, erect, ± downy; lvs oval-oblong, sheathing, pointed, fresh green, strongly pleated-veined. Fls usually 1 (rarely 2) per stem; bract large, lf-like; perianth segments **maroon**, 6–9 cm long, elliptical-lanceolate, the 2 lower outer ones conjoined; lip 3–5 cm long, obovate, bright golden-yellow, **hollow**, like a pouch or blunt slipper, red-spotted inside; column in centre of fl spotted, **without** an anther on top as in most orchids, but with **2 anthers** below it, one on each side; ovary downy. W Yorks **only** vvr; in wds on limestone. Fl 5–6.

White and Red Helleborines (*Cephalanthera*) resemble *Epipactis* in both absence of basal or rosette-lvs, and jointing of fl-lips into 2 segments (though this is less obvious in *Cephalanthera*); but fls much **larger** (2 cm or more long), ± oval, **rarely** widely open, white or rosy-red; ovary **twisted.**

B VU AWI **White Helleborine**, *Cephalanthera damasonium*, erect hairless per herb, 15–50 cm tall; angled stem bears brownish sheathing scales below and oval to elliptical-lanceolate lvs 5–10 cm long above, held at an angle of *c.* 45° to stem, **never** drooping at tips and **not** exceeding the fl-spike. Infl of 3–12 fls, each with a **lf-like bract longer than** the lower fls; fls (**Ba**) **creamy**, 2.0–2.5 cm long, ± **erect**, elliptical, **scarcely opening**; lip with orange blotch on hypochile and a curved, blunt-toothed cordate epichile with orange ridges along it (best seen by easing fl open with a pencil). Ovary hairless, ± erect; perianth segments **blunt**. Eng, calc areas of SE, S central Eng, N to Hereford and Cambridge, o; **rest of Br Isles abs**; in beechwds on calc soils. Fl 6–7.

C VU FPO (Eire) AWI ***Narrow-leaved Helleborine**, *C. longifolia*, like B, but 15–40 cm tall; stem, less angled, bears whitish sheathing scales below, and **long**, **linear**, **lanceolate lvs** above, often folded lengthwise, with **drooping tips**, upper often **longer** than fl-spike. Fls (**Ca**) in a **closer** spike, **much longer** than their bracts, **pure white**, 2 cm long, opening more than in B to show smaller orange spot at lip-base; outer perianth segments **pointed**; ovary held at 45° to stem. Br Isles, wide-scattered but r and declining (lf in E Hants only); Ire r; in beechwds on calc soils, also oakwds on more acid soils. Fl 5–6, earlier than B.

D Sch 8 CR ****Red Helleborine**, *C. rubra*, resembles C in its narrow lvs, but these are shorter, **without** long drooping tips; bracts shorter than the **rosy red**, ± wide-opening fls (**Da**); ovary and upper stem sticky-hairy above; whole plant slightly purple-flushed. Eng: Cotswolds, Chilterns, Hants vr; in beech wds on calc soils. Fl 6–7.

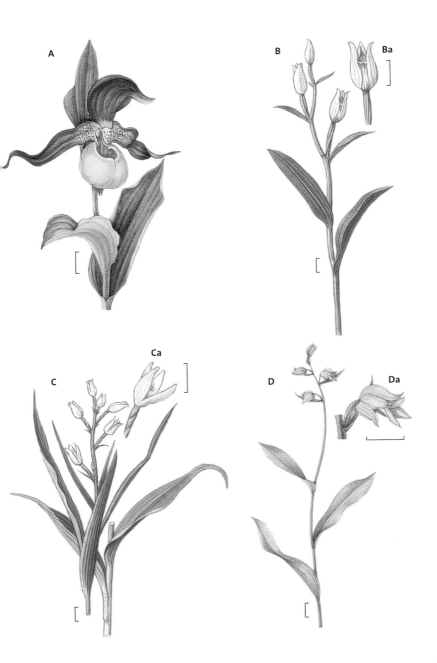

A

B Ba

C Ca

D Da

Lady's-tresses (*Spiranthes*) have small, **downy**, whitish fls in dense, **spirally twisted** spikes, not widely open, with lvs at base of and up stem; lip **frilly-edged**, curved down. **Creeping Lady's-tresses** (*Goodyera*) differs in having **runners**, oval rosette-lvs with conspicuous **cross-veins** between main veins and **untoothed spout-like** lips to fls.

A NT **Autumn Lady's-tresses**, *Spiranthes spiralis*, short (7–20 cm) erect per herb; round stem bears only **bract-like**, **green** scale-lvs; plant unusual in that **lf-rosette** does not **encircle** stem-base, but arises adjacent to it and has **oval**, **blue-green** lvs 2.5–3.0 cm long. Infl a long (3–10 cm), dense, **spirally-twisted**, spike of 7–20 small (4–5 mm long) hardly open, white fls (**Aa**) with **coconut-scent**; bracts equal downy ovaries; perianth segments blunt-lanceolate, downy; lip with **green** centre, **frilled** white edge, curved down. Br Isles: GB, N to Morecambe Bay, IoM, f-lc to S, r to N; Ire, o-lf; on dry calc or sandy gslds, dunes. Fl 8–9.

B WO (NI) FPO (Eire) BAP *****Irish Lady's-tresses**, *S. romanzoffiana*, **stouter** than A, 12–25 cm tall; lvs ± erect, **lanceolate-linear**, in a rosette **around** stem-base; upper stem-lvs small, bract-like, sheathing; infl **short** (2.5–5.0 cm long), **broad** (2 cm wide), **oblong**; fls in **3 spirally-twisted rows**. Fls (**Ba**) larger (8–11 mm long) cream, hawthorn-scented, lowest bract longer than its fl; lip cream with green veins; ovary sticky-hairy, spreading. Br Isles: S Devon vr; W Scot (Coll, Colonsay, Barra, W coast Highlands) vr; W, N Ire, r-vlf; in moist mds, bogs cut away for peat. Fl 7–8.

C **Creeping Lady's-tresses**, *Goodyera repens*, per herb, similar fls to A but **key difference** is **creeping runners** ending in rosettes of **broad-stalked**, oval, **evergreen** lvs with **conspicuous cross-veins** between the main parallel ones; upper stem-lvs bract-like. Stem 10–25 cm tall, with narrow infl of **cream** fls (**Ca**) as in A but hardly opening; perianth segments blunt-oval, sticky-hairy; lip short, narrow, **untoothed**, arched. Eng: Norfolk, Cumbria, Northumberland r; Scot, lf-lc in E and NE, r in W; Ire abs; in ancient Scots pine wds, and (to S, E) in **new** pine plantations. Fl 7–8.

D WO (NI) AWI NT **Bird's-nest Orchid**, *Neottia nidus-avis*, **entirely honey-coloured** erect saprophyte 20–40 cm tall, with brownish sheathing scales (no green lvs) up the stout stem; infl a raceme, 5–20 cm long, loose below; bracts of fls short, scarious; fls (**Da**) with perianth segments forming a loose hood; lip 12 mm long, unspurred, hanging down, divided halfway into 2 oblong, blunt **diverging** lobes. Capsule 12 mm, erect. Br Isles: SE, S and central Eng, f-lc; N to Inverness o-lf; Ire o; in wds, especially beech, in deep lf-litter. Fl 5–7.

E Sch 8 EX ****Ghost Orchid**, *Epipogium aphyllum*, erect saprophyte 10–20 cm tall; stem, ± translucent, pale yellowish, **pink-tinged**, bears a few sheathing brownish scales only, no lvs. Fls (**E**) 1–4, in a v loose spike, each fl about 2 cm across, pendulous, with a bract equal to the slender stalk; perianth segments linear, pale yellow, down-pointed, blunt; lip pointing **upwards**, cordate; ground colour **white**, with **crimson ridges** along it; 2 short side-lobes present; spur 8 mm x 4 mm, **blunt**, upturned, yellowish tinged with red. Eng: Chilterns vr; in beechwds in deep lf-litter, v erratic in its appearances. Fl 7–9. Regarded as **extinct** but may re-occur.

A

Aa

5 mm

B

9 mm

Ba

C

Ca

Da

D

E

A VU *Coralroot Orchid*, *Corallorhiza trifida*, erect, saprophytic herb 7–25 cm tall; **yellow-green**, hairless stem bears long, brownish **sheathing** scales; raceme of 4–12 fls with tiny bracts; fls (**Aa**) with **yellow-green** perianth segments, strap-shaped, curved down; lip 5 mm long, on **lower** side of fl, unspurred, oblong with 2 tiny basal side-lobes, white with crimson spots. GB: N Eng vr; E Scot r; rest of GB and Ire abs; in acid boggy wds, dune slacks. Fl 5–7.

B AWI **Common Twayblade**, *Listera ovata*, erect herb, 20–60 cm; stem green, hairless below, downy above; an opp pair of broad **ovate-elliptical**, unstalked, **strongly-ribbed**, horizontally spreading or ± ascending lvs, 5–15 cm long, on stem **below** its middle but **several cm** above gd level; only a few tiny bract-like lvs above them. Fls (**Ba**) many, usually 15 in a loose raceme, 7–25 cm long; each fl green, with perianth segments forming a hood; lip 10–15 mm long, **yellow-green**, hanging down, forked almost halfway; **spur abs**; capsule ± globular. Br Isles, c except in high mt areas; in wds, mds, gslds, dunes on basic soils. Fl 6–7.

C **Lesser Twayblade**, *L. cordata*, much smaller than B; slender **reddish** stem, only 6–20 cm tall, bears below mid-stem an opp pair of stalkless oval-**triangular** to **cordate** lvs, 1–2 cm long, dark green and **shiny** above, pale green below with **prominent mid-rib**. Some stems bear lvs only; others a loose raceme, 1.5–6.0 cm long, of 3–5 tiny brownish fls (**Ca**); uppermost perianth segments oval, forming a hood; others narrow, forward-pointing, green outside, **reddish** within, 1.5–2.0 mm long; lip **reddish**, hanging down, 3.5–4.0 mm long, v narrow, forked halfway; **spur abs**; capsule (**Cb**) globular. Br Isles: S Eng, Exmoor only, vr; N Eng and N Wales to N Scot, r to S, f-lc to N; Ire r; in pine wds, shaded bogs, hth moors on peaty soils. Fl 6–8.

D EU Sch 8 BAP EN ****Fen Orchid**, *Liparis loeselii*, hairless herb; erect angled stem, 6–20 cm tall, arises from green scaly bulb among wet moss. Lvs normally only **1 opp pair**, arising from **base** of stem, oblong, elliptical, pointed, shiny, **greasy**-looking, yellowish-green, 2.5–8.0 cm long; fls 2–10, in a loose raceme 2–10 cm long; lower bracts as long as fls. Fls (**Da**) yellow-green, perianth segments **linear-lanceolate**, 5–8 mm long, spreading, **incurved**; lip on **upper** side of fl, often vertical, **obovate**, **furrowed**, with **wavy** edges, 5–7 mm long; **spur abs**; capsule erect, narrowed to tip. GB: Norfolk, vr; S Wales, N Devon, r-la (as small form with oval lvs, in dune slacks); in calc mossy fens, moist dune slacks. Fl 6–7.

E FPO (Eire) WO (NI) **Bog Orchid**, *Hammarbya paludosa*, tiny plant 3–12 cm tall; erect hairless stem arises from a tiny bulb among moss; lvs 3–5, one above the other at stem base, only 0.5–1.0 cm long, oval, concave, rounded at tip and with a fringe of tiny green bulbils (**Eb**) from which new plants grow. Raceme 1.5–6.0 cm long, spike-like, of small flat-faced green fls (**Ea**), 7 mm from top to bottom, greenish-yellow, with narrow spreading perianth segments; lip pointed, short, on **upper** side of fl; **spur abs**. Br Isles: Eng, Wales, vr (now extinct C Eng, but New Forest–Dorset vlf); Scot Highlands o-lf; Ire vr; in *Sphagnum*, around pools in valley bogs with some flow of water. Fl 7–9.

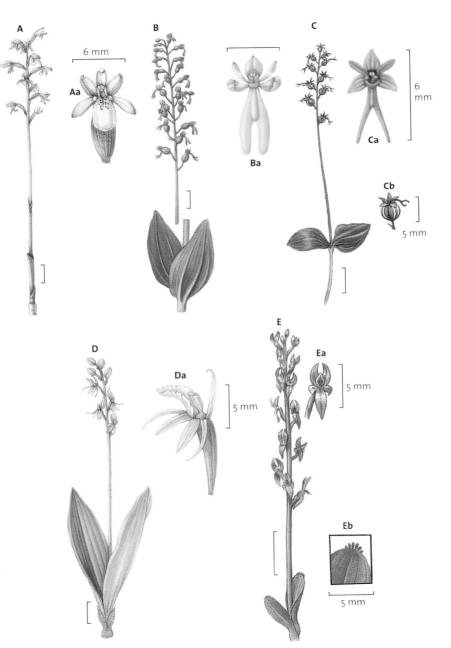

A

Aa

6 mm

B

Ba

C

Ca

6 mm

Cb

5 mm

D

Da

5 mm

E

Ea

5 mm

Eb

5 mm

KEY TO *ORCHIS* & *DACTYLORHIZA*

- Bracts shorter than fls, narrow, membranous (1–2 mm wide) – fl-spike wrapped round with sheathing lvs in bud..
  (***Orchis***) **Key p 534** (below)
- Bracts as long, or longer than, fls, broad (2–6 mm wide), lf-like – fl-spike never enclosed in sheathing lvs in bud ....(***Dactylorhiza***) **Key p 541**

In *Orchis*, fl-spike is **wrapped round** by sheathing lvs in bud; bracts of fls v narrow, **shorter** than fls, **membranous** (not lf-like); lvs arranged in a spiral from basal rosette up stem; fl-lips 3- to 5-lobed, on lower side of fl, not jointed into two parts, but with a stoutish **spur no longer** than lip at rear; 1 erect stalkless anther rises above beak of fl; this anther has, enclosed in a pouch, 2 stalked pollen-masses that give the appearance of 2 'eyes', the beak resembles a 'nose', and the hollow below beak leading to stigma and spur resembles a 'mouth', forming a tiny 'face' within the helmet of the fl; the lip is either divided into 'arms' and 'legs', or resembles a 'skirt', so the whole fl from front resembles a human figure. The tubers are egg-shaped.

KEY TO *ORCHIS*

1. All perianth segments (except lip) forming a helmet to fl – spur curved down **2**
   Two of outer perianth segments erect or spreading – spur ± curved up **6**

2. Lip shallowly three-lobed, mid-lobe no longer than side-lobes, notched at tip – perianth segments of helmet with lengthwise green and purple stripes ....*Orchis morio* (p 536 **F**)
   Lip deeply three-lobed, the much longer central lobe divided again into 2 lobes (often with a tooth between them), so producing 2 'arms' and 2 'legs' like a human figure – perianth segments of helmet without lengthwise green and purple stripes **3**

3. Outer perianth segments of helmet oval, shortly-pointed, much darker than lip **4**
   Outer perianth segments of helmet oval-lanceolate, long-pointed, paler than lip **5**

4. Plant only 8–15 cm – lip only 4–6 mm long, narrow, white with a few flat red spots – 'legs' ± parallel (suggesting a tiny clown's figure) – helmet blackish-purple in bud, paler purple in open fl..*O. ustulata* (p 536 **E**)
   Plant 25–45 cm – lip 12–18 mm long, broad, rosy, with many spots formed of tufts of crimson hairs – 'legs' spreading (but ± variable in shape and colour) – helmet dark reddish-purple-flecked on a green ground...*O. purpurea* (p 535 **B**)

5. All 4 'arms' and 'legs' of lip v narrow, long, curved, red, 1 mm wide – 'body' white – helmet white with ± pink flush and streaks ....*O. simia* (p 536 **C**)
   'Arms' of lip narrow but shorter and broader (1.5 mm wide), 'legs' red, broader (2–3 mm) and shorter, blunt, out-turned – 'body' white red-spotted – helmet ash-grey outside, pink-streaked inside ....*O. militaris* (p 536 **D**)

6. Lvs elliptical-oblong, normally purple-blotched – fls tom-cat-scented at night – lip three-lobed, mid-lobe longest, blunt – spur equalling ovary – widespread ......*O. mascula* (p 535 **A**)
   Lvs linear-lanceolate, unspotted – fls unscented – lip scarcely three-lobed, mid-lobe abs or minute – spur shorter than ovary – CI only ....*O. laxiflora* (p 535 **A.1**)

A AWI **Early-purple Orchid**, *Orchis mascula*, erect hairless herb 20–40 cm tall; lvs shiny dark green, oblong, 5–10 cm long x 2–3 cm wide, usually with dark purple **blotches**, arranged **lengthwise** (blotches **transverse** on lvs in Spotted Orchids, p 542 A, B); lvs in a basal rosette and sheathing up stem, spirally arranged. Stem solid, fleshy; infl a loose raceme 5–15 cm long; fls (**Aa**) bright purple-crimson, tom-cat-scented, especially at night; 2 outer sepals **erect** or **spreading**, rear sepal and 2 petals forming a loose helmet; lip 8–12 mm long x 8–10 mm wide, ± equally, bluntly and shallowly three-lobed, central lobe **longest**, notched, sides of lip ± folded back; lip with paler, purple-dotted, central patch; spur stout, ± curved up, as long as ovary; bracts narrow, purplish. Br Isles, f-lc in wds, gslds on ± base-rich soils. Fl 4–6. **A.1 Loose-flowered Orchid**, *O. laxiflora*, v like A, but lvs never spotted and fls unscented; **key differences** in fl-lip has mid-lobe **much shorter** than side-lobes, or **abs**. **CI only**, lf as native (vr introd SE Eng); in wet mds. Fl 4–6.

B AWI EN *****Lady Orchid**, *O. purpurea*, erect hairless herb, 25–45 cm tall; lvs to 15 cm long x 5 cm wide, oval-oblong, ± blunt-pointed, mostly in a basal rosette, upper lanceolate and sheathing stem, all glossy green, **unspotted**. Fl-spike 5–15 cm long, dense, conical at first then cylindrical; fls (**Ba**) sweetly vanilla-scented, the 3 sepals forming a close **oval**, **blunt** helmet v heavily flecked with dark brownish-purple on a green background, but v variable from pale green (in albino plants) to blackish-purple; petals within helmet narrow. Lip 12–18 mm long, hanging below helmet, white or ± rose-flushed, with 2 'arm'-like side-lobes 2–3 mm wide, and a broad (5 mm wide) central 'body' divided into 2 short **broad**, **blunt**, slightly diverging 'legs' with a tiny tooth between them, the whole lip sprinkled with **tiny tufts of crimson hairs** (rarely wholly white). Spur curved down, cylindrical, blunt, half length of ovary; bracts v tiny (whole fl resembles an early Victorian woman

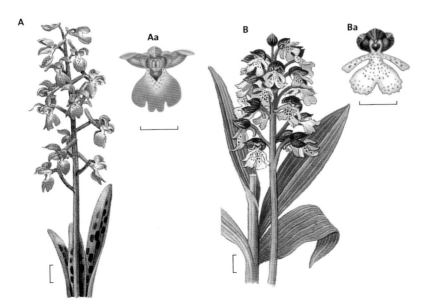

A Aa B Ba

in wide spotted dress and poke-bonnet, hence the name). Eng: Kent f-la; rest of S Eng vr and sporadic only; **rest of Br Isles abs**; in open wds, scrub, gslds on calc soil. Fl 5–6.

C Sch 8 VU ****Monkey Orchid**, *Orchis simia*, has lvs and habit of B, but shorter (15–30 cm), slenderer. In fls (**C**) 3 **long-pointed whitish-grey** sepals form a helmet **nearly as long as lip**, dotted and streaked with pink especially within; lip with narrow white (2 mm wide) 'body' (with tufts of crimson hairs as in B), v narrow (1 mm wide), long, curved crimson 'arms' and 'legs' with a short 'tail' between 'legs'; spur as in B. Eng only: Kent, Chilterns, vr; on calc gslds, open wds. Fl 5–6.

D Sch 8 VU ****Military Orchid**, *O. militaris*, close to B in lvs and habit, but usually less tall; fls (**D**) with helmet of **long-pointed** sepals as in C, but ash-grey, only ± flushed pink or violet; lip nearer to B's, with 'body' **3 mm wide**, white, with tufts of crimson hairs, 'arms' as in C but shorter; 'legs' **much broader** (2–3 mm) than in C, crimson, **blunt**, **diverging widely**, with v short blunt 'tail'; spur as in B and C. Bucks, Suffolk only vr (vvr introd elsewhere); on calc gslds, scrub. Fl 5–6.

E EN ***Burnt Orchid**, *O. ustulata*, small erect herb **8–15 cm** tall; lvs 2.5–6.0 cm long x 0.5–1.0 cm wide, pointed, mostly in a rosette; spike short (2–3 cm), **dense**; fls (**Ea**) with helmet of **blunt**, **oval** sepals only **2–3 mm long**, **blackish-purple** in bud and when young (so tip of spike appears scorched), paler in older fls; lip 4–6 mm long, white with **few**, **flat**, red dots, lobed into short, blunt 'arms' and 'legs' (resembling a tiny clown's suit); scent as of stewed cherries; spur v short. Eng: S, SE, o-lf; N to Durham vr; Scot, Ire abs; on calc gslds with short turf. Fl 5–6 (some populations 7–8 which may be a distinct ssp).

F WO (NI) NT **Green-winged Orchid**, *O. morio*, erect herb 10–30 cm tall; lvs **unspotted**, glossy-green, elliptical-lanceolate, 3–9 cm long x 1–2 cm wide, upper sheathing stem as in other *Orchis* spp, lower in a rosette; spike 2.5–6.0 cm long, rather loose, bracts narrow, purplish. Fls (**Fa**) much as in A, but unscented, with **all 3** sepals in a **close**, **blunt**, **oval** helmet, rich deep **purple-veined** with **strong green stripes**; lip broader than long, the sides ± folded back, **deep purple** with a pale crimson-dotted central patch, and bluntly three-lobed as in A; spur stout, nearly as long as ovary (fls vary to pink or white). Eng, Wales, r, only vla in individual damp mds; SW Scot vr; Ire o; in gslds, mds, on calc or heavy clay soils. Fl 5–6.

G EN ***Man Orchid**, *Aceras anthropophorum*, erect per 20–40 cm; lvs in a basal rosette and up stem, 6–10 cm long, oblong-lanceolate, glossy grey-green, unspotted, pointed, with a few **transverse wrinkles** on upper side of mid-lf. Fl-spike 8–15 cm long, many-fld; fls (**Ga**) with close oval hood 5–6 mm long of yellowish-green sepals, often with a narrow red-brown edge; lip hanging down, 10–12 mm long, yellow, often red or chocolate-edged, **narrow**, with 2 narrow 'arms' and 'legs' but **no 'tail'**; **spur abs**; plant has vanilla smell when bruised. Eng: Kent and Hants N to Chilterns and S Lincs, f-lc in SE, vr to N; **rest of Br Isles abs**; on calc gslds, scrub, open wds. Fl 5–6.

H **Pyramidal Orchid**, *Anacamptis pyramidalis*, erect hairless herb 20–30 cm; lvs grey-green, unspotted, lanceolate, pointed, spirally arranged, in a rosette and up stem. Fls (**Ha**) in a dense, usually short (2–5 cm, sometimes longer) **dome-shaped** to conical spike, **wholly** rich pinkish-red, foxy-scented; 2 outer sepals lanceolate,

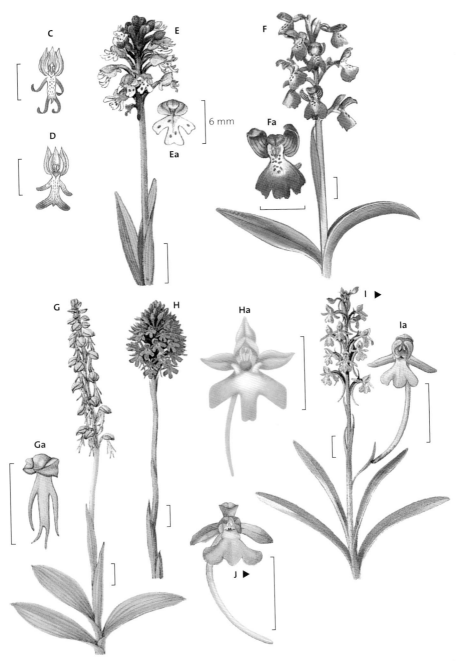

C

D

E

Ea

6 mm

F

Fa

G

Ga

H

Ha

I ▶

Ia

J ▶

spreading; upper sepal and 2 petals forming a small helmet; lip 5–6 mm long, as broad as long, **deeply** three-lobed nearly to base, lobes oblong, ± equal, side-lobes spreading; 2 ± erect **ridges** run out onto lip from its base; spur 12 mm long, ± straight, ± thread-like, pointed, longer than ovary; the two pollen-masses attached to a **common** base, thus always removed together by insects (in *Orchis* spp (p 534) they can be removed separately). Br Isles: GB, f-lc in S Eng, o-lf N to central Scot and Hebrides; Ire f-lc; on calc gslds, dunes. Fl 6–8.

▲ **I Fragrant-orchid**, *Gymnadenia conopsea*, resembles H in its pink long-spurred fls, but taller (to 40 cm), with **glossy green**, **unspotted**, oblong-lanceolate **strongly keeled** lvs, 6–15 cm long x 1–3 cm wide, in **2 dense ranks** up **opp sides** of stem. Fl-spike **cylindrical**, 6–15 cm long, **dense**, with green bracts ± equalling fls; fls (**Ia**) rosy-pink, unspotted, 8–12 mm wide, **sweet-scented**, but with an **acid** or **rancid** overtone; lateral sepals narrow, with rolled-under edges, **pointed**, spreading but **curving down** at a **slight angle**; upper sepal and 2 petals forming a little helmet; lip broader than long, near to that of H, but **without** erect ridges on it, three-lobed about **half-way**; spur narrow-cylindrical, to 20 mm long, twice length of ovary, **curved**, pointed. Br Isles lf-lc; on dry calc gslds. Fl 6–7. **I.1** *G. borealis* (*G. conopsea* ssp *borealis* in the *New Flora* but in the author's opinion a distinct sp), the plants of hill pastures on more acid soils in N and W of GB are more slender, with lips **longer than wide** and only slightly three-lobed; they are carnation-scented like J. Also in bogs in S Hants and E Sussex. Fl 6–7.

▲ **J Marsh Fragrant-orchid**, *G. densiflora* (*G. conopsea* ssp *densiflora* in the *New Flora* but in the author's opinion a distinct sp), often taller and stouter than I, but **key differences** are: fls (**J**) **deep** pink; lip **much broader than long**, with 'shoulders' to the side-lobes; side sepals ± **square-tipped**, **straight**, **horizontally** spreading; spur only a **little** longer than ovary; sweet **carnation**-scent; and fl **later**. Eng, Wales, o but lf; in calc fens, also sometimes on calc gslds. Fl 7–8.

A VU *Musk Orchid*, *Herminium monorchis*, small erect orchid 5–15 cm tall; normally only 2, ± oblong, pointed yellow-green unspotted basal lvs and sometimes 1–2 tiny, bract-like stem-lvs. Fl-spike 2–6 cm long, narrow, dense, cylindrical, many-fld. Fls (**Aa**) **greenish-yellow**, with strong **honey-scent**, 3–4 mm wide, bracts shorter than fls; sepals pointed, forming a **loose** hood; **lip** 3 mm long, **deeply** three-lobed, mid-lobe longer and blunter than side-lobes, **spur abs**. S Eng only, from Somerset and Kent to Cotswolds and Chilterns, r; **abs rest of Br Isles**; on short calc gslds, mostly near coasts. Fl 6–7.

B AWI NT **Greater Butterfly-orchid**, *Platanthera chlorantha*, erect hairless orchid 30–40 cm tall; 2 ± opp, unspotted, elliptical-oblong **blunt** lvs, 10–15 cm long, rise at an angle from near base of stem. Spike 5–20 cm long, loose, conical at first, then cylindrical, with fls (**Ba**) 18–25 mm wide, with long lfy bracts, on long, ± horizontal stalks, v sweetly-scented; side sepals oval-lanceolate, pointed, spreading horizontally; other perianth segments forming a short hood, all **greenish-tinged creamy**; lip **undivided**, strap-shaped, blunt, pointing down, green to greenish-cream, 10–15 mm long; **spur v long** (20–30 mm x 1 mm), often curved downwards and forwards; **pollinia** yellow long (3–4 mm), 4 mm apart at base, **converging above** with tips **close together**. Br Isles o-lf; in wds, scrub, on basic clays and calc soils. Fl 6–7.

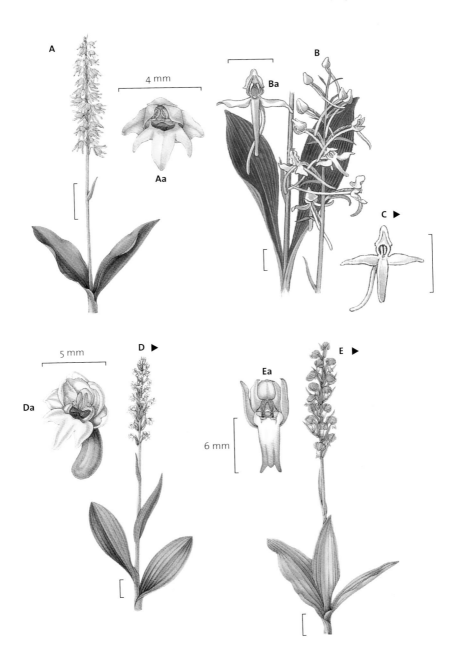

539

▲ **C** VU **Lesser Butterfly-orchid**, *Platanthera bifolia*, close to B p 538, but rather smaller; lvs often shorter (5–8 cm), more oval, especially in bog plants (but more like B in wds); spike narrower in proportion to length; **key differences** from B are: fls (**C**) **narrower** (11–18 mm wide), with less greenish tinge, lip only **6–10 mm** long and proportionately **broader**; **pollinia 2 mm long**, **vertical**, **parallel**, **close together** (**C**); Br Isles, r-lf in open wds on basic or calc soils, mds, fens, acid bogs. More c than B in N GB. Fl 6–7.

▲ **D** FPO (Eire) WO (NI) VU **Small-white Orchid**, *Pseudorchis albida*, small (10–20 cm), slender, erect orchid, with oblong-lanceolate unspotted lvs 2.5–8.0 cm long up the stem, uppermost v narrow and short; spike 3–6 cm, **narrow**, **v dense**, **cylindrical**, of tiny greenish-**creamy** fls (**Da**) 2.0–2.5 mm long x 5 mm wide, with vanilla scent; perianth segments forming a short flat hood, lip 3–4 mm long, three-lobed, curved down, triangular mid-lobe only slightly longer than side-lobes; spur v short (2–3 mm), bluntly conical, curved down. Br Isles: Wales, and N Eng to S Scot, r; N Scot f-lc; Ire r; in mds, gslds in hill or mt areas. Fl 6–7.

▲ **E** VU **Frog Orchid**, *Coeloglossum viride*, short erect orchid, 4–20 cm tall, with basal rosette of blunt oval hairless unspotted lvs, and narrow short lvs up stem; spike short (1.5–7.0 cm long), loose-fld, with green bracts, of which lower are as long as the fls; fls (**Ea**) **greenish** to **brown-purple**, with a close helmet of sepals with darker edges not unlike that of Man Orchid (p 536 G), 3–5 mm long; lip hanging down, 4–6 mm long x 2–3 mm wide, **oblong**, **green-yellow** or brownish, with edge often chocolate, shortly three-toothed at **tip**; spur 2 mm long, bluntly conical, stout and pale. Br Isles, widespread, but r (except in central S Eng lf on chk, and lc in N Eng and N Scot); on calc gslds to S, hill-pastures, dunes, mt rock-ledges to N. Fl 6–8.

A Sch 8 NT **Lizard Orchid*, *Himantoglossum hircinum*, a striking plant, 30–90 cm tall, with stout erect hairless stem; basal lvs oblong-elliptical, in a rosette, 10–20 cm long, withered by time fls open; upper lvs narrower, pointed, clasping, all unspotted, grey-green. Spike 10–30 cm long, many-fld, with lower bracts equal to fls. Fls (**Aa**) with perianth segments forming a close blunt helmet 5–6 mm long, greenish-grey outside, purple-streaked within; lip to **5 cm long**, at first coiled like a **watch-spring**, then **ribbon-like** and **spirally twisted**; three-lobed, mid-lobe v long, narrow, tip notched, side-lobes less than half as long, lobes pale olive-brown, body of lip whitish, spotted with clusters of purple hairs; fls strongly unpleasant-smelling, more tadpole than lizard-shaped; spur blunt, 4 mm long. Br Isles: S Eng, CI, vr, but vlf in a few places; **Wales**, **Scot**, **Ire**, **abs**; on calc gslds, scrub, dunes. Fl 6–7.

B **Dense-flowered Orchid**, *Neotinea maculata* (*N. intacta*), a short (10–30 cm) erect orchid with rosette of oblong basal lvs, **often** with **small brown spots**; upper lvs smaller, clasping. Spike 3–8 cm long, **v dense**, cylindrical. Fls (**Ba**) with helmet 4 mm long, greenish-white or pink; lip pink or white, 4–5 mm long, with 2 narrow pointed side-lobes, and **longer**, **broader**, **notched** mid-lobe, often purple-spotted at base; spur 2 mm, blunt, curved. Br Isles: **GB abs**; central, W Ire vlf; IoM vr-abs; on rocky gslds on limestone, calc dunes. Fl 4–6.

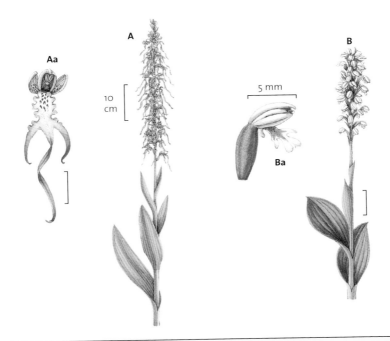

A   **Aa**   **B**   **Ba**

10 cm

5 mm

ID Tips   *Dactylorhiza* **spp**

- **Spotted Orchids** may sometimes have unspotted lvs, but the stems are generally **solid**.

- **Marsh Orchids** may, or may not, have spotted lvs, but stems **hollow** and usually stout (check by gently compressing stem between finger and thumb – do not pick or cut).

- Many *Dactylorhiza* spp have distinctive markings on the lower fl-lip (**labellum**) – there are useful drawings of these markings in the *New Flora*.

- *Dactylorhiza* are a difficult group due to hybrids being frequent between spp and populations being v variable.

- Always try to look at **5 or more** plants in any group and use the **averages** of the characters for all plants when using this key.

KEY TO *DACTYLORHIZA*

1   Stem solid – lvs ± always with transverse blotches – spur usually <2 mm wide at midpoint **2**
Stem hollow (can be felt by compressing – do not pick or cut) – lvs spotted or not – spur usually >2 mm wide at midpoint **3**

2   Lip lobed ± ½ way to base, with 3 deep lobes, mid-lobe usually longest.......................................
....*Dactylorhiza fuchsii* (p 542 **A**)
Lip lobed much less than ½ way, with 3 v shallow lobes, mid-lobe narrower and shorter than rounded, toothed side-lobes ....*D. maculata* (p 542 **B**)

3 Stem-hollow large – lvs erect, keeled, hooded at tips, normally unspotted – lip-edges folded back at sides, perianth segments arched back (so fls appear v narrow) – lips with 2 oblong dotted areas on each side of centre, each surrounded by a dark U-shaped line .... *D. incarnata* (p 544 **D**)

Stem-hollow small – lvs ± spreading, scarcely hooded – lips flat or concave – perianth segments spreading (so fls appear broad) **4**

4 Lvs few (usually <5), narrow (to 2 cm wide only), keeled, with tiny (1 mm) spots near tip, or none – fls few (8–10) – lip three-lobed, mid-lobe much longer than side-lobes .... *D. traunsteineri* (p 544 **F**)

Lvs many (>5), broad (over 2 cm) – fls many in dense spikes **5**

5 Lip with narrow central lobe usually $^1/_3$–$^1/_2$ as long as unlobed basal part, with dark spots and lines mostly in central part but usually some extending to lip margins – lvs either unmarked (parts of Ire only) or heavily spotted, spots mostly >2 mm across .... *D. majalis* (p 544 **G**)

Lip less deeply lobed, with dark spots or lines confined to central part – lvs unmarked or with spots usually <2 mm across or with rings **6**

6 Lip shield- or diamond-shaped, usually <9.5 mm wide – fls deep purple with heavy crimson streaks all over – lvs with spots near tip, or none .... *D. purpurella* (p 544 **E**)

Lip not shield- or diamond-shaped, usually >9.5 mm wide – fls pale pink-purple to rosy-purple – light crimson dots and speckles confined to centre of v shallowly three-lobed, flat or concave lip, which is broader than long, with v short, blunt mid-lobe – lvs normally unspotted (sometimes with ring-spots) ....*D. praetermissa* (p 542 **C**)

A   Common Spotted-orchid, *Dactylorhiza fuchsii*, usually slender, 15–40 cm tall, stem **solid**; rosette-lvs broad-elliptical, blunt, grey-green, with many **transversely** elongated purple spots (sometimes unspotted, especially in white-fld plants); stem-lvs narrow-lanceolate, pointed, merging in size into the lfy bracts of fls above. Fl-spike conical, then cylindrical, to 10 cm long; fls (**Aa**) **pale** pink, with purple streaks and spots (sometimes all pure white); outer sepals **spreading**, upper sepal and 2 petals forming a loose hood; lip 6–8 mm long, **deeply** 3-lobed with **pointed**, well-separated lobes, mid-lobe slightly **longer** than side-lobes; spur 5–8 mm long, conical, usually **<2 mm wide** at midpoint. Br Isles, c in wds, scrub, gslds on calc or ± basic soils. Fl 6–8.

B   Heath Spotted-orchid, *D. maculata*, similar to A but lvs **all** lanceolate, pointed, with ± **circular** purple spots; but **key difference** in the fls (**B**), lip lobed **much less** than ½ way, with 3 v shallow lobes, mid-lobe **narrower and shorter** than rounded, toothed side-lobes; lip markings often more **streak-** or **loop-like** and **less dot-like**. Br Isles, f-lc except in calc or v cultivated districts; in hths, bogs, acid gslds. Fl 6–8.

C   Southern Marsh-orchid, *D. praetermissa*, erect orchid, 20–60 cm tall, with stout **hollow** stem; lvs normally **unspotted** (**ring-spotted** in var *junialis*), 10–20 cm long, widest **below** middle, broad-lanceolate, pointed, in two ± **opp ranks** up stem, **not** sheathing stem, slightly hooded at tip, but hardly keeled lengthwise. Fl-spike dense, conical, 5–10 cm long, the broad (5 mm) lfy bracts **longer** than fls; fls (**Ca**) **pale** pink-purple to rosy-purple, with **spreading** outer sepals; lip ± flat or concave, **over 9.5 mm wide**, broader than long, three-lobed but with side-lobes **v**

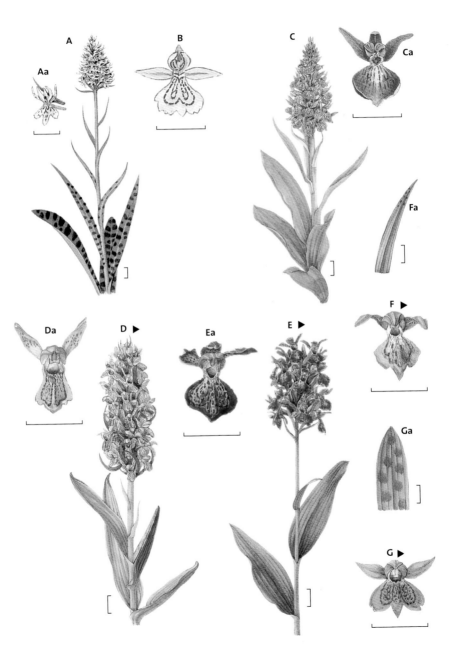

shallow, mid-lobe short and blunt, pattern of **light** crimson dots and speckles **confined to lip centre**; spur v **stout**, conical, ± curved, usually **>2 mm wide** at midpoint. Br Isles: Eng, Wales, c to S, r to N; **Scot**, **Ire**, **abs**; in fens, wet mds on ± calc soils. Fl 6–8.

▲ **D Early Marsh-orchid**, *Dactylorhiza incarnata*, **key differences** from C are: stem-hollow larger (hence more compressible); lvs more **erect**, yellow-green, strongly **keeled**, normally unspotted, tapering to narrow **hooded** tip; fl-spike **narrower** (to 4 cm wide) more cylindrical, **denser**, lower bracts to 3 mm wide, **much** longer than fls. Fls (**Da**) with outer perianth segments ± **erect**, and sides of lip **strongly bent back** in vertical plane, so making fls **v narrow** from front view; lip narrow (5 mm wide) oblong, weakly three-lobed at tip, with a pattern of 2 long **U-shaped purple loops** enclosing purple-dotted **patches**; fls normally flesh-tinted. D is v variable and **several sspp** are recognised: ssp *incarnata* is described as D above, widespread but only o-lf; in calc fens and wet mds; ****ssp *cruenta*, with lvs spotted both sides and fls pale purple, only in Ire, r; *ssp *coccinea*, with brick-red fls, vlc in dune hollows in W of GB and Ire; ssp *pulchella*, like D but with fls purple, f-lc in acid bogs of Br Isles; ****ssp *ochroleuca*, with cream or yellow fls without lip markings, o in fens of E Anglia, Wales; ssp *gemmana*, over 40 cm tall, having lip over 9 mm wide with dots but no lines, r in fens of Norfolk and W Ire. All fl 5–7.

▲ **E Northern Marsh-orchid**, *D. purpurella*, replaces C in N of Br Isles; shorter (10–25 cm) than C; stem with **v narrow hollow** in centre, lvs short, flat (5–10 cm long), with **spots near tip**, or unspotted. Fl-spike short, broad (5–6 cm long x 3–4 cm), dense; fls (**Ea**) with stout spur, **spreading** outer sepals, lip **shield-** or **diamond-shaped**, **under 9.5 mm wide**, deep rich purple with heavy crimson streaks all over. Br Isles: N Wales and N Eng to N Scot, r to S, c to N; S Eng, Hants only, vr; N Ire o-lf; in fens, wet mds. Fl 6–7.

▲ **F** WO (NI) ***Narrow-leaved Marsh-orchid**, *D. traunsteineri* ssp *traunsteineri*, differs from A (p 542) in being shorter (15–20 cm tall); stem **slender**, only slightly hollow; lvs (**Fa**) few (3–5), **v narrow** (1.0–1.5 cm wide), keeled, **linear-lanceolate** with slightly hooded tip, usually with tiny purple spots (1 mm wide) or bars near tip. Bracts tinged reddish-purple **all over**, **without** spots or rings. Fl-spike short, loose, **few-fld**; fls (**F**) red-purple, a little darker than in C (but less dark than in E); side sepals ± spreading; lip 8–10 mm wide, narrowed to base, ± diamond-shaped, flat, but **three-lobed**, the mid-lobe usually **much longer than** side-lobes, marked with a **strong** pattern of dark crimson lines and dots; spur straight, **slender**. Br Isles: central S, E, N Eng, N Wales, r-lf; Ire, Scot vr; in calc fens, dune slacks. Fl 5–6.

F.1 Sch 8 ****Lapland Marsh-orchid**, *D. traunsteineri* ssp *lapponica* (*D. lapponica*), v close to F but usually more heavily spotted lvs with more pronounced hood to lf-tip; fl with heavier markings than F and bracts **green**, tinged reddish-purple **around margins only**, with spots or rings. First recognised 1986; vr in W Scot only. F 6–7.

▲ **G** ****Western (Broad-leaved) Marsh-orchid**, *D. majalis*, resembles C in robust habit (to 60 cm tall), but differs in the more spreading **broad-lanceolate** to oblong lvs (**Ga**), **broadest in middle** and flat-tipped, **heavily transversely purple-blotched** or **ring-spotted** (but unmarked in parts of Ire only), spots mostly **>2 mm** across. Fl-spike **dense**, 5–10 cm long, oblong; fls (**G**) deep magenta to pink-purple, side sepals spreading; lip **v broad** (10–12 mm wide), ± flat, **strongly three-lobed**, with broad, rounded ± notched side-lobes and narrower, pointed mid-lobe usually ⅓–½ **as long as** unlobed basal part; lip with **strong** pattern of crimson loops and dots mainly in central part but some **extending to lip-margins**. Br Isles: Wales, N Scot & Ire only, r in fens, wet mds, dune slacks. Fl 5–7. Several sspp are recognised (see the *New Flora*), including End ssp *cambrensis* in Wales and Scot only.

Bee, Spider and Fly Orchids (*Ophrys*) have a **velvety, insect-like** lower fl-lip, often with a complex pattern of markings; **spur abs**.

A WO (NI) **Bee Orchid**, *Ophrys apifera*, erect orchid, 10–40 cm tall; lvs grey-green, elliptical-oblong, pointed, in a basal rosette and up the hairless stem. Fls (**Aa**) 2–10, large (to 3 cm across), spaced out up stem; sepals 12–15 mm x 5–7 mm, **rosy-pink** or whitish with ± green mid-vein, pointed, oblong-elliptical, spreading; 2 upper petals much shorter, downy, **greenish, square-ended** (or reddish and ± pointed in E Kent); lip 10–15 mm long x 8–10 mm wide, **bee-like**, rich brown, furry, convex, blunt-tipped, with two narrow short pointed side-lobes; lip bears a **U- or W-shaped pale** brown loop bordered within and without with a pale yellow line;

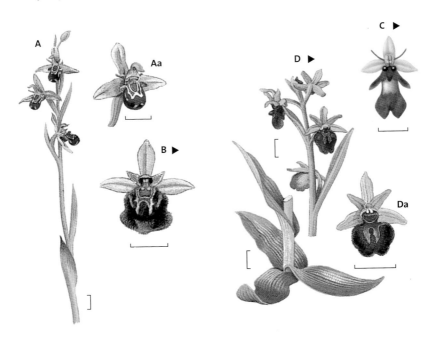

within the U or W is a reddish-orange U-shaped patch; lip ends in a long narrow tooth, but this is curved up underneath, so not visible from above; base of lip has 2 eye-like glossy bosses. Anther large, green, shaped like a duck's head; pollinia **long-stalked**, falling forwards on their stalks onto stigma to cause self-pollination if not removed by insects; **spur abs**. Br Isles: Eng f-lc, especially in S and E; Wales, Ire, r-o; **Scot abs**; on calc gslds, dunes, also disturbed gd, quarries etc. Fl 6–7.
A.1 Wasp Orchid, *O. apifera* var *trollii*, is one of several vars of A; looks v different from A in its **long-pointed**, ± **straight**, yellow lip, with usually **no** U or W pattern, but with **scattered brown blotches** on it. Anther as in A. S Eng, r on calc gsld. Fl 6–7.

▲ **B** Sch 8 VU ****Late Spider-orchid**, *Ophrys fuciflora*, **key difference** from A in the fls (**B**), in which sepals are much as in A, but the 2 upper petals **broad-triangular**, **pink**, downy, **blunt-pointed**; lip **larger** (15–20 mm long x 12–15 mm wide), often **widening** towards the tip, usually with a more complex pattern than in A, often like an H with side-lobes, but sometimes ± as in A; tip of lip, however, **always** has a **cordate, forward-pointing green appendage**. Anther short, pollinia short-stalked, never falling forward. Br Isles: E Kent only vr; on calc gslds. Fl 6–7, earlier than most vars of A, but hybridizes with it (*O.* x *albertiana*) also vr, E Kent only.

▲ **C** AWI VU **Fly Orchid**, *O. insectifera*, usually much taller (20–60 cm), more slender than A, B or D, fls more widely-spaced up stem. Fls (**C**) have **narrower** lip (15 mm x 4–5 mm) with 2 arm-like side-lobes and a deeply notched mid-lobe; lip **sides** turned-down but lip, purple-red to blackish-purple, velvety, with a broad, **oblong shiny blue patch** in the centre, is **straight**, **not convex**, at tip as in A, B and D and has no appendage. Sepals **green** as in D, but only 6–12 mm long; upper petals, **blackish-purple**, **velvety**, **thread-like**, resemble an insect's antennae. Whole lip resembles a fly sitting on a green fl. Anther short; lvs narrower, greener than in A and D. Br Isles: Eng o in S, r N to Cumbria; N Wales, Ire, r; **Scot abs**; in open wds, scrub, gsld on calc soils. Fl 5–6.

▲ **D** Sch 8 ***Early Spider-orchid**, *O. sphegodes*, differs from A in its fls (**Da**) with pale **yellow-green** sepals, **narrower** than in A (4–5 mm); in **brownish-yellow strap-shaped** upper petals **not much shorter** than the sepals; and in furry oval convex lip **without** any appendage, **dark purplish-brown** when fresh (but fading to when older) with a glossy blue *p*-shaped mark on it. Anther short like B; side-lobes of lip sometimes well-developed, but often abs. Br Isles: S Eng only, vr-vlc, mostly near sea; on calc gslds in short turf. Fl 4–5 (or early 6).

ILLUSTRATED GLOSSARY

This glossary includes additional terms, not used in this book, which you will find used frequently in other botanical texts.

Achene   a one-seeded nutlet formed from a single carpel, shed in fruit without opening to release the seed.

Adpressed (or **appressed**)   pressed close to another organ, usually refers to hairs when pressed close to stem or leaves or fruits held close to the stem, as in many species of the Cabbage family.

Adpressed

Aggregate species   a group of very closely related species, in this book treated as one.

Alternate leaves   leaves alternating up a stem, occurring first on one side, then on the other.

Alternate leaves

Annual   a plant completing its life cycle in one year, germinating in autumn or spring, flowering, fruiting and dying by the following autumn. Look for single flowering stems and the absence of any non-flowering shoots (because most of the plant's energy goes into producing flowers).

Anther

Anther   see stamen.

Arrow-shaped leaf   a wide-based leaf, tapering to a point above, with two backward-directed pointed or rounded basal lobes.

Ascending   growing upwards.

Auricle   a lobe or pair of lobes at the base of a leaf.

Arrow-shaped leaf

Awl-shaped leaf   a leaf tapering from a cylindrical base to a fine point.

Axil   the upper angle between a leaf or bract and the stem from which it grows.

Axillary   arising in the axil of a leaf or a bract.

Auricle

Axis (or **rhachis**)   the main stem running up through an inflorescence.

Basal   at the base. In some plants all the leaves occur at the base of the stem, often in a rosette.

Base-rich (or **basic**)   of soils or water, rich in alkaline nutrients (especially calcium salts).

Awl-shaped leaf

Beak   a terminal projection, for example, on a fruit, sometimes formed by a style that is persistent in fruit. Effectively the stalk between the achene and the pappus hairs in the Daisy family.

Axil

Berry   a fleshy fruit containing one or more seeds, which do not have a stony inner coat around each seed (cf drupe).

Biconcave   having concave faces on both sides.

Biconvex   having convex faces on both sides.

Biennial   a plant living for two seasons, normally germinating in the first spring and that year only forming leaves; it flowers and fruits in the second year, and then dies.

Beak

Bifid   forked or cleft in two.

Bisexual   of a flower, with both male (stamens) and female (carpels) parts in the same flower.

Bifid

Blade

Blade   the main part of a flat organ, such as a leaf.

Bog   an area of waterlogged acid peat, usually with vegetation composed of *Sphagnum* moss species (which form the bulk of the peat) and of sedges, low shrubs of the Heath family etc.

Bract   (1) a leaf-like structure, usually green, immediately below a flower and located where the flower-stalk joins the stem.
(2) in Umbellifers (Carrot family), used for the whorl of small leaves at the base of a main umbel.

Bract (1)   (2)

(1) Bracteole

Bracteole   (1) a tiny leaf-like structure, usually green, on a flower-stalk.
(2) in Umbellifers, the whorls of small leaves or bracts at the bases of the secondary or partial umbels.

Bulb   an underground swelling at the base of a plant, eg, in a Daffodil, made up of overlapping layers of scales (which are actually fleshy leaf-bases).

Bulbil   a small bulb, usually developed above ground in a leaf-axil or on a leaf edge, which can fall off and grow into a new plant.

Calyx (pl **calyces**)   the whorl of sepals below an individual flower (used collectively).

Calyx

Calyx-tube   the tube formed when the lower parts of the sepals are fused together.

Capitulum   a dense head of tiny flowers without individual stalks (as in the Daisy and Teasel families), surrounded by an involucre of bracts.

Capsule   a dry fruit that opens into two or more parts (valves), or by holes or a lid, to release the seeds.

Calyx-tube

Carpel   one of the units in the gynoecium (female organs) of a flower. Carpels are either free from one another (as in a Buttercup), or joined together into an ovary (as in Poppy or Flax). Each carpel has a style at its tip.

Casual   a plant that is introduced, occurs sporadically, only persisting in any given population for a short time and has not become naturalised. A casual species is therefore dependent on constant re-introduction.

Capitulum

Catkin   a spike of tiny flowers, usually all of one sex, ie, with either stamen- or carpel-bearing flowers, found mostly on trees and shrubs.

Cell   (1) a cavity, or cavities, within an ovary, in which are attached the ovules.
(2) cavities in an anther, containing the pollen.
(3) the microscopic units of which all plants (and animals) are built up.

Carpel

Character   a feature of a plant that is used to compare and identify plants, for example, fruit shape.

Chlorophyll   a green pigment found in plant cells, which harnesses the sun's energy by complex chemical reactions.

Cladode

Cladode   a leaf-like structure which is really a flattened stem (shown by the fact that it may bear a flower, or flowers, on its surface in the axil of a bract – eg, Butcher's-broom).

Clasping leaf a (usually stalkless) leaf with basal lobes that project backwards and appear to clasp the plant's stem.

Clasping leaf

Claw lower, narrower part of an organ, such as a petal – effectively the petal stalk (cf. limb).

Cleistogamous of flowers, not opening, being self-pollinated during the bud stage.

Claw

Compound leaf a leaf divided into separate leaflets, without any flange joining them together along their common stalk (cf pinnate and palmate).

Cordate heart-shaped (of a leaf, broad below, pointed above, with two rounded basal lobes).

Corm a swollen rounded underground mass of solid tissue at the base of a stem, eg, in a Crocus, not (as in a bulb) composed of layers of scales; each corm is of one year's formation, the next year's growth arising separately on top of it.

Compound leaf

Corolla the petals of a flower (used collectively).

Corolla-tube the tube formed when the petals are fused together below.

Corymb an inflorescence in which the outer flower-stalks are much longer than the inner ones, so that the flowers are all more or less at the same level in a flat-topped cluster (as in an umbel, but in the latter all flower-stalks arise from a single point on the top of the stem).

Cordate

Cotyledon (1) one of the pair of first leaves of a seedling in Dicotyledons, which is one of the two great sub-divisions of the flowering plants. Dicotyledon leaves are usually net-veined. (2) the single first leaf of a seedling in Monocotyledons, which is the other of the two great sub-divisions of the flowering plants. Monocotyledon leaves are usually parallel-veined.

Corolla-tube

Cyme (adj **cymose**) an inflorescence in which the terminal flower opens first, followed in succession by lateral flowers growing from bract-axils lower down the inflorescence-stalk. Cymes may be simple, with the flowers along the side of a single stem, or double (a dichasium).

Deciduous a woody plant that drops its leaves in autumn (in Europe, at least), and which produces new leaves from buds the next spring.

Corymb

Decumbent a stem that lies on the ground but tends to turn upwards at its tip.

Diagnostic of plant characters, character(s) that enable unambiguous identification of a particular species (or subspecies, variety, etc) and distinguish it from other, similar-looking plants.

Cyme

Dichasium a forked cyme, with the oldest flower in the first fork (at the stem-head) and the next oldest flowers terminating the opposite side-branches; successive opposite pairs of branches arise below each terminal flower, and the process is repeated.

Dichasium

Digitate leaf

Dicotyledons   see cotyledon.

Digitate leaf   a palmate leaf, the leaflets of which are narrow and finger-like.

Dioecious plants   those plants with male and female flowers on separate individual plants (only male flowers on some, only female on others).

Disc-floret

Disc   round and flat, shaped like a coin.

Disc-floret   one of the tubular florets in a flower-head in the Daisy and Teasel families etc.

Distant   widely spaced-out.

Drupe   a fleshy fruit, resembling a berry, but with the seed inside enclosed in a hard stony case (like a plum or cherry).

Elliptical

Ellipsoid   the three-dimensional equivalent of elliptical (applied to solid objects such as fruits).

Elliptical   widest in the middle and up to about 3x as long as wide, tapering to the tip and the base (applied to flat objects such as leaves).

Entire

Endemic   only native to one country or area.

Entire   usually of a leaf, without any lobes or teeth along its margins.

Epicalyx

Epicalyx   an extra calyx of small bract-like leaflets below the true calyx.

Epichile   in the Helleborines (part of the Orchid family), the outer segment of the lip of a flower.

Epiphyte   a plant sitting upon another (usually woody) plant, and ± attached to it but, unlike a parasite, obtaining no nourishment from its host.

Erect   upright.

Falls

Falls   the drooping outer perianth segments in an Iris flower.

Family   a natural group of genera, with certain important features of flower, fruit, and/or vegetative structure in common.

Fen   a plant community on base-rich, often alkaline and calcareous, wetland dominated, not by *Sphagnum* mosses and Heath family members (as in a bog), but by Dicotyledonous herbs, Sedges and Grasses, etc.

Fen carr   boggy woodland developed over fen by invasion of shrubs and trees.

Filament

Filament   see stamen.

Floret   a small flower, especially in the flower-head of members of the Daisy and related families; or the individual tiny flower in Grasses, Sedges and Rushes.

Floret

Flush   an area where water (usually ± base-rich) flows or seeps through the ground, making its vegetation often very different from that of the more acid, stagnant or drier surroundings (mostly found in mountains or moorlands).

Follicle   a dry fruit formed from a single carpel, splitting open across the top and then a little down one side.

Follicle

Fruit   the dry or fleshy case, surrounding a plant's seeds, that is formed from the ovary wall.

Garden escape   a cultivated or garden plant that occurs outside of gardens, which may or may not have become naturalised.

Genus (pl. **genera**)   a grouping of species in the same family with important features of flower, fruit, and sometimes vegetative characters in common.

Glabrous   hairless.

Gland (1)

Gland   (1) a sticky tip to a hair, or a sticky spot or small knob on a leaf, petal etc that secretes viscous juice.
(2) the fleshy yellow bracts, oval or crescent-shaped, that alternate with the leafy bracts around the flower-clusters in Spurges.

Glandular hair   a hair with a gland on its tip (through a hand lens, looking like a stalk with a blob on the end).

Gland (2)

Glandular hairs

Glaucous   green strongly tinged blue-grey or whitish.

Globose (or **globular**)   spherical.

Hastate   shaped like a late medieval pike-head known as a 'Halberd', used to describe the base of a leaf.

Heath   an area of vegetation covered with dwarf shrubs of the Heath family, such as Heather (*Calluna*) or Heath (*Erica*), on poor, acid, usually sandy and well-drained soil, or it may be damp (wet-heath).

Hastate

Herb   a non-woody plant dying back each winter to an underground tuber, rhizome etc, or to a basal leaf-rosette, or (if an annual) lasting only one season.

Hybrid   a plant originating from a cross between species in the same genus or, more rarely, between species in different genera. Hybrids are denoted by a multiplication-sign in the plant's scientific name (eg *Crataegus* x *media*).

Inferior ovary

Hypochile   in the Helleborines (part of the Orchid family), the (often cup-shaped) inner segment of the lip of the flower.

Inferior ovary   an ovary below the calyx and/or corolla (ie, the rest of the flower sits on top of it).

Inflorescence   any grouping of flowers on a stem or in a leaf-axil etc.

Internode   the length of stem between nodes.

Internode

Introduced (or **non-native** or **alien**)   a plant brought to a country, region or site by human agency, not necessarily intentionally (as distinct from a native plant that has arrived by natural means) or a plant that has come in by natural means from an area where it is only present as an introduction.

Involucre   a whorl, collar or ruff of bracts at the base of a flower-head, or (less often) of a single flower.

Involucre

Irregular flower   one which is not radially symmetrical, with all petals equal, but only symmetrical on either side of one vertical plane; usually with either one or more petals enlarged as a lip, or with both a top and a bottom lip.

Irregular flower

Keel   (1) of a flower, in the Pea family, the two lower, partly-joined petals which form a shape like the keel of a boat.
(2) a keel-like flange running along the length of one side of a leaf; or the edge of a leaf folded lengthwise or fruit.

Keel (1)   (2)

Kidney-shaped leaf

Lanceolate leaf

Leaflet

Limb

Mucronate

Node

Oblong leaf

Obovate leaf

Kidney-shaped leaf   a leaf with a rounded outline, with the stalk in a notch between two rounded basal lobes.

Labellum   the lip of an Orchid flower.

Laminar   the blade of a leaf (as distinct from the petiole).

Lanceolate leaf   a long narrow leaf, slightly wider below, gradually tapering to its tip (lance-shaped).

Latex   an opaque milky, watery or coloured sap produced by stems or leaves in certain plants when cut.

Leaflets   the separate leaf-blades of a compound pinnate, palmate or trifoliate leaf.

Ligule   the little flap present where the leaf-blade joins its sheathing base in Grasses and Sedges.

Limb   the upper, wider part of an organ such as a petal (cf claw).

Linear leaf   a long, narrow leaf, as in most Grasses.

Lip   the lower (and sometimes also the upper) lobe of the corolla (or sometimes calyx) of an irregular flower.

Lobed   divided but not completely separated, often of leaves, which may be shallowly-lobed or deeply-lobed.

Marsh   an area of vegetation on wet ground on mineral soil (not on peat).

Mealy   with a floury, whitish texture.

Membranous   thin and papery-like, often transparent.

Midrib   the main, central vein of a leaf.

Monoecious   having male and female flowers separate but on the same plant.

Monocotyledons   see cotyledon.

Moor   an area of upland or montane vegetation dominated by heathers and on dry or damp soils.

Mucronate   a leaf or bract (or, possibly, a fruit), otherwise blunt or rounded, with a small, bristle-like point at its tip.

Native   indigenous to a country or region and having arrived by natural means, not introduced, intentionally or unintentionally, by humans.

Naturalised   an introduced plant that has become established by reproducing effectively by seed or spreading vegetatively (cf casual).

Nectary   a structure (knob-like, pouch-like, or tubular) within a flower (sometimes on a petal-base) that secretes the nectar collected by insects.

Node   the point on a stem where a leaf, a pair of leaves, or a whorl of leaves is attached.

Oblong leaf   a leaf about twice or three times as long as broad, and parallel-sided at least in the central part of the leaf-blade.

Obovate leaf   a leaf with its broadest part above the middle, tapering suddenly to its ± blunt tip, and more gradually to its base.

Ochrea (pl **ochreae**)   the tubular, ± translucent, brown or silvery, membranous stipules, often with a fringed tip, in the Dock family.

Opposite leaves   leaves arising in pairs, on opposite sides of the stem.

Oval leaf   a leaf broadest in the middle and about twice as long as broad.

Opposite leaves

Ovary   the carpels of a flower collectively, especially when they are fused together forming a little case of one or several cells in which are attached the ovules. After pollination and fertilization the ovary develops into the fruit, and the ovules into seeds.

Ovate leaf   a leaf ± egg-shaped in outline, broadest at the base and about twice as long as broad.

Ovoid   (of solid objects) ± egg-shaped.

Ovate leaf

Ovules   the tiny bodies attached inside a carpel or ovary, which each contain an egg; when fertilized, the ovules develop into seeds.

Palate   (1) a two-lobed swelling on the lower lip of irregular flowers, such as Toadflaxes and other species in the Figwort Family. (2) a lower lip of an irregular flower.

Palmate leaf

Palmate leaf   a compound leaf with more than three leaflets arising together from the top of the leaf-stalk.

Palmately-lobed leaf   a lobed leaf with the main veins all radiating from the top of the stalk.

Panicle   a branched raceme.

Palmately-lobed leaf

Pappus   a crown of hairs or bristles, sometimes branched like feathers, that sit on top of the tiny fruits of some members of the Daisy and Valerian families (eg 'thistle-down'); formed from the calyx, they may act as parachutes for wind-dispersal of the fruits.

Parallel veins   veins that remain ± the same distance apart along much of the leaf, running from the leaf-base to its tip.

Pappus

Parasite   a plant that derives all, or part of, its nourishment from another plant (the host), by means of suckers attached to the host's roots or stems. Total parasites usually have scale-like tiny leaves and no green chlorophyll; partial parasites have at least some green leaves.

Patent   sticking out ± at right-angles.

Pedicel   a stalk of a single flower.

Pedicel

Peduncle   a stalk of an inflorescence.

Perennial   a plant that lives at least two years, generally flowering each year (cf annual and biennial).

Perfoliate leaf   a leaf (or a pair of conjoined opposite leaves) forming a ring around a stem.

Peduncle

Perianth   the collective term for the calyx (ie sepals) and/or corolla (ie petals) segments or lobes, used when the calyx and corolla are indistinguishable in form and colour, or when there is only one whorl of segments.

Perianth segments   the sepals and/or petals that make up the perianth of a flower (used mostly either when they are indistinguishable, or there is only one whorl).

Perfoliate leaf

Petal   one of the inner whorl of floral leaves (usually white or coloured) that surround the stamens and/or carpels of a flower.

Petiole   a stalk of a leaf.

Petiole

Pinnate leaf

Pinnatifid leaf

Pinnatisect leaf

Raceme

Ray Florets

Rays

Receptacle

Recurved

pH   in simple terms, a measurement of the alkalinity or acidity of soil or water. pH7 is neutral; values above it are alkaline, below it (particularly below pH5 for plants) are acid. pH can be measured roughly with soil-testing kits available at some garden and chemist's shops.

Pinnate leaf   a compound leaf with the separate leaflets arranged along the leaf-stalk, usually in opposite pairs, and often with a terminal leaflet as well.

Pinnate veins   with secondary veins along each side of the mid-rib, running parallel towards the leaf margin, like teeth of a comb, or sometimes re-joining to form a network.

Pinnatifid leaf   a leaf ± deeply cut into pinnately-arranged lobes, but not cut right to its midrib, the lobes remaining connected by at least a narrow flange of leaf-blade bordering the midrib.

Pinnatisect leaf   a leaf ± deeply cut into pinnately-arranged lobes, cut almost to its midrib.

Pollen   minute yellow dust-like granules, containing the male sexual cells, that are produced inside the anthers of stamens. When pollen grains fall on a stigma, they germinate to produce long but microscopic pollen-tubes which carry the male nuclei (the equivalent of sperms in animals) to the eggs to fertilize them.

Pollinium (pl **pollinia**)   a sticky mass of pollen grains cohering together within the anther of an Orchid. The entire pollinia can be transferred by insects to another flower.

Pome   the fleshy but firm fruit of an Apple or Pear, etc, its outer fleshy part, formed from the swollen cup-shaped receptacle of the flower, surrounding the gristly carpels in each of which are one or more seeds (pips).

Prickle   a sharp spine arising from the surface of a shoot or leaf.

Procumbent   lying loosely on the surface of the ground.

Raceme (adj **racemose**)   a ± elongated inflorescence, in which the individual flowers are usually stalked (cf spike). The lowest flower opens first, and then the others in sequence towards the tip, which can continue to form new buds.

Ray-florets (or **ligulate-florets**)   the (usually) strap-shaped spreading florets found in some of the Daisy family, occurring either around the edge of the flower-head (as in a Daisy) or forming the whole of the flower-head (as in a Dandelion).

Rays   the branches of an umbel.

Receptacle   the tip of the flower-stalk, to which all the floral parts (sepals, petals, stamens, carpels) are attached; it may be dome-shaped or conical (as in Buttercups or Mayweeds); cup-shaped and hollowed out, with the carpels within the cup (as in many members of the Rose family); or a flat disc.

Recurved   arched, backwards or downwards in a curve.

Reflexed   sharply (not in a gradual curve) bent backwards, outwards or downwards.

Reflexed

Regular flower   one which is radially symmetrical, where ± any straight line drawn through the centre produces two mirror images.

Rhizome   a creeping underground stem, sometimes fleshy.

Regular flower

Ruderal   a habitat which is dry, generally under-vegetated and often waste ground, usually colonised by pioneering or 'weedy' species of plants.

Runner   a type of stolon, a stem creeping above ground, which can root at its tip, forming a new plant.

Rhizome

Salt-marsh   a marsh where the water is salty, usually from sea water flooding or seeping into it at high tide, but also (though rarely) beside salt springs inland.

Saprophyte   a plant, usually without green chlorophyll, which is not parasitic but nourishes itself on decomposing vegetable humus, usually through partnership with a fungus whose filaments are attached to its roots and spread out through the humus.

Runner

Scale   a structure that is not leaf-like, being papery, brownish or whitish, often ± transparent.

Scarious   dry and papery, often translucent.

Scrub   an area of vegetation dominated by shrubs on any type of soil.

Sepal   one of the outer whorl of floral leaves (usually green) that surround the petals (corolla) of a flower.

Sessile   stalkless.

Sessile

Shrub   a woody perennial sometimes branched from the base (without one main trunk, as in a tree), and of smaller stature than a tree.

Simple leaf   a leaf that is not compound, ie, not divided into leaflets; it may, however, be lobed or toothed.

Spadix   a dense erect spike of florets found in the Arum family, usually with a spathe. Its tip may be club-shaped, without florets (as in Lords-and-Ladies), or bearing florets to the tip (as in Sweet-flag).

Simple leaf

Spathe   a large bract (or pair of bracts) at the base of, wrapping round, or surrounding the inflorescences (especially in bud) of the Arum family, and of some members of the Lily, Daffodil and Iris families.

Species (pl. **species**)   a basic grouping of organisms, ranked below, for example, family and genus.

Spike

Spike   an unbranched raceme without stalks to the individual flowers.

Spoon-shaped leaf   a leaf with a long narrow stalk-like lower part, suddenly widening into an oval, blunt tip (resembling a spoon or paddle).

Spoon-shaped leaf

Spur

Stamen

Standard

Stigma

Stipule

Stolon

Style

Sucker

Superior ovary

Spur   a cylindrical or conical, sometimes curved, hollow projection from the back or base of a petal or sepal, usually containing nectar.

Stamen   one of the individual male organs of a flower, comprising the filament and anther. The anther is the little sac, usually two (or more) -celled, containing the pollen at the top of a stamen; the filament is the stalk.

Staminode   a sterile or rudimentary stamen without a pollen-containing anther.

Standard   (1) the top petal in a flower of members of the Pea family. (2) one of the erect inner perianth segments in an Iris flower.

Stem-leaves   leaves arising from a plant's stem (as distinct from basal leaves).

Sterile   not producing fertile seeds or pollen. A sterile plant has few, shrivelled pollen grains and under-developed fruits. Compare anthers (x20) from a likely sterile hybrid to anthers from another, fertile plant from the same genus. A sterile hybrid anther will not appear well filled with pollen as so many of the grains are shrunken. Under-developed fruits will feel hollow when squeezed.

Stigma   the (usually sticky) surface at the tip of a style, which receives the pollen and where the pollen grains are stimulated to germinate. A style may have one or several stigmas.

Stipule   a leaf-like or scale-like appendage at the base of a leaf-stalk; often a pair of them is present.

Stolon   a runner, not necessarily forming a new plant at its tip. Includes procumbent stems.

Strap-shaped   (of a leaf, petal, etc) flat, parallel-sided and blunt-ended.

Style   the stalk-like structure at the tip of a carpel or on top of an ovary of several joined carpels, that bears on its tip the stigma or stigmas. A persistent style is a style that remains after flowering and is still attached to the end of the fr.

Subspecies   a subdivision of a species, separated for the purposes of plant classification on several conspicuous characters, genetic differences or on ecological or geographical criteria or any combination of these.

Sucker   a shoot arising from the roots of trees or shrubs to produce new plants, often a characteristic of certain plant groups, eg. Willow family.

Superior ovary   an ovary sitting above and within the whorls of stamens, petals and sepals, and which is not fused to them at all.

Taxon (pl **taxa**)   any taxonomic grouping, such as a family, a genus or a subspecies.

Taxonomy   in its narrowest sense, the theory and practice of producing a classification by grouping individuals into species, subspecies etc, and arranging these into larger groups such as families and orders.

Tendril   a climbing filament, usually projecting from the end of a leaf, that coils around other plants or objects for support. Tendrils are generally modified stems, leaves, leaflets or leaf-tips.

Tendril

Tepal   alternative word for perianth segment.

Terminal   at the very end (or tip).

Thallus   a plant body not separated into a distinct stem and leaves, as in Duckweeds (Lemnaceae), and in many lower plants such as Algae, some Liverworts and Lichens, etc.

Thorn   a sharp-pointed woody structure, strictly speaking the tip of a modified branch or stem.

Trifoliate leaf

Trifoliate leaf   a compound leaf with only three leaflets (as in Clovers).

Truncate   (of a leaf or fruit) ± square-ended, as if cut off at right-angles.

Tuber   a swollen part of a root or underground stem, storing food.

Truncate

Umbel   a ± flat-topped inflorescence, with several branches all rising from one point at the top of the main stem; it may be simple (composed of one whorl of unbranched branches only), or compound (when each branch in turn bears a similar 'secondary' or 'partial' umbel).

Umbel

Valve   the part of a fruit, covering the seeds, that opens or falls off.

Variety   a subdivision of a species or subspecies differing by only one to a few characters and genetic differences but generally not by ecological or geographic criteria.

Vascular plant   plants which have vascular tissue, which form channels for conducting water and nutrients. The group includes Conifers, Dicotyledons, Monocotyledons, Ferns, Horsetails, Clubmosses and Quillworts and excludes lower plants, comprising Bryophytes (Mosses, Liverworts, and Hornworts) and, for example, Lichens and Algae.

Valves

Vein   one of the thickened strands or lines visible in a leaf (or petal, etc), which contains the vessels or tubes that conduct water and food (as in the arteries and veins of the human body).

Waste ground   unused or uncultivated, often disturbed ground (as alongside a new road, around an industrial area, on a heap of mine waste or in a quarry, including what are known as brownfield sites).

Wedge-shaped

Wedge-shaped   of a leaf-base, gradually tapering into the leaf-stalk.

Whorled leaves   three or more leaves arising from a stem at the same level or point on it.

Whorled

Wing   one of the two side-petals in members of the Pea family.

Winged stem   a stem with one or more broad flanges along its length.

Woody   hard, not quickly withering and dying like green stems.

Wing

TABLE OF ANCIENT WOODLAND INDICATOR PLANTS (AWIs)
collated by K. Kirby, English Nature (2004)

AWI species have been researched and lists compiled by various authors for the regions of Britain given below.
Species listed in blue are grasses, sedges, rushes, ferns or fern allies not included in this book
(except, in some cases, in the vegetative keys). English names are not included to save space.

South-west = Cornwall, Devon, Somerset, Avon & Dorset
South = Hampshire, Wiltshire, Buckinghamshire, Berkshire & Oxfordshire
South-east = Kent, Surrey, Sussex, London & Hertfordshire
East = Essex, Suffolk, Norfolk, Cambridgeshire, parts of Middlesex & parts of Hertfordshire

	South-west	Dorset	Somerset	South	South-east	East	Lincolnshire	Worcs	Derbyshire	North-east Yorks	Lowland Northumberland	Carmarthen	Angus
Aconitum napellus	✓							✓					
Acer campestre	✓			✓	✓	✓		✓			✓		
Adoxa moschatellina	✓	✓		✓	✓	✓	✓	✓	✓	✓	✓	✓	
Alchemilla filicaulis						✓							
Allium ursinum	✓			✓	✓	✓	✓	✓	✓			✓	✓
Anagallis minima						✓							
Anemone nemorosa	✓	✓	✓	✓	✓	✓	✓	✓	✓		✓		✓
Aquilegia vulgaris	✓	✓		✓	✓			✓	✓	✓			
Arum maculatum											✓		
Athyrium filix-femina								✓					✓
Blechnum spicant	✓			✓	✓	✓		✓					
Brachypodium sylvaticum									✓		✓		
Bromopsis benekenii		✓											
Bromopsis ramosa	✓			✓	✓	✓		✓	✓		✓	✓	
Calamagrostis canescens						✓	✓						
Calamagrostis epigejos	✓			✓	✓	✓	✓						
Campanula latifolia						✓	✓	✓	✓	✓	✓		
Campanula patula								✓					
Campanula trachelium	✓			✓	✓	✓	✓	✓	✓				
Cardamine amara				✓	✓	✓		✓					
Cardamine impatiens								✓	✓				
Carpinus betulus				✓	✓	✓							
Carex acutiformis						✓							
Carex digitata								✓					
Carex elongata								✓					
Carex laevigata	✓	✓		✓	✓	✓	✓	✓	✓			✓	✓
Carex montana								✓					
Carex pallescens	✓	✓		✓	✓	✓	✓	✓	✓				✓
Carex paniculata											✓		
Carex pendula	✓	✓		✓	✓	✓	✓	✓	✓	✓		✓	
Carex remota	✓	✓		✓	✓	✓	✓	✓	✓	✓	✓	✓	
Carex strigosa	✓	✓		✓	✓	✓	✓	✓	✓			✓	
Carex sylvatica	✓	✓		✓	✓	✓	✓	✓	✓	✓	✓	✓	
Cephalanthera damasonium		✓						✓					
Cephalanthera longifolia				✓				✓					
Ceratocapnos claviculata	✓			✓		✓	✓		✓				
Chrysosplenium alternifolium			✓				✓	✓	✓	✓	✓	✓	
Chrysosplenium oppositifolium	✓	✓		✓	✓	✓	✓	✓	✓	✓			✓
Circaea x intermedia									✓		✓		
Cirsium heterophyllum									✓				

	South-west	Dorset	Somerset	South	South-east	East	Lincolnshire	Worcs	Derbyshire	North-east Yorks	Lowland Northumberland	Carmarthen	Angus
Colchicum autumnale	✓		✓	✓				✓					
Conopodium majus	✓			✓	✓	✓	✓		✓				✓
Convallaria majalis	✓			✓	✓	✓	✓	✓	✓	✓		✓	
Cornus sanguinea						✓							
Corylus avellana						✓							
Crataegus laevigata				✓	✓	✓		✓					
Daphne laureola	✓			✓	✓	✓		✓	✓				
Daphne mezereum									✓				
Dipsacus pilosus	✓			✓	✓	✓	✓	✓					
Dryopteris aemula	✓				✓	✓						✓	
Dryopteris affinis	✓			✓	✓			✓	✓				
Dryopteris carthusiana	✓			✓	✓	✓		✓	✓				
Elymus caninus	✓			✓	✓	✓	✓	✓	✓		✓		
Epipactis helleborine	✓	✓		✓	✓	✓		✓			✓		
Epipactis muelleri		✓		✓									
Epipactis phyllanthes		✓											
Epipactis purpurata		✓		✓	✓	✓							
Equisetum sylvaticum	✓			✓	✓	✓	✓	✓	✓		✓		
Equisetum telmateia								✓	✓				
Euonymus europaeus	✓				✓	✓	✓	✓			✓	✓	
Euphorbia amygdaloides	✓	✓	✓	✓	✓	✓		✓					
Festuca altissima								✓	✓		✓		
Festuca gigantea	✓			✓	✓	✓					✓	✓	
Fragaria vesca							✓						
Frangula alnus	✓			✓	✓	✓	✓	✓	✓				
Gagea lutea							✓		✓	✓	✓		
Galanthus nivalis			✓										
Galium odoratum	✓	✓	✓	✓	✓	✓	✓	✓	✓	✓	✓	✓	✓
Geranium robertianum													✓
Geranium sanguineum								✓	✓				
Geranium sylvaticum								✓					
Geum rivale	✓			✓		✓	✓	✓	✓	✓			
Gnaphalium sylvaticum							✓	✓					
Goodyera repens											✓		
Gymnocarpium dryopteris								✓				✓	✓
Helleborus foetidus								✓					
Helleborus viridis	✓			✓	✓	✓		✓	✓				
Holcus mollis	✓			✓	✓								
Hordelymus europaeus				✓	✓		✓	✓	✓	✓	✓		
Hyacinthoides non-scripta	✓			✓	✓	✓	✓				✓		
Hymenophyllum tunbrigense	✓											✓	
Hymenophyllum wilsonii												✓	
Hypericum androsaemum	✓		✓	✓	✓	✓		✓					
Hypericum hirsutum						✓	✓						
Hypericum pulchrum	✓			✓	✓	✓		✓	✓	✓			✓
Hypericum tetrapterum						✓							
Ilex aquifolium	✓			✓	✓	✓			✓				
Iris foetidissima	✓			✓	✓	✓							
Juniperus communis												✓	
Lamiastrum galeobdolon		✓	✓	✓	✓	✓	✓	✓	✓	✓			
Lathraea squamaria	✓	✓	✓	✓	✓	✓	✓	✓	✓	✓	✓	✓	
Lathyrus linifolius	✓	✓		✓	✓	✓	✓	✓	✓				

	South-west	Dorset	Somerset	South	South-east	East	Lincolnshire	Worcs	Derbyshire	North-east Yorks	Lowland Northumberland	Carmarthen	Angus
Lathyrus sylvestris	✓	✓		✓	✓	✓		✓					
Listera ovata								✓					✓
Lonicera periclymenum									✓				
Luzula forsteri	✓	✓		✓	✓			✓					
Luzula pilosa	✓	✓	✓	✓	✓	✓	✓	✓	✓	✓	✓	✓	
Luzula sylvatica	✓	✓	✓	✓	✓	✓	✓	✓	✓	✓			
Lysimachia nemorum	✓	✓		✓	✓	✓	✓	✓	✓				✓
Lysimachia vulgaris								✓					
Lythrum portula						✓							
Maianthemum bifolium						✓	✓						
Malus sylvestris	✓			✓	✓							✓	
Melampyrum cristatum						✓							
Melampyrum pratense	✓			✓	✓	✓	✓	✓	✓			✓	✓
Melampyrum sylvaticum										✓			
Melica nutans								✓	✓		✓		
Melica uniflora	✓	✓	✓	✓	✓	✓	✓	✓	✓	✓	✓	✓	
Melittis melissophyllum	✓	✓										✓	
Mercurialis perennis						✓	✓	✓	✓		✓	✓	✓
Milium effusum	✓	✓	✓	✓	✓	✓	✓	✓	✓			✓	
Moehringia trinervia	✓			✓	✓	✓						✓	
Monotropa hypopitys		✓						✓					
Myosotis sylvatica						✓	✓	✓	✓		✓		✓
Narcissus pseudonarcissus	✓			✓	✓	✓		✓	✓				
Neottia nidus-avis	✓	✓	✓	✓	✓	✓	✓	✓			✓	✓	
Ophioglossum vulgatum						✓							
Ophrys insectifera		✓				✓							
Orchis mascula	✓			✓	✓	✓	✓	✓	✓	✓			
Orchis purpurea					✓								
Oreopteris limbosperma	✓			✓	✓	✓		✓	✓				
Orobanche hederae								✓					
Oxalis acetosella	✓	✓		✓	✓	✓	✓	✓	✓		✓		
Paris quadrifolia	✓	✓	✓	✓	✓	✓	✓	✓	✓	✓	✓	✓	
Phegopteris connectilis	✓										✓		✓
Phyllitis scolopendrium	✓			✓	✓			✓	✓				
Pimpinella major					✓	✓		✓					
Platanthera chlorantha	✓			✓	✓	✓	✓	✓					
Poa nemoralis	✓	✓		✓	✓	✓		✓				✓	✓
Polygonatum multiflorum	✓	✓	✓	✓	✓				✓				
Polygonatum odoratum									✓				
Polypodium vulgare	✓			✓	✓	✓			✓				
Polystichum aculeatum	✓	✓	✓	✓	✓			✓	✓		✓		✓
Polystichum setiferum	✓			✓	✓			✓	✓		✓		
Populus tremula	✓			✓	✓	✓	✓		✓				
Potentilla sterilis	✓			✓	✓	✓	✓		✓				✓
Primula elatior						✓							
Primula vulgaris	✓			✓	✓	✓	✓	✓	✓				✓
Prunus avium	✓			✓	✓	✓	✓	✓	✓				
Prunus padus						✓			✓				
Pulmonaria longifolia	✓			✓									
Pulmonaria obscura						✓							
Pyrola minor									✓				
Pyrus communis						✓							

	South-west	Dorset	Somerset	South	South-east	East	Lincolnshire	Worcs	Derbyshire	North-east Yorks	Lowland Northumberland	Carmarthen	Angus
Quercus petraea	✓			✓	✓	✓	✓	✓					
Radiola linoides					✓	✓							
Ranunculus auricomus	✓		✓	✓	✓	✓	✓	✓	✓		✓		
Rhamnus catharticus									✓			✓	
Ribes nigrum	✓			✓	✓	✓							✓
Ribes rubrum	✓		✓	✓	✓	✓							✓
Ribes spicatum											✓		
Rosa arvensis	✓			✓	✓				✓				
Rubus caesius									✓				
Rubus saxatilis									✓				
Ruscus aculeatus	✓			✓	✓	✓							
Salix aurita								✓					
Salix caprea												✓	
Sanicula europaea	✓	✓		✓	✓	✓		✓	✓			✓	
Scirpus sylvaticus	✓			✓	✓			✓	✓			✓	
Scrophularia nodosa							✓						
Scutellaria minor					✓								
Sedum telephium	✓			✓	✓	✓		✓					
Serratula tinctoria				✓	✓								
Sibthorpia europaea	✓												
Silene dioica													✓
Solidago virgaurea	✓			✓	✓			✓	✓				
Sorbus aucuparia						✓							
Sorbus torminalis	✓	✓	✓	✓	✓	✓	✓	✓	✓	✓		✓	
Stachys officinalis	✓			✓	✓	✓		✓	✓				
Stachys sylvatica													✓
Stellaria holostea							✓	✓	✓				
Stellaria neglecta							✓		✓				
Stellaria nemorum									✓	✓	✓		
Tamus communis	✓			✓	✓	✓			✓				
Tilia cordata	✓		✓	✓	✓	✓	✓	✓	✓	✓	✓	✓	
Tilia platyphyllos								✓	✓				
Trollius europaeus									✓				
Ulmus glabra								✓	✓				
Vaccinium myrtillus	✓			✓	✓				✓				
Valeriana officinalis							✓						
Veronica montana	✓	✓		✓	✓	✓	✓	✓	✓			✓	✓
Viburnum opulus	✓		✓	✓	✓	✓	✓		✓			✓	
Vicia sepium	✓			✓	✓	✓			✓				✓
Vicia sylvatica	✓	✓		✓	✓			✓	✓	✓	✓		
Viola odorata						✓							
Viola palustris	✓			✓	✓				✓				
Viola reichenbachiana	✓		✓	✓	✓	✓	✓	✓	✓	✓		✓	
Viola riviniana							✓	✓					
Wahlenbergia hederacea	✓				✓								

INDEX TO KEYS

INDEX

Both scientific and English names of families and species are included, the latter in italics under generic headings; the few scientific synonyms given are set in brackets. English plant names are always listed under the last word of the name which is often hyphenated.